Physical Metallurgy
and processing of
Intermetallic Compounds

Physical Metallurgy
and processing of
Intermetallic Compounds

Edited by
N. S. Stoloff
Rensselaer Polytechnic Institute
V. K. Sikka
Oak Ridge National Laboratory

CHAPMAN & HALL

THOMSON PUBLISHING
New York • Albany • Bonn • Boston • Cincinnati • Detroit • London • Madrid • Melbourne
Mexico City • Pacific Grove • Paris • San Francisco • Singapore • Tokyo • Toronto • Washington

Cover design: Andrea Meyer, emDASH inc.

Printed in the United States of America

Chapman & Hall
115 Fifth Avenue
New York, NY 10003

Chapman & Hall
2-6 Boundary Row
London SE1 8HN
England

Thomas Nelson Australia
102 Dodds Street
South Melbourne, 3205
Victoria, Australia

Chapman & Hall GmbH
Postfach 100 263
D-69442 Weinheim
Germany

Nelson Canada
1120 Birchmount Road
Scarborough, Ontario
Canada M1K 5G4

International Thomson Publishing Asia
221 Henderson Road #05-10
Henderson Building
Singapore 0315

International Thomson Editores
Campos Eliseos 385, Piso 7
Col. Polanco
11560 Mexico D.F
Mexico

Internation Thomson Publishing–Japan
Hirakawacho-cho Kyowa Building, 3F
1-2-1 Hirakawacho-cho
Chiyoda-ku, 102 Tokyo
Japan

1 2 3 4 5 6 7 8 9 10 XXX 01 00 99 98 97 96

Library of Congress Cataloging-in-Publication Data

Physical metallurgy and processing of intermettalic compounds /
[edited by] Norman S. Stoloff and Vinod K. Sikka.
p. cm.
Includes bibliographical references and indexes.
ISBN 0–412–98971–9
1. Intermetallic compounds. 2. Alloys. 3. Physical metallurgy. I. Stoloff, N.S.
II. Sikka, Vinod K.
TA483.P495 1994
669'.9—dc20 94–18509
CIP

To order this or any other Chapman & Hall book, please contact **International Thomson Publishing, 7625 Empire Drive, Florence, KY 41042.** Phone: (606) 525-6600 or 1-800-842-3636. Fax: (606) 525-7778. e-mail: order@chaphall.com.

For a complete listing of Chapman & Hall titles, send your request to **Chapman & Hall, Dept. BC, 115 Fifth Avenue, New York, NY 10003.**

Contents

II BEHAVIOR OF ALLOY SYSTEMS

III ENVIRONMENTAL EFFECTS

IV PROCESSING

Preface

The attractive physical and mechanical properties of ordered intermetallic alloys have been recognized since early in this century. However, periodic attempts to develop intermetallics for structural applications were unsuccessful, due in major part to the twin handicaps of inadequate low-temperature ductility or toughness, together with poor elevated-temperature creep strength. The discovery, in 1979, by Aoki and Izumi in Japan that small additions of boron caused a dramatic improvement in the ductility of Ni_3Al was a major factor in launching a new wave of fundamental and applied research on intermetallics. Another important factor was the issuance in 1984 of a National Materials Advisory Board reported entitled "Structural Uses for Ductile Ordered Alloys," which identified numerous potential defense-related applications and proposed the launching of a coordinated development program to gather engineering property and processing data. A substantial research effort on titanium aluminides was already underway at the Air Force Materials Laboratory at Wright-Patterson Air Force Base in Ohio and, with Air Force support, at several industrial and university laboratories. Smaller programs also were underway at Oak Ridge National Laboratory, under Department of Energy sponsorship. These research efforts were soon augmented in the United States by funding from Department of Defense agencies such as Office of Naval Research and Air Force Office of Scientific Research, and by the National Science Foundation. Increased research support from several branches of the Department of Energy was funneled to both National Laboratories (Oak Ridge and Los Alamos), as well as to many universities. Research efforts also expanded rapidly in Japan, France, Germany, and later in China. Particularly noteworthy was the launching of a Japanese National Project on Intermetallics, funded by Ministry of International Trade and Industry and by industry, with TiAl and Nb_3Al as the focus of a coordinated effort to develop new high temperature alloys.

As a result of this remarkable increase in activity in the field of intermetallics, numerous conferences, workshops, and symposia on intermetallics were held throughout the world, starting with a biannual symposium on High Temperature Ordered Intermetallic Alloys, held at the Materials Research Society Fall Meeting of 1984. As of the end of 1994, at least eighteen general conference proceedings were in print in English, and others were in preparation.

The editors were approached in 1991 by several publishers to prepare a book on Structural Intermetallics. Because of the explosion in information generated on intermetallics during the last decade, we decided to enlist the aid of experts in a wide variety of fields to help us to carry out this project. We are grateful for their contributions. In planning the scope of the book we took the point of view that no matter how attractive some intermetallics appear to be in the laboratory, the key to successful utilization of these new materials resides in the effective use of modern processing techniques to provide desired microstructures and properties. Accordingly, we have included chapters on powder as well as on melt processing, and other chapters include additional information on processing. Another key feature of this book is its emphasis on aluminides and silicides. These are the classes of intermetallics that have been most intensively studied, and, therefore, are most likely to be successfully applied in commerce. Chapters on creep, fatigue, and fracture also are highlighted in this book because these are the areas where improvement in properties of intermetallics was most clearly needed. One factor that was unrecognized until very recently is the major impact that moisture and hydrogen have on low-temperature fracture behavior of many intermetallics. Therefore, a separate chapter entitled 'Environmental Embrittlement' is devoted to this issue.

It is hoped that this book, with its emphasis on alloy development, processing, mechanical properties, and structural applications will prove to be a basic reference source as well as a valuable adjunct to the conference proceedings mentioned earlier.

N. S. Stoloff
V. K. Sikka

Contributors

Dr. R. R. Bowman
NASA Lewis Research Center
Mail Stop 49-3
21000 Brookpark Rd.
Cleveland, OH 44135

Dr. Ray A. Buchanan
University of Tennessee
434 Dougherty Engineering Building
Knoxville, TN 37996-2200

Dr. S. A. David
Oak Ridge National Laboratory
P.O. Box 2008
Oak Ridge, TN 37831-6095

Professor F. H. Froes
Institute for Materials and Advanced Processes
University of Idaho
Moscow, ID 83843

Professor Randall German
Engineering Science and Mechanical Engineering Dept.
Penn State University
227 Hammond Building
University Park, PA 16802

Professor L. A. Heldt
Department pf Metallurgical and Materials Engineering
Michigan Technological University
Houghton, MI 49931-1295

Dr. Ronald G. Iacocca
Engineering Science and Mechanical Engineering Dept.
Penn State University
227 Hammond Building
University Park, PA 16802

Dr. J. G. Kim
University of Tennessee
434 Dougherty Engineering Building
Knoxville, TN 37996-2200

Dr. K. S. Kumar
Martin Marietta Laboratories
1450 South Rolling Road
Baltimore, MD 21227-3898

Dr. C. T. Liu
Oak Ridge National Laboratory
P. O. Box 2008
Oak Ridge, TN 37831-6114

Dr. M. J. Maloney
Pratt & Whitney Aircraft
P. O. Box 109600
West Palm Beach, FL 33410-9600

Dr. C. McKamey
Oak Ridge National Laboratory
P. O. Box 2008
Oak Ridge, TN 37831-6114

Dr. M. V. Nathal
NASA Lewis Research Center
Mail Stop 49-3
21000 Brookpark Rd.
Cleveland, OH 44135

Dr. M. Nazmy
Senior Scientist and Group Leader
ABB Power Generation Ltd.
5401 Baden
Switzerland

Ronald D. Noebe
NASA Lewis Research Center
Mail Stop 49-3
21000 Brookpark Rd.
Cleveland, OH 44135

Dr. Richard E. Ricker
Corrosion & Wear Group
Metallurgy Division
National Institute of Standards and Technology
Clopper Road
Gaithersburg, MD 20899

Dr. M. L. Santella
Oak Ridge National Laboratory
P. O. Box 2008
Oak Ridge, TN 37831-6115

Professor Erland Schulson
Dartmouth College
Thayer School of Engineering
Tuck Drive
Hanover, NH 03755

Dr. Dilip Shah
Pratt and Whitney Acraft
East Hartford, CT 06108

Dr. Yasuharu Shirai
Kyoto University
Department of Metal Science and Technology
Sakyo-Ku
Kyoto 606
Japan

Dr. V. K. Sikka
Oak Ridge National Laboratory
Metalworking Theory and Practice
Metals and Ceramics Division
P. O. Box 2008
Oak Ridge, TN 37831

Professor N. S. Stoloff
Materials Engineering Department
Rensselaer Polytechnic Institute
Troy, NY 12180-3590

Dr. C. Suryanarayana
Institute for Materials and Advanced Processes
University of Idaho
Moscow, ID 83843

Professor Vaclav Vitek
University of Pennsylvania
Department of Materials Science and Engineering
3231 Walnut
201 LRSM
Philadelphia, PA 19104-6272

Professor Masahara Yamaguchi
Kyoto University
Department of Metal Science and Technology
Sakyo-Ku
Kyoto 606
Japan

Dr. M. Yan
University of Pennsylvania
Department of Materils Science and Engineering
3231 Walnut
201 LRSM
Philadelphia, PA 19104-6272

Physical Metallurgy
and processing of
Intermetallic Compounds

Part I

Basic Properties

Chapter 1

Defect Structures

M. Yamaguchi and Y. Shirai

1. INTRODUCTION

In the early 1900s, many binary combinations of metals that form intermetallic compounds were found. Since then, the mechanical properties of these compounds have been studied mainly from their theoretical implications and requirements to understand the mechanical properties of intermetallic compounds existing as secondary phases in many alloys used as structural materials. As a result of these efforts, various new findings on the lattice defects and properties of intermetallic compounds were made. Among them were discoveries of (a) the complications introduced by heat treatment in the property/composition relationships of some intermetallic compounds, later leading to the finding of ordered structures formed from solid solutions disordered at high temperatures (Kurnakov, Zhemchuznii, and Zasedatelev 1916); (b) the brittleness of intermetallic compounds in general at temperatures lower than 0.7 Tm (Tm: melting temperature) (Tamman and Dahl 1923); (c) the effect of compositional deviation from stoichiometry on hardness and flow stress; (d) the existence of antiphase boundaries and antiphase domains (Smith 1943); (e) the existence of superlattice dislocations traveling in pairs or groups in ordered lattices (Koehler and Seitz 1947; Marcinkowski, Brown, and Fisher 1961; Marcinkowski 1963); (f) the anomalous temperature dependence of hardness of Ni_3Al (Westbrook 1957).

The strength anomaly in Ni_3Al has been studied extensively since its discovery because of its practical implications. Such extensive studies on Ni_3Al and its related compounds result in not only enhancing our understanding of the strength anomaly and deformation behavior of Ni_3Al but also stimulating much new work on other intermetallic compounds and eventually leading to the current boom of research and development of high-temperature structural intermetallics. Recently, an outstanding overview on the origins, status, and future of structural intermetallic compounds was made by Westbrook (1993). This work should be consulted for additional background information on the historical development of intermetallics studies.

In this chapter, we focus our attention on the defect structure of intermetallic compounds and outline the results of studies of point defects and their properties. Particular emphasis is placed upon the intermetallic superlattice compounds that crystallize in typical superlattice structures based on the three principal metallic structures, the f.c.c., b.c.c., and h.c.p. Research and development activities in the intermetallics field have been focused on this class of compounds in particular, because of their potential as high-temperature structural materials. The reader also is referred to the two books on intermetallic compounds edited by Westbrook (1960, 1967), the two reviews on the mechanical properties of intermetallic compounds published in *Progress in Materials Science* (Stoloff and Davies 1966; Yamaguchi and Umakoshi 1990), and the recently published book on intermetallic compounds edited by Westbrook and Fleischer (1994) for general information on the lattice defects and defect structure of intermetallic compounds.

2. CONSTITUTIONAL DEFECTS

In intermetallic compounds, different atomic species occupy differently designated lattice sites and they are usually held together by strong bond. If the intermetallic compound has an extended phase field, its nonstoichiometry may give rise to the existence of constitutional defects such as atoms occupying atomic sites of a sublattice other than its own sublattice (antisite atoms) or vacancies of deficient atomic species (constitutional vacancies). Which of these defects dominates the defect structure of the nonstoichiometric compound depends on the crystal structure, constituent elements, and bonding nature of the compound as well as on which of the constituent elements is deficient and the amount by which the compound is nonstoichiometric. In the following sections, the studies of constitutional defects in intermetallic compounds crystallizing in some typical structures are summarized.

2.1 Superlattice Compounds with b.c.c.-Derivative Structures

2.1.1 B2 Compounds. The B2 structure consists of two simple cubic interpenetrating sublattices, the α and β sublattices (Fig. 1-1) and the general formula of compounds crystallizing in this crystal structure can be given as A_xB_{1-x}, where $x = 0.5$ for the stoichiometric composition. A and B atoms occupy α and β sublattices, respectively. This class of compounds can be classified into two groups with respect to their defect structure. One group consists of compounds in which only antisite defects are formed on both sides of $x > 0.5$ and $x < 0.5$. The other group is made up of compounds in which constitutional vacancies are formed for $x < 0.5$, while only antisite atoms are formed for $x > 0.5$. Compounds such as AgMg and CuZn belong to the former group and those such as NiAl, NiGa, CoAl and CoGa, where Ni and Co correspond to the A atom of the formula of A_xB_{1-x}, belong to the latter group.

The total vacancy concentration, c, which is defined as the number of vacancies relative to the total number of atoms, is obtained from macroscopic density (ρ) measurements and the x-ray density (ρ_x) calculated from lattice constant as follows:

$$c = (\rho_x - \rho)/\rho \tag{1.1}$$

For the perfect B2 structure,

$$\rho_x = 2(x_A m_A + x_B m_B)/a^3 \tag{1.2}$$

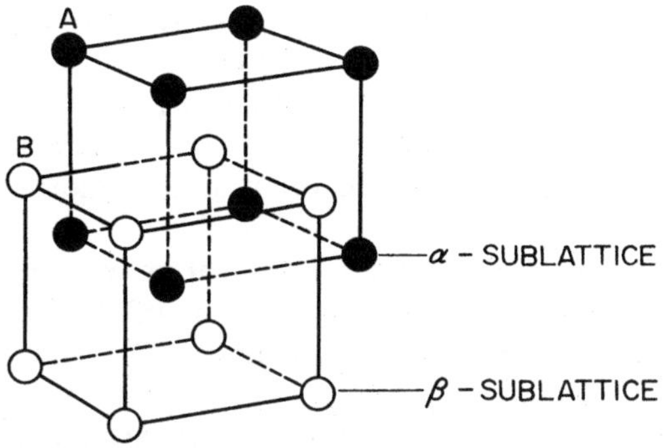

Figure 1-1. B2 structure

where x_i is the atomic concentration of i element and m_i is the mass per atom of i element, which is given by dividing the mass per mole of i element by Avogadro's number.

Both constitutional and thermal vacancies are involved in the concentration given by equation 1.1. The increase in the thermal vacancy concentration can be obtained from thermal expansion measurements (Simmons and Balluffi 1960). If the vacancy concentration at the reference temperature is negligible and the higher order terms containing $\delta l/l$ and $\delta a/a$, which are the incremental changes of the length and lattice parameter between the reference temperature and temperature of interest, T, can be neglected, the thermal vacancy concentration, c_{th}, at T can be given as

$$c_{th} = 3(\delta l/l - \delta a/a) \tag{1.3}$$

which has been commonly used for the determination of thermal vacancies. However, the higher order terms may not be negligible for VIII-IIIA B2 compounds such as NiAl and CoGa with a high vacancy concentration (Neumann and Chang 1979; Van Ommen et al. 1981). The increase in the thermal vacancy concentration can thus be written as (Van Ommen et al. 1981; Van Ommen 1982):

$$\delta c_{\text{th}} = (1 + c)\{(1 + \delta l/l)^3(1 + \delta a/a)^{-3} - 1\} \tag{1.4}$$

where c is the vacancy concentration given by equation 1.1 at the reference temperature. Figure 1-2 shows variations of lattice parameter and macroscopic length as a function of temperature for $Co_{48}Ga_{52}$, $Co_{50}Ga_{50}$ and $Co_{52}Ga_{48}$ (Van Ommen et al. 1981). The figure indicates that no significant difference between $\delta l/l$ and $\delta a/a$ is observed at temperatures lower than about 560°C and thus, as seen from equations 1.1 and 1.2, the thermal vacancy concentration is negligibly small below 560°C, which can be regarded as the onset temperature, T_{th}, of the thermal vacancy formation in CoGa. This indicates that some information on the concentration of constitutional vacancies can be obtained using equation 1.1 and specimens slowly cooled from high temperatures or fully annealed at temperatures lower than T_{th}, where no significant amount of thermal vacancies are formed.

Figure 1-3 shows the vacancy concentration values in NiAl determined by density measurements as a function of Al concentration, x_{Al} (from the review by Chang and Neumann 1982). The solid line is a thermodynamically calculated curve with a very small value of the disorder parameter, that is, a very small concentration of vacancy or antisite atoms at the stoichiometric composition. However, the vacancy concentration is

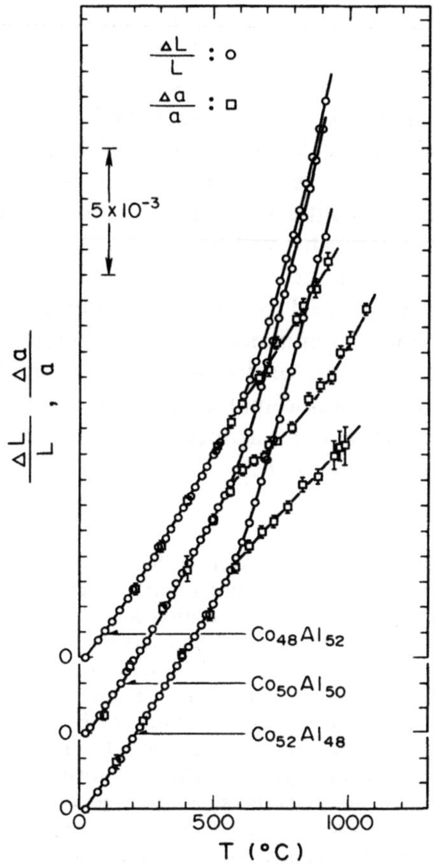

Figure 1-2. Variation of lattice parameter and length with temperature for $Co_{48}Al_{52}$, $Co_{50}Al_{50}$ and $Co_{52}Al_{48}$ (Van Ommen et al. 1981)

seen to be almost zero on the Ni-rich side and the dependence of vacancy concentration on x_{Al} on the Al-rich side is almost linear and its slope is 2. In fact, considering simply that each extra Al atom is compensated by a vacant lattice site of the Ni-sublattice, the concentration of vacancies in a nonstoichiometric compound Ni_xAl_{1-x} ($x < 0.5$) can be given as:

$$c = 1 - 2x = 2x_{Al} - 1 \tag{1.5}$$

Thus, in Ni_xAl_{1-x} with $x < 0.5$, vacancies with a concentration of $1 - 2x$ exist on the Ni-sublattice, while in that with $x > 0.5$ antisite Ni atoms with a concentration of $(2x - 1)/2$ occur on the Al-sublattice. The same trend has been found in CoAl, NiGa, CoGa, and FeAl (for a review, see Chang and Neumann 1982) and recently in RuAl (Fleischer 1993a), although in the case of the former four compounds, in particular, NiGa, CoGa, and FeAl, thermal vacancies are easily quenched-in even in slowly cooled specimens and thereby the vacancy concentration in the slowly cooled specimens if often not zero even on the-transition-metal-rich side ($x > 0.5$) and it gradually increases with increasing Al content even in the composition range of $1 > x > 0.5$. This will be discussed in section 3.

The existence of a large amount of vacancies on the Ni-sublattice in NiAl has been confirmed by positron annihilation measurements.[1] Figure 1-4 shows the dependence of S parameter characterizing the Doppler broadening of the positron-electron annihilation radiation in Ni_xAl_{1-x} compounds (Shimotomai et al. 1985). Since the S parameter increases with increasing the vacancy concentration, the trend of the figure is the same as in Figure 1-3.

---

[1] Positron annihilation measurements. Positrons have the same mass and spin as electrons, but the opposite charge; that is, one positive elementary charge. In conventional positron annihilation experiments, positrons from a positron source, usually ^{22}Na, are injected into the specimen. They are almost immediately thermalized and make a random walk in the specimen. When positron trapping centers such as vacancies exist in the specimen, some of them are trapped at trapping centers during their random walk and then annihilate when they meet electrons. Others annihilate at the interstices of the specimen lattice. The emission of positrons from ^{22}Na and the positron-electron annihilation in the specimen result in the emission of γ rays with different energies. The Doppler broadening of the annihilation γ ray is obtained by measuring the energy of the annihilation γ ray. The S parameter is the area under the energy spectrum in a narrow energy range symmetrical around 511 keV. The time difference between the two different detector signals corresponding to the birth and annihilation γ rays of a positron gives the lifetime of the positron. When some different types of defects that trap positrons exist, the lifetime spectrum $T(t)$ is given as:

$$T(t) = (I_0/\tau_0)\exp(-t/\tau_0) + \sum (I_i/\tau_i)\exp(-t/\tau_i) \tag{1.6}$$

where τ_i is the lifetime of positrons trapped at the i-th type defects and I_i is a measure of the corresponding defect concentration. When no positron-trapping defects exist, τ_0 is identical with the positron lifetime at lattice interstices, τ_l. Thus, analyzing the observed positron lifetime spectrum, various information on the density and properties of positron trapping centers can be obtained. $\tau_i > \tau_l$, since electron density is lower at defects such as vacancies than at interstices. Thus, the average lifetime of positrons given as

$$\tau_{AV} = \sum I_i \tau_i \tag{1.7}$$

increases with increasing the concentration of defects and lifetime at each defect (for reviews on positron annihilation spectroscopy, see Smedskjaer et al. 1983; Triftshäuser 1986; Eldrup 1986; Hautojärvi 1987).

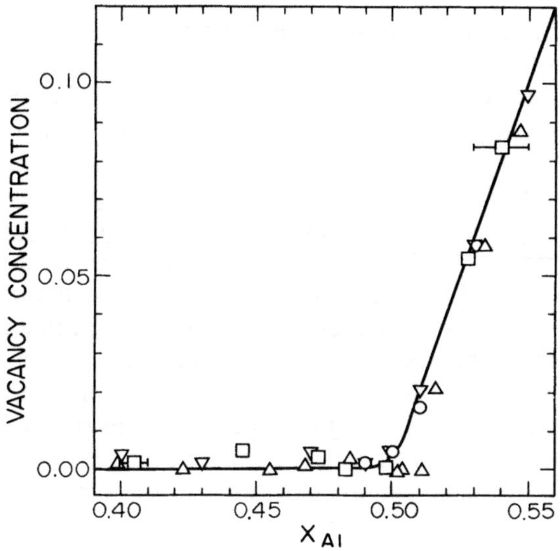

Figure 1-3. Vacancy concentration in NiAl as a function of Al composition (from the review by Chang and Neumann, 1982). The results of the measurements by four different research groups are collectively shown (see the review). The vacancy concentrations were determined by density measurements at room temperature on specimens slowly cooled from high temperatures.

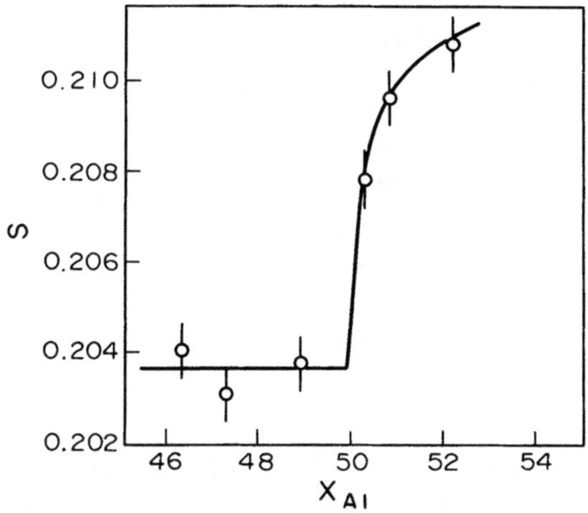

Figure 1-4. Doppler-broadening line shape parameter for positron annihilation in NiAl as a function of Al composition (Shimotomai et al. 1985; reprinted with kind permission World Scientific Publishing Co. Pte Ltd.)

In B2 compounds such as AgMg and CuZn, A or B antisite atoms are believed to be formed on β or α sublattices. Figure 1-5a shows the temperature dependence of the average positron lifetime for several different compounds, including a CuZn compound. The $\tau_{AV} - T$ curves, except that for $CoSi_2$ in Figure 1-5a were obtained by in situ measurements of positron lifetime in specimens thermally equilibrated at temperatures in the range of room temperature to a temperature close to T_m. Positron lifetime in $CoSi_2$ was measured at 100 K using specimens thermally equilibrated at high temperatures and then rapidly cooled to 0°C. The $\tau_{AV} - T$ curves in Figure 1-5b are for specimens which were electron-irradiated and then isochronally annealed. In the figure, τ_{AV} is plotted as a function of isochronal annealing temperature. All the specimens were fully annealed at a high temperature and slowly cooled to room temperature before electron irradiation. The $\tau_{AV} - T$ curves in Figure 1-5a are found to have an S-shape and those for electron irradiated

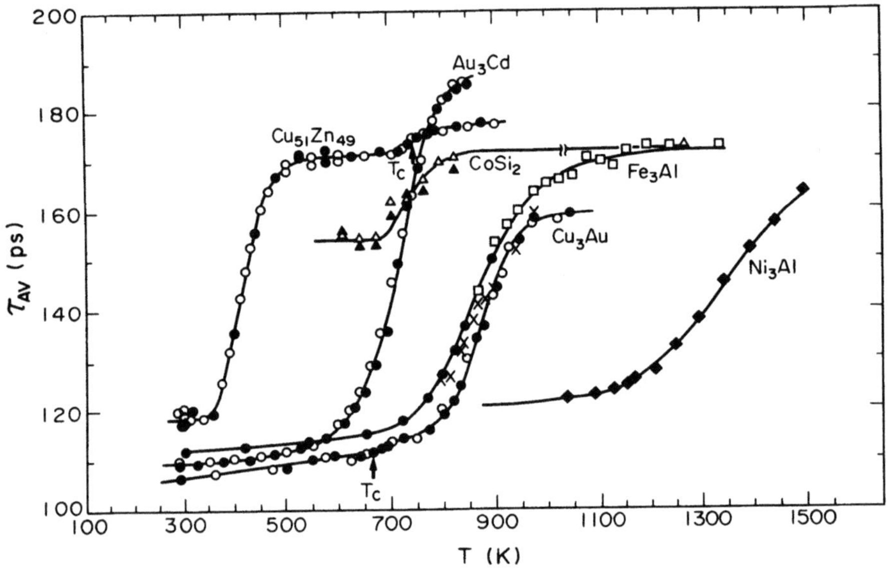

(a)

Figure 1-5a. Temperature dependence of the average positron lifetime in $Cu_{51}Zn_{49}$ (Shirai 1988); Fe_3Al, Ni_3Al (Schaefer et al. 1992), Cu_3Au (Shirai et al. 1989), Au_3Cd and $CoSi_2$ (Ito et al. 1993; Yamaguchi, Shirai and Inui 1993). Different symbols for Fe_3Al: different measuring runs. Open and closed triangles for $CoSi_2$: water quenched and air cooled specimens, respectively; Open and closed circles for $Cu_{51}Zn_{49}$, Cu_3Au and Au_3Cd: heating and cooling runs, respectively; T_C: the order-disorder transition temperature.

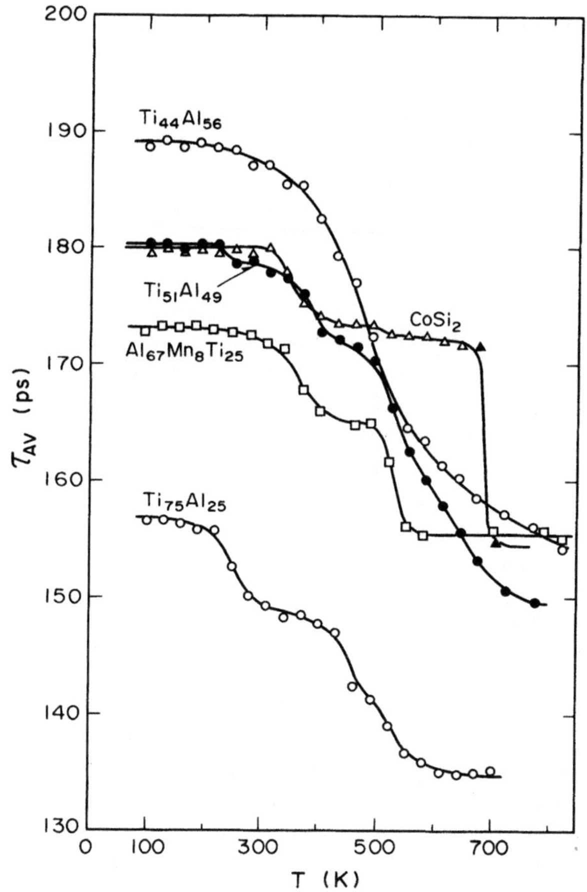

(b)

Figure 1-5b. Annealing temperature dependence of the average positron lifetime in electron irradiated $Ti_{44}Al_{56}$, $Ti_{51}Al_{49}$ (Shirai and Yamaguchi 1992), $Al_{67}Mn_8Ti_{25}$ (Shirai, Masaki, and Yamaguchi 1994), $CoSi_2$ (Ito et al. 1993) and Ti_3Al. Lifetime measurements were made at 100 K or 290 K after each isochronal annealing. Solid triangles for $CoSi_2$ indicate that corresponding specimens were furnace cooled after the corresponding isochronal annealing.

compounds in Figure 1-5b possess an inverse S-shape. The room-temperature positron lifetime in each compound in Figure 1-5a and the lifetime in each compound at fully annealed state in Figure 1-5b are reasonably close to the lifetime values expected for the positron-electron annihilation at lattice interstices in corresponding compounds. This clearly indicates that vacancies are thermally formed at high temperatures or by electron irradiation, otherwise only a very small number of vacancies exist ($< 10^{-6}$, i.e., the detection limit by positron annihilation techniques). If constitutional vacancies exist, whose concentration is usually greater than 10^{-3}, almost all positrons are trapped and annihilate at these vacancies and thus practically no change in positron lifetime should be observed before and after heating, quenching or electron irradiation, which can introduce only a muchsmaller amount of vacancies (about 10^{-4}). It is thus clear that no constitutional vacancies exist in the compounds shown in Figures 1-5a and 1-5b.

Now we return to the case of CuZn. Although Figure 1-5a includes only the result for the Cu-rich $Cu_{51}Zn_{49}$ (Shirai 1988), the same trend was found in Zn-rich Cu_xZn_{1-x} ($x < 0.5$) (Schultz and MacKenzie 1982). We may thus draw the conclusion for CuZn that Cu or Zn antisite atoms are formed on Zn or Cu sites and that no structural vacancies are formed on both sides of stoichiometry. The onset temperature, T_{th}, where an appreciable positron trapping at thermal vacancies occurs in the Cu-rich $Cu_{51}Zn_{49}$ is about 370 K, and thus the relative onset temperature T_{th}/T_m is about 0.33, which is much smaller than the corresponding value (i.e., around 0.6) of many metals. Thermal vacancies are found to be easily formed in this compound, as has been also reported by Chabik and Rozenfeld (1981).

2.1.2 DO_3 Structure. Positron lifetime measurements have been done on Fe_3Al (Schaefer, Würschum and Bub 1992) and Fe_3Si (Würschum et al. 1991). The result for Fe_3Al is presented in Figure 1-5a. The existence of constitutional vacancies is unlikely from the same reason as in CuZn. T_{th} is about 750K. The relative onset temperature, T_{th}/T_m is thus as low as 0.42, similar to CuZn. This suggests that the vacancy formation enthalpy in Fe_3Al is also rather low and the equilibrium vacancy concentration is high. In general, the values of the vacancy formation enthalpy for superlattice compounds with b.c.c.-derivative superlattice structures such as B2 and DO_3 are quite low in comparison to the corresponding values for transition metal constituent elements (Schaefer, Würschum and Bub 1992).

2.2 Superlattice Compounds with f.c.c.-Derivative Structures

2.2.1 Compounds with the $L1_2$ and its derivative structures. In this class of compounds, only antisite atoms are believed to occur without constitutional vacancies on both sides of stoichiometry. This has been confirmed in Ni_3Al, which is one of the most typical $L1_2$ compounds, with a wide range of compositions on both sides of stoichiometry by density measurements (Aoki and Izumi 1975) and in Ni_3Al with Ni-rich compositions by positron annihilation techniques (Wang, Shimotomai, and Doyama 1984; Schaefer et al. 1992, see Figure 1-5a). The large amount of vacancies also was not found in well-ordered Cu_3Au (Kim 1987a; Shirai et al. 1989, see Figure 5a). Some theoretical calculations also predict that only antisite atoms occur in Ni_3Al, without constitutional vacancies on both sides of stoichiometry (Foiles and Daw 1987; Kim 1991). The relative onset temperature of thermal vacancy formation is 0.71 for Ni_3Al (Schaefer et al. 1992) and 0.64 for Cu_3Au (Shirai et al. 1989). This is in sharp contrast to the thermal vacancy formation in compounds with b.c.c.-derivative structures. The equilibrium concentration and formation enthalpy of thermal vacancies in $L1_2$ compounds are close to those of pure transition metal constituents (Schaefer et al. 1992). The effect of defect structure on the mechanical properties of Ni_3Al has been summarized in the reviews by Liu and White (1985) and Liu (1987).

Recently, some information on the defect structure of Al_3Ti with a structure of the $D0_{22}$ type, which is one of the $L1_2$-derivative long-period structures, and its $L1_2$ variations has been obtained by positron annihilation techniques and electrical resistivity measurements. As seen from the variation of the average positron lifetime in an electron-irradiated Mn-modified $L1_2$ trialuminide with annealing temperature (Figure 1-5a), the lifetime value in the fully annealed compound is 156 psec and it increases to 173 psec after electron irradiation (Shirai, Masaki, and Yamaguchi 1994). The average positron lifetimes in Al_3Ti with the $D0_{22}$ structure and various other $L1_2$ variations with Cr, Mn, Fe, Ni, Cu, Pd and Ag were also measured at fully annealed state and found to be reasonably close to the corresponding lifetime value of the Mn-modified $L1_2$ compound. This indicates that constitutional vacancies are unlikely to be formed in the Al_3Ti and related cubic trialuminides. The ternary alloying elements are added to substitute for Al atoms. However, they may replace not only Al but also Ti atoms and, on top of that, a large number of antisite Ti and Al atoms may be formed (Porter et al. 1989; Kogachi, Minamigawa, and Nakahigashi 1992; Kogachi and Kameyama 1993). Such defects are expected to exert a significant influence on the electrical resistivity of this class of materials. In fact, in comparison to the $D0_{22}$ Al_3Ti, its $L1_2$ variations generally exhibit much larger residual resistivities at 4.2 K (Figure 1-6a) and much weaker and often negative temperature dependence (Figure 1-6b). The same trend has been found in the $L1_2$ variations of Al_3Zr (Masaki 1993). Such large residual resistivity and anomalous temperature dependence of resistivity have been partly interpreted in terms of a very small mean free path for electrons, caused by the existence of a large number of antisite and substitutional defects (Mooij 1973; Nishitani et al. 1991, Yamaguchi and Inui 1993). Further work on the

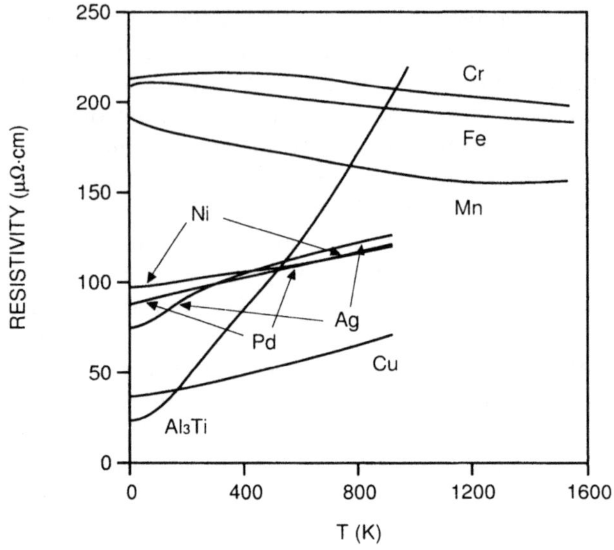

(a)

Figure 1-6a. Electrical resistivity

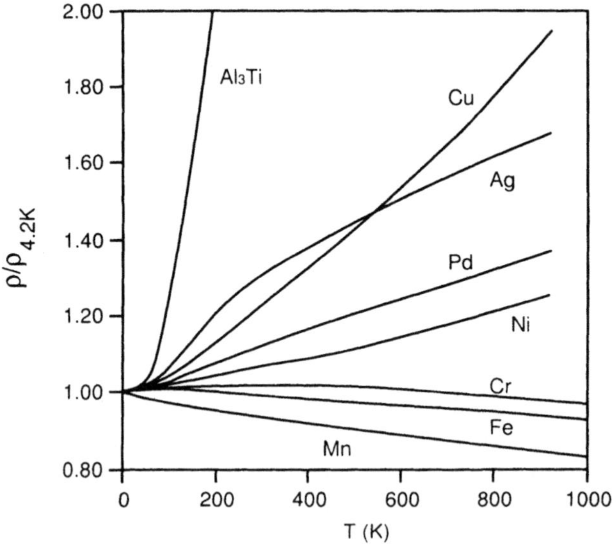

(b)

Figure 1-6b. Electrical resistivity Ratio, $\rho/\rho_{4.2K}$ as a function of temperature for Al_3Ti and its $L1_2$ variations (Nishitani et al. 1991; Masaki 1993). Cr: $Al_{67}Cr_8Ti_{25}$; Mn: $Al_{67}Mn_8Ti_{25}$; Fe: $Al_{67}Fe_7Ti_{26}$; Ni: $Al_{65}Ni_9Ti_{26}$; Cu: $Al_{63}Cu_{11}Ti_{26}$; Pd: $Al_{65}Pd_9Ti_{26}$; Ag: $Al_{64}Ag_9Ti_{27}$.

defect structure and electron band structure in Al_3Ti and its $L1_2$ variations is required for more complete understanding of the results shown in Figures 1-6a and 1-6b. Some relevant studies are presently in progress. Collective information on the $L1_2$ phase field in various Al-Ti-X systems is available in the review by Kumar (1990) and recent papers by Nic, Klansky, and Mikkola (1992) and by Nakayama and Mabuchi (1993).

Figure 1-5 includes a result of measurements of positron lifetime in Au_3Cd as a function of temperature. No large amount of vacancies are found in the fully annealed and ordered Au_3Cd with a structure of the $D0_{23}$ type, also one of the $L1_2$-derivative long-period structures.

2.2.2 $L1_0$ Compounds. In this class of compounds, TiAl has been attracting much attention because of its potential as a new high-performance, high-temperature structural material with a high strength-to-weight ratio. Vujic, Li, and Whang (1988) confirmed by their density measurements that anti-site atoms are predominant defects in Al-rich TiAl compounds. Shirai and Yamaguchi (1992) recently investigated the decaying behavior of vacancies introduced in TiAl compounds containing 49 at%Al, 52 at%Al and 56 at%Al by electron or proton irradiation using positron lifetime spectroscopy, and found that all the irradiated specimens recover to the fully annealed state at about 800 K. Figure 5b includes the results for 49 and 56 at%Al specimens irradiated with 10 MeV electrons to a dose of about 5×10^7 e^-/cm^2 at temperatures lower than 160 K. The TiAl alloys with Ti-rich compositions exhibit a two-phase microstructure composed of TiAl and Ti_3Al phases. However, since the volume fraction of the Ti_3Al phase in the specimen with 49%Al is small, there is no doubt that information on the positron lifetime in the specimen is from the main constituent phase, that is, TiAl which is expected to be Al-rich (Mohandas and Beaven 1991). The average positron lifetime of the two specimens decreases with increasing annealing temperature. This indicates that no constitutional vacancies exist on both sides of stoichiometry and at the same time the equilibrium concentration of thermal vacancies in TiAl is rather small. The $\tau_{AV} - T$ curves of the two TiAl specimens also indicate that the migration of vacancies occurs rather easily in TiAl. The recovery starts around room temperature, indicating that vacancies in TiAl can migrate even around room temperature.

Recently, the defect structure in TiAl has been theoretically investigated based on the internal energies of vacancies and anti-site atoms determined from first-principles calculation by Fu and Yoo (1992), Fu, Ye, and Yoo (1993), and Yoo and Fu (1993). Their calculations give a vacancy concentration at stoichiometry of 10^{-6}–10^{-7} even at 1500 K. However, the antisite defect concentration is predicted to be larger than 10^{-4} above 1000 K. Figure 1-7 shows a result of their calculations showing the

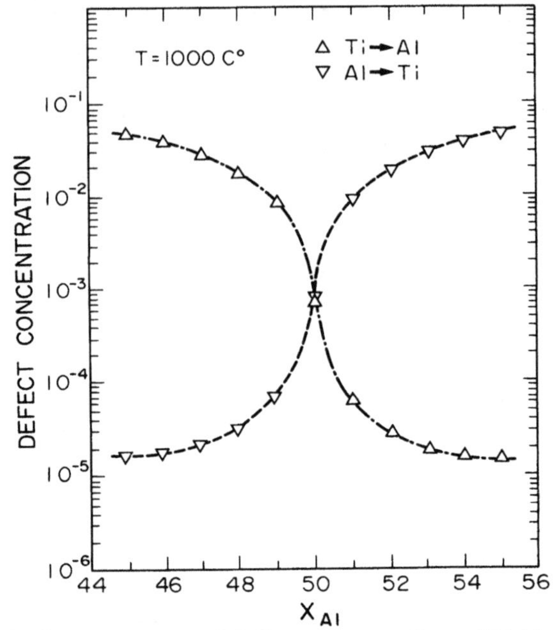

Figure 1-7. Defect structure and defect concentrations of TiAl as a function of Al composition at 1000°C (Yoo and Fu, 1993).

dependence of the defect structure and defect concentrations in TiAl at 1300 K on the atomic concentration of Al. The implications of these results for the mechanical properties, in particular, of TiAl at high temperatures have yet to be clarified.

2.3 Compounds with Other Structures

Studies of the defect structure of intermetallic compounds with crystal structures other than b.c.c.- and f.c.c.-derivative superlattice structures are rather scarce. In this section, we briefly review the recent studies on the defect structure of Ti_3Al with a structure of the $D0_{19}$ type and $CoSi_2$ with a cubic structure of the C1 type. Ti_3Al is one of the three intermetallic compounds occurring in the Ti-Al system and its mechanical properties have been intensively studied since it is a base material of advanced Ti alloys (see the chapter on titanium aluminides), as well as the secondary phase of the two-phase TiAl alloys which often plays a decisive role in the deformation of the two-phase TiAl alloys (Inui, Toda, and Yamaguchi 1993; for a review see Yamaguchi and Inui 1993). As seen from the result of positron lifetime measurements for an electron irradiated Ti_3Al in Figure 1-5b, no constitutional vacancies are likely to be formed in this compound, similar to other intermetallic compounds in the same figure.

$CoSi_2$ is one of the transition metal disilicides that have been receiving much attention not only in Si device applications because of its low electrical resistivity and high-temperature stability (Murarka 1983), but also in the field of high-temperature structural materials since it has a low density, some deformability, and excellent oxidation resistance (Takeuchi, Hashimoto, and Shibuya 1992; Suzuki and Takeuchi 1993; Ito et al. 1992, 1994). As seen from the results shown in Figures 1-5a and b, no large amount of vacancies exists in this compound at low temperatures. Of interest is that the onset temperature of thermal vacancy formation is 675 K and thus T_{th}/T_m is as low as 0.42. The vacancy formation enthalpy is 1.1 eV (Ito et al. 1993) and the ratio of the enthalpy to T_m is close to that for Fe_3Al. This is rather surprising in view of the nature of atomic bonding, which has high covalency as well as metallicity. When $CoSi_2$ is rapidly cooled from high temperatures, a large number of thermal vacancies is quenched-in and their migration, which can occur even at room temperature, results in the formation of secondary defects such as small dislocation loops (Ito et al. 1993; Yamaguchi, Shirai, and Inui 1993; Suzuki and Takeuchi 1993). If the disilicides of transition metals exhibit the same trend as in $CoSi_2$, this finding might have important implications for their Si device applications.

2.4. Uniqueness of VIII-IIIA B2 Compounds in Their Defect Structure

The aforementioned survey of the studies on the defect structure of intermetallic compounds indicates that VIII-IIIA B2 compounds are unique in the formation of constitutional vacancies on the IIIA-element-rich side. This uniqueness has been interpreted in terms of their bonding nature. Recently, Fu and Yoo (1992) calculated the electronic structure and bonding charge density of Ni aluminides as prototypical examples of transition metal aluminides and found that the bonding mechanism of such aluminides can be best described by the combination of charge transfer and strong p(aluminum)-d(transition metals) hybridization effects. Al atoms in transition metal aluminides are electropositive and a depletion of electron density at the Al sites is accompanied by a significant buildup of the directional d-bonding charge along the nearest neighbor Ni-Al direction. On the other hand, Ni atoms are metallically bonded since nonbonding (between Ni-Al) d electrons localized at the Ni sites decrease and delocalized electrons increase at Ni-Ni interstitial regions

(Fu and Yoo 1992). Fox and Tabbernor (1991) also reported that bonding in NiAl is achieved by a similar mechanism based on their low-angle structure factor measurements by high energy electron diffraction.

Thus, a repulsion occurs when electropositive Al atoms substitute for Ni atoms and are brought into nearest neighbors in nonstoichiometric compounds with $x < 0.5$, whereas in those with $x > 0.5$, Ni antisite atoms are in favor of metallic bonding between Ni atoms (Fox and Tabbernor 1991; Fu and Yoo 1992). This explains why Al antisite atoms are not tolerated on the Ni sublattice, but Ni antisite atoms are on the Al sublattice. The uniqueness of VIII-IIIA B2 compounds in their defect structure should thus be attributed to the characteristics of their electronic structure and bonding charge density as well as the atomic size difference.

3. THERMAL DEFECTS

As mentioned in the previous sections, the onset temperature and formation enthalpy of thermal vacancies for intermetallic compounds with b.c.c.-derivative structures are generally low. The equilibrium concentration of thermal vacancies is thus high. Theoretical calculations of equilibrium concentration of vacancies and antisite atoms have been made in these compounds, in particular, those with the B2 structure (Neumann, Chang, and Lee 1976; Bakker and Van Ommen 1978; Chang and Neumann 1982; Kim 1986, 1987b; Fu and Yoo 1992). In this section, we briefly consider the equilibrium concentration of vacancies and antisite atoms in B2 compounds formed between A element of VIII and B element of IIIA with a composition of Ax_ABx_B, where $x_A + x_B = 1$.

Consider first the case of stoichiometric composition. The thermal vacancy concentration begins to increases with increasing temperature at temperatures higher than T_{th} as shown in Figures 1-2 and 1-5a. However, the increase in the vacancy concentration has to be achieved by forming vacancies only on the α (A element) sublattice and at the same time without disturbing the stoichiometric ratio. In order to fulfill the two requirements, three defects, that is, two vacancies on α sites and one antisite atom of A element on a site of the β (B element) sublattices, must be formed simultaneously. Such a set of three defects is called a triple defect (Wasilewski 1968; Neumann et al. 1976). Neumann et al. (1976) derived theoretical expressions for the thermodynamic properties and equilibrium concentration of thermal defects based on the triple defect concept and the Bragg–Williams approximation. The reader is referred to the extensive review by Chang and Neumann (1982) for their calculations. Bakker and Van Ommen (1978) also derived similar expressions and made a comparison of the two results.

Assuming only triple defects are formed, Neumann et al. (1976) derived the following expressions presenting the concentrations of vacancies, c_V and antisite atoms, c_A;

$$c_V^2(c_V - 2\chi) = \alpha^3 \tag{1.8a}$$

$$c_A = (c_V - 2\chi)/2 \tag{1.8b}$$

where α is the disorder parameter given by the vacancy concentration at stoichiometry and χ is the deviation from stoichiometry, $\chi = 0.5 - x_A = x_B - 0.5$. When α is zero, $c_V = 0$ and $c_A = x_A - 0.5$ for $x_A > x_B$ and $c_V = 2x_B - 1$ and $c_A = 0$ for $x_A < x_B$. This is almost the case for NiAl as shown in Figure 1-3 and in agreement with the conventional concept of constitutional defects in VIII-IIIA B2 compounds such as NiAl. However, the disorder parameter α, in general, increases with increasing temperature, that is, triple defects are formed even at stoichiometry, and thus the concentration of vacancies given by equation 1.8a increases monotonically with increasing x_B and temperature as seen in Figure 1-8 showing vacancy concentrations of FeAl as a function of x_{Al} at several different temperatures (Chang et al. 1993). This triple defect mechanism is believed to be

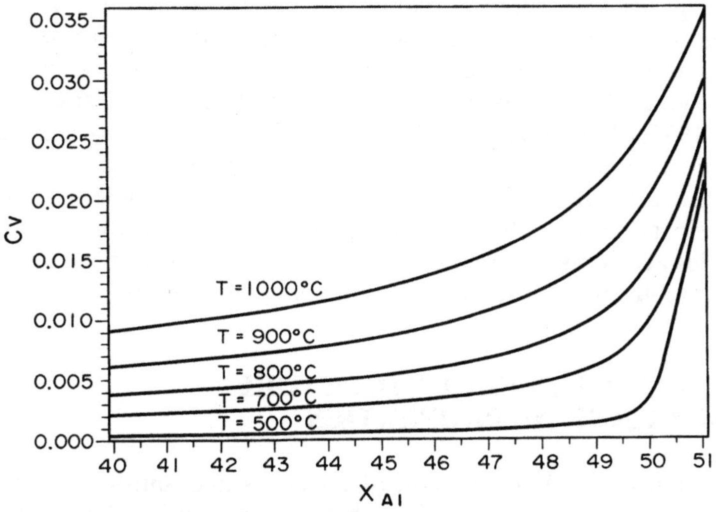

Figure 1-8. Vacancy concentration in FeAl as a function of Al composition at several temperatures (Reprinted from *Intermetallics*, Volume 1, Y. A. Chang, L. M. Pike, C. T. Liu, A. R. Bilbrey, and D. S. Stone, Correlation of the hardness and vacancy concentration in FeAl, pages 107–115, copyright 1993, with kind permission from Elsevier Science Ltd. The Boulevard, Langford Lane, Kidlington OX5 1GB, UK)

responsible for thermal vacancy concentrations in VIII-IIIA B2 compounds that are a factor of a thousand larger than those in pure metals at equivalent temperatures.

Fleischer (1993a) has recently measured the vacancy concentration in Co_xAl_{1-x} with a wide range of x and found increasing vacancy concentrations with increasing Co content for $x > 0.5$. Such a trend of the composition dependence of vacancy concentration may not be interpreted in terms of the triple defect mechanism. In order to explain such a trend under the framework of the triple defect mechanism, the bond energy between α and β sites determining the energy of triple defect formation may need to be modified so that it depends on composition, although it is assumed to be independent of composition and temperature in the triple defect mechanism (Fleischer 1993b).

The vacancy concentration in VIII-IIIA B2 compounds may reach as high as about 10^{-2}, as shown in Figure 1-8. Such a large amount of vacancies may be locally ordered to avoid the same species at the first nearest neighbor distance of their sublattice (Georgopoulos and Cohen 1981). Such local ordering of vacancies may lead to the formation of phases adjacent to the B2 phase of interest in the corresponding VIII-IIIA system and may thus affect the stability of the B2 phase with respect to the phases adjacent to the B2 phase. Upon quenching from high temperatures, high vacancy supersaturations are produced in the VIII-IIIA compounds. The annealing-out of such supersaturated vacancies occurs by the formation of dislocation loops and microvoids and by the climb of dislocations (e.g., Ball and Smallman 1968; Fraser et al. 1975; Yang, Dodd, and Strutt 1972; Parthasarathi and Fraser 1984; Schwab and Gerold 1980). Such secondary defects often play an important role in the deformation of VIII-IIIA B2 compounds. Yang et al. (1972), for example, have reported that dislocation loops formed during the annealing-out of supersaturated vacancies act as dislocation sources and thereby NiAl with such dislocation loops exhibits greater deformability at room temperature.

4. EFFECT OF DEFECT STRUCTURE ON MECHANICAL PROPERTIES

The existence of a large amount of vacancies and antisite atoms exerts a considerable influence on the mechanical properties of B2 compounds. Such defects strengthen the B2 compounds on both sides of stoichiometry at low temperatures, while at high temperatures they soften the compounds because the same defects increase the diffusion rates and even change the diffusion mechanism (Stolwijk, Van Gend, and Bakker 1980;

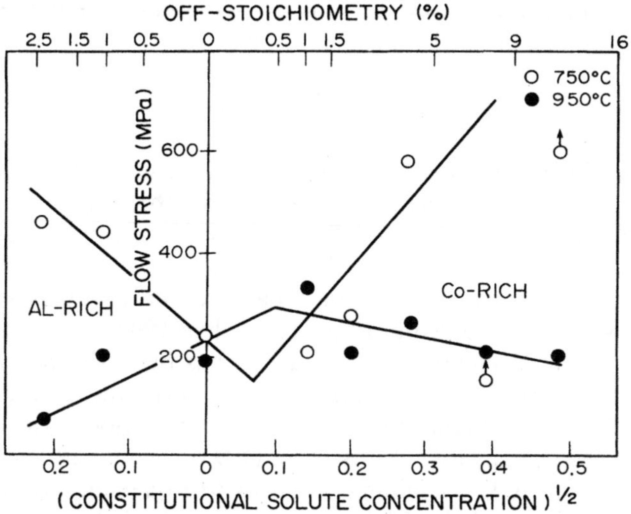

Figure 1-9. Flow stress in compression as a function of off-stoichiometry for CoAl at 750°C and 950°C (Fleischer 1993c)

Kao and Chang 1993). In this section, we briefly consider the defect concentration dependence of the strength of NiAl, CoAl, and FeAl, which have been attracting attention as high-temperature structural materials. The effect of vacancies and antisite atoms on the indentation hardness of NiAl was well established in the 1950s (see the review by Westbrook, 1960). A significantly greater slope of the hardness versus composition curves was observed on the Al-rich side. This suggests that vacancies are more potent hardener than antisite Ni atoms. The elastic constants of NiAl also are considerably affected by supersaturated thermal vacancies. All elastic stiffnesses except the shear constant $(C_{11} - C_{12})/2$ were reported to decrease after quenching (Rusovic and Henig 1980). The reader also is referred to the extensive review by Miracle (1993) for general information on the physical and mechanical properties of NiAl. The recent research and development of NiAl alloys for high-temperature structural applications have been intensively reviewed by Darolia (1991, 1993).

Figure 1-9 shows the flow-stress variations with composition at 750°C ($0.53T_m$) and 950°C ($0.64T_m$) for CoAl (Fleischer 1993c). The flow stress is plotted as a function of square root of defect concentration whose dependence on off-stoichiometry was determined by measurements of x-ray and macrodensities. The low-temperature effect on strength follows a square-root-of-the-concentration relation proposed by Fleischer (1962), which is the most commonly observed in solution hardening. In dilute alloys of metals, the amount of hardening, $\Delta\tau$, is given as:

$$\Delta\tau = G\varepsilon^{3/2}c^{1/2}/\alpha \tag{1.9}$$

where G is shear modulus, α is a material-sensitive number which is about 700, c is solute concentration, ε is a weighted sum of ε_b and ε_g which are the size and modulus misfits given as $d(\ln a)/dc$ and $d(\ln G)/dc$, where a is the lattice parameter. Hardening in CoAl is reported to be consistent in magnitude with conventional size and modulus interactions with vacancies and Co antisite atoms (Fleischer 1993c). The same would be true for other B2 compounds. In fact, it has been shown by Fleischer (1993a) that the results of the flow stress measurements for AgMg by Wood and Westbrook (1962) follow well the square-root-of-the-concentration relation. More recently, Chang et al. (1993) have made microhardness measurements on FeAl containing 40–51 at%Al quenched from temperatures in the range of 500–1000°C and plotted the hardness values as a function of vacancy concentration, c_V, calculated as in Figure 1-8 (Figure 1-10). The square-root-of-the-concentration relation is again followed ex-

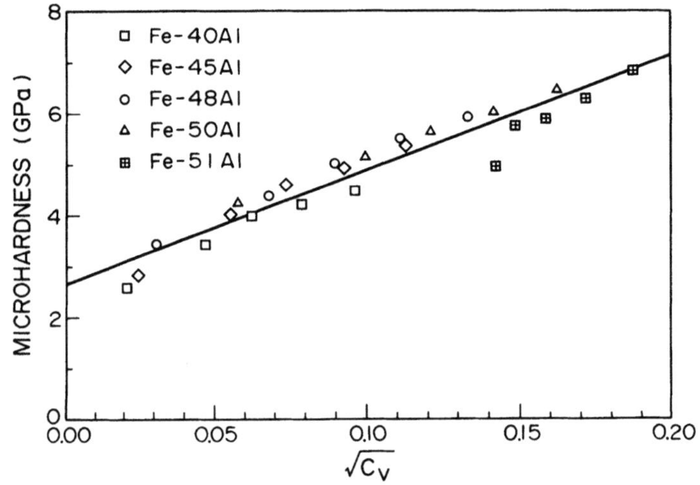

Figure 1-10. Microhardness of FeAl as a function of the square root of vacancy concentration (Reprinted from *Intermetallics*, Volume 1, Y. A. Chang, L. M. Pike, C. T. Liu, A. RF. Bilbrey, and D. S. Stone, Correlation of the hardness and vacancy concentration in FeAl, pages 107–115, copyright 1993, with kind permission from Elsevier Science Lyd, The Boulevard, Langford Lane, Kidlington OX5 1GB, UK)

tremely well. It has been known that the mechanical properties of FeAl is sensitive to annealing temperature and cooling rate (Nagpal and Baker 1990; Crimp, Vedula, and Gaydosh 1987; Gaydosh et al. 1992, Liu and Kumar 1993). This may be interpreted in terms of the quenched-in vacancies and their effects on the mechanical properties of FeAl. The reader also is referred to the recent reviews by Vedula and Stephens (1987) and by Baker and Munroe (1990) for the effects of defect structure on the mechanical properties of B2 compounds and to the review by McKamey et al. (1991) for Fe_3Al where imperfect B2 order is often left after heat treatments.

5. CONCLUDING REMARKS

In this chapter, we have reviewed a wide cross-section of the studies on the defect structure of intermetallic superlattice compounds. VIII-IIIA compounds with the B2 structure are unique in the formation of constitutional vacancies on the IIIA-element-rich side of stoichiometry. The vacancy concentration in these B2 compounds reaches as high as an order of 10^{-2} due to the formation of triple defects. Such a high vacancy concentration exerts serious influences on their mechanical properties, in particular, high-temperature strength and creep resistance. Lack of resistance of the formation of antisite atoms on the VIII element sublattice has been interpreted in terms of electronic reasons as well as the atomic size difference. However, more work is needed to clarify the reasons for the unique defect structure of VIII-IIIA B2 compounds, in particular, those with ternary alloying elements. In fact, some solute atoms are reported to increase the vacancy concentration in CoAl (Fleischer 1993b), although the reason is presently unknown. Improving our understanding of the defect structure would be extremely useful for further development of high-temperature materials based on VIII-IIIA B2 compounds such as NiAl, CoAl and FeAl.

The large amount of vacancies soften the VIII-IIIA B2 compounds at high temperatures because they increase the diffusion rates. The limited high-temperature creep resistance is one of the major reasons which restrict their high-temperature structural applications. However, systematic information on the diffusion coefficients for this class of compounds are scarce, in particular, for those with nonstoichiometric compositions. More diffusion measurements and more work on diffusion mechanisms are required. The reader is referred to the reviews by Bakker et al. (1992), Koiwa (1992), Frohberg and Wever (1992) and Vogl et al. (1992) for the current status of studies on diffusion in intermetallic compounds.

ACKNOWLEDGEMENTS

This work was supported by Grant-in-Aid for Scientific Research on the priority area "Intermetallic Compounds as New High Temperature Structural Materials" from The Ministry of Education, Science, and Culture, Japan, and in part by the research grants from New Energy and Industrial Technology Development Organization and the Research and Development Institute of Metals and Composites for Future Industries.

REFERENCES

AOKI, A., and IZUMI, O., 1975, *Phys. Stat. Sol.* (a), 32: 657–664.

BALL, A., and SMALLMAN, R. E., 1968, *Acta Metall.* 16: 233–241.

BAKER, I., and MUNROE, P. R., 1990, In *High Temperature Aluminides and Intermetallics*, ed. S. H. Whang, C. T. Liu, D. P. Pope, and J. O. Stiegler, 425–452. Warrendale, PA: TMS.

BAKKER, H., and VAN OMMEN, A. H., 1978, *Acta Metall.* 26: 1047–1053.

BAKKER, H., LO CASCIO, D. M. R., and DI, L. M., 1992, In *Ordered Intermetallics—Physical Metallurgy and Mechanical Behaviour*, ed. C. T. Liu, R. W. Cahn, and G. Sauthoff, 433–448. Dordrecht: Kluwer Academic Pub.

CHABIK, S., and ROZENFELD, B., 1981, *Appl. Phys.* 25: 143.

CHANG, Y. A., and NEUMANN, J. P., 1982, *Prog. Solid St. Chem.* 14: 221.

CHANG, Y. A., PIKE, L. M., LIU, C. T., BILBREY, A. R., and STONE, D. S., 1993, *Intermetallics* 1: 107.

CRIMP, M. A., VEDULA, K. M., and GAYDOSH, D. J., 1987, In *High Temperature Ordered Intermetallic Alloys II*, MRS Symp. Proc. 81: 499–504.

DAROLIA, R., 1991, *J. Metals* 43(3): 44.

________, 1993, In *Structural Intermetallics*, ed. R. Darolia, J. J. Lewandowski, C. T. Liu, P. L. Martin, D. B. Miracle, and M. V. Nathal, 495–504, Warrendale, PA: TMS.

ELDRUP, M., 1986, In *Defects in Solids*, ed. A. V. Chadwick, M. Terenzi, p. 145, New York and London: Plenum Press.

FLEISCHER, R. L., 1962, *Acta Metall.* 10: 835–842.

________, 1993a, *Acta Metall.* 41: 863–869; 1197–1205.

________, 1993b, *J. Mat. Res.* 8: 49–58.

________, 1993c, *J. Mat. Res.* 8: 59–67.

FOILES, S. M., and DAW, M S., 1987, *J. Mat. Res.* 2: 5–15.

FOX, A. G., and TABBERNOR, M. A., 1991, *Acta Metall.* 39: 669–678.

FRASER, H. L., LORETTO, M. H., SMALLMAN, R. E., and WASILEWSKI, R. J., 1975, *Phil. Mag. A*32: 873–879.

FROHBERG, G., and WEVER, H., 1992, In *Ordered Intermetallics—Physical Metallurgy and Mechanical Behaviour*, ed. C. T. Liu, R. W. Cahn, and G. Sauthoff, 485–496, Dordrecht: Kluwer Academic Publ.

Fu, C. L., Ye, Y. Y., and Yoo, M. H., 1993, In *High Temperature Ordered Intermetallic Alloys V*, Pittsburgh: MRS Symp. Proc. 288: 21–32.

Fu, C. L., and Yoo, M. H., 1992, *Mater. Chem. Phys*. 32: 25.

Gaydosh, D. J., Draper, S. L., Noebe, R. D., and Nathal, M. V., 1992, *Mat. Sci. Eng*. A150: 7–.

Georgopoulos, P., and Cohen, J. B., 1981, *Acta Metall.* 29: 1535–1551.

Hautojävri, P., 1987, In *Characterization of Defects in Materials*, MRS Symp. Proc. 82: 1.

Inui, H., Tota, Y., and Yamaguchi, M., 1993, *Phil. Mag A* 67: 1315–1332.

Ito, K., Inui, H., Hirano, T., and Yamaguchi, M., 1992, *Mat. Sci. Eng*. A152: 153–159.

Ito, K., Inui, H., Hirano, T., and Yamaguchi, M., 1994, *Acta Metall.* In press.

Ito, Y., Shirai, Y., Yamada, Y., and Yamaguchi, Y., 1993, In *High Temperature Ordered Intermetallic Alloys V*, MRS Symp. Proc. 288: 275–280.

Kao, C. R., and Chang, Y. A., 1993, *Intermetallics* 1: 237–251.

Kim, S. M., 1986, *Phys. Rev. B* 33: 1509–1511.

Kim, S. M., 1987a, *Mater. Sci. Forum* 15–18: 1257; 1987b, *Mater. Sci. Forum* 15–18: 1323.

________, 1991, *J. Mat. Res*. 6: 1455–1460.

Koehler, J. S., and Seitz, F., 1947, *J. Appl. Mech*. 14: A217.

Kogachi, M., and Kameyama, A., 1993, *Scripta Metall.* 29: 1329–1334.

Kogachi, M., Minamigawa, S., and Nakahigashi, K., 1992, *Scripta Metall.* 27: 407–412.

Koiwa, M., 1992, In *Ordered Intermetallics—Physical Metallurgy and Mechanical Behaviour, ed. C. T. Liu, R. W. Cahn, and G. Sauthoff*, 449–464, Dordrecht: Kluwer Academic Publ.

Kumar, K. S., 1990, *International Materials Reviews* 35: 293–.

Kurnakov, N. S., Zhemchuznii, S. F., and Zasedatelev, M., 1916, *J. Inst. Metals* 15: 305.

Liu, C. T., 1987, In *High Temperature Ordered Intermetallic Alloys II*, MRS Symp. Proc., 81: 355–367.

Liu, C. T., and Kumar, K. S., *J. Metals* 45 (5): 38.

Liu, C. T., and White, C. L., 1985, In *High Temperature Ordered Intermetallic Alloys*, MRS Symp. Proc., 39: 365–380.

Marcinkowski, M. J., 1963, In *Electron Microscopy and Strength of Crystals*, ed. G. Thomas and J. Washburn, New York: Interscience Publishers, p. 333.

Marcinkowski, M. J., Brown, N., and Fisher, R. M., 1961, *Acta Metall.* 9: 129–137.

Masaki, K., 1993, Feb., M Sc Thesis, Kyoto University.

McKamey, C. G., Devan, J. H., Tortorelli, P. F., and Sikka, V. K., 1991, *J. Mat. Res*. 6: 1779–1805.

Miracle, D. R., 1993, *Acta Metall.* 41: 649–684.

Mohandas, E., and Beaven, P. A., 1991, *Scripta Metall* 25: 2023–2027.

Mooij, J. H., 1973, *Phys. Stat. Sol.* (a), 17: 521.

MURARKA, S. P., 1983, *Silicides for VLSI Applications*, New York: Academic Publishers.

NAGPAL, P., and BAKER, I., 1990, *Metall. Trans.* 21A: 2281–2282.

NAKAYAMA, Y., and MABUCHI, H., 1993, *Intermetallics* 1: 41–48.

NEUMANN, J. P., CHANG, Y. A., and LEE, C. M., 1976, *Acta Metall.* 24: 593–604.

NEUMANN, J. P., and CHANG, Y. A., 1979, *Z. Metallukde* 70: 118.

NIC, J. P., KLANSKY, J. L., and MIKKOLA, D. E., 1992, *Mat. Sci. Eng.* A152: 132–137.

NISHITANI, S. R., MASAKI, K., SHIRAI, Y., and YAMAGUCHI, M., 1991, *Scripta Metall.* 25: 1921–1925.

PARTHASARATHI, A., and FRASER, H. L., 1984, *Phil. Mag* A, 50: 89–100.

PORTER, W. D., HISATSUNE, K., SPARKS, C. J., OLIVER, W. C., and DHERE, A., 1989, In *High Temperature Ordered Intermetallic Alloys III*, MRS Symp. Proc. 133: 657–668.

RUSOVIC, N., and HENIG, E. T., 1980, *Phys. Stat. Sol.* (a), 57: 529–540.

SCHAEFER, H. E., WURSCHUM, R., and BUB, J., 1992, *Mater. Sci. Forum* 105–110: 439.

SCHULTZ, P. J., and MacKenzie, I. K., 1982, *Positron Annihilation*, ed. Coleman P. G., Sharma S. C., and Diana, L. M., p. 553, North-Holland Publ.

SCHWAB, R., and GEROLD, V., 1980, *Acta Metall.*, 28: 433–441.

SHIMOTOMAI, M., WANG, T.-M., IWATA, T., and DOYAMA, M., 1985, In *Positron Annihilation*, ed. P. C. Jain, R. M. Singru, and K. P. Gopinathan, p. 140, Singapore: World Scientific Publ. Co.

SHIRAI, Y., 1988, *Bulletin of the Japan Institute of Metals*, 27: 869.

SHIRAI, Y., NAKAMURA, F., TAKEUCHI, M., WATANABE, K., and YAMAGUCHI, M., 1989, In *Positron Annihilation*, ed. L. Dorikens-Vanpraet, M. Dorikens, and D. Segers, p. 488, Singapore: World Scientific Publishing Company.

SHIRAI, Y., and YAMAGUCHI, M., 1992, *Mat. Sci. Eng.* A152: 173–181.

SHIRAI Y., MASAKI, K., and YAMAGUCHI, M., 1994, *Intermetallics* 2: 221–224.

SIMMONS, R. O., and BALLUFFI, R. W., 1960, *Phys. Rev.* 117: 52–61.

SMEDSKJAER, L. C., and FLUSS, M. J., 1983, In *Solid State: Nuclear Methods*, ed. J. N. Mundy, S. J. Rothman, M. J. Fluss, and L. C. Smedskjaer, p. 77. Orlando, FL: Academic press.

SMITH, C. S., 1943, *Trans AIME* 152: 144–148.

STEPHENS, J. R., 1987, *High Temperature Ordered Intermetallic Alloys II*, MRS Symp. Proc. 81: 381–395.

STOLOFF, N. S., and DAVIES, R. G., 1966, *Prog. Mater. Sci.* 13: 1–66.

STOLWIJK, N. A., VAN GEND, M., and BAKKER, H., 1980, *Phil. Mag.* A, 42: 783–808.

SUZUKI, K., and TAKEUCHI, S., 1993, *Intermetallics* 1: 21–27.

TAKEUCHI, S., HASHIMOTO, T., and SHIBUYA, T., 1992, *J. Mat. Sci.* 27: 1380.

TAMMANN, G., and DAHL, K., 1923, *Z. Znorg. Chem.* 126: 104.

TRIFSHÄUSER, W., 1986, In *Microscopic Methods in Metals*, ed. U. Gonser, p. 249. Berlin, Heidelberg: Springer-Verlag.

VAN OMMEN, A. H., 1982, *Phys. Stat. Sol.* (a), 72: 273–280.
VAN OMMEN, A. H., WAEGEMAEKERS, A. A. H. J., MOLEMAN, A. C., SCHLATTER, H., and BAKKER, H., 1981, *Acta Metall.*, 29: 123–133.
VOGL, G., KARNER, CH., RANDL, O., and SEPIOL, B., 1992, *Ordered Intermetallics—Physical Metallurgy and Mechanical Behavior*, ed. C. T. Liu, R. W. Cahn, and G. Sauthoff, 497–509. Kluwer Academic Publishers.
VUJIC, D., LI, Z. and WHANG, S. H., 1988, *Metall. Trans. A* 19: 2445–2455.
WANG, T.-M., SHIMOTOMAI, M., and DOYAMA, M., 1984, *J. Phys. F.* 14: 37.
WASILEWSKI, R. J., 1968, *J. Phys. Chem. Solids* 29: 39.
WESTBROOK, J. H., 1957, *Trans AIME* 209: 898.
________, ed., 1960, *Mechanical Properties of Intermetallic Compounds*, New York: John Wiley and Sons.
________, 1967, *Intermetallic Compounds*, New York: John Wiley and Sons.
________, 1993, In *Structural Intermetallics*, ed. R. Darolia, J. J. Lewandowski, C. T. Liu, P. L. Martin, D. B. Miracle, and M. V. Nathal, 1–15. Warrendale, PA: TMS.
WESTBROOK, J. H., and FLEISCHER, R. L., 1994, In *Intermetallic Compounds—Principles and Practice*, New York: John Wiley and Sons, in press.
WOOD, D. L., and WESTBROOK, J. H., 1962, *Trans. AIME* 224: 1024–1037.
WURSCHUM, R., GREINER, W., VALIEV, R. Z., RAPP, M., SIGLE, W., SCHNEEWEISS, O., and SCHAEFER, H. E., 1991, *Scripta Metall.* 25: 2451–2456.
YAMAGUCHI, M., and INUI, H., 1993, In *Structural Intermetallics*, ed. R. Darolia, J. J. Lewandowski, C. T. Liu, P. L. Martin, D. B. Miracle, and M. V. Nathal, 127–142. Warrendale, PA: TMS.
YAMAGUCHI, M. and INUI, H., 1994, In *Intermetallic Compounds—Principles and Practice*, ed. J. H. Westbrook and R. L. Fleischer New York: John Wiley and Sons, in press.
YAMAGUCHI, M., SHIRAI, Y., and INUI, H., 1993, *High Temperature Ordered Intermetallic Alloys V*, MRS Symp. Proc. 288: 131–139.
YAMAGUCHI, M., and UMAKOSHI, Y., 1990, *Prog. Mater. Sci.* 34: 1.
YANG, W., DODD, R. A., and STRUTT, P. R., 1972, *Metall Trans.* 3: 2049–2054.
YOO, M. H., and FU, C. L., 1993, In *Structural Intermetallics*, ed. R. Darolia, J. J. Lewandowski, C. T. Liu, P. L. Martin, D. B. Miracle, and M. V. Nathal, 283–292. Warrendale, PA: TMS.

Chapter 2

Grain Boundary Structure and Chemistry

V. Vitek and M. Yan

ABSTRACT

The importance of understanding the atomic and chemical structure of grain boundaries in intermetallic compounds for a fundamental comprehension of their fracture behavior is the principal theme of this contribution. Since intermetallics are the prime examples of quasi-brittle materials, we first discuss general features of brittle fracture in ductile materials. $L1_2$ intermetallic compounds, in particular Ni_3Al, have been studied most extensively; therefore, we review in detail the present state of our understanding of the atomic and chemical structure of grain boundaries in these alloys and discuss possible reasons for intrinsic brittleness of their grain boundaries. Results of recent atomistic studies of grain boundaries in $L1_2$ alloys are then described. First, we concentrate on comparison of the boundary structures in Ni_3Al and Cu_3Au, the two alloys with the same crystal structure but rather different propensities to ordering. Second, we discuss the effects of temperature and bulk nonstoichiometry on the structure and chemistry of grain boundaries in Ni_3Al. At this point, the most important finding is that significantly different structures are invoked by segregation of nickel and aluminum, respectively, which may be related to the fact that only nickel-rich alloys may be ductilized by boron alloying. Finally, we discuss briefly the structure and properties of grain boundaries in NiAl and assess the possibilities and limits of further research on atomic level behavior of interfaces in intermetallic compounds.

1. INTRODUCTION

In recent years, intermetallic compounds have attracted considerable attention as potential new materials for high-temperature applications. The reason is that they frequently possess many desirable properties, such as low density, high melting temperature, and high corrosion resistance. The atomic mobility is highly restricted in these materials owing to the chemical order and, therefore, metallurgical processes, such as creep, corrosion, and recrystallization, are strongly inhibited (see, e.g., papers in recent symposia on intermetallic compounds (Whang, Liu, Pope, and Stiegler 1990; Izumi 1991; Johnson, Pope, and Stiegler 1991; Liu, Cahn, and Sauthoff 1992). Furthermore, the yield strength of a number of intermetallic alloys increases with increasing temperatures (see reviews by Pope and Ezz 1984; Pope 1990; Yamaguchi and Umakoshi 1990, Dimiduk 1991; Vitek, Sodani, and Cserti 1991).

However, a severe drawback that limits the application of intermetallics is their low fracture toughness, particularly at low temperatures. The fracture that ensues is either transgranular cleavage or intergranular cracking. This is so in spite of the fact that these materials are often reasonably ductile when tested in compression. An example of a material that deforms plastically in compression but cleaves very readily transgranularly when loaded in tension is Al_3Ti converted to $L1_2$ structure by additional alloying by elements such as Fe, Cr, and Mn (George et al. 1990; Kumar and Brown 1992a, b; Morris and Lerf 1991; Wu, Pope, and Vitek 1991a, b). Ni_3Al, which crystallizes in the $L1_2$ structure, is the prime example of an intermetallic fracturing intergranularly. It is ductile in the single-crystalline form when loaded either in tension or compression (e.g., Pope and Ezz 1984; Pope 1990; Vitek et al. 1991; Yamaguchi and Umakoshi 1990; Dimiduk 1991; Liu 1991b), but as a polycrystal it fractures readily along grain boundaries when loaded in tension (e.g., Izumi and Takasugi 1988; Liu 1988; Takasugi and Izumi 1992). At the same time NiAl, which crystallizes in B2 structure, fractures both transgranularly and intergranularly (Chang, Darolia, and Lipsitt 1992; Darolia et al. 1992; Kumar, Mannan, and Visawanadham 1992; Schulson and Barker 1983). Yet, in either case the failure is not totally brittle because even though the overall ductility observed in the fractured sample may be low, there tends to be large amount of inelastic deformation in the region localized around the crack and the work to fracture, extracted from the Griffith equation, is much higher than the true surface energy. Such fracture, which occurs also in other materials, for example in iron-based (b.c.c.) alloys, has been called quasi-brittle (Hack, Chen, and Srolovitz 1989). Thus, a fundamental understanding of the fracture behavior of intermetallic compounds can only be attained in the context of the generic problem of quasi-brittle fracture.

The quasi-brittle behavior combines crack propagation with localized plastic deformation and it is the problem of dislocation emission at the tip of a stationary and/or moving crack that lies in the heart of this problem. In particular, while quasi-brittle materials can be deformed plastically to relatively high strains, the dislocation emission from the crack tip is in these materials inhibited to such an extent that crack propagation can take place. The most widely used model for analyzing the balance between dislocation emission and crack propagation is that of Rice and Thomson (1974), in which these two process are regarded as mutually exclusive. Either dislocation emission evolves at the crack tip and the crack remains stationary, or there is no plastic deformation at the crack tip and the crack starts to propagate. This model has been subsequently further developed and refined (Lin and Thomson 1986; Thomson 1986; Weertman 1978; Bartholomeusz and Wert 1992a, b). An important recent development is inclusion of dislocation core effects via a Peierls–Nabarro type model (Rice 1991; Sun, Rice, and Truskinovsky 1991), which allows for physically more realistic treatment of atomic-level processes involved in the emission of dislocations from the crack tip.

An alternate though not entirely opposite point of view is that local inelastic deformation necessarily accompanies fracture and that the amount of local plastic deformation is a function of the true cohesive energy, γ, so that the critical energy release rate, i.e., the work to fracture, $G_c = 2\gamma + \gamma_p(\gamma)$, where $\gamma_p(\gamma)$ is the inelastic work per unit area of crack face which is a function of γ (Jokl, Vitek, and McMahon 1980; Jokl et al. 1989; Vitek 1991). This work results from an incipient, irreversible shear deformation that occurs simultaneously with the bond breaking and dissipates a significant part of the energy. This assertion is based on the fact that if the local tensile stresses at the crack tip are sufficiently large to break atomic bonds, then the shear stresses, which are of the same order of magnitude at the tip as the tensile stress, must be large enough to shear the lattice locally. Indeed, in quasi-brittle materials the work to fracture is always large compared to true surface energies (Knott 1977, 1987), which suggests that cracking and inelastic deformation are concomitant, rather than mutually exclusive processes.

Notwithstanding, regardless of the details of the models for quasi-brittle cracking, both the cohesive strength of the material and localized plasticity at the crack tip are critical parameters. Thus, when analyzing atomic structures and atomic-level processes affecting the propensity to fracture we always have to consider both these properties and their mutual interaction. In this chapter we are principally concerned with grain boundaries in intermetallic compounds and, therefore, we shall concentrate on relating both the cohesive and deformation properties in the grain-boundary region to the atomic structure of these interfaces. Since the intermetallic compound in which the grain-boundary properties and structure were investigated in most details is Ni_3Al, we focus our discussion on this material but emphasize primarily those structural aspects which may be pertinent for intermetallics in general.

We first review the present state of our understanding of the atomic and chemical structure of grain boundaries in Ni_3Al as well as various theoretical approaches towards understanding the intrinsic brittleness and boron effects which are based either on the atomic- or electronic-structure considerations. Recent atomistic studies of grain boundaries in $L1_2$ alloys carried out both at 0 K and finite temperatures are then discussed in more detail. In particular, we first concentrate on comparison of the boundary structures in Ni_3Al and Cu_3Au, the two alloys with the same crystal structure but rather different propensities to ordering and fracture. Secondly, we discuss the effects of the bulk off-stoichiometry on the structure and chemistry of grain boundaries in Ni_3Al and interpret these results in the context of boron effects. We also summarize the present understanding of the structure of grain boundaries in NiAl and emphasize the recent high resolution electron microscopic results for the $\Sigma = 5$ {310} boundaries in NiAl (Fonda and Luzzi 1993) which reveal nonstoichiometric boundary structures. Finally, we summarize outstanding problems and assess the possibilities and limits of further research on atomic level behavior of interfaces in intermetallics.

2. STRUCTURE, CHEMISTRY AND BRITTLENESS OF GRAIN BOUNDARIES IN $L1_2$ COMPOUNDS: Ni_3Al

It is known that in pure metals, with the exception of iridium (Hecker, Rohr, and Stein 1978; White, Clausing, and Heatherley 1979), grain boundaries are not susceptible to cracking, and it is the segregation of impurities and alloying elements to grain boundaries in disordered alloys that can lead to severe embrittlement (see, e.g., Johnson and Blakeley 1977; Lathanision and Jones 1987; Was and Bruemmer 1989; Pugh 1991). In contrast, grain-boundary brittleness in nickel-based $L1_2$ compounds, as well as in some other compounds possessing this crystal structure (e.g., Co_3Ti), appears to be an intrinsic property of these materials (Liu, White, and Horton 1985; Takasugi et al. 1985; Takasugi and Izumi 1985a, b; Takasugi, Izumi, and Masahashi 1985; Izumi and Takasugi 1988; Liu 1988). This conclusion is valid in spite of the fact that recent experimental observations suggest that the environment, in particular hydrogen, plays an important role in the intergranular brittleness of $L1_2$ alloys (Takasugi, Nagashima, and Izumi 1990; Kimura et al. 1991; Liu 1991a, 1992; Nishimura and Liu 1992; Takasugi, Hanada, and Yoshida 1992). The reason is that these environmental phenomena are again related to some special intrinsic properties of grain boundaries in these materials and are not of the same type as, for example, the hydrogen embrittlement observed in disordered alloys, where it often combines with impurity segregation. At the same time, the intergranular brittleness is not a generic feature of all the $L1_2$ alloys. For example, Cu_3Au, Ni_3Fe, and Ni_3Mn, which also crystallize in the $L1_2$ structure, are not brittle. Hence, the first question that needs to be answered is what the reason is for the intrinsic intergranular brittleness in alloys such as Ni_3Al, and why alloys like Cu_3Au do not exhibit intergranular fracture.

A very important finding associated with the grain-boundary brittleness in intermetallics is that alloying may have striking effect on the ductility of these materials. The most celebrated example is that doping of polycrystalline Ni_3Al by a small amount of boron can substantially increase the ductility of this material (Aoki and Izumi 1979). In subsequent studies it was found that boron segregates to grain boundaries in an interstitial manner and that the beneficial effect of boron takes place only in the nonstoichiometric Ni_3Al rich in Ni (Dimiduk, Weddington, and Lipsitt 1987; George, Liu, and Padgett 1989; Izumi and Takasugi 1988; Liu1988; Liu et al. 1985; Takasugi et al. 1985; Takasugi and Izumi 1985a; Takasugi et al. 1987). In fact, in Co_3Ti the intergranular brittle fracture is suppressed in Co-rich alloys without any additional alloying (Takasugi et al. 1987). This suggests that deviations from stoichiometry toward a surplus of the majority element is at least as important as the effect of boron. Therefore, the second question is: What is the reason for the alloying effects and the effects of off-stoichiometry on brittleness of grain boundaries?

2.1. Cohesion at Grain Boundaries

A natural approach to investigate the source of the intrinsic intergranular brittleness, and of the alloying effects, is to consider the cohesive properties of grain boundaries that are governed by their atomic and electronic structures. Takasugi et al. (1985) and Takasugi and Izumi (1985a) observed that the propensity for grain boundary fracture increases with increasing difference in the electrochemical factors, such as the valency or electronegativity, between the constituent elements A and B forming the A_3B compound. They suggested, therefore, that the low boundary cohesion in ordered compounds may be attributed to a poor atomic bonding across the boundary resulting from these electrochemical differences. Physically, this argument is based on the assumption that the distribution of the electronic charge in ordered alloys will be characterized by *directionality* and *heteropolarity* (Izumi and Takasugi 1988). Owing to the directionality, the A-B bonds will be severely distorted when crossing grain boundaries, thus reducing the p-d hybridization that contributes to the bond strength. At the same time, the heteropolarity causes the B atom to withdraw electrons from the A-A (metal–metal) bonds and thus weakens the strength of these bonds. Hence, both these phenomena may lead to the reduction of the grain-boundary cohesion. Since the effects of both the bond directionality and heteropolarity become more severe when the valency difference or electronegativity difference between A and B species is larger, the boundary strength may be closely related to these electrochemical factors.

Although there are a number of experimental results that support such correlation (Taub et al. 1986; Taub and Briant 1987; Takasugi and Izumi 1991), the physical justification for this argument, which extrapolates from the results of the quantum mechanical cluster calculations (Messmer and Briant 1982; Briant and Messmer 1984; Eberhart and Vvedensky 1988) to metallic solids, is questionable (Sagi and Pettifor 1987; Goodwin, Needs, and Heine 1988, 1990). Moreover, the calculations performed using the embedded atom method (EAM) (Chen, Voter, and Srolovitz 1986; Voter et al. 1989; Chen et al. 1990) indicate that the cohesion energy in Ni_3Al is not significantly different from that in pure Ni, suggesting that the cohesive strength of grain boundaries is not the only reason for the grain-boundary brittleness.

Similarly, as in the case of binary alloys, it can also be speculated that boron directly affects the cohesion at grain boundaries. This hypothesis is supported by the fact that while boron segregates readily to the grain boundaries, it does not segregate strongly to free surfaces (White et al. 1984; Liu et al. 1985; White, Liu, and Padgett 1988). Following the arguments based on the equilibrium thermodynamics (Hirth and Rice 1980; Rice and Wang 1989), this phenomenon implies that boron segregation leads to a decrease of the grain-boundary energy, γ_{gb}, but increase in the surface energy, γ_s. The net effect is thus the increase of the ideal work to fracture, $\gamma_{ideal} = 2\gamma_s - \gamma_{gb}$. The possible strengthening of bonding across grain boundaries has been indicated in tight binding calculations (Masuda-Jindo 1988) as well as in the calculations employing EAM (Chen et al. 1989b; Chen et al. 1990). However, this rationalization does not explain why the beneficial influence of boron on intergranular cohesion is limited to Ni_3Al alloys enriched with Ni, while it has no effect in stoichiometric Ni_3Al or in alloys enriched with aluminum.

An obvious possibility is that the surplus of Ni in the boundary region is an indispensable precursor for the decrease of the grain-boundary brittleness and an important role of boron might be to attract additional Ni via co-segregation. Indeed, atomistic simulation employing EAM has confirmed that boron segregation lowers grain-boundary energy and that co-segregation of Ni and B to grain boundaries is energetically favorable (Chen et al. 1990). If this is the case, it can be speculated that chemical disordering will occur at grain boundaries with surplus of Ni. Nonetheless, the extent of disordering at grain boundaries in Ni_3Al is still a controversial subject (Liu 1991a, b). For example, Schulson and co-workers (Baker and Schulson 1989; Baker et al. 1990; Schulson 1991; Schulson and Baker 1991; Schulson et al. 1991) observed in a boron doped Ni-24%Al alloy disordered regions spreading up to 20 nm away from the boundaries. However, disordered grain-boundary phase of such an extent has not been observed by other authors and disordering on such scale would, indeed, indicate formation of a bulk disordered phase nucleated at the grain boundary. Sass and co-workers (Kung, Rasmussen and Sass 1991, 1992a, b; Kung and Sass 1992) reported compositional order up to the boundary in most boundaries studied and a relatively narrow disordered region near some high angle boundaries; a similar observation has been reported by Mills and co-workers (Mills 1989; Mills and Daw 1991; Mills et al. 1991; Mills, Goods, and Foiles 1992). A the same time, combined HREM and chemical analysis (Krzanowski 1989), as well as FIM/atom probe analysis (Brenner and Ming-Jian 1991; Sieloff, Brenner, and Hua 1989) of grain boundaries in boron doped Ni_3Al suggest an enrichment of boundaries by nickel. On the other hand, the AES observations (George et al. 1989) show no strong tendency for nickel-boron co-segregation. Hence, while nickel segregation is probably an important factor in decreasing the brittleness of grain boundaries in Ni_3Al, the region of disorder is likely to extend only a few lattice spacings away from the boundary.

2.2. Plastic Deformation at and Near Grain Boundaries

An alternative explanation of the effect of alloying on the intrinsic intergranular brittleness in Ni_3Al emerges from the consideration of the plastic deformation associated with propagation of slip across grain boundaries and/or with intergranular fracture. Schulson and co-workers (Schulson et al. 1985a; Baker and Schulson 1989; Baker et al. 1990; Schulson and Baker 1992) proposed that compositional disordering, possibly enhanced by boron segregation, ameliorates the transfer of slip across grain boundaries. This argument was based on the results of experimental measurements of the Hall–Petch relation, which suggest that transfer of slip through grain boundaries is easier in boron-doped Ni_3Al alloys (Schulson et al. 1985b; Schulson and Baker 1991; Schulson et al. 1991). This proposition is supported by the crystallographically based arguments that disordering at grain boundaries relieves certain geometrical constraints upon the transfer of slip through grain boundaries (Frost 1987; King and Yoo 1987; Yoo and King 1988; King, Frost, and Yoo 1991) and thus the slip transfer through strongly ordered grain boundaries may be inherently more difficult than through those that are compositionally disordered. However, in situ observations of the slip transfer through grain boundaries do not show any significant effect of the boron segregation upon the slip transmission and the slip transfer follows the same criteria as in f.c.c. structures (Lee, Robertson and Birnbaum 1992; Lee et al. 1991; Robertson et al. 1992). Similarly, atomistic studies of the interaction between dislocations and grain boundaries do not indicate any pronounced differences between pure f.c.c. metals and $L1_2$ alloys (Pestman and DeHosson 1992; Pestman et al. 1991a, b). Furthermore, in a recent electron microscopic study of grain-boundary dislocations in a boron-doped Ni-24 at %Al alloy it was found that their Burgers vectors corresponded to the DSC lattice of the $L1_2$ structure rather than of the f.c.c. lattice as expected in the case of chemical disorder (Forwood and Bigson 1992).

Nevertheless, the local disordering may play an important role in ductilizing the intergranularly brittle $L1_2$ compounds but its major influence may be on the ability to emit dislocations from the tips of intergranular cracks rather than on the slip transfer through grain boundaries (Sun et al. 1991; Vitek 1991). In general, the chemical disordering will decrease the energy of APB's formed during localized plastic shearing in the vicinity of grain boundaries which then makes the local shear deformation and thus dislocation emission, much easier.

3. ATOMISTIC STUDIES OF THE STRUCTURE AND CHEMISTRY OF GRAIN BOUNDARIES IN $L1_2$ ALLOYS

3.1. Structures of Boundaries at 0 K

A large number of calculations of grain-boundary structures have been made using the molecular statics method. In these calculations the temperature effects are not considered and the calculations effectively corresponds to 0 K. In such calculations the boundary structures are first constructed geometrically (using e.g., the coincidence site lattice analysis) and then relaxed such as to minimize the energy with respect to individual atom positions and relative displacements of the adjoining grains (Sutton 1984; Vitek 1984; Sutton 1990). No compositional changes are considered. Such simulations employed both the pair-potentials representing a stable $L1_2$ structure (Kruisman, Vitek, and DeHosson 1988) and N-body potentials of the EAM and Finnis–Sinclair type (Chen et al. 1986; Chen, Srolovitz, and Voter 1989a; Vitek et al. 1989; Vitek, Ackland and Cserti 1991; Vitek and Chen 1991). As an example Figures 2-1a and b show the structures of the $\Sigma = 73$ (830)/[001] symmetrical tilt boundary for Cu_3Au and Ni_3Al, respectively.[1] In these figures Cu and Ni atoms are depicted as dark circles and Au and Al as white circles. Atoms positioned in two (002) planes along the [001] direction (rotation axis) are shown but not distinguished in these figures.

In general, it has been found in these calculations that significant structural differences exist between crystallographically identical grain boundaries in Ni_3Al and Cu_3Au. In the case of Cu_3Au (low ordering energy), the relaxation in the boundary is such that the atoms in the boundary region can not be uniquely assigned to either the upper or the lower grain. Hence, grain-boundary structures in this alloy are topologically similar to those in pure f.c.c. materials. On the other hand, in the case of Ni_3Al (high ordering energy), the atoms in the grain-boundary region relax much less and they can be regarded as uniquely attached to either the upper or the lower grain. This is visible in the structures shown in Figures 2-1a and b.

Owing to the very small relaxation in grain boundaries in Ni_3Al, there are many distinct columnar cavities at these boundaries because of the mismatch arising from the misoriented $L1_2$ lattices. These columnar cavities may serve as suitable nuclei for the formation of intergranular cracks under external loads and thus render the grain boundary susceptible to cracking (Kruisman et al. 1988; Vitek et al. 1991; Vitek and Chen 1991). Such cavities will not occur, however, in certain special boundaries, for example, in $\Sigma = 3(111)$ symmetrical tilt boundaries which are fully coherent, and these boundaries should not then display any intrinsic brittleness. Indeed, the recent experimental study of the misorientations of the grains along the intergranular fracture path in Ni_3Al shows that the

---

[1] This is the boundary we use as an example throughout this paper but the same structural aspects and properties have been found for other boundaries studied (Min Yan 1992; Min Yan et al. 1992; Vitek et al. 1992).

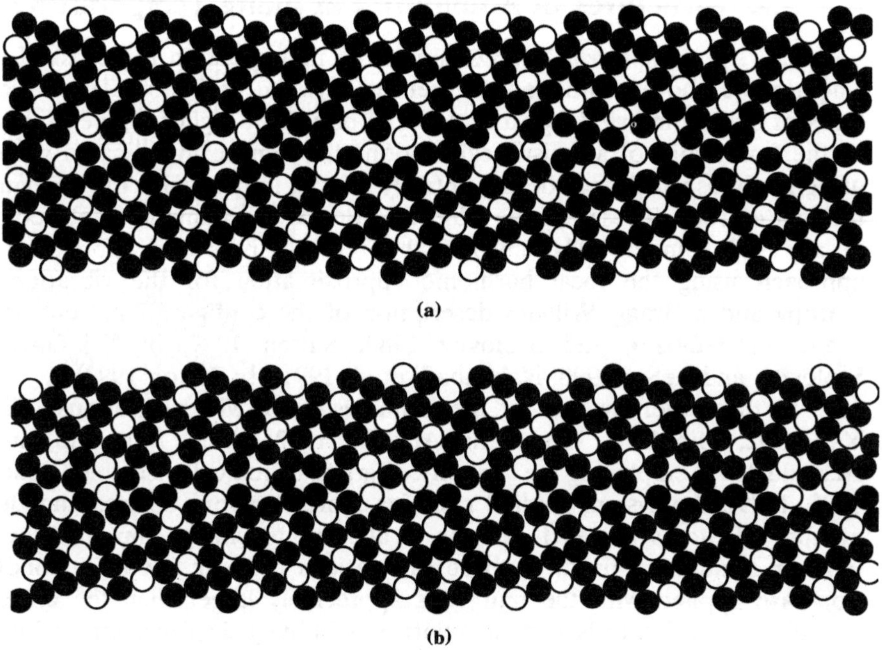

Figure 2-1. The structures of the $\Sigma = 73$ (980)/[001] symmetrical tilt boundary in (a) Cu_3Au and (b) Ni_3Al.

$\Sigma = 3(111)$ twins are avoided (Pope 1992; Lin and Pope 1993). In weakly ordered alloys such as Cu_3Au and, of course, in pure f.c.c. metals, grain boundaries rarely contain such cavities.

However, the columnar cavities may not be the only and possibly not even the most important factor in the intrinsic brittleness of grain boundaries in Ni_3Al. In weakly ordered alloys, such as Cu_3Au, even if micro cracks are formed, they would be rapidly blunted by dislocation emission from the crack tip since the grain-boundary regions are not fully ordered compositionally and the energies of APB's resulting from the related shear displacements are, therefore, small or negligible. On the other hand, in Ni_3Al the dislocation emission accompanying a propagating crack will be difficult since it will lead to the creation of APB's with a high energy owing to the preservation of the ordered $L1_2$ structure up to the boundary plane. Hence, the total work to fracture will be large and the crack propagation difficult in weakly ordered alloys but it will be much lower in strongly ordered alloys (Vitek 1991).

3.2. Structures of Boundaries at Finite Temperatures

3.2.1. Stoichiometric Composition. The first question to be asked is whether the above-mentioned differences between grain-boundary structures of stoichiometric Cu_3Au and Ni_3Al prevail at higher temperatures. In the framework of atomistic calculations this can be investigated when using methods which include the entropy effects, in particular the configurational entropy. These are the Monte Carlo (MC) technique (Metropolis et al. 1953; Binder 1984) and/or the recently proposed approximate approach using the local harmonic approximation for the vibrational entropy and a Bragg–William description of the configurational entropy (LeSar, Najafabadi, and Srolovitz 1989; Sutton 1989a, b; Najafabadi, Srolovitz, and LeSar 1990; Najafabadi et al. 1991a, b; Sutton 1992).

In this paper we present the main results of MC simulations (Min Yan 1992; Min Yan, Vitek, and Ackland 1992; Vitek, Min Yan, and Ackland 1992) employing the "modified" grand canonical ensemble (Foiles 1985, 1989); the system studied contains more than one species and the number of atoms for each species is allowed to change, but the total number of atoms is fixed. In these calculations periodic boundary conditions are applied with the unit cells appreciably larger than the corresponding primitive cells and no relative rigid-body displacements of the two grains either parallel or perpendicular to the boundary plane are allowed. The latter condition is physically justified when considering polycrystals where each grain is constrained by its environment and cannot move freely relative to other grains.

Figure 2-2 shows the equilibrium structure of the Σ73 {830} grain boundary in Cu_3Au at 300 K. Again, only atoms within the distance a[001] along the rotation axis are shown, although the size of the unit cell along this direction used in the calculation was 3a[001]. For each atomic site, i, the average occupancy by Cu has been calculated as τ_{Cu}/τ_{tot}, where τ_{Cu} is

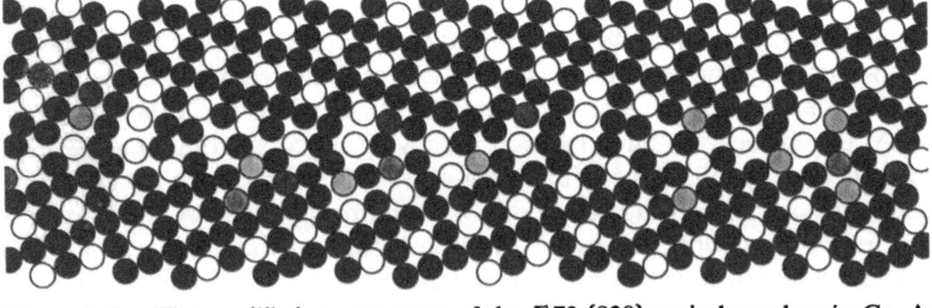

Figure 2-2. The equilibrium structure of the Σ73 {830} grain boundary in Cu_3Au at 300 K.

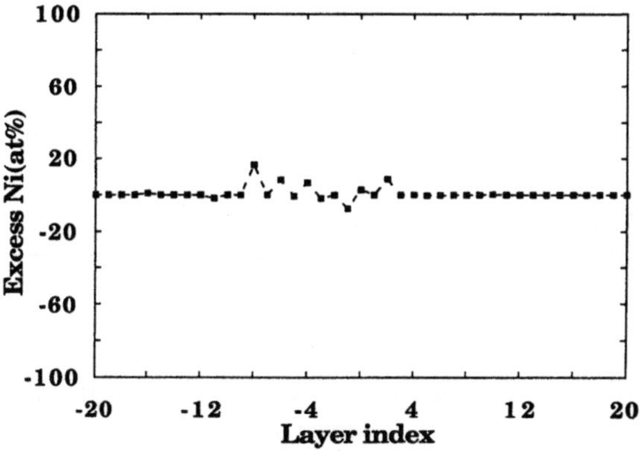

Figure 2-3. Excess concentration of Cu (in units of atomic percent) in different {16 6 0} layers parallel to the Σ73 {830} grain boundary at 300 K.

the number of MC trials for which this site is occupied by Cu among all the MC trials, τ_{tot}. This quantity is depicted in Figure 2-2 by shading of the circles representing individual atoms: The white circles represent the sites occupied by Cu with less than 5% probability, and black circles represent sites occupied by Cu with more than 95% probability. In between these two extremes, circles are shaded by five different grades of gray. The five different levels of shading correspond to occupancies between 95%–80%, 80%–60%, 60%–40%, 40%–20%, 20%5%, respectively.

Comparing this boundary structure with that obtained by the molecular statics calculation at 0 K (Fig. 2-1a), it is seen that the boundary structure at 300 K is not noticeably different topologically but the species redistribute in the boundary region to such an extent that the $L1_2$ unit cells can no longer be identified and the grain boundary becomes chemically disordered. In order to analyze the compositional disorder, the excess concentration of Cu (in units of atomic percent) in different {16 6 0} layers parallel to the boundary is plotted in Figure 2-3. In this picture the layer 0 corresponds to the center of the boundary, layers $-1, -2\ldots$ to the lower grain and layers $+1, +2\ldots$ to the upper grain; note the asymmetry between the upper and lower grain caused by the relative displacement of the grains parallel to the boundary. The compositional disorder is manifested by the positive peaks (indicating surplus of Cu) and the negative peaks (indicating surplus of Au) which imply that even at 300 K, which is much lower than the order-disorder transition temperature of Cu_3Au, there are extensive interchanges between Cu and Au atoms in the grain boundary. Furthermore, the excess Au concentration within the boundary region indicates that compositional disordering is accompanied

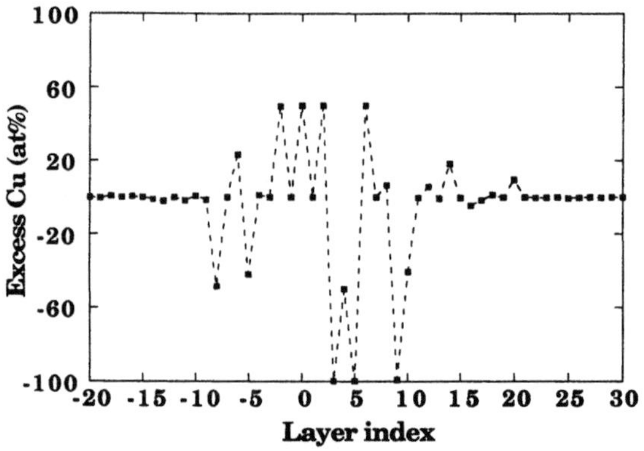

Figure 2-4. Excess concentration of Ni (in units of atomic percent) in different {16 6 0} layers parallel to the Σ73 {830} grain boundary at 600 K.

by the segregation of Au onto the grain boundary. The compositional disordering in the grain-boundary region becomes more extensive as the temperature increases and spreads further into the bulk. Above the bulk order-disorder transition temperature the MC calculation leads to the complete disorder in both the grain boundary and the bulk. However, topologically the boundary structure remains the same even in this case and is very similar to the structure found in pure f.c.c. metals.

The compositional character of grain boundaries in Ni_3Al is quite different than in Cu_3Au. There is no compositional disordering at room temperature and only at 600 K a marginal disordering starts to occur in some boundaries. A more significant compositional disorder is observed only at temperatures close to melting. This is seen in Figure 2-4, where the excess concentration of Ni in different {16 6 0} layers parallel to the Σ73 {830} grain boundary is plotted at this temperature, as well as in Figure 2-5, which shows the atomic structure of the Σ73 {830} grain boundary in the stoichiometric Ni_3Al at 600 K; shading of the circles representing atoms has the same meaning as in Figure 2-1 with Cu replaced by Ni and Au replaced by Al. Comparison of Figures 2-5 and 2-1b shows that no significant topological changes occurred in the boundary due to nickel segregation.

This study demonstrates that the structural differences between boundaries in weakly ordered (Cu_3Au) and strongly ordered (Ni_3Al) $L1_2$ alloys found by molecular statics studies at 0 K prevail and are even more pronounced at finite temperatures. In Cu_3Au the grain-boundary region becomes compositionally disordered well below bulk disordering while in Ni_3Al the $L1_2$ order is disturbed in the boundary region only at tempera-

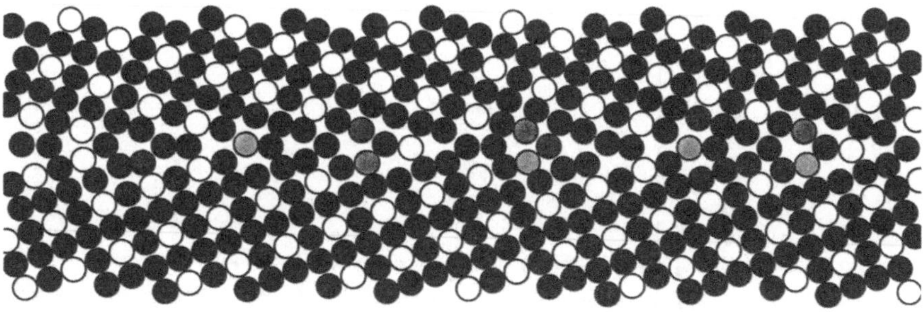

Figure 2-5. The structures of the $\Sigma = 73$ (830)/[001] symmetrical tilt boundary in Ni_3Al at 600 K.

tures close to melting. This finding is in a good agreement with the recent electron microscopic observations. It was, indeed, observed that in Cu_3Au the $\Sigma = 3$ boundaries disorder compositionally at temperatures lower than the bulk order-disorder temperature, though appreciably higher than the room temperature (Tichelaar, Schapink, and Li 1992). However, for such special coherent boundaries disordering must be expected to occur at higher temperatures than at more general boundaries. In Ni_3Al the boundaries have been generally found to be compositionally ordered though a disorder cannot be ruled out in a region extending about 5Å away from the center of the boundary, in particular for off-stoichiometric compositions and in the presence of boron segregation (Mills 1989; Mills and Daw 1991; Mills et al. 1991; Kung et al. 1991; Kung et al. 1992a, b; Kung and Sass 1992; Mills et al. 1992).

3.2.2. Non-stoichiometric Composition. As mentioned above, doping by boron suppresses the intergranular fracture only in nickel-rich Ni_3Al alloys, while in stoichiometric or aluminum rich alloys the effect of boron is not significant. This is the reason for studying the impact of the bulk composition upon structural and chemical features of grain boundaries and several such studies have been made (Najafabadi et al. 1991a; Mills et al. 1992; Min Yan 1992; Min Yan et al. 1992; Najafabadi and Srolovitz 1993; Vitek et al. 1992). In this chapter we present some typical results of the recently performed MC simulations of grain boundaries in nonstoichiometric Ni_3Al. In these studies the starting configurations are the corresponding structures obtained from the static relaxations, except that the bulk composition is adjusted to a chosen off-stoichiometry in the bulk. During the simulation the bulk composition is kept constant by fixing the corresponding difference of chemical potentials between the two species.

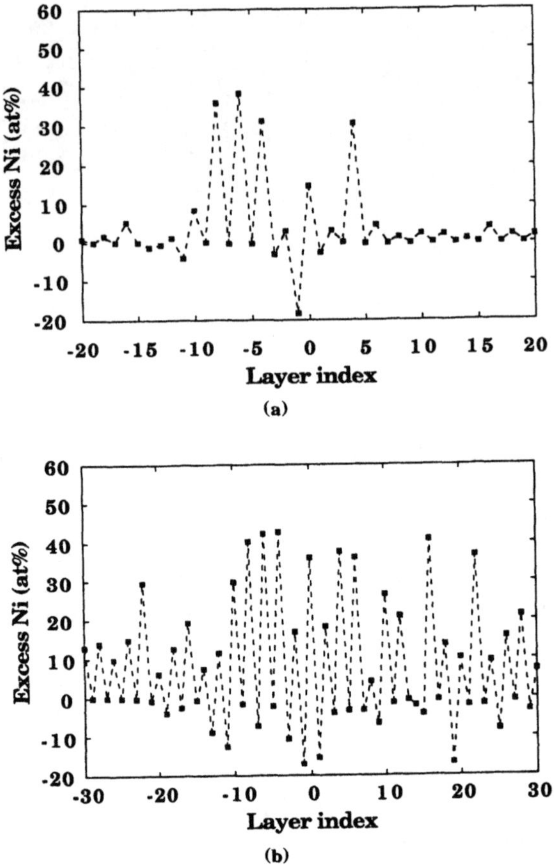

Figure 2-6. Excess concentration of Ni (in units of atomic percent) in different {16 6 0} layers parallel to the Σ73 {830} grain boundary at 600 K at the bulk compositions (a) Ni-24%Al and (b) Ni-20%Al.

Figures 2-6a and b show the excess Ni concentration (in units of atomic percent) on {16 6 0} layers parallel to the Σ73{830} tilt boundary at the bulk compositions Ni-24%Al and Ni-20%Al, respectively, at 600 K. Considerable Ni segregation occurs in the boundary region in both cases and it is, of course, more dramatic in the latter case. At the same time the nickel segregation leads to extensive interchanges between Ni and Al atoms, as manifested by negative peaks in Figure 2-6a, b. Such behavior was also found in MC studies of Mills et al. (1992). This implies that a significant compositional disordering takes place in the vicinity of the

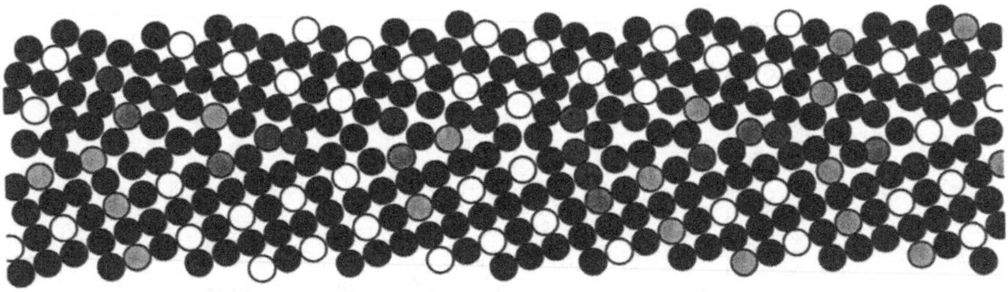

Figure 2-7. The structures of the Σ = 73 (830)/[001] symmetrical tilt boundary in Ni-24%Al at 600 K.

grain boundary as the result of Ni segregation. An extensive "smearing" between Ni and Al atoms takes place on most of the boundary sites and as a consequence, the grain boundary changes even topologically and becomes structurally akin to that in Cu_3Au and/or in an f.c.c. material. This is seen in Figure 2-7, which shows the structure of the Σ73 {830} grain boundary in Ni-24%Al at 600 K; shading of the circles representing atoms has the same meaning as in Figure 2-5.

The situation is, however, very different when the bulk is enriched with Al. Figures 2-8a and b show the excess aluminum concentration on {16 6 0} layers parallel to the Σ73{830} boundary at the bulk compositions Ni-26%Al and Ni-30%Al, respectively, at 600 K. At the bulk composition Ni-26%Al, aluminum segregates strongly to certain atomic sites; these correspond to the four sharp peaks seen in Figure 2-8a. As the bulk concentration of Al increases, aluminum segregation is intensified but is still principally limited to the same atomic sites. This is seen in Figure 2-8b where the main peaks are situated at the same {16 6 0} layers as in Figure 2-8a but their height increased. Such a phenomenon indicates that only some very specific atomic positions in the boundary region can serve as segregation sites for aluminum and it can be speculated that once all these site are filled by Al atoms the grain boundary will become saturated and no further segregation will take place. More importantly, unlike in the case of Ni segregation, the Al segregation does not lead to interchanges between Ni and Al atoms (no negative peaks occur in Figures 2-8a, b) and, therefore, no compositional disordering develops in the vicinity of the boundary plane. On the contrary, a two-dimensional ordered structure is formed in the boundary region in this case. Topologically the boundary structure remains the same as in the stoichiometric case shown in Figure 2-1b.

The atomistic MC calculations show that in nonstoichiometric Ni_3Al alloys both Ni and Al segregate to grain boundaries when in surplus in the

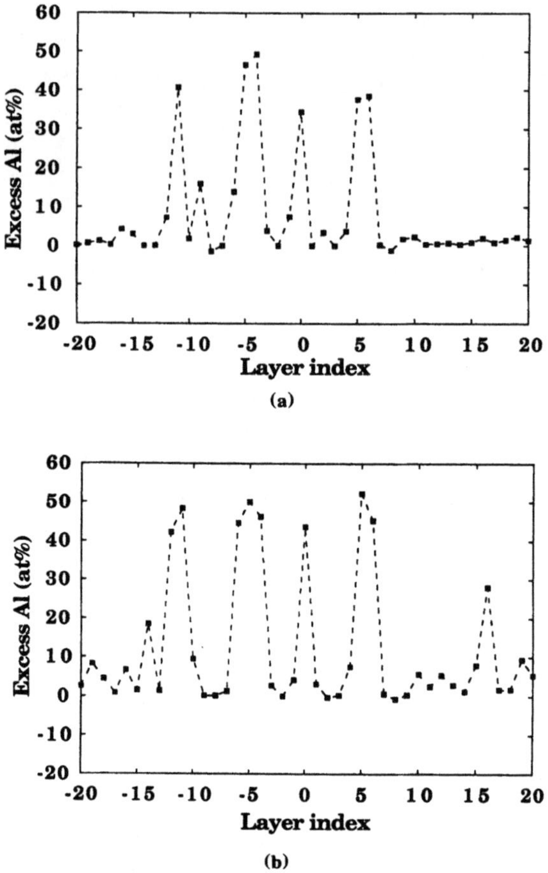

Figure 2-8. Excess concentration of Al (in units of atomic percent) in different {16 6 0} layers parallel to the Σ73 {830} grain boundary at 600 K at the bulk compositions (a) Ni-26%Al and (b) Ni-30%Al.

bulk. However, the impact of the Ni segregation on the boundary structure is very different from that of Al segregation. When nickel segregates, the grain boundary region becomes chemically disordered since additional interchanges of Ni and Al atoms accompany the segregation process and the species "smear" in the boundary region. In contrast, in the case of aluminum segregation, the Al atoms occupy only certain specific sites in the grain-boundary region, and no significant compositional disordering occurs. The selectivity of aluminum segregation is, presumably, related to the fact that Al is a large atom and can only be accommodated at sites associated with a large free volume. Since the segregation of Al is very selective, a two-dimensional ordered Ni-Al compound with a high Al content is formed at the grain boundary. Formation of such ordered structure has also been observed in recent atomistic studies of Najafabadi and Srolovitz (1993), who interpret such structures as narrow regions of NiAl.

The extent of Ni (or Al) segregation varies from boundary to boundary (Min Yan 1992; Min Yan et al. 1992; Vitek et al. 1992) but the compositionally disordered region formed in Ni-rich Ni_3Al is never of macroscopic dimensions but spans only a few lattice spacings. However, this may be sufficient to affect the crack nucleation and propagation along grain boundaries. After nickel segregation the grain-boundary region becomes more akin to that found in weakly ordered alloys and, as discussed above, both the crack nucleation and propagation may be more difficult. In contrast, in Al-rich Ni_3Al the two-dimensional compositionally ordered structure, formed due to the segregation of Al, may render the material even more brittle. This considerable difference between the structural changes invoked by segregation of Ni and Al, respectively, may be the principal reason why only in Ni-rich alloys the intergranular brittleness can be inhibited by alloying with boron.

4. GRAIN BOUNDARIES IN NiAl

NiAl, crystallizing in a b.c.c.-based B2 structure, is the primary candidate as the advanced structural material for high pressure turbine blades because of its relatively low density, excellent oxidation resistance, and high melting temperature (Darolia et al. 1992; Sauthoff 1989). The main problem encountered in the application of this alloy is again the low ductility at room temperature. At this temperature, fracture mode of stoichiometric NiAl is predominantly intergranular (Schulson and Barker 1983; Hahn and Vedula 1989) but, unlike in Ni_3Al, the material brittleness cannot be alleviated by boron alloying in spite of the fact that the alloying suppresses intergranular fracture and the fracture mode changes to transgranular. The reason is that boron is an extremely potent solid solution strengthener and thus any beneficial effect of boron associated with suppression of intergranular cracking is more than offset by the dramatic increase of the yield strength (George and Liu 1990, 1991). Similarly, small deviations from stoichiometry can result in significant hardening of NiAl (Ball and Smallman 1966; Pascoe and Newey 1968).

The grain-boundary brittleness of stoichiometric NiAl again appears to be intrinsic, similarly as in the case of Ni_3Al, since no fracture inducing impurities were detected on the corresponding fracture surfaces by Auger spectroscopy (George and Liu 1990). Hence, the atomic and chemical structure of the grain boundaries in this compound is again likely to provide the clue for understanding the reasons for intergranular fracture. However, the structure of grain boundaries in NiAl has not yet been studied to such an extent as in Ni_3Al.

An important finding attained by atom probe field ion microscopy is that grain boundaries in the stoichiometric NiAl are frequently rich in nickel and the zones depleted of aluminum are narrow, extending about 0.5 nm from the boundary plane. The same experiments also revealed that in boron-doped NiAl, boron segregates to the boundaries (Camus et al. 1988; Miller, Jayram, and Camus 1992). Recently, a high resolution electron microscopic observation of the structure of the $\Sigma = 5$ (310)/[001] tilt boundary has been carried out (Fonda and Luzzi 1993). The atomic structure was analyzed by comparing simulated images generated by the multislice algorithm (e.g., Self and O'Keefe 1988; Spence 1988) for a variety of atomistic (unrelaxed) models. While crystallographically this is a symmetrical tilt boundary, the boundary exhibits an asymmetry produced by a relative rigid body displacement of the adjoining grains parallel to the boundary plane; however, there is no measurable expansion at the boundary. While these features are essential for matching the observed and calculated images, the most important feature is that a very close match is achieved by allowing for changing the stoichiometry of the boundary region. The model structure that leads to the best match is shown schematically in Figure 2-9. In this structure the aluminum sites adjacent to the boundary plane are replaced 50% by vacancies and 50% by nickel antisite defects, while 50% of nickel sites adjacent to the boundary plane contain vacancies. This produces nickel-rich nonstoichiometry in the boundary region that is, in general, compatible with the above-mentioned atom probe studies. Furthermore, the overall nickel enrichment is adjusted to a lower level by formation of nickel vacancies rather than by replacing some nickel atoms by aluminum which is congruent with the observation that aluminum rich nonstoichiometry is accommodated by constitutional vacancies rather than by aluminum antisite defects (Bradley and Taylor 1937). Notwithstanding, a definitive answer whether this proposed structure is the correct minimum energy configuration can only be obtained by combining the image analysis with a full scale atomistic study of the boundary which allows for minimization of the free energy with respect to the atomic positions analogously, as it was done in the recent investigation of the boundary facets produced by segregation of bismuth in a Cu-Bi alloy (Luzzi et al. 1991).

Only very limited atomistic studies of grain boundaries in NiAl have been made using the embedded atom type potentials originally constructed for Ni_3Al. The calculations of Chen et al. (1989) suggest that the

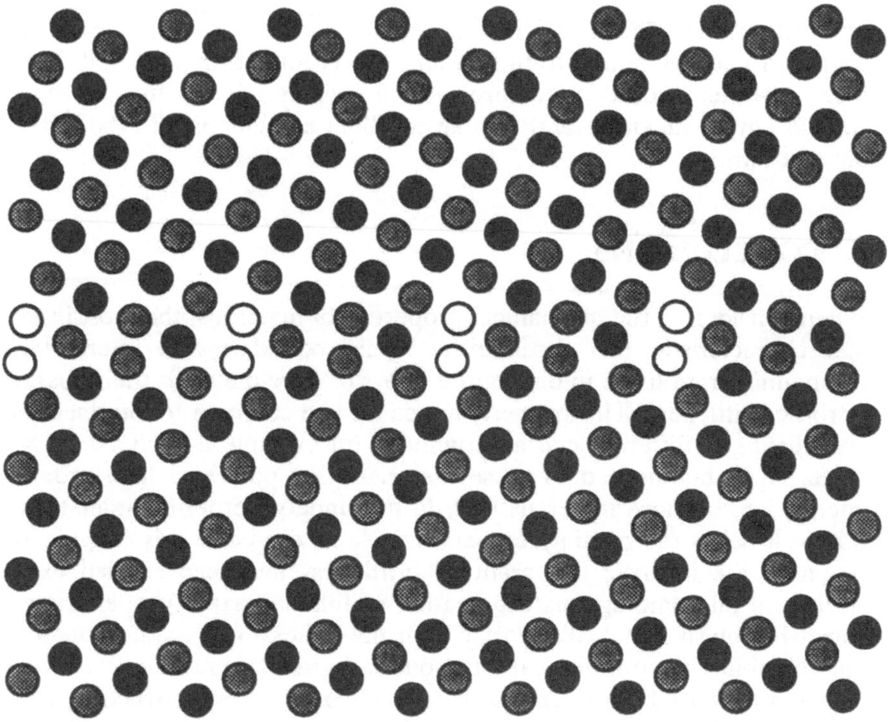

Figure 2-9. Model structure of the $\Sigma = 5$ (310) tilt boundary in NiAl which leads to the best match with the observations of fonda and luzzi (1993). The dark circles represent aluminum atoms, lightly shaded circles nickel atoms and open circles the atomic sites, half of which are occupied by nickel atoms and half by vacancies.

nickel-rich alloy will have a large concentration of nickel antisite defects in the bulk and aluminum vacancies in the grain boundary while aluminum rich alloy will have nickel vacancies both in the bulk and in the grain boundary. Calculations of Petton and Farkas (1991) showed an extensive multiplicity of possible boundary structures and suggest that the lowest energy structures correspond to boundaries with depleted aluminum. These findings of the atomistic studies are in a general agreement with experimental observations (Bradley and Taylor 1937; Camus et al. 1988; Miller et al. 1992). Nonetheless, the lowest energy structure of the $\Sigma = 5$ (310)/[001] boundary calculated by Petton and Farkas (1991) does not reproduce the HREM image observed by Fonda and Luzzi (1993) with acceptable precision. This is not surprising, however, since the structure leading to the best match of images, described above, could not be attained in the molecular statics calculations without choosing the starting configuration very similar to this structure. Recently, Min Yan et al. (1994) found another ordered structure with a surplus of Ni which does not contain any Al vacancies and is in an excellent agreement with the HREM observations.

5. CONCLUSIONS

When considering the mechanical properties of materials, the most important characteristic that distinguishes grain boundaries in intermetallic compounds from those in disordered alloys or pure metals is their possible intrinsic brittleness. This property appears to be common to intermetallics crystallizing in different crystal structures, for example $L1_2$ Ni_3Al, as well as B2 NiAl, but limited to those materials that possess a high ordering energy, i.e., to those materials that do not undergo order-disorder transition well below the melting temperature. As discussed in this chapter, the origin of the intrinsic intergranular brittleness has been studied extensively, but an unambiguous explanation of this important property is still somewhat elusive. It is likely that both the cohesion at grain boundaries and the ability to emit dislocations from intergranular cracks play a crucial role and they are both related to the details of the atomic structure at the boundaries. In particular, it seems to be important whether the stoichiometric order is preserved up to the boundary or whether a chemical disorder evolves in the boundary region. The latter is, presumably, common in alloys with a low ordering energy and may be an important factor in alleviating brittleness in strongly ordered alloys. This would suggest that off-stoichiometry and alloying by elements inducing disorder could improve the ductility of intermetallic compounds. However, in practice the situation is more complex. Alloying and off-stoichiometry may also lead to a significant increase of the yield stress in the bulk, as for example in NiAl, which more than outweighs their beneficial effect on grain-boundary brittleness. Furthermore, off-stoichiometry and associated segregation to grain boundaries do not necessarily lead to chemical disorder. An example is aluminum-rich Ni_3Al, for which the calculations suggest formation of two-dimensional ordered phases at grain boundaries.

Since it is the grain-boundary region which spans only a few lattice spacing that governs the boundary properties, a fundamental comprehension of the mechanical response of these interfaces can only be attained through detailed understanding of their atomic and chemical structure. Thus, the experimental and theoretical approaches attacking this problem are likely to play a major role in future research. On the experimental side, the techniques revealing the atomic and chemical structure of interfaces, such as the high resolution electron microscopy, are the most important. However, these have to be pursued in conjunction with theoretical, atomistic studies of grain boundaries in a synergistic manner.

While atomistic studies have revealed a number of important features of grain boundaries, these findings are mostly qualitative. The reasons are principally twofold. First, the atomistic calculations have all been carried out for relatively special, short-period boundaries, while boundaries in real materials are typically of more general nature. Second, the interatomic forces employed in most atomistic calculations are relatively simple central forces that may not be sufficient for alloys of transition metals. The first limitation can be diminished with the increase of computer power and by developing a more detailed understanding of the relationship between general and short period boundaries. An example of the latter is the structural unit model of grain boundaries (Sutton and Vitek 1983; Sutton and Balluffi 1990). The second shortcoming could, in principle, be resolved by carrying ab initio quantum mechanical calculations (e.g., Freeman 1992; Freeman et al. 1992) but these are usually limited to very simple structures and only very limited applications to grain boundaries in pure elements have been made (Tarnow et al. 1990). The approach that has been dominant in the last few years is the use of the embedded atom method (Daw and Baskes 1984) and of the many-body Finnis–Sinclair type potentials (Finnis and Sinclair 1984). While this development represents a significant advancement over the previously used pair-potentials, it does not include the possible noncentral character of atomic interactions that may be of importance in transition metal alloys in general, and particularly if calculations are to be carried out for intermetallic compounds that do not crystallize in cubic structures, such as aluminum–titanium alloys. The most promising developments in which the noncentral character of interatomic forces is taken into account, and which may be applicable in extensive studies of interfaces, are the bond-order potentials (Pettifor 1989; Pettifor and Aoki 1991) and other tight-binding methods (Sob, Vitek, and Oh 1992). Last, but not least, is the establishment of the link between the atomic and chemical structure of grain boundaries and cohesive and deformation properties of the grain-boundary region. The interplay between the cohesion and the ability to emit dislocations at the crack tips is a general problem in quasi-brittle materials of which intermetallics are prime examples. These studies require a combined atomistic and continuum-mechanics approach and only first steps have been taken in this direction (Rice 1991; Sun et al. 1991).

ACKNOWLEDGEMENTS

This research was supported by the Department of Energy, Office of Basic Energy Sciences, Grant No. DE-FG02-87ER45295.

REFERENCES

AOKI, K., and IZUMI, O., 1979, *J. Japan Inst. Metals* 43: 1190.

BAKER, I., DAROLIA, R., WHITTENBERGER, J. D., and YOO, M. H., eds., 1993, *High-Temperature Ordered Intermetallic Alloys V*, MRS Symp. Proc. Vol. 288.

BAKER, I., and SCHULSON, E. M., 1989, *Scripta Metall.* 23: 1883.

BAKER, I., SCHULSON, E. M., MICHAEL, J. R., and PENNYCOOK, S. J., 1990, *Phil. Mag.* B 62: 659.

BALL, A., and SMALLMAN, R. E., 1966, *Acta Metall.* 14: 1349.

BARTHOLOMEUSZ, M. F., and WERT, J. A., 1992a, *Acta Metall. Mater* 40, 673.

________, 1992b, *J. Mater. Res.* 7: 919.

BINDER, K., 1984, *Application of the Monte Carlo Method in Statistical Physics* (Berlin: Springer).

BRADLEY, A. J., and TAYLOR, A., 1937, *Proc. Roy. Soc. London* A 159: 56.

BRENNER, S. S., and MING-JIAN, H., 1991, *Scripta. Metall. Mater.* 25: 1271.

BRIANT, C. L., and MESSMER, R. P., 1984, *Acta Metall.* 32: 2043.

CAMUS, P. P., BAKER, I., HORTON, J. A., and MILLER, M. K., 1988, *J. Phys. France* 49: C6–329.

CHANG, K. M., DAROLIA, R., and LIPSITT, H. A., 1992, *Acta Metall. Mater* 40: 2727.

CHEN, S. P., SROLOVITZ, D. J., and VOTER, A. F., 1989, *J. Mat. Res.* 4: 62.

CHEN, S. P., VOTER, A. F., ALBERS, R. C., BORING, A.M., and HAY, P. J., 1989, *Scripta Metall.* 23: 217.

________, 1990, *J. Mat. Res.* 5: 955.

CHEN, S. P., VOTER, A. F., BORING, A. M., ALBERS, R. C., and HAY, P. J., 1989, *High-Temperature Ordered Intermetallics Alloys III*, Materials Research Society 133: 149.

CHEN, S. P., VOTER, A. F., and SROLOVITZ, D. J., 1986, *Scripta Metall.* 20: 1389.

DAROLIA, R., LAHRMAN, D. F., FIELD, R. D., DOBBS, J. R., CHANG, K. M., GOLDMAN, E. H., and KONITZER, D. G., 1992, *Ordered Intermetallics—Physical Metallurgy and Mechanical Behaviour*, ed. C. T. Liu, R. W. Cahn, and G. Sauthoff, p. 679. Dordrecht, Kluwer Academic Publishers.

DAW, M. S., and BASKES, M. I., 1984, *Phys. Rev. B.*, 29: 6443.

DIMIDUK, D. M., 1991, *Journal de Physique III*, 1: 1025.

DIMIDUK, D. M., WEDDINGTON, V. L., and LIPSITT, H. A., 1987, *High-Temperature Ordered Intermetallics Alloys II*, MRS Symp. Proc. 81: 221.

EBERHARD, M. E., and VVEDENSKY, D. D., 1988, *Scripta Metall.*, 22: 1183.

FINNIS, M. W., and SINCLAIR, J. E., 1984, *Phil. Mag. A*, 50: 45.

FOILES, S. M., 1985, *Phys. Rev. B*, 32: 7685.

________, 1989, *Phys. Rev. B* 40: 11502.

FONDA, R. W., and LUZZI, D. E., 1993, *Phil. Mag. A*, 68, 1151.

FORWOOD, C. T., and GIBSON, M. A., 1992, *Phil. Mag. A*, 66: 1121.

FREEMAN, A. J., 1992, *Ber. Bunsen Ges. Phys. Chem.*, 96: 1512.

FREEMAN, A. J., XU, J.-H., HONG, T., and LIN, W., 1992, *Ordered Intermetallics—Physical Metallurgy and Mechanical Behaviour*, ed. C. T. Liu, R. W. Cahn, and G. Sauthoff, p. 1. Dordrecht: Kluwer Academic Publishers.

FROST, H. J., 1987, *Acta Metall.*, 35: 519.

GEORGE, E. P., HORTON, J. A., PORTER, W. D.,and SCHNEIBEL, J. H., 1990, *J. Mater. Res.* 5: 1639.

GEORGE, E. P., and LIU, C. T., 1990, *J. Mater. Res.*, 5: 754.

________, 1991, *Alloy Phase Stability and Design*, Materials Research Society 186: 309.

GEORGE, E. P., LIU, C. T., and PADGETT, R. A., 1989, *Scripta. Metall. Mater.*, 23: 979.

GOODWIN, L., NEEDS, R. J., and HEINE, V., 1988, *Phys. Rev. Lett.*, 60: 2050

________, 1990, *J. Phys.: Condens. Matter*, 2: 351.

HACK, J. E., CHEN, S. P., and SROLOVITZ, D. J., 1989, *Acta Metall.*, 37: 1957.

HAHN, K. H., and VEDULA, K., 1989, *Scripta Metall.*, 23: 7.

HECKER, S. S., ROHR, D. L., and STEIN, D. F., 1978, *Metall. Trans. A*, 9: 481.

HIRTH, J. P., and RICE, J. R., 1980, *Metall. Trans. A*, 11: 1501.

IZUMI, O., ed. 1991, *International Symposium on Intermetallic Compounds—Structure and Mechanical Properties—(JIMIS-6)* Sendai, Japan: The Japan Institute of Metals.

IZUMI, O., and TAKASUGI, T., 1988, *J. Mat. Res.* 3:426.

JOHNSON, L. A., POPE, D. P., and STIEGLER, J. O., eds., 1991, *High-Temperature Ordered Intermetallic Alloys IV*, MRS Symp. Proc. 213.

JOHNSON, W. C., and BLAKELEY, J. M., eds., 1977, *Interfacial Segregation*, Metals Park, OH: ASM.

JOKL, M. L., VITEK, V., and MCMAHON, C. J., JR., 1980, *Acta Metall.*, 28: 2479.

JOKL, M. L., VITEK, V., MCMAHON, J., C. J., and BURGERS, P., 1989, *Acta Metall.*, 37: 87.

KIMURA, A., IZUMI, O., IGARASHI, J., MISAWA, T., and TAKASUGI, T., 1991, *International Symposium on Intermetallic Compounds—Structure and Mechanical Properties—(JIMIS-6)*. ed. O. Izumi, p. 737. Sendai: The Japan Institute of Metals.

KING, A. H., FROST, H. J., and YOO, M. H., 1991, *Scripta. Metall. Mater.*, 25: 1249.

KING, A. H., and YOO, M. H., 1987, *Scripta Metall.* 21: 1115.

KNOTT, J. F., 1977, *Fracture 77* 1: 61. Waterloo, Ontario: University of Waterloo Press.

________, 1987, *Atomistics of Fracture*, ed. R. M. Latanision and J. Pickens, p. 209. New York: Plenum Press.

KRUISMAN, J. J., VITEK, V., and DEHOSSON, J. T. M., 1988, *Acta Metall.* 36: 2729.

KRZANOWSKI, J. E., 1989, *Scripta. Metall. Mater.*, 23: 1219.

KUMAR, K. S., and BROWN, S. A., 1992a, *Acta Metall. Mater*. 40: 1923.

________, 1992b, *Phil. Mag. A*, 65: 91.

KUMAR, K. S., MANNA, S. K., and VISAWANADHAM, R. K., 1992, *Acta Metall. Mater*, 40: 1201.

KUNG, H., RASMUSSEN, D. R., and SASS, S. L., 1991, *Scripta. Metall. Mater*., 25: 1277.

________, 1992, *Acta Metall. Mater*., 40: 81.

________, 1992b, *Ordered Intermetallics—Physical Metallurgy and Mechanical Behaviour*, ed. C. T. Liu, R. W. Cahn, and G. Sauthoff, p. 355. Dordrecht: Klewer Academic Publishers.

KUNG, H., and SASS, S. L., 1992, *Acta Metall. Mater* 40: 99.

LATHANISION, R. M., and JONES, R. H., eds., 1987, *Chemistry and Physics of Fracture*, Boston, Martinus Nijhoff.

LEE, T. C., ROBERTSON, I. M., and BIRNBAUM, H. K., 1992, *Acta Metall. Mater*, 40: 2569.

LEE, T. C., SUBRAMANIAN, R., ROBERTSON, I. M., and BIRNBAUM, H. K., 1991, *Scripta. Metall. Mater*., 25: 1265.

LESAR, R., NAJAFABADI, R., and SROLOVITZ, D. J., 1989, *Phys. Rev. Lett*., 63: 624.

LIN, H., and POPE, D. P., 1993, *Acta Metall. Mater*, 41: 553.

LIN, I.-H., and THOMSON, R., 1986, *Acta Metall*. 34: 187.

LIU, C. T., 1988, *Interfacial Structure, Properties and Design*, Materials Research Society 122.

________, 1991a, *International Symposium on Intermetallic Compounds—Structure and Mechanical Properties—(JIMIS-6)*, ed. O. Izumi, p. 703, Sendai: The Japan Institute of Metals.

________, 1991b, *Scripta. Metall. Mater*., 25: 1231.

________, 1992, *Ordered Intermetallics—Physical Metallurgy and Mechanical Behaviour*, ed. C. T. Liu, R. W. Cahn, and G. Sauthoff, p. 321, Dordrecht: Kluwer Academic Publishers.

LIU, C. T., CAHN, R. W., and SAUTHOFF, G., eds., 1992, *Ordered Intermetallics—Physical Metallurgy and Mechanical Behaviour*, Dordrecht: Kluwer Academic Publishers.

LIU, C. T., WHITE, C. L., and HORTON, J. A., 1985, *Acta Metall*, 33: 213.

LUZZI, D. E., MIN YAN, SOB, M., and VITEK, V., 1991, *Phys. Rev. Lett*, 67: 1894.

MASUDA-JINDO, K., 1988, *J. Phys. France*, 49: C5–557.

MESSMER, R. P., and BRIANT, C. L., 1982, *Acta Metall. Mater*, 30: 487.

METROPOLIS, N., ROSENBLUTH, A. W., ROSENBLUTH, N. M., and TELLER, A. H., 1953, *J. Chem. Phys*., 21: 10.

MILLER, M. K., JAYRAM, R., and CAMUS, P. P., 1992, *Scripta. Metall. Mater*., 26: 679.

MILLS, M. J., 1989, *Scripta Metall*, 23: 2061.

MILLS, M. J., and DAW, M. S., 1991, *High Resolution Electron Microscopy of Defects in Materials*, MRS Symp. Proc. 183: 15.

MILLS, M. J., GOODS, S. H., and FOILES, S. M., 1992, *Structure and Properties of Interfaces in Materials*, MRS Symp. Proc. 238: 127.

MILLS, M. J., GOODS, S. H., FOILES, S. M., and WHETSTONE, J. R., 1991, *Scripta. Metall. Mater.*, 25: 1283.

MIN YAN, 1992, Ph.D. Thesis, University of Pennsylvania.

MIN YAN, FONDA, R. W. AND LUZZI, D. E., 1994, *Phil. Mag. A*, to be published.

MIN YAN, VITEK, V., and ACKLAND, G. J., 1992, *Structure and Properties of Interfaces in Materials*, MRS Symp. Proc. 238: 139.

MORRIS, D. G., and LERF, R., 1991, *Phil. Mag. A*, 63: 1195.

NAJAFABADI, R., and SROLOVITZ, D. J., 1993, *High-Temperature Ordered Intermetallic Alloys V*, MRS Symp. Proc., 288: 189.

NAJAFABADI, R., SROLOVITZ, D. J., and LESAR, R., 1990, *J. Mater. Res.* 5: 2662.

NAJAFABADI, R., WANG, H. Y., SROLOVITZ, D. J., and LESAR, R., 1991a, *High-Temperature Ordered Intermetallic Alloys IV*, MRS Symp. Proc. 213: 51.

________, 1991b, *Acta Metall. Mater*, 39: 3071.

NISHIMURA, C., and LIU, C. T., 1992, *Scripta. Metall. Mater.*, 27: 1307.

PASCOE, R. T., and NEWEY, C. W. A., 1968, *Met. Sci. J*, 2: 138.

PESTMAN, B. J., and DEHOSSON, J. T. M., 1992, *Acta Metall. Mater*, 40: 2511.

PESTMAN, B. J., DEHOSSON, J. T. M., VITEK, V., and SCHAPINK, F. W., 1991a, *Phil. Mag. A*, 64: 951.

________, 1991b, *High-Temperature Ordered Intermetallic Alloys IV*, MRS Symp. Proc. 213: 429.

PETTIFOR, D. G., 1989, *Phys. Rev. Lett.*, 63: 2480.

PETTIFOR, D. G., and AOKI, M., 1991, *Phil. Trans. Roy. Soc. London A* 334: 439.

PETTON, G., and FARKAS, D., 1991, *Scripta Metall*, 25: 55.

POPE, D. P., 1990, *High Temperature Aluminides and Intermetallics*, ed. S. H. Whang, C. T. Liu, D.P. Pope, and J. O. Stiegler, p. 51. Warrendale, PA: TMS.

________, 1992, *Ordered Intermetallics—Physical Metallurgy and Mechanical Behaviour*, ed. C. T. Liu, R. W. Cahn, and G. Sauthoff, p. 143. Dordrecht: Kluwer Academic Publishers.

POPE, D. P., and EZZ, S. S., 1984, *Int. Met. Rev*, 25: 233.

PUGH, S. F., 1991, *An Introduction to Grain Boundary Fracture in Metals*, London: The Institute of Metals.

RICE, J. R., 1991, *J. Mech. Phys. Sol*, 40: 239.

RICE, J. R., and THOMSON, R., 1974, *Phil. Mag*, 29: 73.

RICE, J. R., and WANG, J.-S., 1989, *Mat. Sci. Eng. A*, 107: 23.

ROBERTSON, I. M., LEE, T. C., SUBRAMANIAN, R., and BIRNBAUM, H. K., 1992, *Structure and Properties of Interfaces in Materials*, MRS Symp. Proc. 238: 357.

SAGI, M. A., and PETTIFOR, D. G., 1987, *Phil. Mag. Lett*, 56: 245.

SAUTHOFF, G., 1989, *Z. Metallkunde* 80: 337.

SCHULSON, E. M., 1991, *International Symposium on Intermetallic Compounds—Structure and Mechanical Properties—(JIMIS-6)*, ed. O. Izumi, p. 339, Sendai: The Japan Institute of Metals.

SCHULSON, E. M., and BAKER, I., 1991, *Scripta. Metall. Mater*, 25: 1253.

———, 1992, *Ordered Intermetallics—Physical Metallurgy and Mechanical Behaviour*, ed. C. T. Liu, R. W. Cahn, and G. Sauthoff, p. 371, Dordrecht: Kluwer Academic Publishers.

SCHULSON, E. M., and BARKER, D. R., 1983, *Scripta Metall*, 17: 519.

SCHULSON, E. M., WEIHS, T. P., BAKER, I., FROST, H. J., and HORTON, J. A., 1985a, *Scripta Metall*, 19: 1497.

SCHULSON, E. M., WEIHS, T. P., VIENS, D. V., and BAKER, I., 1985b, *Acta Metall*, 33: 1587.

SCHULSON, E. M., XU, Y., MUNROE, P. R., GUHA, S., and BAKER, I., 1991, *Acta Metall. Mater*, 39: 2971.

SELF, P. G., and O'KEEFE, M. A., 1988, *High Resolution Transmission Electron Microscopy and Associated Techniques*, ed. P. R. Buseck, J. M. Cowley and L. Eyring, p. 244, Oxford: Oxford University Press.

SIELOFF, D. N., BRENNER, S. S., and HUA, M. J., 1989, *High-Temperature Ordered Intermetallics Alloys III*, MRS Symp. Proc. 133: 155.

SOB, M., VITEK, V., and OH, Y., 1992, *Computational Methods in Materials Science*, MRS Symp. Proc. 278: 205.

SPENCE, J. C. H., 1988, *Experimental High-Resolution Electron Microscopy*, Oxford: Oxford University Press.

SUN, Y., RICE, J. R., and TRUSKINOVSKY, L., 1991, *High Temperature Ordered Intermetallic Alloys IV*, MRS Symp. Proc. 213: 243.

SUTTON, A. P., 1984, *Int. Met. Rev.*, 29: 377.

SUTTON, A. P., 1989a, *Atomistic Simulations of Materials: Beyond Pair Potentials*, ed. V. Vitek and D. Srolovitz, p. 265. New York: Plenum Press.

———, 1989b, *Philos. Mag. A* 60: 147.

———, 1990, *J. Phys. France*, 51: C1–35.

———, 1992, *Phil Trans Roy Soc London A*, 341: 233.

SUTTON, A. P., and BALLUFFI, R. W., 1990, *Phil. Mag. Lett*, 61: 91.

SUTTON, A. P., and VITEK, V., 1983, *Phil. Trans. Roy. Soc. London A*, 309: 1.

TAKASUGI, T., GEORGE, E. P., POPE, D. P., and IZUMI, O., 1985, *Scripta Metall*, 19: 551.

TAKASUGI, T., HANADA, S., and YOSHIDA, M., 1992, *J. Mat. Res.*, 7: 2739.

TAKASUGI, T., HIRAKAWA, S., IZUMI, O., ONO, S., and WATANABE, S., 1987, *Acta Metall*, 35: 2015.

TAKASUGI, T., and IZUMI, O., 1985a, *Acta Metall*, 33: 1247.

———, 1985b, *Acta Metall*, 33: 39.

———, 1991, *Scripta. Metall. Mater*, 25: 1243.

———, 1992, *Ordered Intermetallics—Physical Metallurgy and Mechanical Behaviour*, ed. C. T. Liu, R. W. Cahn, and G. Sauthoff, p. 391, Dordrecht: Kluwer Academic Publishers.

TAKASUGI, T., IZUMI, O., and MASAHASHI, N., 1985, *Acta Metall.* 33: 1259.

TAKASUGI, T., NAGASHIMA, M., and IZUMI, O., 1990, *Acta Metall. Mater*, 38: 747.

TARNOW, E., DALLOT, P., BRISTOWE, P. D., JOANNOPOULOS, J. D., FRANCIS, G. P., and PAYNE, M. C., 1990, *Phys. Rev. B.*, 42: 3644.

TAUB, A. I., and BRIANT, C. L., 1987, *Acta Metall*, 35: 1597.

TAUB, A. I., BRIANT, C. L., HUANG, S. C., CHANG, K. M., and JACKSON, M. R., 1986, *Scripta. Metall. Mater*, 20: 129.

THOMSON, R., 1986, *Solid State Physics*, ed. H. Ehrenreich and D. Turnbull, 39: 2.

TICHELAAR, F. D., SCHAPINK, F. W., and LI, X., 1992, *Phil. Mag. A*, 65: 913.

VITEK, V., 1984, *Dislocations 1984*, ed. P. Veyssiere, L. Kubin, and J. Castaign, p. 435, Paris: Editions CNRS.

________, 1991, *J. Phys. France III*, 1: 1085.

VITEK, V., ACKLAND, G. J., and CSERTI, J., 1991, *Alloy Phase Stability and Design*, MRS Symp. Proc. 186: 237.

VITEK, V., and CHEN, S. P., 1991, *Scripta. Metall. Mater*, 25: 1237.

VITEK, V., CHEN, S. P., VOTER, A. F., KRUISMAN, J. J., and DEHOSSON, J. T. M., 1989, *Grain Boundary Chemistry and Intergranular Fracture*, Materials Science Forum 46: 237.

VITEK, V., MIN YAN, and ACKLAND, G. J., 1992, *Ordered Intermetallics—Physical Metallurgy and Mechanical Behaviour*, ed. C. T. Liu, R. W. Cahn, and G. Sauthoff, p. 335, Dordrecht: Kluwer Academic Publishers.

VITEK, V., SODANI, Y., and CSERTI, J., 1991, *High-Temperature Ordered Intermetallic Alloys IV*, MRS Symp. Proc. 213: 195.

VOTER, A. F., CHEN, S. P., ALBERS, R. C., BORING, A. M., and HAY, P. J., 1989, *Atomistic Simulations of Materials: Beyond Pair Potentials*, ed. V. Vitek and D. Srolovitz, p. 223, New York: Plenum Press.

WAS, G. S., and BRUEMMER, S. M., eds., 1989, *Grain Boundary Chemistry and Intergranular Fracture*, Materials Science Forum 46.

WEERTMAN, J., 1978, *Acta Metall*, 26: 1731.

WHANG, S. H., LIU, C. T., POPE, D. P., and STIEGLER, J. O., eds., 1990, *High Temperature Aluminides and Intermetallics*, Warrendale, PA: TMS.

WHITE, C. L., CLAUSING, R. E., and HEATHERLEY, L., 1979, *Metall. Trans. A*, 10: 683.

WHITE, C. L., LIU, C. T., and PADGETT, J., R. A., 1988, *Acta Metall*, 36: 2229.

WHITE, C. L., PADGETT, R. A., LIU, C. T., and YALISOVE, S. M., 1984, *Scripta Metall*, 18: 1417.

WU, Z. L., POPE, D. P., and VITEK, V., 1991a, *Scripta Metall.* 24: 2187.

________, 1991b, *Scripta Metall.* 24, 2191.

YAMAGUCHI, M., and UMAKOSHI, Y., 1990, *Prog. Mater. Sci*, 34: 1.

YOO, M. H., and KING, A. H., 1988, *J. Mat. Res*, 3: 848.

Chapter 3

Brittle Fracture and Toughening

E. M. Schulson

ABSTRACT

This chapter reviews the intrinsic and the extrinsic factors that contribute to the brittle fracture of strongly ordered polycrystalline intermetallic compounds, and then considers methods for improving the ductility and the fracture toughness of this class of materials. A high work-hardening rate appears to be common to all brittle intermetallics and may be a major factor in limiting the energy that can be absorbed by plastic flow near the tips of propagating cracks. The incorporation of a ductile, random solid solution within the matrix is an effective toughening method.

1. INTRODUCTION

Brittle fracture or low crack tolerance is the primary barrier to the use of intermetallic compounds in load-bearing applications. This chapter considers the problem. The discussion addresses the factors contributing to this behavior, and then considers possible solutions. The focus is on the tensile ductility and toughness of strongly ordered intermetallic polycrystals. The emphasis is on behavior at room temperature where premature breakage of intermetallic components would prevent service at any temperature.

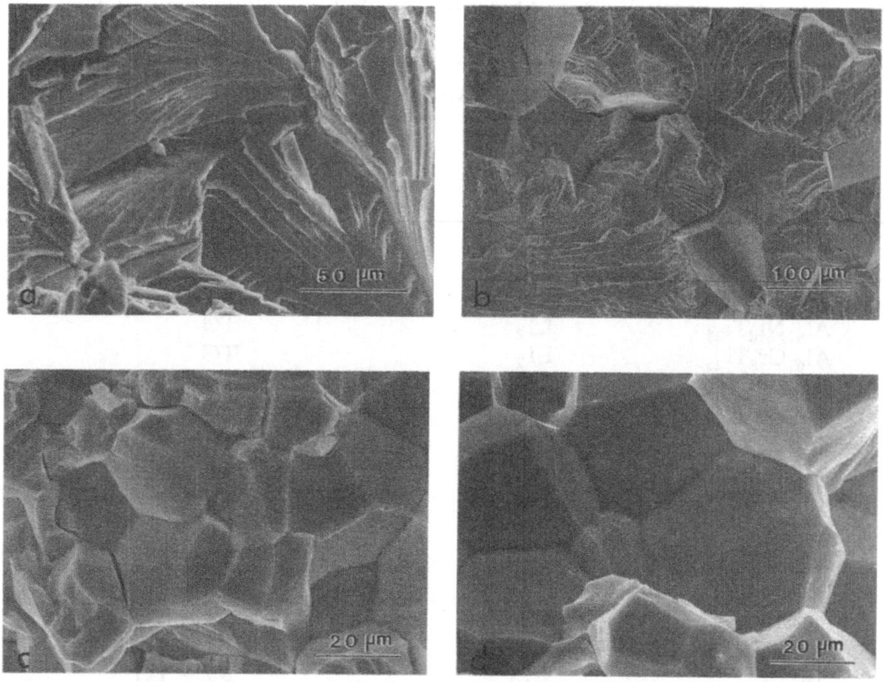

Figure 3-1. SEM fractographs showing typical transgranular, intergranular and mixed modes of brittle fracture. The material is the B2 aluminide FeAl: a) Fe34Al; b) Fe40Al; c) Fe45Al; d) Fe50Al. Note that the transgranular mode increases with increasing deviation from the stoichiometric composition. (Reprinted with permission from *Scripta Metallurgica*, Volume 24, P. Nagpal and I. Baker, 1991, page 2381, Elsevier Science, Ltd. Pergamon Imprint, Oxford, England)

This chapter offers an overview; fuller accounts may be obtained from the literature cited and from chapters in this book on environmental embrittlement, creep, and fatigue.

Brittle fracture of intermetallics under uniaxial tension is marked by the absence of necking and by a fracture path perpendicular to the direction of loading. Occasionally, it can be preceded by significant global plasticity. The process occurs through either intergranular separation or transgranular cleavage (see Figure 3-1), or through a mixture of both modes (see Table 3-1), often with little more resistance than the energy to create new surface. Several factors contribute to the problem.

Table 3-1 Brittle Fracture Modes of Some Polycrystalline Intermetallic Compounds at Room Temperature

Intermetallic	Crystal Structure	Fracture Mode*
Ni_3Al	$L1_2$	IG
Ni_3Si	$L1_2$	IG
Ni_3Ga	$L1_2$	IG
Ni_3Ge	$L1_2$	IG
Pt_3Ga	$L1_2$	IG
Zr_3Al**	$L1_2$	IG
$Al_{67}Ni_8Ti_{25}$	$L1_2$	TG
$Al_{67}Cr_8Ti_{25}$	$L1_2$	TG
Fe_3Al	DO_3	TG
TiAl	$L1_0$	TG
Ti_3Al	DO_{19}	TG
Al_3Ti	DO_{22}	TG
Al_3Nb	DO_{22}	TG
Al_3Zr	DO_{23}	TG
$Mg_{17}Al_{12}$	α-Mn	TG
NiAl	B2	IG + TG
FeAl	B2	IG + TG
Ni30Al20Fe	B2	IG + TG
$MoSi_2$	C_{11b}	IG + TG
$Be_{12}Nb$	$Mn_{12}Th$ (b.c.t.)	IG + TG
Ti_5Si_3	$D8_8$	IG + TG

*IG = intergranular; TG = transgranular cleavage

**In presence of a notch; otherwise Zr_3Al fractures in a ductile mode

2. FACTORS CONTRIBUTING TO BRITTLE BEHAVIOR

The *crystal structure* is the most fundamental factor. Dislocations within an intermetallic with a complex crystal structure have large Burgers' vectors and large Peierls barriers. Consequently, the applied stress necessary for deformation by dislocation slip is greater than that for cracking. Brittle behavior then follows. The lightweight intermetallic $Mg_{17}Al_{12}$, for instance, possesses the α-Mn structure and has 58 atoms per unit cell. The material is so fragile that it cleaves even under the low pressure needed to hold it in a vice (Guha 1987). Indeed, of the thousands of intermetallics listed in Pearson (1967), most have crystal structures that are probably too complicated to allow extensive plastic deformation at room temperature. Fortunately, enough are sufficiently simple (B2, $L1_2$, DO_3, $L1_0$, and DO_{19}) to at least hold the promise of significant ductility.

Even when the crystal structure is relatively simple, intermetallics may not exhibit extensive ductility under tension. For instance, the high melting point alloy NiAl has the B2 or ordered b.c.c. structure, but fractures after 4% to 5% elongation (Rozner and Wasilewski 1966) or less (see reviews by Miracle 1993, Baker 1993). Similarly, the NiAl-based alloy $Ni_{30}Al_{20}Fe$ fractures after about 6% elongation (Guha et al. 1992). The primary reason for the low ductility in these cases is that the compound (Ball and Smallman 1966) and its derivative (Guha et al. 1992) slip on systems of the form $\{110\}\langle 001\rangle$ of which only three (Groves and Kelly 1963) are independent, and this conflicts with the von Mises requirement of five independent modes for extensive, crack-free deformation. Thus, another fundamental factor is the *number of independent deformation modes*.

Yet, when the crystal structure favors slip and the material possesses five independent deformation modes, polycrystals may still exhibit little ductility and fail in a brittle mode. An example is the ferromagnetic order/disorder alloy FeCo. This material slips on the $\{110\}\langle 111\rangle$ system (Yamaguchi, Umakoshi, and Yamane 1982) of which five modes are independent. When fully ordered with the B2 crystal structure, FeCo-2V extends only by about 5% before fracturing transgranularly (Johnston, Davies, and Stoloff 1965). A factor that appears to be important here is the *slip character* (Stoloff and Davies 1966), or the propensity for cross slip. The lower the propensity, the more planar or less wavy is the slip and the greater is the tendency to develop internal stresses to nucleate cracks. Slip is planar in fully ordered FeCo-2V, but is wavy in the disordered state; correspondingly, disordered material is ductile and fractures transgranularly through the nucleation and growth of microvoids following a reduction in area of 90% (Johnston et al. 1965).

When assessing the role of slip character in alloys that undergo an order-disorder transformation, care should be taken to consider the deformation modes as well. In the above case this is not an issue, because both the active slip system and the number of independent modes of deformation are independent of the degree of long-range atomic order. In the DO_{19} alloy Mg_3Cd, on the other hand, these characteristics depend upon the degree of order. In this case ordering actually increases the waviness of slip and promotes both basal and nonbasal slip (Stoloff and Davies 1964). The result is an increase in ductility from around 6% to around 20%. It would not be correct, therefore, to ascribe the increase in ductility of Mg_3Cd solely to a change in its slip character, because the increase in the number of independent deformation modes contributes as well.

Although wavy slip is desirable, planar slip is not necessarily a deleterious characteristic. For instance, the neutron-transparent $L1_2$ line compound Zr_3Al slips in a planar manner on the $\{111\}\langle 110\rangle$ system, but fails in a ductile mode after more than 30% elongation (for review see Schulson 1984). In this case, the slip lines are relatively finely spaced, suggesting that the *slip distribution* is another important factor. Fine slip may account for the thermally induced brittle-to-ductile transition in stoichiometric polycrystals of the lightweight $L1_0$ intermetallic TiAl (Imayev, Kaibyshev, and Salischev 1992) and for the superior (albeit still low) ductility of the polycrystalline form relative to the single crystalline form of the corrosion resistant aluminide FeAl (Gaydosh et al. 1992; Strothers and Vedula 1987). The mechanics are related to the concentration of plastic strain within and at the head of slip bands and the lessening of this and, hence, of the stress available to nucleate cracks by distributing slip on closely spaced planes. The best combination is fine, wavy slip and the worst is coarse, planar slip.

Other factors are related to the grain boundaries, as exemplified by the thermal hardening and corrosion resistant nickel-based $L1_2$ intermetallics Ni_3X [X = Al, Ga, Si, Ge]. These materials can be compressed quite extensively without collapsing (with the possible exception of Ni_3Ge) and single crystals of Ni-rich Ni_3Al can be extended by $\approx$ 20% (Copley and Kear 1967). However, polycrystals fracture intergranularly after only about 1% elongation. *Segregants* are a concern because intergranular fracture in steels and other alloys not possessing a superlattice is usually attributed to a deleterious impurity such as sulphur, phosphorous, or antimony. Indeed, sulphur exacerbates the problem, at least in Ni_3Al (White and Stein 1978), and it can destroy the beneficial effects (described in section 3) of boron (Briant and Taub 1988). Yet, when contaminants are undetectable, the Ni_3X alloys still fracture intergranularly after little or no global deformation (e.g., see Takasugi et al. [1985] for Ni_3Al and Ni_3Si; Fang and Schulson [1992] for Ni_3Ge). Barring the possibility that segregants are harmful at concentrations undetectable by AES (i.e., less than about 1 at%) and excluding hydrogen for the moment, it appears that the grain boundaries themselves are intrinsically fragile. This point is addressed more fully in section 3. Here, suffice it to say that low *boundary cohesion*, although commonly cited to be the reason for the low ductility, is probably not the only, or even the most important factor: brittle fracture of the Ni_3X alloys is always preceded by discontinuous yielding, never by elastic deformation only, suggesting that slip deformation within the vicinity of the grain boundaries is a significant element.

Another important factor is the *external environment* (see chapter on environmental embrittlement for a complete discussion). As already noted, both Ni_3Al and Ni_3Si fail after about 1% elongation. This occurs when the alloys are deformed in moist air. When deformed in dry oxygen, they elongate by about 8–10% (Oliver and White 1987; Liu and Oliver 1991; George, Liu, and Pope 1993). Similarly, the oxidation-resistant DO_3 alloy Fe_3Al fails after about 2% elongation when strained in moist air, but extends by as much as 12% in dry oxygen (Liu et al. 1990). The B2 aluminide Fe36Al elongates by about 18% in dry oxygen compared with about 2% in air (Liu, McKamey, and Lee 1989). In these cases and in all other cases in which the environment is a factor, it has no effect on the yield strength. Environmental embrittlement is a surface effect and probably influences either the nucleation or the inward propagation of cracks, or both. The mechanism appears to involve the weakening of atomic bonds by hydrogen that is made available upon oxidation (of aluminides and silicides) in water vapor (Bond, Robertson, and Birnbaum 1989; Liu 1992). When hydrogen is precharged into otherwise ductile intermetallics, such as Co-rich Co_3Ti (Takasugi and Izumi 1986) and (as discussed in section 3) Ni-rich Ni_3Al doped with boron (Kuruvilla and Stoloff 1985), it also causes embrittlement, particularly at lower strain rates. Although important, *hydrogen embrittlement* is not the fundamental cause of low ductility, because in the absence of hydrogen brittle fracture still occurs, albeit delayed.

When the environment is important, so is the applied *strain rate*. For instance, when the B2 alloy Fe45Al is strained in air at rates below about 10^{-2} s^{-1}, it elongates only by about 3% and fractures mainly intergranularly, independent of strain rate. When strained at higher rates in the same environment, the same material extends a greater amount, reaching around 9% at a rate of 1 s^{-1} (Nagpal and Baker 1991b), Figure 3-2. Fracture still occurs intergranularly, however. The ductile $L1_2$ alloy Co_3Ti (Takasugi and Izumi 1986) shows similar behavior. This effect suggests a competition between hydrogen kinetics and dislocation kinetics, the latter dominating at the higher strain rates.

Strain rate itself is a factor in the general description of the mechanical behavior of intermetallics, as reflected in the *strain rate sensitivity* of the flow stress at elevated temperatures. For instance, it is important in the brittle-to-ductile transition (section 3). However, barring its role in the environmental effect, strain rate appears not to significantly affect low-temperature deformation, at least of the strongly order $L1_2$ alloys (Turner et al. 1989).

Turning to *interstitials*, about all that can be said is that provided such features do not confer overriding benefits to the grain boundaries (as

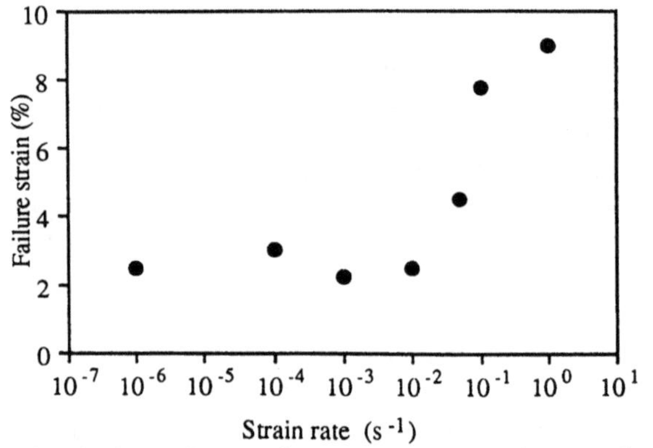

Figure 3-2. Graph of elongation to fracture versus strain rate for the B2 aluminide Fe45Al, loaded in air at room temperature. Note the threshold at $\approx 10^{-1}$ s^{-1}. (Reprinted with permission from *Scripta Metallurgica*, Volume 26, P. Nagpal and I. Baker, 1991, page 2577, Elsevier Science Ltd., Pergamon Imprint, Oxford, England)

does boron to Ni_3Al, section 3), they could impede dislocation slip, thereby raising the flow stress to the point that the nucleation and propagation of cracks might be preferred over global plastic deformation. In fact, both boron and carbon are potent strengtheners of NiAl (George and Liu 1990) and when present at levels as low as 100 ppm by wt. double the yield strength of this material and reduce its already poor ductility to nothing. The more encouraging point to note is that the suppression of interstitial-dislocation interactions in NiAl by rapid cooling from 400°C to room temperature appears to account for an increase to 7% in the elongation of ⟨110⟩ oriented single crystals of stoichiometric composition (Hack et al. 1992). The role of interstitials, particularly in the B2 intermetallics (which have a more open structure), needs further attention.

The *stress state* is another major factor in brittle fracture. One indication of its importance is the beneficial effect of *hydrostatic pressure*. For instance, a pressure of 965 MPa confers an elongation of 22% on Ni_3Al (Zok et al. 1989); similarly, 500 MPa pressure confers an elongation of > 10% on NiAl (Margevicius and Lewandowski 1991). The latter result is particularly interesting because it suggests that non-⟨001⟩ slip occurs under confinement. The pressure-induced ductility reflects the suppression of crack growth by the confining stress.

Notch sensitivity is another indication of the importance of the stress state. Notched Zr_3Al, for instance, fails intergranularly (Schulson 1984), whereas smooth bars of the material, as already noted, are ductile and fail transgranularly. Boron-doped Ni-rich Ni_3Al shows the same behavior

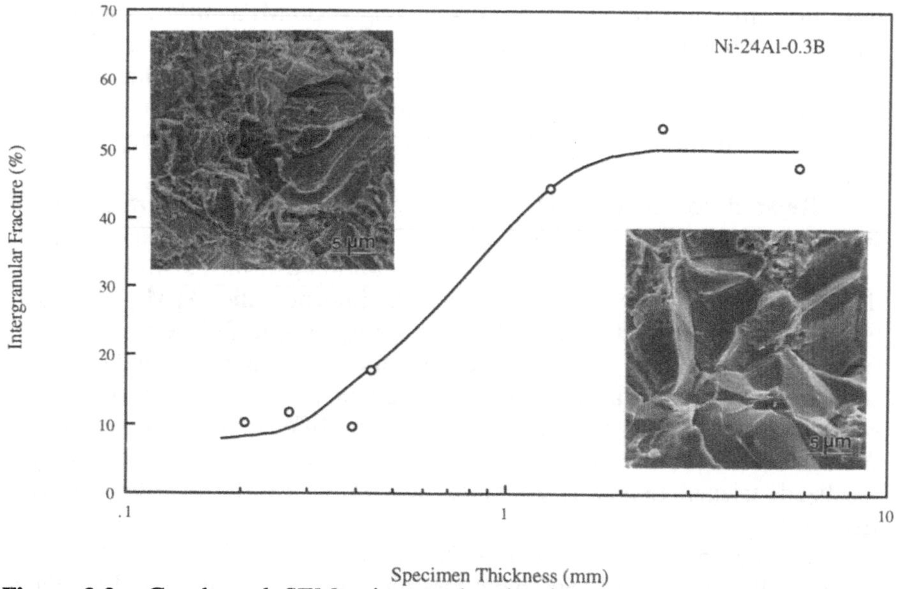

Figure 3-3. Graph and SEM micrographs showing the fracture mode versus thickness of notched, Ni-rich Ni_3Al doped with boron (Ni24Al 0.3B). Note the transition from predominantly transgranular to mixed transgranular and intergranular fracture as the thickness of the section increases to about 1 mm. The intergranular area in the photograph is greater than that averaged over the whole fracture surface. (From Xu and Schulson 1993a)

(Stoloff et al. 1987; Khadkikar, Lewandowski, and Vedula 1989; Xu and Schulson 1993a), Figure 3-3. The effect occurs within thick sections where a loaded notch induces a triaxial state of tension that promotes brittle fracture by suppressing plastic flow. Thin sections are not affected because they cannot support a large through-thickness tensile stress and biaxial tension is not a serious impediment to plastic flow, at least in plastically isotropic polycrystals. Xu and Schulson (1993a) suggested that for relatively sharp notches the ductile-to-brittle transition thickness, t_t, is of the order of twice the plastic zone size under plane stress and may be estimated from the relationship:

$$t_t = \frac{1}{\pi}\left(\frac{K_{IC}}{\sigma_y}\right)^2 \tag{3.1}$$

where σ_y is the yield strength and K_{IC} is the fracture toughness. For the Ni_3Al-based alloy described in Figure 3-3 this relationship predicts a transition thickness of $\simeq$ 1.4 mm (using $K_{IC} = 30$ MPa $\cdot m^{1/2}$, section 3.2, and $\sigma_y = 450$ MPa), in reasonable agreement with experiment. Although little is published on this effect, notch sensitivity is probably characteristic of most of the more strongly ordered intermetallics and may be *the* major impediment to the use of this class of materials in load-bearing applications.

Related to stress state and to notch sensitivity is *surface finish.* Rozner and Wasilewski (1966) isolated this factor during their original research on NiAl, and Kurfman (1965) underlined it in his studies on the deformation of single crystals of the B2 intermetallic AgMg. Schulson (1984) noted the effect on Zr_3Al, whose elongation to failure was reduced from more than 30% to only 1% to 2% upon reducing the quality of the surface by cutting in a lathe. Similarly, Guha et al. (1992) reported that surface quality affects the ductility of Ni30Al20Fe. Like notch sensitivity, the surface effect may originate from the suppression of plastic flow by a localized, triaxial tensile stress state. The effect can be avoided by electropolishing.

Finally, the *elastic factor* is worth noting. This is defined as the ratio of the bulk modulus to the shear modulus, K/G. As suggested by Pugh (1954) and noted by Hecker, Rohr, and Stein (1978), low values of this ratio correlate with the brittle fracture of a variety of f.c.c. and b.c.c. metals. More recently Cottrell (1991) proposed that ductile crystals have a ratio $K/G > 2.5$ and that brittle crystals have a ratio $K/G < 2$. Calculations by Fu (1990) for the brittle $L1_2$ alloy Al_3Sc are consistent with this view, as are some distinctions by Turner, Powers, and Wert (1989) for other $L1_2$ alloys. The paucity of data on the elastic moduli prevents a thorough assessment of this factor.

2.1. Summary

In summary, many intrinsic and extrinsic factors, Table 3-2, contribute to the brittle behavior of intermetallic compounds. The solution for a particular alloy, therefore, might appear to be the systematic elimination of the factors. This approach works to some degree. For instance, as noted by Baker and George (1992), specimens of the B2 alloy Fe40Al with high quality surfaces fracture mainly intergranularly after about 1% elongation in moist air. The addition of a small amount of boron leads to transgranular cleavage after about 4% elongation (Liu and George 1990) and the removal of water vapor then raises the ductility of the same boron-doped alloy to around 18% (Liu et al. 1989). Yet, brittle fracture still occurs. Some other factor or combination of factors must therefore be important, not only to FeAl but to all of the materials noted above. Strength, or the critical resolved shear stress for plastic flow, is probably not a basic one,

Table 3-2 Factors Contributing to the Brittle Behavior of Polycrystals

Intrinsic	Extrinsic
•complex crystal structure	•deleterious segregants
•insufficient independent deformation modes	•interstitials
•planar slip	•moisture in environment
•coarse distribution of slip	•hydrogen
•low bulk modulus/shear modulus ratio	•low strain rate
•low dislocation mobility	•triaxial tensile stress state
•high work hardening rate	•notches
•low emission of dislocations from crack tips	•low quality surface

for even the soft $L1_2$ intermetallic, Al_3Sc, fails by cleavage after negligible elongation (Schneibel and George 1990; Schneibel and Hazzledine 1991). Difficulties in the *emission of dislocations* from crack tips and/or a low *mobility of dislocations* are not easily assessed owing to the paucity of information. These factors may affect the deformation of bulk specimens, but appear not to be a problem in thin foils where in situ TEM deformation studies show that dislocations are profusely emitted from the tips of cracks and rapidly glide across grains; for example, in Ni_3Al (Baker, Schulson, and Horton 1987), in Ni_3Ge (Fang and Schulson 1993), in FeAl (Baker and Horton 1993). Perhaps the *work-hardening rate* is the problem. For strongly ordered polycrystalline intermetallics this property exceeds by a factor of two or more that of ductile metals and alloys (see Table 3-3), including the thermally disorderable and ductile $L1_2$ alloys Cu_3Au, Ni_3Mn and Ni_3Fe (see Schulson 1984). Work hardening may contribute to brittleness by limiting the energy that can be dissipated through plastic work before the local tensile stress reaches the cohesive strength of the atomic bonds. Whatever the origin, the brittle behavior of bulk intermetallics implies that crack propagation dominates crack-tip plasticity. The challenge then is not only to improve our awareness and understanding of all the factors, but to develop practical methods for ameliorating the behavior.

3. METHODS OF IMPROVING DUCTILITY AND TOUGHNESS

Tensile ductility is discussed separately from toughness. To some extent this distinction is artificial, because in strong materials both properties may be viewed as a measure of the inelastic deformation preceding

Table 3-3 Work-Hardening Rate* of Polycrystals† (at axial strain of 0.1)

Material	Work-Hardening Rate (Normalized with respect to the shear modulus, G)	Reference
NiAl	G/15	Dymek et al. (1992)
$FeAl^+$	G/7	Baker and Nagpal (1993)
Zr_3Al	G/10	Schulson (1984)
Ni_3Al	G/12	Weihs et al. (1987)
Al_3Sc	G/15	Schneibel and George (1990)
$Al_{66}Mn_6V_5Ti_{23}$	G/15	Zhang et al. (1990)
$Al_{67}Ni_8Ti_{25}$	G/19	Turner et al. (1989)
Low Carbon Steel	≈ G/50	U.S. Steel (1964)
301 Stainless Steel	G/40	Brickner and Defilippi (1977)
Cu, Al, Ni	G/30–G/40	Feltham and Meakin (1957)
Cu_3Au, Ni_3Mn, Ni_3Fe	G/23–G/38	Schulson (1984)

*For the intermetallics generally obtained from compression tests at room temperature.
†Furnace cooled after annealing.

fracture, the former on a global scale and the latter on a scale local to the vicinity of a propagating crack. However, the correlation between the two properties is not always a direct one. The TiAl-based intermetallics, for instance, appear to show an inverse relationship between ductility and toughness (Chan 1992), albeit for alloys of different compositions. The implication is that in at least some cases different processes may be operating globally from those operating near crack tips. In proceeding with the discussion the unifying theme is the role of the common microstructural elements, including the crystal structure.

3.1. Ductility

*3.1.1. **Crystal Structure**.* In considering the crystal structure the aim is to simplify the atomic arrangement and thereby to shorten the dislocation Burgers' vector and lower the Peierls barrier. Because the crystal structure is a reflection of the atomic bonding, the only practical method to simplification is through *macroalloying.*

More through serendipity than design, successful attempts have been made to simplify to $L1_2$ the structure of several binary trialuminides having either the DO_{22} or the DO_{23} crystal structure. For instance, the structure of Al_3Zr can be changed from DO_{23} to $L1_2$ by macroalloying (typically in the range 5–15 at%) with Cr, Fe, Cu or Ni (Schneibel and

Porter 1989). Similarly, the structure of Al_3Ti can be changed from DO_{22} to $L1_2$ upon macroalloying with Ni, Cu, Zn, Fe, Mn and V (Raman and Schubert 1965; Mikkola et al. 1991; Lerf and Morris 1991). Unfortunately, simplification of this kind does not impart ductility to the materials noted: the alloys still fracture by transgranular cleavage even though they slip on five independent {111}⟨110⟩ systems. Indeed, in the case of the $L1_2$-stabilized alloy $Al_{67}Ni_8Ti_{25}$ cleavage was not even accompanied by significant localized plastic deformation (Turner et al. 1989).

A guide to changing the crystal structure is the electron-to-atom (e/a) ratio. This is defined as the average per atom of the number of electrons beyond the inert gas shells, and was correlated by Dwight and Beck (1959) with the structure of closely packed, binary AB_3 compounds formed by transition metals. The structures build up from ordered AB_3 layers and reflect the sequence of stacking. The correlation is that the stacking sequence changes from purely hexagonal (...ABCBCACAB...) to purely cubic (...ABC...) as the e/a ratio decreases. Liu and Inouye (1978) followed this approach and successfully changed the crystal structure of alloys based upon Co_3V for which $e/a = 8$. By replacing Co ($e/a = 9$) with varying amounts of Fe ($e/a = 8$) and Ni ($e/a = 10$), they changed the structure from the nine-layered complex hexagonal to the three-layered $L1_2$ cubic structure. In this case the critical e/a ratio was found to be 7.888. Correspondingly, they changed the mechanical behavior of the alloys from brittle to ductile, resulting in elongations to failure at room temperature of more than 40%.

3.1.2. Vacancies and Other Point Defects. Although little systematic work has been done to elucidate the role of vacancies and other point defects, it appears from the observations that have been made that it is usually desirable to eliminate them, at least from the B2 compounds. For instance, Crimp, Vedula, and Gaydosh (1987) showed that the tensile elongation at room temperature of polycrystalline Fe40Al increased from almost nothing upon quenching from 1000°C to a few percent upon slowly cooling from the same temperature. Similarly, Schmidt, Nagpal, and Baker (1989) noted that the compressive ductility of Fe45Al was significantly greater upon slowly cooling from a high temperature. Nagpal and Baker (1991c) found that polycrystals of both Fe40Al and Fe45Al could extend by about 3% at room temperature if given a prolonged anneal (5 days) at an intermediate temperature (400°C) to reduce the vacancy concentration to the equilibrium level. Concerning interstitials again, the increase (cited above) in the ductility of NiAl single crystals by rapidly cooling from an intermediate temperature to suppress age-embrittlement (Hack, Brzeski, and Darolin 1992) points to the need to eliminate these defects.

3.1.3. Dislocation Structure. An important point about dislocations, in addition to the length of their Burgers' vector, is the degree to which they are dissociated. This affects the propensity for cross slip and, in turn, the tendency to concentrate stress through dislocation pileups. From this perspective high fault energies and, by implication, undissociated or only narrowly dissociated dislocations are desirable. The only clear example that relates the structure of dislocations in strongly ordered intermetallics to slip character and ductility is the recent work on the core structure of dislocations in NiAl (Mills and Miracle 1993; Tonn, Zhang, and Crimp 1993). As noted above, slip in this material at low temperatures occurs predominantly along ⟨001⟩. Within stoichiometric alloys, the splitting of dislocations with a ⟨001⟩ Burgers' vector has not been observed, implying that the fault energies are very high; correspondingly, slip is wavy and an elongation of up to 4–5% has been measured (Rozner and Wasilewski 1966). Within off-stoichiometric alloys rich in nickel, splitting of the core of such dislocations has been reported (Tonn et al. 1993); correspondingly, slip is more planar and the ductility drops to almost nothing. Thus, even though in the example noted the ductility is limited by an insufficient number of independent slip systems, a tendency towards greater ductility with higher fault energy is apparent.

An opposing view is that low fault energies ease the emission of dislocations from crack tips and thus enhance local plasticity. This is based on the Rice–Thomson (1973) analysis[1] in which shorter, partial dislocations have lower barriers for nucleating slip. It is difficult to assess this view, because too little data are available on true surface energies and elastic constants. It is not clear, therefore, which, if either, extreme of fault energy is the more desirable.

The other important aspect of the dislocation structure is the density of the mobile dislocations. If this were increased, say be predeformation at a low temperature under a confining state of stress, then perhaps some improvement in ductility could be imparted. That preworking can be beneficial has been demonstrated for two B2 intermetallics, FeCo (Stoloff and Dillamore 1970) and NiAl (Schulson 1982). In fact, Hack et al. (1992) showed that the fracture toughness of precrept single crystals of stoichiometric NiAl (about 12 hours at 400°C) is several times higher than that of uncrept material (section 3.2), owing, they suggest, to an increase in the dislocation density to beyond that needed to trap interstitial impurities.

[1]The Rice–Thomson (1973) ratio is defined as Gb/γ where G is the shear modulus, b is the Burgers' vector of the emitted dislocation, and γ is the true surface energy of the crack plane. Ratios greater than ten are taken to indicate high barriers to emission.

3.1.4. Grain Boundaries. In addition to barring deleterious intergranular segregants, one strategy concerning grain boundaries is to modify their structure. This may be done in several ways.

Microalloying or alloying at levels below $\approx$ 1 at% is one method. As alluded to above, a strikingly successful example of this approach is the addition of a small amount of boron to Ni_3Al (Aoki and Izumi 1979; Liu, White, and Horton 1985). Bulk concentrations as small as 0.2 at% impart elongations of around 50% to otherwise brittle polycrystals (see Fig. 3-4a, b, and c). Boron segregates to the grain boundaries and changes the mechanical properties there. Interestingly, boron is effective in improving the ductility only of alloys whose compositions lie on the Ni-rich side of the stoichiometric composition (see Fig. 3-5), even though it segregates in equivalent amounts to the grain boundaries in Al-rich alloys (Briant and Taub 1988). Moderate improvements in ductility have been realized in other alloys susceptible to intergranular fracture, including Ni_3Ga (Taub and Briant 1987; Xu and Schulson 1993b) Ni_3Si (Taub and Briant 1987; Oliver and White 1987; Schulson, Briggs, and Baker 1990) and FeAl

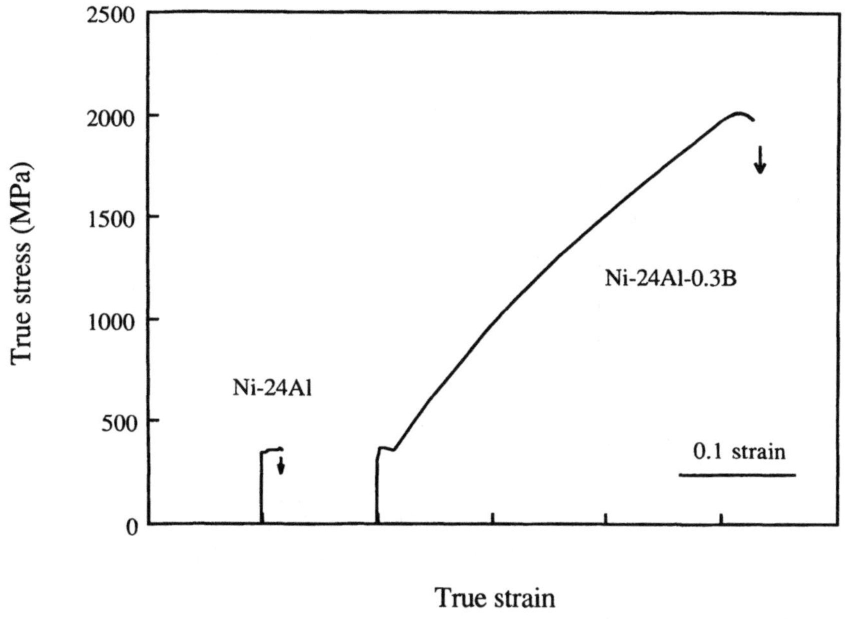

(a)

Figure 3-4. Illustrations showing the effect of boron on the ductility of the $L1_2$ aluminide Ni_3Al:

a) Tensile stress-strain curves. Note the discontinuous yielding in both the doped and undoped alloy. Ni24Al and Ni24Al 0.3B of 18 μm grain size.

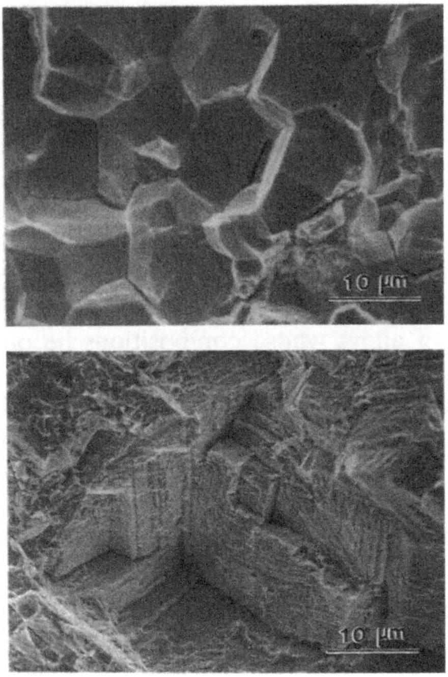

(b)

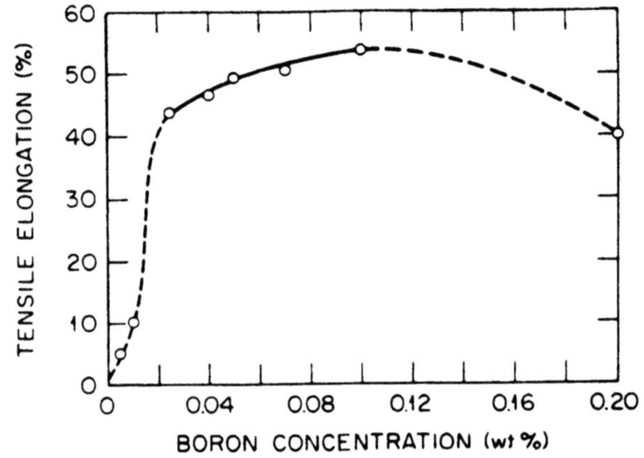

(c)

Figure 3-4. (Continued)

b) SEM fractographs showing the intergranular fracture mode of undoped material and the transgranular mode of the doped alloy. Note the blocky character of the transgranular mode. (Ni24Al and Ni24Al 0.3B)

c) Graph of elongation to fracture versus bulk boron concentration for Ni-rich Ni_3Al (Ni24Al). (Reprinted with permission from *Acta Metallurgica*, Volume 33, C. T. Liu, C. L. White, and J. A. Horton, 1985, pages 213–229, Elsevier Science Ltd., Pergamon Imprint, Oxford, England)

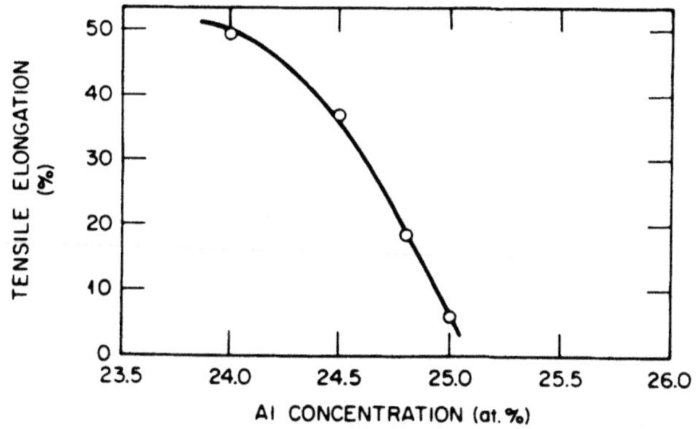

Figure 3-5. Graph of elongation to fracture versus Al concentration in Ni_3Al doped with boron to a bulk concentration of 0.24 at%. Note that the effect of boron decreases to about zero as the stoichiometric composition is approached and exceeded from the Ni-rich side. (Reprinted with permission from *Acta Metallurgica*, Volume 33, C. T. Liu, C. L. White, and J. A. Horton, 1985, pages 213–229, Elsevier Science Ltd., Pergamon Imprint, Oxford, England)

(Crimp et al. 1987; Liu and George 1990). Table 3-4 summarizes the status of microalloying several $L1_2$ and B2 materials. In addition to improving ductility, boron and another interstitial element, carbon, are potent strengthening agents.

The reason for the boron effect in Ni_3Al is still under investigation. The explanations most commonly cited are that boron either raises the grain boundary cohesive strength (proposed by Liu et al. 1985; Takasugi et al. 1985; Taub and Briant 1987), possibly by donating electrons to the boundary atomic bonds (Briant 1990), or enhances slip accommodation at and transmission across the boundaries (proposed by Schulson et al. 1986) by easing the operation of grain-boundary dislocation sources. The first mechanism could increase ductility by allowing tensile stresses greater than the yield stress to be supported; the second would lower the stress concentration on grain boundaries ahead of dislocation pileups and thus suppress crack nucleation in favor of global flow. A more recent explanation is that boron blocks the entry of hydrogen released from moisture in the environment (George et al. 1993). Viewpoints and supporting data/observations on the first two models are given elsewhere (Liu 1991; Schulson and Baker 1991); the third model was recently questioned (Lee and White 1993) because it appears to contradict the occurrence of hydrogen embrittlement at low temperatures. Suffice it to say that the primary evidence supporting the *cohesion model* is a theoretical calculation that indicates the grain boundary cohesive energy increases only slightly in the presence of boron (Chen et al. 1989, 1990). That the

Table 3-4 Effects at Room Temperature of Microalloying Some $L1_2$ and B2 Intermetallics

Intermetallic	Crystal Structure	Element	Effects
Ni-rich Ni_3Al (76 Ni)	$L1_2$	boron, carbon	• increase (from ≤ 2% to 50%) in ductility in air (Aoki & Izumi 1979; Liu et al. 1985) with boron; no effect of carbon (Masahashi et al. 1988) • decrease (by ~ 40%) in contribution of grain boundaries (g.b.) to yield strength (y.s.) (Schulson et al. 1991) with boron
Ni-rich Ni_3Al (76 Ni)	$L1_2$	beryllium	• increase (from ≤ 2% to ≅ 5%) in ductility in vacuum (Takasugi et al. 1986; Masahashi et al. 1988
Ni_3Al (75 Ni)	$L1_2$	boron, carbon	• increase (from ≤ 2% to ≅ 10–30%) in ductility in air with boron (Schulson et al. 1986); no effect of carbon (Masahashi et al. 1988) • decrease (by ~ 40%) in contribution of g.b. to y.s. (Schulson et al. 1986)
Al-rich Ni_3Al (74 Ni)	$L1_2$	boron	• negligible effect on either ductility in air or g.b. contribution to y.s. (Schulson et al. 1991)
Ni-rich Ni_3Ga (76 Ni)	$L1_2$	boron	• increase in ductility in air (from ≤ 2% to ≅ 10%) (Taub et al. 1986; Xu and Schulson 1993b) • decrease (by ~ 40%) in g.b. contribution to y.s. (Xu and Schulson 1993b)
Ni-rich Ni_3Si (77 Ni)	$L1_2$	boron	• increase in ductility in air (< 1% to ~ 3–5%) (Taub et al. 1986; Schulson et al. 1990) • decrease (by ~ 40%) in g.b. contribution to y.s. (Schulson et al. 1990)
Ni-rich Ni_3Ge (76.5 Ni)	$L1_2$	boron	• increase in ductility in air (≪ 1% to 1–2%) (Fang and Schulson 1992) • decrease (by ≈ 20%) in g.b. contribution to y.s. (Fang and Schulson 1992)
Fe-rich FeAl (60 Fe)	B2	boron	• increase in ductility in air (from ≪ 1% to ~ 4%) (Crimp and Vedula 1986; Liu and George 1991)
NiAl (50 Ni)	B2	boron, carbon	• decrease in ductility (from 2 to 3% to ≈ 0%) (George and Liu 1990)
$(Co_{85}Fe_{15})_3V$	$L1_2$	boron	• increase in ductility in air and in water (from ~ 6% to ~ 35%) (Pike and Liu 1992)

increase is small is consistent with a measurement of the cohesion of doped and undoped tilt boundaries in bicrystals of Ni_3Al (Otterbein, Hofmann and Rühle 1993). The primary points supporting the *slip transmission model* are the observations of discontinuous yielding before fracture (Schulson et al. 1986), of dislocation pileups at grain boundaries, and of the transmission of slip across the boundaries ahead of the pileups in both doped and undoped material (Baker et al. 1987), of the reduction in the contribution of the grain boundaries to the yield strength (i.e., a reduction in Hall-Petch (HP) slope) of Ni-rich Ni_3Al when doped with boron but not a reduction in the HP slope of doped but brittle, Al-rich Ni_3Al (Schulson et al. 1986; 1991), and of the recurrence of these effects in the Ni-rich compositions of all of the other members of the Ni_3X family (Ni_3Si: Schulson et al. 1990; Ni_3Ge: Fang and Schulson 1992; Ni_3Ga: Xu and Schulson 1993b). Figure 3-6 illustrates the point about the HP slope. With one caveat, the primary evidence for the *suppression of hydrogen embrittlement* is the improvement, noted above, in the ductility of Ni_3Al (George et al. 1993) and of Ni_3Si (Liu and Oliver 1991) when the materials are strained in a H_2O-free environment. The caveat is that in both materials the improvement was made in alloys sufficiently Ni-rich to contain at equilibrium a second-phase, disordered solid solution (one was seen in Ni_3Si): if present as a film along the boundaries, this phase may have dictated the response of the boron-free material to the environment. Perhaps the reason for the effect of boron on the ductility of Ni_3Al and the other $L1_2$ polycrystals involves to differing degrees the cohesion of the grain boundaries and the ease with which slip is transmitted across them, as well as blocking the entry of hydrogen.

In improving the ductility of Ni-rich Ni_3Al, boron also changes the fracture mode from intergranular to transgranular. The pattern that develops is not a classical dimple pattern, however; that is, not a pattern of the kind seen in ductile metals like copper and stainless steel. Instead, the transgranular mode is characterized more by stepped, blocky features (Figs. 3-3 and 3-4) whose facets appear to be defined by {111} planes (Aoki and Izumi 1979; Liu et al. 1985; Hanada et al. 1987). In other words, although boron imparts extraordinary macroscopic ductility, it appears not to impart fully ductile behavior on a local scale. The other clues to this are the absence of marked necking and, as noted above, notch sensitivity. The reason fully ductile behavior does not develop is not understood, but may be related to the relatively high work-hardening rate of even boron-doped polycrystals (Weihs et al. 1987).

It should be noted that the addition of ≈ 0.1 at% beryllium also improves the ductility of both stoichiometric and Ni-rich Ni_3Al (Masahashi, Takasugi, and Izumi 1988). The improvement is less than that

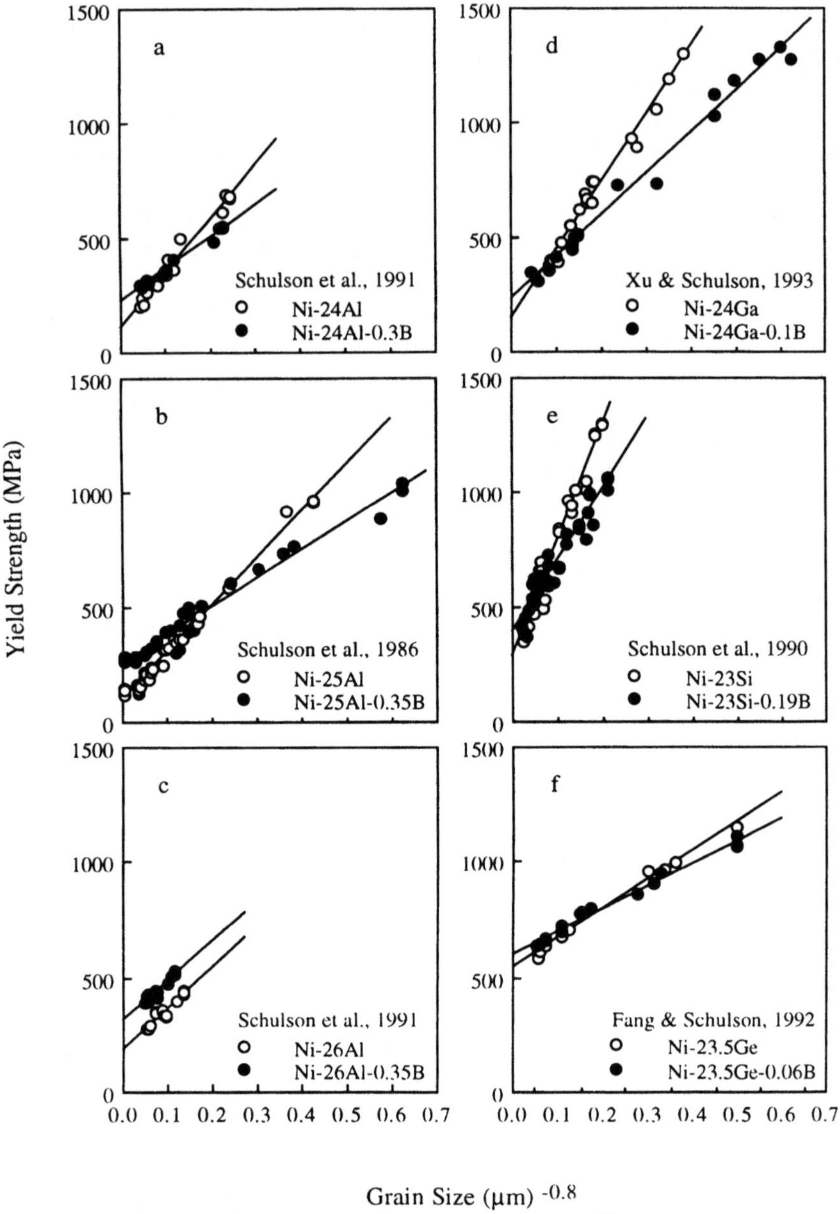

Figure 3-6. Graphs of yield strength versus (grain size)$^{-0.8}$ of Ni_3X (X = Al, Ga, Si, Ge) with and without boron:

a) Ni-rich Ni_3Al d) Ni-rich Ni_3Ga

b) Stoichiometric Ni_3Al e) Ni-rich Ni_3Si

c) Al-rich Ni_3Al f) Ni-rich Ni_3Ge

Note that boron lowers the Hall-Petch slope in cases (a), (b), (d) and (e) by about 4% and by about 20% in case (f); boron does not significantly affect the slope in case (c). (The 0.8 functionality is explained in the references cited.)

imparted by boron (≈ 5% vs. ≈ 50% elongation) and the fracture mode remains intergranular. Beryllium resides substitutionally. Whether it affects the grain-boundary chemistry, the contribution of the grain boundaries to the yield strength and/or their susceptibility to hydrogen is not known.

Another way to modify the grain-boundary structure is to *constitutionally disorder* it. In so doing, this would accommodate slip by increasing the number of interactions between lattice dislocations and grain boundaries, by two times for B2 alloys and by four times for $L1_2$ materials (Frost 1987; King and Yoo 1987). A disordered boundary region would also ease the nucleation of dislocations and enhance slip transmission, and may be a general requirement for tensile ductility (Baker and Schulson 1989). Indeed, grain-boundary disordering may be the reason boron both raises the ductility and lowers the grain-boundary contribution to the yield strength of Ni-rich Ni_3Al, the disordering in this case arising through the co-segregation of nickel, as suggested by Frost (1988). Boron and nickel co-segregation has been observed in Ni-rich Ni_3Al, but not in Al-rich Ni_3Al (Baker et al. 1990); correspondingly, constitutionally disordered grain-boundary regions, Figure 3-7a, b, and c, have been observed in boron-doped, Ni-rich Ni_3Al, but not in boron-doped Al-rich Ni_3Al (Baker et al. 1990; Kung, Rasmussen, and Sass 1992). Although appealing, constitutionally disordered grain boundaries in Ni_3Al have not been seen by all investigators (e.g., Forward and Gibson 1992) owing, perhaps, to differences in impurity level and heat treatment.

The other way to improve the ductility of polycrystals is to design them with *special boundaries*. The idea is to lower the propensity for brittle fracture by eliminating transverse grain boundaries and by forming only low energy/angle boundaries parallel to the applied stress (Hanada et al. 1987; Lin and Pope 1993). This kind of structure, although not eliminating brittle fracture altogether, leads to a 17% elongation of columnar grained, {111}-textured polycrystals of stoichiometric Ni_3Al produced by unidirectional solidification (Hirano 1990). In contrast to the effect of boron, this method is effective in Al-rich Ni_3Al and leads to elongations of around 50% and 25%, respectively, in Ni26Al and Ni27Al when strained along the growth direction (Hirano et al. 1991). The integrity of the low-angle boundaries probably reflects the ease of slip transmission across them and the suppression of stress concentrations large enough to nucleate cracks.

Barring the elimination of the grain boundaries altogether, an alternative approach is to reduce their spacing. This is the classical *grain refinement method*. Although not suppressing brittle fracture, it leads to elongations in Ni-rich NiAl (51/49) as high as 40% at 400°C (Schulson and

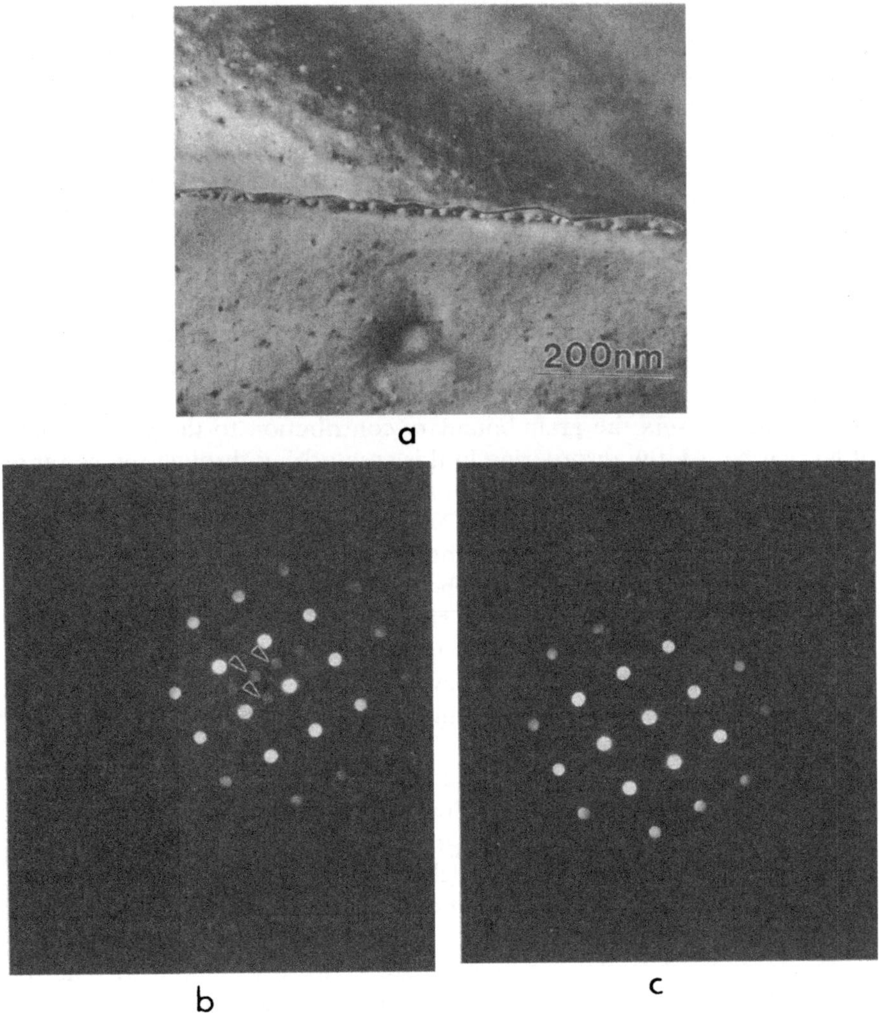

Figure 3-7. a) Bright-field transmission electron micrograph of Ni-rich Ni3Al doped with boron (Ni24Al · 3B). Note the rope-like feature along the grain boundary, bonded to one of the grains. Electron diffraction patterns from; b) the matrix (arrows point to superlattice spots) and c) the grain boundary phase. Note the absence of superlattice reflections from the boundary feature, implying that it is disordered. (From Baker et al. 1990)

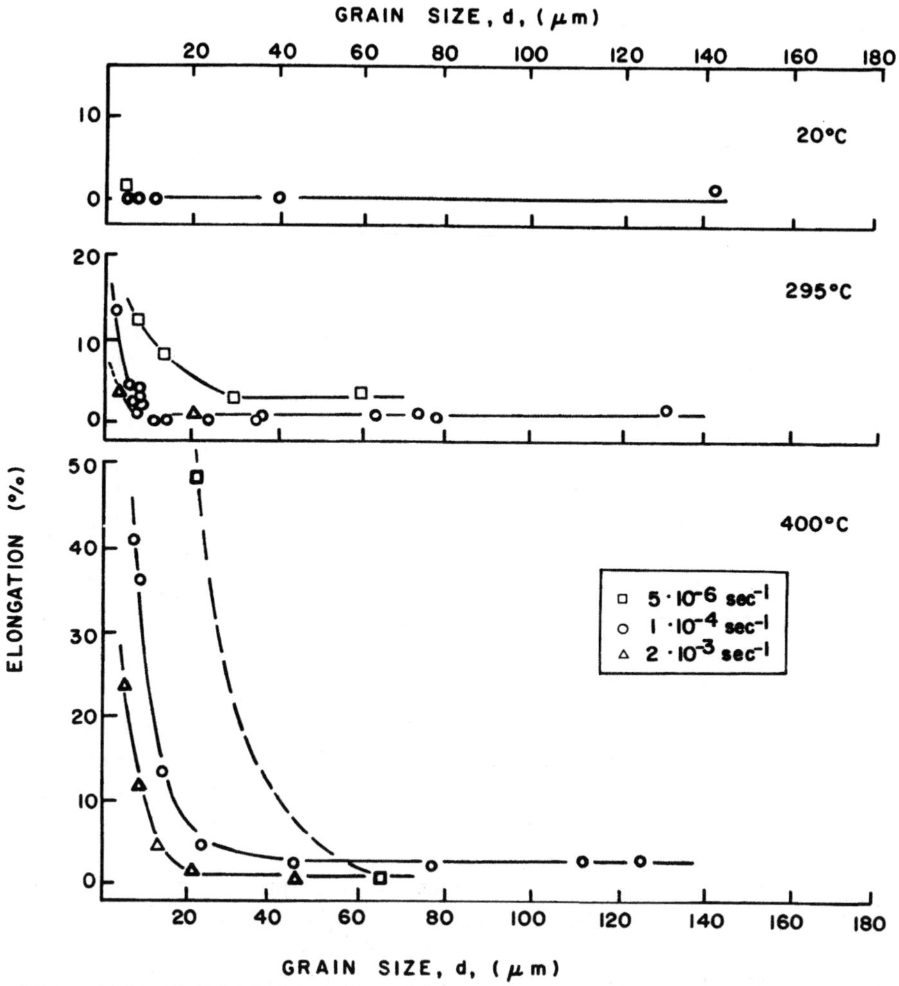

Figure 3-8. Graph of elongation to fracture versus grain size of Ni-rich NiAl (51 Ni). Note the increase in ductility at 295°C and at 400°C upon decreasing the grain size to below a critical size, d_c, and note the increase in d_c upon increasing temperature and decreasing strain rate. (From Schulson 1985)

Barker 1983) and 13% at 295°C (Schulson 1985) when the grains are reduced to below a *critical grain size*, Figure 3-8. At room temperature the effect in NiAl is only marginally evident (Cheng and Cantor 1992), if evident at all, owing possibly to the critical size being smaller than the grain size of the finest grained aggregates so far studied; that is, $\lesssim 1\ \mu m$. The principle is that the grain size governs the crack size and that cracks shorter than the critical size, even though they may nucleate near the macroscopic yield point, may not propagate until the material hardens through global deformation. The ductility is then related to the difference in the stresses to nucleate and to propagate cracks and to the rate of work hardening. Following Cottrell (1958), Schulson (1979) suggested that the critical grain size, d_c, may be expressed by:

$$d_c = \left(\frac{YK_{IC} - k_y}{\sigma_o} \right)^2 \tag{3.2}$$

where Y is a geometrical factor of order unity, K_{IC} is the plane strain fracture toughness, ky is the Hall–Petch slope and σ_o is the Hall–Petch intercept or the lattice friction stress. These parameters may depend upon temperature, strain rate, and deviation from stoichiometry, in which case the critical grain size would change, as shown for NiAl in Figure 3-8 (Schulson 1985). While numerous observations can be rationalized on this basis (see Schulson 1979) and further analysis supports the approach (Chan 1990), the practicality is questionable owing to the apparent critical sizes being smaller than about one micron at room temperature. If nanocrystalline intermetallics could be consolidated free from pores and other defects and in sizes large enough to test, then perhaps (Bohn et al. 1991) this method would show greater utility.

3.1.5. Dispersoids. Seybolt (1966) noted that within materials susceptible to coarse slip, plastic strain could be distributed more homogeneously through the incorporation of a fine dispersion of particles. This could reduce internal stress concentrations and, thus, delay the onset of cracking. The method would probably not change the fracture mode, but it might lead to some improvement in ductility. In addition, it could improve the high-temperature mechanical properties. Experiments, however, reveal conflicting results. Seybolt reported that the addition to Ni-rich NiAl (51/49) of 5% by volume of 0.2–0.3 μm particles of Y_2O_3 actually reduces ductility: the brittle-ductile transition temperature (as measured by bend ductility) increased from about 300°C to 450°C. D. Morris and M. Morris (1991), on the other hand, found that a fine, cellularlike dispersion of ZrB_2 particles of about 0.05 μm raises the room-temperature bend ductility of Fe-rich FeAl (65/35). They attributed this improvement to more homogeneous slip. Similarly, Strothers and Vedula (1987) observed that annealed, slowly cooled and finely grained (5 μm) Fe40Al exhibits in air a relatively high room-temperature elongation of about 7% when the alloy contains a dispersion of Y_2O_3. In both of the latter instances it is not clear whether the improvement is solely related to the particles or whether it is related to a particle-induced microstructural modification such as a reduction in grain size or the development of a favorable crystallographic texture. From a practical point of view, however, the results are encouraging. Seybolt's observations of the opposite effect is puzzling, but may be related to the coarseness of his dispersion. The principle here may be that to be effective the average spacing between particles, whether hard or soft, must be less than the average spacing between the slip bands within the parent material.

3.1.6. Macroalloying and Second Phases. Another method to improve ductility is macroalloying to the point that a ductile, second phase develops within the microstructure. Preferably, this phase should be continuous and should line the grain boundaries. For instance, upon alloying NiAl with Co (Ni31Al18Co) to produce a continuous grain-boundary layer 1–6 μm in thickness, Pank, Nathal, and Koss (1990) observed a bend ductility of 6.6%. Similarly, upon alloying NiAl with Fe (Ni30Al20Fe) and heat treating the alloy to produce a discontinuous grain-boundary film of an austenitic Fe-rich solid solution of 10 μm thickness, Guha et al. (1992) measured a tensile elongation of 6%. They noted also that when the same alloy was heat treated to remove the boundary film and to create instead a dispersion of precipitates within the grains, it failed in tension before any global yielding. Ishida et al. (1991) observed a similar improvement in tensile ductility upon alloying NiAl with sufficient Fe, Co, and Cr to produce an f.c.c. phase at the grain boundaries. Upon alloying Ni-rich Ni_3Al (77 Ni), Chiba, Hanada, and Watanabe (1992) found that additions of so-called γ-formers (i.e., elements which, if present at sufficiently high levels, lead to the formation of a random solid solution as a second phase) promote ductility, possibly through the formation at the grain boundaries of a thin layer the disordered γ-phase. This result reflects the earlier observations by Dimiduk, Weddington and Lipsitt (1987) and Fuchs and Stoloff (1988), who increased the ductility of Ni_3Al by a few percent by alloying with Mn, Cr, or Fe. It is interesting to note that all of the Ni_3Al-based alloys currently undergoing commercial development (Sikka, Mavily, and Anderson 1992) contain sufficient nickel and enough alloying elements to produce a two-phase microstructure the disordered component of which partly lines the grain boundaries. In all of these cases, the second phase probably helps to accommodate/transfer slip and/or to stop/deflect cracks.

Concerning multiphase alloys, Yang, Leake, and Cahn (1992) showed that it is also possible to confer ductility, albeit under compression, by incorporating another intermetallic. In particular, a substantial volume fraction of the relatively ductile phase Ni_3Al in semicoherent epitoxy with NiAl and Ni_2AlTi conferred several percent shortening.

Deviations from the Stoichiometric Composition. Whether deviations from the stoichiometric composition affect ductility is worth noting.

Consider Co_3Ti. The room-temperature ductility of this material (in air at a strain rate of 10^{-4} s^{-1}) increases from 15% to 40% to 60% as the Co concentration increases from 77 at% to 79 at% to 80 at% (Takasugi and Izumi 1985). Correspondingly, the fracture mode changes from primarily intergranular with some transgranular dimple fracture to completely transgranular dimple fracture. The deviations thus appear to be beneficial. However, it would not be correct to conclude that the improvement is caused only by a change in the composition of the $L1_2$ phase. At least part (and possibly all) of the effect is caused by the presence of a disordered Co-Ti solid solution, evidence for which was obtained from X-ray diffraction in the most Co-rich alloy. Whether the disordered phase lines the grain boundaries is not known.

The more common observation is that deviations from stoichiometry within the confines of a single phase do not impart a large increase in ductility or any increase at all, although such deviations generally strengthen intermetallics at low temperatures. For instance, both NiAl (which spans the range 42–55 at%Al at 400°C) and FeAl (36–50 at%Al at 400°C) are brittle across the whole range of composition for which a single phase is in equilibrium. In these materials only slight and opposing effects on ductility are seen upon decreasing the Al concentration on the Al-lean side of stoichiometry (for reviews see Baker and Nagpal 1993, Baker 1993): The ductility in one case (FeAl) increases from almost nothing in air to about 3% ($\leq$ 45 Al), while in the other it decreases from at most about 4% to 5% at the stoichiometric composition (Rozner and Wasilewski 1966) to essentially nothing. Similarly, polycrystalline Ni_3Al (24–26 at%Al at 400°C) remains brittle over its range of composition. The same behavior (Guha 1987) is evident in the α-Mn intermetallic $Mg_{17}Al_{12}$ (40–48 at%Al).

It appears, therefore, that for a single-phase alloy deviation from the stoichiometric composition generally is not an effective method for improving the low-temperature ductility. To be effective the deviation must be great enough and in the right direction to bring out a second phase of a disordered solid solution, as hypothesized by Baker and Schulson (1989).

3.1.7. Brittle Fracture Mode and Ductility. So far, nothing has been said about a possible correlation between ductility and the mode of brittle fracture. This is because the evidence is contradictory. For instance, the small increase in the ductility of FeAl with decreasing Al concentration, noted above, is accompanied by a transition from IG to mixed IG/TG to predominantly TG fracture[2] (Fig. 3-1). Similarly, the ductility of both boron-doped, stoichiometric Ni_3Al (Weihs 1985) and boron-doped, Ni-rich Ni_3Ge (Fang 1992) correlates positively with the degree of transgranular fracture. On the other hand, the moderate increase in the ductility of NiAl with increasing Al concentration on the Al-lean side of stoichiometry, also noted above, is accompanied by a transition from TG to IG fracture. Similarly, the increase (to 9%) in the ductility of Fe_3Al upon the addition of 2 to 6 at%Cr (McKamey et al. 1988) is accompanied by a transition from TG to mixed TG/IG fracture. Thus, although IG and TG are different fracture modes, one mode is not always associated with more ductility than the other.

3.2. Toughness

The discussion now turns to the deformation zone near the tip of a loaded crack and, particularly, to how crack propagation can be impeded. To begin, Table 3-5 summarizes the range of fracture toughness characterizing intermetallics. The data were obtained in air, leaving for future research the question of the degree to which the environment lowers the resistance to crack propagation.

Near the lower end are the structurally complex silicides, $MoSi_2$ (Cllb structure), Ti_5Si_3 ($D8_8$) and Nb_5Si_3 ($D8_m$, $D8_l$ or $D8_8$), as well as some structurally simple $L1_2$- or $L1_2$-stabilized alloys, such as Al_3Sc and $Al_{66}Fe_6V_5Ti_{23}$, and the B2 material CoAl. Single-phase alloys of these intermetallics are as fragile as conventional ceramics and possess fracture toughnesses within the range 1–5 MPa $m^{1/2}$. In these cases the toughness is dominated by the energy to create new surface, with little or no contribution from plastic work.

Near the higher end of the range are FeAl, Ni_3Al and Zr_3Al. Even though the first two of these alloys fracture in a brittle mode, as does Zr_3Al in the presence of a notch, these materials have moderate to good (20 and 56 MPa $m^{1/2}$) fracture toughnesses at room temperature. These values compare favorably with the toughness of some high-strength, structural aluminum alloys and imply that plastic flow localized at the tip of the propagating crack is a significant impediment to propagation. In fact, the measured toughnesses for FeAl and Ni_3Al are many times greater than those derived from theoretical considerations of elastic energy alone (Yoo

[2]There is some variation in the Fe/Al ratio, which marks the IG to TG transition, owing possibly to differences in impurities, processing, and texture (Baker and Nagpal 1993).

Table 3-5 Fracture Toughness at Room Temperature

Intermetallic	Form	K_C (MPa $m^{1/2}$)	Reference
Zr_3Al	polycrystal	50	Rosinger and Wilkens (1980)
Ni_3Al (76 Ni)	polycrystal	20	Rigney and Lewandowski (1992)
Boron-doped Ni_3Al (76 Ni)	polycrystal	> 29	Rigney and Lewandowski (1992)
FeAl (60 Fe)	single crystal	33–56††	Chang et al. (1992)
FeAl (60 Fe)	polycrystal	16–40†	Schneibel and Jenkins (1993)
FeAl (55 Fe)	polycrystal	12–30*	Klein et al. (1993)
NiAl (50 to 60 Ni)	polycrystal	5–7	Vehoff (1992); Kumar et al. (1992)
NiAl (63 Ni)	polycrystal (furnace cooled; or quenched and annealed)	11–13	Kumar et al. (1992)
NiAl (50 Ni)	{110} single crystals	4–7	Vehoff (1992)
NiAl (50 Ni)	{110} single crystals (thermal/mechanically processed	12–17	Hack et al. (1992)
TiAl	polycrystal (equiaxed single phase)	6–10	Vehoff (1992); Déve et al. (1992)
TiAl/Ti_3Al	≈ 5% Ti_3Al, (equiaxed)	12–15	Déve et al. (1992)
TiAl/Ti_3Al	duplex lamellar	20–25	Déve et al. (1992); Chan (1992)
CoAl (40,49 Al)	{110} single crystal	2–3	Chang et al. (1992)
$MoSi_2$	polycrystal	3	Petrovic et al. (1991)
Ti_5Si_3	polycrystal	2–3	Vehoff (1992)
Nb_5Si_3	polycrystal	≈ 3	Nekkanti and Dimiduk (1990)
Al_3Sc	polycrystal	3	George et al. (1990)
$Al_{69}Fe_4V_2Zr_{25}$	polycrystal	2	Schneibel and Porter (1989)
$Al_{66}Fe_6V_5Ti_{23}$	polycrystal	2	George et al. (1989)
Al_3Nb	polycrystal	2	Schneibel et al. (1988)
Nb_3Al	polycrystal	1	Murugesh et al. (1993)

††non-plane strain toughness

†higher value obtained in air at crack growth rate of ≈ 350 $\mu m \cdot s^{-1}$; lower value obtained at 0.1 $\mu m \cdot s^{-1}$.

*higher value obtained in air at a loading rate of 6×10^3 MPa $\cdot$ $m^{1/2}$ s^{-1}; lower value obtained at 0.1 MPa $\cdot$ $m^{1/2}$ s^{-1}.

and Fu 1992), further support for the importance of plastic flow in these cases. The toughness of FeAl increases with loading rate owing to the suppression of the environmental effect (Schneibel and Jenkins 1993; Klein, Nagpal, and Baker 1993). That Ni_3Al and Zr_3Al are as tough as they are and yet are notch-sensitive is curious, but might be reconciled with environmental sensitivity and with its suppression at strain rates as high as those local to the tip of a propagating crack.

Nearer the center of the range is NiAl. Depending upon how this material is processed, its fracture toughness usually ranges from relatively low values of 4–7 MPa $m^{1/2}$, which are close to those calculated from elasticity theory (Yoo and Fu 1992), to higher values of around 13–17 MPa $m^{1/2}$. The lower values are more commonly measured and are obtained from both polycrystals and single crystals. The higher values are also obtained from both polycrystals and single crystals. For instance, tougher polycrystals were obtained by Kumar, Mannon, and Viswanadham (1992) who processed Ni-rich (63 at%) material to have a continuous *grain boundary film* of Ni_3Al about 1–2 μm thick. This result is reminiscent of Russell and Edington's (1972) observation that a two-phase microstructure raises the impact energy of NiAl by about ten times. Interestingly, this improvement is conferred by a phase which itself is brittle as a bulk polycrystal, suggesting that as a boundary layer Ni_3Al absorbs energy either by bridging or by stopping/deflecting cracks in the substrate. The tougher single crystals were obtained by Hack et al. (1992) from experiments on ⟨110⟩-oriented stoichiometric crystals that were homogenized by annealing at 1316°C for 48 hours. In one test, as noted above, Hack crept specimens at 400°C for about 12 hours to increase the *density of mobile dislocations* and thereby obtained a fracture toughness of around 12 MPa $m^{1/2}$. In another experiment, he air-cooled specimens from 400°C to room temperature to *suppress strain-age embrittlement* and obtained values as high as 17 MPa $m^{1/2}$.

TiAl is also near the center of the range. Equiaxed, single-phase material has a relatively low fracture toughness of 6–10 MPa $m^{1/2}$. A small amount (5% by vol.) of second-phase Ti_3Al raises the level to around 12–15 MPa $m^{1/2}$, owing to the dissipation of energy through twinning near the crack-tip zone. Processing to create a refined, *lath-like microstructure* of alternating plates of TiAl and Ti_3Al raises the toughness further to around 20–25 MPa $m^{1/2}$, again owing in part to the absorption of energy through twinning within the TiAl component of the laminate. Microalloying two-phase TiAl/Ti_3Al (55 at%Ti) with silicon ($\simeq$ 0.1 at%) and chromium ($\simeq$ 1 at%) raises the toughness still further, to about 34 MPa $m^{1/2}$ at room temperature (Vehoff 1992).

Perhaps the most effective way to toughen intermetallics is to incorporate a large volume fraction of a *ductile phase*. For instance, Nb particles (20% by vol.) raise the work of fracture of $MoSi_2$ by five times (Lu et al. 1991); also, such particles (40% by vol.) increase the fracture toughness of Nb_3Al by about six times, to $\simeq 6.5$ MPa $m^{1/2}$ (Murugesh, VenKateswara, and Ritchie 1993). Mo fibers (40% by vol.) raise the work of fracture of $MoSi_2$ dispersed with particles of SiC (40% by vol.) by several hundred times (Déve et al. 1992). Similarly, Nb (around 70% by vol.) imparts a toughness of about 21 MPa $m^{1/2}$ to an extruded Nb_5Si_3/Nb composite (Mendiratta et al. 1991). Particles (20–30% by vol.) of a Ni(Si) solid solution hardened with precipitates of Ni_3Si impart a fracture toughness of 13–18 MPa $m^{1/2}$ to a complex nickel-silicide (Li and Schulson 1993); and lamellae of TiNb (20% by vol.) impart to TiAl a fracture toughness above 100 MPa $m^{1/2}$ (Odette et al. 1992). In the last instance, a smaller amount of the ductile phase (10% by vol. TiNb) raises the toughness of the TiAl-based composite to around 30 MPa $m^{1/2}$ (Rao, Odette, and Ritchie 1992). The primary toughening mechanism in these instances is crack bridging by the ductile phase, Figure 3-9, and the attendant dissipation of energy through the plastic deformation of the bridge.

While ductile-phase toughening is promising, two points should be noted. The first is that the large volume fraction of the ductile phase raises questions about the overall balance of properties of the composite. The other point concerns the thermal and the strain-rate response of the toughening phase and the degree to which thermal softening and strain-rate hardening reduce its energy absorbing qualities, say at elevated temperatures or during impact. Preliminary work (Rigney et al. 1991) which showed that the fracture toughness of the Nb_5Si_3/Nb composite (51% Nb

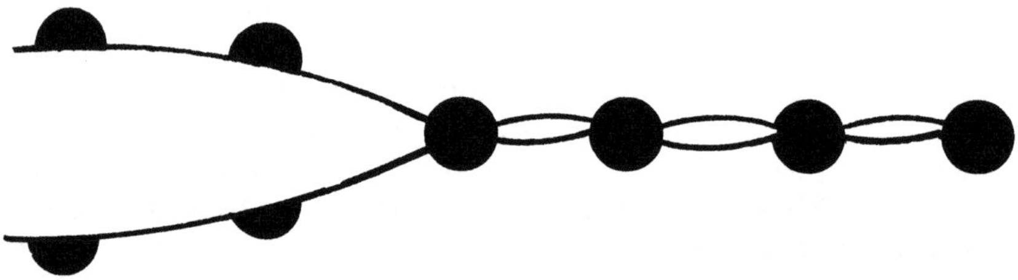

Figure 3-9. Schematic sketch of crack bridging by a ductile (dark) phase. (Reprinted with permission from *Journal of Mechanics and Physics of Solids*, Volume 36, B. Budiansky, J. C. Amazigo, and A. G. Evans, page 167, Elsevier Science Ltd., Pergamon Imprint, Oxford England)

by vol.) decreases with increasing loading rate, from around 25 MPa $m^{1/2}$ at 0.5 MPa $m^{1/2}$ s^{-1} to around 10 MPa $m^{1/2}$ at 200 MPa $m^{1/2}$ s^{-1}.

In comparison, the incorporation of a large volume fraction of *hard particles/fibres* is less effective, at least at room temperature. (This approach might be considered when creep strength is the goal.) For instance, fine particles ($\sim$ 2–3 μm) of SiC (20% by volume) maximize the room-temperature fracture toughness of $MoSi_2$ only at around 8 MPa $m^{1/2}$ from about 3 MPa $m^{1/2}$ (Petrovic et al. 1991). Discontinuous fibres (i.e., whiskers) of Al_2O_3 ($\leq$ 25% by volume) raise the toughness of NiAl of a variety of Ni-rich compositions to only around 10 MPa $m^{1/2}$ from 5–7 MPa $m^{1/2}$ (Kumar et al. 1992), and particles of TiB_2 ($\leq$ 20% by volume) have almost no effect at all on the toughness of this material (Kumar et al. 1992). In fact, in some cases hard particles appear to be detrimental to toughness. For instance, particles of TiB_2 (10 vol%) reduce the fracture toughness of Ni_3Al with and without boron to around 12–13 MPa $m^{1/2}$ from $>$ 30 and 20 MPa $m^{1/2}$, respectively (Rigney and Lewandowski 1992).

Concerning *grain refinement*, the results are mixed. For instance, Reuss and Vehoff (1990) reported higher toughness values at room temperature and above for Ti_5Si_3 of 6 μm grain size ($\approx$ 6 MPa $m^{1/2}$ at 20°C) versus material of 10 to 20 μm grain size ($\approx$ 2 MPa $m^{1/2}$ at 20°). Similarly, Kostrubanic et al. (1991) found a threefold improvement in the toughness of the NiAl-based alloy Ni35Al20Fe upon combining a reduction in grain size (from 28 μm to around 2 μm) with a dispersion (2 vol%) of Y_2O_3. On the other hand, Kumar et al. (1992) observed little effect upon reducing the grain size of Ni-rich (63 at%) NiAl from about 25 to about 12 μm. Similarly, Vehoff (1992) noted that grain refinement does not improve the fracture toughness of NiAl at temperatures where it shows no ductility. Perhaps the discrepancy lies in the range of grain sizes examined and in the possibility that the dependence of toughness on grain size increases with decreasing grain size.

3.3. Brittle-Ductile Transition Temperature (BDTT)

Although the focus of this chapter is on behavior at room temperature, it is instructive to consider whether intermetallics, like some structural materials, undergo a brittle-to-ductile transition above a critical temperature. The comments below are based upon measurements of fracture toughness, few though they are, because it is this property that dictates the ability of a loaded component to tolerate cracklike flaws. Bend ductility may also be used to estimate BDTT. Where such data are available, as for Ni-rich NiAl (51/49) (Seybolt 1966), this measure gives a similar transition temperature.

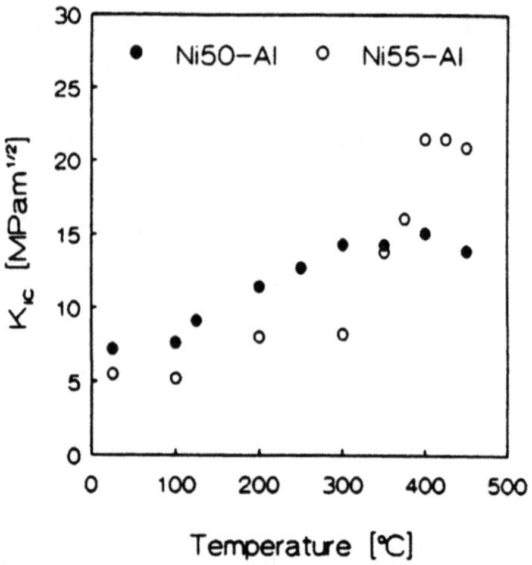

Figure 3-10. Graph of the fracture toughness versus temperature of stoichiometric and Ni-rich NiAl. Note the higher DBTT in the off-stoichiometric alloy. (From Vehoff 1992; reprinted by permission of Kluwer Academic Publishers)

Intermetallic compounds generally become tougher at elevated temperatures. The only case, to the author's knowledge, where the toughness appears to decrease is $MoSi_2$, and this is attributed to the presence at the grain boundaries of SiO_2, which becomes viscous at elevated temperatures. When carbon is added to this material (to reduce SiO_2 to SiC), the toughness rises with increasing temperature (Maloy 1991).

Two kinds of behavior are seen. In one kind, the toughness increases gradually with temperature. For instance, the fracture toughness of the $L1_2$-stabilized trialuminides $Al_{67}Ti_{25}Cr_8$ and $Al_{66}Ti_{25}Mn_9$ rises slowly from about 4 MPa $m^{1/2}$ at room temperature to about 10 MPa $m^{1/2}$ at 700°C (Brown and Kumar 1993); correspondingly, the tensile ductility increases from essentially zero to between about 5% and 10%, and the fracture mode changes from transgranular to intergranular (Kumar and Liu 1993). Similarly, the toughness of the TiAl-based two-phase alloys $TiAl/Ti_3Al$ rises slowly from room temperature to about 700°C (Vehoff 1992). In such cases it is not meaningful to define a transition. In the other kind of behavior, the fracture toughness increases more abruptly over a range of temperature of about 100°C, as in steel. For example, in NiAl this transition occurs from about 150°C to 250°C and leads to roughly a doubling of the resistance to fast crack propagation, as seen in Figure 3-10. This behavior appears to be the more common kind and is described in terms of a brittle-to-ductile transition temperature in Table 3-6.

The BDTT depends upon a number of factors, including the loading rate, the composition, and the scale of the microstructure. For example,

Table 3-6 Brittle-to-Ductile Transition Temperature

Intermetallic	Melting Point (°C)	BDTT (°C)	Reference
NiAl	1638	150–250	Reuss and Vehoff (1990)
TiAl	1460	800–1000	Seetharaman et al. (1993)
Ti_5Si_3	2120	700–900	Reuss and Vehoff (1990)
$MoSi_2$–2%C	≈ 2100	1000–1400	Maloy (1991)
$Be_{12}Nb$	1700	≈ 1000	Henager et al. (1992)
Ti_5Si_3/Ti_3Al		650–850	Vehoff (1992)
NiAl/NiAlNb		600–800	Reuss and Vehoff (1990)

the BDTT's of NiAl and of the Ti_5Si_3/Ti_3Al eutectic composite increases by as much as 100°C and 200°C, respectively, upon increasing the loading rate by three orders of magnitude (Vehoff 1992). Under uniaxial tension TiAl shows a similar dependence (Seetharaman et al. 1993). This kind of response is a measure of increasing strain-rate sensitivity at elevated temperatures and of the onset of thermally activated deformation mechanisms. The BDTT of NiAl increases by about 100°C upon deviating from the stoichiometric (50/50) composition to the Ni-rich side (55/45) of stoichiometry (Vehoff 1992), in keeping with the effect noted by Pascoe and Newey (1968) on the compressive ductility of NiAl. Upon increasing the grain size from about 1 to 3 μm to about 20 to 50 μm, the BDTT of Ti_5Si_3 increases by about 100°C, even though the toughness at temperatures below the transition appears not to change with grain size. The other important factor, which applies to single crystals (of NiAl, at least), is the orientation of the crack plane relative to the loading direction. In NiAl, cleavage along {100} raises the BDTT by about 100°C relative to cleavage along {110} and raises the fracture toughness by a factor of two (Vehoff 1992). (A complicating factor here may be dislocation-solute interactions, in view of the observations on NiAl by Hack et al. (1992), noted previously.)

While the increase in toughness with temperature implies an increasing tolerance for flaws, the improvement is not large. For instance, the fracture toughness of polycrystalline NiAl of stoichiometric composition and of {110}-oriented single crystals increases from about 8 MPa $m^{1/2}$ just below the transition to about 15 MPa $m^{1/2}$ just above (Vehoff 1992). Similarly, the fracture toughness of Ti_5Si_3 increases from about 3 MPa $m^{1/2}$ at 700°C to about 8 MPa $m^{1/2}$ at 1100°C (Vehoff 1992), and that of $MoSi_2$-2%C increases from around 6 to 12 MPa $m^{1/2}$ as the temperature increases from 1000°C to 1400°C (Maloy 1991). The toughness of the structurally more complex tetragonal intermetallic $Be_{12}Nb$ shows perhaps the smallest improvement of all, rising from about 2 MPa $m^{1/2}$ at 500°C to only 4 MPa $m^{1/2}$ at 1000°C (Henager, Jacobson, and Breummer 1992). The toughnesses of the two-phase eutectic composites Ti_5Si_3/Ti_3Al and NiAl/NiAlNb show similar behavior and increase by roughly a factor of two upon exceeding the BDTT (Vehoff 1992).

It appears, therefore, that intermetallics are similar to some structural materials (ferritic steels, for instance) that fail in a brittle mode at low temperatures; namely, that when they can be described in terms of a BDTT, their fracture toughness roughly doubles above this temperature and their BDTT increases with both loading rate and grain size. In another sense, however, intermetallics are different. While the transitions in steel lead to good/very good toughnesses, the transition in the intermetallics studied to date lead only to moderate values. The reason the improvement is not greater reflects the relatively small degree of plastic work accompanying crack propagation in these materials.

4. CONCLUDING REMARKS

In concluding this chapter it is worth considering once again whether the brittle fracture of polycrystalline intermetallics under tension and the solutions so far developed are linked by common elements.

Concerning brittleness, aside from the intrinsic obstacles provided by a complex crystal structure, by an insufficient number of independent deformation modes and by coarse, planar slip, the factor that appears to be common to all strongly ordered intermetallics is a high work-hardening rate. In terms of the deformation near the tips of cracks, the work-hardening rate limits the energy that can be dissipated through plastic flow before the tensile stress reaches the cohesive strength of the atomic bonds. A high hardening rate then favors crack propagation over global plasticity. Of course, a very low work-hardening rate would lead to plastic instability. Thus, for both plastic stability and global ductility an intermediate rate close to that for f.c.c. metals and alloys ($\approx G/30-G/40$) may be required.

Concerning the amelioration of brittle behavior, the element that appears to be common to those, albeit few, cases that exhibit extensive ductility and good fracture toughness is a ductile, second phase comprised of a random solid solution. The Co-rich $L1_2$ material Co_3Ti (80 at%Co), which displays 60% elongation and transgranular ductile fracture in air at room temperature, is characterized by this feature. So, too, are the newly developed $L1_2$ alloys based upon Ni_3Al. The tough silicides and aluminides described in section 3.2 also contain large volume fractions of ductile, solid solutions. Although speculative, the good toughness and ductility of Zr_3Al might also be the result of a small amount of a ductile phase, for this line compound is ductile only on the side of stoichiometry where a disordered solid solution is in equilibrium with the $L1_2$ phase. Finally, there is evidence, albeit controversial, that the beneficial effect of boron on the ductility of Ni-rich Ni_3Al is related to a constitutionally disordered grain-boundary region. Random solid solutions, whether localized at grain boundaries or dispersed as particles, work harden at lower rates than ordered phases and this feature may contribute to their effectiveness in stopping cracks.

Whether a high work-hardening rate is truly fundamental to the brittle behavior of intermetallics and whether a nonordered second phase is a prerequisite to extensive ductility and good toughness are questions that await further research.

5. ACKNOWLEDGMENTS

The author wishes to thank Professor Ian Baker, Professor Robert W. Cahn and Dr. Chain T. Liu for discussions and for helpful comments on the manuscript. The work was sponsored by United States Department of Energy, Grant No. DE-FG02-86ER45260.

REFERENCES

AOKI, K., and IZUMI, O., 1979, *Nippon Kinzoku Gakkaishi*, 43: 1190.

BAKER I., 1993, *J. Mat. Res.*, (submitted).

BAKER, I., and GEORGE, E. P., 1992, *Metals and Materials*, June, 318.

BAKER, I., and HORTON, J. A., 1993, *Phil. Mag. A*, 67: 479.

BAKER, I., and NAGPAL, P., 1993, Proc. Internat. Symp. Structural Intermetallics, September 26–30, Seven Springs, PA.

BAKER, I., and SCHULSON, E. M., 1989, *Scripta Metall.*, 23: 345.

BAKER, I., SCHULSON, E. M., and HORTON, J. A., 1987, *Acta Metall.*, 35: 1533.

BAKER, I., SCHULSON, E. M., MICHAEL, J. R., and PENNYCOOK, S. J., 1990, *Phil. Mag. B*, 62: 659.

BALL, A., and SMALLMAN, R. E., 1966, *Acta Metall.*, 14: 1517.

BOHN, R., HAUBOLD, T., BIRRINGER, R., and GLEITER, H., 1991, *Scripta Metall.*, 25: 811.

BOND, G. M., ROBERTSON, I., and BIRNBAUM, H. K., 1989, *Acta Metall.*, 37: 1407.

BRIANT, C. L., 1990, *Metall. Trans. A.*, 21A: 2339.

BRIANT, C. L., and TAUB, A. I., 1988, *Acta Metall.*, 36: 2761.

BRICKNER, K. G., and DEFILIPPI, J. D., 1977, *Handbook of Stainless Steels*, ed. D. Pechner and I. M. Bernstein, ch. 20, New York: McGraw Hill.

BROWN, S. A., and KUMAR, K. S., 1993, *J. Mat. Res.* (in press)

BUDIANSKY, B., AMAZIGO, J. C., and EVANS, A. G., 1988, *J. Mech. Phys. Solids*, 36, 167.

CHAN, K. S., 1990, *Scripta Met.*, 24: 1725.

________, 1992, *J.O.M.*, May, p. 30.

CHANG, K. M., DAROLIA, R., and LIPSITT, H. A., 1992, *Acta Metall. Mater.* 40: 2727.

CHEN, S. P., STROLOVITZ, D. J., and VOTER, A. F., 1989, *J. Mat. Res.* 4: 62.

CHEN, S. P., VOTER, A. F., ALBERS, P. C., BORING, A. M., and HAY, P. J., 1990, *J. Mat. Res.* 5: 955.

CHIBA, A., HANADA, S., and WATANABE, S., 1992, *Mat. Sci. Eng.* A152: 108.

COPELY, S. M., and KEAR, B. H., 1967, *TMS-AIME* 239: 977.

COTTRELL, A. H., 1958, *TMS-AIME* 212: 192.

________, 1991, *Mat. Sci. Tech.* 7: 981.

CRIMP, M. A., and VEDULA, K. M., 1986, *Mat. Sci. Eng.* 78: 193.

CRIMP, M. A., VEDULA, K. M., and GAYDOSH, D. J., 1987, In *High Temperature Ordered Intermetallic Alloys II*, MRS Symp.Proc 81:499.

DÉVE, H. E., EVANS, A. G., and SHIH, O. S., 1992, *Acta Metall. Mater.* 40: 1259.

DÉVE, H. E., WEBER, C. H., and MALONEY, 1992, *Mat. Sci. Eng.* A153: 668.

DIMIDUK, D. M., WEDDINGTON, V. L., and LIPSITT, H. A., 1987, *Proc. MRS* 81: 221.

DYMEK, K. S., DOLLAR, M., HWANG, S. J., and NASH, P., 1992, *Mat. Sci. Eng.* A152: 160.

DWIGHT, A. E., and BECK, P. A., 1959, *TMS-AIME* 215: 976.

FANG, J., 1992, Ph.D. thesis, Thayer School of Engineering, Dartmouth College.

FANG, J., and SCHULSON, E. M., 1992, *Mat. Sci. Eng.* A152: 138.

________, 1993, *Phil. Mag. A* 67: 1117.

FELTHAM, P., and MEAKIN, J. P., 1957, *Phil. Mag.* 2: 1237.

FORWARD, C. T., and GIBSON, M. A., 1992, *Phil. Mag.* A 66: 1121.

FROST, H. J., 1987, *Acta Metall.* 35: 519.

________, 1988, *Acta Metall.* 36: 2199.

FU, C. L., 1990, *J. Mat. Res.* 5: 971.

FUCHS, G. E., and STOLOFF, N. S., 1988, *Acta Metall.* 36: 2599.

GAYDOSH, D. J., DRAPER, S. L., NEOBE, R. D., and NATHAL, M. V., 1992, *Mat. Sci. Eng.* A150: 7.

GEORGE, E. P., and LIU, T., 1990, *J. Mat. Res.* 5: 754.

GEORGE, E. P., HORTON, J. A., PORTER, W. D., and SCHNEIBEL, J. H., 1990, *J. Mat. Res.* 5: 1639.

GEORGE, E. P., LIU, C. T., and POPE, D. P., 1993, *Scripta Metall.* 28: 857.

GEORGE, E. P., PORTER, W. D., HENSON, H. M., OLIVER, W. C., and OLIVER, B. F., 1989, *J. Mat. Res.* 4: 78.

GROVES, G. W., and KELLY, A., 1963, *Phil. Mag.* 8: 877.

GUHA, S., 1987, MS thesis, Thayer School of Engineering, Dartmouth College.

GUHA, S., BAKER, I., MUNROE, P., and MICHAEL, J. R., 1992, *Mat. Sci. Eng.* A152: 258.

HACK. J. E., BRZESKI, J. M., and DAROLIA, R., 1992, *Scripta Metall.* 27: 1259.

HANADA, S., KIM, M. S., WATANABE, S., and IZUMI, O., 1987, *Scripta Metall.* 21: 277.

HECKER, S. S., ROHR, D. L., and STEIN, D. F., 1978, *Metall. Trans. A* 9A: 481.

HENAGER, C. H., JACOBSON, R. E., and BREUMMER, S. M., 1992, *Mat. Sci. Eng.* A153: 416.

HIRANO, T., 1990, *Acta Metall. Mater.* 38: 2667.

HIRANO, T., CHUNG, S. S., MISHIMA, Y., and SUZUKI, T., 1991, In High Temperature Ordered Intermetallic Alloys IV, MRS Symp. Proc. 213: 635.

IMAYEV, R. M., KAIBYSHEV, O. A., and SALISHCHEV, G. A., 1992, *Acta Metall. Mater.* 40: 589.

ISHIDA, K., KAINUMA, R., UENO, N., and NISHIZAWA, T., 1991, *Metall. Trans. A* 22A: 441.

JOHNSTON, T. L., DAVIES, R. G., and STOLOFF, N. S., 1965, *Phil. Mag.* 12: 305.

KHADKIKAR, P. S., LEWANDOWSKI, J. J., and VEDULA, K. M., 1989, *Metall. Trans. A* 20A: 1247.

KING, A. H., and YOO, M. H., 1987, *Scripta Metall.* 21: 1115.

KLEIN, O., NAGPAL, P., and BAKER, I., 1993, In High Temperature Ordered Intermetallic Alloys V, MRS Symp. Proc. 288: 935.

KOSTRUBANIC, J., KOSS, D. A., LOCCI, I. E., and NATHAL, M., 1991, In High Temperature Ordered Intermetallic Alloys IV, MRS Symp. Proc. 213: 679.

KUMAR, K. S., and LIU, C. T., 1993, J of Metals, 45 (6) 28.

KUMAR, K. S., MANNON, S. K., and VISWANADHAM, R. K., 1992, *Acta Metall. Mater.* 40: 1201.

KUNG, H., RASMUSSEN, D. R., and SASS, S. L., 1992, *Acta Metall. Mater.* 40: 81.

KURFMAN, V. B., 1965, *Acta Metall.* 13: 307.

KURUVILLA, A. K., and STOLOFF, N. S., 1985, *Scripta Metall.* 19: 83.

LEE, K. M., and WHITE, C. L., 1993, *Scripta Metall.* 29: 547.

LERF, R., and MORRIS, D. G., 1991, *Acta Metall. Mater.* 39: 2419.

LI, Z., and SCHULSON, E. M., 1993, In High Temperature Ordered Intermetallic Alloys V, MRS Symp. Proc. 288: 1081.

LIN, H., and POPE, D. P., 1993, *Acta Metall. Mater.* 41: 553.

LIU, C. T., 1991, *Scripta Met.* 25: 1231.

LIU, C. T., CAHN, R. W., and SAUTHOFF, G., Eds., 1992 *Ordered Intermetallics—Physical Metallurgy and Mechanical Behavior*, (Dordrecht: Kluwer Acad. Publ.).

LIU, C. T., and GEORGE, E. P., 1990, *Scripta Metall.* 24: 1285.

LIU, C. T., and INOUYE, H., 1978, *Metall. Trans. A* 10A: 1515.

LIU, C. T., and OLIVER, W. C., 1991, *Scripta Metall.* 25: 1933.

LIU, C. T., MCKAMEY, C. G., and LEE, E. H., 1989, *Scripta Metall.* 24: 385.

LIU, C. T., WHITE, C. L., and HORTON, J. A., 1985, *Acta Metall.* 33: 213.

LU, T. C., EVANS, A. G., HECHT, R. J., and MEHRABIAN, R., 1991, *Acta Metall. Mater.* 39: 1853.

MCKAMEY, C. G., HORTON, J. A., and LIU, C. T., 1988, *Scripta Metall.* 22: 1679.

MALOY, S., 1991, *J. Amer. Cer. Soc.* 74: 27.

MARGEVICIUS, R. W., and LEWANDOWSKI, J. J., 1991, *Acta Metall. Mater.* 41: 485.
MASAHASHI, N., TAKASUGI, T., and IZUMI, O., 1988, *Acta Metall.* 36: 1823.
MENDIRATTA, M. G., LEWANDOWSKI, J. J., and DIMIDUK, D. M., 1991, *Metall. Trans. A* 22A: 1573.
MIKKOLA, D. E., NIC, J. P., ZHANG, S., and MULLIGAN, W. W., 1991, *ISIJ Internat.* 31: 1076.
MILLS, M. J., and MIRACLE, D. B., 1993, *Acta Metall. Mater.* 41: 85.
MIRACLE, D. B., 1993, *Acta Metall. Mater.* 41: 649.
MORRIS, D. G., and MORRIS, M. A., 1991, *Acta Metall. Mater.* 39: 1771.
MURUGESH, L., VENKATESWARA, R., and RITCHIE, R. O., 1993, *Scripta Metall.* 29: 1107.
NAGPAL., P., and BAKER, I., 1991a, *Scripta Metall.* 24: 2381.
________, 1991b, *Scripta Metall.* 25: 2577.
________, 1991c, *Mater. Char.* 27: 167.
NEKKANTI, R. M., and DIMIDUK, D. M., 1990, *Proc. MRS Symp.* 194: 175.
ODETTE, G. R., CHAO, B. L., SHECKHERD, J. W., and LUCAS, G. E., 1992, *Acta Metall. Mater.* 40: 2381.
OLIVER, W. C., and WHITE, C. L., 1987, In *High Temperature Ordered Intermetallic Alloys II*, MRS Symp. Proc. 81: 241.
OTTERBEIN, U., HOFMANN, S., and RÜBLE, M., 1993, In *High Temperature Ordered Intermetallic Alloys V*, MRS Symp. Proc. 288: 183.
PANK, D. R., NATHAL, M. V., and KOSS, D. A., 1990, *J. Mat. Res.* 5: 942.
PASCOE, R. T., and NEWEY, C. W. A., 1968, *Metall. Sci. J.* 2: 138.
PEARSON, W. B., 1967, *Handbook of Lattice Spacings and Structures of Metals and Alloys*, Vol. 2, Oxford: Pergamon Press.
PETROVIC, J. J., HANNELL, R. E., MITCHELL, T. E., WADE, T. E., and MCCLELLAN, K. J., 1991, *Ceram. Eng. Sci. Proc.* 12: 1633.
PIKE, L. M., and LIU, C. T., 1992, *Scripta Metall. Mater.* 27: 1313.
PUGH, S. F., 1954, *Phil. Mag.* 45: 823.
RAMAN, A., and SCHUBERT, K., 1965, *Z. Metallk.* 56: 99.
RAO, K. T., ODETTE, G. R., and RITCHIE, R. O., 1992, *Acta Metall. Mater.* 40: 353.
REUSS, S., and VEHOFF, H., 1990, *Scripta Metall.* 24: 1021.
RICE, J. R., and THOMPSON, R., 1973, *Phil. Mag.* 29: 73.
RIGNEY, J. D., and LEWANDOWSKI, J. J., 1992, *Mat. Sci. Eng.* A158: 31.
RIGNEY, J. D., LEWANDOWSKI, J. J., MATSON, L., MENDIRATTA, M. G., and DIMIDUK, D. M., 1991, In *High Temperature Ordered Intermetallic Alloys IV*, MRS Symp Proc. 213: 1001.
ROSINGER, H. E., and WILKENS, B. J. S., 1980, *J. Nucl. Mater.* 89: 347.
ROZNER, A. G., and WASILEWSKI, R. J., 1966, *J. Inst. Met.* 94: 169.
RUSSELL, K. C., and EDINGTON, J. W., 1972, *Met. Sci. J.* 6: 20.
SCHMIDT, B., NAGPAL, P., and BAKER, I., 1989, In *High Temperature Ordered Intermetallic Alloys III*, MRS Symp. Proc. 133: 755.
SCHNEIBEL, J. H., and GEORGE, E. P., 1990, *Scripta Metall.* 24: 1069.
SCHNEIBEL, J. H., and HAZZLEDINE, P. M., 1991, In *High Temperature Ordered Intermetallic Alloys IV, MRS Symp. Proc.* 213: 323.

SCHNEIBEL, J. H., and JENKINS, M. G., 1993, *Scripta Metall* 28: 389.

SCHNEIBEL, J. H., and PORTER, W. D., 1989, In *High Temperature Ordered Intermetallic Alloys III*, MRS Symp. Proc. 133: 335.

SCHNEIBEL, J. H., BECHER, F. F., and HORTON, J. A., 1988, *J. Mat. Res.* 3: 1272.

SCHULSON, E. M., 1979, *Res. Mech. Letters* 1: 111.

———, 1982, COSAM Program Overview, NASA TN-8300006, Washington, DC, 175.

———, 1984, *Internat. Met. Rev.* 29: 195.

———, 1985, In *High Temperature Ordered Intermetallic Alloys*, MRS Symp. Proc. 39: 193.

SCHULSON, E. M., and BAKER, I., 1991, *Scripta Metall.* 25: 1253.

SCHULSON, E. M., and BAKER, I., 1992, In *Ordered Intermetallics—Physical Metallurgy and Mechanical Behavior*, Ed., C. T. Liu, R. W. Cahn, and G. Sautoff, p. 371, Dordrecht: Kluwer Acad. Publ.

SCHULSON, E. M., and BARKER, D. R., 1983, *Scripta Metall.* 17: 519.

SCHULSON, E. M., BRIGGS, L. J., and BAKER, I., 1990, *Acta Metall. Mater.* 38: 207.

SCHULSON, E. M., WEIHS, T. P., BAKER, I., FROST, H. J., and HORTON, J. A., 1986, *Acta Metall.* 34: 1395.

SCHULSON, E. M., XU, Y., MUNROE, P. R., GUHA, S., and BAKER, I., 1991, *Acta Metall. Mater.* 39: 2971.

SEETHARAMAN, V., SEMIATIN, S. L., LOMBARD, C. M., and FREY, N. D., 1993, *Proc. MRS* 288: 513.

SEYBOLT, A. U., 1966, Trans. ASM. 59: 860.

SIKKA, V. K., MAVILY, J. T., and ANDERSON, K., 1992, *Mat. Sci. Eng.* A153: 712.

STOLOFF, N. S., and DAVIES, R. G., 1964, *TMS-AIME* 57: 247.

———, 1966, *Progress in Materials Science* 13: 1.

STOLOFF, N. S., and DILLAMORE, I. L., 1970, *Ordered Alloys*, eds. B. H. Kear et al. p. 525, Claitor's Publishing Div., Baton Rouge, LA.

STOLOFF, N. S., FUCHS, G. E., KURUVILLA, A. K., and CHOE, S. F., 1987, In *High Temperature Ordered Intermetallic Alloys II*, MRS Symp. Proc. 81: 247.

STROTHERS, S. D., and VEDULA, K. M., 1987, *Prog. Powd. Metall.* 43: 561.

TAKASUGA, T., and IZUMI, O., 1985, *Acta Metall.* 33: 39.

———, 1986, *Acta Metall.* 34: 607.

TAKASUGI, T., GEORGE, E. P., POPE, D. P., and IZUMI, O., 1985, *Scripta Metall.* 19: 551.

TAUB, A. I., and BRIANT, C. L., 1987, *Acta Metall.* 35: 1597.

TAUB, A. I., BRIANT, C. L., HUANG, S. C., CHANG, K. M., and JACKSON, M. R., 1986, *Scripta Metall.* 20: 129.

TONN, S. C., ZHANG, Y., and CRIMP, M. A., 1993, *Mat. Sci. Eng.* (in press)

TURNER, C. D., POWERS, W. O., and WERT, J. A., 1989, *Acta Metall.* 37: 2635.

U.S. STEEL, 1964, "The Making and Shaping of Steel," 8th ed., ed. by H. E. McGannon.

VEHOFF, H., 1992, *Ordered Intermetallics—Physical Metallurgy and Mechanical Behavior*, C. T. Liu et al. (eds.) p. 299, Dordrecht, Kluwer Academic Publishers.

WEIHS, T. P., 1985, MS thesis, Thayer School of Engineering, Dartmouth College.

WEIHS, T. P., ZINOVIEV, V., VIENS, D. V., and SCHULSON, E. M., 1987, *Acta Metall.* 35: 1109.

WHITE, C. L., and STEIN, D. F., 1978, *Metall. Trans. A* 9A: 13.

XU, Y., and SCHULSON, E. M., 1993a, In *High Temperature Ordered Intermetallic Alloys IV*, MRS Symp. Proc. 288: 635.

________, 1993b, unpublished research.

YAMAGUCHI, M., UMAKOSHI, and YAMANE, T., 1982, *Scripta Metall.* 16: 607.

YANG, R., LEAKE, J. A., and CAHN, R. W., 1992, *Mat. Sci. Eng.* A152: 227.

YOO, M. H., and FU, C. L., 1992, *Mat. Sci. Eng.* A153: 470.

ZHANG, S. J., NIC, P., MULLIGAN, W. W., and MIKOLA, D. E, 1990, *Scripta Metall.* 24: 1099.

ZOK, F., EMBURY, J. D., VASUDEVAN, A. K., RICHMOND, O., and HACK, J., 1989, *Scripta Metall.* 23: 1893.

Chapter 4

Creep

Mohamed Y. Nazmy

1. INTRODUCTION

The rotating blades of a gas turbine or a jet engine are subjected to centrifugal stresses and high temperatures, a combination that gives rise to deformation by creep. Hence, blade material requirements include, in addition to satisfactory corrosion and oxidation resistance, high creep resistance, that is, high-temperature capability. The continuous need for ever increasing turbine inlet temperatures, improved efficiency and performance will necessitate both static and rotating components that retain dimensional stability at temperatures exceeding those for current nickel-base superalloys ~ 1100°C (Miller and Chamber 1987). A number of materials of the intermetallic class have the potential to meet the requirements for elevated temperature applications in gas turbines for two main reasons: first, for their inherent higher melting point above that of nickel-base superalloys, and second, for the low density that some of the intermetallics may possess. In fact, the second reason can be a unique advantage of intermetallics over nickel-base superalloys, with respect to the design of the turbine disk. This is because the turbine disks are subjected to radial pull of the blades in addition to the body forces resulting from the centrifugal forces associated with rotation. Hence, for large rotating components, the intermetallics, with their inherent low density, can be of interest in disks design.

In discussing components design with respect to creep loading, one has to consider a number of important aspects. Most critical components are designed to a maximum creep strain on the order of 1% in order to maintain dimensional compatibility. Hence, the life to a specific creep strain and not the life to fracture is the most relevant parameter to assess the creep advantage of certain materials. This point will be addressed later in the discussion of creep behavior of the different types of intermetallics. Another important aspect of creep behavior is the type of deformation, that is, tensile vs. compression. In most applications, the main stress is of a tensile nature. Hence, one has to bear in mind that compressive creep behavior may not necessarily represent actual loading conditions. In many research investigations on intermetallics, compressive creep testing was employed due to either the limited tensile ductility of the material or the availability of material for tensile testing.

In this chapter, the discussion mainly will be concentrated on the description of the creep behavior of monolithic intermetallics with some details on oxide dispersion strengthened or fiber strengthening of intermetallics.

2. CREEP BEHAVIOR OF INTERMETALLICS WITH $L1_2$ STRUCTURE

Ni_3Al is one of the most intensively investigated intermetallics with $L1_2$ structure. This might be due to the fact that Ni_3Al (i.e., γ') is the main strengthening phase in nickel-base superalloys. This is in spite of the fact that the superior resistance to creep of the superalloys with γ f.c.c. solid solution and γ' dual phase is attributed to the nature of dislocations at the γ/γ' interface as they pass from one phase to the other, rather than to the creep properties of the γ' phase itself.

Flinn (1960) reported on the steady-state creep of polycrystalline $Ni_3(Al,Fe)$ in the temperature range of 816–1093°C where the rate of dislocation motion is low compared with the jump rate of atoms. He assumed that in that temperature range the creep is of viscous type, of paired dislocations dissociated on nonglide planes, and the creep was controlled by the thermal diffusion of neighbouring atoms to annihilate the antiphase boundaries (APB's) produced by the motion of the dislocations.

Nicholls and Rawlings (1977) determined the creep rate of the same polycrystalline $Ni_3(Al, Fe)$ over the temperature range of 530–903°C. They divided creep phenomena into two separate mechanisms. The transition between the two mechanisms coincided with the temperature Tp at which the flow stress reaches a peak value during tensile deformation. Since the plastic deformation of Ni_3Al-based components is controlled by cube slip at temperatures above Tp, and by octahedral slip at temperatures below Tp, the creep behavior will differ at temperatures above and below Tp. In the temperature range 680–930°C the creep rate $\dot{\varepsilon}$ was given by the diffusion-controlled mechanism according to the following equation (Weertman 1958):

$$\dot{\varepsilon} = \frac{2\pi D_0 b}{GTk}\sigma^n \exp(-Q/RT) \tag{4.1}$$

where D_0, b, G, k, T, σ, and Q are the preexponent diffusion coefficient, Burgers vector, shear modulus, Boltzmann's constant, absolute temperature, and stress and activation energy, respectively. The exponent n is equal to 3.

At temperatures below 680°C, cross slip from {111} to {001} planes was predicted to be the rate-controlling mechanism from the calculation of activation energies for four possible cross-slip processes. Two different creep behaviors were reported in transient creep tests made on $Ni_3(Al,Cr)$ at low and high temperatures (Thornton, Davies, and Johnson 1970). The transient creep strain ε can usually be presented as a function of time t in the following equation:

$$\varepsilon = A \log t + Bt^m \tag{4.2}$$

where A, B, and m are constants.

The logarithmic term in equation 4.2 is predominant in low temperature transition creep where the creep rate may be controlled by the dislocation mobility. Creep deformation takes place by the exhaustion hardening of mobile dislocations since edge dislocations can move out freely and there is no long-range interaction between the edge and screw components of the dislocations. At 626°C the transient creep strain exhibits the power law with the time dependence being proportional to $1/3$. The creep behavior is controlled by dislocation debris hardening as a result of the generation and interaction of dislocations. The yield stress increases with increasing temperature in the intermediate temperature range while the creep strength of $Ni_3(Al,Fe)$ decreases with an increase in temperature. The rate of recovery also increases with an increase in temperature. This is because the activation-energy of self-diffusion in γ' is not extremely high. Therefore, the lattice resistance for γ' to the motion of dislocations is too small to provide adequate creep strength at the temperatures relevant to high-temperature applications. The phenomenon of inverse creep (i.e., absence of steady state creep of Ni_3Al) was reported in some of the creep tests on alloyed single crystals of Ni_3Al (Anton, Pearson, and Snow 1987). Other researchers have reported inverse creep behavior for $Ni_3(Al,Hf)$ single crystals doped with B at intermediate temperatures and at high stress (Schneibel and Horton 1988; Hazzledine and Schneibel 1989; Hemker and Nix 1989). The creep curve is characterized by a primary region followed by inverse creep without the development of a steady-state part. In the primary creep region, the creep rate slows down due to the exhaustion of screw dislocations on the primary octahedral planes by cross slip. In the creep study on $Ni_3(Al,Hf)$ single crystals doped with B, it was concluded that inverse creep was attributable to the activation of secondary octahedral and cube-slip systems (Hemker and Nix 1991). A model has been proposed for an unlocking mechanism for pinned screw dislocations by the reaction between the primary and secondary dislocations (Hazzledine and Schneibel 1989). Therefore, the creep strength is not expected to increase with increasing temperature in contrast to yield strength, although the inverse creep occurs only in limited orientations. It was recently reported that inverse creep dominates intermediate-temperature creep behavior, and as a consequence the intermediate-temperature creep strength of Ni_3Al is predicted to decrease with increasing temperature in a normal manner (Hemker and Nix 1991).

The inadequate creep strength of Ni_3Al has led to several schemes for its strengthening. Anton et al. (1988) have tried solid solution strengthening of single crystals. They employed 2–8 at%Ta due to its known potent solid-solution strengthening effect. Their creep results showed a stoichiometric effect depending on the test temperature. The same principle of solid-solution strengthening has been applied to directionally solidified (DS) $Ni_3(Al,Ta/Nb)$ (Nazmy and Staubli 1991). The improvement in creep strength in Ni_3Al in both cases was achieved at the expense of the density advantage claimed for this compound. Fujita et al. (1988) also have reported on the improvement of creep properties of Ni_3Al using several alloying elements. The strengthening mechanism in that case was not pure solid-solution strengthening. Funk and Blank (1988) tried to improve the creep strength of Ni_3Al by the formation of Ni_3Al-Mo in situ composite. Once more, the improvement in creep was accompanied by an unacceptable increases in the density of the material. Figures 4-1a and b give a comparison of the creep strength achieved for different alloyed Ni_3Al as compared to DS CM247LC and CMXS4 strong Ni-base superalloys, respectively. Miura et al. (1990) investigated the compression creep behavior of alloyed Ni_3Al single crystals. They reported an exponent of 3–4 for the stress dependence and an activation energy of the order of the diffusion activation energy of the ternary addition. The creep-fatigue

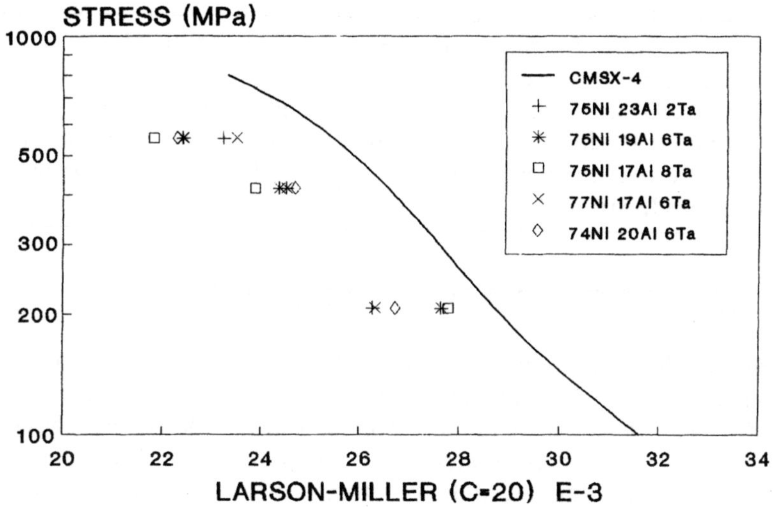

(a)

Figure 4-1a. Comparison of the creep rupture strength between several DS Ni_3Al alloys and DS CM247LC. (Reprinted with permission from *Scripta Metallurgica*, Volume 25, M. Nazmy and M. Staubli, 1991, page 1305, Elsevier Science, Ltd., Pergamon Imprint, Oxford, England)

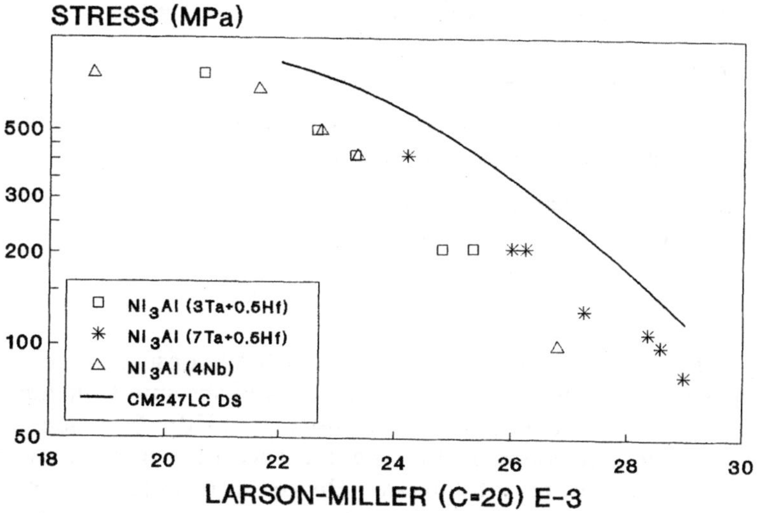

(b)

Figure 4-1b. Comparison of the creep rupture strength between several single crystal Ni_3Al alloys and CSMX-4. (Anton, Pearson, and Snow 1987) phenomenon has been studied on DS and single crystals of Ni_3(Al,B,Hf) by stress-controlled fatigue testing (Bellows et al. 1988). They reported an improvement in the creep-fatigue resistance of Ni_3(Al,B,Hf) over the superalloy UDIMET 115. This improvement is creep-fatigue resistance was attributed to the cyclic work hardening associated with the homogeneous nature of slip since the creep-fatigue resistance is a function of the ultimate tensile strength.

3. CREEP OF B2 AND DO_3 STRUCTURES

Interest in the equiatomic aluminides NiAl, CoAl, and FeAl with the B2 structure and Fe_3Al with DO_3 structure stems from their excellent oxidation resistance, high melting points, and low densities. These characteristics present these materials as candidates for high-temperature applications in jet engines and gas turbines. For these applications, creep resistance is considered to be one of the important properties required for extended use at high temperatures.

The creep behavior of binary single-phase compounds will be presented. Attempts to improve the creep strength of these intermetallics will then be described. The driving factor for these attempts is the low creep resistance of these intermetallics at the temperature range of interest. For example, unalloyed NiAl has much lower creep strength than with the Mar-M200 nickel-base superalloy (Strutt and Kear 1984).

3.1. Creep of Binary Compounds

The creep behavior of B2 binary compounds can be divided into regimes of dislocation creep and diffusional creep.

3.1.1. Dislocation creep. The steady state creep rate $\dot{\varepsilon}$ in this case is usually expressed as in the following equation:

$$\dot{\varepsilon} = A\sigma^n \exp - Q/RT \quad (4.3)$$

where A, σ, n, Q, T and R are microstructural constant, applied stress, stress exponent, activation energy for creep, and absolute temperature and gas constant, respectively (Nix and Ilschner 1980). Equation 4.3 is a simplified form of equation 4.1. The existence of a true steady state is sometimes difficult to substantiate. Hence, it is usually a satisfactory approximation to substitute for it with the minimum creep rate and still get consistent results. For single-phase metals and alloys, dislocation creep can be classified as either of two main types: known as class M, or pure

Table 4-1 Criteria For Classifying Dislocation Creep Behavior

	CLASS M	CLASS A
Controlling Mechanism	Climb	Viscous Glide
n	5	3
Q	Diffusion	Diffusion
Dislocation Structure	Subgrains	Homogeneous
Primary To Transients	Class M	Class A

metal type; and class A, or alloy type. In class M, the dislocation glide is faster than climb and thus creep is controlled by the rate of climb and a dislocation substructure is always observed. In class A creep, viscous dislocation glide is rate controlling. The origin of this viscous glide can be solute atoms or high lattice friction due to long range order. These categories represent limiting conditions and can be distinguished by the criteria listed in Table 4-1 (Mohamed and Langdon 1974; Soliman and Mohamed 1982; Mills, Gibeling, and Nix 1986). The activation energy for creep is equal to that for lattice diffusion in both types. Corrections for the temperature dependence of the elastic modulus frequently are necessary (Nix and Ilschner 1980). There are tests for discriminating between creep mechanisms, whereby sudden changes in the applied stress or strain rate are made and the response is unique to the material and type of test (Langdon and Yavari 1981; Mills et al. 1986; Yaney, Gibeling and Nix 1987). The instantaneous strain increment after a small stress increase would be primarily elastic for class A materials but would have a large plastic component in class M behavior. A summary of the stress exponents and activation energies determined by various researchers (Lawley, Coll, and Cahn 1960; Vandervoort, Mukherjee, and Dorn 1966; Hocking, Strutt, and Dodd, 1971; Bewk, Dodd, and Strutt 1973; Yang and Dodd 1973; Rudy and Sauthoff 1985; Whittenberger 1985, 1986, 1987, 1988; Mannan, Kumar, and Whittenberger 1990; Whittenberger et al. 1990; Nathal 1991) is presented in Table 4-2. This table has been compiled and published by Nathal (1991) in a recent excellent review of the creep of B2 intermetallic compounds. For Nial and CoAl, the reasonable choice for activation energy appears to be close to 310 kJ/mol, which is similar to the activation energy determined in diffusion experiments (Berkowitz, Jaumot, and Nix 1954; Hancock and McDonell 1971; Lutze-Birk and Jacobi 1975;

Table 4-2 Summary of Creep Parameters for B2 Aluminides

At%Al	Grain Size (μm)	T (K)	n	Q (kJ/mol)
		NiAl		
48.25	5–9	1000–1400	6–7.5	313
49.2	15–20	1100–1400	5.75	314
50	12	1200–1300	6	350
50	450	1073–1318	10.2–4.6	283
50	500	1173	4.7	—
50.4	1000	1075–1750	7–3.3	230–290
50	SX[123]	1023–1223	7.7–5.4	—
50	SX	1023–1328	4–4.5	293
50	SX[001]	1000–1300	6	440
		CoAl		
44	8	1100–1400	2.7	200
49.9	10	1100–1400	4.5	345
50	20	1200–1300	3	384
50	SX[123]	1323	2.6	—
		FeAl		
27	70	810–1010	6	370
45.7	40	1300–1400	3.5	410
48.7	40	1100–1200	6.5	470
50	500	1173	4.0	—

Nathal 1991. (Reprinted by permission of Kluwer Academic Publishers.)

Hagel 1977; Shankar and Seige 1978). For FeAl, an activation energy for creep of 450 kJ/mol is much higher than that for diffusion of ~ 300 kJ/mol (Hirano and Hishinuma 1968; Nishida, Yamamoto, and Nagata 1970; Hagel 1977). Correction for the temperature dependence of the elastic modulus is not sufficient to explain the discrepancies observed for FeAl. Nevertheless, Sauthoff (1990) has nicely correlated the creep resistance of Coal, NiAl, and FeAl with their corresponding diffusion coefficients.

For NiAl, the available results on the stress exponent n as well as on the microstructural observations of subgrain formation are consistent with class M behavior (Yang and Dodd 1973; Rudy and Sauthoff 1985;

Whittenberger 1987, 1988; Whittenberger et al. 1990). For CoAl the n values range between 2–5, which indicates a possible tendency towards class A behavior. Actually, most of the data appear to show a transition from class A to M between 1200 K and 1400 K. At 1200 K, inverted primary creep (Whittenberger 1985; Yaney and Nix 1988; Mannan et al. 1990), homogeneous dislocation structures (Hocking, Strutt, and Dodd 1971; Yaney and Nix 1988), and class A response to strain rate changes (Yaney and Nix 1988) have been observed. At 1400 K, normal primary creep and class M response during strain-rate changes are observed (Whittenberger 1985; Yaney and Nix 1988). Dislocation structures during creep have not been reported; nevertheless, subgrains have been observed after extrusion at 1500 K (Yaney, Pelton, and Nix 1986).

FeAl appears to behave in creep by at least two different mechanisms, as shown in Table 4-2. Whittenberger (1983, 1986) studied powder metallurgy FeAl; at high stresses and at temperatures below 1200–1300 K, he reported stress exponents which cluster near 6. Normal primary creep is reported but no subgrains have been observed (Whittenberger 1983, 1986; Rudy and Sauthoff 1985). At low stress levels as 1200 K and 1300 K and at higher temperatures, the creep deformation appears to follow a stress exponent of 3–4 (Whittenberger 1983). In this regime, normal primary creep was observed but a lack of subgrains also was reported (Rudy and Sauthoff 1986; Whittenberger 1986). The evidence of temperature dependence of n values, the absence of the typical steady-state creep conditions, observations of grain growth and dynamic recrystallization during deformation, and an improvement in creep strength with coarse grain sizes all lead to the conclusion that a superposition of both dislocation and grain-boundary mechanisms is contributing to the creep deformation in this regime (Whittenberger 1986; Baker and Gaydosh 1987). Sauthoff and co-workers also have observed such a transition in n values, but at a lower temperature, ~ 1027 K (Rudy and Sauthoff 1985, 1986; Jung, Rudy, and Sauthoff 1987). Their materials were made by casting; hence, the grain-boundary mobility is probably greater than in the powder metallurgy material of Whittenberger, which has oxide particles that restrict grain growth.

3.1.2. Diffusional creep. Time-dependent deformation can take place by stress-assisted vacancy flow at stresses that are too low for dislocation process to be significant. Creep by these diffusional mechanism such as Nabarro–Herring or Coble creep is attributed solely to movement of vacancies from sources to sinks, which are usually grain boundaries of different orientations with respect to the applied stress. These mechanisms are characterized by a stress exponent $n = 1$ and pronounced grain-size dependence, with large-grained materials being more creep resistant as shown in the creep rate ε:

$$\dot{\varepsilon} = A\left(\frac{\sigma\Omega D}{kTd^2}\right) \tag{4.4}$$

where A, Ω, D, k, T, d and σ are dimensionless factor, atomic volume, diffusivity coefficient (through the grain for Nabarro–Herring or along the grain boundaries for Coble creep), Boltzman constant, absolute temperature, grain size (effective diffusion length), and applied stress, respectively.

A clear evidence for diffusional mechanisms in a Ni-20Fe-50Al alloy, namely a stress exponent of 1 has been reported (Jung, Rudy and Sauthoff 1987). For NiAl, the large body of literature on creep does not provide enough support for creep by diffusional mechanisms over large regimes of temperature and stress. In fact, the calculated creep rates on the basis of this type of mechanism are much higher than the actual experimental data, implying that the diffusional mechanisms are somehow suppressed compared to disordered metals (Whittenberger 1987). For CoAl and FeAl, evidence for diffusional mechanisms appear to commence at approximately 0.7 *Tm*. This can be indirectly explained for CoAl through the observation that the stress exponent n has not shown a corresponding increase at higher temperatures (Hocking et al. 1971; Whittenberger 1985; Mannan et al. 1990). In the case of FeAl, the improved creep resistance observed for coarse-grain size supports the presence of a diffusional mechanism (Whittenberger 1986).

3.1.3. Effect of stoichiometry. Literature results on the effect of stoichiometry on the creep rates of NiAl, CoAl, and FeAl have been summarized and discussed as shown in Figure 4-2 (Nathal 1991). The testing temperatures indicated in this figure is approximately 0.7 *Tm*. For NiAl there is a broad range of composition where the creep rate is roughly independent of composition. The largest difference in creep rates within this range of compositions is only a factor of 5 (Pascoe and Newey 1968; Yang and Dodd 1973; Whittenberger 1987; Whittenberger, Kumar and Mannan 1991). The plausible explanation for this effect is the enhanced diffusivity due to the decrease in melting point as one moves from the stoichiometric ratio. These trends as a function of stoichiometry are reversed from those observed at lower temperatures, where point-defect hardening predominates over the effects of diffusion (Jung et al. 1987). For CoAl, the creep dependence on stoichiometry is much stronger, with the highest creep strength close to the equiatomic composition as shown in Figure 4-2 (Whittenberger 1985). The creep behavior of FeAl follows that

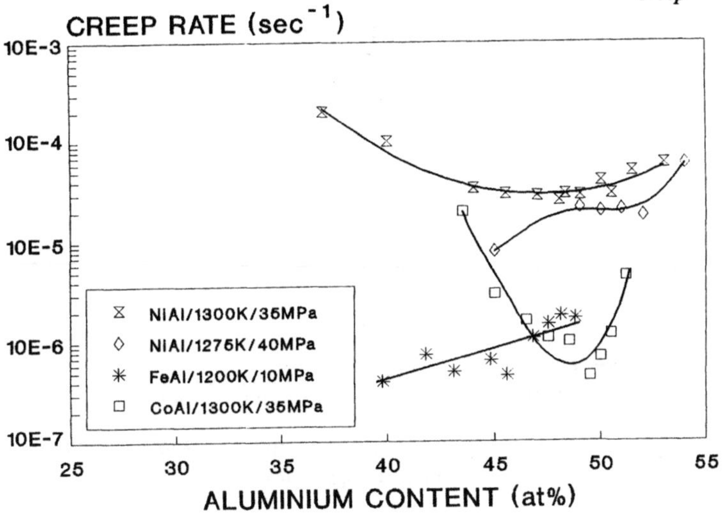

Figure 4-2. Effect of stoichiometry on the creep rate for NiAl, FeAl, and CoAl (Nathal 1991; reprinted by permission of Kluwer Academic Publishers).

of the melting point, where a slight increase in creep rate is observed as Al level is decreased (Whittenberger 1986). The diffusion data for FeAl also follow a weak dependence on Al content (Hirano and Hishinuma 1968; Nishida, Yamamato, and Nagata 1970; Hagel 1977).

3.2. Grain Size Effects

In general, at high temperatures the creep strength decreases with decreasing grain size. This effect has been shown in Ni-1Fe-40Al (Sauthoff 1990). Nevertheless, the studies on NiAl and FeAl have demonstrated that elevated-temperature creep strength can sometimes be improved by decreasing the grain size (Whittenberger 1983, 1986). It is believed that the observed apparent Hall–Petch behavior at high temperatures for the specific case of FeAl is due to precipitates that prevent grain-boundary sliding and hence strengthen the material to various degrees. If the resistance to dislocation creep is improved, for example, by precipitation strengthening, the grain-boundary effects may play an influential role and coarse-grained materials may be desirable.

3.3. Creep of Fe_3Al Intermetallic

The iron aluminide Fe_3Al occurs over the range 18–32 at%aluminum. It exists in the ordered DO_3 structure up to ~ 540°C, an imperfect

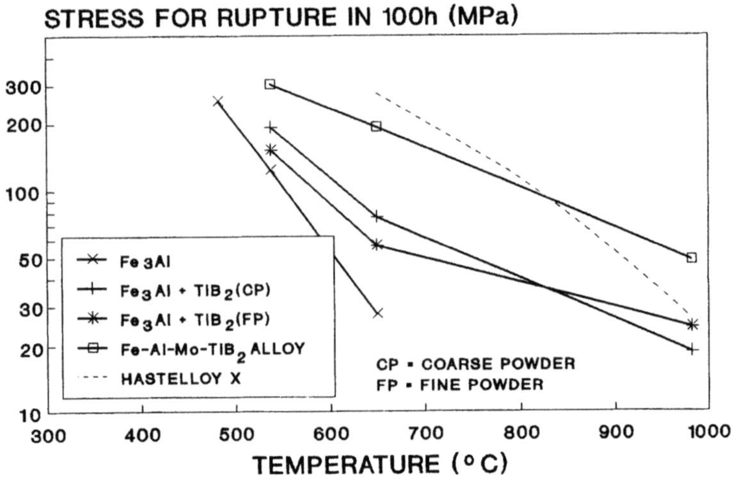

Figure 4-3. Creep properties of Fe_3Al base intermetallics as compared with HASTELLOY-X (McKamey et al. 1991).

ordered B2 structure between 540°C and 760°C, and is disordered above 760°C. The DO_3 structure results from further ordering in the B2 structure on four sublattices instead of two. Fe_3Al exhibits higher strength at elevated temperatures up to 540°C, which is the critical temperature of ordering, than the B2 $(Ni_{0.8}Fe_{0.2})Al$ (McKamey et al. 1991). Chromium is added to Fe_3Al to improve the low-temperature ductility and Mo or Nb is added to enhance its creep strength (McKamey et al. 19991). In fact, Mo tends to increase the critical $DO_3 \rightarrow B2$ temperature. Figure 4-3 shows the creep properties of Fe_3Al-base intermetallics. One can observe in this figure that Mo additions have improved the creep resistance of Fe_3Al almost to the level of Hastelloy X, a well known nickel-base alloy.

3.4. Methods for Increasing Creep Resistance of B2 Structure Intermetallics

In this section various concepts for improving the creep strength, primarily for NiAl, will be discussed. It has been reported that the Al creep resistance of NiAl is inferior to that of the existing nickel-base superalloys (Strutt and Kear 1984). The excellent oxidation resistance at high temperatures of NiAl makes it rewarding from the technological application point of view to investigate the feasibility of applying various strengthening concepts to this intermetallic.

3.4.1. Solid Solution Strengthening. The improvement in compressive creep strength, at 1200 K and a creep rate of 10^{-7} s^{-1}, of NiAl due to solid-solution hardening is illustrated in Figure 4-4 (Rudy and Sauthoff 1985; Jung et al. 1987; Whittenberger 1987; Pathare 1987; Whittenberger et al. 1989; Nathal 1991). The solid-solution alloys show some improvements in strength, but they exhibit a change in the stress exponent in to a value near 3–4. Hence, it appears that these solute additions change the creep mechanism to viscous drag behavior similar to that which occurs when alloying elements are added to pure metals. In a recent study, a transition from class M to class A behavior as a function of applied stress was observed, and the transition appeared to be well described by current theories developed for disordered solid solutions (Mohamed and Langdon 1974; Soliman and Mohamed 1982; Raj, Locci, and Noebe 1991). However, because of this new stress exponent, the creep strength improvements over the binary alloys are only significant at high stresses and this advantage is reduced or eliminated in the lower stress regime. The solid-solution strengthening of FeAl by Ni additions also is shown in Figure 4-4. Solid-solution hardening of CoAl by Fe and Ni also has been reported (Whittenberger 1987). Since both of these binary compounds have tendencies towards class A behavior, it would not be surprising that the ternary alloys would also show similar trends. In summary, it appears that solid-solution strengthening does provide some creep strength improvements over the binary compounds, but this concept is inadequate by

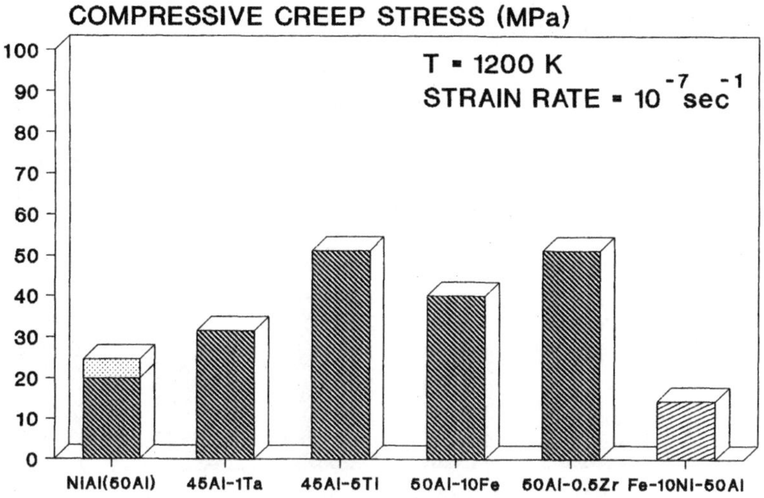

Figure 4-4. Solid solution hardening of polycrystalline NiAl and FeAl in compressive creep. (based on creep data compiled by Nathal 1991)

itself and must eventually be used in combination with other strengthening mechanisms.

3.4.2. Precipitation Strengthening. Marked improvements in creep strength of NiAl by precipitation hardening were first reported by adding Ti to form a two-phase mixture of NiAl and Heusler-phase Ni_2AlTi (Polvani, Tzeng, and Strutt, 1976). Heusler phases and other phases such as Laves (e.g., NiAlTa) can be formed with many ternary additions such as Nb, Ta, Hf, and V. The improvements in compressive creep strength, at 1300 K and creep rate of 10^{-7} s^{-1}, of NiAl, Ti and Ta-containing alloys (Polvani et al. 1976; Rudy and Sauthoff 1985; Jung et al. 1987; Whittenberger 1987; Pathare 1987; Whittenberger et al. 1989; Nathal 1991) are presented in Figure 4.5.

It is evident that these materials are strong; nevertheless, the extrapolation at lower stresses shows the diminishing of these advantages due to a low stress exponent *n*. The reasons for the low stress exponent in these alloys is not clear, since most creep resistant two-phase alloys exhibit significantly higher stress exponents than the matrix phase. An explanation for this behavior may be that the observed *n* values represent a superposition of several deformation mechanisms, including diffusional creep, and/or that coarsening of the precipitate phase results in less strengthening in the low-stress regime. By analogy to the superalloys, optimizing the creep strength requires a balance of the compositions of the two phases, the precipitate-volume fraction, and the size and distribution of the

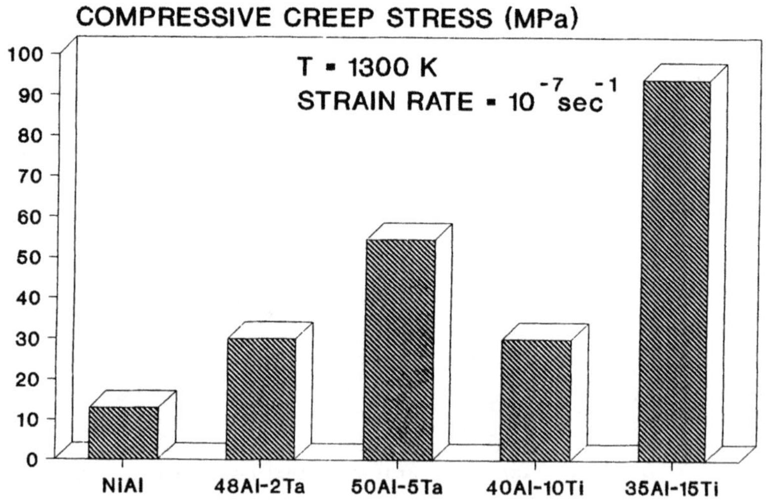

Figure 4-5. Improvement in compressive creep strength of NiAl, Ti, and Ta containing intermetallics. (based on creep data compiled by Nathal 1991).

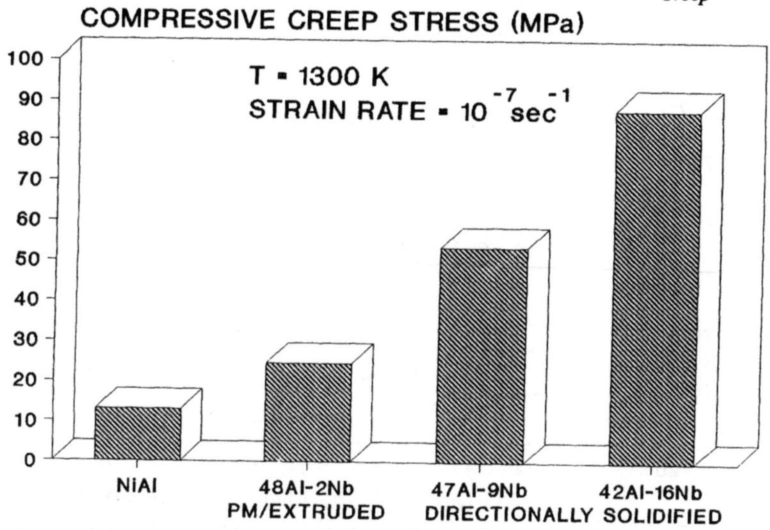

Figure 4-6. Effects of variations in microstructure on the compressive creep strength of NiAl strengthened by Laves phases. (based on creep data compiled by Nathal 1991)

precipitates. Another example of the sensitivity of creep strength to the microstructure in these alloys is shown in Figure 4-6. In this figure alloys with Nb rich Laves phase exhibit similar improvements to the Ta and Ti alloys (Fujita et al. 1988; Pathare 1987; Whittenberger et al. 1989; Whittenberger, Westfall, and Nathal 1989; Oliver, Noebe, and Whittenberger 1991). It has been reported also that by changing the processing of the same alloy from cast and extruded to directionally solidified, both creep strength and the stress exponent were changed (Mills, Gibeling, and Nix 1986). The use of single-crystal technology would also seem to be necessary in order to realize the full benefits of the precipitates. Recently, Darolia (1991) has reported promising tensile creep-rupture properties. His results, not corrected for the density, are presented in Figure 4-7, along with the creep-rupture properties of the superalloy René 80.

3.4.3. Dispersion Strengthening. This section addresses the use of rapid solidification and reaction-milling processes to introduce fine dispersoids in the intermetallic to improve its creep strength. Fine dispersoid particles can be incorporated in NiAl through rapid solidification. By rapid solidification, very fine dispersions of second phases, with diameters on the order of 20–50 nm, can form that are resistant to coarsening due to their very low solubility. For NiAl, both pure elements such as W and Mo

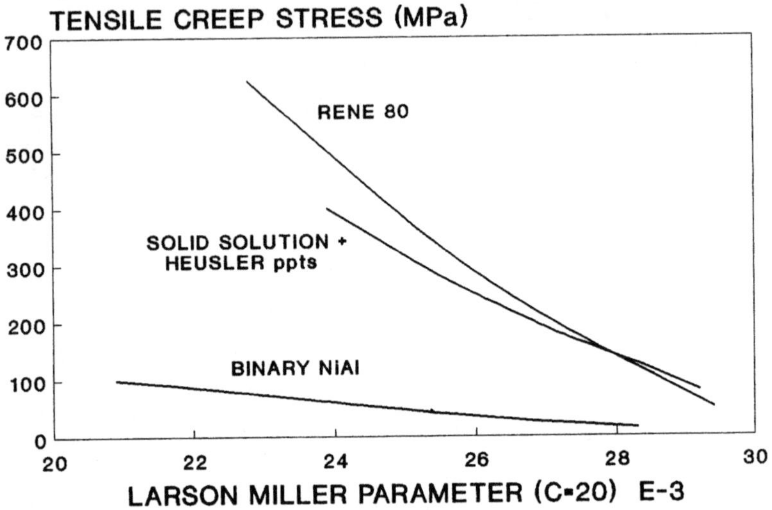

Figure 4-7. Tensile stress rupture properties of NiAl intermetallics compared with superalloy RENÉ 80. (Darolia 1991)

and various carbides or borides are candidates for this type of strengthening. These dispersoids also tend to pin the grain boundaries and result in significant grain refinement (Jha, Ray, and Whittenberger 1989; Locci et al. 1989; Whittenberger, Gaydosh, and Kumar 1990; Whittenberger et al. 1991). The improvements in compressive creep strength, at 1300 K and creep rate of 10^{-7} s^{-1} for different dispersion strengthened NiAl alloys are shown in Figure 4-8. (Whittenberger 1987; Jha et al. 1989; Nathal 1991; Whittenberger et al. 1990; Whittenberger et al. 1991; Whittenberger, Ray, and Jha 1991). It has been reported that TiB_2 exhibited more strengthening effect than W and TiC (Locci et al. 1989; Whittenberger et al. 1990). Nevertheless, the degree of improvement observed may be simply due to grain refinement. The alloys reinforced with HfB_2 and HfC are both considerably stronger than binary NiAl (Jha et al. 1989; Whittenberger et al. 1990; Whittenberger et al. 1991). The mechanism for the improved strength in the Hf-containing dispersoids, is primarily due to the interaction of the dispersoids with mobile dislocations and subgrain boundaries (Jha et al. 1989; Whittenberger et al. 1990; Whittenberger et al. 1991). However, use of rapid solidification to dispersion strengthen NiAl has not provided sufficient strengthening to compete effectively with the superalloys.

In FeAl, strengthening by additions of Fe_6Al_6Zr and ZrB_2 particles has been demonstrated at temperatures up to 1100 K as shown in Figure

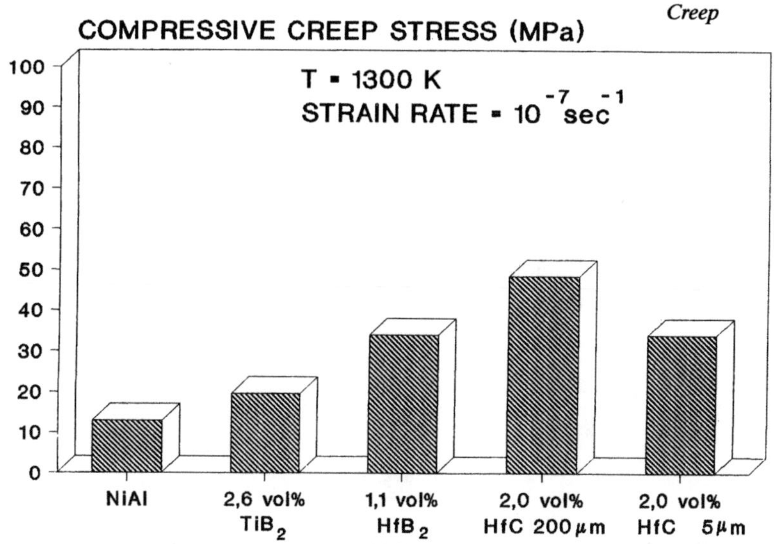

Figure 4-8. The improvements in compressive creep strength for different dispersion strengthened NiAl intermetallics. (based on creep data compiled by Nathal 1991)

4-9 (M. Morris and D. Morris 1990; Nathal 1991). The reaction-milling process produces very fine dispersoids, on the order of 20–50 mm, at relatively high volume fractions of ~ 10% (Whittenberger et al. 1990a, 1990b). These particles are not uniformly distributed but are clustered along prior particle boundaries. The improvements in compressive creep strength, at 1300 K and 10^{-7} s^{-1}, are encouraging and are shown in Figure 4-10. The reasons for the exceptional properties of this type of second-phase reinforcement are not clearly understood.

3.4.4. Strengthening by Composites. This section addresses the strengthening of composites by discontinuous and continuous reinforcements. In the case of NiAl, reinforcement with TiB_2 particulates with 1 μm diameter, produced by an exothermic reaction process known as XD, reveals improvements in creep strength that scale with the amount of reinforcement, as shown in Figure 4-11 (Whittenberger 1987; Whittenberger et al. 1990). It is important to point out that the stress exponents are high, which leads to a creep-strength advantage that is maintained at lower stresses. An increase in the stress exponent usually indicates that the dislocation substructure is refined and stabilized by the second phase when compared to the same matrix without reinforcement. Evidence for this has been provided by transmission electron microscopy, where the

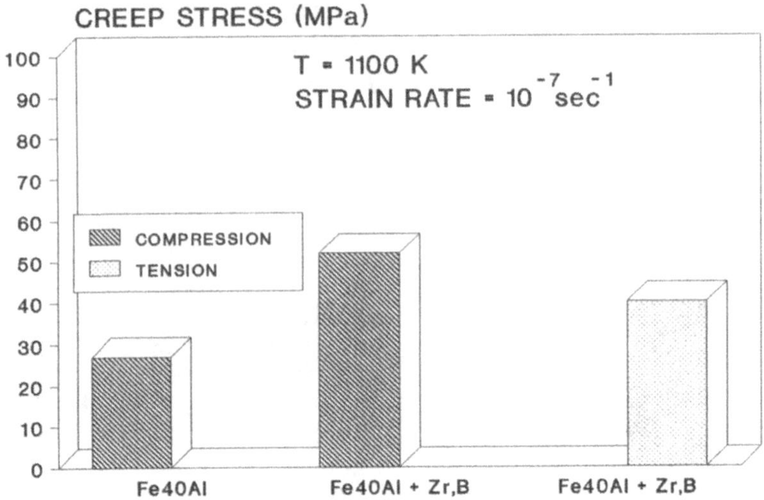

Figure 4-9. Comparison of creep strength under tension and compression for Fe-40Al strengthened by ZrB_2 particles. (based on creep data compiled by Nathal 1991)

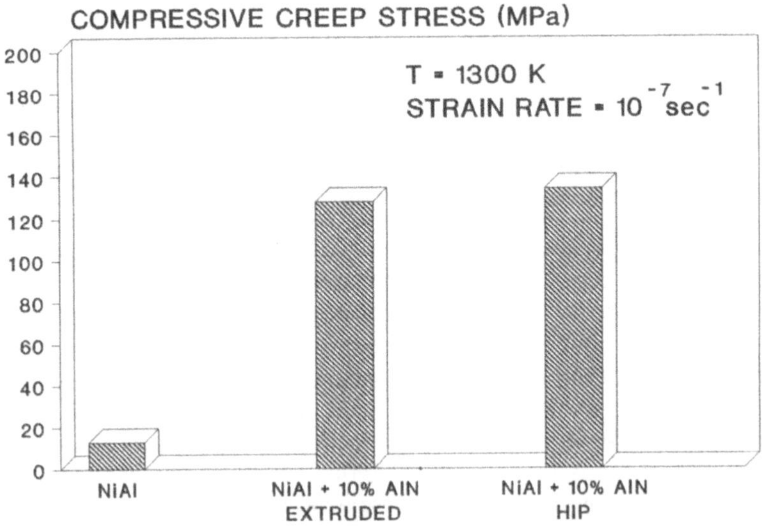

Figure 4-10. Effect of AlN dispersoids added by a reaction milling process on compressive creep strength of NiAl. (based on creep data compiled by Nathal 1991)

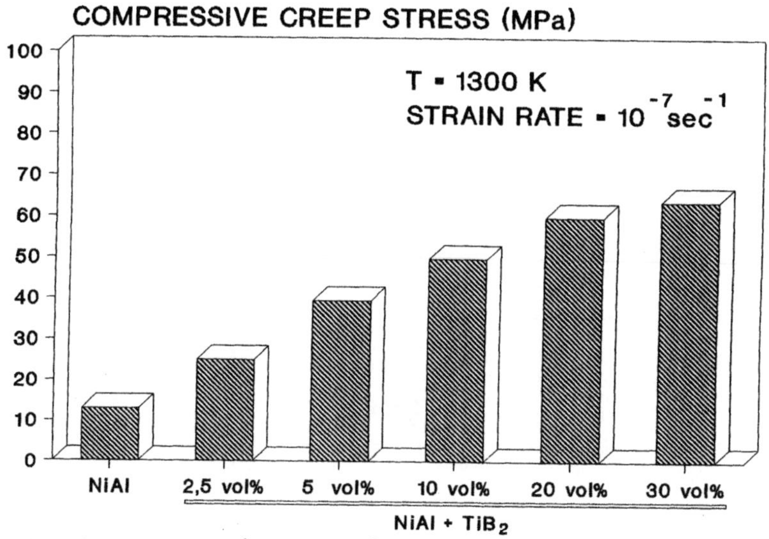

Figure 4-11. Improvements in compressive creep strength of NiAl due to reinforcement with TiB_2 particulate. (based on creep data compiled by Nathal 1991)

creep-deformation structure was characterized by subgrain boundaries that are usually pinned by the particles, in addition to higher dislocation content within the subgrains (Whittenberger et al. 1990). Similar strength improvement have been observed when TiB_2 particles were added to CoAl (Mannan et al. 1990).

Another type of discontinuous reinforcement that has been produced in NiAl is Al_2O_3 whiskers (Whittenberger et al. 1991). These whiskers had an average aspect ratio of approximately 7.4 and were added to mechanical blending at volume fractions up to 25%. The improvements in creep strength, at 1300 K and 10^{-7} s^{-1}, are shown in Figure 4-12. However, these improvements, due to Al_2O_3 whiskers, were not as effective as in the case of TiB_2 particles. The stress exponents for Al_2O_3 whisker-containing material were the same as that of the matrix, which indicates that deformation is controlled by flow in the matrix as predicted by several models of composite strengthening (Kelly and Street 1972). Control of whisker distribution and alignment, and whisker damage during processing, are generally major problems in this type of composite. Hybrid composites containing both TiB_2 particulates and Al_2O_3 whiskers have been made and it has been shown that these strengthening concepts were additive (Nathal 1991). Continuous reinforcement can either be natural, such as directionally solidified eutectics, or artificially fabricated, in order

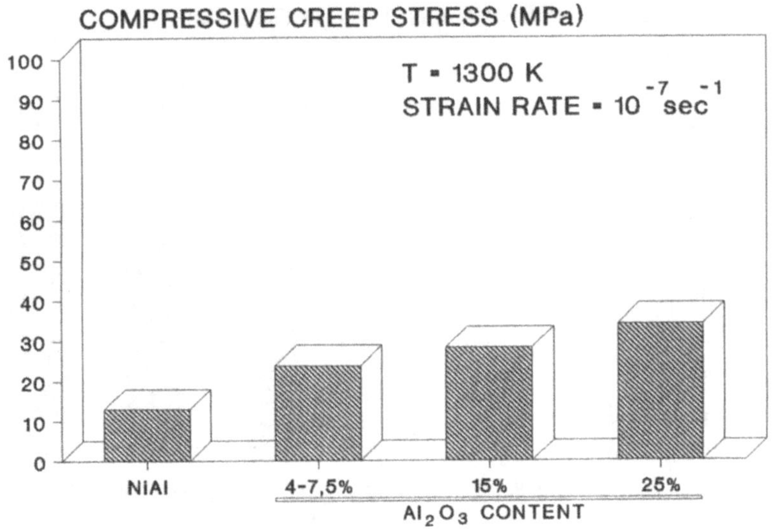

Figure 4-12. Improvements in compressive creep strength of NiAl due to Al_2O_3 whiskers added by mechanical blending. (based on creep data compiled by Nathal 1991)

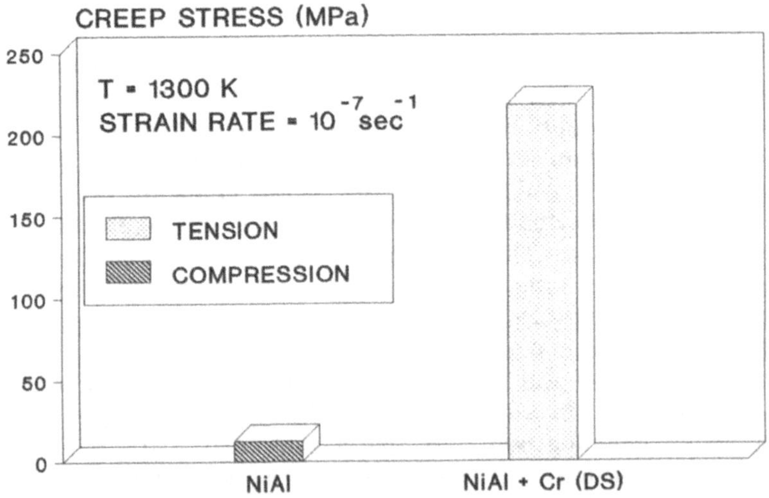

Figure 4-13. Tensile creep strength of DS NiAl-34Cr eutectic. (based on creep data compiled by Nathal 1991)

to use fiber-matrix combinations not achievable through eutectic solidification. Walter and Cline (1970) have shown that an eutectic consisting of α-Cr rods in a NiAl matrix possessed very promising creep properties as shown in Figure 4-13. Similar eutectic microstructures can be produced with Mo, W, Re, and NiAlNb phases and these prove to be advantageous (Walter and Cline 1970; Locci et al. 1989; Darolia 1991; Mason et al. 1991).

4. CREEP BEHAVIOR OF Ti-ALUMINIDES

In this section the creep behavior of three main types of Ti-aluminides will be addressed. The three intermetallic compounds based on the compositions Ti_3Al, TiAl and Al_3Ti crystallize respectively into DO_{19}, $L1_0$ and DO_{22} structures that are all based on geometrically close-packed structures such as h.c.p. and f.c.c. In fact, these intermetallics in the Ti-system are considered to be of great importance for future applications in advanced structures. This is due to their unique combination of attractive physical and mechanical properties, mainly low density, high strength, and good oxidation resistance. The Ti_3Al and TiAl base intermetallics are not very far from application in aeroengine components.

4.1. Creep Behavior of Ti_3Al Base Intermetallic

Ti_3Al is the material that has been investigated over the longest period and on which the development work has proceeded furthest. The research and development program on Ti_3Al has been pursued mainly in the U.S.A. This intermetallic has DO_{19} h.c.p. structure. From the viewpoint of engineering applications, the high strength-to-density ratio of Ti_3Al makes it an attractive material with extensive potential applications for high-temperature components in aircraft engines. The creep behavior of Ti_3Al and Ti_3Al-Nb has been studied by a number of investigations (Mendiratta and Lipsitt 1980; Hayes 1989). At all temperatures and stresses tested, the creep curves of Ti_3Al and Ti_3Al-Nb exhibited primary creep regimes followed by steady-state creep.

The steady-state creep rate, ε, which can be given as a function of the applied stress σ and temperature T, was found to obey the power law (i.e., equation 4.3. Mendiratta and Lipsitt (1980) reported that the creep mechanism of Ti_3Al and Ti_3Al-Nb strongly depends on the deformation temperature and applied stress. At high temperature and high stress, steady-state creep in Ti_3Al may be controlled by dislocation climb since the stress exponent n is about 5, while at low applied stress the n value is

Table 4-3 Summary of Creep Data of Ti_3Al and Ti_3AlNb Compounds

Compound	Temperature (°C)	Stress (MPa)	Activation Energy (kJ/mol)	Stress Exponent
Ti_3Al	> 550	> 138	206	4.3–5
	> 750	< 138	206	2.3–2.5
Ti-26Al-5Nb	650	312–138	190	5.5
	> 650	> 173	285	6.5
Ti-23Al-11Nb	750–650	69–173	259	4.3
Ti-25Al-10Nb-3V-1Mo	650	150–400	305	4.6
	704	100–250	305	4.7
	760	45–150	305	4.0
	870	> 30–60	not det.	3.5
	870	10–30	not det.	2.8

Mendiratta and Lipsitt 1980; Hayes 1989; Wonsuk, Thompson and Williams 1990

about 2.5, and this suggests the importance of another mechanism such as grain-boundary diffusion or grain-boundary sliding. The n value of 6.5 at applied stresses above 173 MPa and of 2.5 below 173 MPa predict also two distinct rate-controlling mechanisms of dislocation climb and grain-boundary sliding for creep in Ti_3Al-Nb. The activation energy for the steady-state creep of Ti_3Al-Nb is higher than that of Ti_3Al, that is, Nb additions improved the creep strength. In fact, Nb additions enhance the oxidation resistance of Ti_3Al (Mendiratta and Lipsitt 1980). Similar creep behavior was reported in Ti_3Al-($21^{wt}/o$ Nb) by Hayes but he confirmed only the dislocation-climb mechanisms at limited test temperatures and stresses as shown in Table 4-3 (Hayes 1989).

Recently, the role of microstructure in elevated-temperature creep behavior of an advanced Ti_3Al-base alloy, Ti-25 at%Al-10Nb-3V-1Mo has been investigated (Wonsuk, Thompson, and Williams 1990). They reported that the β solution-treated colony type (i.e., slow cooled, microstructure), showed superior creep resistance. Study of the stress dependence of steady-state creep rate indicates that increasing temperature caused a gradual decrease in the stress exponent n and a transition mechanism at 870°C, depending on applied stress. They concluded that the creep of Ti-25Al-10Nb-3V-1Mo may be predicted by dislocation-controlled mechanism. A summary of their results on the as received structure also is presented in Table 4-3. In a recent study, the same alloy has been investigated in detail (Hayes 1991). He concluded that steady-state creep of the as-processed Ti-25Al-10Nb-3V-1Mo appears to be governed by two deformation processes, corresponding to a low temperature ($T < 700$°C) and a high temperature ($T > 700°C$) mechanism. The possible creep mechanisms are low-temperature power-law creep controlled by core diffusion below 700°C and high-temperature power-law creep controlled by lattice diffusion above 700°C. β heat treatment increases the apparent creep activation energy and stress exponent over the entire temperature regime evaluated. The results support climb controlled by lattice diffusion coupled with grain-boundary sliding contributing to creep with increasing temperatures. At comparable stresses and temperatures, the steady-state creep rate of the β heat treated Ti-25Al-10Nb-3V-1Mo decreases by approximately an order of magnitude compared to the as-processed material, indicating a significant grain-size dependence. This suggests that the low-temperature ($T < 700°C$) creep deformation of the as-processed Ti-25Al-10Nb-3V-1Mo may result from a Coble creep process coupled with climb controlled power-law creep. The temperature and stress dependence of steady-state creep indicate that creep deformation may lie within a transition regime bounded by Coble creep at low temperature and power-law creep controlled by lattice diffusion at higher temperatures, with a stress dependence reflecting the contribution of both mechanisms. Recently a compilation of creep data on different Ti_3Al-base alloys has been published (Larsen et al. 1990). A Larson-Miller plot of this compilation is shown in Figure 4-14. It can be seen that there is a range of creep properties obtainable in these alloys. The single data point for Ti-25Al-10Nb-3V-1Mo indicates that the microstructural effects are as significant

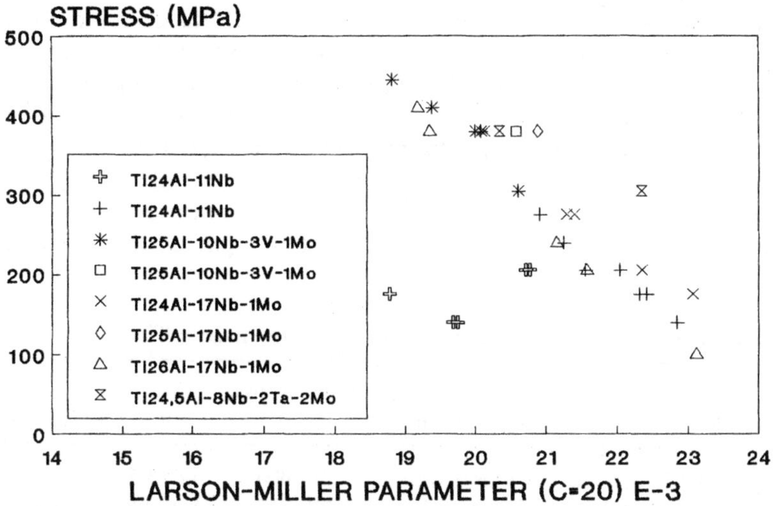

Figure 4-14. Creep rupture strength of different Ti_3Al-base alloys (Larsen et al. 1990).

as compositional effects. Compositionally, increasing the β stabilizer content in the Ti-24Al-17Nb-1Mo and Ti-25Al-17Nb-1Mo alloys has resulted in improved creep resistance.

Environmental effects, mainly due to oxygen and hydrogen, on the properties of Ti_3Al-base alloys are the main problems facing the application of these materials (Larsen et al. 1990; Chan 1992; Chu and Thompson 1992) (see chapter on Environmental Effects in this volume).

4.2. Creep Behavior of γ TiAl-Base Intermetallics

The γ TiAl phase has a wide range of homogeneity (49–66 at%aluminum), which varies depending on the temperature. It has a density of ~ 4 gm/cm^3 and a high modulus of elasticity ~ 170 GPa, which make this material very attractive for applications that require high strength/weight ratio. The creep behavior of single phase γ TiAl with Al contents ranging from 50 to 56 at% has been investigated in compression (Loiseau and Lasalmonie 1984; Takahashi, Nagai, and Oikawa 1989). It has been reported that the creep curves have the normal shape, with a primary stage and a steady-state region (Takahashi et al. 1989).

The stress exponent n and temperature dependence Q of the steady-state creep rate obey the power-law creep equation (4.3). These parameters are given in Table 4-4 for both investigations. The activation energy for creep was found to be higher than that for lattice interdiffusion in the γ TiAl phase, ~ 130–165 kJ/mol (Ouchi, Iijima, and Hirano 1980). This suggests that the high temperature deformation of γ TiAl must involve processes more complex than a simple diffusion-related process.

The creep properties of two phase intermetallics are improved as compared with those of single-phase γ TiAl. It has been shown that the best creep resistance in two-phase α_2/γ structure was obtained in alloys containing aluminum from 45 at% to 48 at% (Mitao, Tsuyama, and Minakuwa 1991). The creep behavior of alloyed two-phase α_2/γ titanium aluminide has been investigated in the temperature range 550–825°C (Martin, Mendiratta, and Lipsitt 1983). They reported that the creep data

Table 4-4 Summary of Creep Data of Single Phase γ TiAl

Alloy	Type of Test	Temp. Range (°C)	n	Q (kJ/mol)
Ti-50 : 56 at% Al*	Compression	750–900	4.6–3.7	—
Ti-53 at% Al**	Compression	727–877	4 ± 0.5	340 ± 20

* Loisau and Lasalmonie 1984

** Takahashi, Nagai, and Oikawa 1989

of the investigated alloys can be expressed in terms of the conventional power law (i.e., equation 4.3). The reported values for the activation energy Q ranged from 300–370 kJ/mol and the stress exponent n was in the range 4–5.5. The creep behavior of two-phase γ TiAl base alloys is strongly dependent on the microstructure of the alloy. This effect has been presented in detail, for Ti-48Al-2Nb-2Cr (Shih et al. 1991). They reported an activation energy of 320 kJ/mol, which suggests that the creep rate-controlling process may be related to the ordering of the TiAl lattice instead of self-diffusion of pure elements. The results of high activation energy value have been confirmed recently for the same alloy (Wheeler, London, and Larsen 1992). In addition, a change in the stress exponent n from a value of 3 at low stresses to a value of 7 at high stresses has also been reported (Wheeler, London, and Larsen 1992). As a matter of fact, Shih et al. (1991) have also shown that the creep properties of two-phase α_2/γ alloy Ti-48Al-2Nb-2Cr are improved for a fully transformed lamellar structure L as compared to a duplex structure D, which consists of primary γ grains in addition to the transformed L grains. Nevertheless, this improvement in creep properties of the L structure depends on the test temperature, below ~ 650°C the D structure is better than the L structure.

A Larson–Miller plot of the creep properties of both structures for Ti-48Al-2Nb-2Cr is given in Figure 4-15. The density-corrected creep properties of IN738LC, a well-known land-base gas turbine blading alloy,

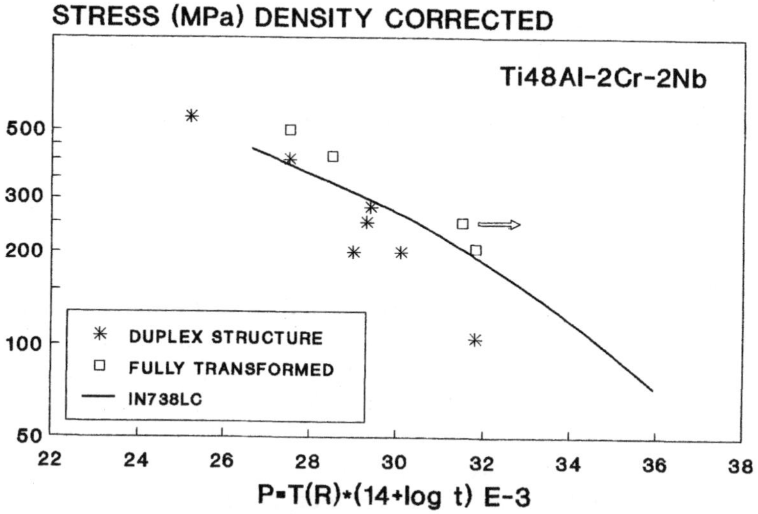

Figure 4-15. Density corrected creep properties of both structures of Ti48Al-2Cr-2Nb as compared with IN738LC (Shih et al. 1991).

also are included on the same figure. The creep deformation of TiB_2-reinforced two-phase γ TiAl was investigated (Kampe, Bryant, and Christodoulou 1991). They reported an activation energy of 340 kJ/mol and a stress exponent of 3–3.5 in the temperature range 800–900°C. Their conclusion is that the mechanisms governing creep deformation are the same in both the unreinforced and particulate-reinforced materials and are independent of the processing route.

The improvement in creep properties, in this type of two-phase γ TiAl alloys, exhibited by the L structure is usually accompanied by lower ductility (Shih et al. 1991). A balance between creep properties and tensile ductility is required for the applications. Huang (1991) has recently reported on the creep of two-phase Ti-47Al-1Cr-1V-2.5Nb alloy. He found that the creep life of this alloy is very sensitive to oxidation. This is in contrast to the observation of Lipsitt and co-workers that after prior oxidation, tensile properties of γ TiAl alloys are not affected (Lipsitt, Schechtmann, and Shafrik 1975). This difference is understandable since creep testing is a long-term test that allows oxygen to diffuse toward the crack tip.

4.3. Creep Behavior of Al_3Ti Intermetallics

This intermetallic compound exhibits attractive properties such as high strength, low density, and high oxidation resistance. The major mode of deformation of Al_3Ti has been reported to be the ordered twinning of the (111) [112]-type which does not disorder the DO_{22} structure (Vasudevan, Wheeler, and Fraser 1989; Whittenberger et al. 1991). At high temperatures, not only the ordered twinning but also slip of the (001) [110]-type becomes operative (Yamaguchi, Umakoshi, and Yamane 1987; Vasudevan et al. 1989). Recently, Whittenberger et al. (1991) have investigated the high-temperature mechanical properties of monolithic and particulate composites of $L1_2$ forms of Al_3Ti. They reported for unreinforced alloys containing Cr and Fe stress exponents of 2.69 and 2.8, respectively, and an activation energy, for the Al_3Ti-Fe alloy form, of 308.6 kJ/mol. They also showed that creep data on these material can be represented by a power law (i.e., equation 4.3). Due to the lack of diffusion data, that is, activation energies for volume, grain boundary or dislocation core on $L1_2$ modified Al_3Ti intermetallics, it was not possible to establish a direct comparison with the calculated activation energy for high-temperature deformation (Edwards, McNelly, and Sherby 1975). Nevertheless, they concluded that for the unreinforced materials that exhibited stress exponents from 2.6 to 2.9, a viscous dislocation glide process is believed to be responsible for creep in these materials. In the case of particulate-reinforced materials, a weakening effect has been reported and was attributed to a dispersion-softening mechanism (Edwards et al. 1975).

5. CREEP BEHAVIOR OF SOME HIGH TEMPERATURE REFRACTORY INTERMETALLICS

Cr_3Si and $MoSi_2$ are two examples of intermetallic compounds with attractive properties suitable for very high temperature, (i.e, > 1400 K) applications. They have A15 and C11 crystal structures, respectively. Recently, the creep properties of a group of refractory intermetallics have been investigated (Anton and Shah 1991). Stress exponents of 3.4 and 2.9 and activation energies of 498 kJ/mol and 356 kJ/mol have been reported for Cr_3Si and $MoSi_2$ at 1473 K, respectively. The cyclic oxidation tests at 1422 K demonstrated the excellent oxidation resistance exhibited by $MoSi_2$, where Cr_3Si was moderately less oxidation-resistant. Nevertheless, both intermetallics did not show any sign of catastrophic oxidation. Recently, Sadananda et al. (1992) investigated the compressive creep behavior of monolithic and fiber-reinforced $MoSi_2$. They reported that the creep strength of MoSi is improved by alloying with WSi_2 and the effect is more pronounced at lower temperatures. The creep strength has been further improved by reinforcement with SiC-whiskers. For monolithic $MoSi_2$, the stress exponent varied between 1.93 to 1.19 at 1373 K and 1473 K, respectively. In the case of SiC fiber-reinforced $MoSi_2$, the stress exponent ranged from 5.2 to 3 at 1373 K and 1573 K, respectively, which is a characteristic value for power-law creep. The activation energies for $MoSi_2$, $MoSi_2$ with 20% SiC, and $MoSi_2$ + WSi_2 alloy, are 433, 596, and 536 kJ/mol, respectively. These values are close to the calculated activation energy for diffusion of Mo in $MoSi_2$, which is 460 kJ/mol. Hence, Sadananda and co-workers have proposed that creep in $MoSi_2$ and its composites occurs by Newtonian viscous deformation involving self-diffusion to power-law creep involving dislocation climb.

ACKNOWLEDGMENTS

I would like to thank M. Staubli for his assistance and stimulating discussions during the preparation of the manuscript. The permission by Dr. M. Nathal to use a number of figures and tables from his publication is greatly appreciated.

REFERENCES

ANTON, D. L., PEARSON, D. D., and SNOW, D. B., 1987, In *High Temperature Ordered Intermetallic Alloys II*, MRS Symp. Proc. 81: 287–295.

ANTON, D. L., and SHAH, D., 1991, In High Temperature Ordered Intermetallic Alloys IV MRS, Symp. Proc. 213: 733–738.

BAKER, I., and GAYDOSH, D. J., 1987, *Metallography* 20: 347.

BELLOWS, R. S., SCHWARZKOPF, E. A., and TIEN, J. K., 1988, *Metall. Trans.* 19A: 479–486.

BERKOWITZ, A. E., JAUMOT, F. E., and NIX, F. C., 1954, *Phys. Rev.* 95: 1185–1189.

BEWK, J., DODD, R. A., and STRUTT, P. R., 1973, *Metall. Trans.* 4A: 159–166.

CHAN, K. S., 1992, *Metall. Trans.* 23A: 497–507.

CHU, W. Y., and THOMPSON, A. W., 1992, *Metall. Trans.* 23A: 1299–1312.

DAROLIA, R. 1991, *J. of Metals* 43(3): 44.

EDWARDS G. R., MCNELLY, T. R., and SHERBY, O. D., 1975, *Phil. Mag. A.* 32: 1245–1264.

FLINN, P. A. 1960, *Trans. Metall. Soc. AIME* 218: 145–154.

FUJITA, K. A., MATSUMOTO, T., NAKAMURA, M., and TAKEDA, Y., 1988, In *High Temperature Ordered Intermetallic Alloys III*, MRS Symp. Proc. 133: 573–578.

FUNK, W., and BLANK, E., 1988, *Metall. Trans.* 19A: 987–998.

HAGEL, H. C. 1977, In *Intermetallic Compounds*, ed. J. H. Westbrook, p. 377. New York: John Wiley & Sons.

HANCOCK, G. F., and MCDONELL, B. R., 1971, *Phys. Stat. Sol.* (a), 4:143–150.

HAYES, R. W., 1989, *Scripta Metall.* 23: 1931–1935.

________, 1991, *Acta Metall.* 39: 569–577.

HAZZLEDINE, P. M., and SCHNEIBEL, J. H., 1989, *Scripta Metall.* 23: 1887–1892.

HEMKER, K. J., and NIX, W. D., 1989, In *High Temperature Ordered Intermetallic Alloys III*, MRS Symp. Proc. 133: 481–486.

________, 1991, *Acta Metall.* 39: 1901–1913.

HIRANO, K., and HISHINMUMA, A., 1968, *Nippon Kinzoku Gakkaishi* 32: 516.

HOCKING, L. A., STRUTT, P. R., and DODD, R. A., 1971, *J. Inst. Metals* 99: 98–101.

HUANG, J., 1991, *Scripta Metall.* 25: 1901–1906.

JHA, S. C., RAY, R., and WHITTENBERGER, J. D., 1989, *Mat. Sci. Eng.* A119: 103–111.

JUNG, I., RUDY, M., and SAUTHOFF, G., 1987, In *High Temperature Ordered Intermetallic Alloys II*, MRS Symp. Proc. 81: 263–274.

KAMPE, S., BRYANT, J., and CHRISTODOULOU, L., 1991, *Metall. Trans.* 22A: 447–454.

KELLY, A., and STREET, K. N., 1972, *Proc. Roy. Soc. London* A328: 283–293.

LANGDON, T. G., and YAVARI, P., 1981, In *Creep and Fracture of Engineering Materials and Structures*, ed. B. Wilshire, and D. R. J. Owen, p. 71. Swansea (UK). Pineridge Press.

LARSEN, J. M., WILLIAMS, K. A., BALSONE, S. J., and STUCKE, M. A., 1990, In *High Temperature Aluminides & Intermetallics*, ed. S. H. Whang, C. T. Liu, D. P. Pope, and J. O. Stiegler, 521–556. Warrendale, PA: TMS.

LAWLEY, A., COLL, J. A., and CAHN, R. W., 1960, *Trans.* TMS-AIME 218: 166–176.

LIPSITT, H., SHECHTMAN, D., and SHAFRIK, R., 1975, *Metall. Trans.* 6A: 1991–1996.

LOCCI, I. E., NOEBE, R. D., MOSER, J. A., LEE, D. S., and NATHAL, M. V., 1989, In *High Temperature Ordered Intermetallic Alloys III*, MRS Symp. Proc. 133: 639–646.

LOISEAU, A., and LASALMONIE, A., 1984, *Mat. Sci. Eng.* 67: 163–168.

LUTZE-BIRK, A., and JACOBI, H., 1975, *Scripta Metall.* 9: 761–765.

MANNAN, S. K., KUMAR, K. S., and WHITTENBERGER, J. D., 1990, *Metall. Trans.* 21A: 2179–2188.

MARTIN, P., MENDIRATTA, M., and LIPSITT, H., 1983, *Metall. Trans.* 14A: 2170–2174.

MASON, D. P., VAN AKEN, D. C., NOEBE, R. D., LOCCI, I. E. and KING, K. L., 1991 In *High Temperature Ordered Intermetallic Alloys IV*, MRS Symp. Proc. 213: 1033–1038.

MCKAMEY, C., DE VAN, J., TORTORELLI, P., and SIKKA, V., 1991, *J. Mat. Res.* 6: 1779–1805.

MENDIRATTA, M. G., and LIPSITT, H. A., 1980, *J. Mat. Sci.* 15: 2985–2990.

MILLER, H. E., and CHAMBER, W. L., 1987, In *Superalloys II*, ed. C. Sims, N. S. Stoloff, and W. Hagel, 27–57. New York: John Wiley & Sons.

MILLS, M. J., GIBELING, J. C., and NIX, W. D., 1986, *Acta Metall.* 34: 915–925.

MITAO, S., TSUYAMA, S., and MINUKAWA, K., 1991, *Microstructure/Property Relationships in Titanium Aluminides and Alloys*, ed. Y.-W. Kim, and R. Boyer, p. 297. Warrendale, PA: TMS.

MIURA, S., TAKEKAWA, T. M., MISHIMA, Y., and SUZUKI, D., 1990, In *High Temperature Ordered Intermetallic Alloys IV*, MRS Symp. Proc. 213: 623–628.

MOHAMED, F. A., and LANDON, T. G., 1974, *Acta Metall.* 22: 779–788.

MORRIS, M. A., and MORRIS, D. G., 1990, *Acta Metall.* 38: 551–559.

NATHAL, M. V., 1991, In *Ordered Intermetallics-Physical Metallurgy and Mechanical Behavior*, ed. C. T. Liu, R. W. Cahn and G. Sauthoff, 541–563, Dordrecht: Kluwer Academic Publishers.

NATHAL, M. V., and EBERT, L. J., 1985, *Metall. Trans.* 16A: 1863–1870.

NAZMY, M., and STAUBLI, M., 1991, *Scripta Metall.* 25: 1305.

NICHOLLS, J. R., and RAWLINGS, R. D., 1977, *J. Mat. Sci.* 12: 2456–2464.

NISHIDA, K., YAMAMOTO, T., and NAGATA, T., 1970, *Nippon Kinzhoku Gakkaishi* 34: 591.

NIX, W. D., and ILSCHNER, B., 1980, In *5th Int. Conf. on Strength of Metals and Alloys*, 3: 1503. New York: Pergamon Press.

OLIVER, B., NOEBE, R. D., and WHITTENBERGER, J. D., 1991, Unpublished research, at University of Tennessee, Knoxville.

OUCHI, K., IIJIMA, Y., and HIRANO, K., 1980, In *Titanium 80 Proc. 4th Int. Conf. on Titanium*, ed. H. Kimura and O. Izumi, p. 559. New York: TMS-AIME.

PASCOE, R. T., and NEWEY, C. W. A., 1968, *Met. Sci. J.* 2: 138–143.

PATHARE, V. M. 1987. Ph.D. thesis, Case Western Reserve University, Cleveland, OH.

POLVANI, R. S., TZENG, W. S., and STRUTT, P. R., 1976, *Metall. Trans.* 7A: 33–40.
RAJ, S. V., LOCCI, I. E., and NOEBE, R. D., 1991, Unpublished research at NASA-Lewis Research Center, Cleveland, OH.
RUDY, M., and SAUTHOFF, G., 1985, In *High Temperature Ordered Intermetallic Alloys*, MRS Symp. Proc. 39: 327–333.
_______, 1986, *Mat. Sci. Eng.* 81: 525–.
SADANANDA, K., FENG, C., JONES, H., and PETROVIC, J. 1992, *Mat. Sci. Eng.* A155: 227–239.
SAUTHOFF, G. 1990. Mechanical Properties of Intermetallics at High Temperatures. In *High Temperature Aluminides and Intermetallics*, ed. S. H. Whang, C. T. Liu, D. P. Pope, and J. O. Stiegler. 329–352. Warrendale: TMS.
SCHNEIBEL, J. H., and HORTON, J. A., 1988, *J. Mat. Res.* 3: 651–655.
SHANKAR, S., and SEIGLE, L. L., 1978, *Metall. Trans.* 9A: 1476–1467.
SHIH, D., HUANG, S. C., SCARR, G., JANGARD, H., and CHESNUTT, J., 1991, In *Microstructure/Property Relationships in Titanium Aluminides and Alloys*, ed. Y.-W. Kim and R. R. Boyer, p. 135. Warrendale, PA: TMS.
SOLIMAN, M. S., and MOHAMED, F. A., 1982, *Mat. Sci. Eng.* 55: 111–119.
STRUTT, P. R., and KEAR, B. H., 1984, In *High Temperature Ordered Intermetallic Alloys* MRS Symp. Proc. 39: 279–292.
TAKAHASHI, T., NAGAI, H., and OIKAWA, H., 1989, *Mat. Sci. Eng.* A114: 13–20.
THORNTON, P. H., DAVIES, R. G., and JOHNSTON, T. L., 1970, *Metall. Trans.* 1: 207–218.
VANDERVOORT, R. R., MUKHERJEE, A. K., and DORN, J. E., 1966, *Trans ASM* 59: 930–944.
VASUDEVAN, V. K., WHEELR, R., and FRASER, H. L., 1989, In High Temperature Ordered Intermetallic Alloys III *MRS Symp. Proc.* 133: 705–710.
WALTER, J. L., and CLINE, H. E., 1970, *Metall. Trans.* 1: 1221–1229.
WEERTMAN, J., 1958, *J. Applied Physics* 29: 1685–1689.
WHEELR, D., LONDON, B., and LARSEN JR., D., 1992, *Scripta Metall.* 26: 939–944.
WHITTENBERGER, J. D., 1983, *Mat. Sci. Eng.* 57: 77–85.
_______, 1985, *Mat. Sci. Eng.* 73: 87–96.
_______, 1986, *Mat. Sci. Eng.* 77: 103–113.
_______, 1987, *J. Mat. Sci.* 22: 394–402.
_______, 1987, *Mat. Sci. Eng.* 85: 91–99.
_______, 1987, *NASA-TM* 101382.
_______, 1988, *J. Mat. Sci.* 23: 235–240.
WHITTENBERGER, J. D., VISWANADHAM, R. K., MANNAN, S. K., and KUMAR, S. K., 1989, *J. Mat. Res.* 4: 1164–1171.
WHITTENBERGER, J. D., WESTFALL, L. J., and NATHAL, M. V. 1989, *Scripta Metall.* 23: 2127–2130.
WHITTENBERGER, J. D., ARZT, E., and LUTON, M. J., 1990a, *J. Mat. Res.* 5: 271–277.
_______, 1990b. *J. Mat. Res.* 5: 2819–2827.
WHITTENBERGER, J. D., GAYDOSH, D. J., and KUMAR, K. S., 1990, *J. Mat. Sci.* 25: 2771–2776.

Whittenberger, J. D., Viswanadham, R. K., Mannan, S. K., and Sprissler, B., 1990, *J. Mat. Sci.* 25: 35–44.

Whittenberger, J. D., Gaydosh, D. J., and Nathal, M. V., 1991, Unpublished Research at NASA-Lewis Research Center, Cleveland, OH.

Whittenberger, J. D., Kumar, K. S., Brown, S., Di Pietro, M., and Farmer, S. C., 1991, In *Light Weight Alloys for Aerospace Applications II*, ed. E. W. Leed and N. J. Kim, p. 327. Warrendale, PA: *TMS*.

Whittenberger, J. D., Kumar, K. S., and Mannan, S. K., 1991, *Materials at High Temperature* 9: 3.

________, 1991, *J. Mat. Sci.* 26: 2015–2022.

Whittenberger, J. D., Noebe, R. D., Cullers, C. L., Kumar, K. S., and Mannan, S. K., 1991, *Metall. Trans.* 22A: 1595–1607.

Whittenberger, J. D., Ray, R., and Jha, S. C., 1991, In *High Temperature Ordered Intermetallic Alloys IV*, MRS Symp. Proc. 213: 581–587.

Whittenberger, J. D., Ray, R., Jha, S., and Draper, S. L., 1991, *Mat. Sci. Eng.* A138: 83–93.

Wonsuk, C., Thompson, A. W., and Williams, J. C., 1990, *Metall. Trans.* 21A: 641–651.

Yamaguchi, M., Umakoshi, Y., and Yamane, T., 1987, *Phil. Mag.* A 55: 301–315.

Yaney, D. L., Gibeling, J. C., and Nix, W. D., 1987, *Acta Metall.* 35: 1391–1400.

Yaney, D. L., and Nix, W. D., 1988, *J. Mat. Sci.* 23: 3088–3098.

Yaney, D. L., Pelton, A. R., and Nix, W. D., 1986, *J. Mat. Sci.*, 21: 2083–2087.

Yang, W. J., and Dodd, R. A., 1973, *Metal Sci. J.* 7: 41–47.

Chapter 5

Fatigue

N. S. Stoloff

1. INTRODUCTION

It was first reported by Boettner, Stoloff and Davies (1966) that long-range order can enhance high-cycle fatigue resistance. However, the lack of tensile ductility manifested by most ordered alloys with potential for structural applications precluded serious attention of researchers to cyclic behavior until very recently. Rather, the emphasis has been upon increasing ductility through alloying or improved processing techniques. Therefore, only a few studies, especially on high-cycle fatigue, were reported in the literature up to about 1987. However, recent progress in producing ductile-ordered alloys has led to renewed interest in fatigue behavior and, where possible to study, to the effects of long-range order on crack initiation and propagation, respectively. The purpose of this chapter is to review the literature on cyclic deformation of ordered alloys, with emphasis upon recent work on Ni_3Al, NiAl, Ti_3Al and Fe_3Al-type alloys.

2. GENERAL FEATURES OF ORDERED SYSTEMS

Direct studies of the influence of long-range order on cyclic behavior can only be carried out in systems that can be disordered by heat treatment. These include Ll_2 superlattices such as Cu_3Au, Ni_3Fe and Ni_3Mn, B2 superlattices such as FeCo-V and CuZn, and DO_{19} superlattices such as

Mg_3Cd. Other, indirect, information on the role of long-range order in fatigue has been provided by studies of Ni_3Al and Ni_3Ge (Ll_2), Fe_3Al (DO_3 or partial B_2) TiAl ($L2_0$) and Ti_3Al (DO_{19}), none of which can be completely disordered by heat treatment. For convenience, cyclic test results on all systems will be described under the headings of high-cycle fatigue and crack propagation, which are usually carried out under stress control, or low-cycle fatigue, which is usually performed under strain control. Under the latter heading will be included also results of cyclic hardening tests. Crack initiation, which may be studied under either stress or strain control, is discussed in a separate section.

3. HIGH-CYCLE FATIGUE

The influence of long-range order on stress-controlled high-cycle fatigue was first established for Ni_3Mn and FeCo-V (Boettner et al. 1966). In both cases, fatigue lives were substantially increased, in spite of a decrease in yield stress with order for FeCo-V. Cu_3Au also was shown to exhibit increased life with order, although the effect was small (Rudolph et al. 1965). The beneficial effects of order in Cu_3Au are maintained at temperatures to 373°C (Gittins 1968). In each case, it was hypothesized that planar slip, brought about by the presence of superlattice dislocations, was responsible for delayed crack initiation. In most alloys studied to date, order favors planar slip by inhibiting cross slip and/or multiple-slip, either of which are generally considered to be necessary for cracks to initiate readily. The one ordered system in which planar slip is favored by *disorder* is Mg_3Cd. Whitehead and Noble (1970) showed that in this system, poor fatigue properties were exhibited by both ordered and disordered material.

Later, high-cycle fatigue studies were extended to other systems: $(Fe,Ni)_3V$ (Ashok et al. 1983), TiAl polycrystals (Sastry and Lipsitt 1977), and Ni_3Al single crystals (Doherty, Giamei, and Kear 1975) as a function of test temperature, and to β brass as a function of ternary alloy content to influence slip planarity (Clark 1966). The results of these studies are summarized in Table 5-1, (Stoloff et al. 1987), together with data for some commercial alloys that are not ordered. It is clear that the ratio of the fatigue limit to yield strength exceeds 0.5 for most of the ordered alloys, while in the disordered condition that ratio decreases significantly. For TiAl, the ratio varied with temperature, ranging from 0.8 at room temperature to 0.5 at 900°C (Sastry and Lipsitt 1977).

Recent work on NiAl polycrystals doped with 0.28at%Fe showed that the endurance ratio at 400°C (below the brittle to ductile transition)

Table 5-1 High Cycle Fatigue Data, R ≅ 0.1

Alloy (a%)	T°C	Environment	$\Delta\sigma 10^6/\sigma_{ys}$	$\Delta\sigma 10^7/\sigma_{ys}$
Fe-24Al-DO_3	25	Air	0.84	0.71
Fe-24Al-B2	25	Air	0.76	0.62
Fe-29Al-DO_3	25	Air	1.02	0.89
Fe-29Al-B2	25	Air	0.67	0.65
Ni-24Al HIP	25	Air	1.79	1.38
Ni-24Al HIP + Ext	25	Air	1.57	1.20
Ni-25Al Cast	25	Air	1.32	1.05
Ni_3Al (crystal)	25	Air	0.72	0.56
LRO-37 $(Fe,Ni)_3V$	25	Air	2.08	—
Nitac 14B	25	Air	1.25	1.12
MarM-2000D (crystal)	25	Vac	0.55	0.47
Waspaloy	25	Air	0.57	0.48
IN617	25	Air	1.53	1.44
Hastelloy C	25	Air	—	1.05
Fe-24Al-DO_3	500	Air	1.08	0.83
Fe-29Al-DO_3	500	Air	1.01	0.83
Fe-24Al-B2	560	Air	0.68	—
Fe-29Al-B2	600	Air	0.95	—
Ni-24Al HIP + Ext	500	Air	0.98	0.92
Ni_3Al crystal	425	Air	0.42	0.33
Ni_3Al crystal	760	Air	0.36	0.28
IC-218	750	Air	0.71	—
IC-218	750	Vac	0.78	—
LRO-1-3 $(Fe,Co,Ni)_3V$	650	Argon	1.28	0.98
LRO-37 $(Fe,Ni)_3V$	400	Argon	1.73	—
Nitac 14B	825	Argon	1.30	1.11
U500	650	Air	1.61	0.47
Waspaloy	800	Air	0.50	0.46
IN718	650	Air	0.56	0.44

Stoloff et al. 1987

was about 1.9 while at 550°C (above the brittle to ductile transition) the ratio was about 1.5 (Matsugi and Stoloff 1993). By comparison, Udimet 700 has an endurance ratio of only 0.6 (Gordon and Unni 1989). The high endurance ratios of intermetallics at both room and elevated temperatures arises from the very rapid cyclic hardening in these alloys, which tends to delay crack initiation. However, many intermetallics (e.g. Ti_3Al, TiAl,

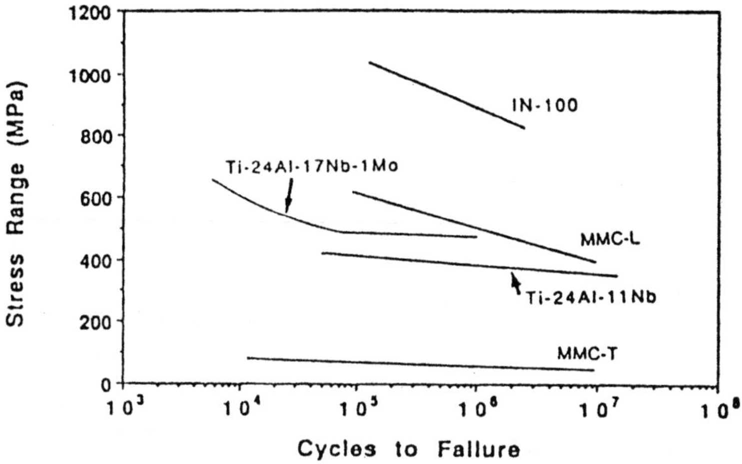

Figure 5-1. High-cycle fatigue of continuous Ti_3Al composites reinforced with SCS 6 fibers, compared to superalloy IN-100, 650°C, 0.2 Hz, R = 0. MMC-L is longitudinal MML-T is transverse orientation (Larsen et al. 1990).

Fe_3Al) display very rapid crack propagation, so that failure may soon follow the initiation of a crack. This may explain the very poor high-cycle fatigue resistance of α_2 alloys compared to IN 100 at 650°C, see Figure 5-1 (Larsen et al 1990). Even when α_2 is reinforced with continuous SiC fibers, IN 100 is superior.

3.1. Ni_3Al Alloys

The influence of Al content on high-cycle fatigue of $Ni_3Al + B$ alloys has been examined. Typical results are shown in Figure 5-2a (Fuchs and Stoloff 1987). The fatigue resistance of Ni-26at%Al is much less than that of Ni-24at%Al, even when normalized for differences in yield stress (see figure 5-2b). Differences in fatigue lives of the two alloys are reflected also in the fracture mode: faceted stage I growth in Ni-24at%Al and intergranular or interdendritic paths in Ni-26at%Al. At 500°C, fatigue lives are nearly unchanged relative to 25°C for Ni-24at%Al, but are considerably higher for Ni-26at%Al. This appears to be a consequence of increased yield stresses with temperature and increased ductility (together with reduced notch sensitivity) for Ni-26at%Al. Ni_3Al single crystals tested at the same stress levels show relatively constant fatigue lives with temperature (Doherty et al. 1975). However, P/M polycrystals reveal a sharp drop in fatigue lives with increasing temperature, Figure 5-3, paralleling a

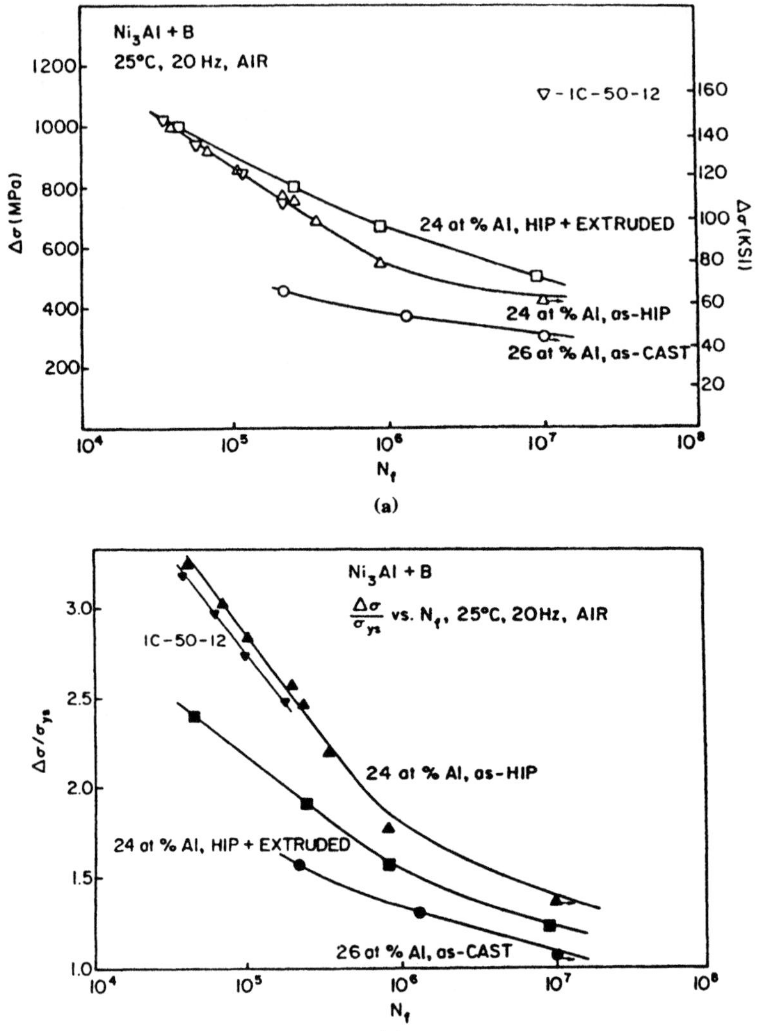

Figure 5-2. High-cycle fatigue of several Ni_3Al + B alloys a) σ-N data b) σ-N data normalized for differences in yield stress (Fuchs and Stoloff 1987; reprinted from *Scripta Metallurgica*, Volume 21 pages 863–868, copyright 1987, with kind permission from Elsevier Science Ltd., The Boulevard, Langford Lane, Kidlington OX5 1GB, UK).

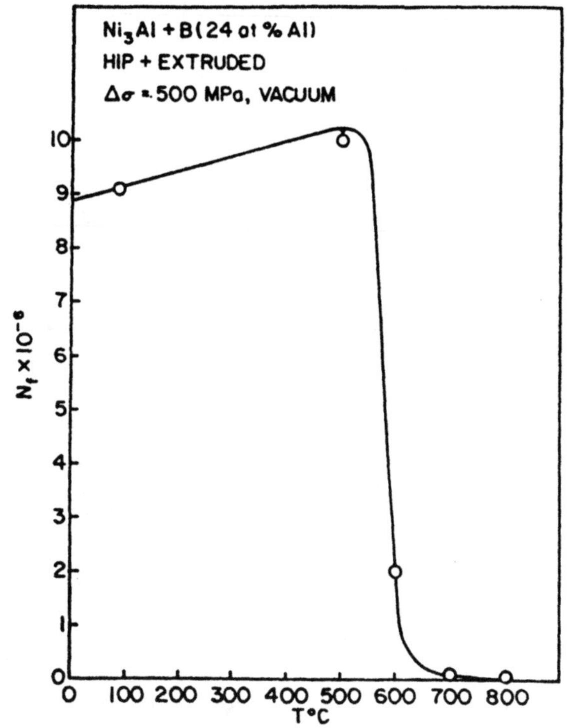

Figure 5-3. High-cycle fatigue life vs. temperature for Ni_3Al + B, $\Delta\sigma = 500$ MPa (Fuchs and Stoloff 1987; reprinted from *Scripta Metallurgica*, Volume 21, pages 863–868, copyright 1987, with kind permission from Elsevier Science Ltd., The Boulevard, Langford Lane, Kidlington OX5 1GB, UK).

drastic loss of tensile ductility in the same temperature regime (Fuchs and Stoloff 1987). Fracture paths in specimens tested at temperatures of 600°C and above revert to intergranular, as in tensile tests. An environmental effect undoubtedly is responsible for the loss of tensile ductility, and is probably responsible for loss of fatigue resistance as well. Note that the applied stress range of $\Delta\sigma = 500$ MPa is above the tensile strength of Ni_3Al + B at temperatures above 500°C.

Additional evidence for an environmental effect has been reported for Ni_3Al alloy IC-218 (see composition in Table 5-2) at 750°C; see Figure 5-4 (Gordon, Unni, and Stoloff 1994). Similarly, hydrogen has been shown to degrade the low temperature HCF resistance of IC-218 (Stoloff 1992).

High-cycle fatigue tests (uniaxial) have been carried out in air on cast Ni_3Al alloy IC-221M; see Figure 5-5 (Gieseke and Sikka 1992). This alloy

Table 5-2 Compositions of Ni_3Al and Fe_3Al Alloys, w%

	Ni	Fe	Al	C	Cr	Zr	Mo	B	Nb	Y
IC-50	bal	—	11.3	—	—	0.6	—	0.2	—	—
IC-218	bal	—	8.5	—	7.8	0.8	—	0.02	—	—
IC-221	bal	—	8.5	—	7.8	1.7	—	0.02	—	—
IC-221M	bal	—	7.95	—	7.74	1.7	1.43	0.008	—	—
IC-396M	bal	—	8.0	—	8.0	0.8	3.0	0.005	—	—
FA-129	—	bal	15.9	0.05	5.5	—	—		1.0	—
FAP-Y	—	bal	8.46	0.03	5.5	0.2	2.0	—	—	0.10

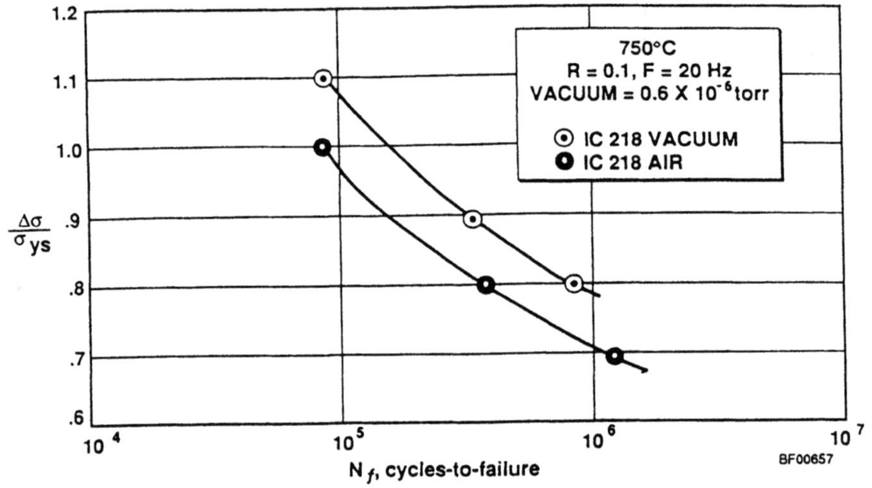

Figure 5-4. High-cycle fatigue data for powder-extruded IC-218 nickel aluminide, 750°C (Gordon, Unni, and Stoloff 1994; reprinted from *Acta Metallurgica*, Volume 36, pages 1381–1387, copyright 1987, with kind permission from Elsevier Science Ltd., The Boulevard, Langford Lane, Kidlington OX5 1GB, UK).

has better fatigue resistance than the Ni-base superalloy IN-713 at 650°C. Higher cooling rates from the melt promoted better fatigue resistance by reducing the volume fraction of micropores, which are the sites of crack initiation. This alloy performed better than a similar alloy, IC-396M, studied by Cheng and Sekhar (1991), due to the large amount and larger size of pores in the latter alloy.

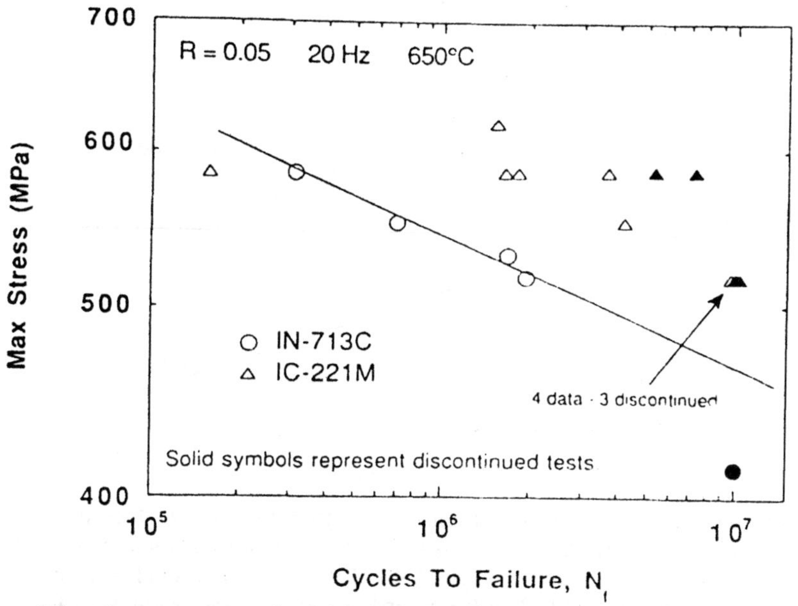

Figure 5-5. A comparison of axial fatigue tests at 650°C on IC-221M cast at 28°C above the nominal pouring temperature to those obtained for as-cast IN-713C (Gieseke and Sikka 1992; reprinted from *Materials Science and Engineering*, Volume A153, pages 520–524, with kind permission from Elsevier Science Ltd., The Boulevard, Langford Lane, Kidlington OX5 1GB, UK).

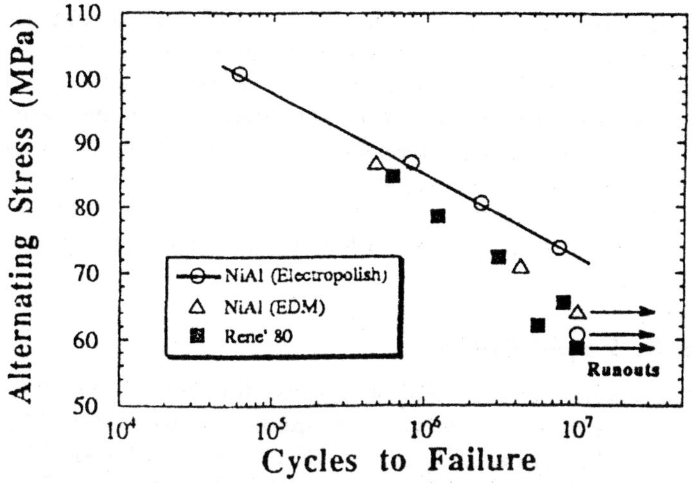

Figure 5-6. The high-cycle fatigue behavior of a Hf-containing NiAl single crystal alloy at 980°C, compared to René 80 (Wright, Maurer, and Bain 1992).

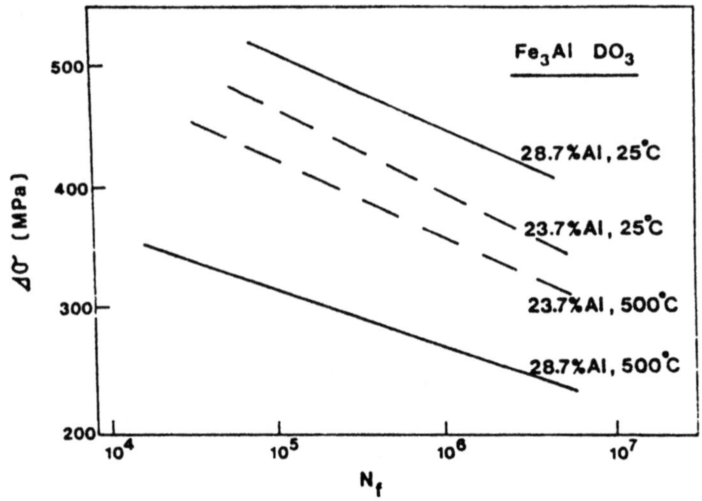

Figure 5-7. Effects of composition and temperature on high-cycle fatigue of Fe_3Al polycrystals at 25°C and 500°C (Fuchs and Stoloff 1988; reprinted from *Acta Metallurgica*, Volume 36, pages 1381–1387, copyright 1988, with permission from Elsevier Science Ltd., The Boulevard, Langford Lane, Kidlington OX5 1GB, UK).

3.2. NiAl Alloys

Considerably fewer studies of high-cycle fatigue have been reported for NiAl and its alloys. At 400°C and 550°C endurance ratios for ternary polycrystals are high (Matsugi and Stoloff 1993), as described above. At higher temperatures, the only published high-cycle fatigue data are for Hf-doped ⟨001⟩ oriented single crystals at 980°C, shown in Figure 5-6 (Wright, Maurer and Bain 1992). Note that fatigue lives are similar to those of René 80, a nickel-base superalloy with comparable creep resistance. The behavior of NiAl appears to be insensitive to surface condition at 980°C, indicating a lack of notch sensitivity. This is in marked contrast to the brittleness and notch sensitivity of similarly oriented crystals at low temperatures.

3.3. Fe_3Al Alloys

High-cycle fatigue (HCF) results for binary Fe_3Al-type alloys reveal an unexpected reversal in properties between Fe-23.7at%Al and Fe-28.7at%Al tested at 25°C and at 500°C (Fuchs and Stoloff 1988). Figure 5-7 shows that the hypostoichiometric alloy is less fatigue resistant at 25°C, but is more resistant at 500°C. This appears to arise from an aging effect, since a two-phase, coherent $\alpha + DO_3$ structure develops in Fe-23.7at%Al at 500°C, but not in Fe-28.7at%Al. Hardness data for Fe-23.7at%Al confirm the conclusion of Inouye (1985) that an aging reaction occurs after exposure at 500°C or 560°C, but not after exposure to 600°C. It has been concluded that enhanced fatigue resistance of the higher Al-containing alloy at 25°C is due to the presence of superlattice dislocations, which cause crack initiation to be more difficultthan in the hypostoichiometric alloy; the latter deforms by the motion of unpaired dislocations.

Recent HCF results for Fe_3Al alloy FA-129 show a thickness effect at 600°C; 16.5 mm plate had better fatigue resistance than a 12.7 mm plate, apparently as a result of a higher yield stress for the former. Extrapolation of the data indicated that the 10^8 cycle fatigue strength should be greater than 120 MPa, which may be compared with about 150 MPa for 304 stainless steel (Gieseke et al. 1993).

4. LOW CYCLE FATIGUE

Low-cycle fatigue (LCF) studies are difficult to carry out with many intermetallics due to lack of appreciable tensile ductility. Nevertheless, the

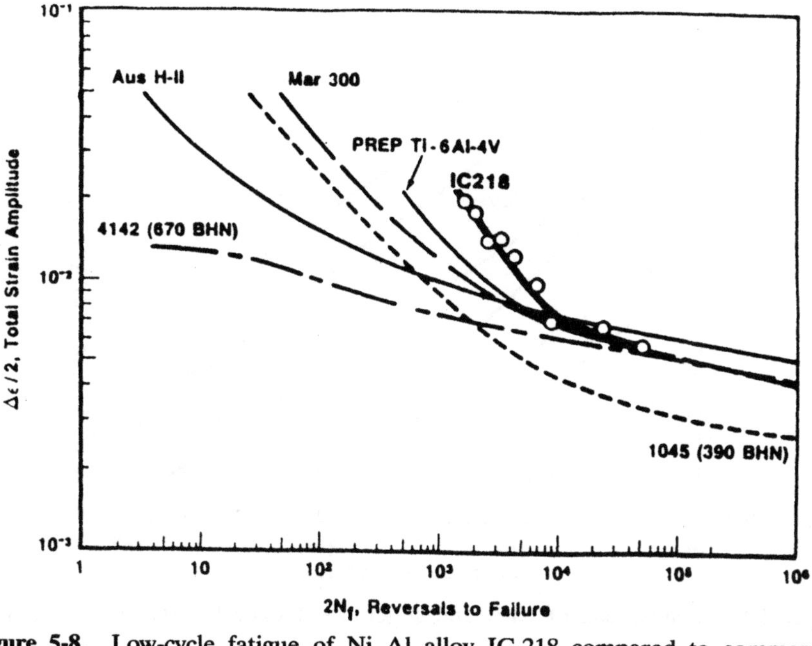

Figure 5-8. Low-cycle fatigue of Ni_3Al alloy IC-218 compared to commercial alloys at 25°C (Gordon and Unni 1991).

LCF behavior of several high-temperature intermetallics, including Ni_3Al, NiAl, Ti_3Al and TiAl, has been reported.

4.1. Ni_3Al Polycrystals

Strain-life data for Ni_3Al alloy IC-218 prepared by hot extrusion of powders, are shown in Figure 5-8 (Gordon and Unni 1991). Note that the LCF performance of IC-218 is superior to that of several other structural alloys, including steels and Ti-6Al-4V, especially at high strain amplitudes. These results were attributed to high tensile ductility of the fine-grained microstructure and good crack-growth resistance of a similar Ni_3Al alloy, IC-221 (Matuszyk et al. 1990).

4.2. NiAl Polycrystals

Recently there have been several reports of LCF behavior of NiAl. The fatigue behavior of powder-processed (P/M) and cast + extruded (C + E) polycrystals of equiatomic NiAl have been compared (Lerch and Noebe 1992; Noebe and Lerch 1992). Two temperature regimes were selected: above and below the ductile-to-brittle transition temperature. At room temperature of Coffin-Manson plot was obtained for C + E material (Noebe and Lerch 1992). The data obeyed the equation:

$$\Delta \epsilon_p = 0.003 N_f^{-0.14} \tag{5.1}$$

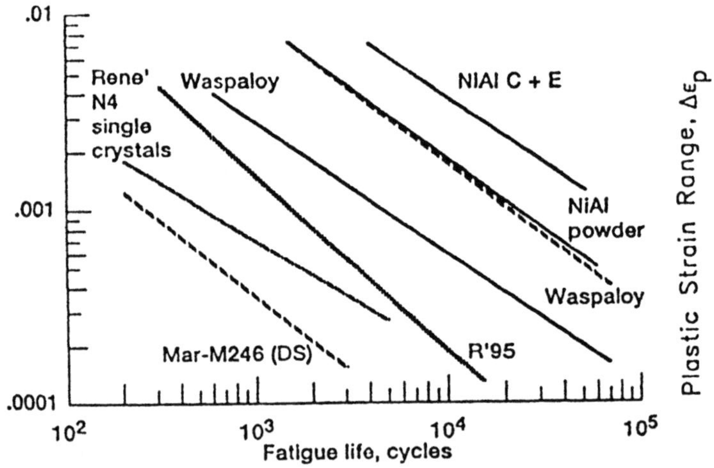

Figure 5-9. Comparison of NiAl and superalloy fatigue life (Noebe and Lerch 1993).

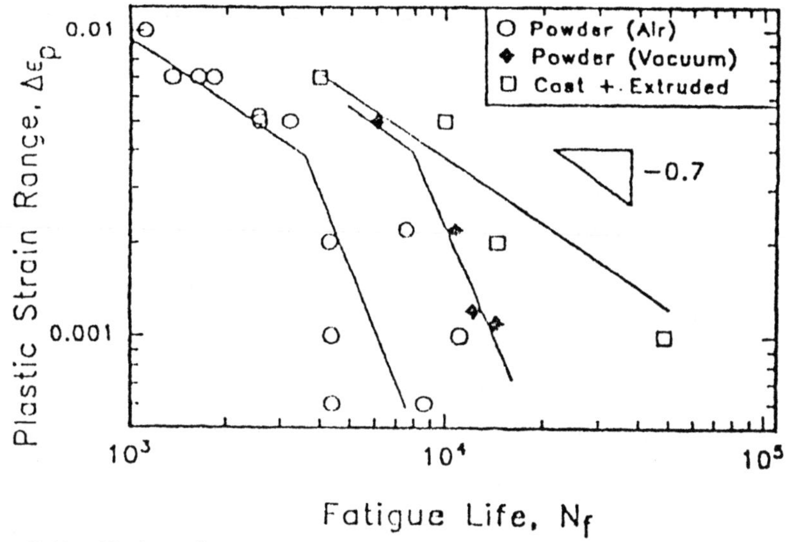

Figure 5-10. Fatigue life of polycrystalline NiAl at 1000°K (Noebe and Lerch 1993).

The exponent is unusually low compared to most structural alloys. The influence of processing technique and environment on the fatigue life at 727°C also was investigated. At low temperatures cyclic hardening is continuous to failure and fracture always initiated at fabrication defects (e.g., inclusions or pores). At 727°C NiAl displayed higher fatigue lives than superalloys on a strain-range basis, see Figure 5-9 (Noebe and Lerch 1993), but had shorter lives on a stress-range basis. Interestingly, C + E material was superior to the P/M alloy based on strain range, but was inferior based on stress range. However, when the P/M samples were tested in strain control at 727°C in 1×10^{-6} torr (MPa) vacuum, lives improved by a factor of three, Figure 5-10 (Noebe and Lerch 1993), so that lives were comparable to those of C + E samples at high strain ranges.

Powder processed Ni-49.5at%Al displays cyclic hardening at 600 K and 700 K that resembles the room temperature behavior: that is, hardening is continuous until final fracture and there is little stable crack growth (Cullers and Antolovich 1992; Cullers and Antolovich 1993). At 1000 K, on the other hand, a maximum in hardening was noted by Noebe and Lerch (1993) after a few cycles, followed by slight cyclic softening until failure. Lives increased substantially between 600 K and 700 K, but considerably less than expected from the large increase in monotonic ductility that occurs over the same temperature range (Cullers and Antolovich 1993). At

1000 K, on the other hand, lives increased by about 100 times relative to room temperature for a plastic strain range of 0.2% (Lerch and Noebe 1992).

4.3. NiAl Single Crystals

Bain, Field, and Lahrman (1991) have studied the cyclic deformation of ⟨001⟩ single crystals of Ni-49.9Al-0.1Mo at 298 K and 1033 K. At the higher temperature, fatigue lives of the single crystals were nearly identical to those for polycrystalline stoichiometric NiAl, and the Coffin-Manson relation was obeyed.

4.4. Surface Film Effects

Thin nickel films applied to extruded Fe-40Al (which deforms by ⟨111⟩ slip) and Ni-30Al-20Fe (which deforms by ⟨100⟩ slip) have been shown to enhance ductility in both monotonic and cyclic tests. Figure 5-11 shows that small amounts of compressive prestrain increase the accumulated strain to failure for the ternary alloy from 18% to 40% while a thin nickel film increased the accumulated strain-to-failure to 34% (Hartfield-Wunsch and Gibala 1991). Combining prestrain and coating increased the strain-to-failure to almost 90%. These results have been explained in terms of mobile ⟨100⟩ dislocations induced by the two techniques. In the binary Fe-40Al alloy, on the other hand, as-extruded material already contains an adequate dislocation density to accommodate deformation. Introducing new dislocations by prestrain has little effect on cyclic deformation. However, surface films in b.c.c.-like materials (such as Fe-40at%Al) may generate mobile nonscrew dislocations that compensate for the reduced mobility of screw dislocations at room temperature and below.

5. CRACK INITIATION

A variety of crack-initiation sites is observed in polycrystalline intermetallics. Crack initiation at 25°C is intergranular in wrought Fe_3Al (Fuchs and Stoloff 1988) and C + E NiAl (Noebe and Lerch 1992). In the latter case, specific crack-initiation sites (e.g., voids) always were observed. The same defects do not influence cracking in this material at 1000 K, probably because of the high ductility above the brittle-to-ductile transition temperature of about 773 K (Noebe and Lerch 1993).

In most other reported cases in polycrystals crack initiation is along slip bands (e.g., $(FeNi)_3V$ and Ni_3Al + B at room temperature and

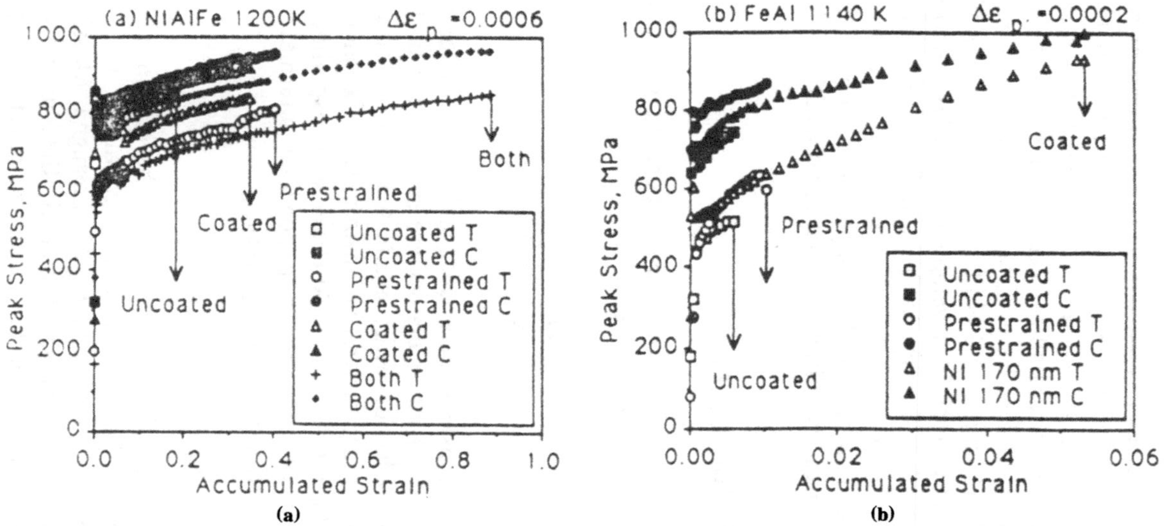

(a) (b)

Figure 5-11. Influence of surface films on cyclic ductility a) Fe-40at%Al; b) Ni-30Al-20Fe (Harfield-Wunsch and Gibala 1991).

Ni_3Al + B and NiAl at elevated temperature (Stoloff et al 1987)). However, some intergranular cracks also were observed in NiAl at 700° K (Cullers and Antolovich 1992). Also, cracks initiated at casting pores in a directionally solidified Ni_3Al alloy (Nazmy and Staubli 1991).

Single crystal studies have shown that persistent slip bands form readily in Ni_3Al (Hsiung and Stoloff 1988) and NiAl (Smith et al. 1992), with intrusions and extrusions developing after a few hundred cycles. The development of surface damage in these and other intermetallics such as FeCo-V and Ni_3Fe closely resembles that seen in conventional metals such as copper and iron, as well as superalloys (Nibkin 1992). However, internal substructures are different in the intermetallics.

6. CRACK PROPAGATION

Crack-growth rates of structural alloys depend upon the stress intensity range, ΔK, according to the relation:

$$\frac{da}{dN} = C\Delta K^m \tag{5.2}$$

The exponent, m, is usually between 2 and 5 for structural alloys. Crack-growth rates of several intermetallics are summarized in Table 5-3. Note the wide range of behavior, which is affected by crystal structure, presence or absence of long-range order, and environment. Exponents as high as 40 have been found for TiAl and Fe_3Al under some experimental conditions. The characteristics of crack growth in individual intermetallic systems will be described in the following paragraphs.

6.1. Ni_3Al Alloys

Detailed studies of crack propagation have been conducted on two aluminides, a Ni_3Al base alloy, IC-221 (Matuszyk et al. 1990; Camus, Duquette, and Stoloff 1990), and Fe_3Al (Castagna, Masziasz, and Stoloff 1993). In both alloy systems environmental effects are very important: for Ni_3Al due to oxygen and for Fe_3Al due to moisture and/or hydrogen. Crack-growth rates for Ni_3Al alloys as well as other Ll_2 alloys are lower than for nickel-base superalloys at room temperature and at 600°C, see Figure 5-12 (Kuruvilla and Stoloff 1987). Also, crack-growth rates increase with increasing temperature in spite of a rise in flow stress over the same temperature range, see Figure 5-13 (Kuruvilla and Stoloff 1987). Crack-tip embrittlement by oxygen seems to be a major factor, although creep

Table 5-3 Crack Growth in Intermetallics

Alloy	Temp. °C	Structure	Environment	Crack Path	$da/dN \Delta K = 40$ MPa (m/cycle)
IC-50 (Ni_3Al + B + Hf)	25	Ll_2	air	TG	10^{-8}
Ni-24Al + B	400	Ll_2	air	IG	6×10^{-5}
Ni-24Al + B	500	Ll_2	vac	TG	5×10^{-5}
Ni-24Al + B	600	Ll_2	vac	TG	8×10^{-7}
FeCo-V (d)	25	b.c.c.	air	TG	2×10^{-9}
FeCo-V (S = 0.7)	25	B2	air	TG	—
Fe-25.05Al	25	B2	air	TG	1×10^{-6}
Fe-25.05Al	25	DO_3	air	TG	7×10^{-6}
Ti_3Al	649	DO_{19}	air	IG	—
Ti_3Al	649	DO_{19}	vac	TG	—
Ni_3Fe (o)	25	Ll_2	air	IG	$> 10^{-4}$
Ni_3Fe (d)	25	f.c.c.	air	IG	$> 10^{-4}$
$(Fe,Ni)_3V$* (d)	25	Ll_2	argon	TG	5×10^{-10}
$(FeNi)_3V$*	600	Ll_2	argon	TG	7×10^{-7}
$(FeNi)_3V$*	25	Ll_2	hydrogen-gas	IG	3×10^{-7}
$(FeNi)_3V$* (d)	25	f.c.c.	hydrogen-gas	TG	3×10^{-7}
α_2-SCS 6(L)	25	DO_{19}	air	—	10^{-8}

*LRO 42
d-disordered
o = ordered
S = degree of order

Source: Stoloff et al. 1987

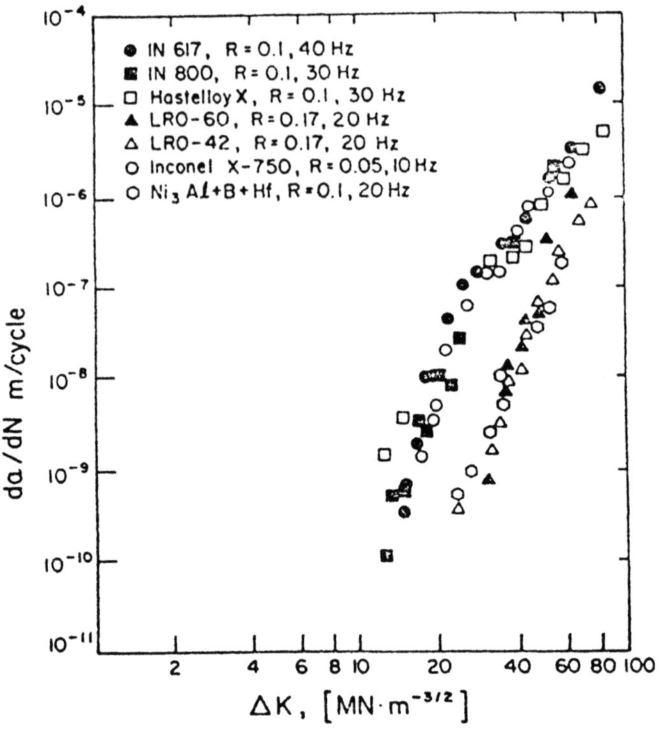

(a)

Figure 5-12. Crack-propagation rates in $(FeNi)_3V$ and Ni_3Al + B + Hf (IC-50) compared to commercial alloys, 25°C (Kuruvilla and Stoloff 1987; reprinted from *Scripta Metallurgica*, Volume 21, pages 873–877, copyright 1987, with kind permission from Elsevier Science Ltd., The Boulevard, Langford Lane, Kidlington OX5 1GB, UK).

processes also influence growth rates at temperatures above about 700°C. Crack-growth rates, da/dN, increase with decreasing frequency in both vacuum and air for alloy IC-221 at 800°C but the rates always are higher in air (Camus et al. 1990). While increased growth rates at low frequency in vacuum are indicative of a creep effect, the higher growth rates in air undoubtedly reflect a contribution of environmental embrittlement. Intergranular fracture occurs in air at all test frequencies, while transgranular fracture is noted in vacuum. Secondary cracking is enhanced at lower

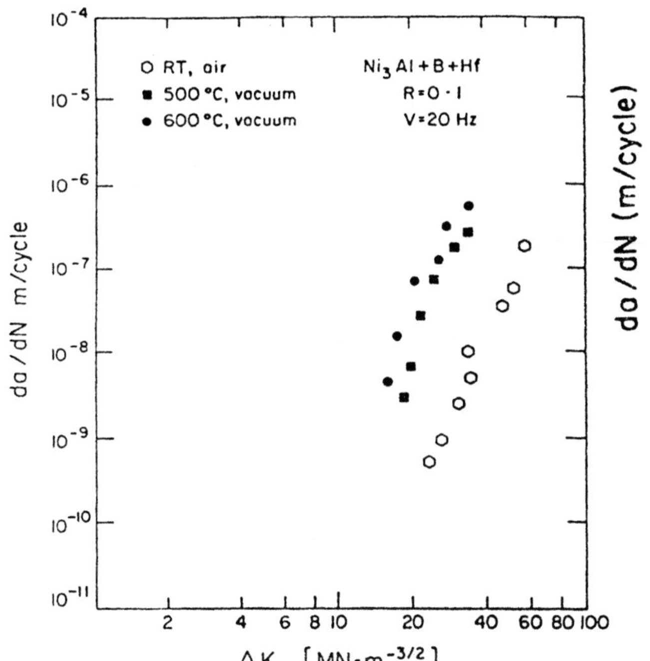

(a)

Figure 5-13. Effect of temperature on crack growth of Ni_3Al + B + Hf. (Kuruvilla and Stoloff 1987; reprinted from *Scripta Metallurgica*, Volume 21, pages 873–877, copyright 1987, with kind permission from Elsevier Science Ltd., The Boulevard, Langford Lane, Kidlington OX5 1GB, UK).

frequencies, suggesting that diffusion of oxygen is required for such cracks to appear.

6.2. Fe_3Al Alloys

Although Fe_3Al alloys are not intended for service above about 500°C, it is instructive to use this system to illustrate the potent effects of ordering and of external environments on the crack growth behavior of ordered intermetallics.

Binary Fe_3Al alloys generally are intermediate in growth rates (in air) between superalloys and $L1_2$ intermetallics, for example, Ni_3Al or $(Fe,Ni)_3V$, at low ΔK levels; at high ΔK, crack growth is more rapid in Fe_3Al (Stoloff et al. 1987).

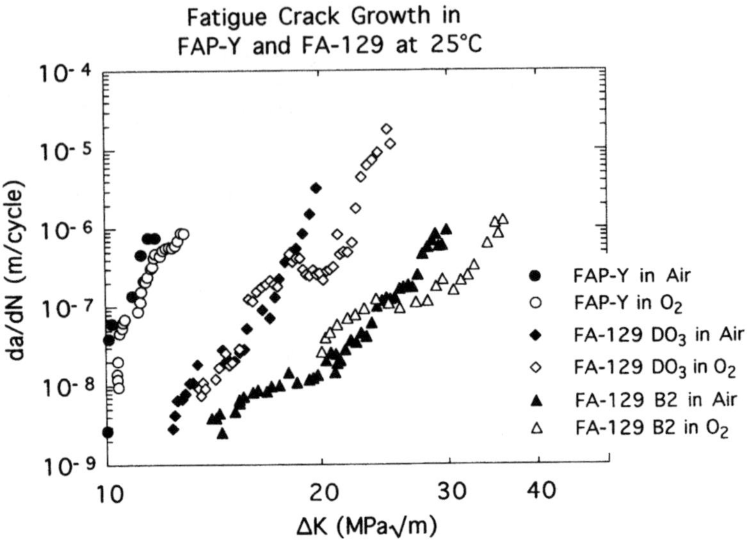

Figure 5-14. Fatigue crack-growth rates in Fe-Al alloys (Castagna and Stoloff 1994).

Recent work has shown that hydrogen or moisture is very detrimental to fracture ductility of iron aluminides (McKamey, Horton, and Liu 1988; Liu, McKamey, and Lee 1990). Details may be found in the chapter on Environmental Embrittlement. Research has extended the study of environmental effects to cyclic loading. Crack-growth rates for an Oak Ridge-developed Fe_3Al alloy (FA-129) (see Table 5-2 for composition), for the partially disordered B2 condition, and for the fully ordered DO_3 condition, are shown in Figure 5-14 (Castagna and Stoloff 1994). Note that lower crack-growth rates are observed in oxygen than air for both ordered conditions. Values for the slopes as well as apparent threshold and critical stress intensities for fracture for each environment and ordered condition may be found in Table 5-4. (Values of ΔK_{TH} and ΔK_c are lower than reported previously by Castagna, Maziasz and Stoloff [1993] because of a machine error in the previous work). In every respect the DO_3 condition is less resistant to cyclic crack growth and fracture, especially air. Hydrogen gas is only slightly more embrittling than air. It is now generally considered that moisture in contact with iron aluminides (Fe_3Al and FeAl) breaks down by chemical reaction at the surface of the sample, thereby liberating hydrogen, which is the specific embrittling agent (Liu et al. 1990).

Table 5-4 Fatigue Crack Growth Data for FA-129 at 25°C

Condition	Fracture Surface	*m*	ΔK_{TH} (MPa$\sqrt{m}$)	ΔK_C (MPa$\sqrt{m}$)
B2 Oxygen	Dimpled, few striations	3.0	20.0	35.9
B2 Air	TG, many striations	6.9	14.5	29.9
B2 H_2 (1 atm)	TG, few striations	8.8	13.8	25.9
DO_3 Oxygen	Dimpled + cleavage	8.5	13.5	25.1
DO_3 Air	Mixed, few striations	11.2	13.3	19.8
DO_3 H_2 (1 atm)	Mixed, no striations	25.1	11.1	17.4

Castagna and Stoloff 1994

Also shown in Figure 5-14 are crack-growth data for an Fe-16at%Al alloy, FAP-Y (see Table 5-2). This alloy, which should be completely disordered, exhibits much more rapid crack growth than FA-129. The reason for this difference is not yet known.

Chromium reportedly improves the ductility of Fe_3Al in monotonic tension tests in air (McKamey et al. 1988), but offers no benefit under cyclic loading conditions. Growth rates are actually higher for FA-129 than for a binary alloy with similar Al content, both tested in the DO_3 condition (Stoloff et al. 1987; Castagna et al. 1993). This unexpected result may be due to the influence of chromium on the repeated formation and rupture of the oxide film on a freshly exposed crack surface.

The influence of temperature on crack growth rates of FA-129 also has been studied (Castagna et al. 1993). Growth rates decrease between 25°C and 150°C in the B2 and DO_3 conditions. However, at higher temperatures growth rates rise again for the DO_3 condition. (Tests of the B2 condition are limited to low temperatures due to the occurrence of DO_3 ordering.) Even at 450°C, crack-growth rates are lower than at room temperature, presumably due to the fact that hydrogen diffuses too rapidly at high temperatures to cause embrittlement. Therefore, it appears that hydrogen embrittlement is maximized near room temperature, as is the case for structural steels.

6.3. Ti_3Al and Ti_3Al Composites

Although crack-growth phenomena in Ti_3Al alloys have been studied extensively, little had been published prior to a series of papers on Ti-24Al-11Nb (α_2) and Ti-25Al-10Nb-3Mo-1V (super α_2). One of the most interesting aspects of crack-growth behavior is the potent effect of microstructure, see Figure 5-15 (Aswath and Suresh 1991). Note that a

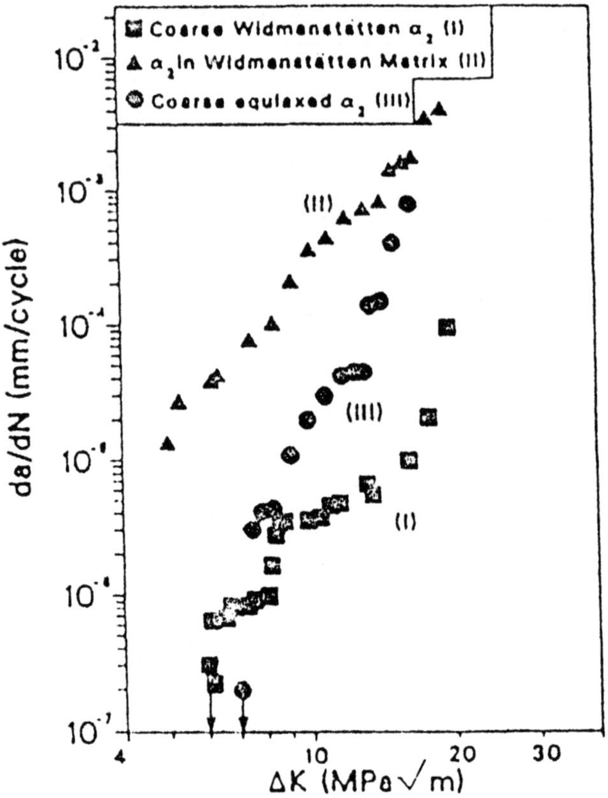

Figure 5-15. Effect of microstructure on crack growth of Ti_3Al at 25°C (Aswath and Suresh 1991).

coarse Widmanstatten microstructure provides the greatest resistance to crack growth. A similar strong effect of microstructure on HCF behavior of super α_2 at 600°C has been reported (Luetjering et al. 1991) and has been attributed to the resulting yield stresses. A bimodal microstructure produces higher yield stresses and longer lives than an equiaxed or lamellar structure. Fatigue crack growth (FCG) resistance improves when the alloy is cycled in vacuum.

Evidence for environmental effects on crack growth of α_2 at both room and elevated temperature has been provided by several groups. There is an effect of frequency on crack-growth rates in laboratory air and in vacuum (10^{-5} torr) for Ti-24Al-11Nb with coarse Widmanstatten α_2 surrounded continuously by transformed β (Williams et al. 1990). This is a

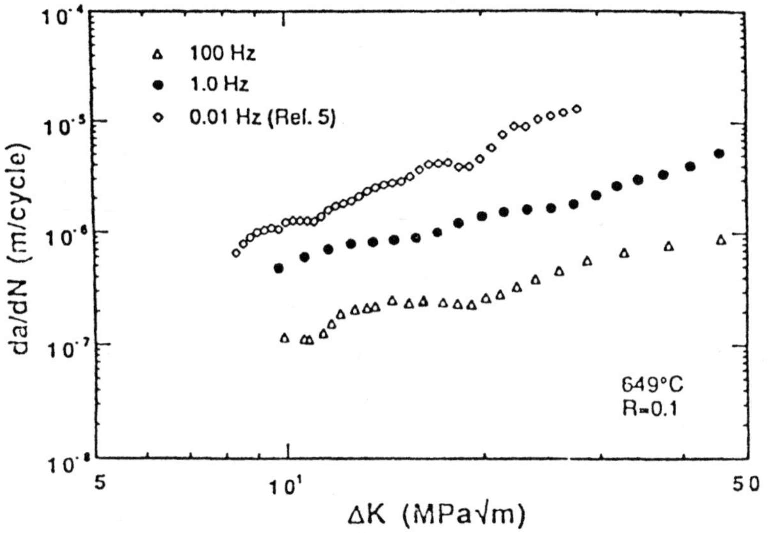

Figure 5-16. Variation of fatigue crack-growth rates of Ti-24-Al-11Nb with frequency at 649°C (Parida and Nicholas 1992).

microstructure with relatively high FCG resistance. Crack growth is much more rapid in air than in vacuum. Similar observations have been made at temperatures in the range 650–800°C (Aswath and Suresh 1989, Balsone et al. 1990). Crack-growth rates of Ti-24-Al-11Nb decrease between 25°C and 250°C, but then increase at higher temperatures. Crack-growth rates in air decreased with increasing frequency, and were further decreased by testing in vacuum. Crack-growth rates for super α_2 are considerably lower in vacuum than in air at 700°C; m increases from 2.3 in vacuum to 4.0 in air (Penton, Cope, and Bowen 1992). Growth rates of Ti-24Al-11Nb increase by about ten times at 649°C when frequency is lowered from 100 Hz to 0.01 Hz, see Figure 5-16 (Parida and Nicholas 1992). In general, fatigue crack-growth rates in α_2 increase with moist environments (probably due to the release of hydrogen), lowered test frequency, increased hold times, and increased test temperature (Wessels, Marquardt, and Krueger 1989).

Small crack behavior in Ti_3Al alloys (Davidson, Campbell, and Page 1991; Ravichandian and Larsen 1992) also have been reported. Long-crack and short-crack data for cast, forged, and rolled super α_2 correlate well with ΔK^*_{eff}, which is defined as:

$$\Delta K^*_{\text{eff}} = \Delta K_i + \Delta K - \Delta K_c \qquad (5.3)$$

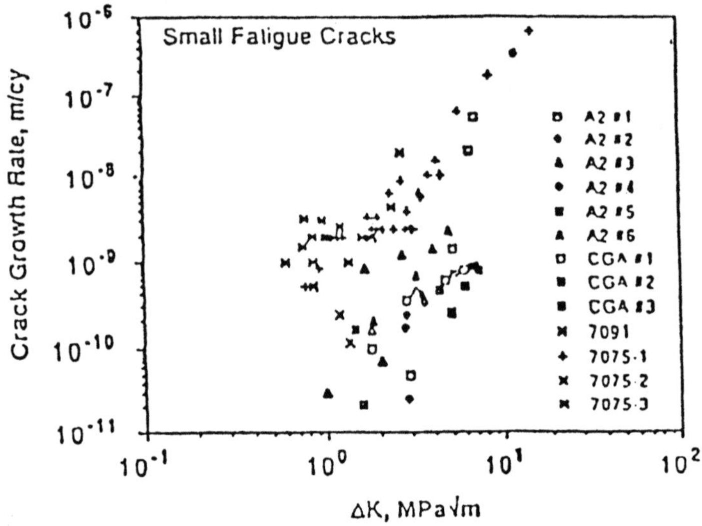

Figure 5-17. Comparison of crack-growth rates for small fatigue cracks in Super Alpha 2 (A2), coarse-grained Astroloy (CGA), and the aluminum alloys 7075-T651 and 7091 (Davidson, Campbell and Page 1991).

where ΔK_i is the value of ΔK^*_{eff} at $\Delta K = 0$ and ΔK_c is the closure level for small cracks and is about $0.4\,\Delta K$ for super α_2. Comparison of the growth rate of small cracks in super α_2 with those in 7075 Al and in the superalloy Astroloy is shown in Figure 5-17 (Davidson et al. 1991). Note that super α_2 has a growth rate slightly higher than for Astroloy but about $10-100\times$ slower than for 7075 Al. Some small cracks arrest in each of the three alloys.

For Ti-24Al-11Nb small cracks develop arbitrary shapes that are influenced by microstructure and texture; the shapes in turn affect crack-growth rates. With a basketweave microstructure as well as with two other microstructures, small and long crack data correlate well when crack closure is accounted for (Ravichandian and Larsen 1992). However, for an aligned colony structure small crack-growth rates were always higher than those for long cracks.

Perhaps the single factor most likely to influence the fatigue crack-growth resistance of titanium aluminides is the use of fibrous reinforcements. Figure 5-18 shows that although the crack growth rates of α_2 and γ are higher than for the P/M Ni-base superalloy IN-100, the longitudinally oriented composite of α_2 (Ti-24Al-11Nb) with continuous SCS-6 SiC fibers is far more crack growth resistant than in IN-100 (Larsen et al.

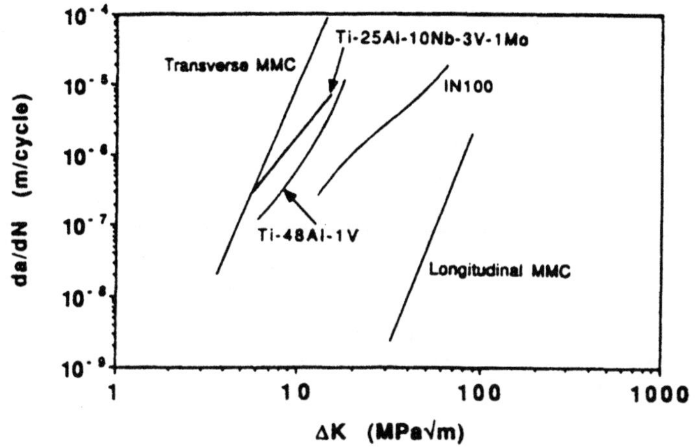

Figure 5-18. Comparison of crack-growth rates of titanium aluminides, α_2 composites, and IN-100; tests at 650°C, R = 0.1, $\nu = 0.2$ Hz except for the composite, for which $\nu = 3.33$ Hz (Larsen et al. 1990).

1990). Unfortunately, the transversely oriented α_2 composite displays the highest growth rate of all. (Note: the specimen geometry differed in that thin tubes of Ti-48Al-1V (γ) were utilized; also, the frequency of tests of the α_2 composite was 3.33 Hz, which would be expected to result in a somewhat lower growth rate than at 0.2 Hz).

6.4. TiAl

Data for TiAl (γ) alloys are even more limited than for Ti_3Al. Due to the low toughness of γ, ΔK values rarely reach 20 MPa$\sqrt{m}$ prior to fracture. Note in Figure 5-18 that the growth rate of Ti-48-Al-1V is comparable to that of super α_2 (Larsen et al. 1990). The influence of microstructure and temperature on fatigue crack-growth behavior of a P/M Ti-48at%Al alloy tested in air has been reported (Soboyejo, Aswath, and Deffeyes 1991). Crack growth at room temperature was slightly slower than for a Ti-6Al-4V alloy. At 700°C, crack growth was even slower, probably due to oxide-induced crack closure.

The constants A and m in Equation 5.2 are extremely sensitive to test conditions for Ti-48-Al-2Mn-2Nb; see Table 5-5 (James and Bowen 1992). The slope m varies from 8.2 for as-cast material tested at 700°C in vacuum to as high as 56.2 for a sample heat treated for 2h at 1200°C and tested at 700°C in air. (This range of slopes is similar to that noted with Fe_3Al alloy FA-129 as a function of long range order and environment, as

Table 5-5 Crack Growth Data for Ti-48Al-2Mn-2Nb

Heat Treatment	Test Conditions	A mm/cycle	*m*
as-cast	800°C/vac.	1.62×10^{-15}	9.4
as-cast	700°C/vac.	3.36×10^{-14}	8.2
as-cast	700°C/air	7.34×10^{-15}	10.0
2h/1200°C	800°C/vac.	7.21×10^{-26}	22.8
2h/1200°C	700°C/air	2.33×10^{-57}	56.2
24h/1200°C	800°C/vac	1.88×10^{-25}	22.8
24h/1200°C	700°C/air	1.47×10^{-31}	29.2

James and Bowen 1992

was shown in Table 5-4 (Castagna and Stoloff 1994). It has been suggested that *static failure modes*, transgranular cleavage of allotriomorphic γ grains at room temperature or transgranular decohesion at 700 or 800°C contribute to high values of m (James and Bowen 1992). Therefore, microstructures containing large amounts of γ grains are not resistant to cyclic loading at elevated temperatures. Rather, a two-phase lamellar α_2/γ microstructure is preferred. Although this alloy displays no change in crack-growth rates between 700°C and 800°C, vacuum decreases the crack-growth rates by about an order of magnitude.

6.5. Ductile-Phase Toughened Composites

With the exception of the α_2-SCS-6 system, crack-growth data have been reported only for composites reinforced with ductile phases; also, data have been published only for tests at room temperature in air. As in the case of monotonic tests, particles do not confer much advantage due to the lack of crack-stopping ability of low aspect ratio reinforcements. For example, subcritical cracks grow at stress intensities as low as 2–3 $\mathrm{MPa}\sqrt{m}$ in $MoSi_2$-Nb composites (Rao, Soboyejo, and Ritchie 1992). For this reason, it has been suggested that increasing the aspect ratio, together with a reduction in particle-matrix interface strength, is necessary for improved crack-growth resistance in these composites (Rao, Odette, and Ritchie 1992).

The most detailed study of crack growth in ductile-phase toughened composites has been carried out in γ-TiAl reinforced with 25 or 50 μm diameter TiNb particles (Rao et al. 1992). As in the case of $MoSi_2$-Nb composites, cracks can propagate subcritically at very low stress intensities, less than $6\mathrm{MPa}\sqrt{m}$; see Figure 5-19 (Rao and Ritchie 1992), compared to a fracture toughness of $25\mathrm{MPa}\sqrt{m}$. However, the Paris slopes, m, (equa-

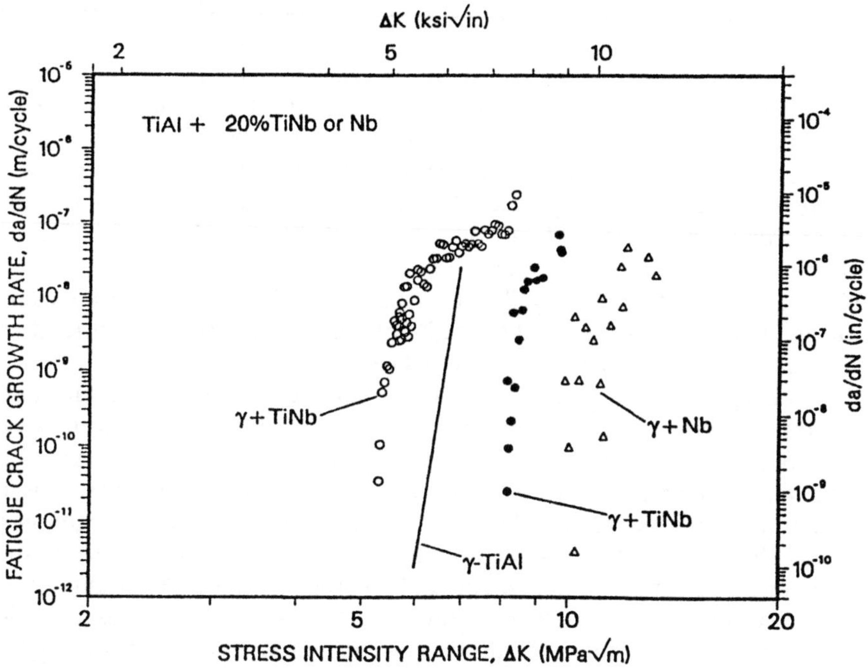

Figure 5-19. Cyclic fatigue-crack propagation rate behavior in ductile-phase toughened γ-TiAl composites (20 vol%), as a function of ductile reinforcement (TiNb vs. Nb), and reinforcement orientation (edge (C-R) vs. face (C-L)), in γ-TiAl composites. Results in unreinforced γ-TiAl (Ti-50 at% Al) are also plotted for comparison (Rao, Odette, and Ritchie 1993).

tion 5.2) are considerably reduced by TiNb, from 29.4 for TiAl to about 9.6 for TiAl-20vol%TiNb. A particularly striking observation was that the crack-growth resistance of monolithic TiAl appears to be diminished by the particles, particularly at near-threshold levels (Rao and Ritchie 1992). Later work showed, however, that Nb particles markedly improved the fatigue crack-growth resistance of TiAl, see Figure 5-19 (Rao, Odette, and Ritchie 1993). The differing effects of the two reinforcements were attributed to different mechanisms of fracture of the two ductile reinforcements during cyclic loading.

7. DISCUSSION

It is clear that crack initiation under stress control is relatively difficultin single-phase $L1_2$ alloys, in spite of the planar slip associated with most intermetallics. Further, crack-growth rates at both 25°C and 600°C are lower than for either single-phase or precipitation-hardened nickel-base alloys, as was shown in Figure 5-12. It is doubtful that planarity of slip is the decisive factor, since slip is planar in both the single-phase intermetallics and in the precipitation-hardened superalloys.

The titanium aluminides are extremely sensitive to microstructure, test temperature, and frequency. The crack-growth rates of both TiAl and Ti_3Al show a high value of m, for example, Table 5-5 (James and Bowen 1992), especially at low temperature and in air. Short crack data for α_2 alloys generally correlate well with long crack data when crack-closure effects in long cracks are subtracted out. However, microstructure does affect short-crack behavior of Ti-24Al-11Nb.

Strong effects of temperature, environment, and microstructure also are found in crack-growth data for Fe_3Al and Ni_3Al alloys. As in the case of titanium aluminides, Fe_3Al alloy FA-129 displays Paris exponents that range from "normal" values of about 3 for B2 material tested in oxygen to as high as 25 for the fully ordered DO_3 condition tested in hydrogen gas, as was shown in Table 5-4. Temperature is extremely important, as the detrimental effects of water vapor and hydrogen on Fe_3Al disappear at 150°C. However, at high temperatures, factors such as decreasing yield stress, creep-fatigue interactions or environmental interaction of oxygen with fatigue cracks can sharply increase crack-growth rates and reduce fatigue lives in Ni_3Al alloys, as shown in Figures 5-3, 5-4, and 5-13.

8. SUMMARY AND CONCLUSIONS

On the basis of fairly limited data for intermetallics, the following conclusions may be drawn:

1. High-cycle fatigue endurance ratios are higher for nickel and titanium aluminides than for superalloys. However, this may be at least in part due to the very high strengths of the superalloys, so that absolute values of endurance limit may be comparable.
2. Cyclic hardening followed by softening occurs in most intermetallics, the only exception being NiAl.
3. Low-cycle fatigue behavior of Ni_3Al and NiAl is superior to that of other high-temperature alloys when expressed on the basis of strain range.
4. Crack-growth rates of Ll_2 alloys are lower than for superalloys. Intermetallics are extremely sensitive to environment, frequency, and temperature.

5. Hydrogen and moisture adversely affect fatigue crack-growth resistance in Ni_3Al, Ti_3Al, and Fe_3Al alloys at low temperatures; oxygen has a similar detrimental effect at high temperatures for Ni_3Al and Ti_3Al alloys.
6. Microstructure strongly influences crack-growth behavior of Ti_3Al alloys.
7. Crack initiation in Ni_3Al and NiAl single crystals occurs after the development of the same pattern of surface damage as seen in conventional alloys.
8. Ductile-phase toughening of intermetallics under monotonic loading does not lead to improved fatigue resistance.
9. Additional research of fatigue of intermetallics needs to be performed in the following areas:
   (a) LCF as a function of temperature, especially for Ni_3Al, Ti_3Al, and TiAl
   (b) environmental effects on crack initiation and growth
   (c) short-crack behavior
   (d) influence of surface modification on fatigue behavior
   (e) all aspects of fatigue behavior of "newer" intermetallics such as $MoSi_2$, Laves phases, Al_3Ti, and Nb_3Al.

REFERENCES

ASHOK, S., KAIN, K., TARTAGLIA, J., and STOLOFF, N. S., 1983, *Metall. Trans. A* 14A: 1997–2003.

ASWATH, P. B. and SURESH, S., 1989, *Mat. Sci. Eng.* A114: L5–L10.

_______, 1991, *Metall. Trans. A.* 22: 817–828.

BAIN, K. R., FIELD, R. D., and LAHRMAN, D. F., 1991, Presented at TMS Fall Meeting, Cincinnati, OH, unpublished.

BALSONE, S. J., MAXWELL, D. C., KHOBAIB, M., and NICHOLAS, T., 1990, *Fatigue 90, III*, Eds., H. Kitaga, and T. Tanaka, 1905–1910, Warley, UK: Materials and Components Engineering Publications, Ltd.

BOETTNER, R. C., STOLOFF, N. S., and DAVIES, R. G., 1966, *Trans. AIME* 236: 131–133.

CAMUS, G., DUQUETTE, D. J., and STOLOFF, N. S., 1990, *J. Mat. Res.* 5: 950–954.

CASTAGNA, A., MAZIASZ, P., and STOLOFF, N. S., 1993, *High Temperature Ordered Intermetallic Alloys V*, MRS Symp. Proc. 288: 1043–1048.

CASTAGNA, A., and STOLOFF, N. S., 1994, In Proceedings Eighth Annual Conference on Fossil Energy Materials, May 10–12, 1994, Oak Ridge, Tennessee, 297–307.

CASTAGNA, A., SHEA, M., and STOLOFF, N. S., 1991, *High Temperature Ordered Intermetallic Alloys IV*, MRS Symp. Proc. 213: 609–616.

CHENG, C. J., and SEKHAR, J. A. 1991, In *High Temperature Ordered Intermetallic Alloys IV*, MRS Symp. Proc., 213: 853–858.

CLARK, H., McI, 1966, Nature, 209: 193.

Cullers, C. L., and Antolovich, S. D., 1992, *Superalloys 1992, Proc. 7 Springs Conf. on Superalloys,* Eds., Antolovich et al., pp. 351–359. Warrendale, PA: TMS.

_______, 1993, in *High Temperature Ordered Intermetallic Alloys V*, MRS Symp. Proc. 288: 531–536.

Davidson, D. I., Campbell, J. B., and Page, R. A., 1991, *Metall. Trans. A.* 22A: 377–391.

Doherty, J. E., Giamei, A. F., and Kear, B. H., 1975, *Metall. Trans. A.* 6A: 2195–2199.

Fuchs, G. E., and Stoloff, N. S., 1987, *Scripta Metall.* 21: 863–868.

_______, 1988, *Acta Metall.* 36: 1381–1387.

Gieseke, B., and Sikka, V. K., 1992, *Mat. Sci. Eng.* A153: 520–524.

Gieseke, B. G., Alexander, D. J., Sikka, V. K., and Baldwin, R. H., 1993, *Scripta Metall.* 29: 129–134.

Gittins, A., 1968, *Met. Sci. J.* 2: 114.

Gordon, D. E., and Unni, C. K., 1991, *J. Mat. Sci.* 26: 6183–6189.

Gordon, D. E., and Unni, C. K., 1989, Presented at TMS-AIME Fall Meeting, Indianapolis, IN.

Gordon, D. E, Unni, C. K., and Stoloff, N. S., 1994, *Fatigue Fract. Engng. Mater. Struct.* 17: 1025–1032.

Hartfield-Wunsch, S. E. and Gibala, R. 1991, in *High Temperature Ordered Intermetallic Alloys IV*, Eds. L. A. Johnson, D. P. Pope, and J. O. Stiegler MRS Symp. Proc. 213: Pittsburgh, 575–580.

Hsiung, L. M., and Stoloff, N. S., 1988, In *High Temperature Ordered Intermetallic Alloys III*, MRS Symp. Proc., 133: 261–267.

Inouye, H., 1985 in *High Temperature Ordered Intermetallic Alloys* Eds. C. C. Koch, C. T. Liu, and N. S. Stoloff, MRS Symp. Proc. 39: 255–262.

James, A. W., and Bowen, P., 1992, *Mat. Sci. and Eng.* 153: 486–492.

Kuruvilla, A. K., and Stoloff, N. S., 1985, *Metall. Trans. A.* 16A, 815.

_______, 1987, *Scripta Metall.* 21: 873–877.

Larsen, J. M., Williams, K. A., Balsone, S. J., and Stucke, M. A., 1990, *High Temperature Aluminides and Intermetallics*, Eds. S. H. Whang, C. T. Liu, D. P. Pope and J. O. Stiegler 521–556. Warrendale, PA: TMS.

Lerch, B. A., and Noebe, R. D., 1992, *HITEMP Review*, NASA Conf. Publ., 10104, v. II, 47-1–47-15.

Liu, C. T., McKamey, C. G., and Lee, E. H., 1990, *Scripta Metall.* 24: 385–390.

Luetjering, G., Proske, G., Albrecht, J., Helm, D. and Daeubler, M., 1991, In *Intermetallic Compounds-Structure and Mechanical Properties*, Ed. O. Izumi, 536–541. Sendai: Japan Inst. of Metals.

Matsugi, K., and Stoloff, N. S., 1993, unpublished.

Matuszyk, W., Camus, G., Duquette, D. J., and Stoloff, N. S., 1990, *Metall. Trans. A.* 21: 2967–2976.

McKamey, C. G., Horton, J. A., and Liu, C. T., 1988, *Scripta Metall.* 22: 1679–1684.

Nazmy, M., and Staubli, M., 1991, *Scripta Metall.* 25: 1305–1308.

NIKBIN, K. M., 1992, *Superalloys* 1992, *Proc. 7 Springs Conf. on Superalloys*, Ed. S. D. Antolovich et al. pp. 637–646. Warrendale, PA: TMS.

NOEBE, R. D., and LERCH, B. A., 1992, *Scripta Metall.* 27: 1161–1166.

________, 1993, *High Temperature Ordered Intermetallic Alloys V*, MRS Symp. Proc. 288: 743–748.

PARIDA, B. K., and NICHOLAS, T., 1992, *Mat. Sci. and Eng.* A153: 493–498.

PENTON, R. J. T., COPE, M. T. and BOWEN, P., 1992, *Mat. Sci. and Eng.* A153: 508–513.

RAO VENKATESWARA, K. T., ODETTE, G. R., and RITCHIE, R. O., 1992, *Acta Metall.* 40: 353–361.

________, 1993, In *Structural Intermetallics*, Eds., R. Darolia, J. J. Lewandowski, C. T. Liu, P. L. Martin, D. B. Miracle, and H. V. Nathal, 829–835. Warrendale, PA: TMS.

RAO VENKATESWARA, K. T., and RITCHIE, R. O., 1992, *Mat. Sci. Eng.* A153: 479–485.

RAO VENKATESWARA, K. T., SOBOYEJO, W. O., and RITCHIE, R. O., 1992, *Metall. Trans. A.* 23A: 2249–2257.

RAVICHANDIAN, K. S., and LARSEN, J. M., 1992, *Mat. Sci. Eng.* A153: 499–507.

RUDOLPH, G., HAASEN, P., MORDIKE, B. L., and NEUMANN, P., 1965, *Proc. First Int. Conf. on Fracture*, Sendai, Japan: 2: 501.

SASTRY, S. M. L., and LIPSITT, H. A., 1977, *Acta. Metall.* 25: 1279–1288.

SMITH, T. R., KALLINGAL, C. G., RAJAN, K., and STOLOFF, N. S., 1992, *Scripta Metall.* 27: 1389–1393.

SOBEYEJO, W. O., ASWATH, P. B., and DEFFEYES, J. E., 1991, *Mat. Sci. Eng.* A138: 95–102.

STOLOFF, N. S., 1992, in *Ordered Intermetallics—Physical Metallurgy and Mechanical Behavior*, Eds. C. T. Liu, R. W. Cahn, and G. Sauthoff, 257–277. Dordrecht: Kluwer Acad. Publ.

STOLOFF, N. S., FUCHS, G. E., KURUVILLA, A. K., and CHOE, S. J., 1987, *High Temperature Ordered Intermetallic Alloys II*, MRS Symp. Proc. 81: 247–261.

WESSELS, J. F., MARQUARDT, B. J., and KRUEGER, D. D., 1989, presented at TMS-AIME Symp. Creep and Fracture of Titanium Aluminides, TMS, Indianapolis, IN.

WHITEHEAD, R. S., and NOBLE, F. W., 1970, *J. Mat. Sci.* 5: 851–861.

WILLIAMS, K. A., BALSONE, S. J., STUCKE, M. A., and LARSEN, J. M., 1990, presented at Aeromat 1990, Long Beach, CA.

WRIGHT, P. K., MAURER, N. and BAIN, K. 1992, General Electric Aircraft Engines, Cincinnati, Ohio, unpublished research.

Part II

Behavior of Alloy Systems

Chapter 6

The Physical and Mechanical Metallurgy of Ni_3Al and Its Alloys

N. S. Stoloff and C. T. Liu

ABSTRACT

The physical and mechanical metallurgy of Ni_3Al and its alloys are reviewed in this chapter. Emphasis is on developments within the past ten years, a period of extremely rapid growth in research on the behavior of this unusually interesting intermetallic compound. This chapter is particularly concerned with solute effects on properties such as strength ductility, fatigue resistance, and environmental stability, subjects which have received much attention. Novel processing techniques, including rapid solidification and powder processing, are shown to have played a key role in providing Ni_3Al-base alloys with improved properties. An assessment of needed areas of research to hasten commercial applications of Ni_3Al alloys concludes the chapter.

1. INTRODUCTION

The ordered intermetallic Ni_3Al has received enormous attention as both a model material and as a potential structural alloy. However, as is the case with most intermetallics, low-temperature ductility and high-temperature creep resistance have been inadequate. An apparent solution to the brittleness of polycrystalline Ni_3Al (single crystals are ductile) had been reported in 1979 by Aoki and Izumi. They noted that small quantities of boron conferred extensive plasticity on Ni_3Al; this work was soon confirmed by Liu and co-workers (1983), Taub and Huang (1984), and many others. The pace of alloy development efforts accelerated rapidly, and was extended to other well-known aluminides in the Fe-Al and Ti-Al systems, as well as to NiAl. Recently, such work has spread to higher melting systems such as Al-Ta and Al-Nb, as well as to beryllides, silicides, and Laves phases. However, the level of effort on Ni_3Al has remained much higher than for any other alloy. Commercialization of Ni_3Al-base alloys for selected applications is now at hand, and research has expanded to Ni_3Al base composites. Consequently, this chapter surveys the rapid progress made in our understanding of the behavior of this very interesting compound, largely focusing upon developments of the past ten years.

2. CRYSTAL STRUCTURE AND PHASE STABILITY

2.1. Order-Disorder Transformation

Most investigators have reported that Ni_3Al is ordered to very near its melting point, although in one case experiments were carried out only to 1000° C (Stockinger and Neumann 1970). Recently, Cahn et al. (1987) reported the results of electrical resistivity and dilatometric studies on Ni_3Al with varying aluminum content in alloys containing iron. They state, on the basis of limited data, that the ordering temperature varies from 1360° C for Ni-22%Al (25° C below the solidus) to 1450° C for stoichiometric Ni_3Al (65° C above the solidus). At 23 at%Al, the ordering temperature and the solidus are identical (based on interpolation of data). These observations require independent confirmation.

2.2. Antiphase Domains

In many A_3B intermetallics, antiphase domains are readily seen, and are usually of uniform size. However, in rapidly quenched Ni_3Al highly anomalous microstructures are seen. For example, in melt-spun ribbons of Ni-22 at%Al, regions of fine domains, all outlined by films of disordered γ, are surrounded by regions of very coarse domains. A mix of fine and coarse domains also is seen in rapidly quenched material containing 500 wppm B (Horton et al. 1983). Horton and Miller (1987) reported that for stoichiometric Ni_3Al, coarse domains were richer in aluminum than

Table 6-1 APB Energy in Ni_3Al, γ_{hkl} (mJm^{-2})

γ_{111}	γ_{010}	$\gamma_{111}/\gamma_{010}$	Method	References
111	90	1.2	Weak-beam TEM	Douin et al., 1986
142	83	1.7	Embedded atom method	Chen et al., 1986
96	28	3.4	Embedded atom method	Baker et al., 1988
175	140	1.25	First principles total-energy calculations	Fu et al., 1992
283	196	1.44	First principles electronic structure	Sluiter et al., 1992

fine domains (25.5 vs 22.5 at%Al) in material prepared by arc-hammer quenching. Cahn et al. (1987) utilizing microanalysis, also concluded that microsegregation was responsible for the difference in domain sizes within grains of rapidly solidified material. The coarse domains are associated with aluminum-rich interdendritic regions in which the ordering temperature is above the solidus. Conversely, fine domains are thought to form only in alloys that freeze from the disordered state. However, Chiba et al. (1991) has reported that fine domains occur in a ternary Ni_3Al alloy containing Pd, Ag, and several other solutes.

Recent experimental values of antiphase boundary (APB) energy determined by transmission electron microscopy (TEM) range from 83–140 mJ/m^2 (Hillier, 1984; Veyssiére, 1984) and by flow stress measurements on γ/γ' alloys a value of 140 mJ/m^2 was deduced (Ardell, 1985). It should be noted that APB energy can be changed by solute additions, and that many elements are highly soluble in Ni_3Al. Recent calculations and experimental data from various sources appear in Table 6-1 and span the range 96–280 mJ/m^2 for γ_{111}.

2.3. Lattice Parameter

The influence of several solutes upon the lattice parameter, a_o, has been reported. In melt-spun ribbon boron, additions to 6 at% increase the lattice parameter from about $3.57 \times 10^{-10} m$ to about $3.60 \times 10^{-10} m$ (Huang, Taub and Chang 1984). No obvious solubility limit is observed within this range. Later work showed a similar large effect of carbon on a_o (Huang et al. 1986). As a result of these large changes in a_o, both boron and carbon confer substantial strengthening upon melt spun Ni_3Al. Following the Mott–Nabarro (1948) model of solid solution strengthening, Huang et al. (1986) have correlated $\Delta\sigma_y/\Delta c$ with $\Delta\varepsilon/\Delta c$ where σ_y is yield stress, (c is concentration and ε is lattice dilation, $\Delta\varepsilon = (a - a_o)/a_o$,

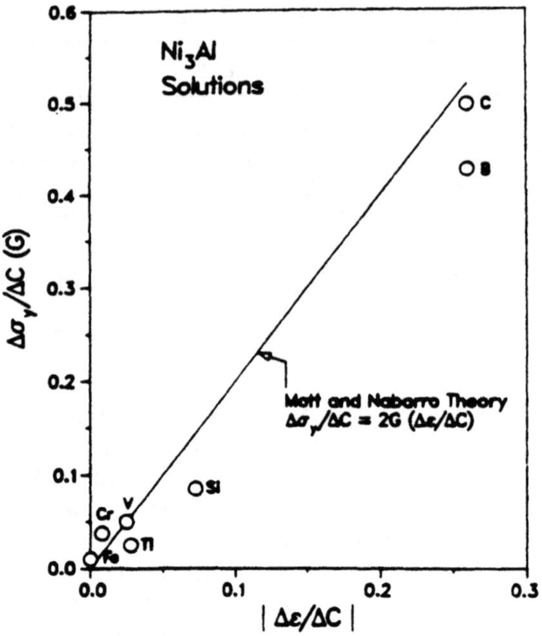

Figure 6-1. Correlation of strengthening potency, $\Delta\sigma_y/\Delta c$, with lattice strain per atom fraction, $\Delta\varepsilon/\Delta C$, for several elements in Ni_3Al. G is the Ni_3Al shear modulus (6.5×10^4 MPa) (Huang et al. 1986).

where a and a_o are atomic radii of solute and solvent, respectively. The value of $\Delta\varepsilon$ is 0.0026/at%B (see also Figure 6-7 for a more extensive compilation of strength vs. lattice parameter). Boron is shown to be a much more potent strengthener than substitutional solutes such as V, Cr, Ti, Fe and Si (see Figure 6-1). Recent work by Baker, Huang and Schulson (1988) indicates a strengthening effect of 387 MPa/at%B, somewhat larger than values previously reported by Liu, White and Horton (1985) and by Huang et al. (1984). The more recent value was corrected for the strengthening effects of grain boundaries, and therefore may be taken as most nearly correct.

Cooling rate also may influence lattice parameter. Lattice parameters measured by different workers are, therefore, difficult to compare.

2.4. Elastic Constants and Specific Heat

The elastic constants of Ni_3Al at room temperature have been tabulated by Yoo (1987) and compared to those of other $L1_2$ and B2

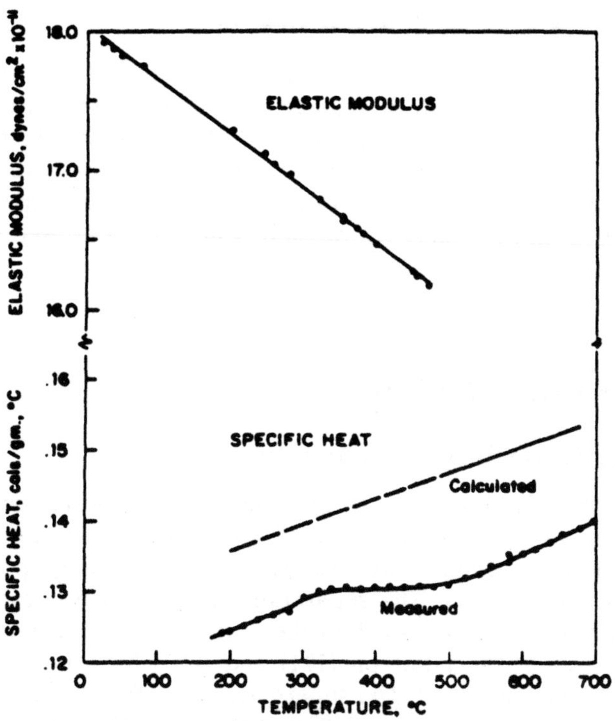

Figure 6-2. Elastic modulus and specific heat of Ni_3Al as a function of temperature (Davies and Stoloff 1965).

compounds. The values of C_{11}, C_{12}, and C_{44} are 2.23, 1.48, and 1.25×10^{11} N/m^2, respectively. Slightly different values were later calculated by Fu (1991). Accordingly, Zener's ratio of shear anisotropy, $A = 2C_{44}/[C_{11} - C_{12}]$, is 3.34 for Ni_3Al, compared to 2.53 for Ni_3Fe and 2.60 for Cu_3Au. In fact, the shear anistropy of Ni_3Al closely resembles that of the B2-type superlattice NiAl, 3.28.

Young's modulus of cast, polycrystalline Ni_3Al between 25° C and 450° C has been measured by a high frequency resonance technique by Davies and Stoloff (1965) (see Figure 6-2). The modulus at room temperature is about the same as that of nickel but the rate of decrease of *E* with temperature is about half that of nickel. Also shown in Figure 6-2 are measured and calculated specific heat data. Note that although there is a plateau in measured specific heat between 300° C and 500° C, there is no comparable anomaly in the modulus data.

2.5. Other Physical Properties

Williams et al. (1985) have reported data for thermal expansion coefficients, electrical and thermal resistivity, and Seebeck coefficients as functions of temperature in high purity Ni-24 at%Al and Ni-25 at%Al (see Table 6-2). Thermal expansion coefficient, α, is not sensitive to stoichiometry. However, earlier work on Ni-23 at%Al indicates an upswing in α at temperatures above 1000 K, possibly due to the onset of sublattice disorder (Stockinger and Neumann 1970).

3. CRYSTAL DEFECTS

3.1. Superlattice Dislocations

Slip in Ni_3Al occurs by the motion of superlattice dislocations comprising two unit dislocations of type $1/2\ \langle 110 \rangle$ on $\{111\}$ at low temperatures. As temperature increases, and depending upon strain rate, $\langle 110 \rangle\ \{100\}$ slip occurs. The separation of the unit dislocations on each plane is inversely proportional to the antiphase boundary (APB) energy. Several values of the latter on $\{111\}$ and $\{001\}$ have been reported (see section 2.1 and Table 6-1). The superpartials are further able to dissociate, and the total energy of a superlattice dislocation, therefore, depends both on the length of the total Burgers vector and on the energies of any faults present. A detailed discussion of fault types on $\{111\}$ in $L1_2$ superlattices may be found elsewhere (Pope and Ezz 1984) so that only the major features of each are described below:

1. Antiphase boundaries. These result from disturbance of unlike nearest neighbors in the slip plane and are relatively high-energy faults.
2. Superlattice intrinsic stacking faults (SISF). These result in a change in stacking sequence but with no nearest neighbor violations. Therefore, these are expected to be low-energy faults, with an estimate of energy as 5–15 mJ/m^2 (Douin, Veyssiére, and Beauchamp 1986).
3. Complex stacking faults (CSF) are produced by shear of type $1/6\ \langle \bar{1}\bar{1}2 \rangle$. This fault vector in $L1_2$ is identical to a stacking fault in the f.c.c. structure, and is expected to have higher energy than APB's. A recent calculation provides a CSF energy of 225 mJ/m^2 (Fu, Ye, and Yoo 1992).
4. Superlattice extrinsic stacking faults (SESF). This fault, similar to the SISF, results in no first nearest neighbor violations, but does change the stacking sequence. Therefore, SESF and SISF energies are expected to be similar. A value of SISF = 40 mJ/m^2 has been reported for Ni_3Al (Fu et al. 1992).

Table 6-2 Smoothed Physical Property Values for Ni-24 and 25 at% Al

	$Ni_{0.75}Al_{0.25}$				$Ni_{0.76}Al_{0.24}$			
Temperature (K)	Electrical Resistivity ($10^{-8}\Omega m$)	Thermal Conductivity (W/m•K)	Absolute Seebeck Coefficient ($\mu V/K$)	Thermal Expansion Coefficient[a] ($10^{-6}K^{-1}$)	Electrical Resistivity ($10^{-8}\Omega m$)	Thermal Conductivity (W/m•K)	Absolute Seebeck Coefficient ($\mu V/K$)	Thermal Expansion Coefficient[a] ($10^{-6}K^{-1}$)
4.2	1.54	—	—	—	9.05	—	—	—
300	32.59	28.85	+5.4	12.5	51.3	21.37	+9.9	12.5
400	39.00	33.72	+4.1	13.0	58.89	24.21	+10.8	13.2
500	44.75	35.13	+1.8	13.5	64.17	26.35	+7.9	13.8
600	50.13	36.18	−0.4	14.0	68.60	28.30	+5.1	14.3
700	55.27	37.07	−2.4	14.6	72.60	30.14	+2.3	14.8
800	60.54	37.27	−4.3	15.2	76.52	31.82	−0.4	15.4
900	66.50	36.86	−5.5	16.0	80.41	33.13	−2.8	16.1
1000	73.02	36.12	−6.3	16.8	84.24	32.70	−4.8	16.8

[a] $(1/L_{300})dL/dT$

Source: Williams et al. 1985

Complex faults have recently been seen by high resolution transmission microscopy (Baluc, Thaler and Mills 1991). SISF are readily seen in Ni_3Al polycrystals deformed in compression between room temperature and 800° C (Veyssiére, Douin, and Beauchamp 1985). Weak beam analysis indicated that the SISF originated at a screw partial with a Burgers vector of $1/2\ \langle 110 \rangle$.

Takeuchi et al. (1973) have pointed out that SISF's do not contribute markedly to the plastic strain during deformation. However, they may act as obstacles to superlattice dislocation motion. Specifically, since their presence depends upon orientation, they may contribute to the high orientation-dependent hardening rates measured in Ni_3Al alloys.

Other complex dislocation dissociations are possible, but only few appear to have important effects on flow behavior of $L1_2$ alloys. For example, when the APB energy on {111} is high, as in Ni_3Al, dissociation by APB

$$[\bar{1}01] \rightarrow [\bar{1}01] + \tfrac{1}{2}[\bar{1}01] \tag{6.1}$$

or dissociation by SISF

$$[\bar{1}01] \rightarrow \tfrac{1}{3}[\bar{2}11] + \tfrac{1}{3}[\bar{1}\bar{1}2] \tag{6.2}$$

is preferred. Although there is no direct evidence of the lattice dissociation, computer simulations of core structures suggest that dissociation by equation 6.2 should occur when APB energy on {111} is high and when the SISF energy is low.

Pair-potentials have been utilized to calculate the core structure of 1/2 [101] superpartials bounding an APB on (111). These calculations demonstrate that the core is confined to the slip plane (Vitek and Pope 1985). The lattice friction (Peierls stress) of this dislocation is very low. It is concluded that the splitting of superpartials on {010} is always energetically favored, but dislocations dissociated in this way are sessile. Dislocations split on {111} on the other hand, are very mobile, but are in a state of higher energy. Thermal activation is required to produce the lower energy configuration. Other calculations of dislocation core structures in $L1_2$ alloys also have been published by this group (Yamaguchi et al. 1982).

3.2. Antisite Defects

In highly ordered intermetallics with AB stoichiometry (e.g., Ni, Al, Co, Al) constitutional vacancies are produced in off-stoichiometric alloys.

However, no such defects have been observed in Ni_3Al and other $L1_2$ alloys. The significance of this observation is that hardening due to changes in aluminum content or the addition of ternary elements arises from intrinsic factors such as size and modulus misfit.

4. ALLOYING

4.1. Solid Solutions

*4.1.1. **Interstitial**.* The principal interstitial solutes that have been studied in Ni_3Al are boron and carbon. It is noteworthy that alloys with up to at least 2000 ppm by wt of boron are single phase; no evidence has been found by TEM for precipitation of borides on grain boundaries at lower nominal boron levels (e.g., 1000 wppm), even though the actual boron concentration at the boundaries must be much higher (Horton 1987). Auger results suggest, in fact, that the boron level at boundaries is of the order of 10 at% in Ni-24 at%Al, doped with 0.05 wt%B. The solubility limit of boron in Ni_3Al is estimated to be 0.3 ± 0.05 wt% (Liu et al. 1985). The lattice strain $1/a\ da/dc$ for boron is 0.13 in bulk samples (Liu et al. 1985), but is determined to be 0.26 in rapidly solidified Ni_3Al (Baker et al. 1988).

Horton and Miller (1987) have utilized atom probe field-ion microscopy to investigate the distribution of boron in rapidly solidified Ni-24 at%Al. Boron was observed to segregate to both grain boundaries and APB's. In addition, a very thin, nonuniformly distributed boron-enriched phase was detected in most grain boundaries. This phase could not be identified, but is possibly disordered f.c.c. with a high aluminum content. Horton and Miller suggested that the necessity for a critical level of about 0.1 at%B to ductilize Ni_3Al might be related to the coverage of grain boundaries by this phase.

Other atom probe results have shown no significant boron clustering in the matrix of a cast Ni-24 at%Al-0.24 at%B alloy (Brenner, Sieloff, and Burke 1986). However, a low degree of boron enrichment at twin boundaries was observed (Horton and Miller 1986). The reason for such discrepancies between groups using the same technique is unknown.

*4.1.2. **Substitutional**.* Many elements dissolve substitutionally in Ni_3Al. A summary of solutes, and site occupancy appears in Table 6-3 (Ochiai, Oya, and Suzuki 1984). Note that the largest number of solutes replace aluminum atoms, but that three of the most extensively studied: iron, hafnium and chromium, reportedly replace both aluminum and

Table 6-3 Substitutional Solutes in Ni_3Al

Ni sites	Al sites	Both sites
Co	Si Zn	Fe
Cu	Ti Zr	Cr
Pd	Mn	Mo
Pt	V	W
Sc	Ge	Hf[a]
	Ga	
	Nb	
	Sb	
	Sn	
	Ta	
	V	

[a]Conflict between Ochiai et al., and Bohn et al.
Source: Ochai et al. 1984 ()Reprinted from *Acta Metallurgica*, Volume 32, pages 289–298, copyright 1984, with kind permission from Elsevier Science Ltd, The Boulevard, Langford Lane, Kidlington OX5 1GB, UK)

nickel. However, there is a disagreement in the literature as to the site occupancy of hafnium, as well as molybdenum and tungsten. For example, X-ray diffraction data indicate that chromium, molybdenum, and tungsten substitute on both sites (Karg, Fornwalt and Kriege 1971), but chemical analysis of extracted precipitates in nickel-base alloys suggest that molybdenum and tungsten substitute mainly for aluminum (Kriege and Sullivan 1968). Additional x-ray work (Aigeltinger et al. 1978) confirms that molybdenum substitutes on aluminum sites. While Ochai et al. (1984) suggest that hafnium can be found on both aluminum and nickel sites, atom probe experiments by Miller and Horton (1987) show a strong preference for aluminum sites. Bentley (1986) has reached the same conclusion from electron channeling experiments. Brenner and co-workers (1986), also using the atom probe, concluded that hafnium is on nickel sites; also in agreement with this position are the results of Rutherford backscattering experiments. There is no apparent reason for this very wide discrepancy in results.

Continuous solid solutions are formed by Ni_3Al with Ni_3Ga, Ni_3Ge and Ni_3Si. Also, according to Masahashi et al. (1986), Ni_3Al and Ni_3Mn form a continuous series of solid solutions at low temperatures. X-ray analysis did not reveal any second phase in Ni-Al-Mn alloys after homogenization at 1323 K and furnace cooling. Dimiduk et al. (1987), on the other hand, report the presence of two phases in arc-melted Ni_3Al containing more than 4 at%Mn, although the same homogenization treatments were used as in the work of Masahashi et al. (1986) and Takasugi et al. (1985).

Iron additions to Ni_3Al have been studied extensively because of several important advantages:

1. solid solution strengthening
2. lower material costs
3. improved hot ductility and fabricability via the formation of an ordered B2 phase.

The solubility limit of iron in Ni_3Al is reported by Dimiduk, Weddington, and Lipsitt (1987) to be near 4 at%, while Horton, Liu, and Santella (1987) place the solubility limit in excess of 10 at%. Further, Horton and co-workers point out that the aluminum/nickel ratio affects the solubility limit, since second phase particles are seen in Ni-20 at%Al-15Fe but not in Ni-17.5 at%Al-15Fe. The two-phase structure consists of a fine distribution of γ in the γ' matrix. A complex cellular structure of small islands of ordered γ' in a web of disordered γ was seen in a 17.5 at%Al-15 at%Fe alloy tensile tested at 600° C.

Titanium additions to Ni_3Al have been widely studied because of the extensive use of titanium in γ/γ' nickel-base superalloys. Lopez and Hancock (1970) suggested that the stacking fault energy of Ni_3Al is increased by the addition of Ti. Others have concluded that titanium raises the APB energy of Ni_3Al. Titanium is known to strengthen Ni_3Al, but there is disagreement as to the effect on peak temperature; Lopez and Hancock report a peak at 400° C, and Thornton et al. (1970) a peak at above 800° C for approximately the same titanium content, 10.5%.

Alloying elements may be expected to influence differently the critical resolved shear stress (CRSS) for slip on different systems. Shah and Duhl (1987) have reported that ternary solutes such as tantalum tend to increase CRSS for octahedral slip, with little influence on CRSS for cube slip. This has the effect of reducing the ductility of cube-oriented single crystals, for which {111} slip is favored.

4.2. Two Phase Alloys

Research efforts on two-phase Ni_3Al-base alloys to be assessed in this section comprise three categories: multiphase alloys, dispersion-strengthened alloys, and composites. The multiphase alloys have reached near commercial status and, therefore, are described in detail in the next section.

4.2.1. Alloy Design of Polycrystalline Ni_3Al Alloys. Mechanical and metallurgical properties of Ni_3Al can be improved by alloying additions.

Recent efforts on alloy design have led to the development of Ni_3Al-base alloys with the following composition range for structural use at elevated temperatures in hostile environments (Liu et al. 1986; Liu 1992):

$$\text{Ni-14 to 18Al-6 to 9Cr-1 to 4Mo-0.1 to 1.5Zr/Hf-0.01 to 0.20B (at\%)}.$$

In these aluminide alloys, chromium at a level of 6% to 9% is added for reducing environmental embrittlement in oxidizing environments at elevated temperatures (section 7.2.2). Zirconium or hafnium additions are most effective in improving the high-temperature strength via solid-solution hardening effects. Molybdenum additions are added for improving strength at ambient and elevated temperatures. Microalloying with boron reduces moisture-induced hydrogen embrittlement and enhances grain-boundary cohesive strength, resulting in sharply increasing the ductility at ambient temperatures. In some cases, moderate amounts (< 20%) of cobalt and iron are added to replace nickel and aluminum and nickel, respectively, in order to further gain in hardening and corrosion resistance. The alloys with optimum properties usually contain 5–15 vol% of the disordered γ phase, which has the beneficial effect of reducing environmental embrittlement in oxidizing atmospheres and improving creep properties at elevated temperatures.

Cast aluminide alloys usually possess a coarse-grain structure that lowers the yield strength at ambient temperatures. The strength can be effectively increased by alloying with molybdenum via solid-solution hardening (Liu 1992). Figure 6-3 compares the tensile properties of a cast Ni_3Al alloy (Ni-16.0Al-8.0Cr-1.7Mo-0.5Zr-0.03, at%) with commercial alloys, such as cast superalloy IN-713C commonly used by high-temperature industry. The strength of the advanced aluminide alloy is comparable to that of IN-713C at room temperature, but it is higher at elevated temperatures because of the significant increase in yield strength with temperature. The aluminide alloys prepared by investment casting showed an excellent fatigue resistance. Limited high-cycle fatigue tests show that the 650° C high-cycle fatigue life of an aluminide alloy (IC-221M: Ni-15.9Al-8.0Cr-0.8Mo-1.0Zr-0.03B, at%) is longer than that of IN-713C by more than two orders of magnitude (Gieseke and Sikka 1992). The cast aluminide alloys usually have a tensile elongation of 10–30% at room and elevated temperatures.

Aluminide alloys with the compositions Ni-16 $\pm$ 2%Al-6 $\pm$ 2%Cr-0.6 $\pm$ 0.5%Zr-0.10 $\pm$ 0.09%B (at%) can be fabricated by cold rolling or hot extrusion. The wrought alloys with a fine-grain structure have excellent strength at ambient temperatures (σ_y > 600 MPa), but lower creep resistance because of grain-boundary sliding (Sikka 1989) at elevated tempera-

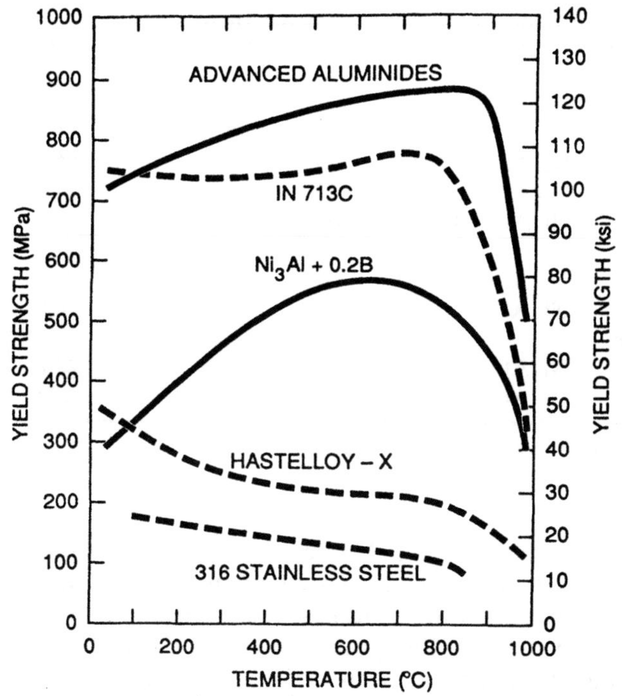

Figure 6-3. Comparison of yield strength of Ni_3Al alloys (advanced aluminide alloy and Ni_3Al + 0.2B) with commercial alloys (IN-713C, Hastelloy X and 316 stainless steel) (Liu et al. 1989; reprinted from *Scripta Metallurgica*, Volume 23, page 877, copyright 1989, with kind permission from Elsevier Science Ltd, The Boulevard, Langford Lane, Kidlington OX5 1GB, UK)

tures (> 700° C). The alloys generally showed good fatigue crack-growth resistance (see section 9.1) (Matuszyk et al. 1990). The fatigue and crack-growth behavior of Ni_3Al alloys are sensitive to test environment (e.g., air vs. vacuum); nevertheless, the aluminide alloys containing chromium show better fatigue-crack resistance when tested in air at elevated temperatures.

4.2.2. Dispersion Strengthened Alloys. Dispersoids of TiB_2, Al_2O_3, ThO_2, and Y_2O_3 have been prepared in an Ni_3Al matrix, but little is known of their properties except for some microhardness data. Baker et al. (1984) studied the influence of titanium and boron additions to rapidly solidified powders of Ni_3Al. The hardness increased from 363 to 413 VHN with TiB_2, but no particles were observed by optical metallography. Mechanically alloyed particles of Ni_3Al with 1% Al_2O_3, hot pressed at 1300° C for 15 min at 28 MPa, resulted in fully dense material with a

Table 6-4 Properties of Dispersion Strengthened Alloys

Processing	Alloy	Temperature C	σ_{ys}(MPa)	σ_{UTS}(Mpa)	$\%\varepsilon_f$
HIP'ed	$Ni_3Al + B/Al_2O_3$	25	474	548	1.2
HIP'ed	$Ni_3Al + B/Y_2O_3$	25	391	464	1.8
HIP'ed	$Ni_3Al + B$	25	265	722	9.9
Reactive Sintered	$Ni_3Al + B$	25	335	729	11.4
HIP'ed	Ni_3Al	25		363	0.0
Reactive Sintered	Ni_3Al	25	—	230	1.0
HIP'ed	$Ni_3Al + B/Al_2O_3$	600	544[a]	617[a]	0.4[a]

[a]Data obtained in 2×10^{-5} torr vacuum.

Source: Bose et al. 1987

uniform dispersion of Al_2O_3. Hardness was approximately twice that of Ni_3Al with no Al_2O_3, but processed identically. However, the Al_2O_3 particles resulted in lower hardness than unalloyed Ni_3Al at 1000° C, probably due to the very fine grain size in the dispersion hardened alloy. Mechanical alloying also has been carried out with 0.5-3.54%Al_2O_3, Y_2O_3 or ThO_2 dispersions in Ni_3Al base alloy powder containing 0.5 at%Hf and 500 wppm B. However, only small hardness increases were noted (Donnelly and Koch 1983).

Later work on $Ni_3Al \pm B$ with Al_2O_3 or Y_2O_3 dispersoids, reactively HIPed (elemental powders reacted under hydrostatic pressure) revealed considerable strengthening at 25° C: 265 MPa yield stress for $Ni_3Al + B$, 391 MPa for $Ni_3Al + B/Y_2O_3$ and 474 MPa for $Ni_3Al + B/Al_2O_3$ (see Table 6-4). As is typical of dispersion strengthened alloys, however, ductility was severely reduced (to less than 2%) for both dispersion-strengthened alloys, compared to 9.9% for the single-phase alloy. Similarly, dispersion-strengthened alloys prepared by mechanical alloying also are strong but brittle (Koch 1988). It has been concluded that the major strengthening effect of oxides in Ni_3Al + B + Hf is in their control of grain size during HIPing. The dispersoids provide a small (~ 13%) additional strengthening contribution by the Orowan mechanism.

4.2.3. Composites. Artificial or mechanically incorporated composites based upon Ni_3Al have received attention in several laboratories, but to date the results have been unsatisfactory in improving strength and/or toughness. Since most composites have been prepared by powder techniques, the reader should consult the chapter by R. M. German for details.

5. YIELDING

5.1. Stoichiometry Effects

Small changes in composition often have a profound effect on the mechanical behavior of intermetallic compounds. Guard and Westbrook (1959) were the first to show that at room temperature and at −196° C the hardness of Ni_3Al increased substantially as aluminum content varied from 25 at%, with a higher rate of hardening on the aluminum-rich side. At 800° C, on the other hand, off-stoichiometric compositions displayed lower hardness than stoichiometric Ni_3Al.

Recently, the influence of aluminum content on yield strength has been reported (Noguchi, Oya, and Suzuki 1981). Again, at 25° C hardening is more rapid on the aluminum-rich side (see Figure 6-4). Note similar behavior in Ni_3Ge. In general, an excess of the more electropositive element leads to more rapid hardening. However, at temperatures above 600 K there is a monotonic increase in flow stress with aluminum content

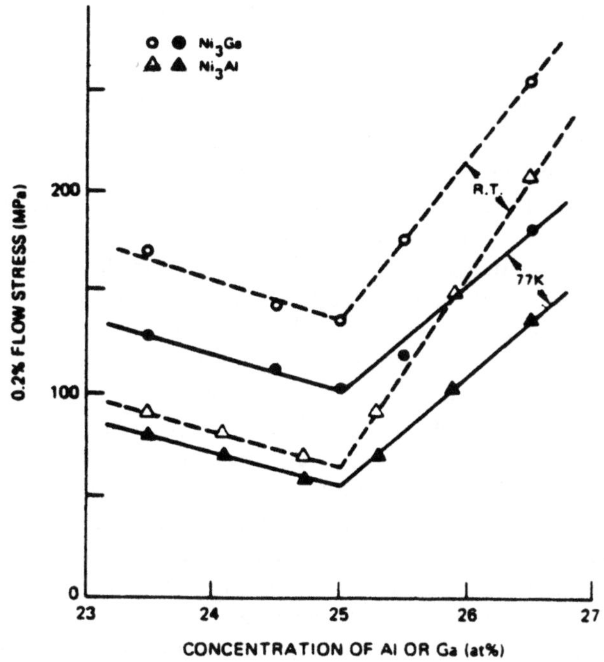

Figure 6-4. Effect of stoichiometry on flow stress of Ni_3Al and Ni_3Ga at 77K and room temperature (Noguchi et al. 1981).

in the range 23.5 to 26.5 at%. This is in sharp contrast to earlier reports of a reduction in flow stress of off-stoichiometric alloys at elevated temperatures (Guard and Westbrook 1959). In the more recent work, the temperature of the flow stress peak, T_p, seems to be unaffected by stoichiometry (Noguchi et al. 1981).

The mechanism for hardening of off-stoichiometric alloys at low temperatures is still not well established, in view of continuing uncertainty about defect populations on either side of stoichiometry, as discussed in section 4.1. As pointed out by Noguchi et al. (1981), the validity of various defect-hardening models should be assessed at very low temperatures, where contributions to strengthening from thermally activated cross slip are reduced. Solution hardening by substitutional atoms on wrong sites seems to be the most likely strengthening mechanism at low temperatures.

5.2. Temperature Dependence and Asymmetry

Certainly the most interesting feature of the plastic deformation of Ni_3Al (and many other $L1_2$ superlattices) is the marked increase in yield and flow stresses with increasing temperature.

The flow stress increase can be detected only at strains higher than those in the microstrain region (Thornton, Davies and Johnston 1970). Further, flow stress increases with temperature are observed in both polycrystals and single crystals. The position and height of the flow stress peak is a function of single crystal orientation, solute additions and, for polycrystals, changes in grain size. In fact, very fine-grained materials (2.9–9.5 μm) show no peak in flow stress (Schulson, Baker, and Frost 1987). Finally, the critical resolved shear stress depends on the sense of the applied stress, being higher in tension for samples oriented near the $[001] - [\bar{1}11]$ side of the unit triangle (Pope and Ezz 1984).

The most widely accepted model for the flow stress peak is based upon an idea Kear and Wilsdorf (1962) put forward to explain the unusually high strain-hardening rate in ordered Cu_3Au. They suggested that screw dislocations of type $\frac{1}{2}[\bar{1}01]$ cross slip from {111} planes, where they are mobile, to {001} planes, where they are sessile. The driving force for cross slip was postulated to be a lower APB energy on {001} compared to {111} planes. More recently, Yoo (1987) has pointed out an additional driving force due to a torque arising from elastic anisotropy. Since cross slip is thermally activated, an increasing number of sessile Kear–Wilsdorf locks are formed as temperature increases from 25° C. While {111} is the primary slip plane at 25° C, cube slip also is observed at 400° C, and at 700° C slip is entirely on {001}. Therefore, massive cube slip renders the Kear–Wilsdorf barrier irrelevant at high temperatures.

Several modifications of the Kear-Wilsdorf model have been suggested in recent years in an effort to explain a significant anisotropy between tensile and compressive flow stresses and the lack of significant strain-rate effects below T_p, the peak temperature. These models, proposed by Takeuchi and Kuramoto (1973), Lall, Chin, and Pope (1979), Pope and Vitek (1987), Paidar, Pope and Vitek (1984), Hirsch (1991) and others are based upon consideration of core configurations of screw superlattice dislocations, and the occurrence of discrete pinning points along $\langle\bar{1}01\rangle$ dislocations moving along the primary {111} slip plane. The details of such models are beyond the scope of this chapter, especially as they are the subject of considerable controversy.

5.3. Solute Effects

With the exception of chromium, all substitutional solute additions to Ni_3Al raise the flow stress, as shown for niobium (Mishima et al. 1986) and silicon (Ochiai et al., 1986) additions in Figures 6.5a and b, respectively. In every case studied the temperature of the flow stress peak, T_p, is reduced; however, titanium additions to 8% (data not shown) seem to have no effect on T_p. In every case the flow stress can be analyzed in terms of the sum of two temperature dependent terms, σ_B and σ_H (see Figure 6-5(b)).

A summary of the hardening effects of numerous solutes at 77 K is shown in Figure 6-6a for transition metal elements and in Figure 6-6b for boron-supgroup elements (Mishima et al. 1986). Note that hafnium and zirconium are the most effective strengtheners among transition metal solutes, while tin, antimony, and indium are the most potent strengthening boron-subgroup elements (Mishima et al. 1986). The rate of hardening for all solutes is approximately linear with lattice parameter change da/dc, as shown in Figure 6-7 (Dimiduk and Rao 1991; Mishima et al. 1986). Other substitutional solutes that have been shown to strengthen polycrystalline Ni_3Al are iron, manganese, and beryllium (Masahashi, Takasugi, and Izumi 1986; Masahashi et al. 1988a). Curwick (1972) has shown that the CRSS for octahedral slip of Ni_3Al single crystals alloyed with molybdenum, titanium, niobium or tungsten was raised, with the effect dependent on temperature and concentration. Similar effects of niobium and tantalum have been shown for a wider range of specimen orientations (Lall et al. 1979; Ezz, Pope, and Paidar 1982). Later work by Heredia and Pope 1989 showed that for orientations in which the tension/compression asymmetry is zero, ternary additions of Hf, Ta or Nb decreased the temperature of the flow stress peak, T_p. Also, the magnitude of the CRSS at T_p tended to increase with increasing solute content and the rate at

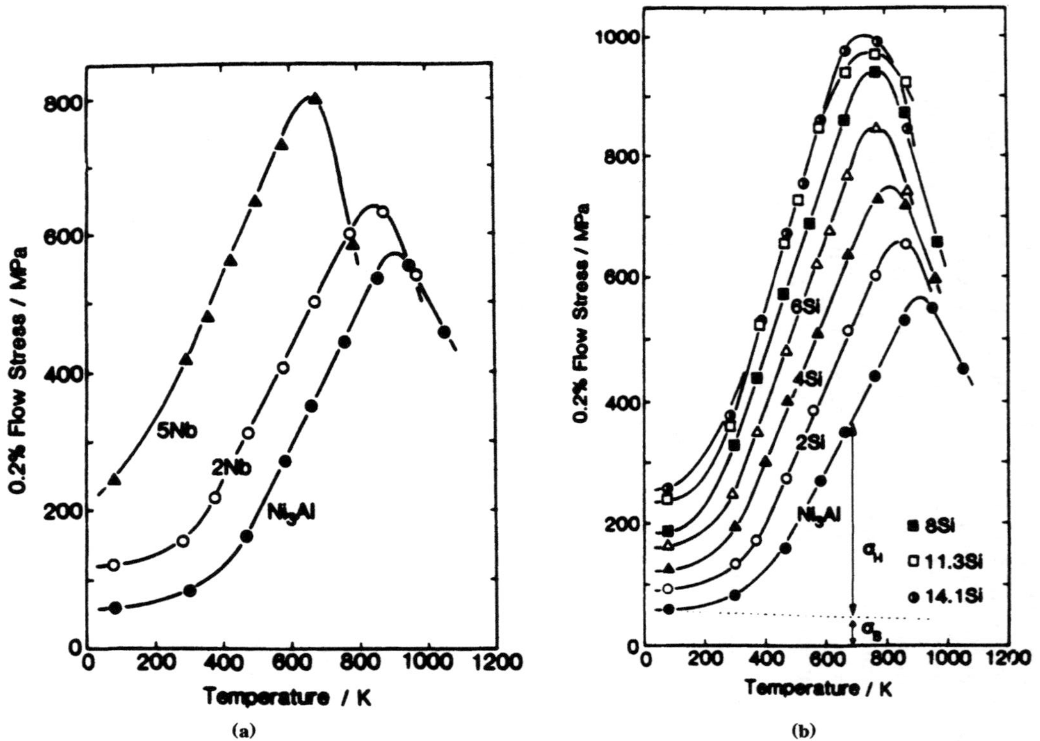

Figure 6-5. Temperature dependence of 0.2% flow stress in Ni_3Al with various solute additions (a) Nb (Mishima et al. 1986) and (b) (Ochiai et al. 1986). The flow stress can be regarded as the sum of two temperature dependent terms, σ_B and σ_H.

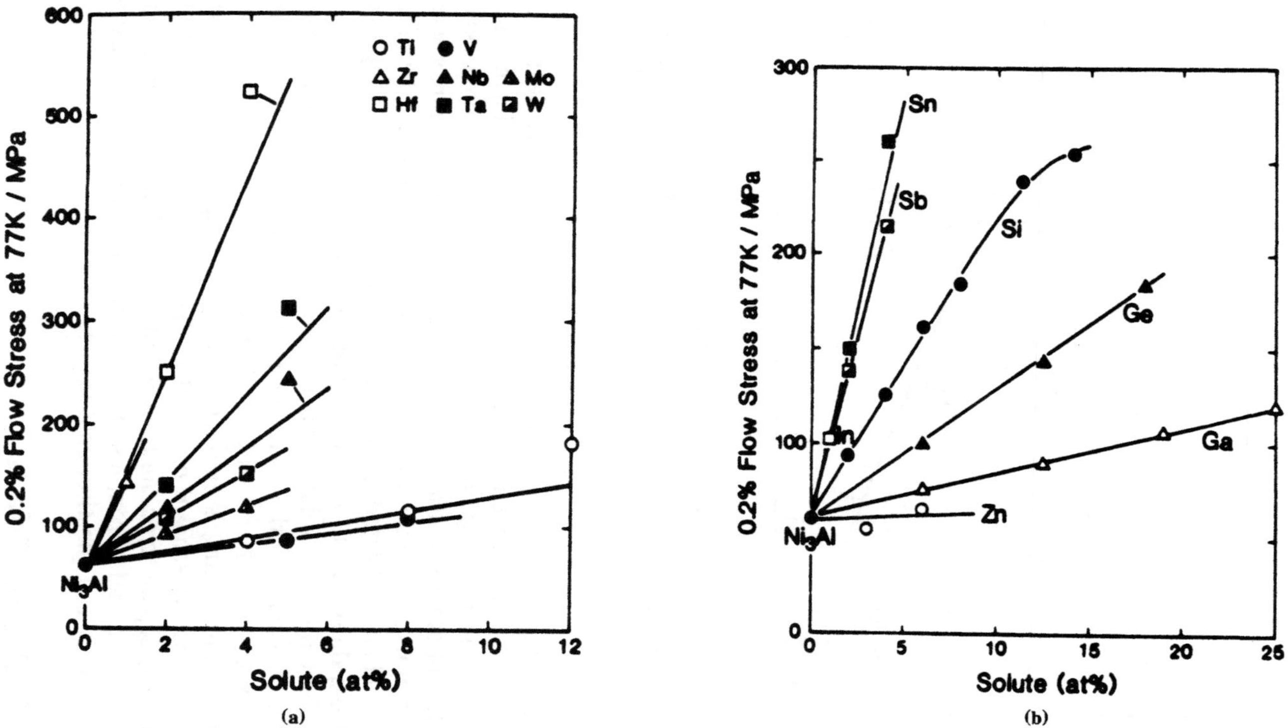

Figure 6-6. Relation between 0.2% flow stress at 77K and solute concentration in ternary Ni_3Al. (a) With addition of transition metal elements (Ochini et al. 1986), (b) with addition of B-subgroup elements (Mishima et al. 1986).

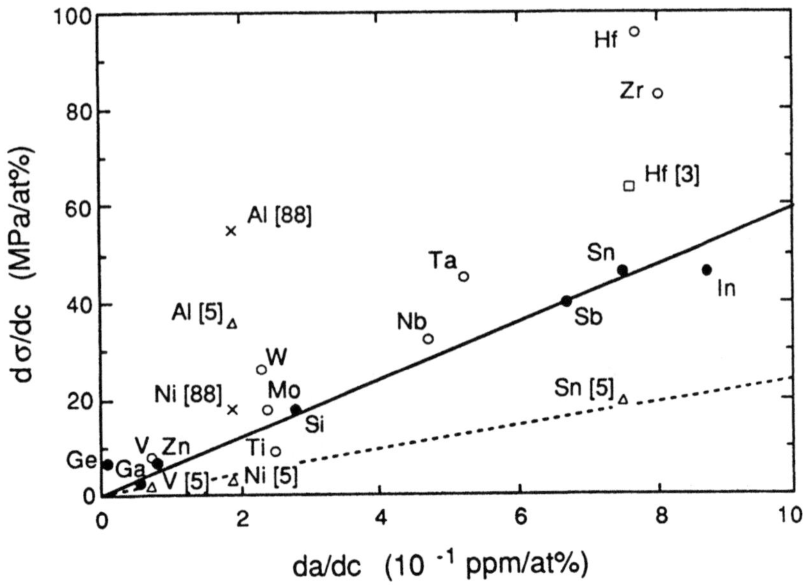

Figure 6-7. Normalized solute-strengthening data at 77K for select solutes in Ni_3Al alloys. Data from [3] and [5] for single crystals (Dimiduk and Rao 1991).

which the strength increased with temperature also increased with addition of solutes. Cube-oriented crystals were strengthened the most by solutes. Finally, ternary additions increased the tension/compression asymmetry for all other orientations. The magnitude of the asymmetry can only be a function of the ease of cross slip from {111} to {001}; therefore, these studies, when completed, should enable the effects of cross slip to be separated from those due to size and modulus misfit.

Most observations of solute hardening have been carried out at room temperature. A study of the hardening effects of boron, carbon and beryllium in Ni-24 at%Al and Ni-25 at%Al as a function of temperature revealed that strengthening is appreciable only at temperatures below the peak in flow stress (Masahashi et al. 1988). The peak occurred at 873 K, independent of dopant type and level. At 77 K, hardening was only effective for carbon and boron, less so for beryllium, which is not surprising, since the latter is a substitutional solute.

5.4. Grain Size Effects

5.4.1. Room Temperature. Grain size effects on yielding in Ni_3Al and Ni_3Al + B have been studied by several groups. Most of the data

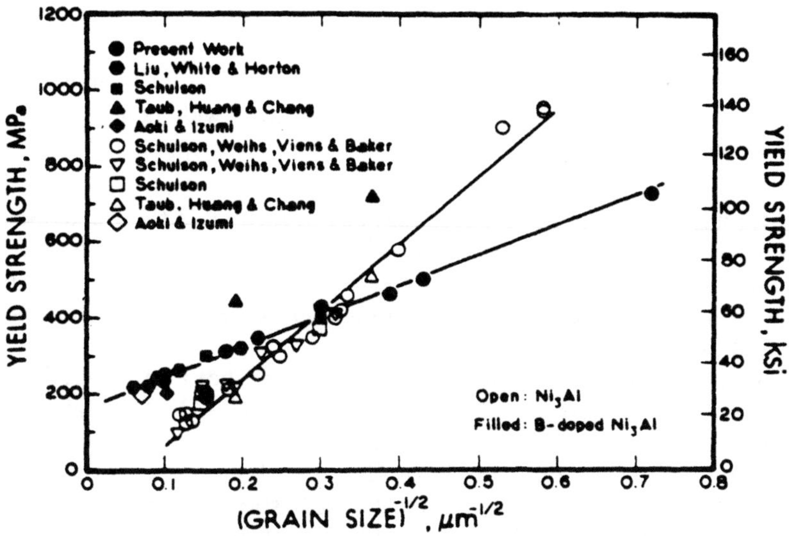

Figure 6-8. Comparison of Hall–Petch relationships for alloys with and without boron (Khadkikar et al. 1987).

reported to date have been obtained at room temperature. Hall–Petch behavior:

$$\sigma_y = \sigma_0 + k_y d^{-\frac{1}{2}} \tag{6.3}$$

is generally obeyed by Ni$_3$Al, and the effect of boron appears to be lowering of the Petch slope, k_y (see Figure 6-8) (Khadkikar, Vedaul, and Shabel 1987). A perplexing feature of the data in Figure 6-8 is that the points extrapolate to a negative intercept at infinite grain size. This clearly cannot be the case. However, as pointed out by Takeyama and Liu (1988a) care must be taken to separately plot results for material of even slightly differing composition (e.g., 24 vs. 25%Al or 0.2 vs. 0.4%B) because of the great sensitivity of Ni$_3$Al to compositional changes. Segregation of data by processing technique also is important, since as-cast material does not fall on the same curve as wrought or powder material (Takeyama and Liu 1988b). When all data are carefully segregated and scrutinized, the following conclusions may be reached (Takeyama and Liu 1988a; Kim et al. 1988):

1. Boron has little or no effect on the Petch slope for Al = 25 at%.
2. Boron reduces the Petch slope by about 25% for Al = 24 at%.

3. Boron increases σ_y; the effect is three times higher for stoichiometric Ni_3Al than for nickel-rich material.
4. The intercept stress σ_0 is negative only for powder-extruded Ni_3Al.
5. flow stresses also obey the Hall–Petch relation, with both σ_y and k_y increasing with strain.

Although these results suggest that boron facilitates the transmission of slip across grain boundaries, the mechanism for such slip accommodation remains unknown. However, it has been suggested that boron may cause a disordered layer to form at grain boundaries (Kung et al. 1991; Baker and Schulson 1989).

There is, in fact, some experimental evidence that a disordered layer exists at grain boundaries in hypostoichiometric Ni_3Al + B. Baker and Schulson (1989) reported that these layers are rather wide and present at many grain boundaries. Kung et al., on the other hand, reports that only a very narrow disordered region exists at a few specially oriented boundaries. While the presence of a disordered layer at grain boundaries would certainly affect the Hall–Petch slope, more research is needed to verify the existence of such layers.

Weihs et al. (1987) have suggested that boron facilitates the motion of dislocations in the grain boundary, but this alone cannot cause yielding. Bond, Robertson, and Birnbaum (1987) have observed by in situ TEM that, in the presence of boron, slip bands impinging on grain boundaries of Ni_3Al lead to plastic deformation instead of fracture near the boundaries; however, the experimental method does not allow measurement of stress levels, so the effects, if any, of boron on the stress to operate dislocation sources, T_c, is unknown. In the absence of direct evidence concerning the influence of boron on slip character, and grain-boundary degree of order, as well as the highly unlikely circumstance of a decrease of T_c by boron segregated to the boundaries, it must be concluded that the Hall–Petch analysis cannot provide a satisfactory explanation of the influence of boron on k_y.

5.4.2. Elevated Temperatures. The temperature at which the yield stress is a maximum in polycrystalline Ni_3Al is located between 600° C and 800° C. Two groups (Hanada, Watanabe, and Izumi 1986; Oya et al.1984) report that the peak temperature is independent of grain size, in the range 4–275 μm and of 12–70 μm, respectively. However, Weihs et al. (1987) report that the maximum stress can be entirely suppressed for material with $d = 2.9$ μm, due to grain boundary sliding and the predominance of cube slip. Takeyama and Liu (1988a) also report a lowering of the maximum temperature from 800° C to 600° C with decreasing grain size in the range 173 to 15.3 μm, consistent with the results of Weihs et al. At temperatures above 700° C, refining grain size weakens Ni_3Al + B, in contrast to low temperature observations.

6. STRAIN HARDENING

$L1_2$ superlattices exhibit high rates of strain hardening relative to disordered f.c.c. metals. Further, a peak in strain hardening rate occurs in $L1_2$ alloys (other than Ni_3Fe) at intermediate temperatures and with a similar orientation dependence to the flow stress peak. Therefore, it has generally been accepted that both the yield stress and strain hardening rate share the common Kear–Wilsdorf mechanism, cross slip of screw dislocations from the primary {111} planes to a plane of type {001}, which is expected to have lower antiphase boundary energy.

There is a transition from (111)[$\bar{1}$01] to (001)[$\bar{1}$10] slip which occurs in Ni_3Al at temperatures near T_p. This observation, together with a calculation by Flinn (1960) that the antiphase boundary energy in Ni_3Al is lower on (001) than on (111) and the Kear–Wilsdorf model of superlattice dislocation cross slip serve as the basis for most modern theories of the flow stress peak.

The following factors promote cross slip of screw dislocations from {111} to {001}:

1. low APB energies on {001} relative to {111} planes
2. high Schmid factors on {001} planes
3. large stress components acting on edge components of the Shockley partials so as to constrict the partials arising from elastic anisotropy
4. elastic anisotropy, giving rise to an interaction torque on the pair of superpartial screw dislocations.

The increased likelihood of cross slip with increasing temperature may result in enhanced strain hardening, either through an exhaustion mechanism (reduction in the number of mobile superlattice dislocations with increased cross slip), or by the formation of APB tubes. Hazzledine and Hirsch (1987) have suggested that tubes are created by cross slip of screw dislocations in the following manner: As a primary screw dislocation moves from its source, jogs are created from intersections with dislocations on other planes. However, cross slip of screw dislocations sometimes occurs in such a way that the following superpartial does not trace exactly the same path as the leading superpartial. As a result, tubes are formed on the jogged dislocation (see Figure 6-9), causing additional work hardening.

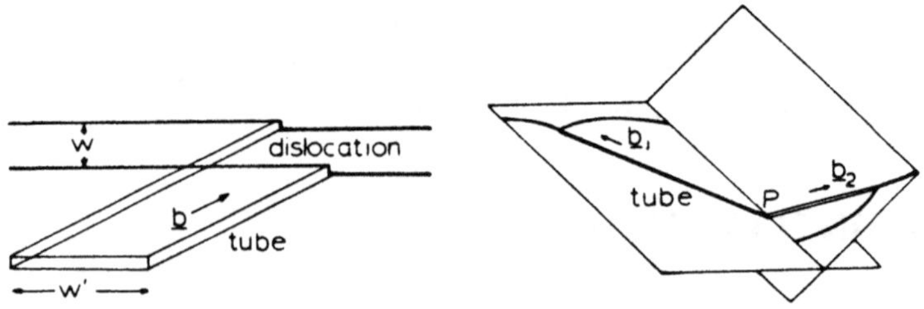

(a) (b)

Figure 6-9. (a) Tube formation at misaligned jogs. (b) Two moving dislocations cutting at P create a single bent tube (Hazzledine and Hirsch 1987).

The size and density of the tubes attached to edge dislocations depends upon the ease of cross slip of screw superpartials, thereby leading to both temperature and orientation-dependent work hardening. Tubes have indeed been observed in Ni_3Al crystals compressed by 4% at 400° C. However, it remains to be demonstrated that APB tubes actually contribute to work hardening.

7. FRACTURE

7.1. Low Temperature Behavior

7.1.1. Intrinsic Grain-Boundary Brittleness? Ni_3Al exhibits {111} ⟨110⟩ slip and has sufficient slip systems for extensive plastic deformation. Single crystals of Ni_3Al are highly ductile, whereas polycrystals are brittle at ambient temperatures. The brittleness originates at grain boundaries. In most metals and alloys, brittle intergranular fracture is associated with strong segregation of harmful impurities, such as sulfur, phosphorus, and oxygen, to grain boundaries (Stein and Heldt 1977), causing embrittlement. In earlier work, trace impurities were also suspected to be a possible cause for embrittlement of grain boundaries in Ni_3Al and other intermetallics. For example, sulfur was found on grain boundaries in polycrystalline Ni_3Al (White et al. 1984).

However, experimental studies of grain-boundary chemistry in sufficiently pure Ni_3Al and other intermetallics indicate that the grain boundaries are basically clean and free from detectable impurities (Liu et al. 1985; Takasugi et al. 1985) by Auger electron spectroscopy. As shown in Figure 6-10, all Auger peaks are associated with nickel and aluminum, and no impurities are detected at Ni_3Al grain boundaries. The

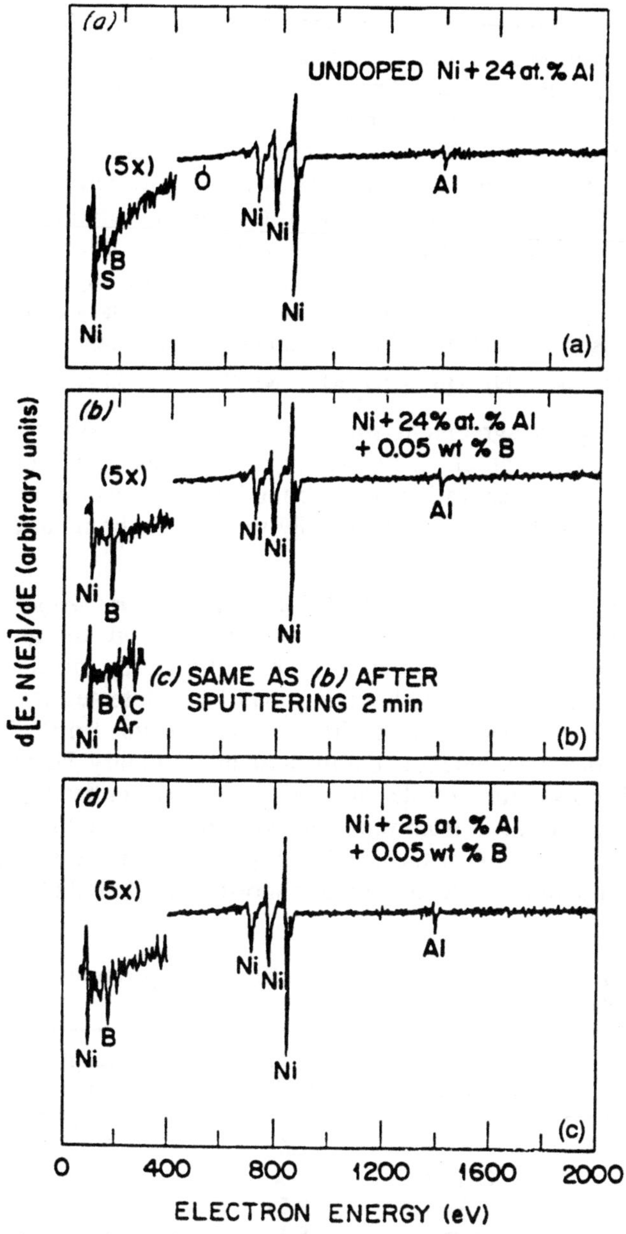

Figure 6-10. Auger spectra from intergranularly fractured region in Ni_3Al specimens (a) undoped Ni-24 at%Al, (b) Ni-24 at%Al doped with 500 wt ppm B, (c) same as (b) after sputtering from 2 min, and (d) Ni-25 at %Al doped with 500 wt ppm B (Reprinted with permission from *Acta Metallurgica*, Volume 33, C. T. Liu, C. L. White, and J. A. Horton, pages 213–229, 1985, Elsevier Science Ltd., Pergamon Imprint, Oxford, England).

aluminide, nevertheless, fractures intergranularly, implying that the grain boundaries may be intrinsically brittle. Similar observations were obtained in other intermetallics, including Ni_3Si, Bi_3Ga,Ni_3Ge, NiAl, and FeAl.

Considerable efforts had been devoted to understanding the "intrinsic" grain-boundary brittleness in ordered intermetallics. A general finding is that the grain-boundary brittleness is associated with ordering energy, electronegativity difference, valency difference, and atomic size difference between A and B atoms constituting binary A_3B ordered intermetallics (Takasugi et al. 1985; Taub 1986, 1987; Farkas and Rangarajan 1987). This correlation appears to correctly rank the grain-boundary cohesive strength of nickel-based $L1_2$ ordered intermetallics in the order

$$Ni_3Fe > Ni_3Mn > Ni_3Al > Ni_3Ga > Ni_3Si > Ni_3Ge,$$

which is in good agreement with experimental observations. Atomistic-simulation (Vitek et al. 1989; Kruisman, Vitalek and Deltosson 1989) calculations and dislocation/grain-boundary interaction analyses (King and Yoo 1987) all support intrinsically weak grain boundaries in strongly ordered intermetallics, such as Ni_3Al.

7.1.2. Environmental Embrittlement as a Major Cause of Grain-Boundary Brittleness. In the foregoing section, the grain boundary in binary Ni_3Al was considered to be intrinsically brittle because no detectable impurities at the boundary had been detected by Auger analyses. In 1992, Liu had made a first attempt to link brittle intergranular fracture with environmental embrittlement in binary Ni_3Si and Ni_3Al. Table 6-5 shows the tensile properties of recrystallized polycrystalline Ni_3Al alloys produced from repeated cold forging and 1000° C annealing. The tensile elongation of the Ni_3Al alloys is sensitive to test environment at room temperature. The aluminides showed only 2.5–2.6% elongation in

Table 6-5 Effect of Test Environment on Room-Temperature Tensile Properties of Ni_3Al Alloys Tested at $3.3 \times 10^{-3} s^{-1}$

Alloy composition (%)	Test environment	Tensile ductility (%)	Yield strength (MPa)	Ultimate tensile strength (MPa)
Ni-24.0Al	Air	2.6	280	333
Ni-24.0Al	Oxygen	7.2	279	439
Ni-23.5Al	Air	2.5	193	230
Ni-23.5Al	Oxygen	8.2	194	351

Source: Liu 1992 (Reprinted from *Scripta Metallurgica*, Volume 27, page 26, copyright 1992, with kind permission from Elsevier Science Ltd, The Boulevard, Langford Lane, Kidlington OX5 1GB, UK)

air but 7.2–8.2% in dry oxygen, an increase in ductility by a factor of ~ 3. These results clearly demonstrate that binary Ni_3Al alloys are prone to environmental embrittlement at room temperature. Thus, an extrinsic factor —environmental embrittlement—is a major cause of low ductility and brittle intergranular fracture in binary Ni_3Al (Liu, 1992), as in the case of many other intermetallics (see chapter on Environmental Embrittlement by N. S. Stoloff).

Environmental embrittlement in aluminides as well as other ordered intermetallics containing reactive elements (such as aluminum, silicon, vanadium, and titanium) is explained by the following chemical reaction (Liu, Lee, and McKamey 1989):

$$2Al + 3H_2O \rightarrow Al_2O_3 + 6H \tag{6.4}$$

Aluminum atoms in the aluminides react with moisture in air, resulting in the generation of atomic hydrogen that penetrates into crack tips and causes brittle grain-boundary crack propagation and premature failure. This embrittlement is similar to hydrogen embrittlement, except that hydrogen here is generated by moisture in air, rather than hydrogen sources such as hydrogen gas or cathodically charged hydrogen. Consistent with equation 6.4, the best ductility is generally obtained in dry oxygen environments. This can be rationalized by considering that the reaction of aluminum atoms with oxygen

$$4Al + 3O_2 \rightarrow 2Al_2O_3 \tag{6.5}$$

competes with the moisture reaction, thereby reducing the generation of atomic hydrogen from equation 6.5. Both specimens tested in air and oxygen showed mainly grain-boundary fracture, with more propensity for transgranular fracture in the specimens tested in oxygen. Note that environmental embrittlement appears not to be the only cause of grain-boundary brittleness in Ni_3Al, because the elimination of the environmental effect by testing in dry oxygen does not lead to extensive ductility and complete suppression of intergranular fracture in Ni_3Al (Liu 1992).

The environmental effect has been also observed in Ni_3Al alloys containing ternary alloy additions. Takasugi et al. (1986) first reported an environmental effect on ductility reduction in beryllium-doped Ni_3Al but not in binary Ni_3Al (24%Al). The Ni_3Al alloy doped with 1 at%Be showed a tensile ductility of 5% in vacuum but 1% in air. On the other hand, no environmental effect was observed in cast binary Ni_3Al, which showed very limited plastic deformation at room temperature. Manganese at a level of 15 at% was added for enhancement of the grain-boundary cohesion in Ni_3Al (Masahashi et al. 1988b). They found that the tensile ductility of $Ni_3(Al, Mn)$ is susceptible to environmental embrittlement at room temperature at different strain rates. For example, the alloy exhibited a ductility of 11% in air but 50% in vacuum at a strain rate of $10^{-3}s^{-1}$. The reduction in ductility is accompanied by a change in fracture mode from ductile transgranular to brittle intergranular. Note that the environmental effect was thought previously to be associated with the ternary alloying additions added to Ni_3Al (rather than in connection with binary Ni_3Al) until the report of a clear environmental effect in fabricated and recrystallized binary Ni_3Al and Ni_3Si.

Most recently, George, Liu, and Pope (1992) have studied the environmental effect in a carefully prepared Ni_3Al alloy with the composition of Ni-22.65% Al-0.26% Zr. In this case, the polycrystalline material was prepared by recrystallization of a cold-worked single crystal. The polycrystalline specimens exhibited a room-temperature tensile ductility of 6–9% in water, 11–13% in air, and 48–51% in oxygen, depending upon orientation with respect to the rolling direction. All specimens showed mainly grain-boundary fracture, including the ductile ones that exhibited in excess of 50% elongation to fracture. These results demonstrate that the grain boundaries in this alloy are severely embrittled in the presence of moisture that reacts with the aluminide to release atomic hydrogen and causes hydrogen embrittlement.

All these results indicate that the grain boundaries in Ni_3Al and its alloys are actually not brittle. The low ductility and brittle grain-boundary fracture in the polycrystalline materials are caused by an extrinsic factor: moisture in air.

7.1.3. The Beneficial Effect of Boron. Boron has been found to be most effective in improving the tensile ductility of polycrystalline Ni_3Al containing less than 25 at%Al (Aoki and Izumi 1979; Liu et al. 1985; Taub et al. 1984) when being tested in air at room temperature. Figure 6-11 is a plot of room temperature tensile ductility in air as a function of boron addition in Ni_3Al (24% Al) (Liu et al. 1985). Microalloying with boron sharply increases the ductility and effectively suppresses intergranular fracture. A tensile ductility higher than 50% in air was achieved in Ni_3Al by microalloying with 0.1 wt%B. The striking effect of boron on the ductility of Ni_3Al, first discovered by Aoki and Izumi (1979), occurs over a wide range of concentration of boron in solid solution (the solubility of boron $\approx$ 1.5 at%). Both Auger microprobe (see Fig. 6-10) and imaging atom probe (Sieloff, Brenner, and Burke 1985) studies had revealed that boron segregates strongly to grain boundaries in Ni_3Al, although the extent of segregation varies from boundary to boundary.

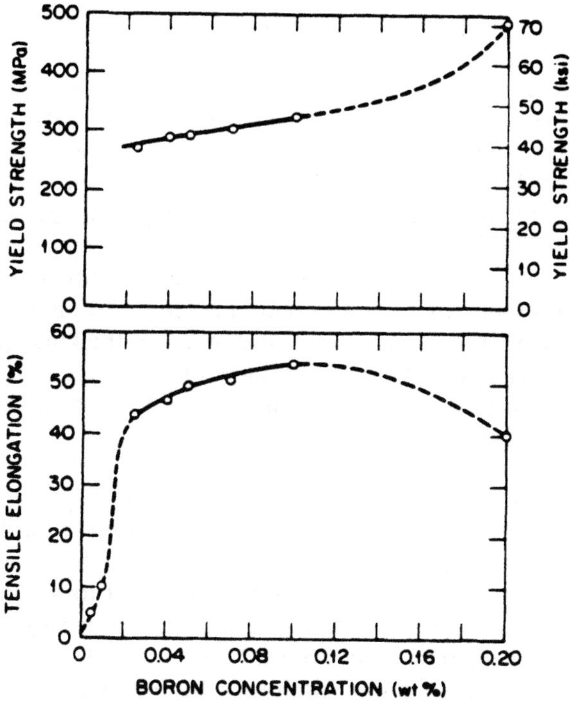

Figure 6-11. Room temperature tensile properties as a function of boron concentration for Ni_3Al (24 at%Al). All specimens recrystallized for 30 min. at 1000° C (Reprinted with permission from *Acta Metallurgica*, Volume 33, C. T. Liu, C. L. White, and J. A. Horton, pages 213–229, 1985, Elsevier Science Ltd., Pergamon Imprint, Oxford, England).

Considerable effort has been devoted to understanding the ductilizing effect of boron in Ni_3Al. As a result, two mechanisms were proposed to explain the beneficial effect of boron: (a) boron-enhanced grain boundary cohesive strength (Liu et al. 1985), and (b) boron-facilitated slip transfer across the grain boundary (Khadkikar et al. 1987). In view of the recent finding of environmental embrittlement (Liu 1992), the ductilizing effect of boron is also related to reduction in moisture-induced hydrogen embrittlement in Ni_3Al. Masahashi et al. (1988a) first reported that the tensile ductility of boron-doped Ni_3Al is insensitive to test environment at room temperature, indicating that boron is effective in eliminating environmental embrittlement in Ni_3Al. Since boron and hydrogen both occupy the same type of sites, it is reasonable to assume that the strong segregation of boron to Ni_3Al grain boundaries would block the diffusion of hydrogen along the boundaries and thus alleviate hydrogen embrittlement. Further studies are required to clarify this point.

An interesting finding in connection with the study of boron in Ni_3Al is the discovery of the alloy-stoichiometry effect. Alloy stoichiometry strongly affects the ductility and fracture behavior of boron-doped Ni_3Al (Liu et al. 1983, 1985). Boron is most effective in improving the ductility and suppressing intergranular fracture in Ni_3Al alloys containing ≤ 24 at%Al. As the aluminum concentration is increased above 24%, the ductility decreases sharply (Fig. 6-11), and the failure mode also changes from transgranular to mixed mode and then to mainly intergranular fracture. Auger (Liu et al. 1985) and particle-tracking autoradiography (Lin et al. 1991) studies reveal that alloy stoichiometry strongly influences boron segregation and grain-boundary chemistry, and thus affects ductility and fracture in boron-doped Ni_3Al.

7.1.4. Other Solutes. Attempts to ductilize Ni_3Al with additions of interstitial carbon have been unsuccessful. However, small additions of beryllium, which is similar in atomic diameter to boron and carbon, can increase the room-temperature ductility of Ni_3Al to about 6% (Takasugi et al. 1986). At elevated temperatures, elongation decreases to zero above 800 K, independent of alloy concentration and level of dopant (Masahashi et al. 1988a). The solubility limit of Be in Ni_3Al is about 0.2 wt% ($\sim$ 1at%). However, even additions above the solubility limit still provided moderate strength with no loss in ductility.

Many solutes other than boron and beryllium have been reported to improve the ductility or fabricability of Ni_3Al: manganese, iron, chromium, hafnium, palladium, and silver, in varying amounts, influence ductility. In the case of chromium, manganese, and iron, there is some question as to whether ductilization occurs in single-phase alloys. According to Takasugi et al. (1985), iron additions in the range 4 to 15 at%, replacing aluminum only, resulted in an increase in room-temperature bend ductility (Fig. 6-12) accompanied by a mixed mode fracture. Later, Horton et al. (1987) reported that the addition of 10 at%Fe to Ni_3Al + B, replacing both nickel and aluminum, produced the most fabricable alloys. The solubility limit of Fe in Ni_3Al is about 15 at%, at which level either a transformed B2 phase (β') or a disordered f.c.c. phase (γ) in the ordered γ' matrix is observed. Above this level of iron, fabrication is increasingly difficult. It is clear from the work of Horton et al. (1987) that iron additions to Ni_3Al + B provided much higher ductility than Fe additions alone. Further, the cold fabricability of a 20Al-10Fe-0.2B alloy was better than that of a 24Al-0.2B alloy, indicating that the effects of boron and iron are additive.

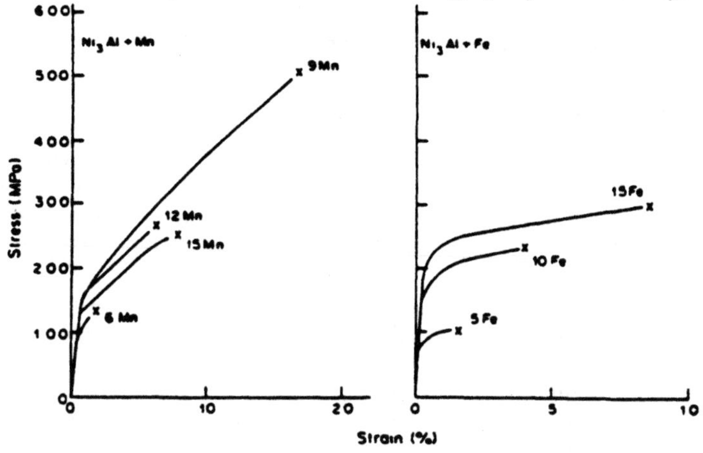

(a) (b)

Figure 6-12. Stress-strain curves of ternary Ni_3Al with (a) Mn and (b) Fe (Reprinted with permission from *Acta Metallurgica,* Volume 33, T. Takasugi, O. Izuma, and N. Masahashi, pages 1259–1269, 1985, Elsevier Science Ltd., Pergamon Imprint, Oxford, England).

Manganese additions to Ni_3Al also improve room-temperature ductility, with a change in fracture mode to transgranular occurring at about 6 at%Mn (Takasugi et al. 1985). Complicating any interpretation of the effect of manganese on ductility is controversy over its solubility in Ni_3Al. Masahashi et al. 1986 state that the Ni-Al-Mn alloy system forms a continuous series of solid solutions at low temperatures. However, Dimiduk et al. (1987) report that at solute levels higher than 4 at%Mn, Ni-Al-Mn alloys are two phase; they associate the ductilizing effect of manganese with the formation of the second phase.

Perhaps of even greater significance is the observation that P/M Ni_3Al + B is notch sensitive in tension, but notch sensitivity is absent in a Ni-15 at%Al, 9% Mn-0.2%B alloy (Fuchs 1986). The latter result is potentially significant for fracture toughness and crack-growth behavior of the aluminides.

Manganese is expected to combine with any sulfur present in Ni_3Al, thereby removing sulfur from grain boundaries to intragranular sites. Liu (1987) has reported that manganese improves the fabricability of B-doped Ni_3Al, so that the ductilizing influence of these two solutes again may be additive.

The addition of 5%Cr had no beneficial effect on the room-temperature bend ductility of Ni_3Al, nor did it change fracture mode (Takasugi et al. 1985). Also, 6%Cr reduced slightly the room-temperature tensile

Table 6-6 Values of the Valency of the Third Element Itself and the Valency Difference Between the Third Element and Constitutive Solvent Atom Substituted by the Third Element.

Third element	Electronic nature	Substitution site	Valency of third element	Valency difference[a] between solvent and third element
Sb	Vb	Al	5	−2
Si	IVb	Al	4	−1
Ge	IVb	Al	4	−1
Al	IIIb	—	3	—
Ga	IIIb	Al	3	0
Cu	Ib	Ni	?(0)	?(+0.25)
Ni	VIII	—	0(0.25)	—
Co	VIII	Ni	0.8(0.72)	−0.8(−0.47)
Fe	VIII	Al	0.2(0.92)	+2.8(+2.08)
Mn	VIIa	Al	0.91(1.35)	+2.1(+1.65)
Cr	VIa	Al	1.2(1.69)	+1.8(+1.31)
Mo	VIa	Al	3	0
W	VIa	Al	3	0
V	Va	Al	1.5(2.19)	+1.5(0.81)
Nb	Va	Al	?	?
Ta	Va	Al	?	?
Ti	IVa	Al	4.2(3.92)	−1.2(−0.02)
Sc	IIIa	Ni	?	?

[a] Defined as (valency of constitutive solvent—valency of ternary element).

Source: Takasugi et al. 1985; reprinted with permission from *Acta Metallurgica*, Volume 33, T. Takasugi, O. Izumi, and N. Masahashi, pages 1259–1269, 1985, Elsevier Science Ltd., Pergamon Imprint, Oxford, England

ductility of B-doped Ni_3Al (Liu 1987). However, 6% to 8%Cr added to B-doped Ni_3Al produces remarkable improvement in ductility in air at elevated temperatures, as described in the next section.

Hafnium also improves the fabricability of B-doped Ni_3Al (Liu 1987), but no systematic data showing the effects have been published.

Takasugi et al. (1985) have examined several hypotheses for improved ductility of Ni_3Al when alloyed with manganese or iron. They concluded that these elements, which have similar electronic bonding to nickel, suppress grain-boundary embrittlement, while other solute additions with dissimilar electronic structures have no beneficial effect. The valency difference between the third element and the solvent atom (Ni or Al) replaced by the solute was taken as a semiquantitative measure of

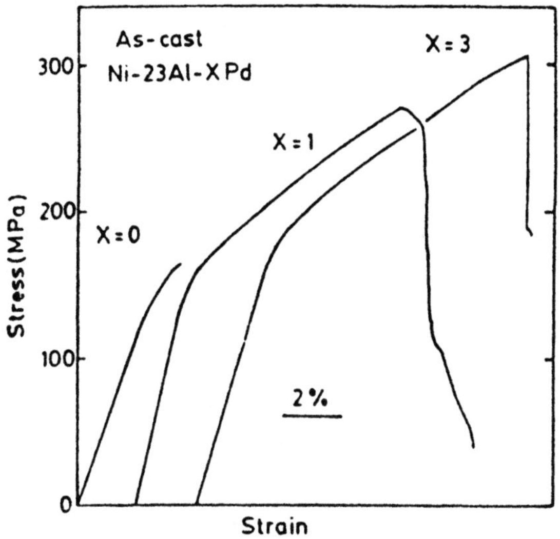

Figure 6-13. Effect of Pd on stress-strain behavior of Ni_3Al (Chiba et al. 1991).

grain boundary strength in ternary alloys. Table 6-6 summarizes these differences (Takasugi et al. 1985). Note that iron is predicted to be the most effective ductilizer, followed by manganese and then chromium.

Chiba et al. (1991) have suggested an alternative explanation for enhanced ductility of Ni-rich Ni_3Al in the presence of elements such as palladium (1 to 2%) as shown in Figure 6-13. Palladium has a strong tendency to occupy nickel sites, independent of composition of the host alloys; in doing so, the ordering energy (APB energy) is hypothesized to decrease. Evidence for reduced APB energy has been provided by TEM examination of thin foils of melt-spun ribbons of Ni-25 at%Al and Ni-23%Al-2%Pd. While a few APB loops can be seen in the stoichiometric alloy, a fine APD structure is seen in the Pd-containing alloy. Similar results have been observed with 0.25% silver or 2% platinum, copper or cobalt additions (Chiba, Hanada, and Watanabe 1992). Chiba et al. (1991) suggest that a lowered APB energy facilitates movement of dislocations in grain-boundary "planes," resulting in disordering in the plane. Alternatively, palladium and silver may affect the degree of environmental embrittlement, but such research has not yet been carried out.

7.1.5. Grain Shape. The ductility of binary Ni_3Al alloys can be substantially improved by control of grain shape. Liu (1987) first reported that the tensile elongation of cast hypostoichiometric Ni_3Al (24 at%Al) increased from 1.2% to 14.1% with a change in grain shape from equiaxed

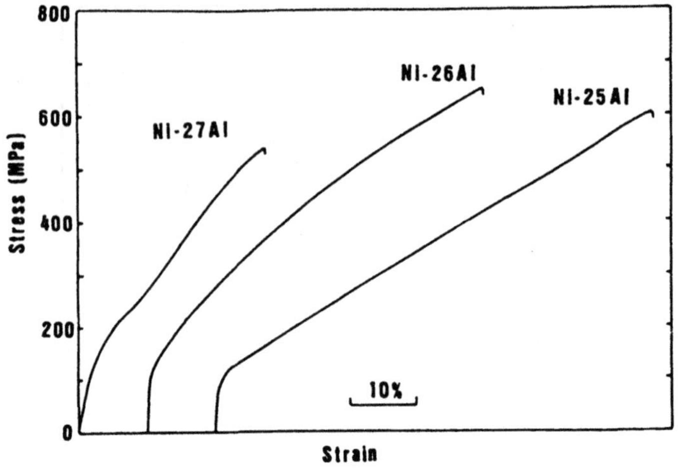

Figure 6-14. Stress-strain curves of the Ni_3Al grown by floating zone-unidirectional solidification at room temperature (Hirano et al. 1991).

to columnar. In this case, the columnar grain structure with a strong ⟨100⟩ texture was produced by a directional levitation zone remelting technique. The beneficial effect of the columnar grain structure tested at the direction parallel to the growth direction is attributed to minimizing the normal stress across the boundaries and, thus, reducing nucleation and propagation of cracks along the boundaries (Liu et al. 1989). Recently, Hirano et al. (1991) reported even higher tensile ductilities of stoichiometric and hyperstoichiometric Ni_3Al alloys with a columnar grain structure with a weak ⟨100⟩ + ⟨111⟩ or ⟨110⟩ + ⟨111⟩ texture tested in the direction parallel to the growth axis, see Figure 6-14. Surprisingly, the same material tested perpendicular to the growth direction also showed a high ductility (28% to 31%) and transgranular fracture. A detailed analysis of grain-boundary chemistry and character (e.g., low angle vs. random boundaries) is required in order to fully understand these interesting results.

7.2. Elevated Temperature Fracture Behavior

*7.2.1. **Single Crystals.*** Cube-oriented single crystals of $Ni_{75}Al_{15}Ti_{10}$, when tested over a range of temperature, show a sharp decrease in ductility, from about 15% at room temperature to about 1% at 1100K (Aoki and Izumi, 1978). The latter also is the temperature of the flow stress peak. Unexpectedly, Heredia and Pope (1989) report that the ductility and fracture stress of crystals of two orientations is increased by B at room temperature. A similar effect was noted for cube-oriented crystals

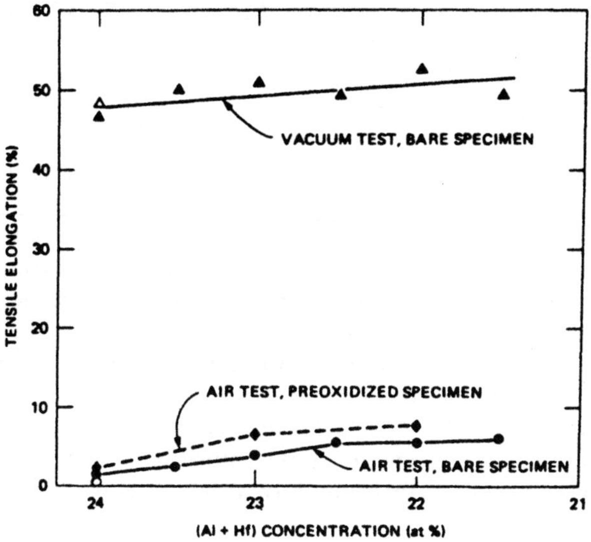

Figure 6-15. Plot of tensile elongation of Ni_3Al alloys as a function of (Al + Hf) concentration for bare specimens tested at 600° C in vacuum or air and for specimens preoxidized at 1100° C/2 h + 850° C/5 h and tested in air. Solid symbols for alloys with 0.5% Hf and open symbols for the alloy without hafnium (Reprinted with permission from *Acta Metallurgica*, Volume 35, C. T. Liu and C. White, pages 643–649, 1987, Elsevier Science Ltd., Pergamon Imprint, Oxford, England)).

only, at elevated temperatures. Therefore, B appears to provide a bulk-ductility effect in addition to the very well documented-grain boundary effect.

*7.2.2. **Polycrystals.*** A ductility minimum is readily apparent also in polycrystalline B-doped Ni_3Al prepared by powder or from the melt and tested in air (Taub et al. 1985). It is clear that the depth of the elevated temperature ductility minimum in polycrystals depends sensitively upon the test environment (Liu and Sikka, 1986). The fracture surface becomes increasingly intergranular as temperature increases to that of the ductility minimum. The minimum is much reduced in severity (and less intergranular cracking occurs) when tests are carried out in 10^{-6} torr vacuum, see Figure 6-15 (Liu and White 1987). Later work showed a steady increase in ductility at 760° C for an alloy containing 0.5Hf and 0.07B as air pressure was reduced from 1 to 10^{-8} torr (Liu 1987). The embrittlement of polycrystalline Ni_3Al in oxygen-bearing environments appears to be largely due to adsorption of oxygen at grain boundaries, thereby lowering cohesion (see chapter on Environment). There is little effect of most solutes on elevated temperature ductility loss; for example, the ductility of Ni-Al-Mn (Mn replacing Al) alloys reaches a maximum at 650 to 800 K (depending upon composition), and dimpled fracture surfaces are seen for all alloys containing at least 9 at%Mn (Masahashi et al. 1986). Essentially, complete ductility loss occurs near 900° C even though tests are conducted in vacuum. Masahashi et al. attribute the low room-temperature ductility of Ni-Al-Mn alloys to hydrogen contained in the air environment and the ductility loss at high temperature to grain-boundary cavitation. Not considered by these authors was the possibility that there is sufficient oxygen in their 1.3×10^{-3} Pa vacuum to cause embrittlement at the strain rate selected for tests, $1.1 \times 10^{-4} s^{-1}$.

Dimiduk et al. (1987) also report that neither manganese nor iron suppress dynamic environmental (air) embrittlement at 600° C. However, in a vacuum of 5×10^{-5} torr, ductility at 600° C is restored in the ternary alloys.

Chromium is the one solute that has been shown to decrease the magnitude of oxygen embrittlement in Ni_3Al (Liu and Sikka 1986). Additions of 6% Cr increase ductility at all temperatures between 25 and 1000° C, but the minimum ductility is still of the order of 10%. However, 8% Cr increases the minimum ductility to 20% elongation, while having the additional benefit of raising room-temperature ductility. It has been suggested that chromium accelerates the formation of a grain boundary oxide that prevents further penetration by oxygen ions (Liu and Sikka 1986). Others have suggested that chromium affects the microstructure and distribution of phases at grain boundaries in cast alloys (Yizhang et al. 1989). Chromium also improves high-temperature fatigue resistance, as described in a later section.

An alternate method of increasing high-temperature ductility is to produce columnar grains by directional solidification (Liu 1987). The number of grain boundaries subject to high-tensile stresses is thus reduced so that some ductility remains even though cohesion is reduced by oxygen.

8. CREEP AND STRESS RUPTURE

One of the major barriers to the development of single-phase Ni_3Al as a structural alloy has been its extremely low creep resistance. This result was initially unexpected because of the usual observation of slowed diffusion and hence reduced creep rate in other alloys when long range order is induced.

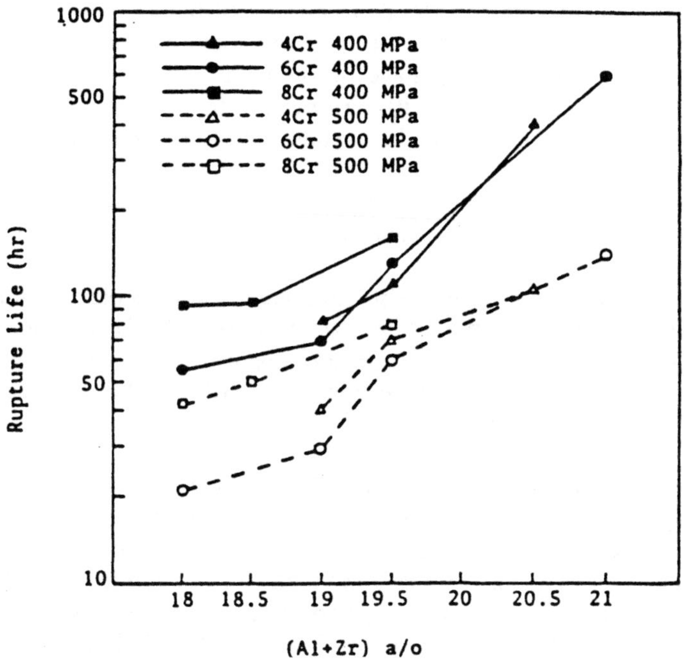

Figure 6-16. Influence of Al + Zr content on rupture life of Ni_3Al, Cr alloys (Hsu et al. 1987).

8.1. Solute Effects

Several solutes have been found to improve the creep resistance of Ni_3Al, most notable being hafnium, chromium, zirconium and tantalum. Hafnium is particularly effective in increasing both yield strength and creep resistance (Schneibel, Peterson, and Liu 1986). Grain-boundary cavitation contributes appreciably to the creep rate of hafnium-containing material tested at 250 MPa in vacuum at 750° C. Improvements in rupture life have been observed for chromium plus zirconium additions in Ni_3Al, see Figure 6-16 (Hsu et al. 1987) implying, therefore, a beneficial effect of these elements on creep rate. This work showed also that single phase γ' was superior to $\gamma' + \beta$ alloys in creep. In tensile tests, on the other hand, the two phase alloys were stronger. Uncertainties in the initial grain sizes complicate an explanation of this behavior. The work of Hsu et al. clearly showed, however, that aluminum content also is important. Rupture lives of alloys with 4, 6 or 8 at%Cr increased with aluminum content (see Fig. 6-16).

Limited results on the creep properties of nickel-iron aluminides at 760° C shows that iron reduces life by more than 10^4 times (Flinn 1960).

Shah and Duhl (1987) have studied the creep behavior of over twenty-five Ni_3Al-type alloys with various solutes. Many were two phase, and some were prepared as single crystals. The following conclusions were drawn from their work:

1. The creep rate of the best multicomponent alloys is almost 100 × less than that of binary Ni_3Al.
2. For any given alloy, creep resistance is greater at the stoichiometric composition, and can be improved by a factor of two with ⟨111⟩ single crystals.
3. The γ' alloys of the strongest (multicomponent) alloys were comparable to those of two-phase nickel-base superalloys such as PWA 1480, but there is no density advantage for the γ' alloys.

Anton, Pearson, and Snow (1987) have shown that the influence of tantalum on creep behavior of Ni_3Al is very dependent upon test temperature. At 760° C and 871° C, tantalum concentrations between 1% and 8% (at a nickel content of 75 at%) weaken Ni_3Al, while at 982° C very pronounced strengthening occurs. Also, at constant Ta = 6%, increasing Ni/(Al + Ta) sharply increases creep life at 760° C. This result is surprising since moving away from stoichiometry should reduce the degree of order and possibly the APB energy, thereby lowering creep resistance. At 871° C and 982° C there is only a small influence of stoichiometry on creep resistance.

8.2. Mechanisms of Creep

Grain-boundary sliding occurs in Ni_3Al at temperatures as low as 400° C (Weihs et al. 1987). Stress-relaxation tests on Ni_3Al at 550° C and 600° C show that sliding occurs in a manner prediced by current theories. Grain-boundary migration and recrystallization during plastic deformation occurs at 600° C (Baker, Viens, and Schulson 1984). Therefore, high creep rates in Ni_3Al may be due to microstructural instabilities.

8.3. Influence of Processing

Several processing techniques may be employed to produce Ni_3Al-base alloys: most important are casting, casting + working, and recrystallization and powder processing. Creep-rupture properties of Ni_3Al alloys such as IC-218 are very dependent on grain size (Sikka 1987). By far the best creep properties are achieved with large-grained castings, which display up to five times the creep strength of powder product. This strong effect of grain size is related to the diffusional creep range used for testing of these alloys.

9. FATIGUE

9.1. High-Cycle Fatigue

The earliest study of high-cycle fatigue behavior of Ni_3Al was carried out on nearly single-phase single crystals by Doherty, Giamei, and Kear (1975). It was shown that the fatigue life was invariant with temperature to 800° C, possibly due to the rise in flow stress over the same temperature range; a small drop in life was observed at 1040° C. Crystals displayed crack initiation at defects or inclusions, but cracks propagated along slip planes.

Later work, on high-cycle fatigue of boron-doped polycrystals, emphasized the importance of grain boundaries as crack-initiation sites

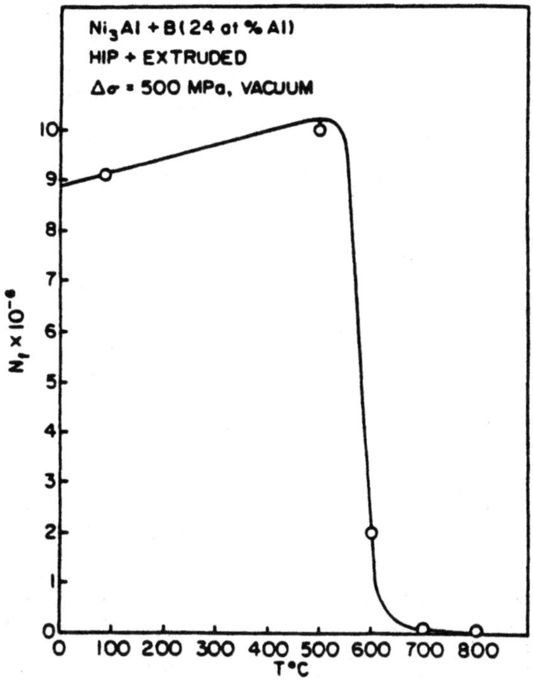

Figure 6-17. Effect of temperature on high cycle fatigue life of HIP + extruded Ni-24% Al + B at constant stress range of $\Delta\sigma = 500$MPa (Stoloff et al. 1987).

Table 6-7 Compositions of Ductile, Ni_3Al-Based Alloys

	ALLOY DESIGNATION (WT%)			
Element	IC-50	IC-218LZr	IC-221M	IC-396M
Al	11.3	8.69	7.98	8.0
Cr	—	8.08	7.74	8.0
B	0.02	0.02	0.008	0.005
Zr	0.6	0.20	1.70	0.8
Mo	—	—	1.43	3.0
Ni	88.08	83.01	81.15	Bal.

Source: Alexander and Sikka, 1992; Sikka et al., 1992

(Stoloff et al. 1987). However, crack propagation is by stage I (slip-band) cracking. The overload zone is again intergranular. The fatigue resistance at both 25° C and 500° C is decreased substantially as aluminum content increases from 24–26 at%, although part of the loss in fatigue life may be attributed to the cast, coarse-grained structure of the higher aluminum-containing alloy compared to the P/M 24 at% alloy. Particularly striking was the observation of sharply decreased fatigue life at $\Delta\sigma = 500$ MPa at temperatures above 500° C for HIP + extruded Ni-24% Al + B (see Fig. 6-17) (Stoloff et al. 1987). This sudden drop in fatigue resistance in moderate vacuum parallels a similar loss in ductility of such alloys when tested in tension in air. It may be concluded that time-dependent fatigue life is much more sensitive to low oxygen levels in the atmosphere than is tensile ductility. Manganese-containing alloys display a similar catastrophic loss in fatigue resistance at elevated temperatures.

Recently, high-cycle fatigue data have been reported for several cast Ni_3Al alloys developed at Oak Ridge National Laboratory (Gieseke and Sikka 1992). The results show that one alloy, IC-221 M (Table 6-7) has fatigue strength exceeding that of the nickel-base superalloy In-713 C.

9.2. Crack Growth

Although several studies of fatigue crack growth in intermetallic compounds, including Ni_3Al, have been reported in recent years, little is known about mechanisms. Crack-growth rates in Ni_3Al + B and the isomorphous compounds $(Fe, Ni)_3V$ are low at room temperature and at 600° C compared to those of commercial alloys (see Fig. 6-18) (Matuszyk et al. 1990). Although growth rates increase significantly for both ordered alloys as temperature rises to 500° C and above, there still is an improve-

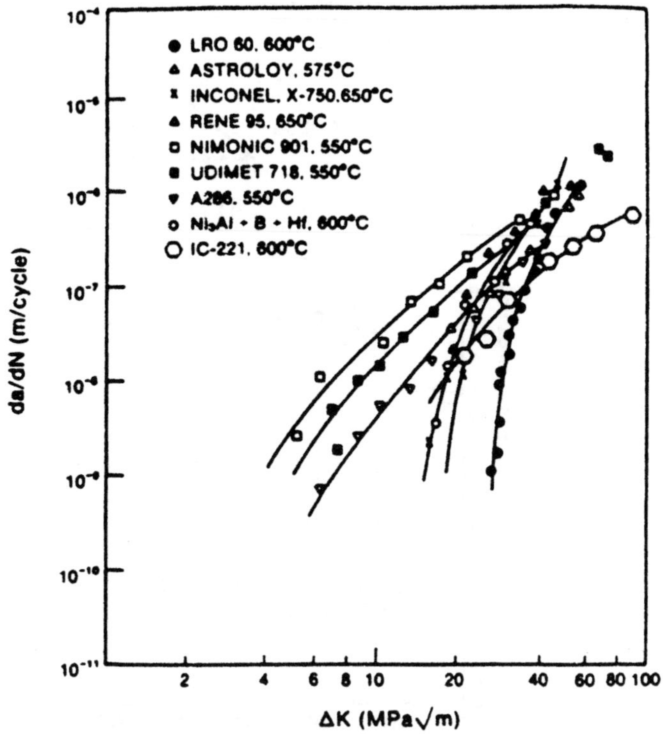

Figure 6-18. Comparative crack propagation rates as a function of ΔK for IC-221 and a number of commercial and experimental alloys between 550° C and 650° C (Matuszyk et al. 1990).

ment over commercial alloys. The increase in growth rates of Ni_3Al with temperature occurs even though yield stress rises significantly in the same range, suggesting that the moderate vacuum (10^{-5} torr) environment contains sufficient oxygen to facilitate cracking. Crack-growth rates obtained in air increase about ten times between 399° C and 538° C. Chromium-containing alloys such as IC-221 are somewhat more crack-growth resistant at 600° C.

9.3. Low-Cycle Fatigue

Several recent studies of cyclic strain hardening and crack initiation and growth under strain-controlled cycling have been reported. At room temperature, cyclic hardening is asymmetric (see Fig. 6-19). and the difference increases with increasing strain amplitude (Hsiung and Stoloff

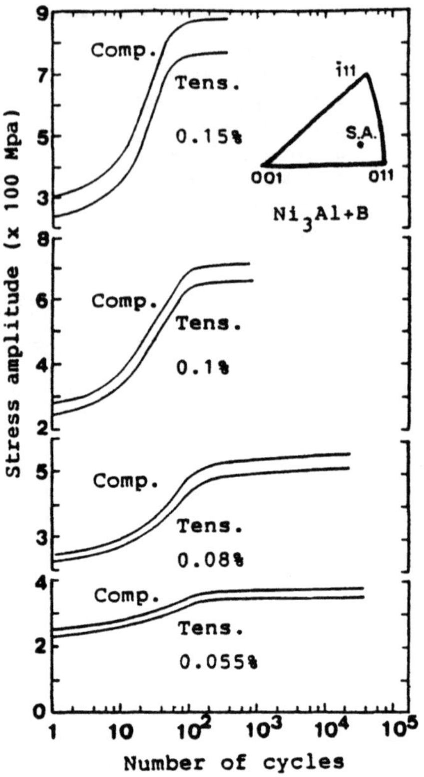

Figure 6-19. Cyclic hardening curves for [$\bar{2}$ 13 17] oriented Ni_3Al + B single crystals fatigued with various strain amplitudes (Hsiung and Stoloff 1988).

1988). A minimum plastic strain of 1×10^{-4} is needed to observe this effect. This phenomenon also had been reported by Bonda, Pope, and Laird (1987a, b), who showed that the nature of the asymmetry between tension and compression is a function of crystal orientation. At low temperatures the asymmetry follows the predictions of the model of Paidar et al. 1984. At 700° C, however, cyclic tensile stresses were higher than compressive stresses for all orientations, perhaps due to the occurrence of massive cube slip and dislocation climb at this temperature. An asymmetric flow stress also is noted in fine-grained polycrystals prepared by P/M techniques and tested at 600° C (Dowling and Gibala 1987). Environment plays a significant role in the accumulation of plastic strain prior to failure under strain-controlled cycling. There is an approximately 10 × decrease in strain to failure in air vs. vacuum in fine grained polycrystals tested at 600° C (Dowling and Gibala, 1987).

10. FABRICATION

Prior to the discovery of the boron ductilization effect, the mechanical properties of Ni_3Al had to be studied via compression tests on polycrystals or by use of single crystals. In recent years, rapid solidification also has been utilized by several laboratories to prepare material for metallurgical studies. Due to the limitation of the boron effect on ductility with respect to aluminum content and elevated temperatures, the mechanical fabrication of aluminides remains a nontrivial task. In this section, the major techniques utilized to prepare aluminide materials are discussed. The choice in large part will be dependent on desired grain size. Powder processing produces much finer grained product than any casting operation. Fine-grain size, however, is advantageous only for low temperature strength and high temperature (> 800° C) ductility. A detailed review of processing techniques for Ni_3Al alloys of industrial interest has recently been published (Sikka, Mavity, and Anderson 1992); see Table 6-6 for a summary of compositions of several significant alloys, as well as section 4.2.1 for a description of alloy design principles.

10.1. Melting and Casting

A variety of melting techniques have been utilized for Ni_3Al alloys. Argon induction melting (AIM), vacuum induction melting (VIM), electroslag remelting (ESR), and vacuum arc double electrode remelting (VADER) have been used to produce aluminide ingots (Liu and Sikka 1986). However, high temperature forming of castings is very sensitive to composition, thermal treatment prior to forming and strain rate; hot forging of castings is generally unsuccessful.

Air induction melted heats of 500 lb each of IC-50 (Ni-23.5 at% Al-0.25Hf-0.2B) and IC-218 (Ni-17 at% Al-8Cr-1Zr) have been centrifugally cast into 305 cm (120 in.) long tubes, with 12.7 cm (5 in.) outside diameter and 2.54 cm (1 in.) wall thickness.

Direct casting of 1 to 2 mm thick sheet has been accomplished by the twin roller casting process (Sikka 1987). This sheet has been cold rolled, annealed at 1050° C, cleaned and reannealed at 1050° C to reproduce material with good surface quality and equiaxed grains through the thickness.

Bar and rods also can be directly cast from liquid (Sikka 1987). A starting stub is used to withdraw a small amount of liquid in a nozzle. The solidified metal on the stub is withdrawn and a new batch of liquid solidifies on the stub. The process is repeated until continuous formation of bar is achieved. Cold swaging or cold rolling, with intermediate anneals, is then used to produce the final size.

10.2. Working of Cast Ingots

Fine-grained, hypostoichiometric Ni_3Al + B cast in small heats can subsequently be cold rolled into sheet without difficulty, provided that intermediate anneals are performed at 1000–1050° C. However, efforts to cold roll larger ingots, which typically exhibit a coarse grain structure, have largely been unsuccessful. Similarly, hot fabrication by rolling and forging is difficult because of the limited ductility of coarse-grained castings at elevated temperatures.

Extrusion of AIM, VIM, and ESR billets has been carried out in a container at 1100° C. Reduction ratios of at least 5/1 are needed to recrystallize the cast structure (Sikka 1987).

Fine-grained (~ 10 μm) ingots produced either by extrusion or by powder metallurgy (P/M) techniques also can be isothermally forged to near net shape at 1100° C at a strain rate of 0.5/min.

10.3. Powder Metallurgy

Powders can be consolidated either by hot pressing, reactive sintering, HIP, HIP + extrusion or by direct extrusion of powders. HIPing is usually carried out at 1100–1150° C for 3 h, with a pressure of 120 MPa. HIPing lead to 100% density with grain size ~ 10–15 μm for ternary and quarternary alloys. Direct extrusion with a reduction ratio of 8/1 results in 100% density. Billets up to 20 cm (8 in.) diameter, compacted under atmospheric pressure (CAP) generally contain large noninterconnected porosity, but can be further consolidated by extrusion at 1100° C in a stainless steel can. Complex shapes made by CAP can be further consolidated by HIPing. Wright and Flinn (1986) report that extruded powder generally has higher strength and ductility than HIP material. Further details can be found in the chapter on Powder Processing by R. M. German.

10.4. Rapid Solidification

Rapid solidification of Ni_3Al and its alloys has been studied by many investigators. The principal effects of rapid solidification in this alloy, as in other systems, are decreased grain size, increased homogeneity, and extension of solid solubility (Baker et al. 1986). The methods used to rapidly solidify Ni_3Al include melt spinning to produce ribbons, in rotating water spinning to produce wires, by Rapid Solidification Rate (RSR) processing, gas atomization to prepare powders, and by using an arc hammer apparatus to produce splats. Rapidly solidified microstructures also have been prepared as a result of various welding processes. With all of these techniques a cooling rate in the range 10^5–10^7 K/s may be achieved.

Considerable strengthening of Ni_3Al alloys results from rapid solidification, due to a combination of high dislocation density, the presence of the martensite, and a fine grain size. Very fine antiphase domains also are seen, as in Ni-Al-Cr (Huang et al. 1985). Martensite alone is not responsible for strengthening, since nickel-rich Ni_3Al is almost as hard as the stoichiometric alloy, even though only the latter contains martensite. Ductility of binary rapidly quenched Ni_3Al remains very low, independent of composition. As in the case of conventionally processed Ni_3Al, considerable ductility results from doping of Ni_3Al with small quantities of boron prior to quenching.

Other macroalloying additions to rapidly solidified Ni_3Al include zirconium, vanadium, niobium, titanium, molybdenum, copper, iron, and cobalt (Inoue, Tomioku, and Masumoto 1983; Inoue et al. 1985). The mechanical properties of wire and microstructures of ribbons of similar thickness ($\sim$ 100 μm) were reported. Alloys containing chromium, iron, cobalt, manuganese or silicon contained 30–55 nm APD which did not preferentially lie along any planes, indicating anisotropic APD energy. High room-temperature yield strengths, from 3900–590 MPa depending upon grain size, were observed. The latter were in the range 1.8–4.9 um. Rapidly solidified alloys containing manganese showed the largest elongation, 27%, but the lowest strengths, 260 MPa; other alloys displayed elongation from 4% (with Fe) to 10% (with Co). However, upon annealing at 800–1150 K, large decreases in yield strength and a drop in ductility to less than 2% were noted. Accompanying reduced ductility was a change in fracture mode from transgranular to intergranular.

Alloys containing zirconium, titanium, vanadium, copper or niobium revealed fewer widely spaced APD on cooling, and were extremely brittle. Some of these alloys also showed higher APB energies which were anisotropic (APB's were observed on cube planes). In the case of silicon, increasing solute (above 12 at%) increased APD size and reduced ductility (Inoue et al. 1983).

11. WELDABILITY

Ni_3Al and its alloys can be welded with some difficulty. The principal variables in the weldability of these alloys are alloy composition and welding speed. A B level of 200 ppm is optimum for welding of binary Ni_3Al, but at welding speeds above 13 mm/s, significant weld-cracking problems exist (Santella, David, and Horton 1987). Even at low speeds (3.4 mm) and with 400° C preheat alloy IC-15 (500-ppm B) showed cracking after electron-beam (EB) or gas tungsten arc (GTA) welding.

Figure 6-20. Photograph of investment cast turbocharger rotor of Ni_3Al alloy IC-396 (Ni-16.4 at%Al-8.0%Cr-15%Mo-0.5%Zr doped with 300 wt ppm B). Casting was carried out at Precision Castparts Corporation (PCC) (Sikka 1990).

Iron additions significantly improve the high-speed weldability of Ni_3Al containing 200-ppm B, (IC-103) but are ineffective in improving the behavior of material with 500-ppm B (IC-25), see Figure 6-20 (Santella et al. 1985). Most cracks in IC-25 were confined to the heat-affected zone (HAZ) and were always intergranular (190). In both alloys, the fusion zone consisted of a γ' matrix with an interdendritically distributed second phase identified as β NiAl (David et al. 1985). More information on weldability of Ni_3Al may be found in Chapter 16 by S. A. David and M. K. Santella.

12. STRUCTURAL APPLICATIONS

Industrial interest in Ni_3Al-based alloys is high at the present time (Sikka 1989; Patten 1990). This is because the alloys possess a unique combination of properties including high strength and good oxidation and corrosion resistance at elevated temperatures, and relatively low density as compared with many Ni-base superalloys for high-temperature applications. A detailed description of the industrial interest in Ni_3Al alloys is given in the paper by Sikka et al. (1992). Here, only a brief description of these applications is outlined below.

1. Automotive turbochargers—Turbocharger rotors in diesel trucks are the near-term application of Ni_3Al alloys. The aluminide alloy IC-221M (see Table 6-7) with good castability has been selected for this application, and aluminide rotors are successfully produced by investment casting (see Fig. 6-20) (Sikka 1990). This castable aluminide alloy is expected to replace IN-713C with the major benefits of the improved fatigue life and potential for lower cost.
2. High-temperature dies and molds—The high-temperature oxidation resistance together with excellent strength during deformation at high strain rates make Ni_3Al alloys attractive as die material for isothermal forging and mold material for glass processing. At present, IC-218 and IC-221M alloys are being evaluated for these applications.
3. Furnace fixtures for heat treatments of auto parts—Because of resistance to carburizing atmospheres, General Motors Corporation is currently evaluating the use of IC-221M as fixture material for heat treatment of auto parts in carburizing furnaces.
4. Rollers for steel slab heating furnaces—The high-temperature strength and oxidation resistance of Ni_3Al alloys can produce significant savings in energy costs by not requiring water cooling and in material costs by extending the life four to six times over the current material in use.
5. Hydroturbines—Nickel aluminides, which offer excellent vibration cavitation resistance in water, are being considered for either repair of current hydroturbine rotors or even replacement of the entire units with these alloys.
6. Cutting tools—Nickel-aluminide-bonded tungsten-carbide composites have better low- and high-temperature strength and cutting properties than Co-bonded tungsten-carbide materials.
7. Piston and valves for automobile diesel engines—Ni_3Al alloys are suitable for piston and valve applications because of their good wear resistance and their capability of developing a thermal barrier by high-temperature oxidation treatments.
8. Turbine blades and valves for jet engines—Directionally solidified Ni_3Al alloy with the composition of Ni-16.3Al-8.2Mo-0.26B (at.%) has superior strength and creep resistance at temperatures above 1000° C in oxidizing environments. Hence, this alloy has the potential to be used for turbine blades and valves for jet engines (Han et al. 1992).

13. CONCLUDING REMARKS

It is clear that considerable progress has been made in understanding yield phenomena, alloy development, and improving fracture resistance of Ni_3Al. Environmental embrittlement has emerged as a crucial factor, and the effects of solute additions on fracture behavior must now be reexamined in the light of their effects on environmental sensitivity. Fatigue and creep data remain sparse, but there is now good understanding of crack-initiation phenomena under cyclic loading. Clearly, much work remains to be done in improving creep behavior before Ni_3Al-base alloys can be seriously considered for turbine blades. Nevertheless, it has been shown that creep behavior can be enhanced greatly by hafnium, chromium, and zirconium. Near-term commercial applications are likely in applications less extreme than gas turbines. Processing of monolithic alloys, as well as components, remain a key area for future research, as is the case for other intermetallics.

ACKNOWLEDGMENTS

This research was sponsored by the Division of Materials Sciences, U.S. Department of Energy under contract DE-AC05-84OR21400 with Martin Marietta Energy Systems, Inc. and by the National Science Foundation under Grant No. DMR-8911975. We wish to thank Connie Dowker and Lou Pyatt for manuscript preparation.

REFERENCES

AIGELTINGER, E. H., BATES, S. R., GOULD, R. W., HREN, J. J., and RHINES F. N., 1978, *In Proc. Int. Conf. on Rapid Solidification Processing, Principles and Technology*, Vol. 29, Claitors Publ., Baton Rouge, La.

ALEXANDER, D. J. and SIKKA, V. K. 1992, *Mat. Sci. Eng.*, A152: 114–119.

ANTON, D. L., PEARSON, D. D. and SNOW, D. B., 1987, In *High Temperature Ordered Intermetallic Alloys II* MRS Symp. Proc., 81: 287–295.

ARDELL, A. J., 1985, *Metall. Trans. A* 16A: 2131–2165.

AOKI, K. and IZUMI, O., 1979, *J. of Mat. Sci.* 14: 1800–1806.

________, 1979, *J. Japan Inst. Met.* 43: 1190–1196.

BAKER, I., ICHISHITA, F. S. SUPERNANT, V. A., and SCHULSON, E. M., 1984 *Metallography* 17: 299.

BAKER, I., VIENS D. V., and SCHULSON, E. M. 1984, *Scripta Metall.* 18: 237–240.

BAKER, I. SCHULSON, E. M. and STOLOFF N. S., 1986, In *Mechanical Behavior of Rapidly Solidified Materials*, 257–276. TMS-AIME, Warrendale, PA: 257–276.

BAKER, I., HUANG, B. and SCHULSON, E. M. 1988, *Acta Metall.* 36: 493–500.

BAKER, I., and SCHULSON, E. M., 1989, *Scripta Metall.* 23: 1883.

BALUC, N., KARN THALER, H. P. and MILLS, M. J. 1991, *Phil. Mag.* A64: 327.

BENTLEY, J., Aug. 1986, In *Proc. 44th General Meeting of Electron Microscope Soc. of America*. Albuquerque, NM, 704.

BOND, G. M., ROBERTSON, I. M. and BIRNBAUM, H. K. 1987, Abstract, *Journal of Metals* 39 (7): 64, presented at TMS-AIME Fall Meeting, Cincinnati, OH.

BONDA, N. R., POPE D. P. and LAIRD, C. 1987a, *Acta Metall.* 35: 2371–2383.

________, 1987b *Acta Metall.* 35: 2385–2392.

BOSE, A., MOORE, B. GERMAN, R. M., and STOLOFF, N. S, 1987, In *Proc. Conf. on PM Aerospace Materials*, MPR Publ. Services, Ltd. Shrewsbury, England, 4.1–4.7.

BRENNER, S. S., SIELOFF, D. and BURKE, M. G., *J. de Physique*, Colloque Supl. 47: C2, 215–225.

CAHN, R. W., SIEMERS, P. A., GEIGER, J. E., and BARDHAN, P., 1987, *Acta Metall.* 35: 2737–2751.

CHEN, S. P., VOTER, A. F., and SROLOVITZ, D. J., 1986, *Scripta Metall.* 20: 1389–1394.

CHINA, A. HANADA, S., and WATOMABE, S., *Acta Metall.*, 39, 1799 (1991)

CHIBA, A., HANADA, S. and WABANABE, S., 1991, In *Intermetallic Compounds, Structure and Mechanical Properties*, JIMIS-6, ed. O. Izumi, 445–450. Sendai, Japan: Japan Inst. Metals.

CHIBA, A., HANADA, S., and WATANABE, S., 1992, *Mat. Sci. Eng.* A152: 108–113.

CURWICK, L. R., 1972, Ph.D. thesis, University of Minnesota.

DAVID, S. A., JEMIAN, W. A., LIU, C. T., and HORTON, J. A., *Welding J.* 64(1): 22s–28s.

DAVIES, R. G., and STOLOFF, N. S. 1965, *Trans. TMS-AIME* 233: 714–719.

DIMIDUK, D. M., WEDDINGTON, V. L. and LIPSITT, H. A. 1987, In *High Temperature Ordered Intermetallic Alloys II*, MRS Symp. Proc. 81: 221–228.

DIMIDUK, D. and RAO, S. 1991, In *High Temperature Ordered Intermetallic Alloys IV*, MRS Symp. Proc. 213: 499–513.

DONNELLY, S. G. and KOCH, C. C., 1983, ECUT *Quarterly Reports, Sept.* 1, 1983 *to Sept.* 1, 1986, North Carolina State University.

DOHERTY, J. E., GIAMEI, A. F. and KEAR, B. H., 1975, *Metall. Trans. A.* 6A: 2195–2199.

DOUIN, J., VEYSSIÉRE, P., and BEAUCHAMP, P., 1986, *Phil. Mag. A.*, 54: 375–393.

DOWLING, W. E., and GIBALA, R., 1987 Journal of Metals 39(11); abstract for TMS-AIME Annual Meeting, Phoenix Arizona Jan. 24–28, 1988.

EZZ, S., POPE, D. P. and PAIDAR, V., 1982, *Acta Metall.* 30: 921–926.

FARKAS, D. and RANGARAJAN, V., 1987, *Acta Metall.* 35: 353–365.

FLINN, P., 1960, *Trans. TMS-AIME* 218: 145–154.

FU, C. L., YE, Y. Y. and YOO, M. H., 1992, In *High Temperature Ordered Intermetallics V*, MRS Symp. Proc., 288: 21–32.

FUCHS, G. E., 1986, Ph.D. thesis, Rensselaer Polytechnic Institute, Troy, NY.

GEORGE, E. P., LIU, C. T. and POPE, D. P., 1992, *Scripta Metall.*. 27: 365–370.

GIESEKE, B., and SIKKA, V. K., 1992, *Mat. Sci. Eng.* A153: 520–524.

GUARD, R. W., and WESTBROOK, J. H., 1959, *Trans. TMS-AIME* 215: 807–814.

HAN, Y. F., LI, S. H., MA, S. and TAN, Y. N., Paper presented at First Pacific Rim Int. Conf. on Advanced Mat. and Processing Hongshou, China, June 23–27, 1992.

HANADA, S., WATANABE, S., and IZUMI, O., 1986, *J. Mat. Sci.* 21, 302.

HAZZELDINE, P. and HIRSCH, P. 1987, In *High Temperature Ordered Intermetallic Alloys II*, MRS Symp. Proc., 81: 75–86.

HEREDIA, F. and POPE, D. P., 1989, In *High Temperature Ordered Intermetallic Alloys III*, MRS Symp. Proc., 133: 287–291.

HILLIER, G. S., 1984, Ph.D. thesis, Cambridge University.

HIRANO, T., CHUNG, S. -S., MISHIMA, Y. and SUZUKI, T., 1991, In *High Temperature Ordered Intermetallic Alloys IV*, MRS Symp. Proc., 213: 635–640.

HIRSCH, P. B., 1991, In *Ordered Intermetallics—Physical Metallurgy and Mechanical Behavior*, ed. C. T. Liu, R. W. Cahn, and G. Sauthoff 197–216. Dordrecht: Kluwer Acad. Publ.

HORTON, J. A., 1987, Oak Ridge National Laboratory, unpublished.

HORTON, J. A., KOCH, C. C., and LIU, C. T., 1983, *Proc. 41st Ann. Mtg., Electron Microscopic Soc. of America*, Vol. 248, San Francisco Press.

HORTON, J. A., LIU, C. T. and SANTELLA, M. L., 1987, *Metall. Trans. A.* 18A: 1265–1277.

HORTON, J. A., AND MILLER, M. K., 1986, *J. de Physique*, Colloque Supl. 47: C2, 209–214.

________, 1987, *Acta Metall.*, 35: 133–141.

HSIUNG, L. M., and STOLOFF, N. S., 1988, *Proc. ICSMA 8, Tampere, Finland*, 683–689. ed. P. O. Kettunen et al., Oxford: Pergamon.

HSU, S. E., HSU, N. S., TONG, C. H., MA, C. Y., and LEE, S. Y., 1987, In *High Temperature Ordered Intermetallic Alloys II*, MRS Symp. Proc., 81: 507–512.

HUANG, S. C., TAUB, A. I. and CHANG, K. M., 1984, *Acta Metall.* 32: 1703–1707.

HUANG, S. C., CHANG, K. M., HALL, E. L., and LAFORCE, R. P., 1985, In *High Temperature Ordered Intermetallic Alloys*, MRS Symp. Proc., 39: 125–132.

HUANG, S. C., TAUB, A. I. and CHANG, K. M., 1986. *J. Mat. Res.* 1: 60–67.

INOUE, A., TOMIOKU, H., and MASUMOTO, T., 1983, *Metall. Trans. A* 14A: 1367–1377.

INOUE, A. L., MASUMOTO, T., and TOMIOKA, H., 1985, *J. Mat. Sci.* 19: 377–380.

KARG, A. V., FORNWALT, D. E., and KRIEGE, O. H., 1971, *J. Inst. Met.* 99: 301–305.

KEAR, B. H., and WILSDORF, H. G. F., 1962, *Trans. TMS-AIME* 224: 382–386.

KHADKIKAR, P. S., VEDULA, K., and SHABEL, B. S., 1987, *Metall. Trans. A.* 18A: 425–428.

KIM, M. S., HANADA, S., WATANABE, S., and IZUMI, O., 1988, *Trans. Japan Inst. Met.* 29: 274–283.

KING, A. H., and YOO, M. H., 1987, In *High Temperature Ordered Intermetallic Alloys II*, MRS Symp. Proc., 81: 99–104.

KOCH, C. 1988, North Carolina State University, Final Report, ORNL Subcontract No. 19X-43368C.

KRIEGE, O. H., and SULLIVAN, C. P., 1968, *Trans. ASM* 61: 278–282.

KRUISMAN, J. J., VITEK, V., and DELTOSSON J. Th. M., 1989, *Acta Metall.* 36: 2729–2741.

KUNG, H. RASMUSSEN, D. R. and SASS, S. L. 1991, *Scripta. Metall.*, 25: 1277–1281.

LALL, C., CHIN S. and POPE, D. P., 1979, *Metall Trans. A.*, 10A: 1323–1332.

LIN, D. L., CHEN, D., and LIN. H., 1991, *Acta Metall.* 39: 523–528.

LIU, C. T., 1987, In *High Temperature Ordered Intermetallic Alloys II*, MRS Symp. Proc., 81: 355–368.

LIU, C. T. and C. C. KOCH, p 42–1-19 in Technical Aspects of Critical Materials use by the steel industry, Vol II B, NBSIR 83-2679-2, NBS, June 1983.

LIU, C. T., LEE, E. H. and MCKAMEY, C. G., 1989, *Scripta Metall.* 23: 875–880.

LIU, C. T. and SIKKA, V. K., 1986, *J. Metals* 38(5):19–21.

LIU, C. T., and WHITE, C., 1987, *Acta Metall.* 35:643–649.

LIU, C. T., WHITE, C. L., and HORTON, J. A., 1985, *Acta Metall.* 33:213–229.

________, April 1992. U.S. Patent No. 5, 108, 700.

________, 1992, *Scripta Metall.* 27: 25–28.

LOPEZ, J. A., and HANCOCK, G. F., 1970, *Phys. Stat. Sol.(a)* 2: 469–474.

MASAHASHI, N., TAKASUGI, T., and IZUMI, O., 1986, *J. Mat. Sci.* 21: 2599–2607.

________, 1988b, *Metall. Trans.* 19A: 353–358.

________, 1988a, *Acta Metall.* 36: 1823–1836.

MATUSZYK, W., CAMUS, G., DUQUETTE, D. J., and STOLOFF, N. S., 1990, *Metall, Trans. A* 21A: 2967–2976.

MILLER, M. K., and HORTON, J. A., 1987, In *High-Temperature Ordered Intermetallic Alloys II*, MRS Symp. Proc., 81: 117–126.

MISHIMA, Y., OCHIAI, S., YODOGAWA, M., and SUZUKI, T., 1986, *Trans. Japan Inst. Met.* 27: 41–50.

MOTT, N. F., and NABARRO, F. R. N., 1948, In *Report Conf. Strength Sol. Phys. Soc., London*, 1–9.

NOGUCHI, O., OYA, Y. and SUZUKI, T., 1981, *Metall. Trans. A* 12A: 1647–1653.

OCHIAI, S., OYA, Y., and SUZUKI, T., 1984, *Acta Metall.* 32: 289–298.

OCHIAI, S., MISHIMA, Y., YODOGAWA, M., and SUZUKI, T., 1986, *Trans. Japan Inst. Met.* 27: 32–40.

OYA, Y., MISHIMA, Y., YAMADA, K., and SUZUKI, T., 1984, *J. Iron and Steel Inst. of Japan* 70: 80.

PAIDAR, V., POPE, D. P., and VITEK, V., 1984, *Acta Metall.* 32: 435–448.

PATTEN, J. W., 1990 in Proc. "High Temperature Aluminides and Intermetallic." ed. S. H. Whang, C. T. Liu, D. P. Pope and J. O. Stiegler, 493–504, Warrendale PA: TMS.

POPE, D. P., and EZZ, S. S., 1984, *Int. Met. Rev.* 23: 136–167.

POPE, D. P., and VITEK, V., 1987, In *High Temperature Ordered Intermetallic Alloys II*, MRS Symp. Proc., 81: 183–192.

SANTELLA, M. L., DAVID, S. A., and WHITE, C. L., 1985, In *High Temperature Ordered Intermetallic Alloys*, MRS Symp. Proc., 39: 495–503.

SANTELLA, M. L., DAVID, S. A., and HORTON, J. A. 1987, In *Advances in Welding Science and Technology*, ASM, Metals Park, OH, 629–633.

SCHNEIBEL, J. H., PETERSON, G. F., AND LIU, C. T., 1986, *J. Mat. Res.* 1: 68–72.

SCHULSON, E. M., BAKER, I., and FROST, H. J., 1987, In *High Temperature Ordered Intermetallic Alloys II*, MRS Symp. Proc., 81: 195–205.

SHAH, D. M., and DUHL D. N., 1987, In *High Temperature Ordered Intermetallic Alloys II*, MRS Symp. Proc., 81: 411–417.

SIELOFF, D. D., BRENNER, S. S., and BURKE, M. G., 1987, In *High Temperature Ordered Intermetallic Alloys II*, MRS Symp. Proc., 81: 87–97.

SIKKA, V. K., 1987, In *High Temperature Ordered Intermetallic Alloys II*, MRS Symp. Proc., 81: 1487–1494.

________, 1989, *Mal. Manuf. Proc.*, 4: 1–24.

SIKKA, V. K., 1990, In Proc. "High Temperature Aluminides and Intermetallics, ed. S. H. Whang, C. T. Liu, D. P. Pope and J. O. Stiegler, 505–520, Warrendale, PA: TMS.

SIKKA, V. K., MAVITY, J. T., AND ANDERSON, K., 1992, *Mat. Sci. Eng.* A153: 712–721.

SLUITER, M., TURCHI, P.E.A., PINSKI, F. J., and STOCKS, G. M., 1992, *Mat. Sci. Eng.* A152: 1–8.

STEIN, D. F., and HELDT, L. A., 1977, In *Interfacial Segregation*, eds. W. C. Johnson and J. M. Blakeley, 239–260. Metals Park, OH: ASM.

STOCKINGER, C. T., and NEUMANN, J. P., 1970, *J. Appl. Crystall.* 3, 32.

STOLOFF, N. S., FUCHS, G. E., KURUVILLA, A. K., and CHOE, S. J., 1987, In *High Temperature Ordered Intermetallic Alloys II*, MRS Symp. Proc., 81: 247–261.

TAKASUGI, T., IZUMI, O., and MASAHASHI, N., 1985, *Acta Metall.* 33: 1259–1269.

TAKASUGI, T., MASAHASHI, N., and IZUMI, O., 1986, *Scripta Metall.* 20: 1317–1321.

TAKEUCHI, S., and KURAMOTO, E., 1973, *Acta Metall.* 21: 415–424.

TAKEUCHI, S., KURAMOTO, E., YAMAMOTO, T., and TAOKA, T., 1973, *Japanese J. of Appl. Phys.* 12: 1486.

TAKEYAMA, M. and LIU, C. T., 1988a, *Acta Metall.* 36: 1241–1250.

________, 1988b. *J. Mater. Res.* 3: 665–674.

TAUB, A. I. and BRIANT, C. L., 1987, *Acta Metall.* 35, 1597–1603.

TAUB, A. I., BRIANT, C. L., HUANG, S. C., CHANG, K. M., and JACKSON, M. R., 1986. *Scripta Metall.* 20: 129–134.

TAUB, A. L., HUANG, S. C., AND CHANG, K.M. *Metall. Trans. A* 1984, 15A 399–402.

TAUB, A. L., CHANG, K.M., and LIU, C. T., 1986, *Scripta Metall.* 20: 1613–1618.

TAUB, A., HUANG, S. C., and CHANG, K. M., 1985 In *High Temperature Ordered Intermetallic Alloys*, MRS Symp. Proc., 39: 221–228.

THORNTON, P. H., DAVIES, R. G., and JOHNSTON, T. L., 1970, *Metall. Trans.* 1: 207–218.

VEYSSIÉRE, P., 1984, *Phil. Mag.* 50A, 189–203.

VEYSSIÉRE, P., DOUIN, J., and BEAUCHAMP, P., 1985, *Phil. Mag. A* 51: 469–483.

VITEK, V., and POPE, D. P., 1985, In *High Temperature Ordered Intermetallic Alloys*,MRS Symp. Proc., 39: 205–212.

VITEK, V., CHEN, S. P., VOTER, A. F., KRUISMAN, J. J., and DEHOSSON, J. Th. M., 1989, In *Grain Boundary Chemistry and Intergranular Fracture, Mat. Sci. Forum*. 46: 237.

WEIHS, T. P., ZINOVIEV, V., VIENS, D. V., and SCHULSON, E. M., *Acta Metall.* 35: 1109–1118.

WHITE, C. L., PADGETT, R. A., LIU, C. T., and YALISOVE, S. M., 1984, *Scripta Metall.* 18: 1417–1422.

WILLIAMS, R. K., GRAVES, R. S., WEAVER, F. J., and MCELROY, D. L., 1985, In *High Temperature Ordered Intermetallic Alloys*, MRS Symp. Proc. 39: 505–512.

WRIGHT, R. N., and FLINN, J. E., 1986 "Consolidation and Properties of Rapidly Solidified Nickel Aluminide Powders," presented at ASM/AIME Fall meeting, Lake Buena Vista, FL. Oct. 2–4, 1986 (unpublished).

YAMAGUCHI, M., PAIDAR, V., POPE, D. P., and VITEK, V., 1982, *Phil. Mag. A* 45: 867–882.

YIZHANG, Z., TIANXIANG, Z., YINGJIE, T., BINGDA, Z., YAOXIA, Z., AND ZHUANGQI, H. In *High Temperature Ordered Intermetallic Alloys II*, MRS Symp. Proc. 133: 555–559.

YOO, M. H., 1987, In *High Temperature Ordered Intermetallic Alloys II*, MRS Symp. Proc. 81: 207–212.

Chapter 7

The Physical and Mechanical Metallurgy of NiAl

Ronald D. Noebe, Randy R. Bowman, and Michael V. Nathal

ABSTRACT

Considerable research has been performed on NiAl over the last decade, with an exponential increase in effort occurring over the last few years. This is due to interest in this material for electronic, catalytic, coating, and especially high-temperature structural applications. This report uses this wealth of new information to develop a comprehensive description of the properties and processing of NiAl and NiAl-based materials. Emphasis is placed on the controlling fracture and deformation mechanisms of single and polycrystalline NiAl and its alloys over the entire range of temperatures for which data are available. Creep, fatigue, and environmental resistance of this material are discussed. In addition, issues surrounding alloy design, development of NiAl-based composites, and materials processing are addressed.

1. INTRODUCTION

As with many intermetallic alloys, NiAl was originally studied as a potential structural material because of its high melting temperature, hardness, and chemical stability. Early investigations were exploratory in nature, designed to determine whether NiAl held promise as a high-temperature refractory compound (Wachtell 1952). These were followed by studies in the early 1960s that concentrated on the effects of processing and other metallurgical variables on mechanical behavior (Wood et al. 1960–1964). By the mid-1960s NiAl was identified as a possible leading-edge material for a superalloy turbine vane. However, no solution was found for the low-temperature brittleness of this compound, and by the end of the 1960s government and industrial interest in NiAl had faded. At this point, research shifted primarily to universities; between 1970 and the mid-1980s, there was a very small, but steady, effort to investigate the oxidation behavior, mechanical properties, and deformation mechanisms of NiAl.

Then, in the mid- to late-1980s, research on NiAl exploded on several fronts. The physics/surface science community became intrigued with the surface properties and catalytic behavior of NiAl. This work was motivated by the fact that NiAl is one of the few intermetallic systems known to have a stable, well-defined surface structure on an atomic level (Franchy, Wuttig, and Ibach 1987; Castro et al. 1991). Because of NiAl's high melting point, excellent thermal stability, and serendipitous lattice match with GaAs compounds, the electronics industry began to take a serious look at using NiAl as a buried interconnect in electronic components (Sands 1988; Chambers and Loebs 1990; Joo et al. 1992). Also, research on NiAl is again centered on its possible use as a high-temperature structural material. The renewed driving forces for this general application have been generic government aeropropulsion programs (Stephens 1988; Doychak 1992) and industrial development of NiAl as a turbine engine material (Darolia 1991; Darolia et al. 1992).

Current researchers face the same problems that hindered widespread acceptance of NiAl as a structural material in the early 1960s: namely, poor creep resistance and inadequate low-temperature toughness and ductility. However, these barriers are disappearing as the properties of NiAl and our understanding of its behavior improve.

This chapter reviews the physical and mechanical metallurgy of NiAl-based materials, concentrating on the effects of processing, alloying, and microstructural modification on behavior. Directions for improving low-temperature toughness and ductility are described, and successful strategies for enhancing creep resistance are reviewed. References are made primarily to the most recent studies in each area because in-depth reviews of the NiAl literature have recently been published (Noebe, Bowman, and Nathal 1993; Miracle 1993). All alloy compositions are given in atomic percent.

2. PHYSICAL METALLURGY

NiAl is an ordered intermetallic compound that crystallizes in a primitive cubic CsCl structure that exists over the composition range of 45 to almost 60 at%Ni. Although the phase diagram by Singleton, Murray, and Nash (1986) indicates that stoichiometric Ni-50Al melts congruently at 1911 K, more recent determinations place the melting temperature T_m of binary stoichiometric NiAl closer to 1955 K (Walston and Darolia 1993). The original lower value for T_m could be attributed to the steep drop-off in melting temperature with deviations from stoichiometry or to unintentional additions of ternary elements. NiAl not only has the highest melting temperature of any compound in the Ni-Al system but also is the most stable, as evident from its very large negative heat of formation, −72 kJ/mol at the stoichiometric composition (Henig and Lukas 1975). This high degree of thermodynamic stability and the existence of a wide phase field makes NiAl relatively easy to fabricate in a range of forms from fine homogeneous powders to single crystals.

Properties such as lattice parameter and density have been thoroughly studied as a method to infer the type of defect structure that occurs in the NiAl lattice, and consequently, significant data have been generated for these properties (Fig. 7-1). Because Ni is a smaller, but heavier, atom than Al, increasing the Ni content by substituting Ni on Al sites should decrease the lattice parameter and increase the density, consistent with data for alloys containing greater than 50 at%Ni (Fig. 7-1). However, the behavior of Al-rich alloys does not continue to follow this trend. Instead, the lattice parameter decreases, and the decrease in density is more rapid than would be expected from the replacement of Ni atoms by Al. Consequently, the deviation in stoichiometry of Al-rich alloys is accommodated by the creation of vacancies on the Ni-lattice sites instead of by substitutional defects as in Ni-rich alloys. This process can result in extremely large numbers of constitutional vacancies. For example, a Ni-55Al intermetallic would contain a constitutional vacancy concentration of almost 10%.

Another property that has been extensively studied is the elastic behavior of NiAl. The single-crystal elastic constants for NiAl, which have been determined as a function of temperature, cooling rate (thermal vacancy concentration), and stoichiometry, have been summarized by Rusovic and Warlimont (1977). Overall, the elastic properties of NiAl are anisotropic, $A = 3.28$ (Wasilewski 1966), and the degree of anisotropy depends mildly on temperature and strongly on stoichiometry. Figure 7-2a shows the dynamic Young's modulus E for near-stoichiometric NiAl as a function of orientation—demonstrating the mild, but linear, dependence of temperature on modulus and the slightly decreasing anisotropy with increasing temperature. Good agreement exists between the temperature dependence of modulus for ⟨100⟩ crystals that was determined by Wasilewski in 1966 and by Walston and Darolia in 1993. Also, minor alloying additions have relatively little effect on the dynamic Young's modulus of ⟨001⟩ single crystals (Walston and Darolia 1993).

The elastic properties of polycrystalline NiAl have also been determined in some detail. Young's modulus is very dependent on processing technique and temperature but relatively insensitive to stoichiometry (Rusovic and Warlimont 1979; Harmouche and Wolfenden 1987; Moose 1991). The effect of processing route on the temperature dependence of the Young's modulus for near-stoichiometric NiAl is demonstrated in Figure 7-2b. Extrusion produces a material that has a higher modulus and a slightly different temperature dependence compared with materials produced from hot-pressed prealloyed powders or cast and homogenized ingots. By comparison with the single-crystal data, the effect of processing

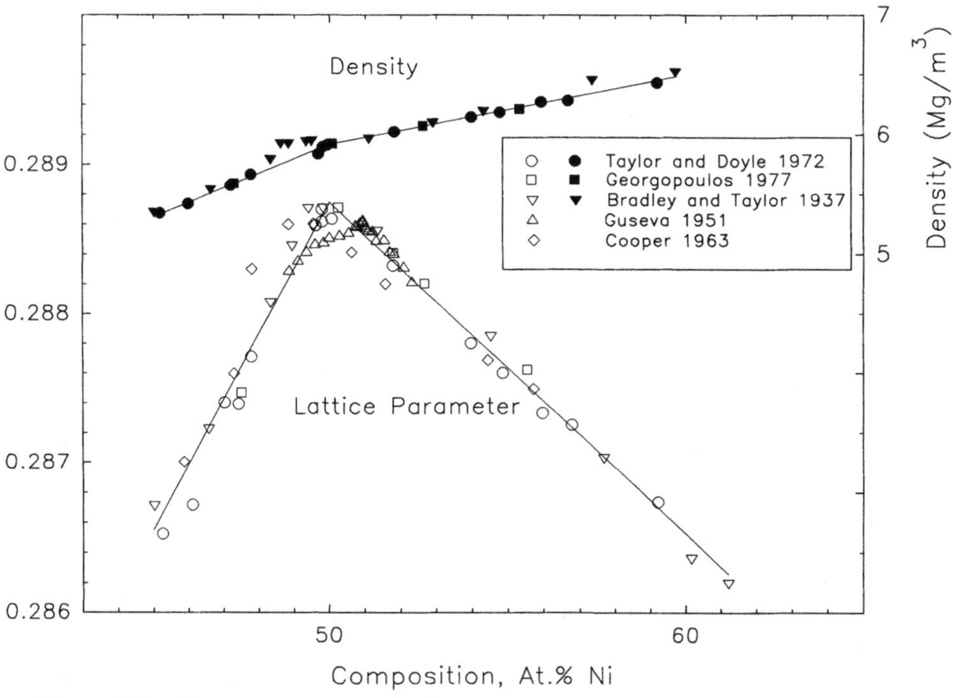

Figure 7-1. Room-temperature lattice parameter and density of NiAl as a function of stoichiometry.

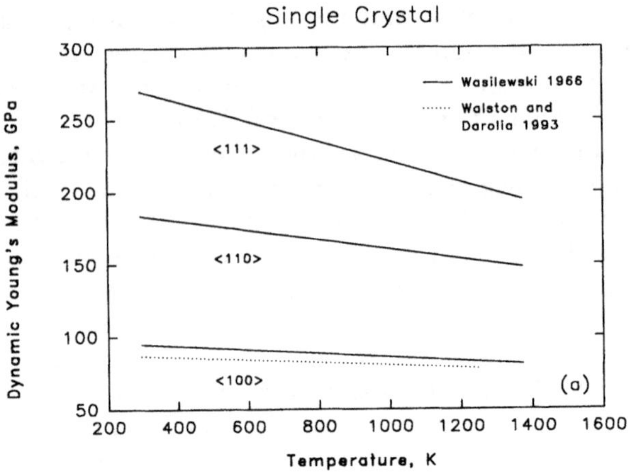

(a)

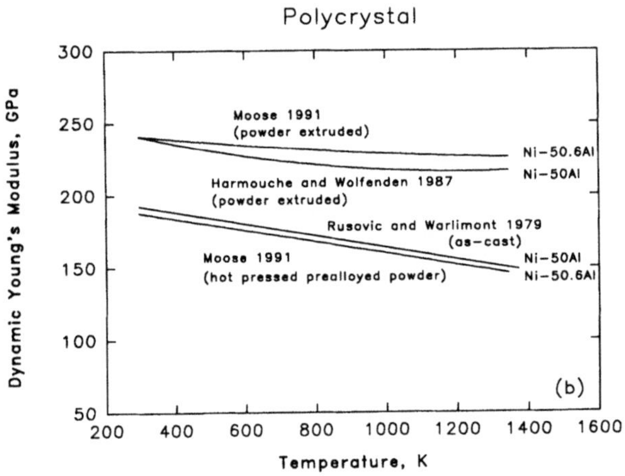

(b)

Figure 7-2. (a) Dynamic Young's modulus for single-crystal NiAl as a function of orientation and temperature. (b) Effect of processing and temperature on dynamic Young's modulus of polycrystalline NiAl.

on Young's modulus can be rationalized in terms of crystallographic texture. The extruded material probably exhibits a preferred ⟨111⟩ orientation because this is the most commonly observed texture in extruded NiAl (Khadkikar, Michal, and Vedula 1990; Beiler et al. 1992). Cast or hot-pressed material would not be expected to have a strong preferred orientation and would, therefore, have a lower modulus in comparison to the higher modulus for ⟨111⟩-oriented materials.

Table 7-1 Physical Properties of NiAl, NiAl Alloys, and an Advanced Superalloy

Property	Temperature, K	NiAl	NiAl alloys[a]	Advanced superalloy
Bonding	—	Covalent/metallic	Covalent/metallic	Metallic
Melting point, K	—	1955[b]	1883 to 1949	1663
Matrix lattice parameter, Å	Room	2.887	2.888 to 2.900	3.580
Matrix/ppt. lattice mismatch, percent		N/A	1 to 6 (β')	0 to −0.5 (γ')
Density, g/cm^3		5.9	up to 6.30	8.60
Young's modulus, polycrystal, GPa		188	188	205
Young's modulus, ⟨001⟩, GPa		88	88	130
Anisotropy factor		3.25	3.25	2.72
Shear modulus, polycrystal, GPa		71.5	71.5	74
Poisson's ratio		0.313	0.313	0.380
Thermal expansion, $10^{-6}/K$	873	13.2	13.7	13.5
Specific heat, $J/g \cdot K$		0.64	0.61 to 0.64	0.46
Thermal diffusivity, cm^2/sec		0.22	0.10 to 0.22	0.033
Thermal conductivity, $W/m \cdot K$		76	35 to 76	15
Electrical resistivity, $\mu\omega \cdot cm$	Room	8 to 10	10 to 30	120 to 140

[a] NiAl alloys containing primarily β' precipitates and less than 5 at% alloying addition.

[b] General Electric Aircraft Engines Differential Thermal Analysis data show that the melting point of NiAl is about 40 K higher than shown in literature data.

Source: Walston and Darolia 1993

Most other physical properties have not been characterized to the same degree as lattice parameter, density, and elastic modulus. Yet, once researchers realized that the high thermal conductivity of NiAl was one of its major advantages over superalloys, additional characterization of NiAl alloys has taken place (Walston and Darolia 1993).

Table 7-1 summarizes the physical properties of binary NiAl and creep resistant NiAl alloys containing Hf, Re, or Ti and compares these properties to those of a typical Ni-base superalloy. For high-temperature structural applications, NiAl would have the greatest advantage over superalloys in terms of its thermal conductivity and density. Because of these differences in properties, replacing superalloy high-pressure turbine blades with ones made from NiAl would reduce turbine rotor weight by 40% and the "hot spot" temperature by 50 K (Darolia et al. 1992). Also, direct replacement of superalloy components with a NiAl alloy would be straightforward because the moduli and thermal expansion characteristics are similar.

3. FLOW AND FRACTURE

3.1. Plasticity and Operative Slip Systems

The operative slip systems in NiAl single crystals and polycrystalline material have been rigorously investigated and described in detail (Noebe et al. 1993; Miracle 1993). In general, NiAl exhibits two significantly different types of slip behavior depending on crystal orientation. The dominant slip vector for "soft" single-crystal orientations and polycrystalline material is ⟨001⟩. But, if the loading direction is along [001], known as the "hard" single-crystal orientation, the operative slip vector is ⟨111⟩ at low and intermediate temperatures and a combination of ⟨110⟩ and ⟨100⟩ at elevated temperatures (Table 7-2). Soft orientations include all non-⟨001⟩ loading directions where ⟨100⟩ slip dominates. Test orientations close to [001] are hard orientations because ⟨001⟩ Burgers vectors have a zero or near-zero resolved shear stress.

By all indications, NiAl obeys Schmid's law and deforms by a ⟨001⟩ slip vector in either {100} or {110} planes for all but near-[001] crystal orientations (Lahrman, Field, and Darolia 1993a). Ball and Smallman (1966a, b) were the first to make a complete slip system determination for NiAl by identifying a ⟨001⟩ slip vector and {110} slip plane in all soft orientations and at all temperatures investigated (300 K to 1273 K). They also observed cross slip or pencil glide on orthogonal {110} planes. In addition to ⟨001⟩{110} slip, Wasilewski, Butler, and Hanlon (1967)

Table 7-2 Observed Slip Systems in Uniaxially Deformed NiAl

Temperature range (K)	Slip vector	Slip plane
	POLYCRYSTALS	
300 to 1200[a]	⟨100⟩	{011}, {001}
	SINGLE CRYSTALS "SOFT" ORIENTATIONS (NON - [100])	
77 to 1300	⟨100⟩	{011}, {001}
	"HARD" ORIENTATIONS - [001]	
77 to 600	⟨111⟩	{112}, {011}, or {123}
600 to 1372	⟨110⟩	{011}
300 to 1300	⟨100⟩	{011}

[a] Less than five independent slip systems.

observed duplex cube slip, ⟨001⟩{100}, in [110]-oriented single crystals (as have Field, Lahrman, and Darolia 1991a). Cube slip also was seen by Loretto and Wasilewski (1971) in [112] crystals deformed between 77 K and 1053 K. In soft-orientation single crystals, only ⟨001⟩ slip is observed because of the nondissociated, compact structure of the ⟨001⟩ dislocation core (Mills and Miracle 1993), making ⟨001⟩ dislocations much more mobile than dislocations with any other slip vector.

Consistent with deformation studies on soft-orientation single crystals, investigators have reported the operation of a ⟨001⟩ slip vector in polycrystalline NiAl gliding on either {110} or {100} planes (Bowman et al. 1992; Nagpal and Baker 1992; Cotton, Noebe, and Kaufman 1993b). Isolated dislocation segments with Burgers vectors other than ⟨001⟩ have been identified in as-extruded NiAl (Lloyd and Loretto 1970; Dollar et al. 1992). However, the presence of non-⟨001⟩ dislocations in these polycrystalline studies does not indicate the operation of an alternate deformation mechanism. Dislocations such as ⟨110⟩ are probably formed by interactions between ⟨001⟩ dislocations because of the extensive deformation that occurs during the extrusion process (Baker and Schulson 1984). Furthermore, non-⟨001⟩ dislocations are not observed in as-cast materials or after room-temperature deformation of cast alloys (Cotton et al. 1993b).

The operation of a ⟨100⟩ slip vector on planes other than {001} and {011} has been observed under special conditions of constrained flow. For example, Miracle (1991) observed a (210) slip plane in one of the many deformed bicrystals he analyzed. Also, Wunderlich, Machon, and Sauthoff (1992) observed ⟨100⟩{310} slip in NiAl within the plastic zone produced by a blunted crack. This crack was initiated in the NiNbAl phase after high-temperature deformation of a NiAl/NiAlNb alloy. Although observations of atypical slip systems are ardently reported, these results need to be tempered by the realization that even in the above studies ⟨100⟩{011} slip was the dominant deformation mechanism.

Deformation of single crystals oriented along [001] is a special case in terms of operative slip systems because the resolved shear stress for ⟨100⟩ slip approaches zero. As a result, deformation occurs primarily by non-⟨001⟩ dislocation processes. This gives rise to a high yield stress at low temperatures (Wasilewski et al. 1967), as well as enhanced creep strength at elevated temperatures for [001] crystals compared to non-[001] crystals (Forbes et al. 1993). Consequently, significant effort has been spent on experimental analyses of the operative deformation mechanisms in hard-orientation crystals.

From 77 K to approximately 600 K, the primary slip vector in hard-orientation crystals is ⟨111⟩, with the most likely slip plane being {112} (Pascoe and Newey 1968a; Kim and Gibala 1991; Veyssiére and Noebe 1992). Previous reports of deformation by ⟨001⟩ dislocations in [001] crystals over this temperature range (Fraser, Loretto, and Smallman 1973) have been attributed to an unstable sample geometry, which invariably leads to kinking (Bowman, Noebe, and Darolia 1989). At higher temperatures (> 600 K, which is above the brittle-to-ductile transition temperature (BDTT) for [001] crystals), the operative slip vector changes, and deformation occurs by nonconservative motion of ⟨001⟩ and ⟨011⟩ dislocations (Kim and Gibala 1991; Field et al. 1991b).

3.2. Yield Strength

Although yield stress is highly sensitive to many metallurgical variables, all studies on single-crystal and polycrystalline material concur that with increasing temperature, yield strength decreases or remains constant over certain temperature regimes (Figs. 7-3a and b). In general, the flow stress of NiAl is similar to that of body-centered cubic metals, exhibiting a strong temperature dependence at low absolute temperatures that is attributed to a large Peierls stress. At intermediate temperatures, there is usually a plateau where yield stress is only mildly dependent on temperature; this is followed by a further drop in strength at elevated temperatures.

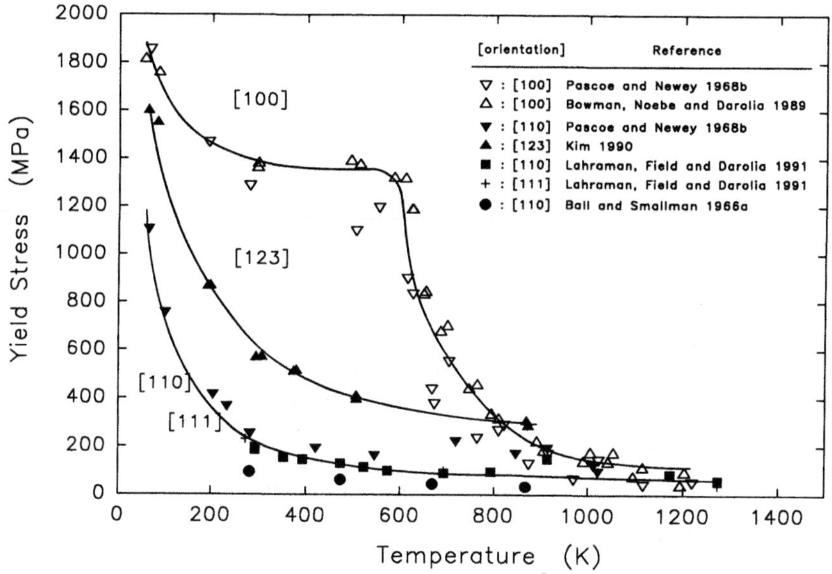

(a)

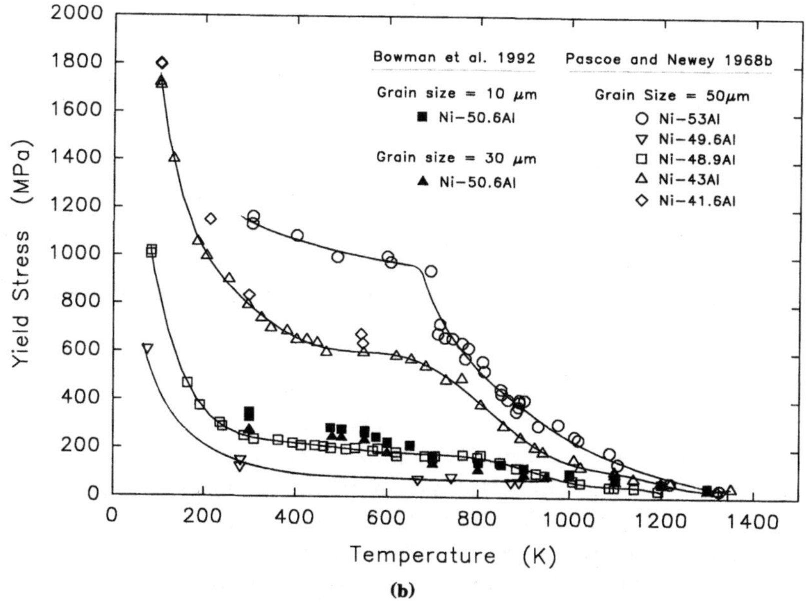

(b)

Figure 7-3. Yield stress as a function of temperature for (a) several different single-crystal orientations and (b) polycrystalline NiAl as a function of strain rate, stoichiometry, and temperature.

The yield strength of NiAl single crystals has a very strong dependence on orientation (Fig. 7-3a). Single crystals with soft orientations exhibit a yield stress σ_y versus temperature T relation that is distinct from that for [001] crystals but very similar to that for near-stoichiometric polycrystalline NiAl. At lower temperatures, single crystals of NiAl that are loaded along $\langle 001 \rangle$ directions exhibit yield stresses several times higher than those for other orientations and are less strongly dependent on temperature than soft orientations are. At temperatures above 600 K, the yield strength of [001] crystals becomes very sensitive to temperature, with a significant reduction in σ_y occurring over a relatively narrow temperature range. Within this temperature regime, the slip vector in hard-orientation crystals begins to change from $\langle 111 \rangle$ to $\langle 001 \rangle$ and $\langle 110 \rangle$ (Kim and Gibala 1991). Above 1000 K, where bulk diffusional processes begin to dominate, hard-orientation single crystals have yield strengths similar to soft-orientation single crystals and polycrystalline NiAl.

Figure 7-3b shows typical polycrystalline NiAl yield-stress behavior as a function of temperature, stoichiometry, and grain size. Although the values of the yield stress and the shape of the σ_y versus temperature curves depend on composition and strain rate, in all cases the yield stress decreases with increasing temperature. The behavior of near-stoichiometric polycrystalline NiAl resembles that of the soft-orientation, single-crystal material. With greater deviations from stoichiometry, polycrystalline material has considerably higher strength, comparable to that of stoichiometric [001] crystals, even though slip is expected to occur by $\langle 001 \rangle$ dislocations in all polycrystalline NiAl alloys.

Not only is the yield strength for NiAl anisotropic, but the strain-rate sensitivity is anisotropic also, with cube-oriented crystals displaying a greater sensitivity to strain rate than soft orientations at intermediate temperatures (500 K to 1000 K) (Pascoe and Newey 1968b; Lahrman et al. 1991). The greater sensitivity of [001] crystals to strain rate is due to the decomposed core structure of $\langle 110 \rangle$ dislocations, which cannot glide conservatively in hard-orientation crystals and the zero critical-resolved shear stress on $\langle 001 \rangle$ dislocations, making it necessary for both types of dislocations to rely on thermal activation to assist their motion (Mills et al. 1993). On the other hand, deformation of soft-orientation single crystals is controlled by $\langle 001 \rangle$ dislocations that have a compact core structure. Consequently, the yield stress of soft-orientation single crystals is relatively insensitive to strain rate below the brittle-ductile-transition temperature (BDTT). The BDTT itself is only mildly dependent on strain rate for soft-orientation single crystals. (Lahrman et al. 1991).

In polycrystalline NiAl, strain rate has almost no effect on yield strength below 600 K, with a strain-rate sensitivity $m < 0.01$. Strain rate has a moderate effect at temperatures between 600 K and 1000 K ($m \approx 0.06$) and a significant effect on yield strength above 1000 K in the creep deformation regime ($m \approx 0.18$) (Bowman et al. 1992). However, the BDTT for polycrystalline NiAl is strongly affected by strain rate. A three orders of magnitude increase in strain-rate results in approximately a 200 K increase in BDTT (Noebe, Cullers, and Bowman 1992). A similar rate dependence of the BDTT is observed in [001] single crystals (Lahrman et al. 1991).

Because σ_y is strongly dependent on temperature, it is useful to represent the yield stress data in an Arrhenius form where the slope of the curve is proportional to the activation energy for deformation. Consequently, changes in slope on an Arrhenius plot usually indicate a change in the deformation mechanism. When the polycrystalline NiAl yield strength data of Bowman et al. (1992) from Figure 7-3b are replotted in an Arrhenius form, three distinct deformation regimes are observed (Fig. 7-4). The lower temperature ($\sim$ 550 K) discontinuity in Figure 7-4, or change in deformation mechanism, is of particular interest because it coincides with the BDTT in NiAl. This change in slope in the Arrhenius plot reinforces the concept that additional deformation mechanisms are operating, accounting for the large-scale plasticity in NiAl above the BDTT. A change in slope of an Arrhenius plot of yield stress is also observed at the BDTT of Zr- and Re-doped polycrystalline NiAl alloys and hard-orientation single crystals (Noebe et al. 1990; Bowman et al. 1989). The actual deformation mechanisms responsible for this change in behavior are discussed in section 3.3. Finally, the change in slope beginning at region III is due to typical creep-deformation processes.

In NiAl, significant deviations are possible from the stoichiometric composition without altering the basic crystal structure of the intermetallic. This in turn, has a significant effect on mechanical behavior (as reviewed previously by Vedula and Khadkikar 1990). Although an increase in yield stress is observed for both Ni- and Al-rich alloys at low temperatures, the magnitude of the strengthening effect is not equivalent. From a previous compilation of low-temperature data, the average hardening rate for Ni-rich alloys was 120 MPa/at%, whereas that for Al-rich NiAl was approximately 350 MPa/at% (Noebe et al. 1993). The greater hardening rate for Al-rich alloys suggests that Ni vacancies provide a greater resistance to dislocation motion than antisite atoms. Regardless of whether the material is Ni-rich or Al-rich, the effects of nonstoichiometry on strength become negligible around 1000 K (Fig. 7-3b), and a reversal in strength occurs at higher temperatures such that stoichiometric alloys become stronger than nonstoichiometric compositions (Ball and Smallman 1966a; Whittenberger et al. 1991).

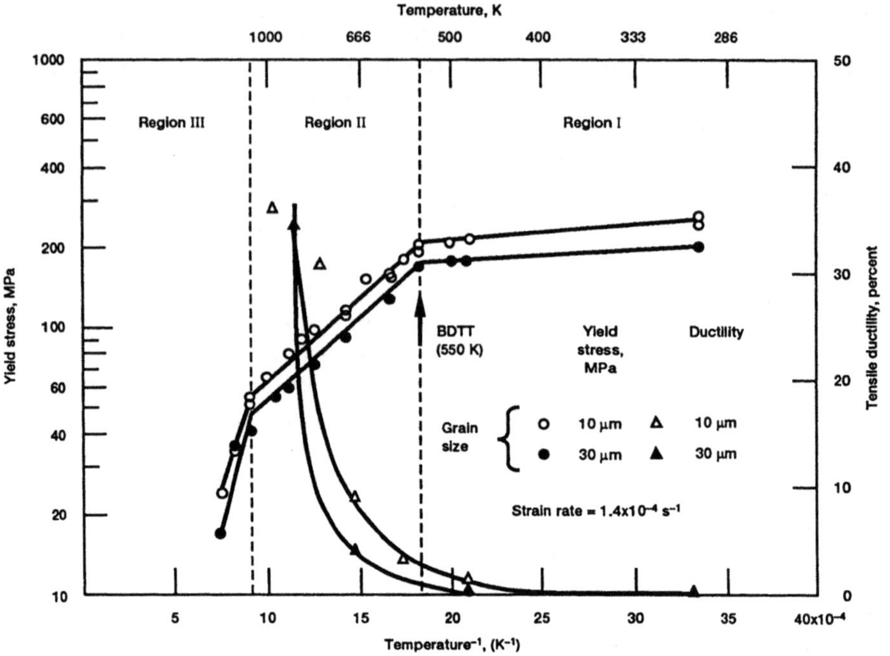

Figure 7-4. Yield stress and tensile ductility of polycrystalline NiAl as an inverse function of temperature. The brittle-to-ductile transition temperature (BDTT) defined by the dramatic increase in tensile ductility corresponds to a change in activation energy for plastic flow beginning at region II (Bowman et al. 1992). Strain rate, $1.4 \times 10^{-4}\ s^{-1}$.

Vacancies significantly influence the flow properties of NiAl as evident from the high hardening rates in Al-rich alloys, where constitutional vacancies are formed on the Ni sublattice. Another type of vacancy defect that can exist in NiAl is a thermal vacancy introduced by rapid quenching from elevated temperatures. Therefore, cooling rate becomes another important variable to be considered before testing NiAl. In polycrystals, a fifty-fold increase in cooling rate from temperatures above 1000 K can increase the compressive-yield stress almost 30% for near-stoichiometric binary NiAl. However, when the material is doped with minor alloying additions such as 500 ppm Zr, there is no dependence of cooling rate on strength, though the solute addition itself has a significant effect (Bowman et al. 1992). Similarly, deviations from stoichiometry reduce the sensitivity of binary NiAl to cooling-rate effects (Nagpal and

Baker 1990a). Cooling rate has a similar effect on the yield strength of NiAl single crystals (Weaver, Kaufman, and Noebe 1993).

Substitutional and interstitial elements also significantly affect the strength of NiAl. However, published single-crystal data are relatively scarce, and much of the available information is for hard orientations. Additions of Cr (Field et al. 1991c), V (Darolia et al. 1989), and Zr (Noebe et al. 1989) do not have any demonstrated hardening effect on [001]-oriented single crystals at low temperatures but do strengthen NiAl at temperatures above 600 K to 700 K, where a change in slip mode is known to occur. Systematic study of the influence of Mo, Ga, and Fe on the yield strength of ⟨110⟩-oriented, single-crystal NiAl revealed a range of strengthening behaviors at room temperature (Darolia, Lahrman, and Field 1992). These researchers determined that Mo is a potent solid-solution hardening agent but that it has a very limited solubility in NiAl. Ga has a mild strengthening effect, whereas Fe, at levels of less than 1 at%, slightly decreases the yield stress. Preliminary testing of high-purity single crystals also indicates that interstitial levels can significantly influence flow stress (Weaver et al. 1993). The critical resolved shear stress reported for high-purity, low-interstitial, [123]-oriented NiAl is only 57 MPa compared to 79 MPa for a commercial-purity material that received an identical thermal treatment.

Solid-solution alloying data for polycrystalline NiAl are more abundant than for single-crystal material. Ternary additions to polycrystalline NiAl have included Be, B, and C (George and Liu 1990); Cr and Cu (Cotton et al. 1993a, d); Y, Mo, and La (Graham 1984); Fe and Ga (Noebe and Behbehani 1992); and Zr (Bowman et al. 1992). Figure 7-5 shows the relation between the hardening rates of these elements in polycrystalline NiAl and atom size. In all cases the flow strength of NiAl was enhanced by the presence of solutes, and in general the hardening rate was proportional to solute size, though exceptions do exist. For example, Ga and Mo have very similar Goldschmidt or atomic radii but result in significantly different hardening rates when they are added to NiAl.

No acceptable theory of solid-solution strengthening can explain all aspects of hardening in intermetallic alloys, even for the relatively simple B2 compounds (Fleischer 1993). For example, substituting a particular ternary addition such as Cu for Ni in NiAl will result in significantly different hardening characteristics than if the element were substituted for Al (Cotton et al. 1993d). A further complication occurs if the alloying addition changes the overall stoichiometry of the intermetallic or if it is added to a nonstoichiometric base. For example, Figure 7-6 demonstrates the influence of Cr and Al levels on the hardness of NiAl. Chromium has a

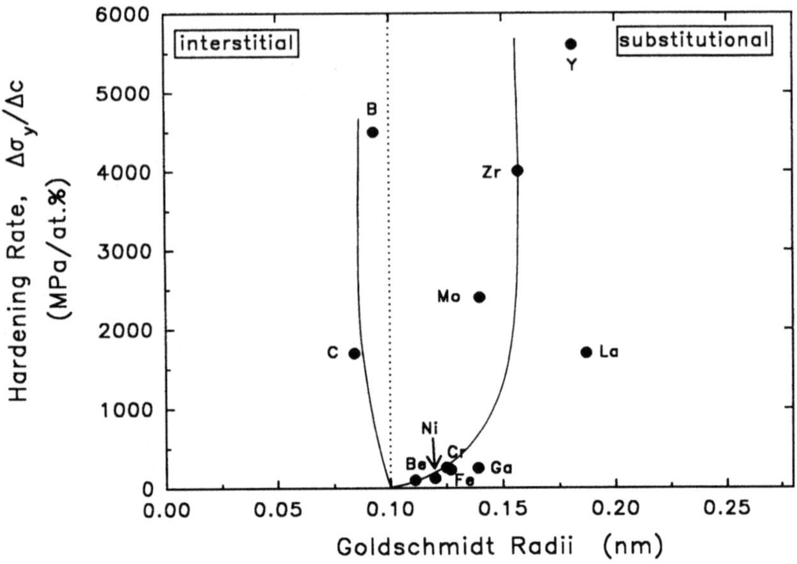

Figure 7-5. Relation between hardening rate and element size for various alloying additions in NiAl.

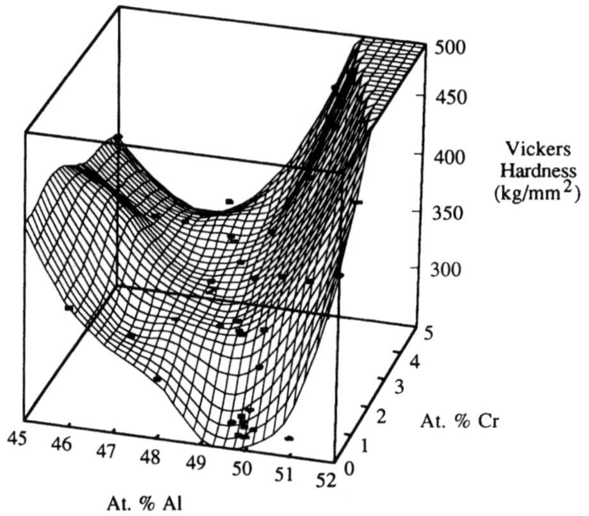

Figure 7-6. Combined effect of stoichiometry and Cr additions on the hardness of Cr-doped B2 NiAl alloys (Reprinted with permission from *Intermetallics*, J. D. Cotton, R. D. Noebe, and M. J. Kaufmann, pages 3–20, 1993, Elsevier Science Ltd., Oxford, England).

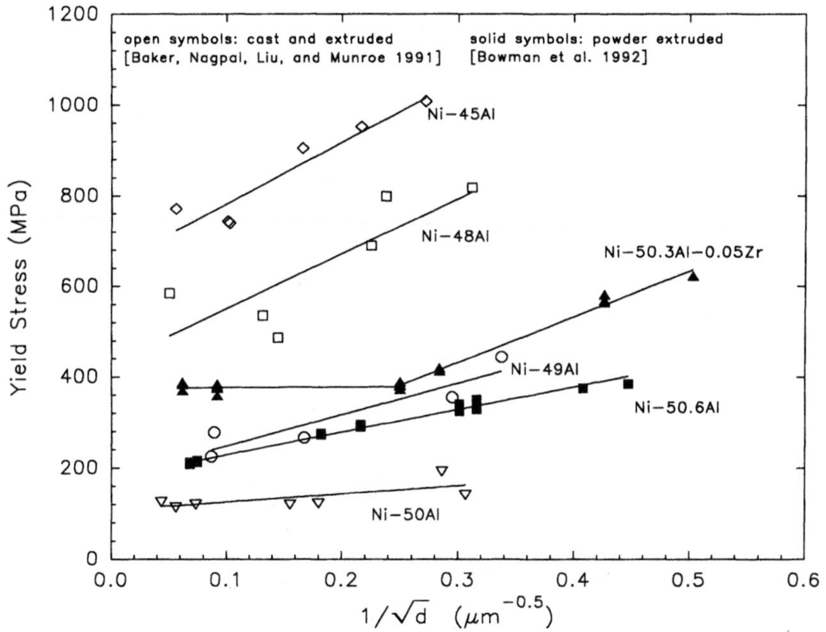

Figure 7-7. Effect of grain size, stoichiometry, and alloying additions on the room-temperature yield stress of NiAl alloys.

moderate strengthening effect on near-stoichiometric NiAl up to its solubility limit of about 2 at%, but even more significant strengthening is observed on either side of stoichiometry because of constitutional defects. Consequently, defect structures in ternary alloys need to be characterized so that addition and cancellation rules for the various defect-hardening mechanisms can be defined. The potential for defect pairing or clustering also has to be accounted for if solid-solution hardening models are to be effective (Cotton et al. 1993d).

Finally, grain size also influences the room-temperature yield strength of NiAl, but not in a straightforward manner. Grain-size effects are complicated by the additional variables of alloy stoichiometry and third-element additions (Fig. 7-7). The relation between alloy stoichiometry and Hall–Petch parameter (slope of the curves in Fig. 7-7) clearly shows that the influence of grain size on yield stress is more significant with increasing deviation from stoichiometry (Nagpal et al. 1991). Consequently, yield stress is essentially independent of grain size for Ni-50Al, whereas the yield stress of even slightly nonstoichiometric alloys is strongly influenced by grain size.

The effect of grain size on yield strength is further complicated when alloying additions are involved, as demonstrated in Figure 7-7. When NiAl was alloyed with 0.05 at%Zr, the yield stress was independent of grain size for grain sizes between 16 μm to 280 μm. However, it was strongly dependent on grain size for grains less than 16 μm because of the enrichment of Zr at the grain boundaries (Zeller, Noebe, and Locci 1990). In the region where the yield stress was independent of grain size, the mean free path of the dislocations was suspected to be less than the average grain diameter because of solid-solution effects.

In summary, many metallurgical parameters influence the flow strength of NiAl. However, the magnitude of the strengthening effect typically depends on other factors in a usually complicated manner. As previously noted, the manner in which temperature affects flow stress is sensitive to alloy composition and crystal orientation. The sensitivity of yield stress to strain rate depends on the temperature range being investigated and on the orientation for single crystals. The effect of cooling rate (thermal vacancy concentration) is not well understood but is mainly a concern for binary near-stoichiometric NiAl, but not ternary alloys. Several interstitial and substitutional alloying additions significantly affect flow strength even when they are present at rather low levels. In fact, these elements may have a significant effect on strength at concentrations that are not readily detectable by present analytical techniques. Finally, the amount of strengthening due to grain refinement is sensitive to both stoichiometry and alloying additions. Even with all of these variables accounted for, our present understanding of the flow behavior of NiAl alloys is still suspect because our present knowledge may be clouded by as-of-yet unknown interactions.

3.3. Ductility and Fracture

Cube-oriented Ni-50Al single crystals exhibit essentially zero plastic strain to failure at room temperature but undergo a sharp brittle-to-ductile transition at temperatures just above 600 K (Bowman et al. 1989; Darolia et al. 1992; Takasugi, Watanabe, and Hanada 1992). This BDTT corresponds to the temperature that [001] crystals begin to undergo a steep decrease in yield stress with increasing temperature (Fig. 7-3a) because the deformation mechanism changes from ⟨111⟩ slip to climb of ⟨100⟩ and ⟨110⟩ dislocations. Ni-rich cube-oriented Ni-40Al crystals undergo a similar brittle-to-ductile transition but at approximately 1000 K (Noebe, Misra, and Gibala 1991).

Previously, it appeared that soft-orientation, single-crystal NiAl also possessed very limited room-temperature tensile ductility. Near-stoichiometric, soft-orientation single crystals have been routinely reported to exhibit room-temperature tensile elongations on the order of 1% (Lahrman et al. 1993b; Takasugi et al. 1993). On occasion, however, room-temperature tensile ductilities on the order of 5% to 7% have been measured in binary NiAl soft-orientation single crystals without any obvious mechanism for the increased tensile ductility (Field et al. 1993; Brzeski et al. 1993). Similarly, an intriguing effect of microalloying additions on the tensile properties of soft-orientation, single-crystal nickel aluminides has been observed (Darolia et al. 1992). When near-stoichiometric, ⟨110⟩ single crystals were doped with approximately 1000 ppm of Fe, Mo, or Ga, the room-temperature tensile elongation increased from approximately 1% to upwards of 6% (Fig. 7-8). From Figure 8, it is obvious that the peak in ductility occurs at very small alloying additions, and as the level of dopant exceeds 0.5 at% the benefits to ductility are lost. This ductilizing effect is possibly due to gettering of interstitial contaminants, which would be consistent with recent observations on the fracture toughness of NiAl that will be discussed shortly.

Soft-orientation single crystals also undergo a fairly abrupt brittle-to-ductile transition at temperatures slightly below those of polycrystals. Depending on the specific orientations, the BDTT's range from 475 K to

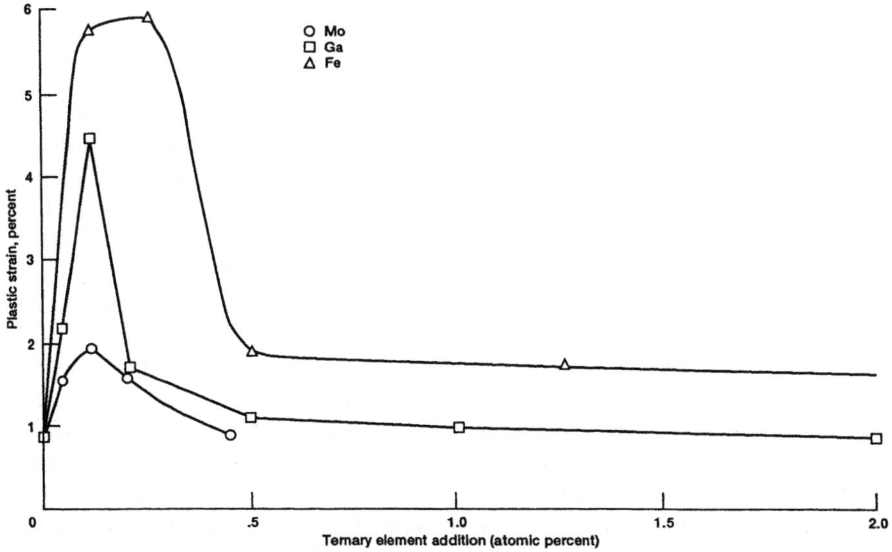

Figure 7-8. Room-temperature tensile ductility of ⟨110⟩ single-crystal NiAl microalloyed with Fe, Ga, or Mo (Reprinted with permission from *Scripta Metallurgica et Materialia*, Volume 26, R. Darolia, D. F. Lahrman, and R. D. Field, pages 1007–1012, 1992, Elsevier Science Ltd., Oxford, England).

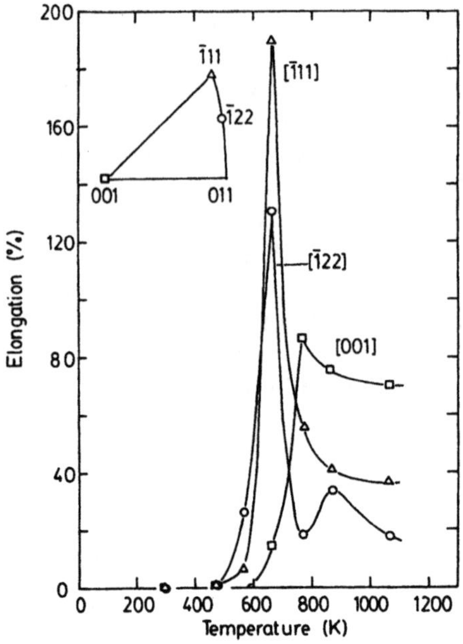

Figure 7-9. Effect of temperature on the tensile ductility of NiAl single crystals (Takasugi, Watanabe, and Hanada 1992).

525 K (Lahrman et al. 1991; Takasugi et al. 1993). Just above the BDTT, anomalously large tensile elongations (greater than 120%) were observed for soft-orientation single crystals; ductility decreased to around 50% at higher temperatures because of the onset of necking (Fig. 7-9). The anomalously large elongations at intermediate temperatures have been attributed to a balance between work hardening due to glide and relaxation processes due to climb resulting in a high necking resistance (Takasugi et al. 1993).

Originally, the room-temperature fracture toughness of single-crystal NiAl was measured in the range of 7 to 12 $MPa\sqrt{m}$ when the notch is cut perpendicular to ⟨100⟩-oriented crystals and from 4 to 6 $MPa\sqrt{m}$ when the notch is cut perpendicular to ⟨110⟩-oriented crystals (Chang, Darolia, and Lipsitt 1992; Vehoff 1993). The fracture toughness of the soft-orientation single crystals agrees perfectly with fracture-toughness values for binary, single-phase, polycrystalline NiAl, which have been measured between 4 and 7 $MPa\sqrt{m}$. Furthermore, the fracture toughness of polycrystalline NiAl is essentially independent of grain size, stoichiometry, or processing technique (Reuss and Vehoff 1990a; Rigney and Lewandowski 1992a; Kumar, Mannan, and Viswanadham 1992).

However, it was recently discovered (Hack, Brzeski, and Darolia 1992) that the fracture toughness of commercial-purity, single-crystal NiAl is extremely sensitive to heat treatment and cooling rate, as determined from double cantilever beam specimens with a notch plane perpendicular to ⟨110⟩. For example, single-crystal samples that were rapidly air cooled to room temperature from 1573 K had a fracture toughness of almost 16 $MPa\sqrt{m}$, but when they were reannealed at 473 K and slowly cooled to room temperature, the fracture toughness dropped to about 3 $MPa\sqrt{m}$ (Hack et al. 1992). Comparable heat treatments had no effect on the toughness of zone-refined, low-interstitial, single-crystal NiAl when it was tested in four-point bending with the crack plane normal to ⟨100⟩. The fracture toughness of this high-purity NiAl was 10 to 12 $MPa\sqrt{m}$ independent of heat treatment (Johnson et al. 1993). Furthermore, experiments in biaxial bending and tension indicate that zone-refined, low-interstitial NiAl single crystals have a much greater room-temperature ductility and fracture strength than commercial-purity material does (DeMarco and Ardell 1993; Johnson et al. 1993). These observations are consistent with a strain-aging phenomenon in NiAl.

This behavior would be similar to that observed in mild steels where carbon is responsible for pinning the mobile dislocation density, causing brittle fracture. Thus, the inherent ductility and fracture toughness of single-crystal NiAl may be much greater than originally determined. However, considerable work is still necessary to understand and sort out the various metallurgical phenomena. For instance, higher toughness values in NiAl are always associated with an increased percentage of transient fracture along higher index planes, such as {511}, before transitioning to {110} planes (Darolia, Chang, and Hack 1993). But the mechanism for this transient fracture behavior and the roles that composition and heat treatment play are still unclear.

Polycrystalline alloys have an additional complicating factor in the presence of grain boundaries. Because NiAl deforms by a ⟨001⟩ slip vector, there are only three independent slip systems available for deformation; no extra independent systems are provided by cross slip (Ball and Smallman 1966b). Because this is less than the five independent deformation modes considered necessary for extensive, uniform, crack-free deformation of a polycrystalline aggregate, polycrystalline NiAl has little potential for significant room-temperature ductility. Experimental evidence to date supports this view. In room-temperature studies of NiAl, the reported tensile ductilities have ranged from zero to a maximum of about

4% (Rozner and Wasilewski 1966; Hahn and Vedula 1989). Furthermore, intergranular fracture in both tensile and compression specimens at low temperatures appears to confirm that the limited ductility of binary, near-stoichiometric NiAl is the result of incompatible shape changes of neighboring grains. These incompatible shape changes are due, in turn, to an insufficient number of independent slip systems (Ball and Smallman 1966a; Noebe et al. 1991).

However, intergranular fracture can result from many different sources. Intergranular fracture of NiAl due to impurity segregation at the grain boundaries has been ruled out through in-situ Auger electron spectroscopy studies (George and Liu 1990; Zeller, Noebe, and Locci 1990). Examination of cast plus extruded alloys and powder-extruded materials after various thermal treatments revealed that the grain boundaries in NiAl were clean and free from measurable impurity contamination including O and C. Intergranular fracture also could arise if the grain boundaries were intrinsically weak because of their structure, as was once thought to be the case in Ni_3Al (Chaki 1991). However, grain-boundary structure simulations for NiAl indicate that there are no periodic structural defects present at the grain boundaries in near-stoichiometric NiAl that would cause the grain boundaries to be inherently weak (Vitek and Chen 1991). The grain-boundary modeling results indicate the possible presence of slight discontinuities in the grain-boundary structure of nonstoichiometric alloys; however, nonstoichiometric NiAl fractures in a transgranular manner (Nagpal and Baker 1991). Finally, Ni_3Al has limited ductility in polycrystalline form because of intergranular fracture aggravated by environmental effects when tested in room-temperature air (George, Liu, and Pope 1992). However, in comparable testing of NiAl in air, vacuum, and various gas environments, environment had no impact on the tensile ductility of polycrystalline NiAl (personal communication with C. T. Liu, 1993, Oak Ridge National Laboratory, Oak Ridge, TN) or single-crystal material (Lahrman et al. 1993b). Elimination of these other factors suggests that grain-boundary incompatibility due to an insufficient number of independent slip systems is the primary factor responsible for the observed intergranular fracture and limited tensile ductility of near-stoichiometric NiAl.

Both intergranular and transgranular fracture modes have been observed in alloyed NiAl. Intergranular fracture initiation in NiAl depends on the operation of at least localized plastic flow to nucleate a critical defect for fracture, whereas transgranular fracture initiation and propagation generally occur before yielding. This behavior is summarized in the fracture map shown in Figure 7-10 (Noebe 1994). Low-yield-strength binary NiAl alloys exhibit limited tensile ductility that leads to the forma-

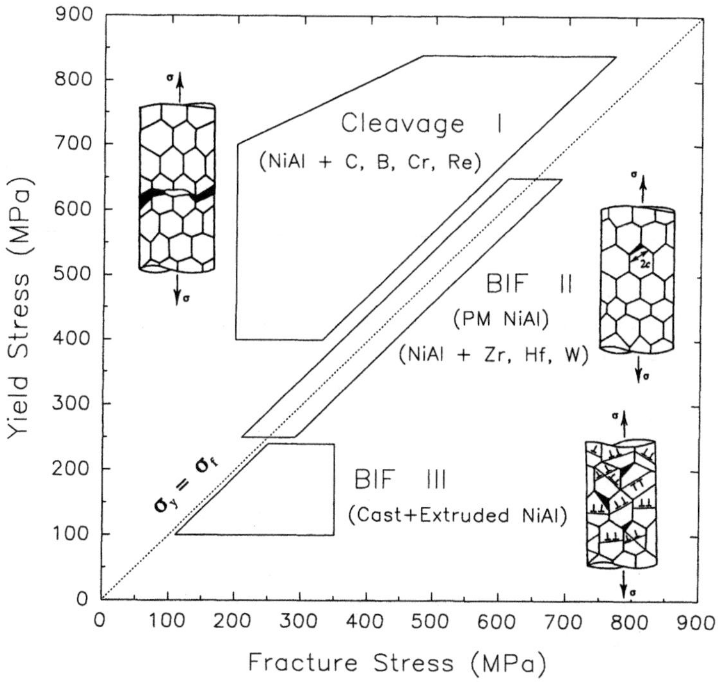

Figure 7-10. Relation between alloying additions, room-temperature tensile properties, and fracture initiation mechanisms in NiAl alloys. Brittle intergranular fracture, BIF; powder metallurgy, PM.

tion of intergranular microcracks because of incompatible plastic deformation at the grain boundaries. This results in a fracture mechanism defined as brittle intergranular fracture III (BIF III) by Gandhi and Ashby (1979). In this region, fracture is intergranular or mixed mode in nature, occurring after some measurable plastic deformation, but initiating intergranularly. In higher strength alloys, usually powder-processed material or microalloyed NiAl, fracture occurs at the macroscopic yield strength of the material, but no significant tensile ductility is achieved (BIF II). Localized or microscopic yielding is responsible for initiating fracture in the material; but because the stresses are higher, the critical stress for crack propagation has already been reached, and fracture occurs at the first defect formed in the material. Finally, in some NiAl materials, such as those alloyed with Re, B, C, or Cr, fracture occurs without any prior plasticity in a completely brittle manner. In such cases, fracture almost always initiates and propagates in a transgranular fashion (cleavage I). Similarly, nonstoichiometric binary NiAl alloys do not exhibit room-temperature ductility and fail in a predominantly transgranular manner (Nagpal and Baker 1991).

Consequently, the origin of fracture in polycrystalline NiAl depends significantly on composition. In low-strength, stoichiometric, binary NiAl alloys, grain boundaries become the sites of Griffith defects after slip is activated. In alloyed materials of moderate strength, local yielding is responsible for initiating intergranular fracture. In contrast, high-strength NiAl-based materials fail transgranularly, with fracture originating at some internal or surface defect.

Although limited tensile ductility can be achieved in low-yield-strength NiAl alloys at room temperature, this is not the brittle-to-ductile transition temperature for polycrystalline alloys. Instead, a dramatic increase in ductility corresponding to the BDTT is observed in the range of 550 K to 700 K (Noebe et al. 1992). Not only is the BDTT defined by a substantial increase in tensile ductility, but a significant and concurrent increase in fracture strength (Fig. 7-11) and fracture toughness (Fig. 7-12) are observed. The fracture toughness of polycrystalline NiAl increases abruptly as temperature increases, with the transition to ductile behavior and high-toughness values occurring between 550 K and 650 K. In this intermediate temperature regime, the fracture toughness of NiAl ranges from 20 to nearly 50 MPa$\sqrt{m}$ depending on microstructure and grain size (Reuss and Vehoff 1990a). The increasing toughness of NiAl with temperature corresponds to the observed increase in tensile fracture strength with temperature and would be expected because of the linear dependence between the critical-stress intensity factor K_{IC} and fracture stress for a constant flaw size.

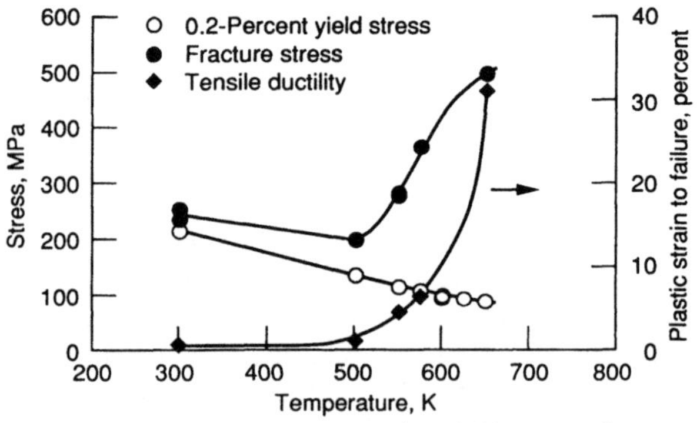

Figure 7-11. Effect of test temperature on the yield stress, fracture stress, and tensile elongation of powder-extruded NiAl (Noebe et al. 1991).

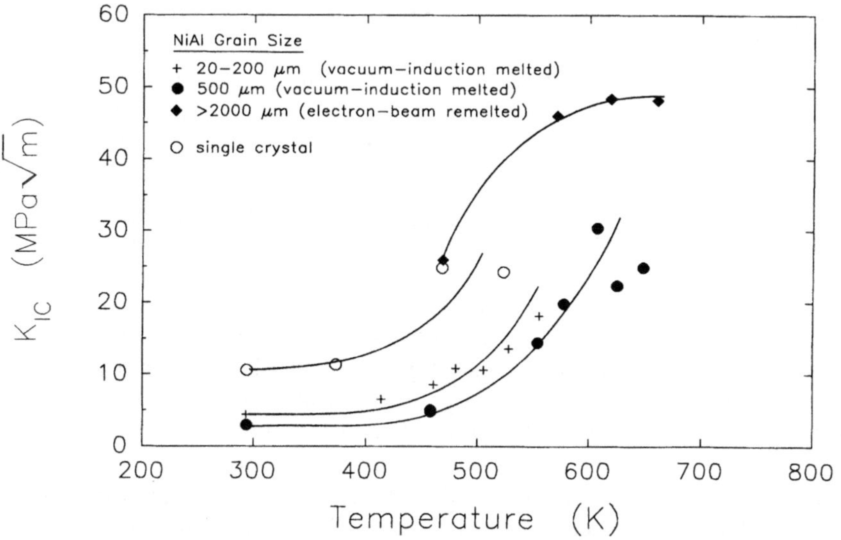

Figure 7-12. Fracture toughness of NiAl as a function of temperature. Alloys were produced by several different techniques, resulting in materials with a range of grain sizes (Reuss and Vehoff 1990b).

For polycrystalline material, these changes in behavior have been attributed to the onset of localized dislocation climb processes that occur within the vicinity of the grain boundaries (Noebe 1994). These dislocation climb processes result in five independent deformation mechanisms, permitting the accommodation of strains across the grain boundaries. In situ annealing studies have verified the onset of significant dislocation climb activity near grain boundaries in NiAl beginning at the BDTT, correlating with the beginning of region II deformation in Figure 7-4 (Bowman et al. 1992). The significance of dislocation climb in a material such as NiAl was demonstrated by Groves and Kelly (1969). They showed that the combination of glide and climb of dislocations with ⟨100⟩ Burgers vectors will result in five independent deformation mechanisms. This condition would satisfy the Von Mises criterion allowing for extensive plasticity in polycrystalline NiAl by relieving stresses at the grain boundaries and other sites of extensive stress concentration. The lowest BDTT observed in polycrystalline NiAl, $\sim 0.30T_m$, falls within the temperature regime for which thermally activated deformation mechanisms occur by short-circuit diffusion. For example, climb could be restricted to grain-boundary regions and still be very influential in relieving compatibility stresses. Typical dislocation glide processes would continue to dominate the deformation behavior of the grain interiors until temperatures of approximately $0.5T_m$ are reached. This description is consistent with in-situ transmission electron microscope annealing studies on NiAl that have qualitatively determined the temperatures for which dislocation climb by short circuit and bulk diffusion processes becomes significant (Bowman et al. 1992). This description of the deformation behavior of polycrystalline NiAl near the brittle-to-ductile transition is also consistent with the observed effects of strain rate and alloying additions on the BDTT (Noebe et al. 1992; Noebe 1994).

Dislocation climb processes driven by short circuit diffusion along dislocation cores could similarly be used to explain the brittle-to-ductile transition in [001]-oriented single crystals. The BDTT for hard-orientation crystals corresponds to the change in deformation mechanism from ⟨111⟩ slip to deformation by ⟨100⟩ and ⟨110⟩ dislocations. Because the resolved shear stress on the ⟨001⟩ dislocations would be small and the ⟨110⟩ dislocations have a decomposed core structure and can only move by nonconservative processes (Mills et al. 1993), a significant amount of thermal activation would be necessary before these dislocations could accommodate imposed strains on [001] crystals.

The mechanism responsible for the brittle-to-ductile transition in soft-orientation single crystals is not obvious. For soft-orientation single crystals, the BDTT occurs as low as $0.25T_m$ and is relatively insensitive to strain rate (Lahrman et al. 1991). This lower BDTT could be due to enhanced cross slip leading to slip homogenization or due to the unlocking of dislocations from point defects or impurities. This last mechanism is consistent with the strain-aging effects recently observed in single-crystal NiAl (Hack et al. 1992). However, the onset of thermally activated deformation processes may still play an important role in the relaxation of stress concentrations that may occur at surface notches or internal defects such as pores and shrinkage cavities. Dislocation glide alone (generally on a single-slip system) would not be sufficient to accommodate the complex stress state near these types of defects in single crystals. Furthermore, single crystals (Takasugi et al. 1993), like polycrystalline NiAl, undergo a dramatic change in ductility at intermediate temperatures where thermally activated deformation processes become active.

4. ALLOY DESIGN

Figure 7-13 shows several basic possibilities for alloying NiAl. The first is adding elements with high solubility in NiAl. Because NiAl is isostructural with several other intermetallic compounds and has a wide phase field, Mn, Cu, and group VIII elements are generally quite soluble. This family

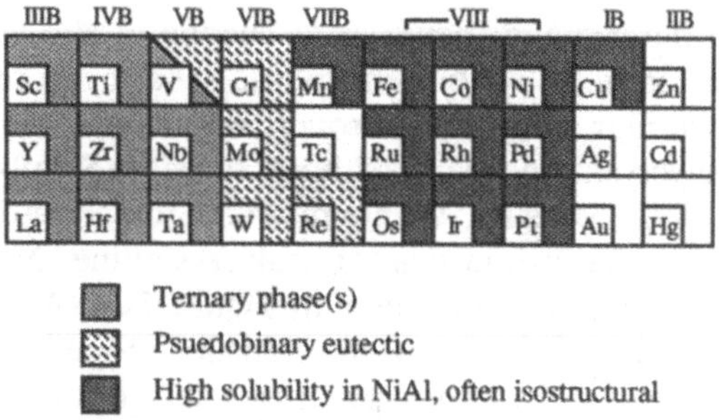

Figure 7-13. Portion of the periodic table of elements illustrating general alloying behavior of ternary additions to NiAl (Cotton 1991).

of elements offers a considerable alloying potential that is mainly unexplored at this time. The refractory metal group VIB elements plus V and Re exhibit very little solubility in NiAl and form no ternary intermetallic phases, with the exception of V in Ni-rich alloys. Instead, these elements form pseudobinary eutectic systems with stoichiometric NiAl and, therefore, may both strengthen NiAl at high temperatures and act as ductile reinforcing agents at low temperatures (Johnson et al. 1993). Finally, group IIIB, IVB, and VB elements have very limited solubility in NiAl and form at least one ternary-ordered intermetallic compound with Ni and Al. The most common of these intermetallics are the Heusler phases, Ni_2AlX, and the Laves phases, NiAlX. These ordered phases are significantly more brittle than NiAl and would, therefore, preclude the development of a ductile-phase toughened material. Nevertheless, NiAl-Heusler and NiAl-Laves alloys are being actively investigated because of their superior creep resistance. In fact, some of the most creep-resistant NiAl-based alloys produced to date are single-crystal NiAl-Heusler compounds (Darolia et al. 1992; Darolia 1993).

Alloy design for creep strength will be discussed in detail in section 5. The remainder of this passage addresses the most common approaches for increasing the ductility and/or toughness of NiAl at ambient temperatures and presents an overview of their relative success.

4.1. Effect of Microalloying Additions on Properties

As discussed in section 3.3 and demonstrated in Figure 7-9, microalloying additions of Fe, Ga, and Mo at the 0.1 to 0.2 at% level consistently and significantly increase the room-temperature tensile ductility of soft-orientation, single-crystal NiAl. The mechanism for this increase in ductility has not been determined; however, a change in slip vector or even an obvious modification in dislocation morphology has been ruled out (Darolia et al. 1992). A similar alloying approach with these elements has been attempted on polycrystalline NiAl without success (Schulson 1982; Noebe and Behbehani 1992; Matsugi, Wenman, and Stoloff 1992). The temperature-dependent tensile properties of polycrystalline NiAl microalloyed with 0.1 at%Fe, Ga, and Zr are shown in Figure 7-14. At this level, Fe and Ga have no beneficial effect on the tensile properties of NiAl: they nominally decrease the room-temperature tensile ductility and slightly increase the BDTT (by approximately 25 K). At a level of 0.28 at%Fe, the BDTT of the microalloyed material is increased even further to approximately 730 K (Matsugi et al. 1992). Moreover, additions of up to 1 at%Fe had no effect on the room-temperature fracture toughness of polycrystalline NiAl (Schneibel, Jenkins, and Maziasz 1993).

Whereas microalloying additions of Fe, Ga, and Mo have not improved the fracture toughness or tensile ductility of polycrystalline NiAl, other additions, such as Zr and B, have had an extremely strong, but detrimental, effect on ductility (Bowman et al. 1992; George and Liu 1990). For example, Figure 7-14 demonstrates the effect of Zr on the tensile properties of NiAl when present at a level of only 0.1 at%. Zirconium not only has a significant influence on the yield strength, but its presence nearly doubles the BDTT of NiAl. Small additions of Mo, La, Y, Re, and Hf to NiAl have also been observed to increase the yield strength, decrease the ductility, and increase the BDTT (Schulson 1982; Mason et al. 1991). Alloying additions of W and Be have no beneficial effect on ductility, though they also do not seem to be detrimental (Mason et al. 1991; George and Liu 1990). This is probably due to the extremely low hardening rate of Be (Fig. 7-5) and the very low solubility of W in NiAl (Locci et al. 1989). Finally, the interstitial element C has an adverse effect on the room-temperature ductility of NiAl (George and Liu 1990), though other interstitial elements such as O and N do not seem to be equally detrimental to tensile properties (Noebe and Garg 1994). Consequently, most alloying elements studied to date seem to reduce the already limited tensile ductility of polycrystalline NiAl, and no additions (except the addition of Mo, Ga, and Fe to single-crystal material) have been identified as being beneficial for ductility.

The major shortcoming in attempting to improve the tensile properties of polycrystalline NiAl through microalloying additions is that the additions are so small that a change in deformation behavior to a slip system that results in five independent deformation modes is not expected. Therefore, the grain-boundary compatibility problem, which is a limiting factor in the low-temperature ductility of polycrystalline NiAl, is not

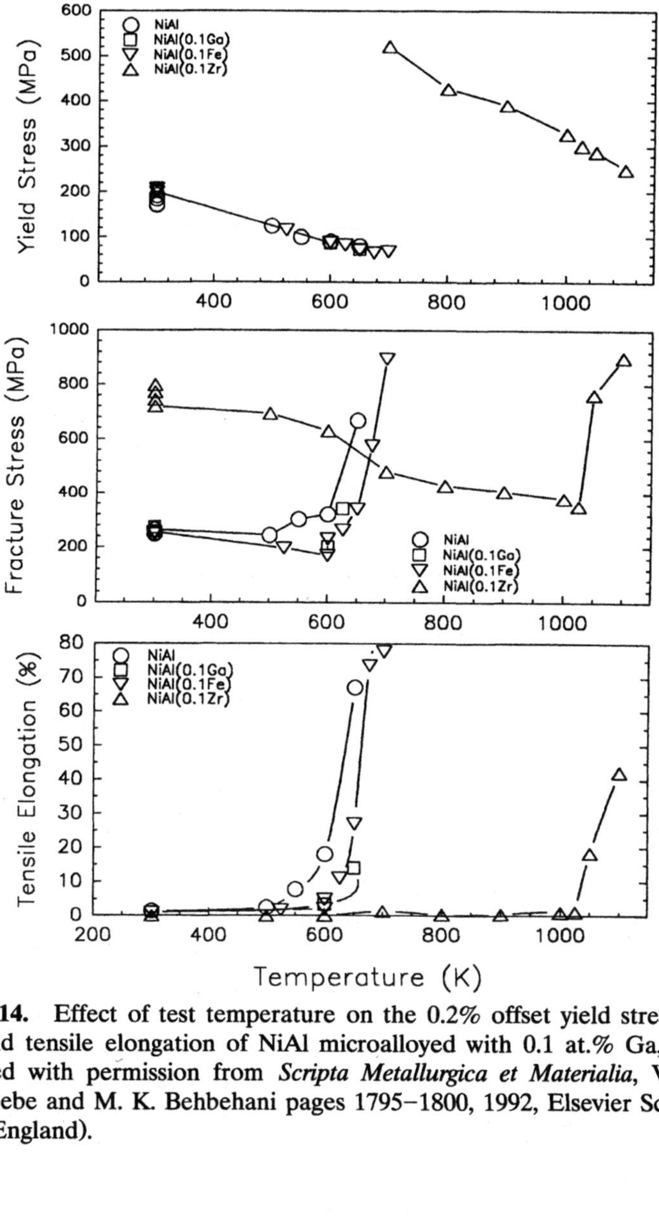

Figure 7-14. Effect of test temperature on the 0.2% offset yield stress, fracture stress, and tensile elongation of NiAl microalloyed with 0.1 at.% Ga, Fe, or Zr (Reprinted with permission from *Scripta Metallurgica et Materialia*, Volume 27, R. D. Noebe and M. K. Behbehani pages 1795–1800, 1992, Elsevier Science Ltd., Oxford, England).

resolved. Conversely, most microalloying additions have had very dramatic, but deleterious, effects on the mechanical properties of polycrystalline NiAl. In single-crystal NiAl, microalloying additions are suspected of acting as gettering agents for some interstitial impurity. Therefore, the alloying addition does not alter the behavior of the single-crystal but getters a detrimental impurity in the intermetallic, permitting NiAl to display its inherent properties. The similarity in the room-temperature tensile properties of high-purity, low-interstitial NiAl and microalloyed single crystals supports this idea with both materials exhibiting 4–5% tensile elongations, in comparison to 1% or less for commercially pure NiAl (Lahrman et al. 1993a; Johnson et al. 1993).

4.2. Effect of Macroalloying Additions on Slip Character

Another strategy for improving the ductility and toughness of polycrystalline NiAl is to add ternary macroalloying elements to enhance or modify the slip processes (Law and Blackburn 1987; Cotton et al. 1993b). The general approach has been to identify elements that should lower the ordering energy of NiAl, thus making ⟨111⟩ slip easier. Cr, Mn, and V are reasonable choices for promoting ⟨111⟩ slip in NiAl, based on calculations resulting from interatomic potential models (Hong and Freeman 1991). These models demonstrate that up to 70% reductions in antiphase boundary energy are possible when alloying additions of at least 17 at% are substituted for Ni or Al. In reality, such large alloying additions are not possible. The solubility of Cr in NiAl is approximately 1 to 2 at% on either sublattice (Cotton, Noebe, and Kaufman 1993c). The solubility of V is about 5 to 12 at% when substituted for Al and is essentially zero when substituted for Ni (Cotton et al. 1991). According to calculations, V substituted for Ni would be the preferred substitution method for lowering the antiphase boundary energy.

In spite of the low solubility for Cr in NiAl, but in apparent agreement with the concept just presented, ⟨111⟩ slip has been reported in polycrystalline NiAl alloyed with approximately 5 at % Cr or Mn (Law and Blackburn 1987). Although the operation of ⟨111⟩ slip satisfies the requirement for generalized polycrystalline plasticity, no tensile ductility was reported in these materials at low temperatures. However, a change in fracture mode from intergranular to transgranular cleavage was observed. According to Figure 7-10, a change in fracture mode would not be unexpected because the yield strength of the alloy was greatly affected by the alloying additions.

Nevertheless, this result prompted more detailed work on NiAl(Cr) alloys by Cotton et al. (1993a–c). In this more recent work, analysis of over 2,000 dislocation segments in deformed NiAl(Cr) alloys covering 16 different ternary alloy compositions failed to identify any significant ⟨111⟩ dislocation activity. To eliminate any chance that processing or chemistry effects could influence slip behavior, Cotton et al. (1993b) also examined a piece of the original casting studied by Law and Blackburn. Again, no evidence of ⟨111⟩ dislocations was observed in the as-cast or deformed material. Therefore, in spite Law and Blackburn's contention in 1987, Cr additions to NiAl do not promote ⟨111⟩ slip.

Iron would be another obvious alloying addition for slip alteration because B2 FeAl alloys deform by ⟨111⟩ slip (Crimp and Vedula 1991). However, single-phase β-Ni-Fe-Al alloys like binary NiAl exhibit only 0–2.5% tensile ductility at room temperature (Guha, Munroe, and Baker 1991; Russell et al. 1991; Raj, Locci, and Noebe 1992). The observed slip vector in these Ni-Fe-Al alloys was ⟨100⟩ so that the ternary alloys were still slip-system limited. The slip behavior of a series of polycrystalline (Ni,Fe)-40Al alloys was also examined by Patrick et al. (1991), who concluded that ⟨100⟩ slip is dominant in B2-phase alloys containing up to 40 at%Fe. Despite the lack of tensile ductility, ternary Ni-Fe-Al alloys may exhibit very high toughness levels, approaching 25 MPa$\sqrt{m}$ in fine-grained material (Kostrubanic et al. 1991).

Other studies that are usually referenced to support the operation of ⟨111⟩ slip in NiAl due to Cr or V additions were performed on [001]-oriented single crystals (Field et al. 1991c; Miracle, Russell, and Law 1989). Because ⟨111⟩ is already the preferred slip vector in this crystal orientation, no conclusion about the influence of alloying on slip mode can be drawn from these studies. Therefore, alloying of NiAl to alter the operative slip vector has been unsuccessful to date. Not only is the slip system left unchanged, but the BDTT of NiAl macroalloyed with Cr (Cotton et al. 1993b) and Fe (Raj et al. 1992) is significantly higher than that of binary NiAl. However, little effort has been expended to date on NiAl(Mn) alloys. Because of the much greater solubility of Mn in NiAl in comparison to either Cr or V (Chakrabarti 1977), more extensive investigation of Mn on the slip behavior of NiAl is warranted.

4.3. Microstructural Modification

Instead of trying to alter the inherent properties of the intermetallic, an alternate approach to improving the ductility and toughness of NiAl is to modify the microstructure of the material. The idea is to improve the mechanical behavior of NiAl through extrinsic toughening mechanisms. This approach includes microstructural variations from grain size refinement to the development of multiphase microstructures.

Grain refinement is one of the oldest techniques recognized for improving the ductility and toughness of metals (Cottrell 1958). Based on this approach, Schulson (1981) suggested that grain refinement could also be used to increase the tensile ductility of brittle intermetallics like NiAl. However, grain size has a critical effect on the ductility of NiAl only within the 550 K to 750 K window, which is generally defined as the BDTT for polycrystalline binary NiAl (Schulson and Barker 1983; Schulson 1985). Even though room-temperature tensile ductility was not achieved in these original tests by Schulson and Barker (1983), the results are often quoted to support a critical grain size for room-temperature NiAl ductility. However, further testing of extruded NiAl castings or powder demonstrated that room-temperature tensile ductility is essentially independent of grain size (Nagpal and Baker 1990b; Noebe et al. 1991), at least within the range of grain sizes that can be realistically achieved by conventional processing methods. In addition, the room-temperature fracture toughness of Ni-45Al is independent of grain size in the range of 20–2000 μm (Rigney and Lewandowski 1992a).

Chan (1990a) developed a quantitative model based on a critical J-integral approach that accurately predicts the dependence of tensile ductility on grain size for semibrittle materials. It is clear from the application of Chan's model to NiAl that grain size affects ductility in markedly different ways depending on the temperature range (Noebe et al. 1991). Above the BDTT, there is a strong dependence of tensile ductility on grain size for NiAl because of the higher fracture toughness; however, at room temperature there is very little dependence of ductility on grain size for grain sizes greater than 1 μm. Although room-temperature ductility is predicted to slightly increase when the grain size decreases below
1 μm (Noebe 1994), tensile ductility is predicted to be only about 5% for a grain size of approximately 0.1 μm. Because it would be extremely difficult to keep such a fine grain size from coarsening during processing and service exposure, grain-size refinement alone does not seem to be a practical method for significantly improving the ductility of NiAl at room temperature.

An alternative approach to increasing ductility is to develop two-phase microstructures where toughness is enhanced through crack interactions with a ductile second phase. Toughness would be improved by such processes as crack bridging, crack blunting, or crack deflection. An example of an intermetallic system that has been successfully toughened by such an approach is Ti_3Al, which was toughened by using a Ti-24Al-11Nb or similar two-phase composition (Chan 1990b). As yet, a successful counterpart to this system has not been identified for NiAl, though significant work is ongoing in this area. Because the development of two-phase NiAl-based alloys has been reviewed previously (Noebe et al. 1991), only a brief discussion centering on more recent advances is presented.

One has only to look at the binary Ni-Al phase diagram to find a model system for ductile reinforcement of NiAl. Within the single-phase

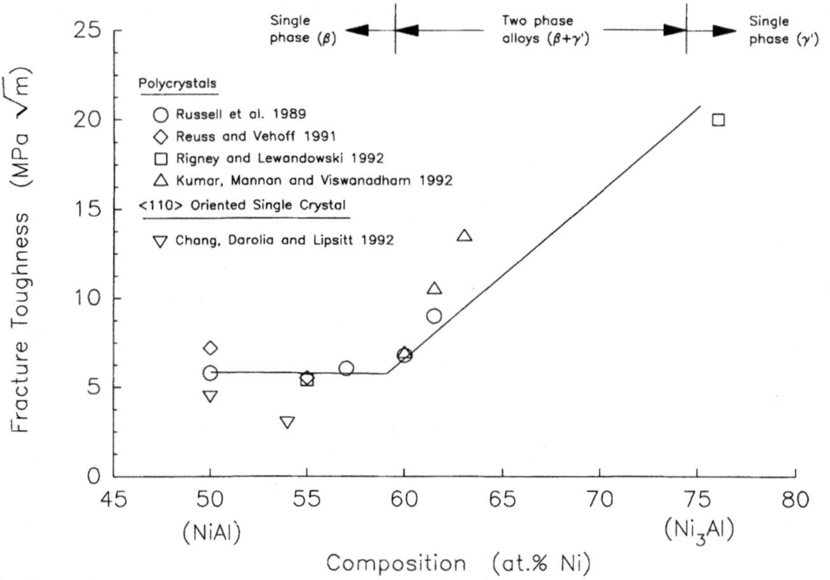

Figure 7-15. Room-temperature fracture toughness of Ni-Al alloys as a function of stoichiometry.

regime, deviations from stoichiometry have no apparent effect on the toughness of NiAl. However, the fracture toughness of polycrystalline NiAl can be increased by going to two-phase, Ni-rich compositions (Fig. 7-15). There are several possible toughening mechanisms responsible for the increased toughness in this regime. A toughness of approximately 9 $MPa\sqrt{m}$ has been reported for quenched single-phase β-alloys with 61.5 at% Ni because of martensitic-transformation toughening (Kumar et al. 1992). However, because of the small volume increase during the NiAl to martensite transformation (Chakravorty and Wayman 1976), only a small increase in toughness results from this mechanism. The toughness and ductility of Ni-rich Ni-Al alloys can be increased best by the formation of a two-phase microstructure. In directionally solidified materials, room-temperature tensile ductilities of 10% have been achieved in a binary Ni-30Al alloy consisting of aligned γ' rods in a NiAl matrix (Noebe et al. 1991). Conventionally processed, two-phase, polycrystalline Ni-Al alloys consist of a Ni_3Al necklace structure surrounding β-grains, with fracture toughness increasing roughly proportional to the amount of γ' phase present (Fig. 7-15). Fracture-toughness values as high as 25 $MPa\sqrt{m}$ also have been measured in two-phase ($\gamma' + \beta$) Ni-Al-Co alloys (Russell et al. 1989). The increased toughness and ductility of these two-phase alloys can be attributed to typical ductile-phase toughening mechanisms such as crack bridging and crack blunting as well as to an additional mechanism known as slip transfer (Misra, Noebe and Gibala 1993). During this last process, dislocation activity in the ductile phase initiates slip in the β-phase, thus increasing the plasticity of the overall alloy.

Brittleness in NiAl can be traced to a dislocation-initiation problem as opposed to a dislocation-mobility problem. This was originally demonstrated in surface-film softening studies (Noebe and Gibala 1986). Other studies, including prestrain experiments on single-crystal NiAl (Hack et al. 1992) and pressurization experiments on polycrystalline NiAl (Margevicius and Lewandowski 1993), also illustrate that the lack of tensile ductility in NiAl is, in part, due to dislocation-initiation problems. Consequently, slip-transfer mechanisms could increase the inherent toughness of NiAl because they directly address the problem of slip initiation.

However, a two-phase alloy can be less tough than NiAl if the reinforcing phase has poor toughness, especially when the brittle second phase is of significant volume fraction and strongly bonded. For example, toughness values similar or less than that of binary NiAl have been observed in two-phase NiAl/NiAlNb alloys (Reuss and Vehoff 1990a) and in NiAl/NiAlTa eutectics (Johnson et al. 1993). In these alloys, cracks run easily through the brittle-reinforcing phase and the solid-solution-strengthened NiAl phase.

In certain instances, limited toughening attributed to crack-deflection mechanisms has been observed in NiAl alloys because of the presence of brittle second phases, though this is generally an inefficient way of increasing toughness. Kumar et al. (1992) found that 1-μm-diameter TiB_2 particulates had essentially no effect on the room-temperature fracture toughness of NiAl but that Al_2O_3 whisker reinforcement increased toughness 50% when present in volume fractions of 15 to 25 vol%. Rigney and Lewandowski (1992b) noted that in two out of three heats of NiAl/TiB_2 particulate-reinforced material, no statistically significant increase in toughness was achieved. In one heat, toughness increased 40%, but the increase in toughness was attributed to the clustering of TiB_2 particles, leading to a "fiberlike" toughening response.

Fiberlike or semicontinuous reinforcements, if brittle but weakly bonded or tough and strongly bonded, provide greater toughening than particulates because cracks cannot circumvent the reinforcement as easily. Therefore, a natural extension of high-aspect-ratio reinforcements would be continuous-fiber-reinforced NiAl alloys. Somewhat similar, naturally occurring materials can also be produced by eutectic solidification processing. For example, when a NiAl-9Mo eutectic is arc melted it has a fracture toughness of approximately 9 MPa$\sqrt{m}$ (Subramanian et al. 1990). But after directional solidification, the eutectic composite consists of ~ 11 vol% of 1-μm-diameter rods in a single-crystal NiAl matrix and has a fracture toughness of 15 MPa$\sqrt{m}$ because of crack bridging by the ductile Mo ligaments (Darolia et al. 1992). However, increasing the Mo content beyond the eutectic composition does not produce additional toughening (Johnson et al. 1992). Directionally solidified eutectics containing larger volume fractions of other refractory-metal phases demonstrate similar or higher levels of toughness. For example, NiAl-Cr eutectics have a toughness of approximately 18 MPa$\sqrt{m}$ (Darolia et al. 1992), whereas a fracture toughness of 31 MPa$\sqrt{m}$ has been measured in a NiAl-V eutectic alloy (Johnson et al. 1992).

5. CREEP

Adequate creep resistance is another critical requirement for the extended use of NiAl in elevated-temperature structural applications. Consequently, high-temperature mechanical properties have been studied extensively. This has led to significant progress in developing creep-resistant NiAl alloys in recent years. To help the reader fully appreciate and understand the work performed in this area, the basic creep mechanisms in NiAl are described; then the various strategies used for improving high-temperature strength are analyzed.

5.1. Creep Mechanisms in NiAl

Creep behavior in B2 aluminides, which is analogous to that for metals and alloys, can be analyzed accordingly. The second stage, or steady-state creep rate $\dot{\varepsilon}$, is usually expressed as a form of the Dorn equation:

$$\dot{\varepsilon} = A(\sigma/E)^n \exp(-Q/RT) \qquad (7.1)$$

where σ is the applied stress, E is the Young's modulus, n is the stress exponent, Q is the activation energy for creep, R is the gas constant, T is the absolute temperature, and A is a constant that takes into account such variables as microstructure and stacking-fault or antiphase-boundary energy. The values for n and Q depend on the operative deformation mechanisms within a given temperature and stress regime and thus, are useful indicators of deformation mode. Creep is most commonly controlled by diffusion; hence the value of Q for creep is often similar to that for diffusion Q_D. Therefore, knowledge of diffusion behavior also is desirable as an aid to understanding creep processes.

Diffusion data for NiAl have been compiled and discussed previously (Noebe et al. 1993; Miracle 1993). Unfortunately, no new experimental data have been generated for a number of years. Additional work is required to determine the activation energy and mechanisms of diffusion at lower temperatures because almost no data exist for diffusion in NiAl below 1200 K. From previous high temperature studies (Hancock and McDonnell 1971; Lutze-Birk and Jacobi 1975; Berkowitz, Jaumot, and Nix 1954), it appears that the activation energy for diffusion is a maximum at Ni-50Al, consistent with the diffusion-coefficient minima at the same composition. Although there is considerable scatter among the various diffusion studies, critical analysis of the available data for NiAl indicates that the most likely range for Q_D is on the order of 250 to 300 kJ/mol.

Table 7-3 summarizes the stress exponents and activation energies for the high-temperature deformation of NiAl. The average value for the activation energy of creep is approximately 315 kJ/mol. However, the data in Table 7-3 have not been corrected for the temperature dependence of the elastic modulus, which would reduce Q by 20 to 30 kJ/mol, thus bringing Q well within the range of data for bulk diffusion.

Dislocation creep in single-phase metals and alloys can usually be classified as one of two main types; class M, or pure metal creep, and class A, or alloy-type behavior (Mohamed and Langdon 1974). Class M creep is characterized by glide being much faster than climb; thus creep is controlled by the rate of dislocation climb past substructural obstacles. Class A creep is often called viscous-glide-controlled creep because the glide of dislocations is restricted by solute atoms or perhaps by a high lattice-friction stress because of long-range order. This reduced glide mobility is the limiting creep process, whereas climb can occur readily. These two types of behavior can be distinguished by several criteria including the stress exponent, the shape of the primary creep curve, the formation of a dislocation substructure, and the response of the material to stress or strain rate transients.

Figure 7-16 summarizes measured and interpolated creep data for binary NiAl at 1175 K. Most of the data fall within reasonable agreement, with no more than a factor of 5 difference in creep rate at a given stress level—except for the strengths reported by Vandervoort, Mukherjee, and Dorn (1966), which are abnormally weak for no known reason. The alloys in Figure 7-16 also have similar stress exponents, about 4.5 to 6, except for a transition to a lower stress exponent at the lowest stress levels. Examination of the stress exponents in Table 7-3 for materials with a wide variety of grain sizes, including single crystals, reveals that between 1100 K to

Table 7-3 Summary of Creep Parameters For NiAl

Al content, at.%	Grain size, μm	Temperature, K	Stress exponent, n	Activation energy for creep, Q, kJ/mol	Reference
48.25	5 to 9	100 to 1400	6.0 to 7.5	313	Whittenberger (1988)
44 to 50.6	15 to 20	1100 to 1400	5.75	314	Whittenberger (1987)
50	12	1200 to 1300	6	350	Whittenberger et al. (1990)
50	450	1073 to 1318	10.2 to 4.6	283	Yang and Dodd (1973)
50	500	1173	4.7	—	Rudy and Sauthoff (1985)
50.4	1000	1075 to 1750	7.0 to 3.3	230 to 290	Vandervoort, Mukherjee, and Dorn (1966)
50	SX [123]	1023 to 1223	7.7 to 5.4	—	Hocking, Strutt, and Dodd (1971)
50	SX [multiple]	1023 to 1328	4.0 to 4.5	293	Bevk, Dodd, and Strutt (1973)
50	SX [001]	1000 to 1300	6	440	Whittenberger and Noebe (1991)
50	SX [001]	1123 to 1473	3 to 7	305	Forbes et al. (1993)
50	SX [223]	1123 to 1473	3 to 7	330	Forbes et al. (1993)

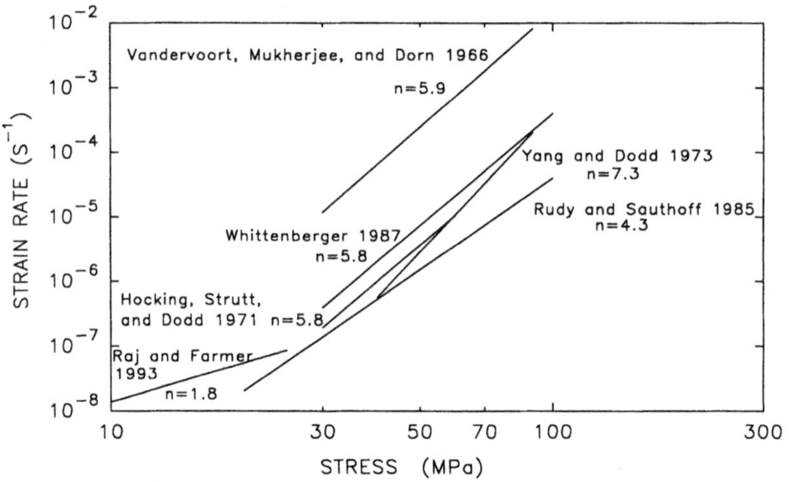

Figure 7-16. Measured or interpolated creep behavior of near-stoichiometric NiAl at 1175 K.

1400 K, the values for n cluster between 5 and 7. Class A materials are usually characterized by a stress exponent of 3, whereas class M behavior is typically characterized by a stress exponent of 5. Although a stress exponent of 7 is higher than is characteristic for class M behavior, such high values have been observed in several class M materials (Ashby 1972). Observations of subgrain formation after high-temperature deformation (Kanne, Strutt, and Dodd 1969; Yang and Dodd 1973); operation of normal primary creep behavior under both constant load and constant crosshead speed tests (Strutt, Dodd, and Rowe 1970; Whittenberger 1987, 1988); strain rate transient test results (Yaney and Nix 1988; Forbes et al. 1993); and strengthening via a reduction in grain size (Whittenberger 1988) are all consistent with class M behavior. Consequently, high-temperature creep in binary NiAl is apparently climb controlled over most stresses and temperatures.

Forbes et al. (1993) argue that creep deformation in single-crystal NiAl is a mixture of both class A and M behavior, with the relative proportions of each type dependent on crystal orientation and the stress/temperature regime. Both Forbes et al. (1993) and Vandervoort et al. (1966) show a gradual decrease in stress exponent, approaching $n \simeq 3$ at high temperatures (1400 K to 1500 K), which is consistent with this idea. However, it remains to be determined whether the viscous glide contribution is significant in cases other than soft-orientation single crystals, where subgrain formation is suppressed because of the limited availability of slip systems.

At stresses that are too low for dislocation processes to be significant, time-dependent deformation also can occur by stress-assisted vacancy flow. Creep by these diffusional mechanisms, such as Nabarro–Herring or Coble creep, is attributed solely to the movement of vacancies from sources to sinks, which are usually grain boundaries that have different orientations than that of the applied stress. These mechanisms are characterized by a stress exponent of $n = 1$ and a clear dependence of creep strength on grain size, with large-grained materials being more creep resistant.

Evidence for the operation of diffusional-creep mechanisms in B2 alloys was reported by Rudy and Sauthoff (1985), who observed $n = 1$ for their experiments with a Ni-20Fe-50Al alloy. Some recent data at very low stress levels for NiAl revealed that the stress exponent was between 1.6 and 2.1 (Raj and Farmer 1993), which indicates the operation of a grain-boundary-assisted deformation mechanism. In addition, there is evidence of diffusional-creep processes in binary NiAl at temperatures above 1300 K and at low strain rates, where grain growth during creep testing resulted in coarser grained material with higher strengths (Whittenberger 1987). Therefore, the limited data generated to date indicate that diffusional creep may occur in polycrystalline NiAl at $T > 1200$ K and at very low stresses.

One final point about the creep behavior of binary NiAl is worth noting. Unlike most NiAl properties, creep rate is roughly independent of stoichiometry for a broad range in composition, between about 45 and 51 at%Al (Yang and Dodd 1973; Whittenberger 1987; Whittenberger, Kumar, and Mannan 1991a). The largest difference in creep rates within this range of compositions is about a factor of 5. Only at very low and high Al contents are nonstoichiometric alloys noticeably weaker (Whittenberger et al. 1991; Whittenberger et al. 1991a). This loss in strength is most easily explained by the lower melting points of these compositions, which in turn implies a higher diffusivity. The diffusivity data for NiAl (Hancock and McDonnell 1971; Lutze-Birk and Jacobi 1975; Berkowitz, Jaumot, and Nix 1954) actually imply a more significant effect on properties than is observed. Nevertheless, these trends in strength as a function of stoichiometry are reversed from those observed at lower temperatures, where defect hardening predominates (Pascoe and Newey 1968b).

5.2. Design of Creep-Resistant NiAl-Based Alloys

As progressive alloying changes are made from pure Ni to a Ni-base superalloy, creep rate decreases by about eight orders of magnitude. The formation of a second phase is one of the major reasons for this improvement (Nathal, Diaz, and Miner 1989). Comparable improvements in creep

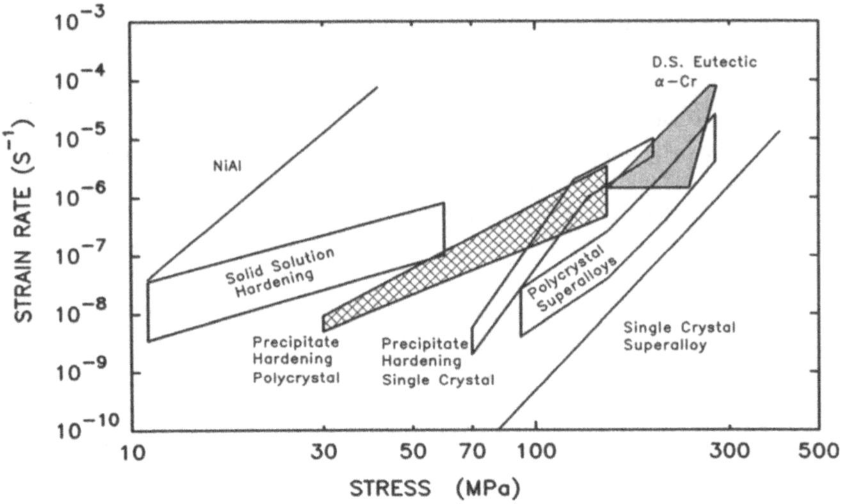

Figure 7-17. Effects of alloying on the creep behavior of NiAl at 1300 K. Both solid-solution alloys and precipitation-hardened NiAl-based materials are included in the figure.

strength by similar alloying strategies are necessary for nickel aluminides to compete effectively with current superalloys. To measure progress in this direction, Figure 7-17 and 7-18 compare the 1300 K creep strengths of NiAl-based materials (which were generated primarily in compression) to the tensile-creep response of a first-generation, single-crystal superalloy (Nathal and Ebert 1985) and a range of polycrystalline superalloys typically used in engine applications, including IN100 and Udimet 700. The boxed regions in these figures represent the range in creep strengths observed for a given strengthening mechanism, whereas the angle of the box indicates the resulting stress exponents. More detailed comparisons of the various strengthening mechanisms in NiAl are given in a previous review (Nathal 1992).

Figure 7-17 summarizes the effects of alloying on the creep behavior of NiAl. The effects of solid-solution hardening by Fe, Nb, Ta, Ti, and Zr on creep strength were studied (Jung, Rudy, and Sauthoff 1987; Pathare 1988; Whittenberger et al. 1989; unpublished research by J. D. Whittenberger and R. D. Noebe, 1991, NASA Lewis Research Center, Cleveland, OH). All the solid-solution alloys were stronger than binary NiAl, but their stress exponents decreased to values near 3 to 4. Apparently, these solute additions altered the creep process in NiAl to a viscous drag mechanism, in a manner very similar to that which occurs when

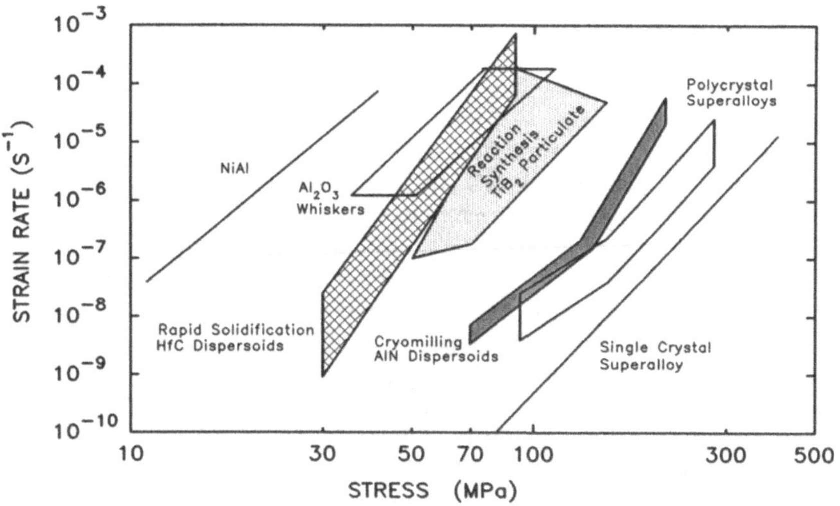

Figure 7-18. Effect of composite strengthening and second-phase reinforcement on the creep behavior of NiAl at 1300 K.

alloying elements are added to pure metals. In one recent study, a transition from class M to class A behavior was observed in the ternary B2 alloy Ni-20Fe-30Al as a function of applied stress (Raj et al. 1992). This transition appears to be well described by current theories developed for disordered solid solutions (Mohamed and Langdon 1974). Although solid-solution hardening does improve creep strength over binary NiAl, it is inadequate by itself and must be used in combination with other strengthening mechanisms to compete with superalloys.

The use of precipitation hardening to significantly improve the creep strength of NiAl was first demonstrated by Polvani, Tzeng, and Strutt (1976). They added Ti to NiAl, forming a two-phase mixture of NiAl and the Heusler phase Ni_2AlTi, resulting in a substantial increase in creep strength. Additional Heusler phases and other intermetallic compounds such as Laves (e.g., NiAlX) can be formed with ternary additions such as Nb, Ta, Hf, Zr, and V. The range of creep properties observed for precipitation-hardened polycrystalline alloys (Polvani et al. 1976; Pathare 1988; Whittenberger et al. 1989) are shown in Figure 7-17. These materials are reasonably strong, but the advantage diminishes when the data are extrapolated to low stresses because of the low-stress exponent. It is not entirely clear why these alloys exhibit such low *n* values, because most creep-resistant, two-phase alloys exhibit significantly higher stress exponents than the matrix phase does. In most cases, the microstructures of the ternary alloys were probably not optimized. For example, if the second phase is too coarse, effective strengthening would not be expected. An equally valid explanation may be that the observed n values represent a superposition of several deformation mechanisms. By analogy to the superalloys, optimizing the creep strength requires a balance between the compositions of the two phases, the precipitate-volume fraction, and the size and distribution of the precipitates.

One well-known strategy for improving creep strength is to eliminate grain boundaries by directional solidification. Yet surprisingly, binary NiAl, single crystals have little or no creep strength advantages over polycrystalline material (Whittenberger et al. 1991). However, a significant advantage in strength is obtained through single-crystal processing of alloyed NiAl. Figure 7-17 compares the creep response of the Heusler-containing NiAl alloy (containing 1 at% Hf) in both polycrystalline (Whittenberger et al. 1991) and single-crystal form (Locci et al. 1993), the polycrystalline data falling within the band for precipitation hardening of polycrystals. The single-crystal version is not only stronger than the poly-crystalline equivalent, but it displays a significantly higher stress exponent, presumably because grain-boundary deformation mechanisms have been eliminated. Figure 7-19 summarizes the progress made in developing creep-resistant NiAl alloys (Darolia et al. 1992; Darolia 1993). Unlike Figures 7-17 and 7-18, which rely heavily on compressive creep properties, Figure 7-19 plots tensile rupture stress for single-crystal NiAl alloys versus the Larson-Miller Parameter, which takes into account both time to

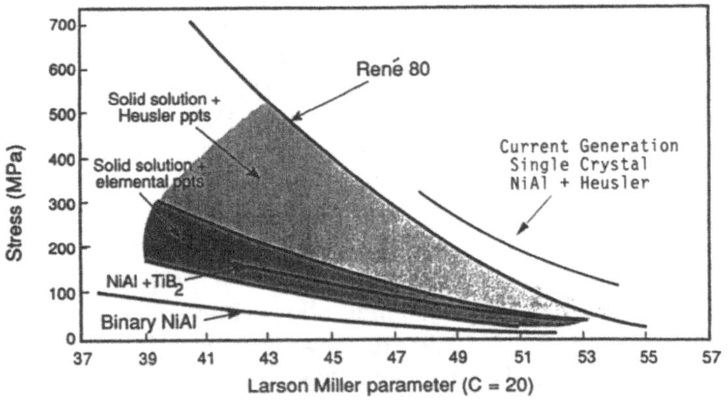

Figure 7-19. Tensile stress rupture properties of single-crystal NiAl alloys and a $NiAl/TiB_2$ composite compared with the superalloy René80. The Larson-Miller parameter is calculated as $(T_R/1000)(\log t + 20)$, where T_R is the absolute temperature in degrees rankine and t is the time to rupture in hours (after Darolia et al. [1992] and Darolia [1993]).

failure and temperature. Through alloying strategies similar to the development of superalloys, NiAl alloys have been produced with greater than polycrystalline René 80 strength levels. The strength advantage of NiAl over superalloys is even more significant when the data are normalized with respect to density.

Anisotropy in creep strength is another factor that may be exploited with single crystals. Certain orientations may have higher creep strengths that could be used to advantage in applications such as turbine blades. During experiments with single-phase binary compounds at temperatures above approximately 1050 K, some investigators have observed significantly higher creep strengths for hard-orientation crystals (Strutt et al. 1970; Forbes et al. 1993). However, other investigators have not observed significant anisotropy (Pascoe and Newey 1968b; Whittenberger et al. 1991).

One reason for these discrepancies is the basic definition of strength used in the various studies (personal communication with K. R. Forbes, 1993, Stanford University, Stanford, CA). For example, above 1200 K the yield strengths of [001] single crystals, soft-orientation single crystals, and polycrystalline NiAl are all similar (Fig. 7-3); but there is a large difference in strength at large strains because of differences in work-hardening rates. At high temperatures, soft-orientation single crystals work soften appreciably, polycrystals reach a steady-state stress level just beyond the yield point, and hard-orientation single crystals strain harden considerably. Therefore, although the yield strength of all these forms of binary NiAl are nearly equivalent, the steady-state creep strengths determined at large strains ($> 10\%$) are significantly different. At 1300 K and a given stress level, the steady-state creep rate for polycrystalline NiAl is at least an order of magnitude lower than for [223]-oriented single crystals, and [001] single crystals have a minimum creep rate about an order of magnitude lower than that of polycrystalline NiAl (personal communication with K. R. Forbes, 1993; Stanford University, Stanford, CA). This difference in creep strength is due to the degree that a dislocation substructure is formed in the various materials. Forbes et al. (1993) reported that a dislocation substructure formed during creep of hard-orientation single crystals but not during creep of soft crystals regardless of whether the samples were tested in tension or compression. The formation of a dislocation substructure along with the reduced mobility of $\langle 011 \rangle$ dislocations that operate in hard-orientation crystals can explain the increased creep strength of the hard-orientation NiAl single crystals over soft orientations or polycrystalline materials (Forbes and Nix 1993).

A variation on single-crystal processing that incorporates a second phase for strengthening is to form an in-situ composite by directionally solidifying eutectic alloys. In their early work, Walter and Cline (1970) showed that a eutectic consisting of α-Cr rods in a NiAl matrix has promising creep properties (Fig. 7-17). However, good creep strength is only maintained as long as the Cr rods are continuous in nature—not just short fibers aligned parallel to the growth direction (Johnson et al. 1992). In addition, rod-type and lamellar eutectic microstructures can be produced in NiAl by directionally solidifying NiAl containing W, Mo, Re, V, NiAlNb, NiAlTa, or other intermetallic phases; and these microstructures may significantly increase the creep resistance of NiAl. For example, in NiAl/NiAlNb alloys, changing the processing from casting plus extrusion to directional solidification significantly improved both the creep strength and the stress exponent, even when the second phase was only partially aligned (Whittenberger, Westfall, and Nathal 1989; Whittenberger et al. 1992).

The effects of inert second phases (other than precipitates) on the creep strength of NiAl are summarized in Figure 7-18. Rapid solidification, one method for fabricating two-component alloys, can produce very fine dispersions of second-phase particles, with diameters ranging from 20–50 nm. These particles usually resist coarsening because of their very low solubility in the matrix. For NiAl, pure elements, such as W and Mo, and various carbides or borides are candidates for this type of strategy. Experiments have determined that TiC and TiB_2 dispersoids (Whittenberger, Gaydosh, and Kumar 1990) have very little strengthening effect but dispersions of HfB_2 and HfC resulted in considerable improvements in creep strength in comparison to binary NiAl (Whittenberger et al. 1991; Whittenberger, Ray, and Jha 1992). These studies also revealed that HfC-strengthened, coarse-grained NiAl was more creep resistant than the finer grained material.

The common mechanism for improved creep resistance of dispersion-strengthened alloys is related primarily to the interaction of the dispersoids with mobile dislocations and subgrain boundaries. However, it is probably not a coincidence that the two most effective dispersoids contained Hf—a very potent solid-solution strengthening agent. To date, the use of rapid solidification to dispersion strengthen NiAl has not provided sufficient creep resistance for NiAl to compete effectively with superalloys (Figure 7-18). However, its strength may be improved by optimizing the dispersoid volume fractions and devising the thermomechanical processing schedules needed to produce grain structures similar to those that have been proven successful in the oxide-dispersion strengthening of Ni-base alloys.

An unusual but effective example of a creep-resistant composite is AlN dispersion-reinforced NiAl, produced by milling NiAl powder in liquid nitrogen followed by high-temperature consolidation (Whittenberger, Arzt, and Luton 1990a, b). This process produces very fine AlN dispersoids, on the order of 50 nm, at relatively high volume fractions of 10 vol% or greater. The particles are not uniformly distributed but clustered along prior particle boundaries surrounding particle-free areas. Creep strengths approaching those of superalloys were obtained with these composites (Fig. 7-18). After correcting for density, the deformation resistance of a typical superalloy and NiAl/AlN are nearly equivalent. It is interesting to note that the consolidation process has relatively little effect on the strength of these materials, at least in compression. Properties of extruded versus hot isostatically pressed (HIP'ed) NiAl/AlN were roughly equivalent (Whittenberger et al. 1990b). The reasons for the exceptional strength produced by this type of second-phase reinforcement are currently being investigated. Nevertheless, cryomilling of NiAl has proven to be a consistently reproducible method for generating NiAl/AlN composites with very high creep strength (Whittenberger and Luton 1992; Hebsur et al. 1993) as well as excellent oxidation resistance (Lowell, Barrett, and Whittenberger 1990).

Another approach to strengthening metals is the use of discontinuous reinforcements. These reinforcements are typically larger in size and present in higher concentrations than those found in dispersion-strengthened materials. One example is Al_2O_3 whisker-reinforced NiAl (Whittenberger, Mannan, and Kumar 1989; Whittenberger et al. 1991b). The whiskers, which had an average aspect ratio of approximately 7.5, were added by mechanical blending at volume fractions ranging from 0 to 25 vol%. Some improvement in creep resistance is seen in Figure 7-18, but it is not very significant. The stress exponents of the NiAl/Al_2O_3 composites were about the same as that of the matrix material. This indicates that deformation is controlled by flow in the matrix, as predicted by several models of composite strengthening (Kelly and Street 1972). These models predict that creep strength may be further improved by increasing the aspect ratio of the whiskers. Because the aspect ratio of the whiskers was reduced during processing due to fracture (Whittenberger et al. 1991b), higher strength whiskers will also be needed. Controlling whisker distribution and alignment and limiting whisker damage during processing are major concerns with this type of composite.

Another type of discontinuous, reinforced composite that has been characterized fairly extensively is NiAl containing 1-μ-diameter TiB_2 particles produced by an exothermic reaction process (Viswanadham et al. 1989; Wang and Arsenault 1990, 1991). These composites have improved strength that scales with the amount of reinforcement (Whittenberger et al. 1990). In addition, the stress exponents are all high, indicating that the dislocation substructure was refined and stabilized by the second-phase reinforcement. Evidence for this contention has been provided by transmission electron microscopy, where subgrain boundaries pinned by the particles, in combination with a high dislocation content within the subgrains, were characteristic of the deformation structure (Whittenberger et al. 1990). TiB_2-reinforced composites are stronger than Al_2O_3-whisker-containing materials but still fall short of the Ni-base superalloys (Fig. 7-18). As an additional comparison, the tensile rupture strengths of a TiB_2-dispersed alloy (Kumar et al. 1992) are also included in Figure 7-19. This gives a further indication of how far particulate-reinforced composites are from competing with superalloys or even single-crystal NiAl-based alloys.

Finally, tests of hybrid composites containing both TiB_2 particulates and Al_2O_3 whiskers have demonstrated that these strengthening mechanisms are additive (Whittenberger et al. 1991b). It should be noted, however, that combining creep-strengthening mechanisms is not always effective. For example, adding TiB_2 particles to a Heusler-phase-(Ni_2AlTi) reinforced NiAl alloy does not improve creep strength (Kumar and Whittenberger 1991).

Various strategies to improve the creep resistance of NiAl have been attempted, although none have been fully optimized. It is clear from these efforts that the poor creep strength exhibited by binary NiAl may be overcome by conventional metallurgical techniques. Of the processes investigated, both solid-solution strengthening and precipitation hardening are relatively ineffective when used with polycrystalline material; however, precipitation-hardened single crystals show extreme promise, exhibiting stress-rupture properties superior to many superalloy systems. Rapid solidification combined with alloying has generally resulted in only small improvements in strength, whereas Al_2O_3 whiskers and TiB_2 particulates have provided larger, but still insufficient, advances in creep resistance. Finally, NiAl/AlN composites and alloyed single crystals exhibit some of the best creep properties to date, with specific strengths comparable to first-generation single-crystal superalloys.

6. FATIGUE

In many structural applications, materials are subjected to repeated loading or vibration. In complex machinery, such as gas turbine engines, cyclic loading is compounded with the additional effects of creep at high temperatures and environment. Both creep processes and environment have been proven to adversely affect the fatigue behavior of most intermetallic systems studied previously (Stoloff, Smith, and Castagna 1993). Furthermore, even though fatigue is a complicated subject and a critical property for many applications, it is usually one of the last properties to be studied with a new material system. This was the case for NiAl, with fatigue studies beginning in about 1991. Fatigue investigations are a natural extension of the multitude of monotonic-deformation studies performed on NiAl, which have led to a significant understanding of the deformation and fracture behavior of this material (section 3). The recent number of fatigue studies also indirectly indicates the maturity of this intermetallic system because substantial quantities of reproducible material are usually necessary for a detailed fatigue investigation. Both single-crystal and polycrystalline NiAl at room and elevated temperatures have been studied under zero-tension or fully reversed loading conditions.

Stress- and strain-controlled fatigue tests of single-crystal NiAl were performed on [001]-oriented Ni-49.9Al-0.1 Mo crystals at room temperature and 1033 K (Bain, Field, and Lahrman 1991). At room temperature, the single-crystal NiAl failed elastically because of the high stress necessary for plastic flow in [001]-oriented crystals. At elevated temperatures, the crystals followed the expected Coffin–Manson strain life behavior, typical of conventional Ni-based superalloys, with fatigue lives very similar to those of polycrystalline NiAl (Fig. 7-20).

Room-temperature, total-strain-controlled fatigue testing of soft-orientation single crystals revealed substantial cyclic ductility in [121]-oriented crystals (Smith et al. 1992). The samples hardened continuously during cycling for at least the first 400 cycles, generally reaching stresses 50% greater than the initial yield strength (Smith et al. 1992). Failure of these single-crystal NiAl specimens was attributed to the formation of cracks at slip extrusion-intrusions. This type of crack-initiation mechanisms appears to be common to many intermetallic systems (Stoloff et al. 1993).

High-cycle fatigue (HCF) testing was performed on a high-strength, Hf-containing NiAl single crystal alloy with creep-rupture properties equivalent to that of René 80 (Darolia 1993). At 1253 K the high-cycle fatigue properties of the NiAl alloy were similar to those of the Ni-base superalloy. In addition, the material was not sensitive to surface finish because fatigue properties were not degraded when the alloy was tested in the as-electrodischarge-machined condition in comparison to electropolished samples. Preliminary tests indicate that the HCF behavior of this material at room temperature is also quite good (personal communication with R. Darolia, 1992, General Electric Aircraft Engines, Evendale, OH).

Low-cycle fatigue (LCF) testing of polycrystalline NiAl has been performed at room temperature, near the monotonic BDTT, and at 1000 K. During room-temperature LCF testing, NiAl continuously work hardened during cyclic straining, eventually reaching stress levels that were around 60% greater than the monotonic fracture strength (Noebe and

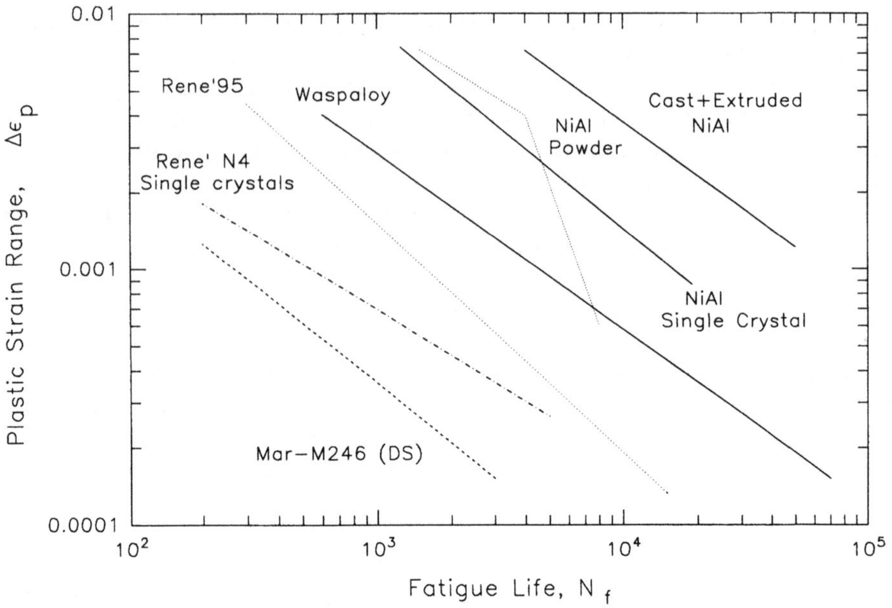

Figure 7-20. Fatigue life at nominally 1000 K for NiAl in comparison to Ni-base superalloys and single-crystal NiAl on a plastic strain range basis (Lerch and Noebe 1994).

Lerch 1992; Edwards and Gibala 1993). The cyclic stress-strain response for NiAl was similar to that of body-centered cubic metals and other B2 alloys, and the fatigue life on a plastic-strain-range basis was considerably greater than that of comparable B2 intermetallics (Noebe and Lerch 1992). The plastic-strain-range/fatigue-life relation appears to be independent of grain size for cast and extruded NiAl but exhibits a much shallower slope (−0.14) than usually observed for metallic materials (−0.6) (Noebe and Lerch 1992; Edwards 1993). This shallow slope is representative of the brittle fracture behavior of NiAl at room temperature, with fracture initiating at prior defects in the material once a critical stress level was achieved (Lerch and Noebe 1993). In addition, high-cycle fatigue testing of hot-pressed flat sheet specimens of NiAl and NiAl/Al_2O_3 fiber composites was conducted at 0.3 Hz and $R = 0$ loading conditions under displacement control between prescribed stress ranges (Bowman 1992a). Consistent with other room-temperature studies, long fatigue lives were attained for NiAl when the stresses were kept just below the monotonic fracture stress of the samples. The fatigue tests on both the monolithic and composite samples were interrupted after 100,000 cycles without a measurable loss in compliance in either material.

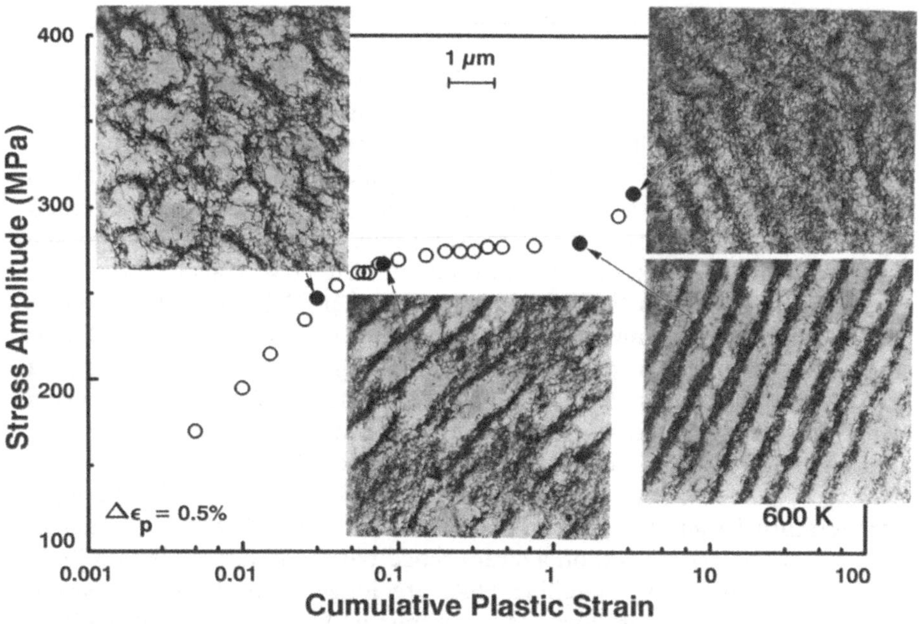

Figure 7-21. Cyclic hardening curve for polycrystalline NiAl at 600 K and a plastic strain range, $\Delta \varepsilon_p$, of 0.5%, and the dislocation microstructures observed at various stages of fatigue life (courtesy of C. L. Cullers).

The cyclic deformation behavior of NiAl was also investigated between 600 K and 700 K, representing temperatures just above and below the monotonic BDTT of a powder-extruded alloy (Cullers and Antolovich 1992). Even though monotonic properties change dramatically over this temperature range (Figs. 7-11 and 7-12), a more gradual transition in fatigue behavior was noted. Increasing the temperature from 600 K to 700 K improved fatigue lives slightly and decreased stress response levels. However, the cyclic hardening behavior was similar over this temperature range as was the manner in which the dislocation structure developed. In this temperature range, the material work hardened significantly for the first few cycles, reached a plateau, and finally exhibited an additional region of work hardening very near the end of life. This behavior was traced to the dislocation structure that evolved during fatigue cycling (Cullers, Antolovich, and Noebe 1993). During the initial period of work hardening, the dislocation structure formed into a loose cellular array (Fig. 7-21). These cells developed into well-aligned veins or walls of dense dislocation tangles leading to a period of stress saturation. This was followed, near the end of the sample's life, by a collapse of the vein structure leading to the second period of accelerated work hardening and failure of the sample.

LCF testing of polycrystalline NiAl has also been performed well above the monotonic BDTT (Lerch and Noebe 1994; Noebe and Lerch 1993). At 1000 K, both processing route and environment affected fatigue life. The fatigue lives of NiAl samples, produced by HIP'ing prealloyed powders, were about a factor of 3 less than those of cast and extruded NiAl because of the lower flow stress and thus, a lower crack-driving force in the extruded material. An environmental effect was also observed. A factor of 3 increase in fatigue life occurred when HIP'ed powder samples were tested in vacuum instead of air. This increase in fatigue life is indicative of an environmentally assisted fatigue-damage mechanism at elevated temperatures. At 1000 K, NiAl exhibits a typical Coffin–Manson strain-life behavior; however, below a plastic strain range of about 0.3%, the fatigue life of HIP'ed NiAl is less than expected because of additional damage by creep cavitation processes (Fig. 7-20). Consequently, the high-temperature fatigue behavior of NiAl is influenced by both environmental and creep processes. Yet, on a strain-range basis, the fatigue lives of both single-crystal and polycrystalline NiAl are longer than those of conventional Ni-based superalloys at a nominal temperature of 1000 K (Fig. 7-20). This advantage is due partly to the higher ductility of NiAl. The disadvantage is that the fatigue lives of binary NiAl would be shorter than those of Ni-based superalloys on a stress-range basis because of the higher strength of superalloys at elevated temperatures (Lerch and Noebe 1993).

The poor LCF performance of the related B2 compounds Fe-40Al and Ni-20Fe-30Al (Hartfield-Wunsch and Gibala 1991) first raised concerns about the viability of any B2 alloy as a structural material because of fatigue considerations. However, initial results for both single-crystal and polycrystalline NiAl indicate that its fatigue life is generally equivalent or superior to other intermetallic compounds and other structural materials such as Ni-base superalloys. Consequently, fatigue behavior does not appear to be an area of immediate concern, though research will be continuing in this area for some time. Of particular interest will be additional fatigue studies on some of the stronger alloys and composites that tend to be less defect tolerant.

7. ENVIRONMENTAL RESISTANCE

The excellent oxidation resistance of NiAl-based materials is well known and has been exploited for many years in the form of coatings for Ni-base

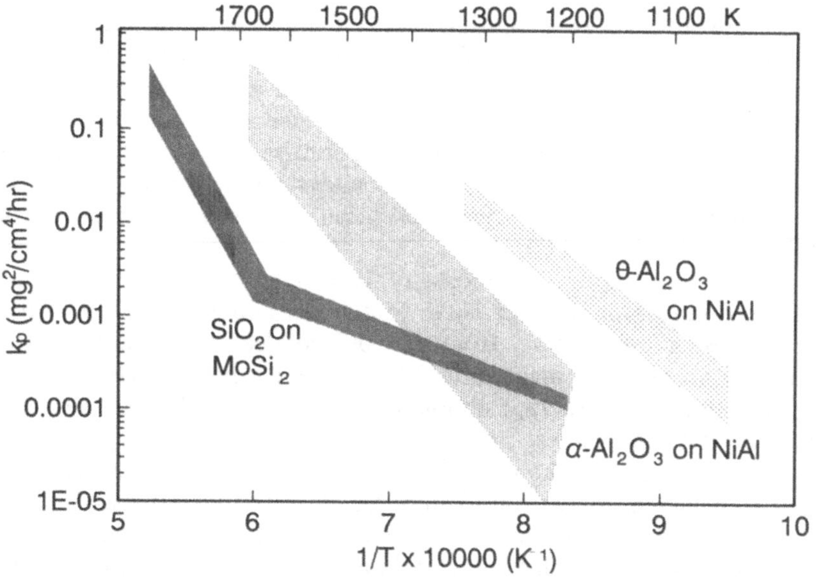

Figure 7-22. Range of observed parabolic rate constants, k_p, for Al_2O_3 formation on NiAl and for SiO_2 formation on $MoSi_2$ as a function of temperature (Nesbitt and Lowell 1993).

superalloys (Rhys-Jones 1989; Wood and Goldman 1987). The oxidation resistance of NiAl alloys is due to the easy formation and slow growth rate of protective Al_2O_3 scales. At low temperatures, γ-Al_2O_3 is the commonly observed oxide (Doychak, Smialek, and Mitchell 1989; Brumm and Grabke 1992). At temperatures near 1200 K, θ-Al_2O_3 appears to be the predominant oxide in mature scales, but at higher temperatures, α-Al_2O_3 is formed (Rybicki and Smialek 1989). Figure 7-22 compares the parabolic growth rate constants k_p for the higher temperature α-phase and the faster growing, lower temperature θ-phase Al_2O_3 scales. The growth rate of the lower temperature oxides, γ-Al_2O_3 and θ-Al_2O_3, are similar, but the growth rates are higher than that of α-Al_2O_3, such that oxidation at approximately 1200 K is faster than at 1300 K (Rybicki and Smialek 1989; Brumm and Grabke 1992).

At temperatures greater than 1300 K, $MoSi_2$ has better oxidation resistance because SiO_2 scales on $MoSi_2$ grow more slowly than Al_2O_3 scales grow on aluminide intermetallics (Fig. 7-22) (Nesbitt and Lowell 1993). However, in the low-temperature range of 650 K to 775 K, $MoSi_2$ is plagued by a rapid, catastrophic mode of oxidation known as "pest," which can quickly reduce a sample to a pile of powder (Chou and Nieh 1993).

NiAl is not affected by a pesting reaction under ambient environments, although it can be forced to undergo a mild pesting reaction by testing at very low partial pressures of oxygen (Grabke et al. 1991). Therefore, alumina-forming alloys such as NiAl exhibit the most oxidation-resistant behavior over the broadest range of temperatures in comparison to other metallic or intermetallic materials. NiAl alloys can be further protected at very high temperatures by the application of thermal barrier coatings (Miller and Doychak 1992).

Apparently, alloy stoichiometry has only a minor influence on isothermal oxidation rates of NiAl. Hutchings and Loretto (1978) found that increasing the Al level from 42–50% reduces the oxidation rate at 1173 K by an order of magnitude. More recent work (Doychak, Smialek, and Barrett 1989) shows that although oxidation rates differed by a factor of 10, the drop in rate was not proportional to Al level over the range of alloys studied. Instead, the oxidation rate at temperatures between 1273 K and 1673 K increased by close to an order of magnitude after a 3% drop in Al content from the stoichiometric composition. Oxidation rates were then relatively constant with slight fluctuations for Ni-Al alloys containing 47 to 37 at%Al.

Several studies have concentrated on the mechanisms of oxidation in NiAl in terms of mass transport phenomenon. Examination of tracer diffusion data (Young and DeWit 1986) and scale morphologies (Doychak et al. 1989) indicates that the transition oxides grow primarily by outward cation diffusion. At higher temperatures, the mature α-Al_2O_3 oxide also grows initially by outward Al diffusion (Prescott et al. 1992) with the contribution of inward oxygen transport increasing during later stages of oxidation because of a strong influence from short-circuit diffusion paths (Jedlinski and Borchardt 1991; Nicolas-Chaubet, Huntz, and Millot 1991). The oxidation behavior for NiAl is in contrast to that for MCrAl alloys, where inward-growing scales are typically formed (Reddy, Smialek, and Cooper 1982).

Cyclic oxidation testing between high and ambient temperatures is significantly more severe than isothermal oxidation and more closely approximates actual service conditions. During the cooling portion of the cycle, the difference in thermal expansion coefficients between metal and oxide produces high stresses at the oxide-metal interface that can lead to scale spallation. Spalling occurs randomly over the surface of the sample, exposing either fresh metal or fractured oxide scales to the next thermal cycle. Continued cycling causes repetitive selective oxidation of Al and, therefore, an overall loss of Al from the alloy. Cyclic lifetimes can be very short in comparison to those for isothermal conditions because the Al consumption rate is inversely proportional to the oxide thickness, and spalling keeps the oxide scale thin.

Three factors strongly affect the cyclic oxidation behavior of NiAl: temperature, alloy stoichiometry, and reactive element additions. Increased temperatures result in more rapid oxide growth and, therefore, a greater volume of oxide spalling during cooling (Nesbitt and Lowell 1993). Cyclic oxidation life decreases by about an order of magnitude for each 100 K-increase in temperature above 1400 K (Doychak et al. 1989). Ni-Al alloys are also significantly more sensitive to stoichiometry under cyclic conditions than under isothermal tests. Ideally, for long-term cyclic conditions, NiAl alloys should contain at least 45 at%Al. For even a few cycles, the composition should contain greater than 40 at%Al to avoid transformation of the matrix to martensite and formation of a less protective Ni-containing oxide (Smialek and Lowell 1974; Doychak et al. 1989).

Finally, it is well known that rare-earth or oxygen-active dopants such as Y, Hf, Zr, Ce, and La have a very large and favorable effect on the cyclic-oxidation resistance of MCrAl, NiAl, and superalloy materials (Smialek and Meier 1987). For example, the excellent cyclic-oxidation resistance of NiAl compared with conventional oxidation-resistant alloys and other Al-forming intermetallics (Fig. 7-23) is only achieved when reactive elements such as Zr are added. The enhanced oxidation resistance of NiAl is achieved when approximately 0.1 at% of the reactive

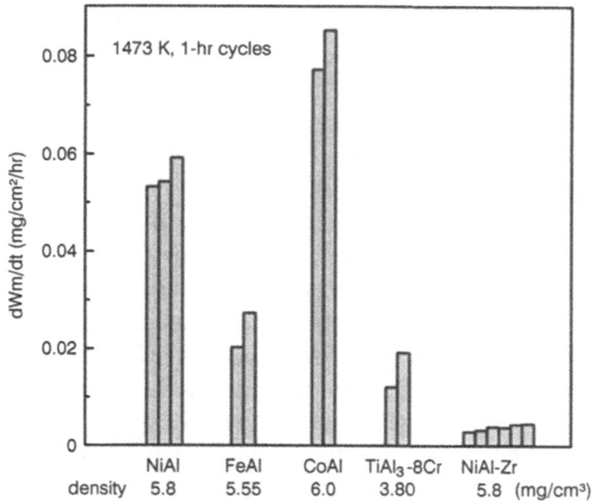

Figure 7-23. Rates of weight loss due to cyclic oxidation at 1473 K for several Al_2O_3-forming alloys (Nesbitt and Lowell 1993).

element is in solution (Barrett 1988), is added in the form of oxide dispersions, or is incorporated in the surface by ion implantation (Jedlinski and Mrowec 1987). These additions slightly decrease the isothermal-oxide growth rate (Rybicki and Smialek 1989; Jedlinski and Mrowec 1987) by suppressing oxygen diffusion through the scale (Jedlinski, Borchardt, and Mrowec 1992). But the main benefit of these additions is a dramatic increase in scale adherence during cyclic oxidation testing (Doychak et al. 1989).

Oxide adherence has been attributed to several causes including oxide "pegging," elimination of voids at the interface, increased scale plasticity, and changes in the chemical bond between the oxide and metal (Smialek and Browning 1985). However, the most likely effect of reactive elemental additions is the role of these elements in preventing sulfur segregation to the oxide/metal interface as previously demonstrated in Ni-base superalloys (Smialek 1991; McVay et al. 1992). Sulphur segregation would reduce the surface energy, adversely reducing the scale adherence. Sulphur also prefers to segregate to free surfaces, favoring the formation of voids under the oxide scale and accelerating their growth to cavities (Grabke, Wiemer, and Viefhaus 1991).

The behavior of NiAl(Zr) alloys under isothermal and cyclic oxidation conditions has been sufficiently characterized that life prediction can be made with reasonable accuracy (Doychak et al. 1989; Nesbitt and Vinarcik 1991). Useful lives are usually defined as the time needed to obtain a given amount of metal recession, a given amount of weight change (typically 5 to 10 mg/cm^2), or a transition to a less protective oxide. In the case of NiAl, severe cyclic-oxidation exposure eventually leads to Al depletion and subsequent formation of NiO and $NiAl_2O_4$, which can be used as a criterion to define life. Therefore, the Al reserve of an alloy, which is critical for extended oxidative life, makes stoichiometric alloys superior to Ni-rich materials.

Another environmental concern, hot corrosion due to molten sulfate deposits, has been studied because of the importance of this type of attack in marine, industrial, and aircraft gas turbine applications. This problem, which is most prominent at intermediate temperatures (900 K to 1300 K), can result from several different mechanisms (Pettit and Giggins 1987). For example, molten deposits containing Na_2SO_4 can cause fluxing reactions that dissolve protective Al_2O_3 oxides. In addition, the formation of sulfides beneath the deposits, and subsequent oxidation of these sulfides, can be quite detrimental to surface integrity (Kaufman 1969).

Binary NiAl is not particularly resistant to hot corrosion, with corrosion rates far exceeding the oxidation rate at equivalent temperatures (Ellis 1987; Godlewski et al. 1989). This occurs because Al depletion is much faster in sulfur-containing environments than in air, causing the less resistant Ni-containing oxides to form early during exposure. Chromium was found to increase the hot corrosion resistance of NiAl, and an additional improvement (albeit marginal) was noted when both Y and Cr were added to NiAl (McCarron, Lindblad, and Chatterji 1976). When Ni-Al was macroalloyed with Cr in sufficient quantity to form MCrAl-based alloys in the $\gamma + \gamma' + \beta$ phase field, hot corrosion resistance was very good (Kaufman 1969; Godlewski et al. 1989). However, Cr additions adversely affect the cyclic oxidation performance of NiAl alloys in air (Smeggil 1991; Grabke, Brumm, and Steinhorst 1992).

Overall, NiAl alloys have questionable sulphidation resistance and will probably need to be macroalloyed or coated if such conditions are encountered during use. NiAl alloys have good isothermal oxidation resistance. However, only when alloyed with reactive elements such as Zr does NiAl exhibit the exceptional cyclic oxidation resistance it is known for. NiAl(Zr) alloys exhibit long-term cyclic oxidation resistance to approximately 1500 K (Nesbitt and Lowell 1993), which exceeds the capabilities of essentially all other Al_2O_3-forming alloys.

8. NiAl-FIBER COMPOSITES

Although individual alloying strategies have successfully improved the ductility, fracture toughness, oxidation resistance, or creep strength of NiAl, no alloy composition has yet been developed that has an acceptable balance of properties in either polycrystalline or single-crystal form. Thus far, alloys that show significant improvements in high-temperature properties have lower toughness or ductility than binary NiAl. Similarly, when attempts are made to optimize low-temperature properties, creep strength is usually sacrificed. This dilemma is the driving force behind the research activities into fiber-reinforced NiAl-based composites, which are under increased scrutiny as a possible means for achieving a balance of properties that is presently unobtainable in monolithic NiAl.

Three categories of NiAl-based composites have received the majority of interest: (a) NiAl reinforced with continuous fibers of either W, Mo, or Al_2O_3; (b) NiAl containing a fine dispersion of TiB_2 or AlN particles; and (c) multi-phase, in-situ NiAl composites. The mechanism for improving strength in continuous-fiber-reinforced composites differs from that in either the particulate or two-phase systems. In continuous-fiber composites, strengthening occurs when load is transferred from the relatively weak matrix to the high-strength fibers, rather than when the motion of dislocations is impeded as generally occurs in the other types of materials.

The classification of a particular system as a "true" composite is often subjective, but for the purposes of this discussion, only artificial continuous-fiber-reinforced systems are considered because of their distinctive strengthening mechanism and unique problems.

To substantially improve strength in a fiber composite, the reinforcing fiber must possess high strength and have an elastic modulus significantly greater than that of the matrix. In addition, a strong bond must be present between the matrix and fiber to allow load transfer from the matrix to the fiber. Conversely, for materials where improved toughness is the primary goal, such as ceramics, a weak interface is desirable. In such cases, the matrix/fiber interface retards crack propagation by debonding in the wake of an advancing crack. Unfortunately, binary NiAl as a polycrystalline composite-matrix material suffers from both low high-temperature strength and poor low-temperature toughness. Attempting to both strengthen and toughen NiAl by fiber reinforcement makes the task of engineering this composite system extremely challenging. Following the example of SiC/Ti aluminide matrix composites, an intermediate level of bonding can provide a combination of strength, toughness, and fatigue crack-growth resistance (Larsen, Revelos, and Gambone 1992). However, the bond strength in this system is low enough that transverse properties are a major concern.

The choice of reinforcing fiber for a NiAl-based composite is limited by the requirements of high elevated-temperature strength, high-modulus, low-density, environmental resistance, and compatibility with the matrix. The last requirement refers to mechanical suitability, similar coefficients of thermal expansion (CTE), and chemical compatibility with the matrix. In addition, the fiber must be readily available in quantities sufficient for developmental studies. At present, single-crystal Al_2O_3 fibers, typically 125 μm in diameter, are the best choice for meeting these requirements (Misra 1988; Noebe, Bowman, and Eldridge 1991). Polycrystalline Al_2O_3 fibers have been used to reinforce NiAl (Anton and Shah 1992), although the fibers suffer from a greater CTE mismatch with NiAl than single-crystal fibers, as well as a loss of creep strength at high temperatures. However, the polycrystalline fibers have such a small diameter (approximately 12 μm) that residual stresses in the matrix should be significantly reduced compared with those of the larger single-crystal Al_2O_3 fibers (Lu et al. 1991).

There are few other fiber systems that are available for use in metal or intermetallic matrix composites. Strengthening in combination with toughening has been observed in NiAl composites reinforced with W (Bowman 1992b) or Mo (Bowman et al. 1991). Because of greater ductility, the use of either W or Mo as a reinforcing fiber has successfully reduced

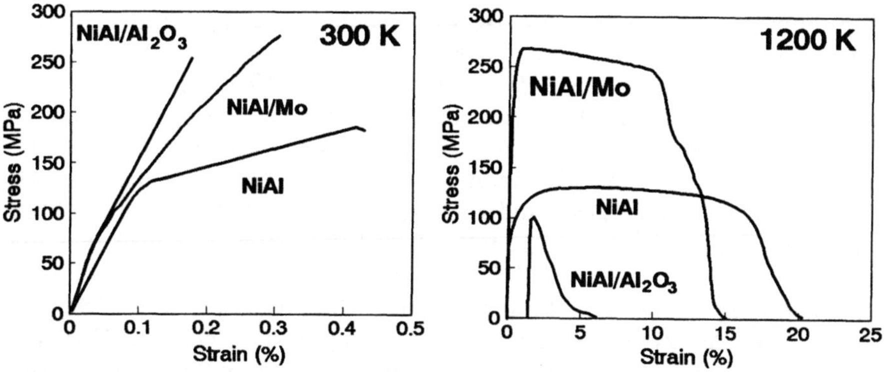

Figure 7-24. Tensile stress-strain curves for NiAl, $NiAl/Al_2O_3$, and NiAl/Mo composites at 300 and 1200 K (Bowman 1992a).

the susceptibility of the fiber to processing-related strength loss. Therefore, even greater strength improvements have been noted in these systems than in $NiAl/Al_2O_3$ (Fig. 7-24). These fibers are also chemically compatible with the matrix. They do, however, suffer from even greater CTE mismatch than the Al_2O_3 fibers do (α of NiAl = 16×10^{-6}/K, Al_2O_3 = 9×10^{-6}/K, Mo = 4.8×10^{-6}/K, W = 4.5×10^{-6}/K). In addition, because of their poor high-temperature oxidation resistance and high densities, W and Mo are not considered reasonable reinforcing candidates for NiAl-based aerospace materials.

SiC fibers have proven successful in Ti-based composites (Larsen et al. 1992) but are an unacceptable choice for NiAl because of the large CTE mismatch ($\alpha_{SiC} = 5.8 \times 10^{-6}$/K) and severe chemical incompatibility at high temperatures (Chou and Nieh 1991). Because beryllides such as Nb_2Be_{17} have a CTE almost identical to that of NiAl, they would make a good reinforcing fiber; unfortunately they are neither chemically compatible with NiAl nor available in fiber form (Misra 1991).

Significant toughening has been observed in NiAl composites reinforced with stainless steel tubes (Nardone 1992). The stainless-steel-tube-reinforced system has demonstrated the proof-of-concept of using ductile reinforcements in a brittle matrix to improve toughness and tensile properties while maintaining a high-fatigue limit (Nardone 1992; Bannister et al. 1992). Nevertheless, this particular system would not be suitable for high-temperature applications because of the extensive solubility of Fe in NiAl.

Because all other available continuous-fiber reinforcements have been eliminated for various reasons, single-crystal sapphire is presently the only potential fiber system for use with NiAl. Consequently, preliminary tensile, creep, fatigue, and cyclic oxidation properties for this system were generated (Bowman 1992a, 1992b; Doychak et al. 1992; Jeng, Yang, and Amato 1992). Unfortunately, current composite processing schemes reduce fiber strength by 50% and decrease the fracture strain of the single-crystal Al_2O_3 fibers by about a factor of 4 (Draper, Gaydosh, and Chulya 1991; Draper, Locci, and Eldridge 1992). Such reductions in fiber properties limit the current strength of the composite to a level that the present generation of NiAl/Al_2O_3 composites do not compete with current high-temperature structural materials. The ultimate tensile strengths of the NiAl/Al_2O_3 composites tested by Jeng et al. (1992) and Bowman (1992a) were only comparable to the strength of monolithic NiAl at both room and elevated temperatures. In addition, the tensile curves in Figure 7-24 indicate that the toughness of NiAl/Al_2O_3 is less than that of monolithic NiAl. The reduced toughness or failure strain of the NiAl/Al_2O_3 composite can be attributed to the brittle nature of the fibers. Because of damage during processing, Al_2O_3 fibers have failure strains on the order of 0.18% in tension (Draper et al. 1991). In the case of NiAl-based composites, the relatively low matrix ductility is incapable of tolerating flaws that are created from the fiber failures. Consequently, the composite fails at the onset of fiber fracture. Maintenance of as-received fiber strength is essential for the success of these composites.

Still preliminary results indicate that the uniaxial creep strength of NiAl/Al_2O_3 composites is quite good (Bowman 1992a). Creep strengths similar to NiAl(1Hf) single crystals have been measured at 1300 K. Also, NiAl/Al_2O_3 composites have excellent fatigue resistance as long as the applied strains during fatigue loading are below the fracture strains of the composite (Bowman 1992a).

Individually good cyclic oxidation resistance of the fiber and matrix does not necessarily guarantee an oxidation-resistant composite. When the bond strength between the NiAl and Al_2O_3 is weak, the cyclic oxidation of the composite is worse than for the matrix-only material because of oxidation along the matrix/fiber interface (Doychak et al. 1992). This is demonstrated in Figure 7-25 where, for the first 150 cycles, the weight gain of the composite is much greater than for the unreinforced matrix because of additional oxidation along the fiber matrix interface. Shortly thereafter the sample fails catastrophically. When the bond is strong, the oxidation resistance of the composite is comparable to that of the matrix-only material. Figure 7-25 also illustrates the importance of reactive element additions for achieving the best cyclic oxidation resistance in NiAl alloys. The oxidation resistance of a NiAl(Zr)/Al_2O_3 composite containing

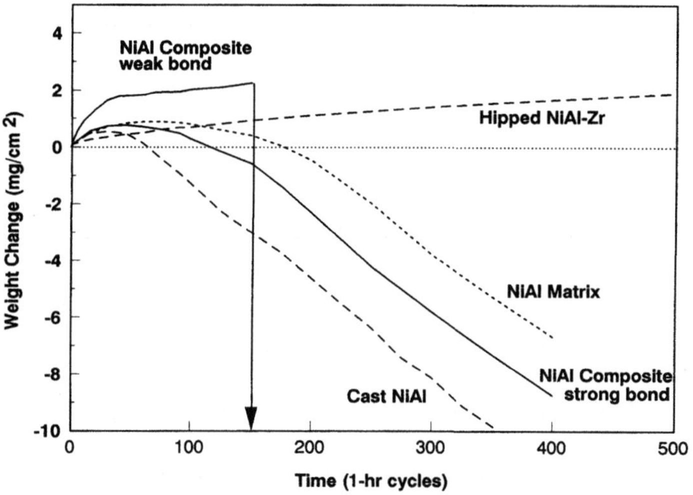

Figure 7-25. Cyclic oxidation behavior of $NiAl/Al_2O_3$ composites with either strongly bonded or weakly bonded fibers in comparison to various monolithic NiAl materials for 1 hr. cycles at 1473 K (Bowman 1992a).

strongly bonded fibers is expected to be similar to that of NiAl(Zr); however, the composite does not survive thermal cycling because of the buildup of residual stresses in the more brittle Zr-doped matrix (Noebe et al. 1991).

If processing-induced fiber-strength degradation can be eliminated in this system, the primary obstacle to the use of Al_2O_3 fibers in NiAl will be the difference in the CTE between NiAl and c-axis-oriented, single-crystal Al_2O_3. This fiber/matrix CTE mismatch leads to the development of high residual stresses in the composite after various thermal excursions such as cooling from the processing consolidation temperature and subsequent thermal cycling. For a strongly bonded $NiAl/Al_2O_3$ system with 25 vol% fibers, an elastic-plastic model (unpublished research by S. M. Arnold, 1992, NASA Lewis Research Center, Cleveland, OH) predicts that the thermally generated residual stresses induced by cooling from the processing temperature will be 150 MPa axial, −175 MPa radial, and 200 MPa tangential, resulting in an effective stress of approximately 350 MPa at the fiber/matrix interface. Although some refinements in the current models are needed to account for creep and viscous flow, the results indicate that stresses in excess of the yield and fracture stresses of the matrix can be generated in this system. In qualitative agreement with the model, matrix cracking has been observed in strongly bonded $NiAl/Al_2O_3$

composites after only a few thermal cycles because of the inability of the matrix to accommodate the imposed strains (Bowman 1992a).

As a result, a substantial effort is under way to identify a suitable coating for the Al_2O_3 fibers to act as either a compliant or compensating layer. The compliant layer concept uses an interfacial layer that increases or maintains the ductility of the matrix/fiber interface but does not necessarily reduce the residual stresses within the system. A compensating fiber coating is designed not only to act as a compliant layer but also to reduce the residual stresses by having a CTE that is higher than that of the matrix. To date, no single material has been identified that can meet either of these requirements while being chemically compatible with the constituent fiber and matrix.

Consequently, continuous-fiber composite technology in the NiAl system is a long way from maturity. The problem is twofold: first, a lack of an adequate fiber or processing technique that does not degrade fiber properties, and second, an unoptimized matrix. It is unlikely that a single type of fiber will be able to increase both high-temperature strength and low-temperature toughness of a NiAl matrix while still meeting all other property requirements. Instead, optimal alloying of NiAl will be necessary to solve one of these material deficiencies; then, a suitable fiber with a better CTE match can be used to overcome the remaining problem.

9. PROCESSING AND FABRICATION

Compared with that of most intermetallics, the processing of NiAl is relatively easy in spite of a high melting temperature and limited room-temperature ductility. Processing is greatly facilitated by a wide single-phase field, congruent melting point, single-phase microstructure, and high ductility above the BDTT. Therefore, NiAl can be routinely fabricated into polycrystalline, single-crystal, or composite form via a variety of processing routes. Numerous conventional processing techniques have been employed, including powder metallurgy (PM), casting and extrusion, directional solidification, and some less orthodox techniques such as mechanical alloying and reaction synthesis. Even though NiAl has a low fracture toughness at room temperature, machining complex geometries from the as-fabricated material is possible with standard techniques such as grinding, abrasive machining, and electrodischarge machining (EDM). Water-jet cutting is an ideal method for machining thin composite shapes that contain nonconductive fibers, such as Al_2O_3, which are not amenable to EDM.

9.1. Powder Metallurgy

In addition to being both economical and reproducible, powder metallurgy techniques may be used to fabricate shapes not readily possible by other techniques. The ability to obtain near-net-shape components significantly limits the amount of post-processing machining necessary. Consequently, polycrystalline NiAl is usually fabricated via conventional PM techniques or the extrusion of small castings (Table 7-4), and NiAl-based composites are fabricated almost exclusively by PM-related processes at the present time.

Usually, NiAl powder processing takes advantage of prealloyed powders that are gas or vacuum atomized in helium or argon. Several commercial vendors using pilot plant facilities routinely produce high-quality, low-interstitial, spherical NiAl powder in heats of 20–45 kg, which sell for about $45.00/kg. A greater demand for powder in the future should drive the cost down as larger heats are produced. Because of the increased chemical uniformity of the final product, prealloyed powders are preferred to material that is reaction synthesized and milled. Cold compaction of the resulting NiAl powders to high densities is not feasible because of the inherent brittleness of the intermetallic. Instead, the prealloyed powders are typically consolidated by extrusion, HIP, vacuum hot pressing, thermal spray techniques, or some combination of these (Noebe and Locci 1990; German and Iacocca 1993).

Although powder processing has several advantages, PM techniques are susceptible to problems with bulk composition or elevated impurity levels. The most common impurity found in powder-processed material is oxygen. Because small powders have large surface-area-to-volume ratios, oxygen contamination presents a concern, although the effect of interstitial oxygen on the mechanical properties of NiAl is probably not a serious problem (Noebe and Garg 1994). In addition, the oxide scales on the powder particles result in prior particle boundaries after consolidation and can be potential sites for failure initiation. Extrusion of NiAl powder breaks up these boundaries into fine oxide particles (Zeller et al. 1990), thus reducing their impact on mechanical properties. Prior particle boundaries would be of greater concern in HIP'ed, hot-pressed, or thermally sprayed NiAl materials. Interstitials such as C, N, and S are typically present at the same levels in prealloyed NiAl powders as in wrought castings of NiAl (Bowman et al. 1992), although a "dirty" consolidation atmosphere can elevate the level of these impurities (Pickens et al. 1989).

Of equal or more concern is the issue of stoichiometry. High-temperature atomization and thermal spray processes can change the final composition of NiAl if one constituent, such as Al, is evaporated preferen-

Table 7-4 Effect of Processing on Room-Temperature Tensile Properties of Polycrystalline NiAl

Al content, at.%	Process	Grain size, μm	Yield stress, MPa	Ultimate tensile strength, MPa	Ductility, percent	Reference
				POWDER METALLURGY		
50.6	PM + extrude	5	370	405	.8	Noebe et al. (1991)
50.6	PM + extrude	10 to 33	290 to 350[a]	—	0	Bowman et al. (1992)
49.6	PM + hot press	35	120	175	.42	Bowman (1992a)
				WROUGHT PROCESS		
50.0	As-cast	150	—	100	0	Grala (1960)
49.0	Cast + extrude	5 to 140	—	234 to 538	0	Schulson (1982)
50.5		—	167 to 193	≃ 220	3.5 to 4.8	Rozner and Wasilewski (1966)
49.4		11 to 16	—	490	0	Hahn and Vedula (1989)
50.3		11 to 16	245	335	2.1	Hahn and Vedula (1989)
50.0		30	154	229	2.2	George and Liu (1990)
50.3		13 to 270	≃ 110	200 to 300	.5 to 2.3	Nagpal and Baker (1990b)
50.0		40	175	229	.45	Margevicius and Lewandowski (1991)
50.0		18	180	257	1.4	Noebe and Behbehani (1992)

[a] Compressive yield stress.

tially (German and Iacocca 1993). Therefore, although powder processing offers many advantages, it is very easy to upset the stoichiometry of the powder or to introduce higher impurity levels during intermediate or final powder-consolidation steps, resulting in a material with a higher yield stress and a greater tendency for completely brittle behavior. However, room-temperature tensile ductility can be obtained from powder-extruded or even hot-pressed NiAl powders (Table 7-4), although tensile ductility is generally achieved more consistently from extruded ingots because the chemistry is more easily and reliably controlled.

As indicated in Table 7-4, NiAl has been produced by extrusion of either powder or cast material encased in steel cans. In powder extrusion, both densification and deformation occur in one step with a very short process cycle time (Roberts and Ferguson 1991). Although the number of slip systems in NiAl does not allow for generalized plasticity in polycrystalline material at room temperature (Ball and Smallman 1966b), considerable reductions in area without cracking are possible by extrusion because of additional elevated-temperature deformation mechanisms and the hydrostatic compressive stress on the intermetallic due the presence of a canning material. During extrusion of NiAl, the cross-sectional area of the material is typically reduced from 7:1 to 16:1 at temperatures between 1100 K to 1400 K, resulting in grain sizes ranging from 3–100 μm. Lower extrusion temperatures result in finer grain sizes that scale linearly with the extrusion temperature (Zeller et al. 1990). Most extruded NiAl alloys have a recrystallized and equiaxed grain structure with a weak $\langle 111 \rangle$ fiber texture (Khadkikar et al. 1990), although extrusion of NiAl at temperatures less than about 1100 K produces a partially recrystallized material with a preferred $\langle 110 \rangle$ orientation (Bieler et al. 1992). The relation between processing variables, resulting microstructure, and mechanical properties for extruded NiAl alloys has been reviewed previously (Baker and Nagpal 1993).

Consolidation of NiAl powders by techniques other than extrusion has also been studied. Powder rolling, which has been demonstrated as a viable consolidation process, produces a strong $\langle 111 \rangle$ texture in the plane of the sheet (K. Bowman et al. 1993). The parameters for densifying NiAl powders by HIP and by uniaxial deformation have been studied experimentally as well as by modeling and are, therefore, relatively well understood (Panda, Lagraff, and Raj 1988; Wright, Knibloe, and Noebe 1991). Finally, less conventional consolidation techniques such as reactive hot pressing and powder injection molding have been investigated (Alman and Stoloff 1991). Reactive hot pressing of elemental powders is preferably performed by adding prealloyed NiAl powder to the compact to dilute the exothermic reaction. However, complete consolidation is achieved only when the reactive-sintered compact is HIP'ed to eliminate porosity.

9.2. Single Crystals

Melt processing of stoichiometric single crystals is, in principle, relatively easy because of the congruent melting temperature and high ordering energy of NiAl. The high melting point does, however, push conventional furnace and crucible materials to their limit. Because the melting point of NiAl is near 1955 K (Walston and Darolia 1993), furnace temperatures are nearly 300 K higher than those used to produce superalloys. Mold and core reactivity as well as structural integrity are already serious concerns in the production of single-crystal superalloys. Such concerns are magnified in NiAl single-crystal production because higher temperatures are required. Mold reactivity issues were first observed during a study of Zr and Hf additions to NiAl single crystals (Locci and Noebe 1989; Locci et al. 1991). During transmission-electron-microscope analysis of these alloys, an unexpected G-phase ($Ni_{16}X_6Si_7$, X = Hf or Zr) was observed. Because Si was not an intentional alloying addition, Locci et al. determined that the contamination originated from the mold material used during the directional solidification process. Si contamination of commercially produced single crystals is still a problem, though the effect of Si on mechanical properties has not been discerned (Hack et al. 1992).

In spite of the high temperatures involved, NiAl single crystals have been successfully produced by a Bridgman melt-processing technique, originally in only small quantities used primarily in slip system studies (McDonnell et al. 1967). The commercial development of NiAl single crystals is presently being led by General Electric Aircraft Engines (GEAE) in an attempt to achieve the high-temperature creep strengths necessary to compete with current superalloys. Using a modified Bridgman process, single crystals measuring 10.2 cm by 2.5 cm by 3.8 cm have been routinely produced from binary NiAl as well as from more highly alloyed compositions (Darolia et al. 1992). These ingots have been used to characterize the physical and mechanical properties of single-crystal NiAl and NiAl alloys in an effort to identify a suitable composition for turbine blade applications.

In addition to the Bridgman technique, melt processes such as the Czochralski and edge-defined, film-fed growth process have been used successfully to produce NiAl single crystals and eutectics (Goldman 1993). Large NiAl single crystals can also be grown by containerless float-zone

Figure 7-26. Small, high-pressure turbine blade machined from a single-crystal NiAl ingot (courtesy of R. Darolia, General Electric Aircraft Engines, Cincinnati, OH).

technology. This procedure has an advantage over other melt-processing techniques because no mold is involved. In addition, zone refining a chemical system open to the atmosphere can be used as a method to further purify the ingots (Reviere, Oliver, and Bruns 1989). Binary NiAl alloys with low Si and very low levels of interstitial elements have been produced by this process (Johnson et al. 1993).

Although these techniques are adequate for producing ingots for characterization, the fabrication of actual single-crystal turbine blades is more involved. Starting with single-crystal ingots and using a combination of EDM, electrochemical machining, and grinding processes, GEAE has successfully fabricated small single-crystal turbine blades (Fig. 7-26) and vanes (Darolia 1991). More complicated, single-crystal blade designs that contain intricate internal cooling passages would require sophisticated core technology. However, the cracking that can occur as the molten NiAl solidifies around the ceramic core is a serious concern. Such core-related problems are a major source of low yields in superalloys and are expected to be worse with NiAl because of its lower ductility and toughness.

To produce complex, thin-walled NiAl-alloy turbine airfoils, GEAE considered an alternative approach from that used with superalloys (Darolia 1991). The new approach involves machining internal cooling passages into each half of a single-crystal preform that has been split by wire-EDM. The two halves are then bonded back together before final machining of the airfoil contours. Initial results with NiAl (Darolia 1993) show that single-crystal bond joints, with properties equivalent to the base metal without forming a grain boundary, may be possible. Also, diffusion brazing of NiAl using self-generated filler metal has been used successfully to bond polycrystalline NiAl alloys (Moore and Kalinowski 1993) and would work even better with single-crystal components. Therefore, the fabrication of intricate turbine blades from single-crystal NiAl seems completely feasible. The major hurdle is identifying the proper alloy composition.

9.3. Continuous-Fiber Composites

Fabrication techniques for continuous-fiber-reinforced NiAl are being investigated by several research groups. Possible processing routes include the powder-cloth process (Pickens et al. 1989), foil/fiber or tape casting, thermal spray processes (Kern, Kaczmarek, and Janczak 1992), and pressure casting (Nourbakhsh et al. 1991). Of these techniques, powder-cloth has been used most often for generating NiAl matrix composites for mechanical property characterization. In the powder-cloth process, the matrix material is processed into flexible clothlike sheets by combining matrix powders with a fugitive organic binder. Likewise, fiber mats are produced by winding the fibers on a drum and applying another organic binder. The composite panel is assembled by stacking alternating layers of matrix cloth and fiber mats. This assemblage is consolidated by vacuum hot pressing to drive off the binders followed by HIP to ensure complete densification of the composite. One interesting feature of the powder-cloth technique is that by altering the binders, or by eliminating them completely, either a strong or weak matrix/fiber bond can be created in the NiAl/Al_2O_3 system (Bowman 1992a). This offers a convenient method for investigating the effect of strong versus weak interfaces on the mechanical properties of the composite without altering the constituents or coating the fibers.

In tape casting, a binder/powder slurry is spread as a thin sheet over the fiber mats. These matrix/fiber mats are then consolidated in a manner similar to the powder-cloth process. In foil/fiber production, alternating layers of thin foils of matrix material and mats of the reinforcing fiber are consolidated. Alternatively, the fibers can be placed within grooves machined into the thin foils—no binders are necessary with this technique. The foil/fiber process has not been attempted with NiAl because very thin sheets of matrix material are not readily available. During thermal spray processing, matrix/fiber mats are fabricated by depositing the matrix material onto fiber mats by either plasma or arc-spray techniques. The matrix/fiber mats are then consolidated in the normal fashion. Noebe and Locci (1990) have presented an overview of these processes.

All the fabrication techniques mentioned above usually result in some damage of the sapphire fibers, reducing the composite strength. Fiber damage can consist of either fiber fragmentation or surface roughening, which severely degrades the fracture stress of brittle fibers such as Al_2O_3. In addition, consolidation can be challenging when large-volume fractions of fibers are used because of the difficultyin promoting material flow between the small interfiber regions. At a critical fiber-volume fraction, the interfiber spacing becomes so small that the composite cannot be completely densified at reasonable temperatures and times. In addition, with fabrication techniques that require volatile binders, trapped gases and incomplete binder burn-off can elevate impurity levels. A major drawback to the thermal spray techniques is that during matrix deposition the high-speed molten matrix droplets can severely damage the fiber on impact. Molten metal infiltration, which is physically less damaging to the fiber than other processes, is one of the few techniques that is suitable for use with fine-diameter fibers; however, it comes with its own set of problems. These include inadequate control of fiber spacing, incomplete infiltration of the fibers, improper wetting of the fibers, or severe reaction between the fiber and matrix.

Without adequate control, these processing problems can be aggravated to a point where the composite properties are inferior to those of the monolithic alloy (German and Iacocca 1993). Consequently, development of continuous-fiber-reinforced composites is trailing that of other types of NiAl-based materials. And unless a better fiber system is found, no significant work will continue in this area.

9.4. General Considerations

Although NiAl processing is generally straightforward and relatively easy, it is still difficult to yield reproducible mechanical properties. As discussed in previous sections, many intrinsic low-temperature properties of NiAl are extremely sensitive to metallurgical variables such as stoichiometry, interstitial content, lattice defects, and solute concentrations. This sensitivity indicates that processing parameters, such as charge composition and cooling rate, are critical to mechanical properties. Imprecise control over processing parameters, which can lead to a wide range of material properties, has been a major contributor to the disparity in the mechanical behavior of nominally identical materials reported over the past several decades. For instance, although yield stress is one of the most common mechanical properties, it is one of the most inconsistent properties reported for NiAl (Table 7-4).

Part of the problem is that after processing, the material can vary from the target composition by as much as $\pm 1\%$Al. Chemistry variations can arise from inaccuracies in the initial charge or from changes in chemistry that occur during processing, such as vaporization of Al. What is equally frustrating is that chemical analyses for major constituents such as Ni and Al are usually no more precise than ± 0.5 at.% (Whittenberger and Luton 1992). In a preliminary study in which identical NiAl samples were sent to several chemical laboratories and research facilities, an almost 1 at.% range in Al content was reported by the various groups (unpublished research by D. F. Johnson and R. D. Noebe, 1993, NASA Lewis Research Center, Cleveland, OH). Yet, the properties of Ni-50Al compared with Ni-49Al or Ni-51Al are as different as a pure metal is to a solid-solution alloy. Minor deviations from stoichiometry have a direct effect on the lattice parameter, thermodynamic properties, diffusion, and mechanical properties such as yield stress, fracture stress, and ductility. For example, an alloy that is 1 at.% Ni-rich exhibits a yield stress 120 MPa greater than stoichiometric NiAl, whereas a Ni-51Al alloy has a yield stress nearly 350 MPa greater than that of Ni-50Al. Because routine chemical analyses cannot generally detect variations of 0.5 at.% for major elements, undetectable deviations from stoichiometry can contribute to apparent discrepancies in the properties of nominally identical materials. Detecting trace metallic impurities is even more difficult,though these impurities also can have an overwhelming effect on properties.

In addition to the bulk chemistry, trace levels of nonmetallic impurities such as C can profoundly affect the properties of NiAl alloys. Impurities such as C dramatically increase the flow stress of NiAl alloys (George and Liu 1990) and possibly degrade properties such as fracture toughness (Hack et al. 1992). Also, very minute amounts of S are probably responsible for degrading cyclic-oxidation resistance (Smialek and Browning 1985). In many cases, the concentration of interstitials necessary to significantly affect properties is less than the resolution of analysis techniques. Concerns over accuracy in interstitial level measurements were recently confirmed in a round-robin chemical analyses program. Using a variety of techniques but identical samples, eleven laboratories measured C, S, O, and N in directionally solidified NiAl (Parker, Johnstone, and Oligee 1991). The measured composition differed considerably from group to group. For example, the range in reported C levels varied from 12–335 ppm. Because the estimated hardening rate of C in NiAl is about 1750 MPa/at.% (Fig. 7-5), the possible range in flow stress due to such a compositional variation is 200 MPa. As with bulk chemistries, dramatic variations in properties may be observed in alloys that have impurity levels that are identical within analytical resolution.

Therefore, although a substantial body of data exists for NiAl, most studies do not report processing parameters or chemistry in sufficientdetail to allow for meaningful comparisons. Such problems have placed a premium on stringent control over processing to ensure the consistency necessary to successfully study this material. As the effect of processing variables on properties and improvements in process technology emerge, a more consistent view of the properties of NiAl will appear.

10. CONCLUDING REMARKS

Our knowledge of the behavior of NiAl has come a long way since the first exploratory studies of the late 1950s and early 1960s. Most of this understanding and progress have come about in only the last few years, with more papers published on NiAl in the last three years than during the entire previous history of this material. At this currently intense rate of research, some theories presented in this chapter may be out of date by the time it is published. Yet, all the previous work on NiAl lays a solid foundation for future studies so that we can build upon lessons of the past. One of the most critical of these lessons is to never underestimate the effect of processing and chemistry on properties.

We also know that the environmental durability of NiAl alloys, which is very good, is so well understood that it can be reliably predicted or modeled. The creep strength of NiAl can be improved to levels equivalent to Ni-base superalloys through alloying combined with single-crystal processing or through compositing. Fatigue behavior even at this early stage of investigation appears to be better than expected, and the inherent fracture toughness and ductility of single crystals may be much higher than originally believed.

This optimistic outlook needs to be tempered by the reality that impact behavior, which may be a formidable problem, has not yet been addressed. Also, the poor low-temperature ductility and toughness of most

NiAl alloys still need to be dealt with. Nevertheless, with the significant research effort being invested in NiAl, it is only a matter of time until this intermetallic material is used in some practical application or spin-off. However, some degree of patience is necessary and justified. For instance, Ni-base superalloys and Al-Li alloys for aerospace applications were developed over a period of several decades at an intensity greater than what NiAl is presently experiencing.

ACKNOWLEDGMENTS

The authors acknowledge the useful discussions on NiAl that have occurred over the years with Ron Gibala, Dan Whittenberger, Ram Darolia, Jim Cotton, and the many others who have worked in this area.

REFERENCES

ALMAN, D. E., and STOLOFF, N. S., 1991, Powder Fabrication of Monolithic and Composite NiAl. *Int. J. Powder Metal.* 27(1): 29–41.

ANTON, D. L., and SHAH, D. M., 1992, Mechanical Evaluation of FP Alumina Reinforced NiAl Composites. In *Intermetallic Matrix Composites II*, MRS Symp.Proc. 273: 157–164.

ASHBY, M. F., 1972, A First Report on Deformation-Mechanism Maps. *Acta Metall.* 20: 887–897.

BAIN, K. R., FIELD, R. D., and LAHRMAN, D. F., 1991, Fatigue Behavior of NiAl Single Crystals. Presented at the 1991 TMS Fall Meeting, Oct. 21, Cincinnati, OH.

BAKER, I., and NAGPAL, P., 1993, Processing of Iron and Nickel Aluminides via Hot Extrusion. *Processing and Fabrication of Advanced Materials for High Temperature Applications* eds., T. S. Srivatsan and R. A. Ravi, The Minerals, Metals and Materials Society, Warrendale, PA.

BAKER, I., NAGPAL, P., LIU, F., and MUNROE, P. R., 1991, The Effect of Grain Size on the Yield Strength of FeAl and NiAl Alloys. *Acta. Metall. Mater.* 39: 1637–1644.

BAKER, I., and SCHULSON, E. M., 1984, The Structure of Extruded NiAl. *Metall. Trans. A.* 15A: 1129–1136.

BALL, A., and SMALLMAN, R. E., 1966a, The Deformation Properties and Electron Microscopy Studies of the Intermetallic Compound NiAl. *Acta Metall.* 14: 1349–1355.

________, 1966b, The Operative Slip System and General Plasticity of NiAl-II. *Acta Metall.* 14: 1517–1526.

BANNISTER, M. K., SPEARING, S. M., LOFVANDER, J. P. A., and DE GRAEF, M., 1992, Fatigue of Extruded Steel/NiAl Composites. In *Intermetallic Matrix Composites II*, *MRS Symp. Proc.* 273: 177–182.

BARRETT, C. A., 1988, Effect of 0.1 at.% Zirconium on the Cyclic Oxidation Behavior of β-NiAl. *Oxid. Met.* 30: 361–390.

BERKOWITZ, A. E., JAUMOT, F. E., and NIX, F. C., 1954, Diffusion of Co^{60} in Some Ni-Al Alloys Containing Excess Vacancies. *Phys. Rev.* 95: 1185–1189.

BEVK, J., DODD, R. A., and STUTT, P. R., 1973, The Orientation Dependence of Deformation Mode and Structure in Stoichiometric NiAl Single Crystals Deformed by High Temperature Steady-State Creep. *Metall. Trans.* 4: 159–166.

BIELER, T. R., NOEBE, R. D., WHITTENBERGER, J. D., and LUTON, M. J., 1992, Extrusion Textures in NiAl and Reaction Milled NiAl/AlN Composites. *Intermetallic Matrix Composites II*, *MRS Symp. Proc.* 273: 165–170.

BOWMAN, K. J., NOEBE, R. D., JENNY, J. R., and KIM, S. T., 1993, Texture in Hot-Worked B2-Structure Aluminides. *Mater. Sci. Eng. A.* A160: 201–208.

BOWMAN, R. R., 1992a, Influence of Interfacial Characteristics on the Mechanical Properties of Continuous Fiber Reinforced NiAl Composites. In *Intermetallic Matrix Composites II*, *MRS Symp. Proc.* 273: 145–155.

________, 1992b, Influence of Interfacial Bond Strength on the Mechanical Properties of Continuous Fiber Reinforced NiAl Composites. HITEMP Review—1992: Advanced High Temperature Engine Materials Technology Program, NASA CP-10104, pp. 38-1–38-15.

BOWMAN, R. R., NOEBE, R. D., and DAROLIA, R., 1989, Mechanical Properties and Deformation Mechanisms of NiAl. HITEMP Review—1989: Advanced High Temperature Engine Materials Technology Program, NASA CP-10039, pp. 47-1–47-15.

BOWMAN, R. R., NOEBE, R. D., DOYCHAK, J., CRANDALL, K. S., and LOCCI, I. E., 1991, Effect of Interfacial Properties on the Mechanical Behavior of NiAl Based Composites. HITEMP Review—1991: Advanced High Temperature Engine Materials Technology Program, NASA CP-10082, pp. 43-1–43-14.

BOWMAN, R. R., NOEBE, R. D., RAJ, S. V., and LOCCI, I. E., 1992, Correlation of Deformation Mechanisms With the Tensile and Compressive Behavior of NiAl and NiAl(Zr) Intermetallic Alloys. *Metall. Trans. A.* 23A: 1493–1508.

BRADLEY, A. J., and TAYLOR, A., 1937, An X-ray Analysis of the Nickel-Aluminum System. *Proc. Roy. Soc.* (London), A159: 66–72.

BRUMM, M. W., and GRABKE, H. J., 1992, Oxidation Behavior of NiAl-I. Phase Transformations in the Alumina Scale During Oxidation of NiAl and NiAl-Cr Alloys. *Corros. Sci.* 33: 1677–1690.

BRZESKI, J. M., HACK, J. E., DAROLIA, R., and FIELD, R. D., 1993, Strain Aging Embrittlement of the Ordered Intermetallic Compound NiAl. *Mat. Sci. Eng.* A170: 11–18.

CASTRO, G. R., ISERN, H., SCHNEIDER, U., STOCKER, M., and WANDELT, K., 1991, Two-Dimensional Phase Transition of Adsorbed Xenon on NiAl(110) and Al(110). *J. Vac. Sci. Technol. A* 9: 1676–1679.

CHAKI, T. K., 1991, Mechanism of Boron-Induced Strengthening of Grain Boundaries in Ni_3Al. *Philos. Mag. Lett.* 63: 123–126.

CHAKRABARTI, D. J., 1977, Phase Stability in Ternary Systems of Transition Elements With Aluminum. *Metall. Trans. B* 8B: 121–123.

CHAKRAVORTY, S., and WAYMAN, C. M., 1976, The Thermoelastic Martensitic Transformation in β'Ni-Al Alloys: I. Crystallography and Morphology. *Metall. Trans. A* 7A: 555–568.

CHAMBERS, S. A., and LOEBS, V. A., 1990, Schottky Barrier Height and Thermal Stability of the NiAl/n-Ge/GaAs(001) Interface. *J. Vac. Sci. Technol A* 8: 2074–2078.

CHAN, K. S., 1990a, Theoretical Analysis of Grain Size Effects on Tensile Ductility. *Scripta. Metall. Mater.* 24: 1725–1730.

_______, 1990b. Fracture and Toughening Mechanisms in an Alpha-2 Titanium Aluminide Alloy. *Metall. Trans. A* 21A: 2687–2699.

CHANG, K.-M., DAROLIA, R., and LIPSITT, H. A., 1992, Cleavage Fracture in B2 Aluminides. *Acta. Metall. Mater.* 40: 2727–2737.

CHOU, T. C., and NIEH, T. G., 1991, Interfacial Reactions of SiC with NiAl. *Scripta. Metall. Mater.* 25: 2059–2064.

_______, 1993. Mechanism of $MoSi_2$ Pest During Low Temperature Oxidation. *J. Mat. Res.* 8: 214–226.

COOPER, M. J., 1963, An Investigation of the Ordering of the Phases CoAl and NiAl. *Phil. Mag.* 8: 805–810.

COTTON, J. D., 1991, The Influence of Chromium on Structure and Mechanical Properties of B2 Nickel Aluminide Alloys. Ph.D. thesis, The University of Florida, Gainesville, FL. (Also NASA CR-189124, 1992.)

COTTON, J. D., KAUFMAN, M. J., and NOEBE, R. D., 1991, Constitution of Pseudobinary Hypoeutectic β-NiAl + α-V Alloys. *Scripta. Metall. Mater.* 25: 1827–1832.

COTTON, J. D., NOEBE, R. D., and KAUFMAN, M. J., 1993a, Chromium-Bearing NiAl Intermetallic Alloys: Part I. Microstructure and Mechanical Properties. *Intermetallics* 1: 3–20.

_______, 1993b, Chromium-Bearing NiAl Intermetallic Alloys: Part II. Slip Systems. *Intermetallics* 1: 117–126.

_______, 1993c, NiAl-Rich Portion of the NiAl-Cr Pseudobinary Eutectic System. *J. Phase Equil.* 14(5): 579–582.

_______, 1993d, Ternary Alloying Effects In Polycrystalline β-NiAl. In International Symposium on Structural Intermetallics, eds. R. Darolia, J. J. Lewandowski, C. T. Liu, P. L. Martin, D. B. Miracle, and M. V. Nathal, pp. 513–522. Warrendale, PA: The Minerals, Metals and Materials Society.

COTTRELL, A. H., 1958, Theory of Brittle Fracture in Steel and Similar Metals. *Trans. Metall. Soc. AIME* 212: 192–205.

CRIMP, M. A., and VEDULA, K., 1991, Room-Temperature Deformation of Single-Crystal B2 Fe-Al Alloys: the Effect of Stoichiometry and Cooling Rate. *Phil. Mag. A* 63: 559–570.

CULLERS, C. L., and ANTOLOVICH, S. D., 1992, Low Cycle Fatigue Behavior of NiAl Deformed Near the Brittle-to-Ductile Transition Temperature. In *Superalloys 1992*, eds. S. D. Antolovich, R. W. Stusrud, R. A. MacKay, D. L. Anton, T. Khan, R. D. Kissinger and D. L. Klarstrom, pp. 351–359. Warrendale, PA: The Minerals, Metals and Materials Society.

CULLERS, C. L., ANTOLOVICH, S. D., and NOEBE, R. D., 1993, Deformation Behavior of Polycrystalline NiAl Cyclically Deformed Near the Brittle-to-Ductile Transition Temperature. In *High Temperature Ordered Intermetallic Alloys V*, MRS Symp. Proceedings 288: 531–536.

DAROLIA, R., 1991, NiAl Alloys for High-Temperature Structural Applications. *J. Met*. 43(3): 44–49.

________, 1993, Structural Applications of NiAl. *International Symposium on Structural Intermetallics*, eds. R. Darolia, J. J. Lewandowski, C. T. Liu, P. L. Martin, D. B. Miracle and M. V. Nathal, pp. 495–504. Warrendale, PA: The Minerals, Metals and Materials Society.

DAROLIA, R., CHANG, K.-M., and HACK, J. E., 1993, Observation of High Index {511} Type Fracture Planes and Their Influence on Toughness in NiAl Single-Crystals. *Intermetallics* 1: 65–78.

DAROLIA, R., LAHRMAN, D., and FIELD, R. D., 1992, The Effect of Iron, Gallium and Molybdenum on the Room Temperature Tensile Ductility of NiAl. *Scripta. Metall. Mater*. 26: 1007–1012.

DAROLIA, R., LAHRMAN, D. F., FIELD, R. D., DOBBS, J. R., CHAN, K. M., GOLDMAN, E. H., and KONITZER, D. G., 1992, Overview of NiAl Alloys For High Temperature Applications. *Ordered Intermetallics—Physical Metallurgy and Mechanical Behavior*, eds. C. T. Liu, R. W. Cahn and G. Sauthoff, pp. 679–698. Dordrecht: Kluwer Academic Publishers.

DAROLIA, R., LAHRMAN, D. F., FIELD, R. D., and FREEMAN, A. J., 1989, Alloy Modeling and Experimental Correlation for Ductility Enhancement in NiAl. *High Temperature Ordered Intermetallic Alloys III*, MRS Symp. Proc. 133: 113–118.

DEMARCO, H. K., and ARDELL, A. J., 1993, Mechanical Behavior of Monocrystalline NiAl Using a Miniaturized Disk-Bend Test. *High Temperature Ordered Intermetallic Alloys V*, MRS Symp. Proc. 288: 641–646.

DOLLAR, M., DYMEK S., HWANG, S. J., and NASH P., 1992, The Occurrence of ⟨110⟩ Slip in NiAl. *Scripta. Metall. Mater*. 26: 29–34.

DOYCHAK, J., 1992, Metal- and Intermetallic-Matrix Composites for Aerospace Propulsion and Power Systems. *J. Met*. 44: 6: 46–51.

DOYCHAK, J., NESBITT, J. A., NOEBE, R. D., and BOWMAN, R. R., 1992, Oxidation of Al_2O_3 Continuous Fiber-Reinforced/NiAl Composites. *Oxid. Met*. 38: 45–72.

DOYCHAK, J., SMIALEK, J. L., and BARRETT, C. A., 1989, The Oxidation of Ni-rich Ni-Al Intermetallics. *Oxidation of High Temperature Intermetallics*, eds. T. Grobstein and J. Doychak, pp. 41–55. Warrendale, PA: The Minerals, Metals and Materials Society.

DOYCHAK, J., SMIALEK, J. L., and MITCHELL, T. E., 1989, Transient Oxidation of Single-Crystal β-NiAl. *Metall. Trans. A* A20: 499–518.

DRAPER, S. L., GAYDOSH, D. J., and CHULYA, A., 1991, Tensile Behavior of Alumina-Reinforced FeAl and FeCrAl. *HITEMP Review—1991: Advanced High Temperature Engine Materials Technology Program*, NASA CP-10104, pp. 42-1–42-14.

DRAPER, S. L., LOCCI, I. E., and ELDRIDGE, J. L., 1992, Al_2O_3 Fiber Strength Degradation in MMC's and IMC's. *HITEMP Review—1992: Advanced High Temperature Engine Materials Technology Program*, NASA CP-10104, pp. 17-1–17-13.

EDWARDS, K. M., 1993, Low-Cycle Fatigue Behavior of Polycrystalline NiAl at Room Temperature. M.S. thesis, The University of Michigan, Ann Arbor, MI.

EDWARDS, K. M., and GIBALA, R., 1993, Low-Cycle Fatigue Behavior of Polycrystalline NiAl at Room Temperature. In *High Temperature Ordered Intermetallic Alloys V*, MRS Symp. Proc. 288: 665–670.

ELLIS, D. L., 1987, Hot Corrosion of the B2 Nickel Aluminides. M.S. thesis, Case Western Reserve University, Cleveland, OH.

FIELD, R. D., LAHRMAN, D. F., and DAROLIA, R., 1991a, Room Temperature Deformation in "Soft" Orientation NiAl Single Crystals. *High Temperature Ordered Intermetallic Alloys IV*, MRS Symp. Proc. 213: 255–260.

________, 1991b, Slip Systems in ⟨001⟩ Oriented NiAl Single Crystals. *Acta. Metall. Mater.* 39: 2951–2959.

________, 1991c, The Effect of Alloying on Slip Systems in ⟨001⟩ Oriented NiAl Single Crystals. *Acta. Metall. Mater.* 39: 2961–2969.

________, 1993, A Mechanistic Study of the Microalloying Effect in NiAl Base Alloys. *High Temperature Ordered Intermetallic Alloys V*, MRS Symp. Proc. 288: 423–428.

FLEISCHER, R. L., 1993, Solid Solution Strengthening of Intermetallic Compounds. *High Temperature Ordered Intermetallic Alloys V*, MRS Symp. Proc., 288: 165–170.

FORBES, K. R., GLATZEL, U., DAROLIA, R., and NIX, W. D., 1993, High Temperature Deformation of Single Crystals of NiAl. In *High Temperature Ordered Intermetallic Alloys V*, MRS Symp. Proc. 288: 45–58.

FORBES, K. R., and NIX, W. D., 1993, Mobility and Substructure Controlled Creep Deformation of [100] Oriented NiAl Crystals. Creep and Fracture of Engineering Materials and Structures, eds. B. Wilshire and R. W. Evans, The Institute of Materials, London, pp. 265–274.

FRANCHY, R., WUTTIG, M., and IBACH, H., 1987, The Adsorption of Sulfur, Carbon Monoxide and Oxygen in NiAl(111). *Surf. Sci.* 189–190: 438–447.

FRASER, H. L., SMALLMAN, R. E., and LORETTO, M. H., 1973, The Plastic Deformation of NiAl Single Crystals Between 300 K and 1050 K: I. Experimental Evidence on the Role of Kinking and Uniform Deformation in Crystals Compressed Along ⟨001⟩. *Phil. Mag.* 28: 651–665.

GANDHI, C., and ASHBY, M. F., 1979, Fracture-Mechanism Maps for Materials Which Cleave: F.C.C., B.C.C. and H.C.P. Metals and Ceramics. *Acta Metall.* 27: 1565–1602.

GEORGE, E. P., and LIU, C. T., 1990, Brittle Fracture and Grain Boundary Chemistry of Micro-Alloyed NiAl. *J. Mat. Res.* 5: 754–762.

GEORGE, E. P., LIU, C. T., and POPE, D. P., 1992, Environmental Embrittlement: The Major Cause of Room-Temperature Brittleness in Polycrystalline Ni_3Al. *Scr. Metall. Mater.* 27: 365–370.

GEORGOPOULOS, P., and COHEN, J. B., 1977, The Defect Structure and Debye Waller Factors vs. Composition in $\beta Ni_{1\pm x}Al_{1\pm x}$. *Scr. Metall.* 11: 147–150.

GERMAN, R. M., and IACOCCA, R. G., 1993, Powder Metallurgy of Intermetallics. In *Processing and Fabrication of Advanced Materials for High Temperature Applications*, eds. T. S. Srivatsan and R. A. Ravi, Warrendale, PA: The Minerals, Metals and Materials Society.

GODLEWSKI, K., GODLEWSKI, E., MROWEC, S., and DANIELEWSKI, M., 1989, Sulphide Corrosion of Pure and Chromium-Modified β-NiAl Intermetallic Compound at High Temperatures. *Mat. Sci. Eng.* A120: 105–109.

GOLDMAN, E. H., 1993, Single Crystal Processing of Intermetallics for Structural Applications. In *High Temperature Ordered Intermetallic Alloys. V*, MRS Symp. Proc. 288: 83–94.

GRABKE, H. J., BRUMM, M., and STEINHORST, M., 1992, Development of Oxidation Resistant High Temperature Intermetallics. *Materials Science and Technology* 8: 339–344.

GRABKE, H. J., STEINHORST, M., BRUMM, M., and WIEMER, D., 1991, Oxidation and Intergranular Disintegration of the Aluminides NiAl and $NbAl_3$ and Phases in the System Nb-Ni-Al. *Oxid. Met.* 35: 199–222.

GRABKE, H. J., WIEMER, D., and VIEFHAUS, H., 1991, Segregation of Sulfur During Growth of Oxide Scales. *Appl. Surf. Sci.* 47(3): 243–250.

GRAHAM, R. B., 1984, The Effect of Temperature, Composition and Grain Size on the Mechanical Properties of NiAl. M.E. thesis, Dartmouth College, Hanover, NH.

GRALA, E. M., 1960, Investigations of NiAl and Ni_3Al. *Mechanical Properties of Intermetallic Compounds*, ed. J. H. Westbrook, pp. 358–402. New York: John Wiley & Sons.

GROVES, G. W., and KELLY, A., 1969, Change of Shape Due to Dislocation Climb. *Phil. Mag.* 19: 977–986.

GUHA, S., MUNROE, P. R., and BAKER, I., 1991, Room Temperature Deformation Behavior of Multiphase Ni-20 at.% Al-30 at.% Fe and its Constituent Phases. *Mater. Sci. Eng. A* A131: 27–37.

GUSEVA, L. N., 1951, On the Nature of the β-phase in the System Ni-Al. *Dokaldy Akad. Nauk. U.S.S.R.* 77: 415–418.

HACK, J. E., BRZESKI, J. M., and DAROLIA, R., 1992, Evidence of Inherent Ductility in Single Crystals of the Ordered Intermetallic Compound NiAl. *Scripta. Metall. Mater.* 27: 1259–1263.

HAHN, K. H., and VEDULA, K., 1989, Room Temperature Tensile Ductility in Polycrystalline B2 NiAl. *Scripta. Metall.* 23: 7–12.

HANCOCK, G. F., and MCDONNELL, B. R., 1971, Diffusion in the Intermetallic Compound NiAl. *Phys. Status Solidi.* 4: 143–150.

HARMOUCHE, M. R., and WOLFENDEN, A., 1987, Temperature and Composition Dependence of Young's Modulus in Polycrystalline B2 Ni-Al. *J. Test. Eval.* 15: 101–104.

HARTFIELD-WUNSCH, S. E., and GIBALA, R., 1991, Cyclic Deformation of B2 Aluminides. In *High Temperature Ordered Intermetallic Alloys IV*, MRS Symp. Proc. 213: 575–580.

HEBSUR, M. G., WHITTENBERGER, J. D., DICKERSON, R. M., and AIKIN, B. J. M., 1993, Microstructure/Property Relations in AlN and Al_2O_3 Particulate Strengthened NiAl. In *High Temperature Ordered Intermetallic Alloys V*, MRS Symp. Proc. 288: 1111–1116.

HENIG, E.-T., and LUKAS, H. L., 1975, Kalorimetrische Bestimmung der Bildungsenthalpie und die Beschrzeibung der Fehlordnung der Geordneten β-Phase $(Ni, Cu)_{1-x}Al_x$. *Z. Metallk.* 66: 98–106.

HOCKING, L. A., STRUTT, P. R., and DODD, R. A., 1971, Comparison of Steady-State Compression Creep Behavior in Stoichiometric CoAl and NiAl Single Crystals Between 850 and 1050 °C. *J. Inst. Met.* 99: 98–101.

HONG, T., and FREEMAN, A. J., 1991, Effect of Antiphase Boundaries on the Electronic Structure and Bonding Character of Intermetallic Systems: NiAl. *Phys. Rev. B* 43: 6446–6458.

HUTCHINGS, R., and LORETTO, M. H., 1978, Compositional Dependence of Oxidation Rates of NiAl and CoAl. *Metall. Sci.* 12: 503–510.

JEDLINSKI, J., and BORCHARDT, G., 1991, On the Oxidation Mechanism of Alumina Formers. *Oxid. Met.* 36: 317–337.

JEDLINSKI, J., BORCHARDT, G., and MROWEC, S., 1992, Transport Properties of Alumina Scales on the β-NiAl Intermetallic. *Solid State Ionics* 50: 67–74.

JEDLINSKI, J., and MROWEC, S., 1987, The Influence of Implanted Yttrium on the Oxidation Behavior of β-NiAl. *Mat. Sci. Eng.* 87: 281–287.

JENG, S. M., YANG, J.-M., and AMATO, R. M., 1992, Mechanical Properties and Deformation Mechanisms of an Al_2O_3 Fiber-Reinforced NiAl Matrix Composite. *Intermetallic Matrix Composites II*, MRS Symp. Proc. 273: 217–225.

JOHNSON, D. R., JOSLIN, S. M., OLIVER, B. F., NOEBE, R. D., and WHITTENBERGER, J. D., 1992, Intermetallic/Metallic Polyphase In Situ Composites. In *Intermetallic Matrix Composites II*, MRS Symp. Proc. 273: 87–92.

________, 1993, High-Purity NiAl Single Crystals and Composites by Containerless Automated Processing. In *First International Conference on Processing Materials for Properties*, eds. H. Henein and T. Oki, pp. 865–870. Warrendale, PA: The Minerals, Metals and Materials Society.

JOO, G. C., KALMAN, Z., TSAKALAKOS, T., and CHEN, S. P., 1992, On the Growth of NiAl Intermetallics on III-V Semiconductors. *NanoStruct. Mater.* 1: 213–218.

JUNG, I., RUDY, M., and SAUTHOFF, G., 1987, Creep in Ternary B2 Aluminides and Other Intermetallic Phases. In *High Temperature Ordered Intermetallic Alloys II*, MRS Symp. Proc. 81: 263–274.

KANNE, W. R., STRUTT, P. R., and DODD, R. A., 1969, Nature of Slip Line Formation During Creep in Stoichiometric NiAl at Temperatures Between 475° and 775° C. *Transactions Metall. Soc. AIME*, 245: 1259–1267.

KAUFMAN, M., 1969, Hot Corrosion in Nickel, Cobalt and Nickel-Aluminide-Base Alloys. *Transactions of ASM* 62: 590–606.

KELLY, A., and STREET, K. N., 1972, Creep of Discontinuous Fibre Composites II. Theory for the Steady State. *Proc. Roy. Soc. London A*. 328(1573): 283–293.

KERN, H., KACZMAREK, R., and JANCZAK, J., 1992, Thermally Sprayed Fiber-Reinforced MMCs. *J. Therm. Spray Technol.* 1(2): 137–142.

KHADKIKAR, P. S., MICHAL, G. M., and VEDULA, K., 1990, Preferred Orientations in Extruded Nickel and Iron Aluminides. *Metall. Trans. A* 21A: 279–288.

KIM, J. T., 1990, On the Slip Behavior and Surface Film Effects in B2 Ordered NiAl Single Crystals. Ph.D. thesis, University of Michigan, Ann Arbor, MI.

KIM, J. T., and GIBALA, R., 1991, Slip Transition in [001] Oriented NiAl at High Temperatures. In *High Temperature Ordered Intermetallic Alloys IV*, MRS Symp. Proc. 213: 261–266.

KOSTRUBANIC, J., KOSS, D. A., LOCCI, I. E., and NATHAL, M., 1991, On Improving the Fracture Toughness of a NiAl-Based Alloy by Mechanical Alloying. *High Temperature Ordered Intermetallic Alloys IV*, MRS Symp. Proc. 213: 679–684.

KUMAR, K. S., DAROLIA, R., LAHRMAN, D. F., and MANNAN, S. K., 1992, Tensile Creep Response of an NiAl-TiB_2 Particulate Composite. *Scripta. Metall. Mater*. 26: 1001–1006.

KUMAR, K. S., MANNAN, S. K., and VISWANADHAM, R. K., 1992, Fracture Toughness of NiAl and NiAl-Based Composites. *Acta. Metall. Mater*. 40: 1201–1222.

KUMAR, K. S., and WHITTENBERGER, J. D., 1991, Intermetallic Matrix Composites via XD™ Synthesis. *Proceedings of the International Symposium on Intermetallic Compounds—Structure and Mechanical Properties*, ed. O. Izumi, pp. 927–934. Sendai, Japan: Japan Institute of Metals.

LAHRMAN, D. F., FIELD, R. D., and DAROLIA, R., 1991, In The Effect of Strain Rate on the Mechanical Properties of Single Crystal NiAl. *High Temperature Ordered Intermetallic Alloys IV*, MRS Symp. Proc. 213: 603–607.

________, 1993a, Effect of Crystallographic Orientation on the Mechanical Properties of a NiAl + Fe Alloy. In *High Temperature Ordered Intermetallic Alloys V*, MRS Symp. Proc. 288: 679–684.

________, 1993b, Evaluation of Environmental Effects on the Room Temperature Tensile Properties of Single Crystal Stoichiometric NiAl and NiAl Alloys. *Scripta. Metall. Mater*. 28: 709–714.

LARSEN, J. M., REVELOS, W. C., and GAMBONE, M. L., 1992, An Overview of Potential Titanium Aluminide Composites in Aerospace Applications. In *Intermetallic Matrix Composites II*, MRS Symp. Proc. 273: 3–16.

LAW, C. C., and BLACKBURN, M. J., 1987, Rapidly Solidified Lightweight Durable Disk Material. Final Technical Report AFWAL-TR-87-4102, Pratt and Whitney Aircraft Co., West Palm Beach, FL.

LERCH, B. A., AND NOEBE, R. D., 1993, Low Cycle Fatigue Behavior of Polycrystalline NiAl at 300 and 1000 K. NASA TM-105987.

LERCH, B. A., AND NOEBE, R. D. 1994, Low-Cycle Fatigue Behavior of Polycrystalline NiAl at 1000k. *Metall. Trans. A*. 25A: 309–319.

LLOYD, C. H., and LORETTO, M. H., 1970, Dislocations in Extruded β'-NiAl. *Phys. Status Solidi*. 39: 163–170.

LOCCI, I. E., DICKERSON, R., BOWMAN, R., WHITTENBERGER, D., and NATHAL, M. V., 1993, Microstructure and Mechanical Properties of Cast, Homogenized and Aged NiAl Single Crystal Containing Hf. In *High Temperature Ordered Intermetallic Alloys V*, MRS Symp. Proc. 288: 686–692.

LOCCI, I. E., and NOEBE, R. D., 1989, Characterization of Second Phase Particles in NiAl-Containing Trace Additions of Zr. *Proceedings of the 47th Annual Meeting of the Electron Microscopy Society of America*, ed. G. W. Bailey, pp. 308–309. San Francisco, CA: San Francisco Press.

LOCCI, I. E., NOEBE, R. D., BOWMAN, R. R., MINER, R. V., NATHAL, M. V., and DAROLIA, R., 1991, Microstructure and Mechanical Properties of a Single Crystal NiAl Alloy With Zr or Hf Rich G-Phase Precipitates. In *High Temperature Ordered Intermetallic Alloys IV*, MRS Symp. Proc. 213: 1013–1018.

LOCCI, I. E., NOEBE, R. D., MOSER, J. A., LEE, D. S., and NATHAL, M., 1989, Processing and Microstructure of Melt Spun NiAl Alloys. In *High Temperature Ordered Intermetallic Alloys III*, MRS Symp. Proc. 133: 639–646.

LORETTO, M. H., and WASILEWSKI, R. J., 1971, Slip Systems in NiAl Single Crystals at 300 K and 77 K. *Phil. Mag*. 23: 1311–1328.

LOWELL, C. E., BARRETT, C. A., and WHITTENBERGER, J. D., 1990, Cyclic Oxidation Resistance of a Reaction Milled NiAl-AlN Composite. *Intermetallic Matrix Composites*, MRS Symp. Proc. 194: 355–360.

LU, T. C., YANG, J., SUO, Z., EVANS, A. G., HECHT, R. and MEHRABIAN, R., 1991, Matrix Cracking in Intermetallic Composites Caused by Thermal Expansion Mismatch. *Acta. Metall. Mater*. 39: 1883–1890.

LUTZE-BIRK, A., AND JACOBI, H., 1975, Diffusion of ^{114m}In in NiAl. *Scripta. Metall*. 9: 761–765.

MARGEVICIUS, R. W., and LEWANDOWSKI, J. J., 1991, The Effects of Hydrostatic Pressure on the Mechanical Behavior of NiAl. *Scripta. Metall. Mater*. 25: 2017–2022.

________, 1993, Pressure-Induced Dislocations and Subsequent Flow in NiAl. *Acta. Metall. Mater*. 41: 485–496.

MASON, D. P., VAN AKEN, D. C., NOEBE, R. D., LOCCI, I. E., and KING, K. L., 1991, Microstructure and Mechanical Properties of Near Eutectic β-NiAl Plus α-Re Alloys Produced by Rapid Solidification and Extrusion. In *High Temperature Ordered Intermetallic Alloys IV*, MRS Symp. Proc. 213: 1033–1038.

MATSUGI, K., WENMAN, D. W., and STOLOFF, N. S., 1992, Observation of the Brittle to Ductile Transition Temperature in he Iron Microalloyed Polycrystalline NiAl Intermetallic Compound. *Scripta. Metall. Mater*. 27: 1633–1638.

MCCARRON, R. L., LINDBLAD, R. L., and CHATTERJI, D., 1976, Environmental Resistance of Pure and Alloyed γ-Ni_3Al and β-NiAl. *Corrosion* 32: 476–481.

McDonnell, B. R., Pascoe, R. T., Hancock, G. F., and Newey, C. W. A., 1967, The Growth of Single Crystals of the Intermediate Phases NiAl and Ni_3Al. *J. Mater. Sci.* 2: 365–370.

McVay, R. V., Williams, P., Meier, G. H., Pettit, F. S., and Smialek, J. L., 1992, Oxidation of Low Sulfur Single Crystal Nickel-Base Superalloys. *Superalloys 1992*, eds. S. D. Antolovich, R. W. Stusrud, R. A. MacKay, D. L. Anton, T. Khan, R. D. Kissinger and D. L. Klarstrom, pp. 807–816. Warrendale, PA: The Minerals, Metals and Materials Society.

Miller, R. A., and Doychak, J., 1992, Plasma-Sprayed Ceramic Thermal Barrier Coatings for Smooth Intermetallic Alloys. *Journal of Therm. Spray Technol.* 1 (3): 211–214.

Mills, M. J., Daw, M. S., Foiles, S. M., and Miracle, D. B., 1993, HRTEM Observation and EAM Calculation of Dislocation Cores in NiAl. In *High Temperature Ordered Intermetallic Alloys V*, MRS Symp. Proc. 288: 257–262.

Mills, M. J., and Miracle, D. B., 1993, The Structure of $a\langle 100\rangle$ and $a\langle 110\rangle$ Dislocation Cores in NiAl. *Acta. Metall. Mater.* 41: 85–95.

Miracle, D. B., 1991, Deformation in NiAl Bicrystals. *Acta. Metall. Mater.* 39: 1457–1468.

———, 1993, The Physical and Mechanical Properties of NiAl. In *Acta. Metall. Mater.* 41: 649–684.

Miracle, D. B., Russell, S., and Law, C. C., 1989, Slip System Modification in NiAl. In *High Temperature Ordered Intermetallic Alloys III*, MRS Symp. Proc. 133: 225–230.

Misra, A. K., 1988, Thermodynamic Analysis of Compatibility of Several Reinforcement Materials with Beta Phase NiAl Alloys. NASA CR-4171.

———, 1991, Preliminary Studies on $NiAl/Nb_2Be_{17}$ Reaction and Effectiveness of BeO as an Interfacial Barrier. *Metall. Trans. A.* 22A: 2535–2538.

Misra, A., Noebe, R. D., and Gibala, R., 1993, The Role of Interfaces in the Room Temperature Mechanical Behavior of Directionally Solidified $\beta + \gamma$-$Ni_{70}Al_{30}$ and $\beta + (\gamma + \gamma)$-$Ni_{50}Fe_{30}Al_{20}$ Alloys. *High Temperature Ordered Intermetallic Alloys V*, MRS Symp. Proc. 288: 483–488.

Mohamed, F. A., and Langdon, T. G., 1974, The Transition from Dislocation Climb to Viscous Glide in Creep of Solid Solution Alloys. *Acta Metall.* 22: 779–788.

Moore, T. J., and Kalinowski, M., 1993, Diffusion Brazing NiAl With Self-Generated Filler Metal. In *High Temperature Ordered Intermetallic Alloys V*, MRS Symp. Proc. 288: 1173–1178.

Moose, C. A., 1991, Interfacial Shear Studies of Sapphire Fiber-Reinforced Nickel Aluminide Matrix Composites. M.S. thesis, Pennsylvania State University, University Park, PA.

Nagpal, P., and Baker, I., 1990a, Effect of Cooling Rate on Hardness of FeAl and NiAl. *Metall. Trans. A* 21A: 2281–2282.

———, 1990b, The Effect of Grain Size on the Room-Temperature Ductility of NiAl. *Scr. Metall. Mater.* 24: 2381–2384.

________, 1991, Room Temperature Fracture of FeAl and NiAl. *Mater. Character.* 27: 167–173.

________, 1992, Dislocation Arrangements in Polycrystalline NiAl After Room Temperature Deformation. *J. Mater. Lett.* 11: 1209–1210.

NAGPAL, P., BAKER, I., LIU, F., and MUNROE, P. R., 1991, Room Temperature Strength and Fracture of FeAl and NiAl. In *High Temperature Ordered Intermetallic Alloys IV*, MRS Symp. Proc. 213: 533–538.

NARDONE, V. C., 1992, Fracture Behavior of Stainless Steel-Toughened NiAl Composite Plate. *Metall. Trans. A*. 23A: 563–572.

NATHAL, M. V., 1992, Creep Deformation of B2 Aluminides. In *Ordered Metallics —Physical Metallurgy and Mechanical Behavior*, eds. C. T. Liu, R. W. Cahn and G. Sauthoff, pp. 541–563. Dordrecht: Kluwer Academic Publishers.

NATHAL, M. V., DIAZ, J. O., and MINER, R. V., 1989, High Temperature Creep Behavior of Single Crystal Gamma Prime and Gamma Prime Alloys. In *High Temperature Ordered Intermetallic Alloys III*, MRS Symp. Proc. 133: 269–274.

NATHAL, M. V., and EBERT, L. J., 1985, Elevated Temperature Creep-Rupture Behavior of the Single Crystal Nickel-Base Superalloy NASAIR 100. *Metall. Trans. A*. 16A: 427–439.

NESBITT, J. A., and LOWELL, C. E., 1993, High Temperature Oxidation of Intermetallics. In *High Temperature Ordered Intermetallic Alloys V*, MRS Symp. Proc. 288: 107–118.

NESBITT, J. A., and VINARCIK, E. J., 1991, Predicting the Oxidative Lifetime of β NiAl-Zr Alloys. *Damage and Oxidation Protection in High Temperature Composites*, eds. G. K. Haritos and O. O. Ochoa, pp. 9–22. New York: The American Society of Mechanical Engineers.

NICOLAS-CHAUBET, D., HUNTZ, A. M., and MILLOT, F., 1991, Electrochemical Method for the Investigation of Transport Properties of Alumina Scales Formed by Oxidation. Part II: Scale Formed on a β-NiAl Alloy. *J. Mat. Sci.* 26: 6119–6126.

NOEBE, R. D., 1994, The Effect of Various Metallurgical Parameters on the Flow and Fracture Behavior of Polycrystalline NiAl Near the Brittle-To-Ductile Transition. NASA TM 106534.

NOEBE, R. D., and BEHBEHANI, M. K., 1992, The Effect of Microalloying Additions on the Tensile Properties of Polycrystalline NiAl. *Scripta. Metall. Mater.* 27: 1795–1800.

NOEBE, R. D., BOWMAN, R. R., CULLERS, C. L., and RAJ, S. V., 1990, Flow and Fracture Behavior of Binary NiAl With Prospects for Future Alloy Development. HITEMP Review—*1990:* Advanced High Temperature Engine Materials Technology Program, *NASA CP*-10051, pp. 20-1–20-19.

________, 1991, Flow and Fracture Behavior of NiAl in Relation to the Brittle-to-Ductile Transition Temperature. In *High Temperature Ordered Intermetallic Alloys IV*, MRS Symp. Proc. 213: 589–596.

NOEBE, R. D., BOWMAN, R. R., and ELDRIDGE, J. I., 1991, Initial Evaluation of Continuous Fiber Reinforced NiAl Composites. In *Intermetallic Matrix Composites*. MRS Symp. Proc. 194: 323–331.

NOEBE, R. D., BOWMAN, R. R., LOCCI, I. E., and RAJ, S. V., 1989, Effect of Zr Additions on the Microstructure and Mechanical Behavior of NiAl. *HITEMP Review—1989: Advanced High Temperature Engine Materials Technology Program*. NASA CP-10039, pp. 48-1–48-15.

NOEBE, R. D., BOWMAN, R. R., and NATHAL, M. V., 1993, Review of the Physical and Mechanical Properties of the B2 Compound NiAl. *Inter. Mater. Rev.* 38: 193–232.

NOEBE, R. D., CULLERS, C. L., and BOWMAN, R. R., 1992, The Effect of Strain Rate and Temperature on the Tensile Properties of NiAl. *J. Mat. Res.*, 7: 605–612.

NOEBE, R. D., and GARG, A., 1994, Characterization of Nitrogen-Doped NiAl. *Scr. Metall. Mater*. 30: 815–820.

NOEBE, R. D., and GIBALA, R., 1986, Surface Oxide Softening of Single Crystal NiAl. *Scripta. Metall.* 20: 1635–1639.

NOEBE, R. D., and LERCH, B. A., 1992, Room Temperature Cyclic Deformation Behavior of Cast and Extruded NiAl. *Scripta. Metall. Mater*. 27: 1161–1166.

———, 1993, Low Cycle Fatigue Behavior of Polycrystalline NiAl at 1000 K. In *High Temperature Ordered Intermetallic Alloys V*, MRS Symp. Proc. 288: 743–748.

NOEBE, R. D., and LOCCI, I. E., 1990, Metal Matrix Composites, Rapid Solidification. In *International Encyclopedia of Composites, Vol. 3*, ed. S. M. Lee, pp. 294–313. New York: VCH Publishers.

NOEBE, R. D., MISRA, A., and GIBALA, R., 1991, Plastic Flow and Fracture of B2 NiAl-Based Intermetallic Alloys Containing a Ductile Second Phase. *Iron Steel Inst. Japan. Int*. 31: 1172–1185.

NOURBAKHSH, S., SAHIN, O., RHEE, W. H., and MARGOLIN, H., 1991, Pressure Casting of a Zirconia-Toughened Alumina Fiber-Reinforced NiAl Composite. *Metall. Trans. A* 22A: 3059–3064.

PANDA, P. C., LAGRAFF, J., and RAJ, R., 1988, Shear Deformation and Compaction of Nickel Aluminide Powders at Elevated Temperatures. *Acta Metall.* 36: 1929–1939.

PARKER, D. J., JOHNSTONE, W. H., and OLIGEE, R. M., 1991, Determining Carbon, Sulfur, Oxygen, and Nitrogen in NiAl—Assessment of Techniques Through Interlaboratory Chemical Analysis Round-Robin. General Electric Technical Momorandum, TM-91-451, Evendale, OH.

PASCOE, R. T., and NEWEY, C. W. A., 1968a, Deformation Modes of the Intermediate Phase NiAl. *Phys. Status Solidi*. 29: 357–366.

———, 1968b, The Mechanical Behavior of the Intermediate Phase NiAl. *Metall. Sci. J.* 2: 138–143.

PATHARE, V. M., 1988, Processing, Physical Metallurgy and Creep of NiAl + Ta and NiAl + Nb Alloys. In Ph.D. thesis, Case Western Reserve University, Cleveland, OH. (Also, NASA CR-182113, 1988.)

PATRICK, D. K., CHANG, K.-M., MIRACLE, D. B., and LIPSITT, H. A., 1991, Burgers Vector Transition in Fe-Al-Ni Alloys. In *High Temperature Ordered Intermetallic Alloys IV*, MRS Symp. Proc. 213: 267–272.

PETTIT, F. S., and GIGGINS, G. S., 1987, Hot Corrosion. *Superalloys II*, eds. C. T. Sims, N. S. Stoloff, and W. C. Hagel, pp. 327–358. New York: John Wiley & Sons.

PICKENS, J. W., NOEBE, R. D., WATSON, G. K., BRINDLEY, P. K., and DRAPER, S. L., 1989, Fabrication of Intermetallic Matrix Composites by the Powder Cloth Process. NASA TM-102060.

POLVANI, R. S., TZENG, W. S., and STRUTT, P. R., 1976, High Temperature Creep in a Semicoherent NiAl-Ni_2AlTi Alloy. *Metall. Trans. A* 7A: 33–40.

PRESCOTT, R., MITCHELL, D. F., SPROULE, G. I., and GRAHAM, M. J., 1992, Transport in α-Al_2O_3 Scales on Fe-Al and Ni-Al Alloys at 1100°C. *Solid State Ionics* 53: 229–237.

RAJ, S. V., and FARMER, S. C., 1993, Observation of a New Creep Regime in Polycrystalline Ni-50(at.%)Al Intermetallic Alloy. In *High Temperature Ordered Intermetallic Alloys V*, MRS Symp. Proc. 288: 647–652.

RAJ, S. V., LOCCI, I. E., and NOEBE, R. D., 1992, Deformation Behavior of a Ni-30Al-20Fe-0.05Zr Intermetallic Alloy in the Temperature Range 300 to 1300 K. *Metall. Trans. A* 23A: 1705–1718.

REDDY, K. P. R., SMIALEK, J. L., and COOPER, A. R., 1982, ^{18}O Tracer Studies of Al_2O_3 Scale Formation on NiCrAl Alloys. *Oxid. Met.* 17: 429–449.

REUSS, S., and VEHOFF, H., 1990a, Temperature Dependence of the Fracture Toughness of Single Phase and Two Phase Intermetallics. *Scr. Metall. Mater.* 24: 1021–1026.

________, (as reported by G. Sauthoff) 1990b, Mechanical Properties of Intermetallics at High Temperatures. *High Temperature Aluminides and Intermetallics*, eds. S. H. Whang, C. T. Liu, D. P. Pope and J. O. Stiegler, p. 344. Warrendale, PA: TMS.

REVIERE, R. D., OLIVER, B. F., and BRUNS, D. D., 1989, Computer Controlled Containerless Processing of High Temperature Intermetallic Compounds. In *Materials and Manufacturing Processes* 4(1): 103–131.

RHYS-JONES, T. N., 1989, Coatings for Blade and Vane Applications in Gas Turbines. *Corr. Sci.* 29: 623–646.

RIGNEY, J. D., and LEWANDOWSKI, J. J., 1992a, Fracture Toughness of Monolithic Nickel Aluminide Intermetallics. *Mat. Sci. Eng.* A149: 143–151.

________, 1992b. Effects of Reinforcement Size and Distribution on Fracture Toughness of Composite Nickel Aluminide Intermetallics. *Mater. Sci. Eng.* A158: 31–45.

ROBERTS, P. R., and FERGUSON, B. L., 1991, Extrusion of Metal Powders. *Int. Mater. Rev.* 36: 62–79.

ROZNER, A. G., and WASILEWSKI, R. J., 1966, Tensile Properties of NiAl and NiTi. *J. Inst. Met.* 94: 169–175.

RUDY, M., and SAUTHOFF, G., 1985, Creep Behavior of the Ordered Intermetallic (Fe, Ni)Al Phase. In *High Temperature Ordered Intermetallic Alloys*, MRS Symp. Proc. 39: 327–333.

RUSOVIC, N., and WARLIMONT, H., 1977, The Elastic Behavior of β_2-NiAl. *Phys. Stat. Solidi.*, (*A*) 44: 609–619.

———, 1979, Young's Modulus of β_2-NiAl Alloys. *Phys. Stat. Solidi.*, (*A*) 53: 283–288.

RUSSELL, S. M., LAW, C. C., BLACKBURN, M. J., CLAPP, P. C., and PEASE, D. M., 1989, Lightweight Disk Alloy Development. Report PWA-FR-19577-8, West Palm Beach, FL: Pratt & Whitney.

———, 1991, Lightweight Disk Alloy Development. Report PWA-FR-21413, West Palm Beach, FL: Pratt & Whitney.

RYBICKI, G. C., and SMIALEK, J. L., 1989, Effect of the θ-α-Al_2O_3 Transformation on the Oxidation Behavior of β-NiAl + Zr. *Oxid. Met.* 31: 275–304.

SANDS, T., 1988, Stability and Epitaxy of NiAl and Related Intermetallic Films on III-V Compound Semiconductors. *Appl. Phys. Lett.* 52: 197–199.

SCHNEIBEL, J. H., JENKINS, M. G., and MAZIASZ, P. J., 1993, Crack Propagation in NiAl and FeAl. In *High Temperature Ordered Intermetallic Alloys V*, MRS Symp. Proc. 288: 549–554.

SCHULSON, E. M., 1981, Comments on the Brittle to Ductile Transition of Long-Range Ordered Alloys. *Res. Mech. Lett.* 1: 111–114.

———, 1982, The Strength and Ductility of Polycrystalline NiAl in Tension. *COSAM* (*Conservation of Strategic Aerospace Materials*) Program Overview, ed. J. R. Stephens, pp. 175–182. NASA TM-83006.

———, 1985, The Effects of Grain Size on the Flow and Fracture of Long-Range Ordered Alloys. In *High Temperature Ordered Intermetallic Alloys*, MRS Symp. Proc. 39: 193–204.

SCHULSON, E. M., and BARKER, D. R., 1983, A Brittle to Ductile Transition in NiAl of a Critical Grain Size. *Scripta. Metall.* 17: 519–522.

SINGLETON, M. F., MURRAY, J. L., and NASH, P., 1986, NiAl. In *Binary Alloy Phase Diagrams*, ed. T. B. Massalski, pp. 140–143. Metals Park, OH: American Society for Metals.

SMEGGIL, J. G., 1991, The Effect of Chromium on the High Temperature Oxidation Resistance of Ni:Al. *Surf. Coat. Technol.* 46: 143–153.

SMIALEK, J. L., 1991, The Effect of Sulfur Content on Al_2O_3 Scale Adhesion. *Microscopy of Oxidation*, eds. M. J. Bennett and G. W. Lorimer, pp. 258–270. London, UK: The Institute of Metals.

SMIALEK, J. L., and BROWNING, R., 1985, Current Viewpoints on Oxide Adherence Mechanisms. NASA TM-87168.

SMIALEK, J. L., and LOWELL, C. F., 1974, Effects of Diffusion on Aluminum Depletion and Degradation of NiAl Coatings. *J. Electrochem. Soc.* 121: 800–805.

SMIALEK, J. L., and MEIER, G. H., 1987, High-Temperature Oxidation. In *Superalloys II*, eds. C. T. Sims, N. S. Stoloff, and W. C. Hagel, pp. 293–326. New York: John Wiley & Sons.

SMITH, T. R., KALLINGAL, C. G., RAJAN, K., and STOLOFF, N. S., 1992, Strain-Controlled Fatigue of NiAl Crystals at Room Temperature. *Scripta. Metall. Mater.* 27: 1389–1393.

STEPHENS, J. T., 1988, High Temperature Metal Matrix Composites for Future Aerospace Systems. AIAA Paper 88-3059.

STOLOFF, N. S., SMITH, T. R., and CASTAGNA, A., 1993, Fatigue Behavior of Intermetallic Compounds and Their Composites. In *High Temperature Ordered Intermetallic Alloys V*, MRS Symp. Proc. 288: 59–70.

STRUTT, P. R., DODD, R. A., and ROWE, G. M., 1970, Creep in Stoichiometric Beta-NiAl. In *Strength of Metals and Alloys, Vol. III*, ASM International, Metals Park, OH, pp. 1057–1061.

SUBRAMANIAN, P. R., MENDIRATTA, M. G., MIRACLE, D. B., and DIMIDUK, D. M., 1990, Microstructures and Mechanical Properties of NiAl + Mo In Situ Eutectic Composites. In *Intermetallic Matrix Composites, MRS Symp. Proc.* 194: 147–154.

TAKASUGI, T., KISHINO, J., and HANADA, S., 1993, Anomalous Elongation Behavior of Stoichiometric NiAl Single Crystals at Intermediate Temperatures. *Acta. Metall. Mater.* 41: 1009–1020.

TAKASUGI, T., WATANABE, S., and HANADA, S., 1992, The Temperature and Orientation Dependence of Tensile Deformation and Fracture in NiAl Single Crystals. *Mater. Sci. Eng.* A149: 183–193.

TAYLOR, A., and DOYLE, N. J., 1972, Further Studies on the Nickel-Aluminum System. I. The β-NiAl and δ-Ni_2Al_3 Phase Fields. *J. Appl. Cryst.* 5: 201–209.

VANDERVOORT, R. R., MUKHERJEE, A. K., and DORN, J. E., 1966, Elevated-Temperature Deformation Mechanisms in β'-NiAl. *Trans. ASM* 59: 930–944.

VEDULA, K., and KHADKIKAR, P. S., 1990, Effect of Stoichiometry on Low Temperature Mechanical Properties. In *High Temperature Aluminides and Intermetallics*, eds. S. H. Whang, C. T. Liu, D. P. Pope, and J. O. Stiegler, pp. 197–217. Warrendale, PA: The Minerals, Metals and Materials Society.

VEHOFF, H., 1993, Fracture and Toughness of Intermetallics. *High Temperature Ordered Intermetallic Alloys V*, MRS Symp. Proc. 288: 71–82.

VEYSSÍERE, P., and NOEBE, R. D., 1992, Weak-Beam Study of ⟨111⟩ Superlattice Dislocations in NiAl. *Phil. Mag.* 65: 1–13.

VISWANADHAM, R. K., MANNAN, S. K., KUMAR, K. S., and WOLFENDEN, A., 1989, Elastic Modulus of NiAl-TiB_2 Composites in the Temperature Range 300 to 1273 K. *J. Mat. Sci. Lett.* 8: 409–410.

VITEK, V., and CHEN, S. P., 1991, Modeling of Grain Boundary Structure and Properties in Intermetallic Compounds. *Scripta. Metall. Mater.* 25: 1237–1242.

WACHTELL, R. L., 1952, An Investigation of Various Properties of NiAl. In WADC TR 52-291.

WALSTON, W. S., and DAROLIA, R., 1993, Effect of Alloying on Physical Properties of NiAl. In *High Temperature Ordered Intermetallic Alloys V*, MRS Symp. Proc. 288: 237–242.

WALTER, J. L., and CLINE, H. E., 1970, The Effect of Solidification Rate on Structure and High Temperature Strength of the Eutectic NiAl-Cr. *Metall. Trans.* 1: 1221–1229.

WANG, L., and ARSENAULT, R. J., 1990, Microstructure of TiB_2-NiAl. *Mat. Sci. Eng.* A127: 91–98.

________, 1991, Interfaces in XD Processed TiB_2/NiAl Composites. *Metall. Trans. A* 22A: 3013–3018.

WASILEWSKI, R. J., 1966, Elastic Constants and Young's Modulus of NiAl. *Trans. Metall. Soc. AIME* 36: 455–457.

WASILEWSKI, R. J., BUTLER, S. R., and HANLON, J. E., 1967, Plastic Deformation of Single-Crystal NiAl. *Trans. Metall. Soc. AIME* 239: 1357–1364.

WEAVER, M. L., KAUFMAN, M. J., and Noebe, R. D., 1993, The Effects of Alloy Purity on the Mechanical Behavior of Soft Oriented NiAl Single Crystals. *Scripta. Metall. Mater.* 29: 1113–1118.

WHITTENBERGER, J. D., 1987, Effect of Composition and Grain Size on Slow Plastic Flow Properties of NiAl Between 1200 to 1400 K. *J. Mater. Sci.* 22: 394–402.

________, 1988, The Influence of Grain Size and Composition on 1000 to 1400 K Slow Plastic Flow Properties of NiAl. *J. Mater. Sci.* 23: 235–240.

WHITTENBERGER, J. D., ARZT, E., and LUTON, M. J., 1990a, Preliminary Investigation of a NiAl Composite Prepared by Cryomilling. *J. Mater. Res.* 5: 271–277.

________, 1990b, 1300 K Compressive Properties of a Reaction Milled NiAl-AlN Composite. *J. Mater. Res.* 5: 2819–2827.

WHITTENBERGER, J. D., GAYDOSH, D. J., and KUMAR, K. S., 1990, 1300 K Compressive Properties of Several Dispersion Strengthened NiAl Materials. *J. Mater. Sci.* 25: 2771–2776.

WHITTENBERGER, J. D., KUMAR, K. S., and MANNAN, S. K., 1991a, 1000 to 1300 K Slow Plastic Compression Properties of Al-Deficient NiAl. *J. Mater. Sci.* 26: 2015–2022.

________, 1991b, 1200 and 1300 K Slow Plastic Compression Properties of Ni-50Al Composites. *Mater. High Temp.* 9: 3–12.

WHITTENBERGER, J. D., and LUTON, M. J., 1992, Elevated Temperature Compressive Properties of Reaction Milled NiAl-AlN and Zr-doped NiAl-AlN Composites. *J. Mat. Res.* 7: 2724–2732.

WHITTENBERGER, J. D., MANNAN, S. K., and KUMAR, K. S., 1989, 1100 to 1300 K Slow Plastic Compression Properties of Ni-38.5Al Composites. *Scripta. Metall.* 23: 2055–2060.

WHITTENBERGER, J. D., NATHAL, M. V., RAJ, S. V., and PATHARE, V. M., 1991, Slow Strain Rate 1200–1400 K Compressive Properties of NiAl-Hf. *Mat. Lett.* 11: 267–272.

WHITTENBERGER, J. D., NOEBE, R. D., CULLERS, C. L., KUMAR, K. S., and MANNAN, S. K., 1991, 1000 to 1200 K Time-Dependent Compressive Deformation of Single-Crystalline and Polycrystalline B2 Ni-40Al. *Metall. Trans. A* 22A: 1595–1607.

WHITTENBERGER, J. D., RAY, R., and JHA, S. C., 1992, Influence of Grain Size on the Creep Behavior of HfC-Dispersed NiAl. *Mater. Sci. Eng.* A151: 137–146.

WHITTENBERGER, J. D., RAY, R., JHA, S. C., and DRAPER, S., 1991, 1000–1300 K Slow Strain Rate Properties of NiAl Containing Dispersed TiB_2 and HfB_2. *Mat. Sci. Eng.* A138: 83–93.

WHITTENBERGER, J. D., REVIERE, R., NOEBE, R. D., and OLIVER, B. F., 1992, Compressive Strength of Directionally Solidified NiAl-NiAlNb Intermetallics at 1200 and 1300 K. *Scr. Metall. Mater.* 26: 987–992.

WHITTENBERGER, J. D., VISWANADHAM, R. K., MANNAN, S. K., and KUMAR, K. S., 1989, 1200 to 1400 K Slow Strain Rate Compressive Behavior of Small Grain Size NiAl/Ni_2AlTi Alloys and NiAl/Ni_2AlTi-TiB_2 Composites. *J. Mat. Res.* 4: 1164–1171.

WHITTENBERGER, J. D., VISWANADHAM, R. K., MANNAN, S. K., and SPRISSLER, B., 1990, Elevated Temperature Slow Plastic Deformation of NiAl-TiB_2 Particulate Composites at 1200 and 1300 K. *J. Mater. Sci.* 25: 35–44.

WHITTENBERGER, J. D., WESTFALL, L. J., and NATHAL, M. V., 1989, Compressive Strength of a B2 Matrix NiAl-Nb Intermetallic at 1200 and 1300 *Scripta. Metall.* 23: 2127–2130.

WOOD, D. L., GRENOBLE, H. E., and WESTBROOK, J. H., 1960–1964, Effect of Basic Physical Parameters on Engineering Properties of Intermetallic Compounds. Report WADD TR-60-184, Pts. I-VII, Wright-Patterson AFB, OH.

WOOD, J. H., and GOLDMAN, E. H., 1987, Protective Coatings. In *Superalloys II*, eds. C. T. Sims, N. S. Stoloff, and W. C. Hagel, pp. 359–384, John Wiley & Sons, New York.

WRIGHT, R. N., KNIBLOE, J. R., and NOEBE, R. D., 1991, Consolidation of NiAl Powders Using Hot Isostatic Pressing. *Mat. Sci. Eng.* A141: 79–83.

WUNDERLICH, W., MACHON, L., and SAUTHOFF, G., 1992, Dislocation Analysis at Crack Tips at Phase Boundaries in an Intermetallic NiAl-NbNiAl Alloy. *Z. Metallkd.* 83: 679–684.

YANEY, D. L., and NIX, W. D., 1988, Mechanisms at Elevated-Temperature Deformation in the B2 Aluminides NiAl and CoAl. *J. Mat. Sci.* 23: 3088–3098.

YANG, W. J., and DODD, R. A., 1973, Steady-State Creep and Associated Microstructures in Stoichiometric and Nonstoichiometric Polycrystalline NiAl. *Metal Science Journal* 7: 41–47.

YOUNG, E. W. A., and DEWIT, J. H. W., 1986, The Use of ^{18}O Tracer and Rutherford Backscattering Spectrometry to Study the Oxidation Mechanism of NiAl. *Solid State Ionics* 16: 39–46.

ZELLER, M. V., NOEBE, R. D., and LOCCI, I. E., 1990, Grain Boundary Segregation Studies of NiAl and NiAl(Zr) Using Auger Electron Spectroscopy. *HITEMP Review—1990: Advanced High Temperature Engine Materials Technology Program*, NASA CP-10051, pp. 21-1–21-17.

Chapter 8

Titanium Aluminides

F. H. Froes and C. Suryanarayana

ABSTRACT

Excellent elevated-temperature properties and low density make the titanium aluminides attractive candidates for use in both engine and airframe applications particularly in the aerospace industry. The challenge to the materials scientist is to maintain these characteristics while building in "forgiveness." The basic Ti-Al-phase diagram and crystal structure of both the Ti_3Al and TiAl phases are reviewed, followed by a consideration of chemistry, processing, microstructure, deformation/fracture, and mechanical property relationships in monolithic material. Synthesis/processing methods are presented, including ingot and powder-metallurgy approaches and use of hydrogen as a temporary alloying element. Composite concepts as a method to enhance not only "forgiveness" but also elevated-temperature behavior are discussed. Consideration of present and projected applications of both monolithic and composite material are then presented. It is concluded that while the titanium aluminides in monolithic form can be used now in nondemanding applications, much further research and development is required before this material class can be used in critical applications, especially in composite concepts.

1. INTRODUCTION

The low density of the ordered intermetallic titanium aluminides, based on Ti_xAl ($x = 1$ or 3), in combination with a raft of mechanical properties

Table 8-1 Properties of Titanium Aluminides, Titanium-Base Conventional Alloys and Superalloys

Property	Ti-Base	Ti_3Al-Base	TiAl-Base	Superalloys
Density, g/cm^3	4.5	4.1–4.7	3.7–3.9	8.3
Modulus, GPa	96–100	100–145	160–176	206
Yield Strength, MPa	380–1150	700–990	400–650	—
Tensile Strength, MPa	480–1200	800–1140	450–800	—
Creep Limit, °C	600	760	1000	1090
Oxidation Limit, °C	600	650	900	1090
Ductility, %atRT	20	2–10	1–4	3–5
Ductility, %atHT	High	10–20	10–60	10–20
Structure	A3/A2	DO_{19}	$L1_0$	f.c.c./$L1_2$

Source: Lipsitt 1985

that have been enhanced as a result of extensive studies, are candidates for applications in advanced aerospace and terrestrial components both in monolithic and composite concepts (Froes, Eylon, and Bomberger 1985; Lipsitt 1985; Miracle, Anton, and Graves 1992; Kim 1989, 1991; Kim and Froes 1990; Liu, Froes, and Stiegler 1990). Over the past 25 years, as a result of the extensive work carried out on these systems, the titanium aluminides can be considered, along with the iron and nickel aluminides, as relatively mature materials, particularly in comparison with the more exotic intermetallics discussed in other chapters of this book. In order to illustrate the attraction of the monolithic titanium aluminides, a comparison with conventional titanium alloys and superalloys is given in Table 8-1 (Lipsitt 1985). Elevated temperature properties such as creep behavior are very attractive, especially on a density-normalized basis. However, ambient temperature "forgiveness" (ductility, fracture toughness, fatigue-crack-growth rate) is a concern, and low oxidation resistance and susceptibility to hydrogen pickup at elevated temperatures are below desired levels; environmental effects are covered in another chapter.

To circumvent some of the concerns with the titanium aluminides, composite concepts are being actively pursued (Froes 1988a, b; Anton et al. 1990), to increase "forgiveness" and enhance elevated-temperature performance, particularly modulus. These materials do exhibit enhancement in certain properties, but transverse behavior and cost are concerns.

This chapter will review both monolithic and composite titanium aluminides with consideration being given to both the science and technology (applications) of these systems. The text is, in part, an update of three earlier reviews of this subject by the same authors (Froes, Suryanarayana, and Eliezer 1991, 1992; Froes 1993) and other recent reviews (Beddoes, Wallace, and de Malherbe 1992; Ward 1993).

2. PHASE DIAGRAMS

Alloys based on TiAl have been under development since 1955 and those based on Ti_3Al since 1960, but the strong effect of interstitial elements (oxygen and hydrogen) has hampered determination of an accurate Ti-Al-phase diagram. Using earlier literature (Kornilov et al. 1965) and thermodynamic modeling, Murray (1987) proposed a comprehensive binary Ti-Al-phase diagram that includes $Ti_3Al(\alpha_2)$, TiAl(γ), $TiAl_2$, δ and $TiAl_3$ intermediate phases, and α(Ti), β(Ti) and (α-Al) terminal solid solutions. However, neither the invariant reaction temperatures nor some phase boundaries, such as $\alpha_2/\alpha_2 + \gamma/\gamma$, have been accurately defined. Further, Murray predicts that a Ti-50 at%Al[1] alloy will solidify as cubic dendrites of β surrounded by γ formed according to the peritectic reaction $L + \beta \rightarrow \gamma$. At lower Al contents and at lower temperatures, the α phase forms from the β and γ phases by a peritectoid reaction, with subsequent α decomposition taking place eutectoidally to a lamellar $\alpha_2 + \gamma$ structure. In fact recent results from both ingot metallurgy and rapid solidification routes indicate that around 50%Al, the α phase forms directly from the liquid state according to the peritectic reaction $L + \beta \rightarrow \alpha$; the presently accepted version of the binary Ti-Al-phase diagram is shown in Figure 8-1a (Rowe and Huang 1988). The ternary Ti-Al-Nb-phase diagram at 900°C is illustrated in Figure 8-1b (Rowe 1993a).

Kattner, Lin, and Chang (1992) have recently assessed the binary Ti-Al-phase diagram using three different analytical thermodynamic descriptions. They showed that the calculated phase diagram agrees well with the critically evaluated experimental data from the literature.

The Ti_3Al phase undergoes a disorder-order transformation in the range of 22% to 39% Al and congruently at 1180°C. The ordered α_2 phase has a hexagonal structure with $a = 0.5782$ nm and $c = 0.4629$ nm. The TiAl(γ) phase, stable from 49% to 66% Al is considered to be ordered up to the melting point, although this is not certain because antiphase boundaries were observed in a rapidly solidified Ti-60% Al alloy. The γ phase has an ordered face-centered tetragonal structure with $a = 0.4005$ nm and $c = 0.4070$ nm, and a tetragonality of only 1.02 at the equiatomic position; this increases to 1.03 at higher Al contents.

[1]Throughout the text all compositions are expressed in at% unless otherwise stated.

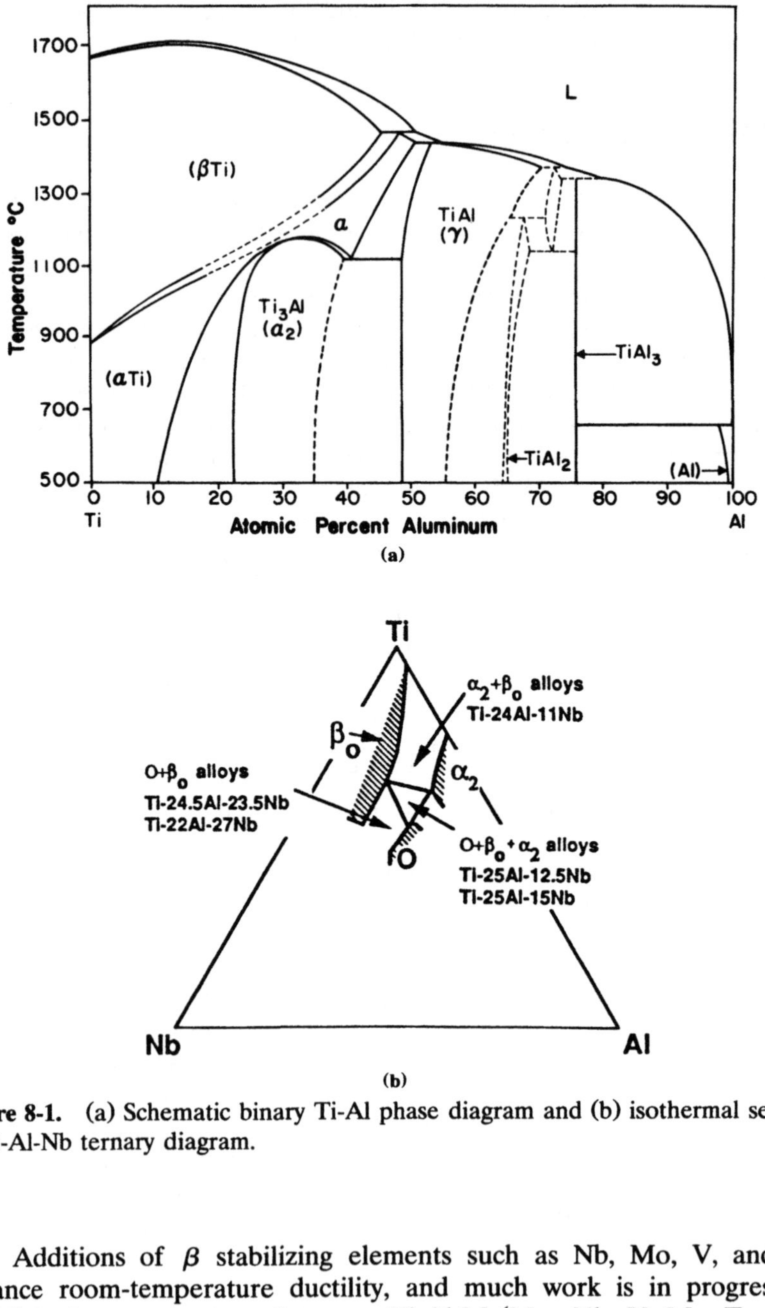

Figure 8-1. (a) Schematic binary Ti-Al phase diagram and (b) isothermal section of Ti-Al-Nb ternary diagram.

Additions of β stabilizing elements such as Nb, Mo, V, and Ta enhance room-temperature ductility, and much work is in progress to establish the constitution of ternary Ti-Al-M (M = Nb, V, Mo, Ta, Mn, etc.) systems (Banerjee et al. 1989; Kaufman et al. 1989; Suryanarayana

Table 8-2 Crystal Structure Data of the Stable and Metastable Phases in Titanium-Rich Ti-Al and Ti-Al-Nb Alloys

Phase	Strukturbericht Symbol	Crystal Structure	Space Group	Pearson Symbol	Lattice Parameters (nm) a	b	c
Ti (α)	A3	h.c.p.	$P6_3/mmc$	hP2	0.29508	—	0.46885
Ti (β)	A2	b.c.c.	Im3m	cI2	0.33132	—	—
$Ti_3Al(\alpha_2)$	DO_{19}	ordered hexagonal	$P6_3/mmc$	hP8	0.5782	—	0.4629
$TiAl(\gamma)$	$L1_0$	ordered face-centered tetragonal	P4/mmm	tP4	0.4005	—	0.4070
B2*	B2	ordered b.c.c.	$Pm\bar{3}m$	cP2	0.322	—	—
—	$B8_2^{@}$	hexagonal	$P6_3/mmc$	—	0.45803	—	0.55204
$Ti_2AlNb(O)$		orthorhombic	Cmcm	—	0.45	0.588	0.960
					0.62	0.94	0.47
					0.596	0.986	0.467
					0.636	0.969	0.479
					0.609	0.957	0.467
Ti_5Al_2Nb	T	ordered tetragronal			0.65		0.66
ω**	—	hexagonal	P6/mmm	—	0.46	—	0.58
ω''**	—	trigonal	$P\bar{3}m1$	—	0.4555	—	0.5542

*observed only in ternary Ti-Al-M alloys where M is a β-stabilizing element

@low-temperature equilibrium phase in Ti-Al-Nb alloys

**transitional phases formed during decomposition of the B2 phase

Source: Froes et al. 1992

and Lee 1992), with emphasis on Ti-Al-Nb. In Ti-Al-Nb alloys containing 5 to 17%Nb, a ductile ordered b.c.c. phase (B2) forms on quenching from high temperatures; this also occurs in alloys containing other β stabilizing elements. Alloys quenched from the high-temperature β phase (both in the binary and the ternary alloys) undergo a variety of martensitic transformations. Similarly, alloys containing the B2 phase form the ω and ω-related phases on cooling to room temperature from elevated temperatures > 1100°C. Table 8-2 presents the crystal structure data of the stable and metastable phases detected to date in titanium-rich binary Ti-Al and ternary Ti-Al-Nb systems.

An equilibrium ordered orthorhombic (O) phase is detected in Ti-Al-Nb alloys with an approximate stoichiometry of Ti_2AlNb (Banerjee et al. 1988). The crystal structure and lattice parameters have been determined by X-ray diffraction, electron diffraction, and neutron diffraction techniques and the lattice parameters are listed in Table 8-2. It has been shown (Hsiung, Cai, and Wadley 1992a, b) that the isothermal $\beta \rightarrow \alpha_2$ transformation takes place through a complex reaction sequence $\beta \rightarrow B2 \rightarrow T \rightarrow O \rightarrow \alpha_2$ at temperatures < 900°C. The T phase has been shown to have an ordered tetragonal structure with lattice parameters a = 0.65 nm and c/a = 1.02 (Hsiung and Wadley 1992a). This T phase is suggested to result from a slight tetragonal distortion of a DO_3-type cubic-ordered phase.

3. SPECIFIC SYSTEMS

3.1. Ti_3Al (α_2)

*3.1.1. **Chemistry***. Semicommercial and experimental α_2 alloys developed to the present time are two-phase ($\alpha_2 + \beta$/B2) with compositions based on Ti-(23-25)Al-(10-30)Nb. Alloy compositions with current engineering significance are Ti-24Al-11Nb, Ti-25Al-10Nb-3V-1Mo, Ti-25Al-17Nb-1Mo, and Ti-23.5Al-24Nb, Ti-(21.8-28.5)Al-(17-30)Nb, Ti-30Al-20Nb, Ti-(21.5-22.0)Al-(20-23)Nb-(1-5)V; the latter two compositions are of the orthorhombic (three phase α_2, orthorhombic and B2) type.

Generally, increasing Nb content enhances the majority of the material properties with the exception of creep performance. Nb, which substitutes for Ti atoms and increases the number of slip systems, may be replaced by specific elements for improved strength (Mo, Ta, Cr), creep resistance (Mo), and oxidation resistance (Ta, Mo). Good elevated-temperature strength, creep resistance, and environmental resistance are favored by high Al, a moderately high β stabilizer addition, and low V.

For the orthorhombic alloys, quaternary additions of V enhance strength, creep resistance, and ambient temperature ductility (Rowe 1993b).

Care should be exercised in evaluating early work because of variations in composition, processing, microstructure, and testing procedures. Of particular importance is oxygen level, which raises the ductile-to-brittle transition temperature of titanium alloys. Reduction of oxygen content from 0.186 to 0.081 wt% for the Ti-24Al-11Nb alloy results in an increase in tensile elongation from 1.1% up to 4.8%, perhaps due to increased basal slip as the oxygen level decreases (Koss et al. 1990). Additions of Y (discussed elsewhere) and B (up to 0.5%) can be used to control grain size. Some interesting additions may have been prematurely abandoned; for example, Zr, which when added at the 7–14% level showed attractive levels of strength and ductility.

3.1.2. Processing / Microstructure Development. Microstructural features that can be varied by thermomechanical processing (TMP) include the β grain morphology and distribution, primary α_2 grain size and volume fraction, and secondary α_2 plate morphology and thickness (Kim and Froes 1990; Liu et al. 1990; Ward 1993). β processing generally results in elongated Widmanstätten α_2 in large primary β grains, while α-β processing results in more equiaxed primary α grains. Both microstructure and mechanical properties follow the behavior of more conventional titanium alloys (Froes et al. 1985), with the caveat that because of their ordered nature the Ti_3Al alloys are very brittle and fracture by a cleavage mechanism at ambient temperatures (Ward et al. 1989; Ward 1993). An indication of the variety of microstructures that can be achieved is shown in Figure 8-2 (Rowe 1990).

Until recently, transformed β microstructures with elongated plate-like α_2 have been considered to be the optimum microstructure because this structure was considered necessary to achieve the best balance of room-temperature ductility and elevated-temperature properties (Larsen et al. 1990). In the past few years, more equiaxed α_2 morphologies have been investigated because of their potential for increased room-temperature ductility (Larsen et al. 1990). A small α_2 grain size enhances strength, and generally a refined slip length improves ductility levels, using either grain size, the allotropic change, or precipitation of second-phase particles. A comprehensive study of the effect of thermomechanical processing on the microstructure of a Ti-24Al-14Nb-3V-0.5Mo alloy indicated that a fine-equiaxed primary α resulted (Fig. 8-3) in up to 26% room-temperature ductility in the as-aged condition; this ductility level decreased to 9.4% after aging due to the precipitation of secondary α_2 laths (Jia et al. 1992).

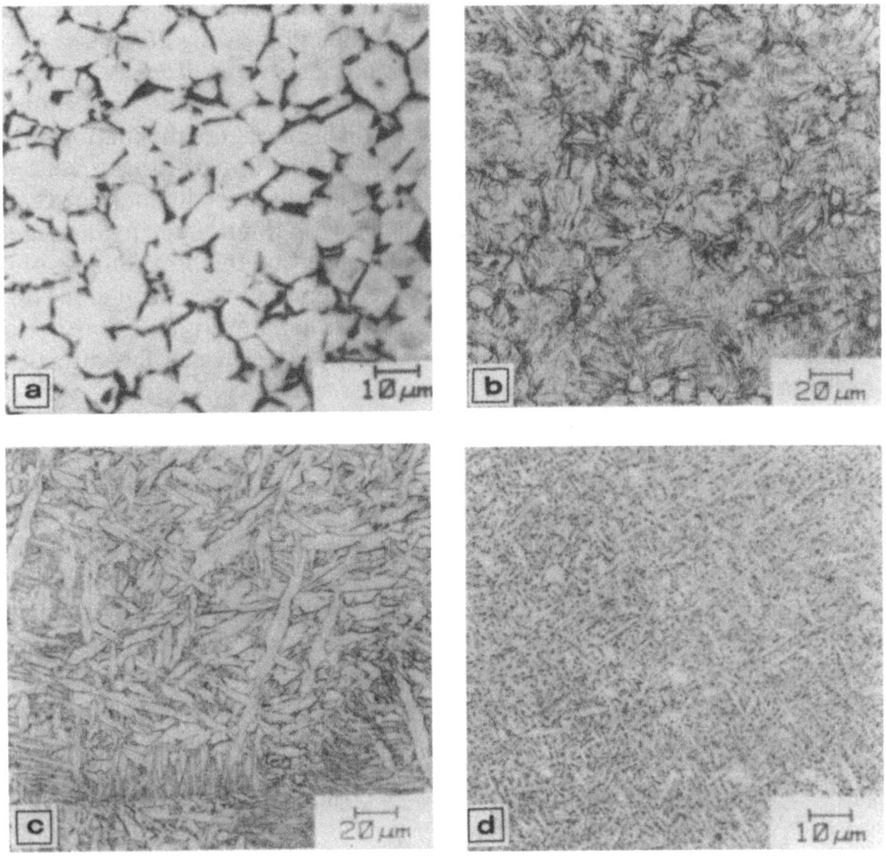

Figure 8-2. Microstructures developed in a Ti-25Al-7.5Nb (α_2) alloy by transforming from the β phase. (a) equiaxed microstructure with a high volume fraction of primary α_2 phase, (b) low volume fraction of primary α_2 phase in a Widmanstätten transformed matrix, (c) coarse Widmanstätten transformed microstructure and (d) fine Widmanstätten transformed microstructure.

Microstructural control has been basically ignored until recently as a method to control ductility. Rather emphasis was placed on (a) adjusting the amount of β phase by chemistry and heat treatment, or (b) modifying the basic deformation mechanisms of the DO_{19} lattice of the α_2 (Ward et al. 1989). This appears to be a serious omission (see section 3.1.3).

The O phase forms as equiaxed grains (Hsiung and Wadley 1992c) or laths coexisting with the T phase [Hsiung and Wadley 1992b, Hsiung et al. 1992b]. In alloys with a Nb content close to 12%, the metastable O phase

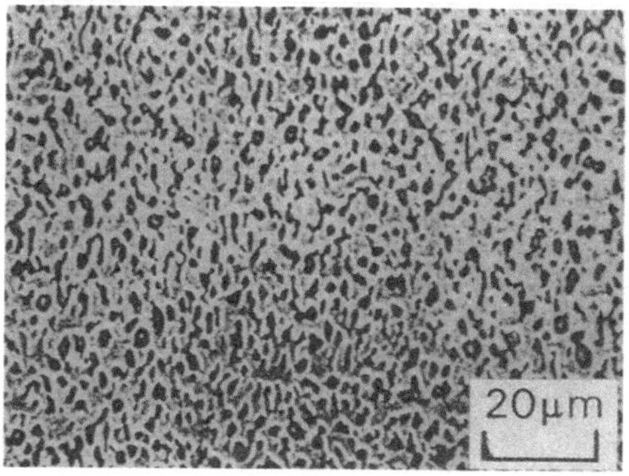

Figure 8-3. Fine microstructure in specially processed Ti-24Al-14Nb-3V-0.5Mo, exhibiting 26% room temperature ductility.

forms from the intermediate α_2 phase by congruent ordering (Bendersky 1992). Higher Nb contents (> 15%) lead to direct formation of the O phase from the B2 phase (Muraleedharan et al. 1992a, b). In alloys close to the Ti_2AlNb composition, the transformation is very sluggish but nevertheless massive, without phase separation when multiphase fields are crossed. The transformation proceeds by copious nucleation of the O-phase regions which grow uniformly outward until impingement occurs.

Since the O phase has been found to form at different temperatures and to coexist with other phases, orientation relationships have been determined as follows:

$$[001]_O//\langle 110\rangle_{B2} \quad \text{and} \quad (110)_O//(111)_{B2} \text{ (Bendersky 1992)} \quad (8.1)$$

$$(\bar{1}\bar{1}0)_T//(001)_O \quad \text{and} \quad [1\bar{1}0]_T \wedge [010]_O = [001]_T \wedge [100]_O = 4.3^\circ$$

$$\text{(Hsiung et al. 1992b; Hsiung and Wadley 1992b)} \quad (8.2)$$

$$(001)_O//(0001)_{\alpha_2} \quad \text{and} \quad [100]_O \wedge [11\bar{2}0]_{\alpha_2} = [110]_O \wedge [01\bar{1}0]_{\alpha_2} = 1.6^\circ$$

$$\text{(Hsiung et al. 1992b; Hsiung and Wadley 1992c)} \quad (8.3)$$

The "O" alloys studied to the present time consist of a fine Widmansätten "O" + B2 structure (Rowe 1993a), with microstructural variations again being a fertile area for additional studies, an example being in situ aligned laths of "O" phase (Rowe, Siemers, and Larsen 1992) which could lead to tailoring of high fracture toughness in specific orientations.

3.1.3. Deformation and Fracture. The limited ductility of Ti_3Al-based alloys at low temperatures is attributed to the planarity of slip and the absence of slip in planes parallel to and/or inclined to the hexagonal axis, (i.e., lack of an adequate number of independent slip systems). The crystal structure of Ti_3Al is shown in Figure 8-4a and the possible slip planes and directions are shown in Figure 8-4b. Three types of dislocations are possible:

1. ⟨a⟩ type with a Burger's vector of the type $1/6\langle 11\bar{2}0\rangle$ moving on basal (0001), prism $\{10\bar{1}0\}$ or pyramidal $\{20\bar{2}1\}$ planes,
2. ⟨c⟩ type with a Burger's vector of [0001] on second order $\{11\bar{2}0\}$ prism planes, and
3. ⟨c + a⟩ type with a Burger's vector of $\langle 11\bar{2}6\rangle$ on $\{11\bar{2}1\}$ or $\{20\bar{2}1\}$ pyramidal planes.

However, on deformation at room and intermediate temperatures, only ⟨a⟩ type dislocations are observed in Ti_3Al alloys. The absence of c-component deformation results in generation of local stresses resulting in brittle cleavage and poor ductility. Deformation-induced twinning can produce the ⟨c⟩-type dislocations; but this mode of deformation is not possible at room temperature without causing disordering. Occurrence of microtwinning (Lipsitt 1985) and presence of ⟨c⟩ and ⟨c + a⟩-type dislocations in samples deformed at high temperatures (Gogia, Banerjee, and Nandy 1990) have been suggested as responsible for the increased ductil-

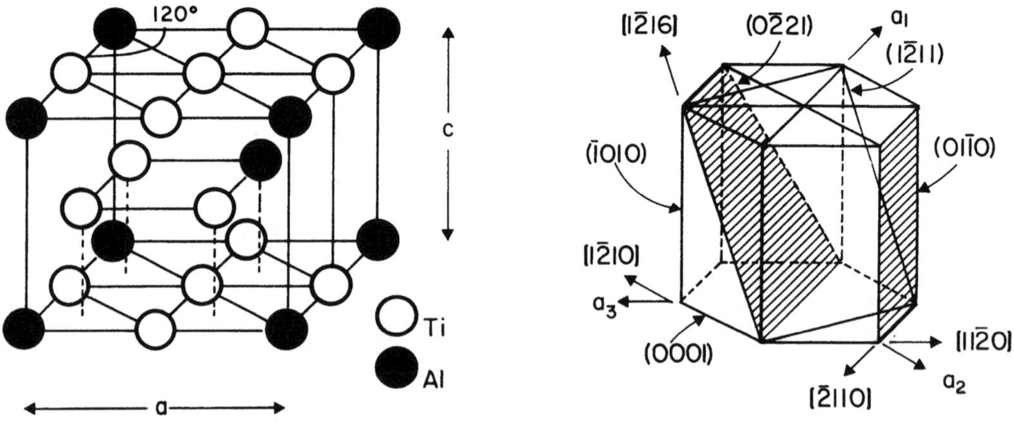

Figure 8-4. (a) Crystal structure of Ti_3Al (α_2) phase and (b) possible slip planes and slip vectors in the structure.

ity. ⟨c⟩ and ⟨c + a⟩-type dislocations have also been reported in samples deformed at room temperature; thus, it is not clear whether the increased ductility at high temperatures is the result of increased ⟨c + a⟩ dislocation activity.

The mechanism by which the ductility is improved with increasing Nb content is unclear. Since the ⟨c + a⟩ dislocation activity decreases with increasing temperature, a possibility is that Nb reduces the slip length by refining the microstructure and increasing the amount of the retained β phase. The $1/6\langle 11\bar{2}0\rangle$ superpartials could then transfer to the β phase to become $1/2\langle 111\rangle$ dislocations according to the Burger's relationship $(0001)\alpha_2//(110)_\beta$ and $[11\bar{2}0]\alpha_2//[111]_\beta$.

A detailed study of the deformation behavior of the B2 phase in Ti-24Al-11Nb alloy showed that it deformed inhomogeneously by ⟨111⟩ slip on {110}, {123} and {112} planes. The slip-band structure is complex, consisting of segments of heavily pinned edge dislocations with periodic cross slip of screw components onto secondary slip planes, with isolated observations of [100] slip (Gogia et al. 1990; Banerjee, Gogia, and Nandy 1990).

At room and intermediate temperatures, fracture in this alloy class occurs by cleavage. Primary α_2 particles have been observed to crack during a tensile test giving an effective drop in load-bearing area. Even at higher temperatures, when the alloy exhibits some ductility, completely ductile fracture has not been observed; instead intergranular cracking takes place accommodating strain across grain boundaries.

The fracture behavior in these alloys is related to slip behavior in the α_2 phase and also to the presence of the β/B2 phase. The tensile properties of Ti_3Al-Nb alloys are critically dependent on the shape, distribution, and volume fractions of the α_2 and B2 phases (Koss et al. 1990; Ward et al. 1989; Gogia et al. 1990; Ward 1993), with boundary strengthening playing a dominant role (Ward et al. 1989). Widmanstätten secondary α_2 resulted in better ductility than microstructures with an aligned α_2 morphology, with aligned colonies acting as single units. Increasing the primary α_2 volume fraction (up to 30%) in a fine Widmanstätten microstructure in the Ti-25Al-10Nb-3V-1Mo alloy decreases the yield strength of the alloy due to a decrease in volume fraction of fine matrix. The ductility depends on both the distribution and volume fraction of the α_2 phase and peaks at 30–35% α_2, a trend that was attributed to the cracks which formed in the relatively large primary α_2. This behavior apparently does not parallel that in conventional titanium alloys in which increasing the amount of equiaxed primary α enhances ductility. However, very recent work (Jia et al. 1992) has clearly shown that if an optimum amount of equiaxed α_2 grains is produced by thermomechanical processing, room-temperature ductility can be dramatically improved (to 26% elongation), removing the inconsistency with the behavior of conventional titanium alloys. Grain boundary α_2 allows easy crack propagation and increases strain incompatibility at the boundary.

In other work on the Ti-24Al-11Nb alloy the maximum elongation of close to 10% in $\alpha_2 + \beta$ processed material occurred at 30–35 vol% of the α_2 phase. Interestingly, this corresponds to the minimum in yield strength for fine α_2 plates with primary α_2 present (Ward et al. 1989). Continuation of this work (Koss et al. 1990) confirmed that $\alpha_2 + \beta$-processed material exhibited higher ductilities than β-processed material. It was suggested that easy cracking in the primary α_2 regions dominated with some evidence of cracking along α_2/β interfaces (Koss et al. 1990); however, microcrack propagation across adjacent transformed β-regions was difficult. Generally the authors concluded that the room-temperature deformation of Ti_3Al-based materials could be understood primarily on the basis of slip within the α_2. Optimum creep behavior was suggested to be associated with a basketweave α_2 microstructure with a minimum of β phase.

Crack propagation within the α_2 phase is relatively easy, again due to the difficulty of the $\langle c \rangle$ component slip. When a crack is initiated on the basal plane and activates co-planar$\langle a \rangle$ slip, large normal stresses near the crack tip are generated, which relax only when $\langle c + a \rangle$ slip is activated, and without which the cracks continue to grow in the α_2. The effectiveness of tough β phase in blunting and bridging cracks increases with temperature and this, at least partially, explains the increased ductility at higher temperatures.

The finer scale structure obtained when stochiometric α_2 is alloyed with β stabilizers such as Nb, V, or Mo, in combination with β regions forming a continuous layer surrounding α_2 grains enhances ductility (Koss et al. 1990; Ward et al. 1989). However, much further work is required to fully optimize the microstructures for a combination of mechanical properties, the recent high ductility (26% room-temperature elongation) with an optimum amount of fine equiaxed primary α_2 emphasizing this point (Jia et al. 1992).

The ordered orthorhombic "O" phase can be considered as arising from the α_2 (ordered h.c.p.) structure by a small orthorhombic distortion. This implies that the three equivalent $\langle 11\bar{2}0 \rangle$ directions of the α_2 phase are changed slightly in magnitude with respect to each other, as well as with regard to the APB configurations associated with them. Similarly, the $\langle 11\bar{2}0 \rangle$ vectors are no longer equivalent in the O phase (Koss et al. 1990, Banerjee, Rowe, and Hall 1991).

Both [100] and [110] dislocations (the equivalent $\langle 11\bar{2}0 \rangle$ dislocations in α_2) are active in the O phase and junctions between these different types of dislocations dominate the microstructure (Banerjee et al. 1991). Cross slip to the (001) plane was found to occur with dislocation dissociation on this plane.

However, in contrast to the behavior of the α_2 phase, extensive ⟨c⟩ component slip (equivalent to $\langle 11\bar{2}0 \rangle$ in α_2) is seen in the O phase and these dislocations are arranged in well-defined slip bands. Both [114] and [102] Burger's vectors are active and dissociated on their slip planes. Thus, the orthorhombic phase, despite its ternary ordered structure and lower symmetry than α_2, deforms by at least as many slip systems and shows extensive ⟨c⟩ component slip.

3.1.4. Mechanical Properties. The mechanical properties of monolithic α_2 alloys, which have been summarized in an excellent overview article (Larsen et al. 1990), depend on composition and the microstructure developed by thermomechanical processing (Kim and Froes 1990; Liu et al. 1990; Rowe 1990; Larsen et al. 1990; Koss et al. 1990, Ward et al. 1989; Ward 1993). The compositions evaluated span 0-30 equivalent Nb content with the majority of the mechanical properties improving with increasing Nb content, with the caveat that Nb may reduce creep resistance (Kim and Froes 1990; Liu et al. 1990), although recent work suggests that richer Nb-containing alloys containing the orthorhombic (O) phase may be useful for elevated-temperature applications (Rowe 1990; Graves, Smith, and Rhodes 1992). Selected mechanical properties of α_2 alloys are listed in Table 8-3 (Fores et al. 1991, 1992; Froes 1993), with the two-phase alloys exhibiting up to a doubling of strength due predominantly to boundary strengthening, with long range order, solid solution, and texture effects also contributing (Ward et al. 1989). The specially thermomechanically processed Ti-24Al-14Nb-3V-0.5Mo (Jia et al. 1992) alloy exhibits high ductility, a result attributed to fine grains, decreased strain incompatibility between α_2 and B2, and an optimum α_2 size/spacing. The orthorhombic alloys rely on the presence of α_2, the orthorhombic phase, and B2 for enhanced ductility (Graves et al. 1992). For this latter alloy class, additions of V enhance strength and ductility (Rowe 1993b).

Increasing the Al level and the amount of β stabilizers both result in increased yield strengths. A fine Widmanstätten microstructure with a small amount of primary α_2 grains exhibits better ductility than microstructures with a coarse Widmanstätten microstructure, or an aligned acicular α_2 morphology (Kim and Froes 1990; Liu et al. 1990).

The fatigue properties of titanium alloys are strongly influenced by microstructure with high-ductility alloys being optimum under LCF conditions, as anticipated from work on conventional titanium alloys (Froes et al. 1985). The low ductility exhibited in material with Widmanstätten α plates in $\alpha + \beta$ alloys is responsible for low elevated-temperature LCF

Table 8-3 Selected Room Temperature Mechanical Properties and Elevated Creep Rupture Life of α_2

Alloys	YS, (MPa)	UTS, (MPa)	El., (%)	K_{IC}, (MPa$\sqrt{m}$)	Creep[a] Rupture
Ti-25Al	538	538	0.3		
Ti-24Al-11Nb	787	824	0.7		44.7
	761	967	4.8		
Ti-24Al-14Nb	831	977	2.1		59.5
Ti-24Al-14Nb-3V-0.5Mo			26.0[b]		
	1034	797	9.4[c]		
Ti-25Al-10Nb-3V-1Mo	825	1042	2.2	13.5	360
Ti-24.5Al-17Nb	952	1010	5.8	28.3	62
	705	940	10.0		
Ti-25Al-17Nb-1Mo	989	1133	3.4	20.9	476
Ti-15Al-22.5Nb	860	963	6.7	42.3	0.9
Ti-22Al-23Nb[d]	863	1077	5.6		
Ti-22Al-27Nb	1000		5.0	30.0	
Ti-22Al-20Nb-5V	900	1161	18.8		
	1092	1308	8.8		

[a] hours at 650°C/380 MPa
[b] specially processed, unaged
[c] specially processed, aged
[d] foil
Source: Froes et al. 1991; Froes et al. 1992; Froes 1993

strength. Early data suggests that the fatigue-crack growth rate (FCGR) is relatively insensitive to microstructure although the coarse Widmanstätten microstructure exhibits the slowest FCGR at low-stress intensities. Fracture toughness appears to depend on microstructure as well as alloy composition but the precise relationship is yet to be defined (Kim and Froes 1990). The decreased toughness of the Ti-25Al-10Nb-3V-1Mo alloy may result from an increase in Mo level in conjunction with a higher strength level. The higher (17%) Nb levels appear to have the capability of combining toughness with strength and creep behavior (Larsen et al. 1990). The Ti-15Al-22.5Nb alloy has the highest fracture toughness reported for an α_2-type alloy but the creep rupture is poor. The extremely low Charpy impact-strength levels of the titanium aluminides (2–3 Joules compared to ~ 25 Joules for conventional titanium alloys) is a major cause for concern in applications requiring foreign-object-damage resistance such as blades (Larsen et al. 1990).

For the alloys listed in Table 8-3, microstructural effects on creep can be as significant as compositional variations (Larsen et al. 1990). Although compositional effects are not yet clear, a recent detailed investigation on microstructure/creep behavior of Ti-25Al-10Nb-3V-1Mo has shown that the colony-type microstructure is characterized by enhanced creep resistance over other microstructures. Creep resistance of Ti-25Al-10Nb-3V-1Mo is raised by a factor of ten in the steady-state regime over the conventional titanium alloy Ti-1100 and two orders of magnitude over Ti-6Al-2Sn-4Zr-2Sn-Si (Ti-6242S). However, 0.4% creep strain is reached within 2 h. The elements Si and Zr appear to improve creep resistance, but the most significant increase is attained by increasing Al content to 25% and limiting β-stabilizing elements to about 12%. However, the Ti-24.5Al-17Nb-1Mo alloy exhibits rupture life superior to other α_2 alloys (Kim and Froes 1990). A significant trade-off between creep behavior on one hand, and ambient temperature ductility and fracture toughness on the other, generally occurs in these alloys; however, the ternary orthorhombic alloys exhibit less creep degradation at high fracture-toughness levels than the α_2 type alloys (Rowe 1993b, c). A comparison of mechanical properties with conventional materials is given in section 6.1.

3.2. TiAl (γ)

3.2.1. Chemistry. γ-TiAl alloys introduced to the present time are in the range Ti-(46-52)Al-(1-10)M, with M being at least one element from V, Cr, Mn, Nb, Ta, W, and Mo. These alloys can be divided into single phase (γ) alloys and two-phase ($\gamma + \alpha_2$) alloys. Single-phase γ alloys contain third alloying elements such as Nb or Ta that promote strengthening and additionally enhance oxidation resistance. The role of third alloying elements in two-phase alloys is to raise ductility (V, Cr, Mn), oxidation resistance (Nb, Ta), or combined properties. Additions such as B and Mg can markedly enhance hot workability (Pu et al. 1992); particulary the Mg is suggested to raise the critical stress for cracking. Also, as with the α_2 alloys, interstitial oxygen content is critical; for example, in a two-phase Ti-48%Al alloys the tensile elongation increased from 1.9% to 2.7% with an oxygen reduction from 0.08% to 0.03%

3.2.2. Processing / Microstructure Development. The microstructure of the nominally γ alloys can be single-phase γ or, in slightly leaner compositions, two-phase $\gamma + \alpha_2$. Further, by appropriate thermomechanical processing the morphology of the phases can be adjusted to be either lamellar or equiaxed, or a mixture of both morphologies.

The lamellar structure can lead to an effective refinement of the microstructure and improved ductility either directly or in combination with fine, recrystallized γ grains. Improved ductility occurs at about 10 vol% α_2; when the α_2 phase content exceeds 20 vol%, ductility can be degraded. This is consistent with the fact that α_2 becomes increasingly more brittle with increasing Al content over 25% and the α_2 plates have a composition of approximately 35% Al.

Currently, two-phase alloys appear to offer the greatest potential for development of a useable material. Here great flexibility exists in controlling the ratio of lamellar to equiaxed γ phase (Kim 1989, 1991) along with their distribution. A lamellar volume fraction of about 30% gives a reasonable combination of properties with good high-temperature creep resistance and acceptable tensile strength and useable ductility. The optimum structure for enhanced ductility has been suggested to consist of adjacent regions of lamellar and equiaxed morphologies (Froes et al. 1991, 1992). TMP refines the microstructure when processing is conducted such that both the α and γ grains are recrystallized in the $(\alpha + \gamma)$ phase field. Grain morphology varies considerably depending on composition, solution treatment temperature and time, cooling rate, and stabilization temperature and time (Kim 1989). Presently, four classes of microstructures have been defined and are receiving considerable attention (Kim 1989, 1992; Kim and Dimiduk 1991, Pu et al. 1992). Examples of these microstructures for a Ti-46.5Al-2.5V-1Cr alloy are shown in Figure 8-5 (Pu et al. 1992).

The grain size decreases with reduced Al level and with additions of W, V, Mn, and Cr. The number of annealing twins in the γ phase increases as Al content decreases or when Mn or V levels are increased while Cr additions increase the volume fraction of the lamellar structure. Results from material with extremely small nanoscale grains exhibit high room-temperature elongation up to 5%, clearly an area worthy of further study.

3.2.3. Deformation and Fracture. The deformation behavior of γ alloys is anisotropic and emphasizes the tetragonality (Fig. 8-6), albeit low (Whang and Hahn 1990). The $L1_0$ alloys show an anomalous increase in flow stress as temperature increases, with an asymmetry of the peak position, similar to other ordered intermetallics with the $L1_2$ and B2 structures. This has been shown in single crystal $Ti_{45}Al_{55}$ and polycrystalline $Ti_{40}Al_{55}Nb_5$, $Ti_{35}Al_{55}V_{10}$ alloys. The dislocation behavior was studied during the mid-70s, but only recently has detailed identification and behavior of the dislocations been carried out. The ordered nature of TiAl (successive (002) planes are made up entirely of Ti and Al atoms) and its slight tetragonality have important implications on the available Burger's vectors and the different stacking faults produced by partial dislocations.

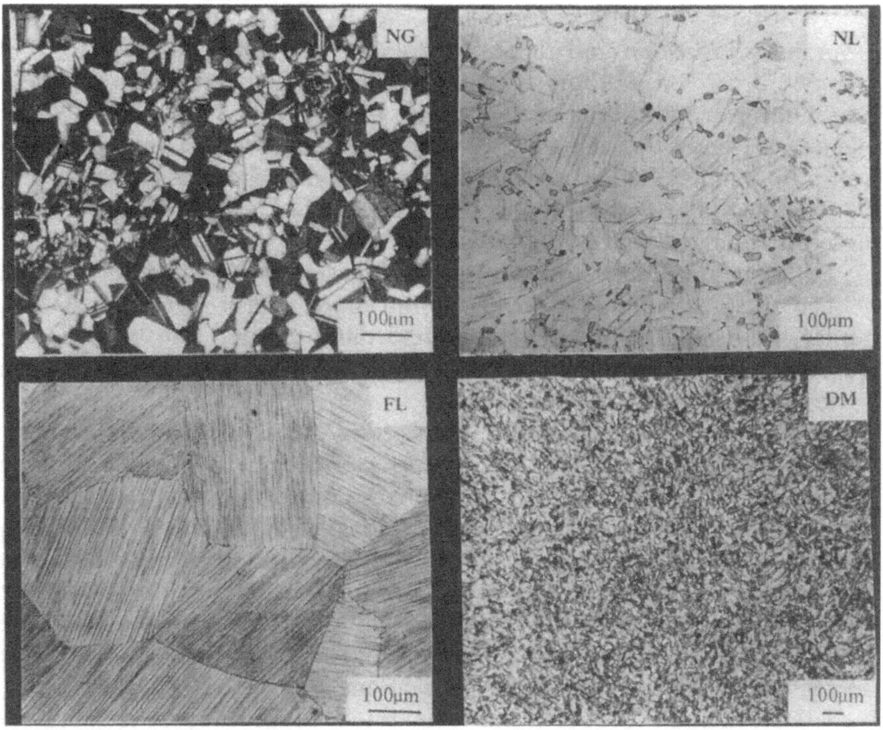

Figure 8-5. Variety of microstructures in Ti-47.5Al-2.5V-1Cr (γ), see Table 8-4.

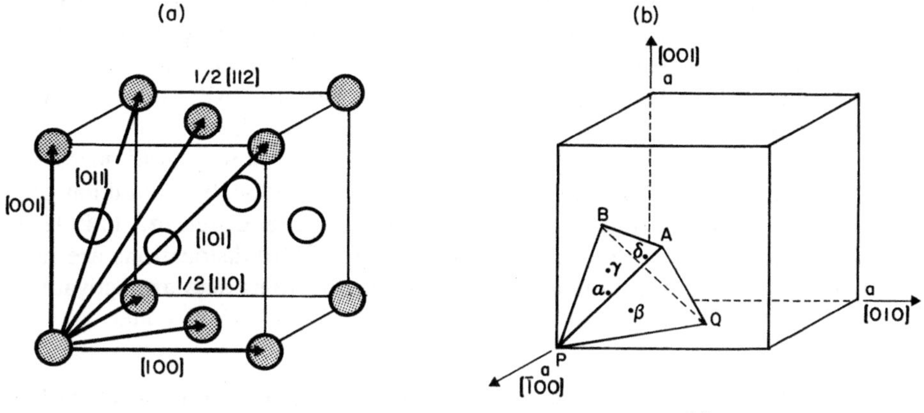

(a) (b)

Figure 8-6. (a) Crystal structure and Burgers vectors in the TiAl (γ) phase and (b) Modified Thompson tetrahedron for the $L1_0$ structure.

On the basis of energetic considerations, 1/2⟨110]{111}[2] is the most favorable slip system and these dislocations have been observed in TiAl samples deformed at room temperature and higher temperatures; these were, however, not observed in samples deformed at 77 K. On the other hand [011] and [101] slip destroys the ordered structure and creates a series of complex faulted structures, implying that slip of these types may be feasible only if dislocations move in pairs so as not to disturb the ordered structure, that is, superdislocations.

The deformation modes in the TiAl-base alloys seem to depend on Al content (i.e., whether a small amount of α_2 phase is present or not), ternary alloying additions, temperature, and purity. The temperature dependence of deformation mode and dislocation structures have been reviewed by Whang and Hahn (1990).

Only ⟨011] and 1/2⟨112] dislocations were observed in samples deformed at 77 K and 1/2⟨110] dislocations were not seen, while all three types were observed on deforming the samples at room temperature. Dissociation behavior of these dislocations and formation of stacking faults and antiphase boundaries have been investigated by the weak-beam electron microscopy technique because of the short fault widths. Although 1/2⟨110] dislocations have been observed, they do not seem to dissociate into Shockley partials. Increasing the purity of the alloy appears to increase the density of 1/2⟨110] dislocations and the ductility.

Deformation of TiAl in the temperature range of room temperature to 600°C (for single crystals and 800°C for polycrystals) is accompanied by strain hardening. Hence, the 1/2⟨110] dislocations are curly and tangled due to massive cross slip during deformation. The ⟨011] dislocations appear linear at 800°C and dissociate into one superpartial and two Shockley partials and the APB lies on a (100) plane, while the APB on room-temperature deformation lies on (111); ⟨101] dislocations and their dissociation have also been observed. Further ⟨101] dislocations cross slip onto other {111} planes; evidence also exists that the APB between two superpartials lies on a {100} plane instead of a {111} plane.

At temperatures > 800°C, TiAl and its alloys show a rapid decrease in strength, probably due to lattice softening and dislocation annihilation. The dislocation morphology at this stage is quite different; 1/2⟨112] dislocations are long and stretched and the 1/2⟨110] dislocations are not tangled or curled. Further, new ⟨100]-type dislocations also are activated.

---

[2] We will use the mixed notation ⟨hkl] to indicate a set of equivalent crystallographic directions which are generated by permutations of $\pm$h and $\pm$k; only a sign change is allowed for l.

These observations suggest that the positive temperature dependence of yield stress may be explained on the basis of either (a) $1/2\langle 110](111\}$ slip dislocations becoming entangled at an intermediate temperature, or (b) a cross-slip model involving $\langle 101]$ and $1/2\langle 112]$ dislocations, or (c) dislocation–dislocation pinning.

Twinning has been found to help in increasing the ductility of TiAl alloys with $\{111\}\langle 11\bar{2}]$-type twins being most commonly observed. The tendency of increased twinning above the ductile-brittle temperature suggests that it is responsible for the increased ductility.

A comprehensive study of the deformation and fracture behavior of two-phase γ-TiAl indicated the strong influence of microstructure on both monotonic and cyclic loading conditions (Kim 1993). Ductility depends on grain size, being limited by strain-incompatibility-induced microcracks. The fracture toughness, highest with the lamellar structures, increases with high plastic anisotropy and bridge-ligament formation. The author rationalized the inverse ductility/toughness relationship in terms of the general tensile yielding and near-tip local plasticity at the crack-initiation toughness. For both the duplex (DM) and fully lamellar (FL) microstructures, plastic deformation is followed by cleavage-like fracture. The DM material fails transgranularly at ambient temperatures and intergranularly at elevated temperatures. The FL material also exhibits transgranular fracture, with a reduction in grain size favoring increased intergranular fracture.

3.2.4. Mechanical Properties. Alloys based on the TiAl(γ) composition have higher elastic modulus, lower density, enhanced elevated-temperature capabilities, and are less likely to ignite than Ti_3Al (α_2) alloys. However, generally room-temperature ductility and fracture resistance can only be classed as "poor" (Larsen et al. 1990). General trends are for a slight decrease in stoichiometry, into the two phase $\gamma + \alpha_2$ region, to increase ductility (Cahn 1989), with small (1–3%) V, Mn, and Cr additions enhancing ductility (Fleischer, Dimiduk, and Lipsitt 1989; Kim 1989, 1993), or Nb, Ta, and Mn improving strength and oxidation resistance.

The Ti-52%Al alloy demonstrates the lowest hardness value at room temperature independent of the TMP treatment. At 1000°C, however, the strength tends to gradually decrease with increasing Al levels. Tensile strength and hardness vary in the same fashion with changes in aluminum content. The room-temperature (RT) tensile elongation is maximum at approximately Ti-48%Al.

The microstructural variation shown in Figure 8-5 results in quite different combinations of mechanical properties, Table 8-4 (Pu et al. 1992). There is a reverse relationship between ductility and fracture

Table 8-4 Microstructure and Mechanical Properties in Ti-46.5Al-2.5V-1Cr TiAl-Type Titanium Aluminide

	FL	NL	DM	NG
YS, MPa	360	430	440–450	387
UTS, MPa	400	480	505–538	468
El., %	0.5	2.3	3.3–4.8	1.7
K_{IC}, MPa$\sqrt{m}$	21	17	12	17

FL: Fully Lamellar
NL: Near Lamellar
DM: Duplex
NG: Near Gamma

Source: Pu 1992

toughness with the duplex microstructure with the finest grain size exhibiting the highest ductility but the lowest fracture toughness, and vice versa for the large-grained, fully lamellar structure. It has been suggested that the best combination of properties may well occur with an "engineered" (Kim and Dimiduk 1991) microstructure of reduced lamellar grain size (Fig. 8-7).

A nanostructure-scale microstructure can increase ductility to 5% and at the same time decrease the ductile-to-brittle transition temperature by as much as 200°C (Froes et al. 1991).

Low-cycle fatigue experiments suggest that fine grain sizes raise the fatigue life at temperatures below 800°C (Froes et al. 1991, 1992). High-cycle fatigue is optimized with a duplex microstructure at low temperatures, while at higher temperatures (> 800°C) a lamellar structure gives a longer fatigue life. Fatigue-crack growth rates, which are slowest for a fully lamellar structure, are more rapid than those for superalloys, even when density normalized. Impact resistance, like fracture toughness, is highest for a fully lamellar structure (~ 3 Joules) (Froes et al. 1991, 1992).

Creep properties are improved beyond the behavior of superalloys when normalized by density, but are strongly influenced by alloy chemistry and TMP. For example, increased Al content and additions of W or C increase creep resistance. In the single-phase γ region (Ti-53.4%Al) there appears to be no influence of grain morphology on creep behavior. However, at leaner Al contents (Ti-50%Al) a small amount of second phase was observed along with enhanced creep behavior and a clear improvement in creep behavior with increasing grain size. This latter work suggested that creep in TiAl is not simply diffusion controlled as it is in

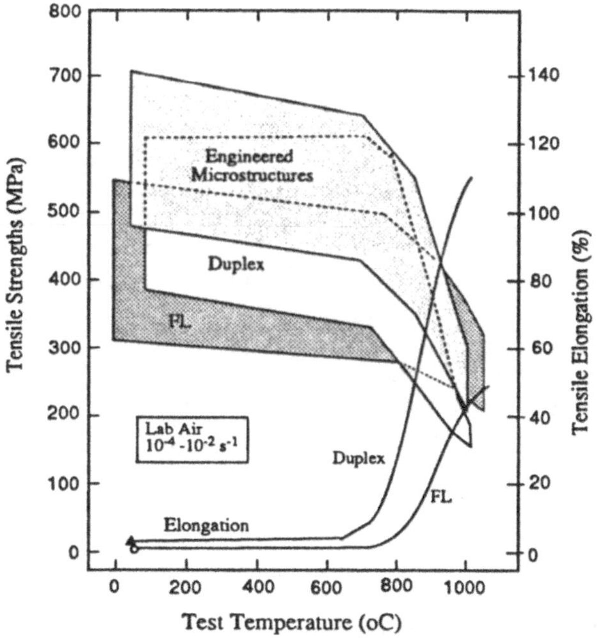

Figure 8-7. Tensile properties of two-phase γ alloys in three different conditions, shown as bands at temperatures up to 1000°C. The upper limit of each band is the highest UTS, and the lower limit the lowest YS.

normal metals. Increasing the volume fraction of the lamellar structure, particularly with serrated grain boundaries, enhances creep properties but lowers ductility. The level of creep strain occurring due to elongation on initial loading and primary creep is a concern since it can exceed projected design levels for maximum creep strain in the part (Larsen et al. 1990).

4. SYNTHESIS / PROCESSING

4.1. Ingot and Casting

Melting, solidification, and processing of the α_2 compositions, while not easy, are an extension of the melting parameters used for conventional terminal titanium alloys. However, the extremely low ductility of γ compositions has necessitated the use of innovative practices, including much smaller than normal ingots for the γ compositions (200 kg) and the exercise of great care in cooling of the ingots to avoid cracking due to thermal stresses. Subsequent processing of γ ingots to various mill product forms has then been carried out with a great deal of caution: higher temperatures, smaller reductions, and increased conditioning compared to conventional alloys.

Near net shape (NNS) cast components are particularly attractive for titanium aluminides, especially the γ compositions, since difficult working operations are eliminated. For the α_2 composition, casting parameters are close to those for the Ti-6Al-4V alloy, despite some flow problems. The use of the TCP method (see below) will allow refinement of the microstructure of the α_2 composition at least, but no reported studies have yet been carried out in this area.

A fruitful area for further study is the use of small additions of yttrium or other rare earth elements to enhance the workability and ductility of both the α_2 and γ titanium aluminides. Small amounts of Y for example (< 100 ppm), added in a highly controlled fashion, should lead to numerous advantages, including increased metal yield, reduced cost, improved inspectability, and overall mechanical-behavior enhancement.

4.2. Powder Metallurgy

The powder metallurgy (PM) approach is used in general terms to either reduce cost or enhance behavior, and well-established, cost-effective techniques that have been developed for the production of NNS conventional titanium alloys, such as Ti-6Al-4V, are readily adapted for use with both families of aluminides (Froes and Eylon 1990a; Froes and Suryanarayana 1993). Techniques that allow improved properties such as rapid solidification (Jones 1982; Suryanarayana, Froes, and Rowe 1991; Arantharaman and Suryanarayana 1987) and mechanical alloying (Suryanarayana and Froes 1993) can also be used with the titanium aluminides. Additionally, a number of other PM methods, such as reactive processing, have been applied to these alloys.

4.2.1. Conventional PM. Because of the extreme difficulty in fabricating the γ compositions, a blended elemental (BE) approach followed by extrusion to bar product was used. Further optimization is required with consideration given to homogeneity, the effect of remnant chloride/voids, and the various phases which have been detected.

Early work using γ compositions and prealloyed (PA) powders was partially successful with problems due to porosity, coarse grains, and cracks. More recently, parts from both families of titanium aluminides have been produced utilizing the plasma rotating electrode process (PREP) and gas atomization (GA) (Moll, Yolton, and McTiernan 1990) methods (see applications section). The mechanical properties of compacts were equivalent to those attained by ingot techniques.

4.2.2. Rapid Solidification. Rapid solidification (RS) processing of the titanium aluminides offers the potential of improved ductility by disordering, grain refinement, and deoxidation of the matrix, and enhanced elevated temperature properties due to a dispersion of fine, thermodynamically stable second-phase particles (Froes and Eylon 1986; Suryanarayana et al. 1991).

α_2 ALLOYS. Rapid solidification of Ti_3Al-Nb alloys with rare earth additions such as Er result in a refinement in the grain size and a dispersion of Er_2O_3 particles, Figure 8-8 (Kaufman et al. 1989). Some increases in room-temperature ductility and elevated temperature performance occur. However, in general it must be concluded that in comparison

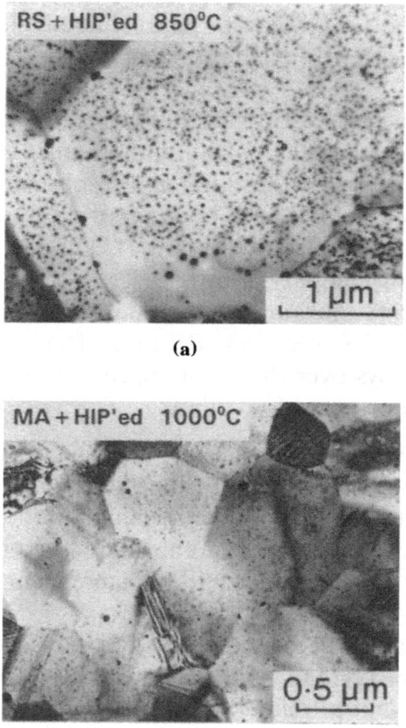

(b)

Figure 8-8. Microstructure of (a) RS and (b) MA Ti_3Al + Er, HIP'd at 850°C and 1000°C, respectively.

with aluminum alloys in which a dispersion of as much as 25–30 vol% can be obtained, with resulting significantly enhanced elevated temperature performance, the 1–3 vol% rare earth oxide obtained in RS Ti_3Al alloys (and Ti-based alloys in general) is quite disappointing (Suryanarayana et al. 1990).

γ ALLOYS. The γ phase forms by a peritectic reaction, therefore RS can lead to a large improvement in alloy homogeneity and formation of metastable phases and microstructures. RS TiAl alloys often contain metastable α_2, and evidence of metastable disordered TiAl has been observed (Graves et al. 1987; Mishurda et al. 1989; Shih, Scarr, and Chesnutt 1989). Studies of the solidification of RS Ti-50%Al powder indicated some dendrites with sixfold symmetry (Graves et al. 1987) that was initially thought to be due to undercooling of the melt; however, this has now been shown to be the equilibrium solidification path for some TiAl alloys.

The homogeneous and fine microstructure of RS alloys leads to improvements in the ductility of γ alloys at room temperature. Further, alloys in the composition range of approximately 48–60%Al processed by RS showed a consistent advantage over equivalent IM material (Huang 1989).

Ternary additions of Co, Cr, Fe, Ge, Mn, Ni, V and W have been made to RS TiAl-based alloys (Hall and Huang 1989). Co, Fe, Ni and W additions result in the greatest hardening observed in TiAl alloys, and additions of Cr, Mn, and V have been reported to improve the room-temperature ductility of RS Ti-48%Al alloys (Hall and Huang 1989) to 3.5%, up from $< 1\%$ in the binary alloy.

Thus it appears that, in combination with the near-net shape advantages offered by PM fabrication, RS may offer some advantages for the processing of γ alloys over the ingot approach; presently the same cannot be said for RS processing of α_2 alloys.

4.2.3. Mechanical Alloying. The mechanical alloying (MA) powder metallurgy approach allows even further departures from equilibrium than RS, including extended solubility, production of novel crystalline and amorphous phases, refinement of the microstructure and dispersoids of second-phase particles (Froes and Suryanarayana 1993; Suryanarayana and Froes 1993).

Mechanical alloying of a number of α_2 alloys including Ti-24Al-11Nb and Ti-25Al-10Nb-3V-1Mo, with and without rare earth additions, resulted in a refinement in grain size, especially when a large amount of the B2 phase was present (Ti-25Al-10Nb-3V-1Mo), which persisted even after exposure at 1090°C for 6 h. Further, little dispersoid coarsening occurred, and there were no dispersoid free zones close to grain boundaries as observed in RS alloys, Figure 8-8 (Suryanarayana and Froes 1993). However, MA conditions were clearly not optimum in this work, with mixed grain sizes and relatively coarse rare-earth oxide particles, and further studies are necessary.

Amorphous and nanoscale grains have been obtained by MA in both α_2 and γ compositions. In both systems the nanoscale grains were maintained after compaction, for example the nanostructured grains shown in Figure 8-9 are present in the Ti-24Al-11Nb and Ti-55Al alloys after HIP'ing at 975°C, 207 MPa, for 2 hours (Suryanarayana et al. 1993). Mechanical alloying has also been used in conjunction with thermochemical processing to synthesise the TiAl composition from a mixture of Al_3Ti and TiH_2 (Suryanarayana, Sundaresan, and Froes 1992).

4.2.4. Spray Deposition. In spray deposition, finely divided molten metal droplets impinge on a substrate before they completely solidify, allowing direct formation of near-net shapes rather than having to go through the intermediate stage of powder production. This process also offers some advantage in terms of solidification rates, but these tend to be slower than for RS processes.

Preliminary work has shown that spray deposition can be used for the titanium aluminides with appropriate equipment modifications to accommodate the reactivity of molten titanium (Fig. 8-10a); the product of spray deposited Ti-25Al-10Nb-3V-1Mo is shown in Figure 8-10b (Froes 1993).

4.2.5. Other Powder Metallurgy Techniques. The reaction sintering and hot isostatic pressing of prereacted powders were studied as potential synthesis methods for TiAl starting from blended elemental powders. While the reaction sintering was unsuccessful, the prereacted approach gave a fully homogeneous TiAl matrix, with a dispersion of alumina particles, and properties at least as good as those obtained from a prealloyed approach. This could well lead to a reduced-cost PM processing route for TiAl.

4.3. Thermochemical Processing (TCP)

In the thermochemical processing (TCP) technique hydrogen is added as a temporary alloying element to titanium-based materials to enhance processability and refine the final microstructure, the latter characteristic leading to significantly improved mechanical properties, including those which are crack-initiation related, such as fatigue (Froes and Eylon 1990b; Froes et al. 1990b). This process is particularly useful for refinement of the

(a)

(b)

Figure 8-9. Nanostructured grains in (a) Ti-24Al-11Nb and (b) Ti-55Al compacts, mechanically alloyed and HIP'ed at 975°C, 207 MPa, 2 hours.

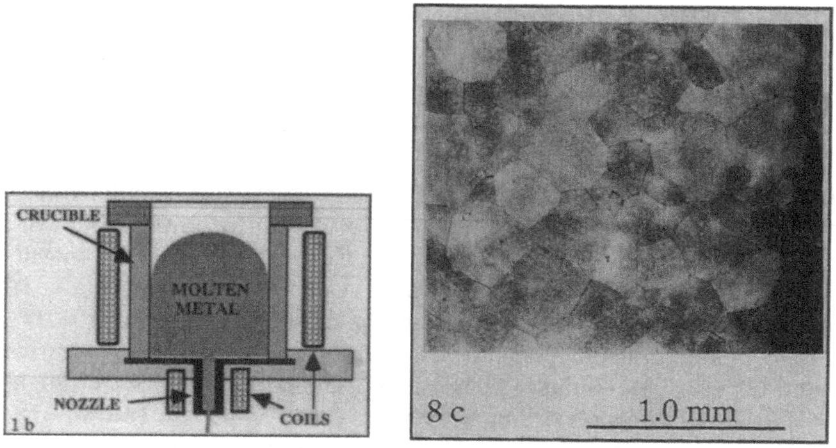

(a) (b)

Figure 8-10. (a) Schematic of spray deposition equipment, and (b) resulting microstructure in Ti-25Al-10Nb-3V-1Mo near substrate.

microstructure with near-net shape (NNS) fabrication methods, such as castings or powder metallurgy, where the working operation normally used to reduce the scale of the microstructure in wrought product cannot be used.

Preliminary studies on the applicability of TCP to the α_2 alloy Ti-24Al-11Nb demonstrated that plasma rotating electrode process (PREP) powder could be successfully compacted by hot isostatically pressing (HIP'ing) at 100°C, or more, lower temperatures with hydrogen containing powder than with normal powder. Additionally, a significantly refined microstructure was obtained. Full evaluation of the advantages to be gained with TCP of the α_2-type alloys remains to be studied; however, the TCP technique has been used on a semicommercial basis in the Soviet Union to process α_2 and other difficult to fabricate titanium alloys.

The face-centered tetragonal γ ($L1_0$ structure) has virtually no solubility for hydrogen under generally applied temperatures and pressures (Froes and Eylon 1990b). However, stabilization of the α_2 phase using RS (Graves et al. 1987) can allow hydrogen to enter the structure of a nominal TiAl composition, thus opening up the possibility of using TCP with this composition. Preliminary work has also demonstrated that hydrogen can be used to assist in the formation of TiAl by mechanical alloying (Suryanarayana et al. 1992). Clearly much further work is required to define the potential of TCP with the γ-type alloys.

4.4. Designer Microstructures

One method of improving "forgiveness" of ceramic materials is by dispersion of a ductile phase in the brittle ceramic matrix. This concept has recently been applied to the γ titanium aluminides by blending Ti-36%Al PREP powder with either powders of commercially pure titanium or Ti-50%Nb. No improvement in compressive strain or toughness was recorded, perhaps because of microporosity, but metallographic work indicated that the second-phase ductile particles arrested cracks propagating through the γ matrix. Other work using Ti-6Al-4V (wt%), a β titanium alloy, or a niobium-containing alloy powder at the 10–20% volume level in a γ matrix resulted in increased toughness when excessive reaction between the ductile phase and the matrix was minimized; a Ti-30%Nb ductile phase working very well. Clearly this is a fruitful area for further exploration with developments in the ceramics area serving as useful guidelines to follow.

5. COMPOSITE CONCEPTS

5.1. Concept

In a composite concept, a second phase, which can have a particulate, short fiber, continuous fiber, or lamellar type configuration (Chawla 1987) is introduced into the matrix to enhance physical and mechanical behavior. Generally, the type of reinforcement that has been used with the intermetallic titanium aluminides has been continuous fibers, although considerable attention has also been given to a novel "in situ" concept using the γ alloys. Because of the low ambient-temperature ductility of the γ compositions, almost all work with continuous fibers has been on α_2 compositions of Nb-equivalent ranging between 10% and 30%. However, preliminary work is in progress on γ compositions reinforced with SiC fibers, produced by chemical vapor deposition, using coatings such as Hf, Ta, and Zr. The major attributes of the continuous fiber approach are that elevated temperature strength and stiffness are enhanced, and at the same time the fibers can provide increased toughness by at least three mechanisms (Chawla 1987), all of which involve additional work:

- plastic deformation of the matrix
- fiber pull-out
- presence of weak interfaces/fiber separation, and deflection of the crack.

This latter attribute of the MMC approach could well play a major role in application of titanium aluminides using "damage tolerant" designs in which fracture toughness and fatigue crack growth rate are more important than ductility (Dimiduk and Miracle 1989); some cast superalloys in service have ductility levels of less than 2% at room temperature.

5.2. Continuous Fibers

The advantages to be gained by embedding ceramic fibers in a titanium matrix have received considerable attention (Smith and Froes 1984). A density-normalized plot of the SiC/Ti_3Al + Nb data versus wrought superalloy data (René 41 and Hastelloy X) indicating the attraction of the intermetallic system is shown in Figure 8-11.

*5.2.1. **Composite Fabrication.*** There are a number of techniques that have been used for composite fabrication, including hot isostatic pressing, vacuum hot pressing, arc spraying, plasma spraying, powder cloth technique, woven fiber mat, and electron beam vapor deposition (Ward-Close and Partridge 1990).

*5.2.2. **Composite Characteristics.*** The desirable characteristics of a reinforcing fiber include compatible coefficient of thermal expansion (CTE) with the matrix, chemical compatibility with the matrix even after extended elevated temperature exposures, low density, high modulus of elasticity, high strength at elevated temperature, good oxidation resistance, 50–250 μm diameter continuous monofilament, and mass production feasibility (including affordable cost).

The first two items depend on the behavior of both the fiber and the matrix, whereas the latter six of these characteristics are attributes of the

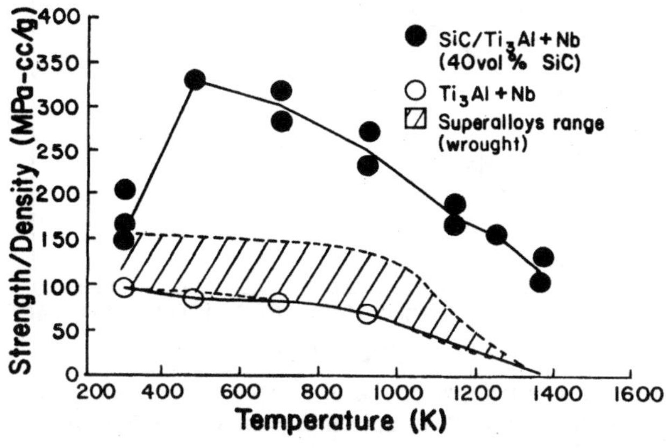

Figure 8-11. Comparison of σ/ρ values of SiC/Ti_3Al + Nb, Ti_3Al + Nb and wrought René 41 and Hastelloy X superalloys for a range of temperatures.

fiber itself. Silicon carbide fibers with the required last six characteristics, are currently available from at least three sources (one commercial), and a number of other fibers (such as Al_2O_3) are in various stages of development. However, the compatibility issues of CTE mismatch and chemical interaction are not yet resolved.

Measured CTE values for two fibers (SiC [SCS-6] and Al_2O_3) and two representative matrices (α_2 and γ) are shown in Figure 8-12. The CTE for the aluminides is more than twice that of the SCS-6 fiber, while the CTE of the Al_2O_3 fiber more closely approaches that of the potential matrix materials. This CTE difference means that upon cool-down after consolidation, significant stresses occur which can degrade behavior.

Chemical reaction between the fiber and the matrix may be the major reason for less than ROM for many of the mechanical properties in titanium matrix composites (Fig. 8-13), with increasing thickness of the complex reaction zone being related to a general degradation in mechanical properties. This relationship has led to a considerable amount of work on fundamental thermodynamic studies of chemical compatibility using

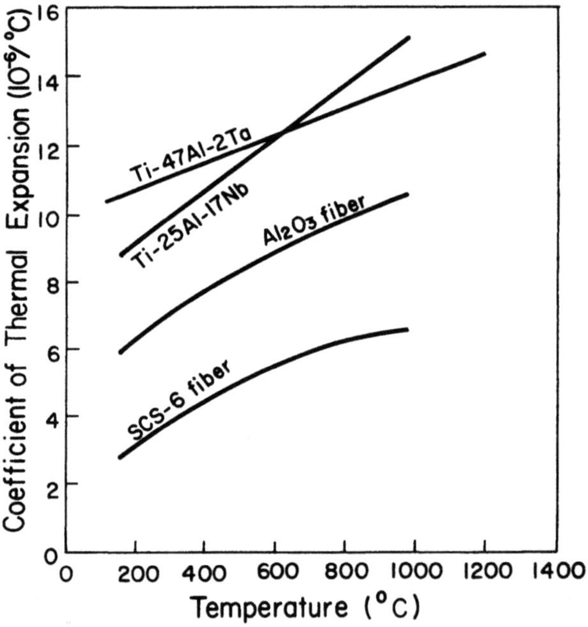

Figure 8-12. The coefficients of thermal expansion of fiber reinforcements and titanium aluminide matrix alloys determined at various temperatures by thermomechanical analysis.

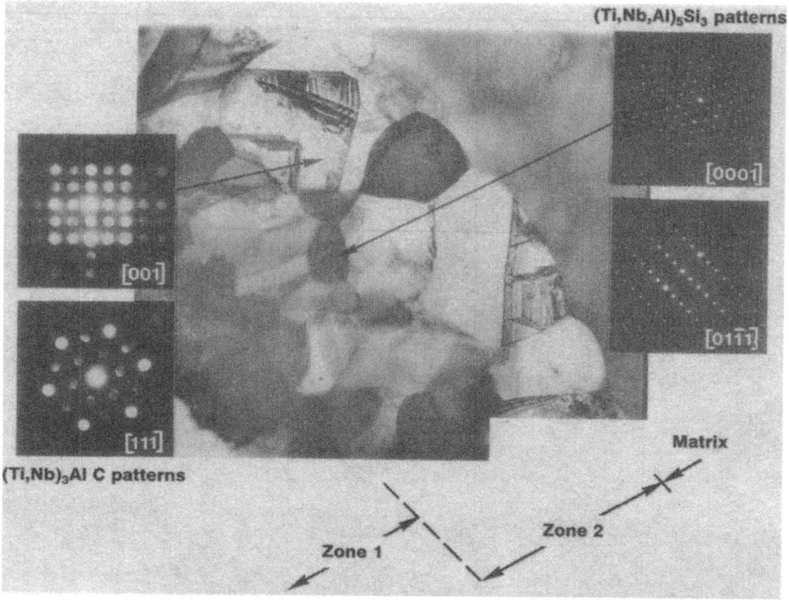

Figure 8-13. High magnification micrograph of the reaction zone. Diffraction patterns on left are from the (Ti,Nb)$_3$AlC phase, which are seen as faulted grains in the micrograph. Patterns on the right are from the (Ti, Nb, Al)$_5$Si$_3$ particles, one of which is indicated with an arrow.

phase diagrams where minimal reaction of the titanium aluminides with Al_2O_3, TiB_2, CaO, Ti_5Si_3, Y_2O_3, rare earth oxides and sulfides, and TiN has been reported. Factors such as trace elements and crystal structure clearly play a role in explaining discrepancies between various studies.

In studies of actual titanium-matrix fiber composites it has been demonstrated that less reaction takes place between fibers such as SCS-6 and Ti_3Al, than with the conventional alloy Ti-6Al-4V (wt%) (Smith and Froes 1984); and that with increasing Nb addition to the Ti_3Al the reaction zone width decreases. Considerable reaction also occurs between TiAl and SCS-6 particularly with the higher fabrication temperature used with this alloy. No reaction was reported between Al_2O_3 fibers and a Ti-45Al-2Ta (γ) matrix, which in combination with the lower CTE mismatch makes this an interesting system; surface flaws on the Al_2O_3 may necessitate a coating for mechanical compatibility with the matrix.

Comprehensive studies using fiber push-out techniques indicate that the room-temperature chemical bonding (119 MPa shear stress) is more than twice as high as the frictional bonding due to CTE differences

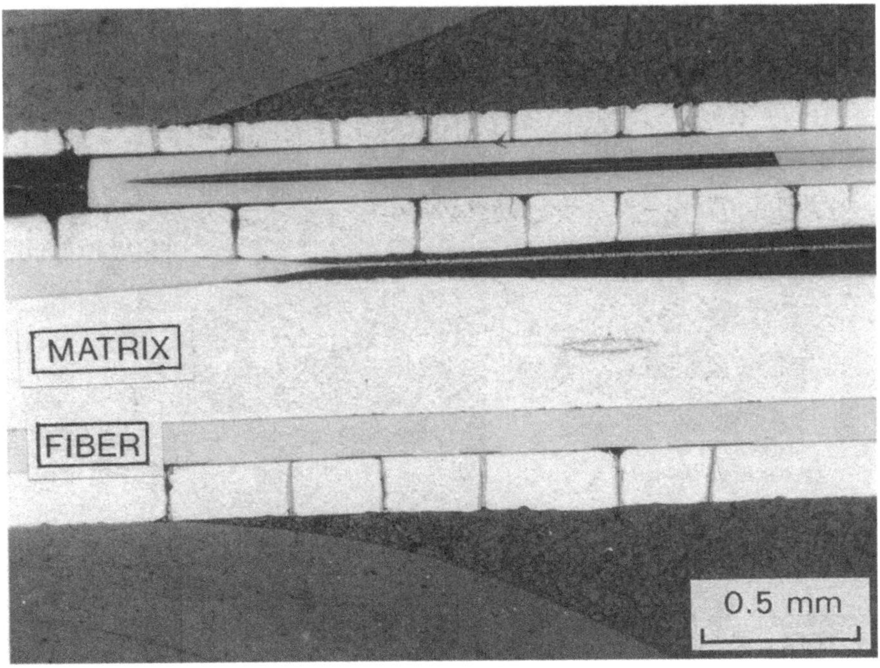

Figure 8-14. Internal transverse matrix cracking in a Ti-24Al-11Nb/SCS-6 composite after experiencing 500 cycles in the temperature range of 150°C to 815°C.

(48 MPa) in a Ti-24Al-11Nb/SCS-6 composite; no significant reduction in strength was noted, with debonding occurring at the remaining coating/reaction zone interface.

The major problem with fiber/matrix incompatibility is illustrated under conditions of purely thermal fatigue, which led to residual stresses from moduli and CTE mismatches and extensive matrix damage in a Ti-24Al-11Nb/SCS-6 composite (Fig. 8-14 (Russ 1990)). Clearly, with the brittle matrices involved, not only is a diffusion barrier necessary to limit interfacial chemical reactions, but also compliant layer coatings to accommodate CTE differences; finite element analysis studies are in progress to better define requirements in this area (Upadhyaya et al. 1993).

Very recent work with "orthorhombic" composites (Ti-22Al-23Nb/SCS-6) has indicated enhanced behavior compared to Ti-24Al-11Nb composites; particularly transverse tensile strength and the lack of a reduction in tensile strength after thermal cycling in an air environment (Smith et al. 1992; Graves et al. 1992). These improvements were related to the reduced reaction between the fiber and the matrix and the change in phase proportions (Ti-24Al-11Nb consists of 90 vol%α_2, 10 vol%β, the Ti-22Al-23Nb 40 vol%α_2, 60% mixture of orthorhombic plus β).

Based on thermodynamic data coatings of TiC, TiB_2 and CaO were applied to Sigma SiC and α-Al_2O_3 fibers prior to insertion into a γ (Ti-47Al-2Ta) matrix. Little or no reaction with TiB_2, CaO or straight—Al_2O_3 was noted. However, TiC was found to react with the SiC. A study of Pt + W coating on SCS-6 fibers in a γ (Ti-48Al-1V) matrix indicated virtually no reaction, suggesting that this combination may help to solve both the CTE and reactivity problems. Many other coatings are possible; however, the ultimate test will be to investigate thermal cycle effects, and the mechanical properties of the composite. Another, but perhaps even more challenging solution is to develop a more "forgiving" matrix (composition and/or microstructure) while maintaining good elevated temperature properties; it could well be that the high fracture-toughness α_2 alloy (42.3 MPa$\sqrt{m}$, Table 8-3) could be optimized as a matrix for "forgiveness" and environmental resistance, letting the fibers bear the stresses—the concept used in polymeric matrix composites (Chawla 1987).

At the present time there is only one fully commercial source of SiC fiber partially optimized for use in a titanium matrix—the 150 μm SCS-6 fiber, which is produced by chemical vapor deposition with a coating of carbon-rich nonstoichiometric amorphous SiC on top of amorphous carbon, with a total coating thickness of about 2.5 μm. This coating blunts the intrinsic surface flaws resulting from CVD, and greatly reduces the chemical interaction between the fiber and the matrix. A second 100 μm SiC-based fiber is rapidly approaching commercial production status, with three alternate surfaces being offered—uncoated, structured carbon-coated and titanium diboride-coated. Coated SiC fibers are also available in limited quantities from the CIS. Other fibers such as B_4C/B, Borsic and B-fibers appear unsatisfactory for use in a titanium matrix (Smith and Froes 1984). However, another alternative which is also being evaluated is Al_2O_3 fibers.

5.2.3. Mechanical Properties. The mechanical properties of SCS-6/(Ti-24Al-11Nb) have recently been reviewed in detail (Larsen et al. 1990). Longitudinal strengths are generally close to ROM values with the behavior being controlled by the fibers, but the transverse properties are significantly lower with the strength being controlled by the matrix capability, creep-mirrored tensile behavior with the rupture stress in the longitudinal direction being an order of magnitude better than at right angles to the fiber direction. Fatigue-crack initiation behavior appeared to be controlled by the matrix strain-life characteristics; with notches degrading life

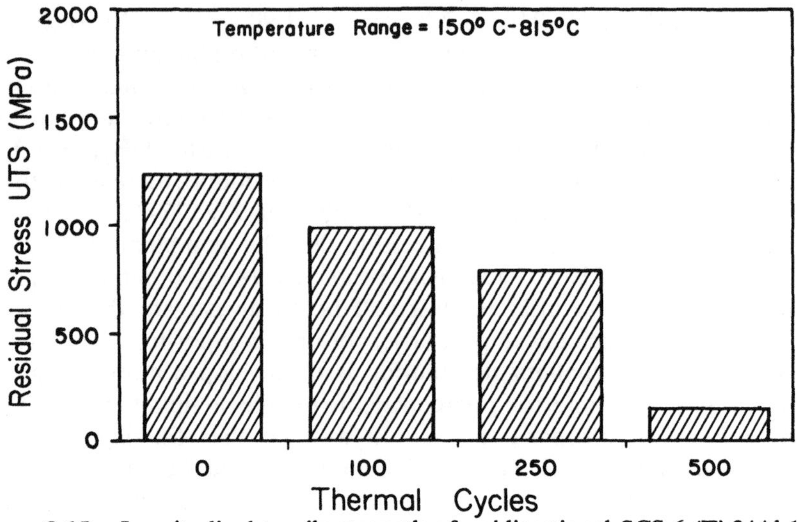

Figure 8-15. Longitudinal tensile strength of unidirectional SCS-6/Ti-24Al-11Nb composite versus number of prior 150°C to 815°C thermal cycles in air.

in the longitudinal direction, but not in the (much lower) transverse direction. This suggests that under transverse loading the fibers act as potent notches. Under cycling loading cracks propagated five orders of magnitude faster in the fiber direction than orthogonal to the fibers. The stress intensity at fracture in the transverse direction ranged from 14–19 MPa$\sqrt{m}$, close to the fracture toughness of monolithic material. However, in the longitudinal direction substantial toughening occurred, attributable to crack bridging, with toughness values ranging from 110–150 MPa$\sqrt{m}$ at room temperature. Thermal fatigue is a great concern since cycling in air can reduce initial strength by almost 90%, Figure 8-15 (Russ 1990). This degradation has been attributed to heavily oxidized surface cracks (Smith and Revlos 1990), which do not occur in vacuum. Thus, while the composite α_2 material does demonstrate enhanced behavior in some areas, several concerns remain, including transverse properties, environmental resistance, and designing with a material of limited ductility.

Some of the problems mentioned above appear to be solved at least to an extent in the "orthorhombic" composites (Smith et al. 1992; Graves et al. 1992). A reduction in the reaction zone and change in the phase proportions resulted in enhanced transverse strength and removal of the thermal fatigue debit compared to the Ti-24Al-11Nb.

The even lower inherent "forgiveness" of the γ compositions has precluded the development of meaningful mechanical property data from composite concepts to date.

5.3. Exothermic Dispersion (XD^{TM})

Production of a dispersion of thermodynamically stable "in situ" second-phase particles, using an exothermic reaction (Christodoulou and Brupbacher 1990), has been evaluated with a γ matrix. In this process the interface is free of reaction products or externally developed contamination such as an oxide layer which should avoid easy decohesion of the reinforcing phase, and early failure. The microstructure of the XD^{TM} γ containing the TiB_2 phase can be controlled both in terms of matrix grain size and morphology (lamellar or equiaxed), and the shape of the reinforcing particles, resulting in material with enhanced room-temperature and elevated temperature strength. The best room-temperature ductility is obtained from an equiaxed matrix morphology in combination with particulate reinforcement ($\sim 1\%$ elongation), while a lamellar/short fiber combination optimizes strength, creep resistance and fracture toughness (17 $MPa\sqrt{m}$). Various mill products are available including forgings, extrusions, investment casting, powder metallurgy, and rolled sheet (Christodoulou and Brupbacher 1990) and secondary fabrication techniques such as superplastic forming, welding, and brazing/diffusion bonding have been demonstrated (Christodoulou and Brupbacher 1990). Clearly the XD^{TM} approach is worthy of further development, perhaps using some of the knowledge developed in other theoretical (Christman, Needleman, and Suresh 1989) and experimental studies regarding the effects of second-phase reinforcements in γ on mechanical behavior.

6. TECHNOLOGY

6.1. Mechanical Property Comparisons

Comparisons of monolithic and composite Ti_3Al materials and monolithic TiAl with conventional materials are shown in Figure 8-16 (Larsen et al. 1990).

The ductility of α_2-based alloys has been increased particularly with additions of Nb (Rowe 1990). However, Charpy impact toughness values are low, raising questions surrounding impact resistance (e.g., foreign object damage) (Larsen et al. 1990). Surface cracking can occur after long time exposure to static or cyclic loading, reducing overall ductility. However, fatigue life can be enhanced with the use of protective coatings. The recently developed α_2 + "O" Ti-24-Al-17Nb-1Mo composition exhibits an enhanced combination of elevated temperature strength and creep performance with room-temperature ductility and fracture toughness among the best obtained for α_2-type compositions. A further attraction of this

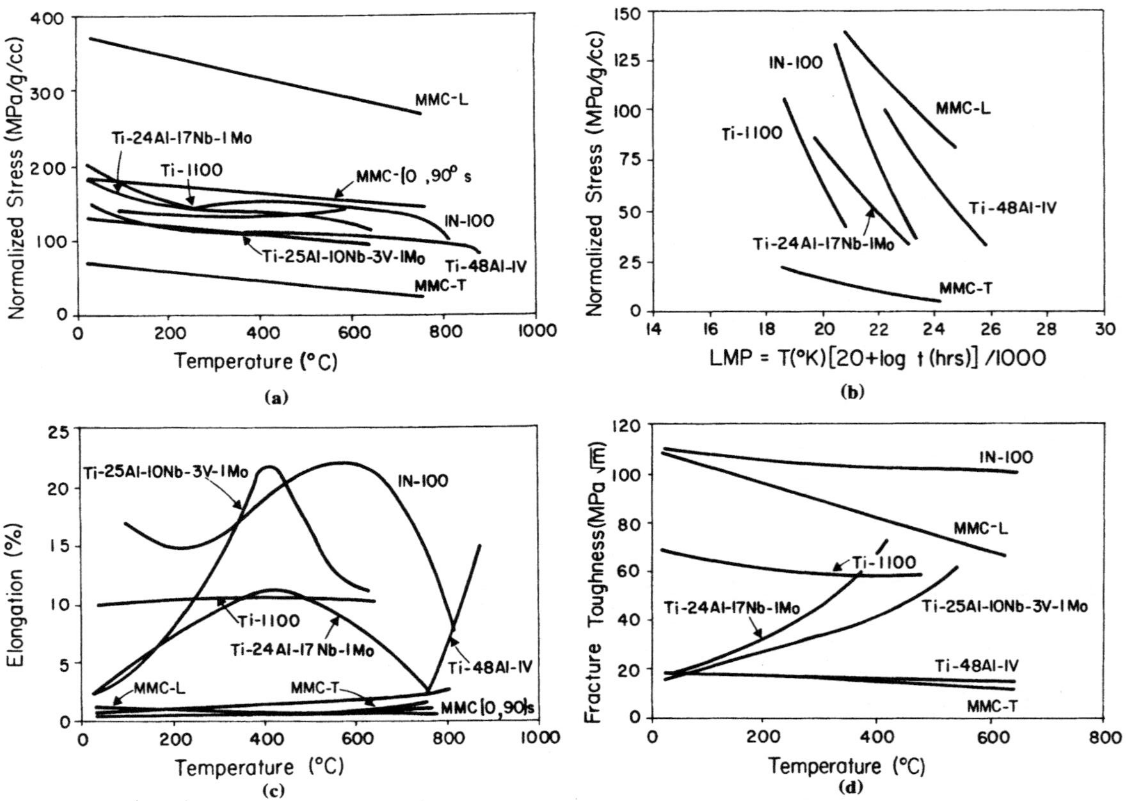

Figure 8-16. Comparison amongst titanium-rich alloys, titanium aluminides, superalloys and titanium aluminide composites: (a) yield stress/density versus temperature, (b) tensile elongation versus temperature, (c) Larson–Miller parameter/density for rupture life and (d) fracture toughness versus temperature.

composition is the increase in fracture toughness with temperature, from 19 MPa$\sqrt{m}$ at RT to 50 MPa$\sqrt{m}$ at 300°C (Larsen et al. 1990). Furthermore, much is still to be done in optimizing the microstructure.

The ordered orthorthombic ("O") alloys, based on Ti_2NbAl, have higher strength-toughness combinations than current α_2-type alloys (Rowe 1992, 1993a, b). They also offer the opportunity to simultaneously enhance both creep behavior and fracture toughness.

The α_2 composites demonstrate clear advantages over the monolithic alloy in the longitudinal direction with excellent elevated temperature strength and stiffness and a significant enhancement in fracture and fatigue-crack growth resistance (Larsen et al. 1990). However, the transverse properties are significantly lower with the strength being controlled by the matrix capability. Thermal fatigue is a great concern since cycling in air can reduce the strength by almost 90% compared to the initial value, a degradation attributed to heavily oxidized surface cracks, which do not occur in vacuum (Russ 1990). Some of these problems mentioned above appear to be solved at least to an extent in the "O" composites (Smith et al. 1992; Graves et al. 1992). A reduction in the reaction zone and change in the phase proportions resulted in enhanced transverse strength and removal of the thermal-fatigue debit compared to the α_2 alloys. However, other matrices, such as Ti-15Al-22.5Nb with a fracture toughness of 42.3 MPa$\sqrt{m}$ and RT elongation of 6.7%, and Ti-24Al-14Nb-3V-0.5Mo (Jia et al. 1992) with an unaged RT elongation of 26% deserve attention, the latter alloy having a further advantage of lower density because of the reduced Nb content.

The monolithic γ alloys combine attractive elevated temperature strength, creep performance, and environmental resistance with high stiffness and low density. Recent enhancement in ambient temperature ductility, up to almost 5% elongation (Pu et al. 1992), has significantly increased interest in this alloy class as a structural material. However, low impact resistance may well preclude use in applications such as turbine blades (Larsen et al. 1990). Thus, this class of materials may be restricted to use in noncritical aerospace applications and terrestrial use such as automobile components.

6.2. Commercialization and Applications

It can take thirty years or more from the research stage to get a new material into full-scale use in demanding flying applications (Dogan 1988; Dogan, Wagner, and Beaven 1991; Froes et al. 1990a); and this is with "forgiving" materials. With "nonforgiving" materials such as the monolithic and composite titanium aluminides, the time period for commercialization will clearly be extended further, even with creative design philosophies that emphasize toughness rather than ductility, use anisotropic behavior to advantage, and use a concurrent engineering approach. Further, the high initial cost of most emerging new materials means that total cost, that is, acquisition and ownership cost, must be used as a criterion for material selection.

The higher operating temperatures offered by use of the titanium aluminides can be seen in terms of the enhancement in relative specific thrust in Figure 8-17, and from a cost viewpoint in that a 100°F increase in operating temperature could result in a decrease of about $1M per military aircraft in life-cycle costs (Dogan 1988; Dogan et al. 1991). In fact, it has been stated that "titanium [aluminide] based composites... are key to the achievement of the goals of the U.S. Integrated High Performance Turbine Engine Technology (IHPTET) advanced engine program" (Dix 1990), a sentiment echoed in other countries in developing advanced engines (Driver 1990). The titanium aluminides are also contenders for use on transatmospheric vehicles such as the U.S. National Aerospace Plane (NASP) and on similar vehicles under consideration in other parts of the world (Ronald 1989; McConnell 1990). Here applications could include skin surfaces, internal structural parts, rocket nozzles, and engine parts (McConnell 1990). An example of a honeycomb structure fabricated

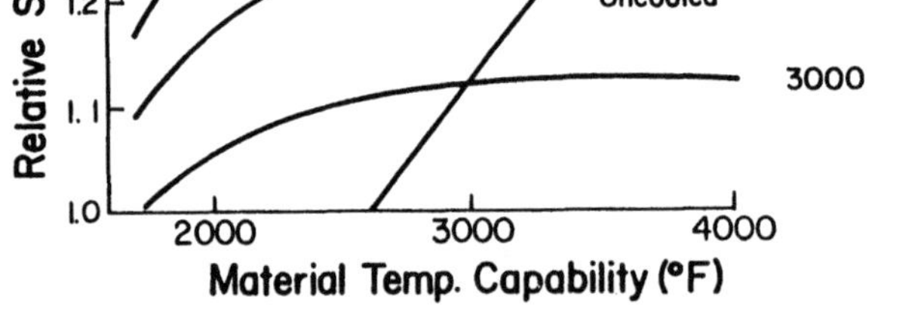

Figure 8-17. Effect of material operating temperature capability, turbine inlet temperature and cooling requirements on relative specific thrust of turbojet and turbofan engines.

Figure 8-18. Honeycomb panel constructed of Ti-24Al-11Nb face sheets and Ti-3Al-2-1/2V core.

using an enhanced diffusion-bonding process is shown in Figure 8-18 (Hoffman, Bird, and Bales 1989).

The primary (ingot production, casting, ingot breakdown, powder metallurgy) and secondary (forming, machining, joining, chemical milling, etc.) fabrication of monolithic α_2 seems to be a (more difficult) extension of conventional titanium processing. In some cases, forming characteristics as good as conventional titanium alloys (e.g, Ti-6Al-4V) can be obtained with careful processing; for example over 1000% elongation under superplastic forming conditions (Zhong 1992). A number of innovative processing techniques such as cold rolling of foil offer interesting opportunities.

Monolithic γ is considerably more difficult to work than α_2 and may require NNS processes such as castings, powder metallurgy (Moll et al. 1990) or isothermal forging. There are indications, however, that control of chemistry and grain size can enhance hot workability and superplasticity ($>$ 400% tensile elongation). Further, by use of innovative preheating and cooling cycles γ alloys such as Ti-48Al-2Nb-2Cr can be successfully welded using electron beam methods, see Figure 8-19 (Fores et al. 1993).

Composite α_2 material can be fabricated by a number of techniques, with incompatibility problems (CTE differences and chemical reaction)

Figure 8-19. Solid state crack in electron beam welded Ti-48Al-2Nb-2Cr (γ) alloy, occurring when preheating and cooling cycles are nonoptimum.

having been at least partially solved with use of orthorhombic matrices. Assuming these concerns are resolved, cost may dictate use. It has been suggested that the current cost of α_2 MMC's ($4–5 K/lb) must be reduced to < $1 K/lb for extensive application to occur (Williams and Johnson 1990). Plasma spray approaches appear to offer the greatest potential for acceptable quality/cost product (Moll et al. 1990; Williams and Johnson 1990); however, when we add compliance/diffusion barrier coatings to current concepts it is clear that this cost goal is a major challenge.

To date there are no flying applications of the titanium aluminides; however, the γ composition may see use in terrestrial components including a turbocharger rotor; see Figure 8-20 (Nishiyama et al. 1990) and a piston head for a diesel engine, Figure 8-21 (Moxson 1993). The reduced weight of the cast TiAl rotor compared to a superalloy resulted in a more positive response, and the increased toughness compared to ceramics gave a much more durable product. Investigation of binary compositions revealed that a Ti-47%Al alloy had the best combination of room tempera-

Figure 8-20. A TiAl cast turbocharger rotor.

Figure 8-21. TiAl piston head for a diesel engine, fabricated using plasma rotating electrode process powder.

ture and 900°C properties (> 2% elongation at room temperature). Studies of ternary additions were then followed by successful testing of a TiAl component.

A number of Ti_3Al and TiAl jet engine components have been produced successfully using casting and ingot approaches (Lipsitt 1985), although, according to the open literature, none have yet flown; however, it is likely that noncritical parts will fly soon. The components illustrated in Figure 8-22 were both produced using a powder metallurgy approach (Moll et al. 1990). The parts shown in Figure 8-23 are a high pressure stator inner support ring, in which use of Ti_3Al saves 43% in weight compared to a superalloy part, a high pressure compressor casing section (Lipsitt 1985); and a T700 compressor case cast from γ is shown in Figure 8-24 (Froes 1993; Froes et al. 1993). Other parts produced in various parts of the world include PM and forged TiAl low pressure turbine blades (Fig. 8-25) and turbine baffle and vane support rings (Fig. 8-26).

The reinforced aluminides offer even greater potential for weight savings (Dix 1990; Driver 1990). This is the situation for components such as complex-shaped blades, but even more so in ax-symmetrical parts such as rings, castings, and discs where the filament direction can be oriented

(a)

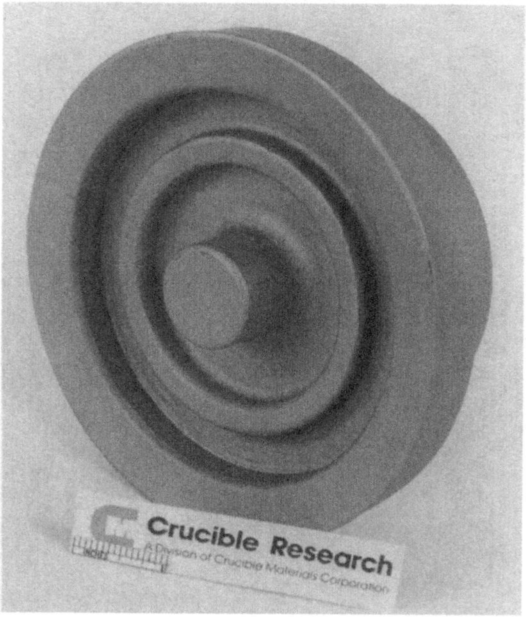

(b)

Figure 8-22. (a) Jet engine compressor rotor produced by HIP of Ti_3Al-Nb PREP powder, and (b) machined preform produced by HIP of gas atomized TiAl alloy powder. Both produced using the Ceramic Mold process.

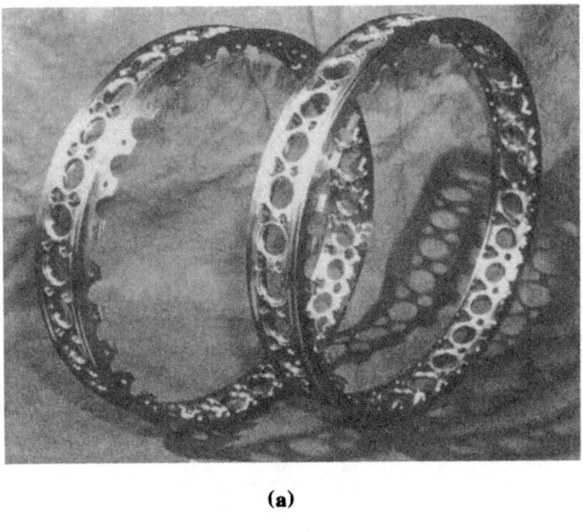

(a)

(b)

Figure 8-23. (a) High pressure turbine stator inner support rings for a GE demonstrator engine, run for 25 h and (b) high pressure compressor casing section, P & W F100 engine, run for 65 h.

Figure 8-24. Cast TiAl T700 compressor case.

so that the large hoop stresses are accommodated by these filaments. The dramatic weight saving that can be achieved (up to 75%) by replacing a conventional disc and spacer assembly with a titanium aluminide reinforced ring is shown in Figure 8-27 (Driver 1990).

7. REMAINING CHALLENGES AND CONCLUSIONS

Because of their low density and attractive high-temperature behavior the monolithic and composite titanium aluminides could replace superalloys in many elevated-temperature terrestrial, airframe, engine, and missile applications. However, the great importance attached to structural integrity in advanced engine and airframe designs means that reliability and reproducibility are receiving increasing emphasis, and the low levels of "forgiveness" in the titanium aluminides is of major concern. At the present time, there are no commercial applications of the titanium aluminides, although many components have been fabricated from both the α_2 and γ alloys and

Figure 8-25. Turbine blade produced from TiAl using a powder metallurgy and hot forging approach.

have performed quite satisfactorily in ground-based tests. It is likely that we will see terrestrial and flying applications of the titanium aluminides in the near future, initially in noncritical components.

Much progress has been made but there are many remaining roadblocks and challenges to implementation. These include a better definition of an optimum Ti_3Al (α_2, α_2 + "O" or "O") monolithic composition/microstructure, combining acceptable ductility and creep and other mechanical properties, which could be anywhere in the range of 10–25%Nb equivalent and includes full definition of interstitial effects, particularly oxygen. There is an equal or greater challenge in defining an optimum γ monolithic composition/microstructure, combining acceptable ductility and creep and other mechanical properties, in which an acceptable balance of ambient temperature ductility and toughness is obtained. A good understanding of chemistry-processing-microstructure-mechanical property relationships for both α_2 and γ alloys was virtually ignored until recently; a serious omission. This area of research requires considerably more work, including nanostructure material.

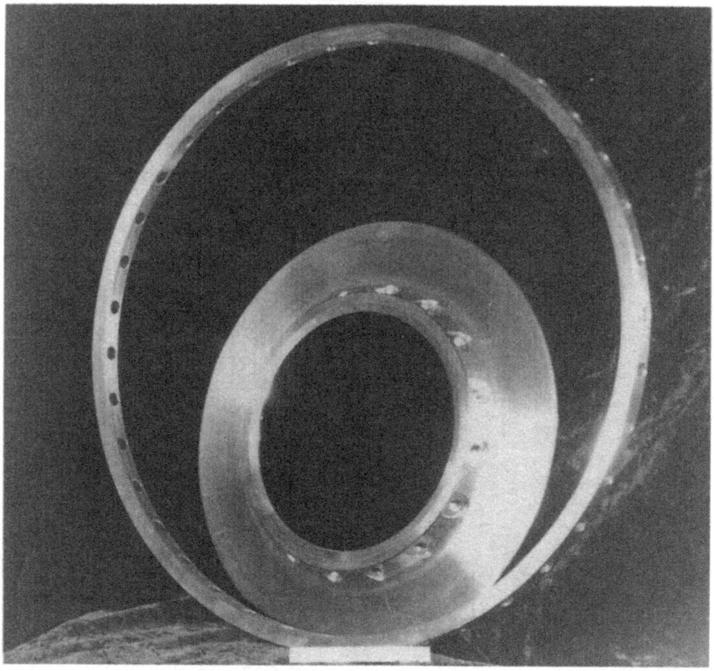

Figure 8-26. A 430 mm vane support ring, and a 300 mm turbine baffle produced from Ti_3Al-based alloys.

There is much further work to be done to define the best matrix for α_2/MMC material, with ductility and environmental issues tending to favor Nb-equivalent compositions in the range 15–25%. However, the use of chemistry and microstructure variations to influence properties is not well understood and requires considerably more study. There are now at least three sources of continuous SiC fibers, although only one is fully commercialized. This has led to the availability of improved quality, cost-effective, reproducible, compatible fibers/coatings (CTE and chemical reactivity) of the right diameter (generally thinner than the present ~ 150 μm SCS-6 fiber) for various applications. This situation has been helped by progress in development of (micromechanical) modeling capabilities for composites and advances in innovative design with these new "brittle" and anisotropic (MMC) materials.

Widespread application of components produced from monolithic- and composite-Ti_xAl materials is still to come. It will require achievement of a high confidence level with designers that these materials will behave in a controlled, reproducible, and cost-affordable manner.

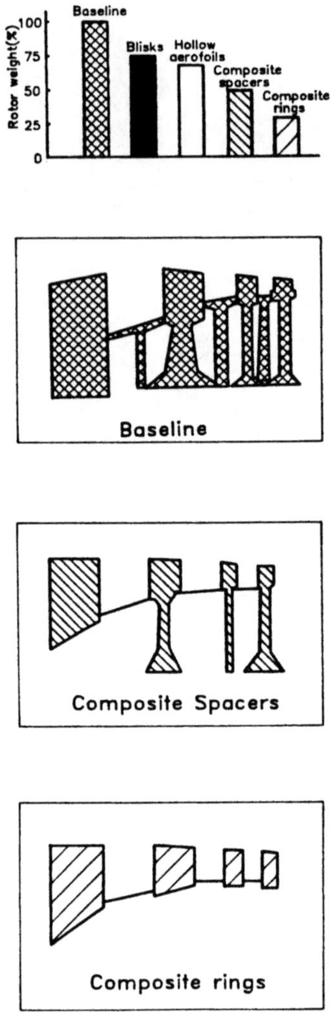

Figure 8-27. The weight savings which can be achieved (up to 75%) by replacing a conventional disc and spacer assembly with a titanium aluminide composite ring.

8. ACKNOWLEDGEMENTS

In compiling this paper the authors acknowledge the help provided by T. Bales, S. Balsone, P. Bania, P. K. Brindley, Y.-W. Kim, J. M. Larsen, C.-G. Li, H. A. Lipsitt, R. A. Mackay, I. S. Polkin, T. M. F. Ronald, Z.

Zhong and D. Zou. Also thanks are given to Mrs. Kandy Nelson, Mrs. Susan L. Goetz, and Miss Aimee Keller for assistance in manuscript preparation.

REFERENCES

ANANTHARAMAN, T. R., and SURYANARAYANA, C., 1987, *Rapidly Solidified Metals—A Technological Overview.* Aedermannsdorf, Switzerland: Trans Tech Publications.

Annual Report on Japanese Basic Technologies for Future Industries, High-Performance Materials for Severe Environments, Nov. 1992, Toranomon Hirai Bldg. 17-7 Toranomon 3-Chome, Minato-Ku, Tokyo 105, Japan.

ANTON, D. L., MARTIN, P. L., MIRACLE, D. B., and MCMEEKING, R. ed., 1990, *Intermetallic Matrix Composites*, Pittsburgh, PA: MRS.

ANTON, D. L., SHAH, D. M., DUHL, D. N. and GIAMEI, A. F., 1989, *JOM* 41(9): 12.

BALSONE, S. J., 1989, In *Oxidation of High Temperature Intermetallics*, ed. T. Grobstein and J. Doychak, p. 219. Warrendale, PA: TMS.

BALSONE, S. J., NICHOLAS, T., MAXWELL, D. C. and KHOBAIB, M., 1990, In *Elevated Temperature Crack Growth*, ed. S. Mall and T. Nicholas, p. 87. New York, NY: Amer. Soc. Mech. Eng..

BANERJEE, D., GOGIA, A. K., and NANDY, T. K., 1990, *Metall. Trans.* 21A: 627.

BANERJEE, D., GOGIA, A. K., NANDY, T. K., and JOSHI, V. A., 1988, *Acta Metall.* 36: 871.

BANERJEE, D., NANDY, T. K., GOGIA, A. K., and MURALEEDHARAN, K., 1989, In *Proc. Sixth World Conf. on Titanium*, ed. P. Lacombe, R. Tricot, and G. Beranger, p. 1091, Les Editions de Physique, Les Ulis Cedex.

BANERJEE, D., ROWE, R. G., and HALL, E. L., 1991, In *High-Temperature Ordered Intermetallic Alloys IV*, MRS Symp. Proc. 213: p. 285. Pittsburgh, PA: MRS.

BEDDOES, J. C., WALLACE, W., and DEMALHERBE, M. C., 1992, *Mats. and Man. Proc.* 7: 527.

BENDERSKY, L. A., 1992, In *Advancements in Synthesis and Processes*, ed. F. H. Froes, W. Wallace, R. A. Cull, and E. Struckholt, p. M346. Covina, CA: SAMPE.

CAHN, R. W., 1989, *Metals, Materials and Processes* 1: 1.

CHAWLA, K. K., 1987, *Composite Materials—Science and Technology*, New York, NY: Springer-Verlag.

CHRISTMAN, T., NEEDLEMAN, A., and SURESH, S., 1989, *Acta Metall.* 37: 3029.

CHRISTODOULOU, L., and BRUPBACHER, J. M., 1990, *Materials Edge* 7: 29.

Congress of the U.S., Office of Technology Assessment, "Advanced Materials by Design," (June 1988).

DIMIDUK, D. M., and MIRACLE, D. B., 1989, In *High-Temperature Ordered Intermetallics III*, MRS Symp. Proc. 133: 349.

DIX, D., July 16, 1990, Defense News: 7.

DOGAN, B., 1988, In *Intermetallic Alloys: Deformation, Mechanical and Fracture Behavior*, GKSS-Forschungszentrum Geesthacht GmBH, Germany, Report 88/E/53.

DOGAN, B., WAGNER, R., and BEAVEN, P. A., 1991, *Scripta Metall. et Mater.* 25: 773.

DRIVER, D., 1990, In *High Temperature Materials for Power Engineering*, ed. E. Bachelet et al., p. 883. Dordrecht: Kluwer Academic Publishers.

FLEISCHER, R. L., DIMIDUK, D. M., and LIPSITT, H. A., 1989, *Ann. Rev. Mater. Sci.* 19: 231.

FLEISCHER, R. L., and TAUB, A. I., 1989, *JOM* 41(9): 8.

FROES, F. H., 1988a, *Materials Edge* 5: 19.

_______, 1988b, In *Space Age Metals Technology*, ed. F. H. Froes and R. A. Cull, p. 1. Covina, CA: SAMPE.

_______, 1989, *Materials Edge* 6: 17.

_______, 1990a, In *Advanced Materials-Outlook and Information Requirements*, ed. Louis J. Sousa, and Charles A. Sorrell, 1990, p. 41. Washington, DC: USBOM.

_______, 1990b, In *P / M in Aerospace and Defense Technologies*, ed. F. H. Froes, p. 23. Princeton, NJ: MPIF.

_______, June 1993, In *High Performance Materials for Severe Environments*, R & D Institute of Metals and Composites for Future Industries (RIMCOF), Japan Industrial Technology Association (JITA), Ministry of International Trade and Industry (MITI), Japan.

FROES, F. H., DEBARBADILLO, J. J., and SURYANARAYANA, C., 1990a, In *Structural Applications of Mechanical Alloying*, ed. F. H. Froes and J. J. deBarbadillo, p. 1, Materials Park, OH: ASM.

FROES, F. H., and EYLON, D., (ed) 1986, *Titanium-Rapid Solidification Technology*, Warrendale, PA: TMS.

_______, 1990a, *Internat. Mater. Rev.* 35: 162.

_______, 1990b, In *Hydrogen Effects on Materials Behavior*, ed. N. R. Moody and A. W. Thompson, p. 261. Warrendale, PA: TMS.

FROES, F. H., EYLON, D., and BOMBERGER, H. B., ed. 1985, *Titanium Technology: Present Status and Future Tends*, Dayton, OH: Titanium Development Association.

FROES, F. H., EYLON, D., and SURYANARAYANA, C., 1990b, *JOM* 42(3): 26.

FROES, F. H., KIM, Y.-W., and HEHMANN, F., 1987, *JOM* 39(8): 14.

FROES, F. H., and PICKENS, J. R., 1984, *JOM* 36(1): 14.

FROES, F. H., and ROWE, R. G., 1986, In *Rapidly Solidified Alloys and Their Mechanical and Magnetic Properties*, MRS Symp. Proc. 58: 309. Pittsburgh, PA: MRS.

_______, 1989, In *Proc. Sixth World Conf. on Titanium*, ed. P. Lacombe, R. Tricot, and G. Beranger, p. 801: Les Editions de Physique, Les Ulis Cedex.

FROES, F. H., and SURYANARAYANA, C., 1993, In *Reviews in Particulate Materials*, ed. A. Bose, R. German, and A. Lawley, p. 233 Princeton, NJ: MPIF.

FROES, F. H., SURYANARAYANA, C., and ELIEZER, D., 1991, *ISIJ Int.* 31: 1235.

———, 1992, *J. Mater. Sci.*, 27: 5113.

FROES, F. H., TSAKIROPOULOS, P., SURYANARAYANA, C., and BAESLACK, W., 1993, In *Light Materials for Transportation Systems*, ed. N. Kim, p. 27, Pohang, Korea: Center for Advanced Aerospace Materials.

GOGIA, A. K., BANERJEE, D., and NANDY, T. K., 1990, *Metall. Trans.* 21A: 609.

GRAVES, J. A., PEREPEZKO, J. H., WARD, C. H., and FROES, F. H., 1987, *Scripta Metall.* 21: 567.

GRAVES, J. A., SMITH, P. R., and RHODES, C. G., 1992, In *Intermetallic Composites II*, ed. D. B. Miracle, D. L. Anton, and J. A. Graves, MRS Symp. Proc. 273: 31.

HALL, E. L., and HUANG, S. C., 1989, In *High-Temperature Ordered Intermetallics III*, MRS Symp. Proc. 133: 693. Pittsburgh, PA: MRS.

———, 1990–91, Work in Progress. GE Corporate R & D Center, Schenectady, NY.

HOFFMAN, K. K., BIRD, R. K., and BALES, T. T., 1989, In *Light-Weight Alloys for Aerospace Applications*, ed. E. H. Chia, and N. J. Kim, p. 481. Warrendale, PA: TMS.

HSIUNG, L. M., CAI, W., and WADLEY, H. N. G., 1992a, *Mat. Sci. Eng.* A152: 295.

———, 1992b, *Acta Metall. et Mater.* 40: 3035.

HSIUNG, L. M., and WADLEY, H. N. G., 1992a, *Scripta Metall. et Mater.* 26: 35.

———, 1992b, *Scripta Metall. et Mater.* 26: 1071.

———, 1992c, *Scripta Metall. et Mater.* 27: 605.

HUANG, S. C., 1989, General Electric Corporate Research and Development Center, Schenectady, NY, Private Communication.

JIA, T., WANG, B., ZOU, D. X., MA, H., and ZHONG, Z. Y., 1992, *Mat. Sci. Eng.* A 152: 317.

JOHNSON, L. A., POPE, D. P., and STIEGLER, J. O., ed. 1991, *High-Temperature Ordered Intermetallic Alloys IV*, MRS Symp. Proc. 213.

JONES, H., 1982, *Rapid Solidification of Metals and Alloys*. Monograph No. 8. London, UK: Institution of Metallurgists.

KATTNER, U. R., LIN, J.-C., and CHANG, Y. A., 1992, *Met. Trans.* 23A: 2081.

KAUFMAN, M. J., BRODERICK, T. F., WARD, C. H., KIM, J. K., ROWE, R. G., and FROES, F. H., 1989, In *Proc. Sixth World Conf. on Titanium*, ed. P. Lacombe, R. Tricot, and G. Beranger, p. 985 Les Ulis Cedex: Les Editions de Physique.

KIM, Y.-W., 1989, *JOM* 41(7): 24.

———, 1991, In *High Temperature Ordered Intermetallic Alloys IV*, MRS Symp. Proc. 213: 177.

———, 1992, *Acta Metall.* 40: 1121.

———, 1993, In Proc. of Int. Workshop on *Ordered Intermetallics*. Hangzhou, PRC: To be published in the proceedings.

KIM, Y.-W., and DIMIDUK, D. M., 1991, *JOM* 43 (8): 40.

KIM, Y.-W., and FROES, F. H., 1990, In *High Temperature Aluminides and Intermetallics*, ed. S. H. Whang, C. T. Liu, D. P. Pope, and J. O. Stiegler, p. 485. Warrendale, PA: TMS.

KOCH, C. C., LIU, C. T., and STOLOFF, N. S., ed., 1985, *High-Temperature Ordered Intermetallic Alloys*, MRS Symp. Proc. 39 Pittsburgh, PA: MRS.

KORNILOV, I. I., PYLAEVA, E. N., VOLKOVA, M. A., KRIPYAKEVICH, P. I., and MAEKIV, V. I., 1965, *Dokl. Akad. Nauk SSSR* 161: 843.

KOSS, D. A., BANERJEE, D., LUKASAK, D. A., and GOGIA, A. K. 1990, In *High-Temperature Aluminides and Intermetallics*, ed., S. H. Whang, C. T. Liu, D. P. Pope, and J. O. Stiegler, p. 175. Warrendale, PA: TMS.

LARSEN, J. M., WILLIAMS, K. A., BALSONE, S. J., and STUCKE, M. A., 1990, In *High-Temperature Aluminides and Intermetallics*, eds., S. H. Whang, C. T. Liu, D. P. Pope, and J. O. Stiegler, p. 521. Warendale, PA: TMS.

LIPSITT, H. A., 1985, In *High Temperature Ordered Intermetallic Alloys*, MRS Symp. Proc. 39: p. 351, Pittsburgh, PA: MRS.

LIU, C. T., FROES, F. H., and STIEGLER, J. O., 1990, *Metals Handbook*, p. 913, Materials Park, OH: ASM.

LIU, C. T., TAUB, A. I., STOLOFF, N. S., and KOCH, C. C. ed., 1989, High-Temperature Ordered *Intermetallics III*. MRS Symp. Proc. 133 Pittsburgh, PA: MRS.

MACKAY, R. BRINDLEY, P. K., and FROES, F. H., 1991, *JOM* 43(5): 23.

MCCONNELL, V. P., Nov./Dec., 1990, *Advanced Composites*: 37.

MIRACLE, D. B., ANTON, D. L., and GRAVES, J. A., eds., 1992, In *Intermetallic Composites II*, MRS Symp. Proc. 273.

MISHURDA, J. C., PEREPEZKO, J. H., GRAVES, J. A., and FROES, F. H., 1989, In *Proc. Sixth World Conf. on Titanium*, ed. P. Lacombe, R. Tricot, and G. Beranger, p. 1127, Les Ulis Cedex: Les Editions de Physique.

MOLL, J. H., YOLTON, C. F., and MCTIERNAN, B. J., 1990, *Int. J. of PM* 26: 149.

MOXSON, V., May 1993, ADMA Corporation, Cleveland, OH, Private Communication.

MOZER, B., BENDERSKY, L. A., BOETTINGER, W. J., and ROWE, R. G., 1990, *Scripta Met. et Mater.* 24: 2363.

MURAELEDHARAN, K., GOGIA, A. K., NANDY, T. K., BANERJEE, D., and LELE, S., 1992a, *Met. Trans.* 23A: 401.

________, 1992b, *Met. Trans.* 23A: 417.

MURRAY, J. L., 1987, *Phase Diagrams of Binary Titanium Alloys*, p. 12, Materials Park, OH: ASM.

NISHIYAMA, Y., MIYASHITA, T., ISOBE, S., and NODA, T., 1990, In *High-Temperature Aluminides and Intermetallics*, ed., S. H. Whang, C. T. Liu, D. P. Pope, and J. O. Stiegler, p. 557. Warrendale, PA: TMS.

PU, Z. J., SHI, J.-D., ZOU, D. X., and ZHONG, Z. Y., 1992, In *International Workshop on Intermetallics*, Hongzhou, China. Published by National Natural Science Foundation of China, Beijing, p. 66. T. L. Lin, and C. T. Liu.

RONALD, T. M. F., 1989, *Ad. Mats. and Proc.* 135(5): 29.

ROWE, R. G., 1990, In *High-Temperature Aluminides and Intermetallics*, eds., S. H. Whang, C. T. Liu, D. P. Pope, and J. O. Stiegler, p. 375. Warrendale, PA: TMS.

________, 1991, In *Microstructure / Property Relationships in Titanium Aluminides and Alloys*, ed. Y.-W. Kim, and R. R. Boyer, P. 387. Warrendale, PA: TMS.

________, 1993a, In *Proc. Seventh World Conference on Titanium*, ed. F. H. Froes and I. Caplan, p. 1259. Warrendale, PA: TMS.

________, 1993b, In *Proc. of the Seventh World Titanium Conference*, ed. F. H. Froes, and I. Caplan, p. 343. Warrendale, PA: TMS.

________, 1993c, In *Synthesis, Processing and Modelling of Advanced Materials*, ed. F. H. Froes, and T. Khan, p. 61. Aedermannsdorf, Switzerland: Trans Tech Publ.

ROWE, R. G., and HUANG, S. C., 1988, *Israel J. Tech*. 24: 255.

ROWE, R. G., KONITZER, D. G., WOODFIELD, A. P., and CHESUT, J. C., 1991, In *High Temperature Ordered Intermetallic Alloys IV*, MRS Symp. Proc. 213: 703.

ROWE, R. G., SIEMERS, P. A., and LARSEN, M., 1992, In *Advancements in Synthesis and Processes*, ed. F. H. Froes, W. Wallace, R. A., Cull, and E. Struckholt, p. M171. Covina, CA: SAMPE.

RUSS, S. M., 1990, *Met. Trans*. 21A: 1595.

SHIH, D. S., SCARR, G. K., and CHESNUTT, J. C., 1989, In *High-Temperature Ordered Intermetallics III*, MRS Symp. Proc. 133: 167.

SMITH, P. R., and FROES, F. H., 1984, *JOM* 36(3): 19.

SMITH, P. R., GRAVES, J. A., and RHODES, C. G., 1992, In *Intermetallic Composites, II*, ed. D. B. Miracle, D. L. Anton, and J. A. Graves, MRS Symp Proc. 273: 43.

SMITH, P. R., and REVELOS, W. C., 1990, In *Fatigue '90*, ed. H. Kitagawa, and T. Tanaka, P. 1711. Birmingham, UK: Materials and Components Eng. Publ. Ltd.

SMITH, P. R., RHODES, C. G., and REVELOS, W. C., 1990, In *Interfaces in Metal-Ceramic Composites*, ed. R. Y. Lin, R. J. Arsenault, G. P. Martins, and S. G. Fishman, p. 35. Warrendale, PA: TMS.

STOLOFF, N. S., KOCH, C. C., LIU, C. T., and IZUMI, O. eds., 1987, *High-Temperature Ordered Intermetallic Alloys II*, MRS Symp. Proc. 81.

SURYANARAYANA, C., and FROES, F. H., 1990, *JOM* 42(3): 22.

________, 1993, *Advanced Mater*. 5: 96.

SURYANARAYANA, C., FROES, F. H., KRISHNAMURTHY, S., and KIM, Y.-W., 1990, *Int. J. of PM*, 26: 117.

SURYANARAYANA, C., FROES, F. H., and ROWE, R. G., 1991, *Internat. Mater. Rev*., 36: 85.

SURYANARAYANA, C., KORTH, G. E., CHEN, G.-H., FREFER, A., and FROES, F. H., 1993, *Nanostructured Mater*. 2: 527.

SURYANARAYANA, C., and LEE, D. S., 1992, *Scripta Metall. et Mater*. 26: 919.

SURYANARAYANA, C. SUNDARESAN, R., and FROES, F. H., 1992, *Mat. Sci. Eng*. A150: 117.

UPADHYAYA, D., BLACKKETTER, D. M., and FROES, F. H., 1993, *Proc. of the Seventh World Titanium Conf*., ed. F. H. Froes, and I. Caplan. p. 2537. Warrendale, PA: TMS.

WARD, C. H., 1993, *Internat. Mater. Rev.* 38: 79.
WARD, C. H., WILLIAMS, J. C., THOMPSON, A. W., ROSENTHAL, D. G., and FROES, F. H., 1989, In *Proc. Sixth World Conf. On Titanium*, ed. P. Lacombe, R. Tricot, and G. Beranger, p. 1103. Les Ulis Cedex: Les Editions de Physique.
WARD-CLOSE, C. M., and PARTRIDGE, P. G., 1990, *J. Mat. Sci.* 25: 4315.
WESTWOOD, A. R. C., 1988, *Metall. Trans. B* 19B: 155.
WHANG, S. H., and HAHN, Y. D., 1990, In *High Temperature Aluminides and Intermetallics*, ed. S. H. Whang, C. T. Liu, D. P. Pope, and J. O. Stiegler, p. 91, Warrendale, PA: TMS.
WHANG, S. H., LIU, C. T., POPE, D. P., and STIEGLER, J. O., eds. 1990, *High-Temperature Aluminides and Intermetallics*, Warrendale, PA: TMS.
WILLIAMS, J. C., and JOHNSON, L. A., Dec. 1990, Private Communication, GE/Aircraft Engines, Evendale, OH.
ZHONG, Z., 1992, Private Communications, CISRI, Beijing, China.

Chapter 9

Iron Aluminides

C. G. McKamey

1. INTRODUCTION

Iron aluminides have been of interest since the 1930s when the excellent corrosion resistance of compositions with more than about 18 at% Al was first noted (DeVan 1989; Ziegler 1932). These alloys offer relatively low material cost, conservation of strategic elements, and a lower density than stainless steels. Their tensile strength also compares favorably with many ferritic and austenitic steels. These property advantages have led to the consideration of iron-aluminum alloys for many applications including those listed in Table 9-1 (McKamey et al. 1991; Nachman and Buehler 1956; Sikka et al. 1993). However, limited ductility at ambient temperatures and a sharp drop in strength above 600° C have been major deterrents to their acceptance for many structural applications. More recent studies have demonstrated that improved engineering ductility (to 10–15% in Fe_3Al) can be achieved in wrought Fe_3Al-based iron aluminide alloys through control of composition and microstructure (Bordeau 1987; Culbertson and Kortovich 1986; Sikka 1991b; Sikka et al. 1993). Accompanying this improvement has been an increased understanding of the causes for ambient temperature embrittlement in this system (Liu et al. 1989, 1990; McKamey and Liu 1992). Because of these advances, iron-aluminide alloys (especially those of up to 50 at% Al) are again being considered for structural uses, especially for applications where their excellent corrosion resistance can be exploited.

Table 9-1 Potential Applications for Iron Aluminides

Application	Component System
heating elements	toasters, stoves, ovens, cigarette lighters, dryers
wrapping wire	insulation wrapping for investment-casting molds
regenerator discs	automotive gas-turbine engines
hot-gas filters	coal-gasification system
piping, tubing	coal-gasification system, automotive systems
tooling	dies for superplastic forming of titanium-based alloys
shields	superheater and reheater tubes in coal-fired power plants
automotive	exhaust manifolds, catalytic converters, exhaust support hangers
other	components needing high-temperature sulfidation and oxidation resistance

Source: Nachman and Buehler 1956, Sikka et al. 1993.

In past years, several efforts have been devoted to understanding and improving the metallurgical properties of iron aluminides, with the aim of producing strong, ductile, corrosion-resistant materials for structural applications. These studies have resulted in significant contributions to the understanding of the fabrication and mechanical properties of iron aluminides. General reviews of the literature on iron aluminides are available (Baker and Munroe 1990, Hardwick and Wallwork 1978; Holladay 1961; Lepkowski and Holladay 1957; McKamey et al. 1991; Prakash et al. 1991; Tomaszewicz and Wallwork 1978). Although iron and aluminum form several intermetallic phases, this chapter will describe only the iron-rich iron-aluminide compositions that have the most potential for structural applications. These include the FeAl and Fe_3Al ordered phases, as well as compositions down to about 15 at% Al which exist in disordered solid solutions.

2. PHASE RELATIONS IN THE IRON-ALUMINUM SYSTEM

The presently accepted phase relationships in the Fe-Al system are shown in Figure 9-1 (Massalski 1986). For structural applications, the most important phases in this system are those on the iron-rich half of the phase diagram up to FeAl. A disordered solid solution of aluminum in iron (α) is stable at lower aluminum concentrations (up to approximately 18–20 at%). Above this composition, two ordered phases ($D0_3$ and B2)

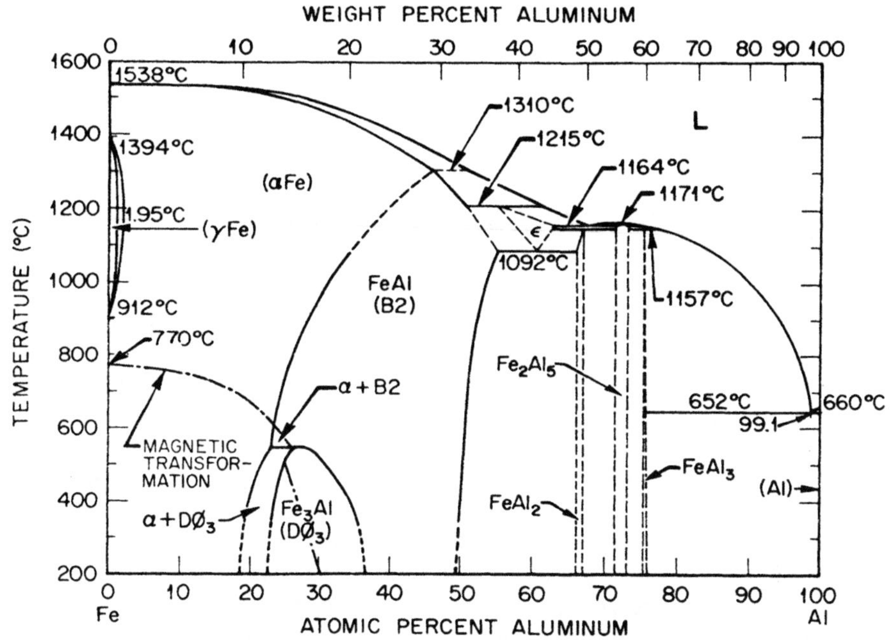

Figure 9-1. Iron-aluminum phase diagram (after Massalski, T. B., 1986, *Binary Alloy Phase Diagrams*. Materials Park, OH: ASM, Int.; Okamoto, H., and Beck, P. A., 1971, *Metall. Trans.* 2: 569–574; Oki, K., Hasaka, M., and Eguchi, T., 1973, *Jap. J. Appl. Phys.* 12: 1522–1530; Swann, P. R., Duff, W. R., Fisher, R. M., 1969, *Trans. Met. Soc. AIME* 245: 851–853).

exist, both possessing a wide compositional range of stability. Near Fe_3Al the following equilibrium phases have been confirmed (Allen 1975; Okamoto and Beck 1971; Oki et al. 1973; Swann et al. 1969): the disordered solid solution (α); an Fe_3Al with an imperfectly ordered B2 structure; an ordered Fe_3Al with the $D0_3$ structure; and the two phase regions, $\alpha + D0_3$ and α + B2. An important note for understanding mechanical property variations in alloys of 23–37% Al is that the Fe_3Al composition can possess the B2 structure above the B2-$D0_3$ ordering temperature (T_c), and that some B2 phase can be retained in material at room temperature by quenching from higher temperatures where the B2 is stable. The transition temperature (T_c) between the B2 and $D0_3$ structures at the stoichiometric Fe_3Al composition is approximately 550° C. With increasing aluminum concentration (at room temperature), the B2 phase becomes more stable, until the $D0_3$ structure ceases to exist above approximately 36–37 at% Al. The B2 phase has a wide range of solubility, existing in what can be described as "imperfect" order at temperatures of about 550° C to 700° C in the Fe_3Al phase and increasing in stability up to the highly ordered FeAl composition.

3. CHARACTERISTIC IRON ALUMINIDE MICROSTRUCTURES

The ordered $D0_3$ (BiF_3-type) and B2 (CsCl-type) crystal structures, the unit cells for which are presented in Figure 9-2 (Davies 1964), are body-centered cubic (b.c.c.) derivative structures. The B2 superlattice can be described as consisting of two interpenetrating simple-cubic sublattices with iron occupying one sublattice and aluminum the other, giving an AB stoichiometry with lattice parameter a_0. The $D0_3$ superlattice is composed of eight B2 superlattices stacked together, but with alternating iron and aluminum body-centered atoms, positioned in such a way as to maximize the spacing (and thereby minimize the energy) between aluminum atoms. This structure can be viewed as composed of four nonequivalent interpenetrating face-centered cubic sublattices, with a lattice parameter of $a'_0 = 2a_0$, giving an A_3B stoichiometry.

Antiphase domains (APD's) are ordered regions that nucleate homogeneously below the ordering transition temperature and grow on sublattices of the previously disordered material (Beeler 1967). Individual

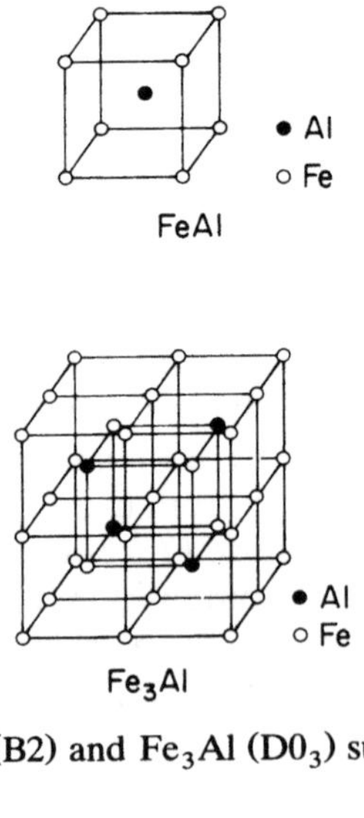

Figure 9-2. Unit cell of FeAl (B2) and Fe_3Al ($D0_3$) superlattices.

domains nucleate on different sublattice sites and grow until they impinge, creating a planar ordering fault at the interface. This ordering fault is known as an antiphase domain boundary (APDB), or more commonly APB. The stability and morphology of APB's depend on many variables including composition and temperature. In the B2 phase field, only one type of domain exists, with an APB fault vector that consists of an $1/2a_0$ $\langle 111 \rangle$ displacement on {110} planes (Marcinkowski and Brown 1962). Formation of the B2 superlattice cannot be suppressed by quenching from above the disorder-to-B2 transition. Near the Fe_3Al composition, B2 domains exist above T_c, but cooling below the B2-$D0_3$ transition temperature into the $D0_3$ phase results in the formation of $D0_3$ domains within the former B2 domains creating a finer domain structure (Marcinkowski and Brown 1962). In this case, the domain structure consists of large domains with a characteristic APB vector of $1/4a'_0\langle 111 \rangle$ characteristic of the B2 ordered structure (a'_0 being the lattice parameter of the $D0_3$ phase), and within these large domains, smaller domains with a characteristic APB fault vector of $1/2a'_0\langle 100 \rangle$indicative of the $D0_3$ structure. These fault vectors correspond to nearest neighbor (NN) and next-nearest neighbor (NNN) bonding, characteristic of the B2 and $D0_3$ phases, respectively.

The APB's in iron aluminides have no preference for particular lattice planes, and therefore exhibit a wavy, isotropic appearance that can be imaged in transmission electron microscopy (TEM) by using the appropriate diffraction conditions (Marcinkowski and Brown 1961, 1962; Ray et al. 1970; Saburi et al. 1972). Imaging with a $\langle 111 \rangle$ superlattice diffraction vector will give rise to contrast from both the nearest-neighbor (NNAPB) and the next-nearest-neighbor APB (NNNAPB) faults (Figure 9-3), enclosing the B2 and $D0_3$ antiphase domains, respectively, while a superlattice diffraction vector of $\langle 002 \rangle$ or $\langle 222 \rangle$ will give rise to contrast only from NNAPB faults.

4. DEFORMATION MECHANISMS

Deformation behavior in iron aluminides is dependent on composition, temperature, and the presence or absence of ordered structures (Leamy and Kayser 1969a). The low aluminum alloys, outside the ordered regions of the phase diagram, deform by movement of $a_0\langle 111 \rangle$ dislocations, similar to b.c.c iron (Justusson, Zackay, and Morgan 1957). Deformation behavior of the higher aluminum compositions, which contain the $D0_3$ and/or B2 ordered phases, is characteristic of most other ordered alloys. In stoichiometric Fe_3A1, the major slip system is {110} $\langle 111 \rangle$, and theoretically the deformation behavior should be dominated by

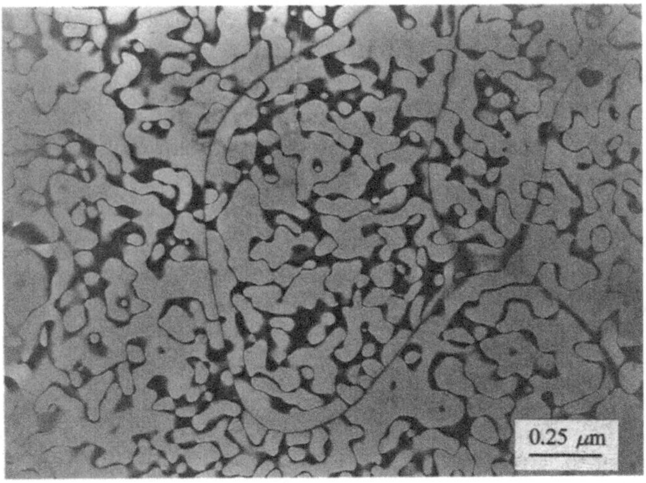

Figure 9-3. Transmission electron microscope image of antiphase domain boundaries in Fe_3Al. Imaging was done using a ⟨111⟩ superlattice diffraction vector. The large wavy boundaries are NNAPBs characteristic of the B2 order; the smaller domains inside the B2 domains are $D0_3$ enclosed by NNNAPBs (Courtesy of J. A. Horton, Oak Ridge National Laboratory).

the motion of superlattice dislocations consisting of fourfold dislocations in the $D0_3$ structure (Fig. 9-4a) or dislocation pairs in the higher temperature B2 structure (Fig. 9-5a), each enclosing faults corresponding to their respective APB types (Ehlers and Mendiratta 1982; Leamy and Kayser 1969; Leamy, Kayser, and Marcinkowski 1969a; Marcinkowski and Brown 1961). These multiple dislocations are necessary to make a perfect lattice translation in the ordered structures, as described by Marcinkowski and Brown (1961) for the $D0_3$ superdislocation. Calculations have shown that the APB energies associated with both B2 and $D0_3$ superdislocation configurations are very low, resulting in large extensions of the individual dislocations (see Fig. 9-6) and allowing for the existence of imperfect configurations (Marcinkowski and Brown 1961, 1962). Therefore, although two- and fourfold dislocations have been observed in TEM, during deformation these dislocations often travel as unit dislocations, leaving ribbons of APB's behind on their slip planes (as shown in Fig. 9-6b). Screw dislocations, producing wavy slip, have also been observed (Crawford, Ray, and Cockayne 1973; Crawford 1976; Crawford and Ray 1977; Ray, Crawford, and Cockayne 1970).

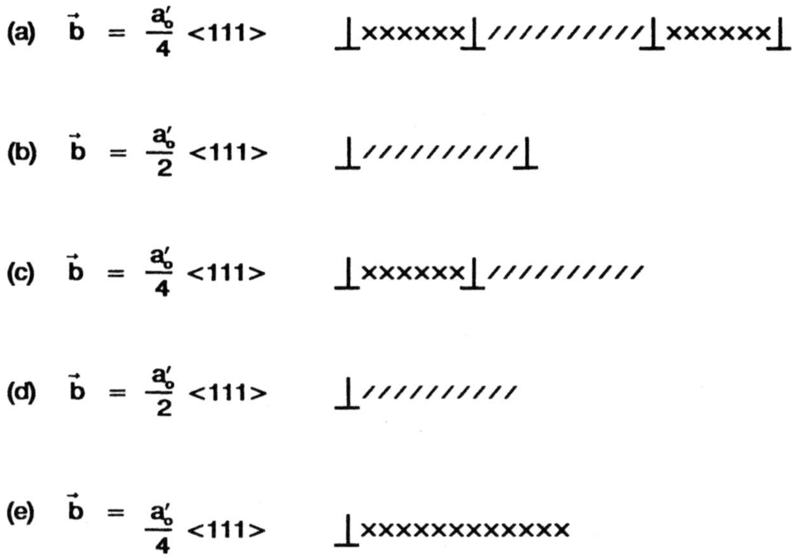

Figure 9-4. Schematic illustration of possible $D0_3$ superlattice dislocations and their imperfect variants. a'_0 is the lattice parameter for the $D0_3$ cell (after Ehlers, S. K., and Mendiratta, M. G., 1982. *Tensile Behavior of Fe*-31 *at% Al Alloy*, AFWAL-TR-82-4089. Wright Patterson Air Force Base, OH: Air Force Wright Aeronautical Laboratories; Leamy, H. J., and Kayser, F. X., 1969a. *Phys. Stat. Solidi* 34: 765–80; Leamy, H. J., Kayser, F. X., and Marcinkowski, M. J., 1969b. *Phil. Mag.* 20: 763–77; Marcinkowski, M. J., and Brown, N., 1961. *Acta Metall.* 9: 764–86; McKamey, C. G., Horton, J. A., and Liu, C. T., 1987, MRS Symp. Proc. 81: 321–327.

Dislocations in the $D0_3$ ordered phase consist of four $1/4a'_0\langle 111\rangle$ dislocations (where a'_0 is the lattice parameter of the $D0_3$ unit cell) connected by two types of APB (Fig. 9-4a)(Ehlers and Mendiratta 1982; Leamy and Kayser 1969; Leamy et al. 1969a; Saburi, Yamauchi, and Nenno 1972). The outer pairs of dislocations are connected by a $1/4a'_0$ $\langle 111\rangle$ APB. These pairs are similar to the B2 dislocation configuration, being affected mainly by nearest-neighbor interaction energies. The inner pair of dislocations is connected by a $1/2a'_0\langle 100\rangle$ APB and is affected by second-nearest neighbor interaction energies. All four superpartial dislocations are required to maintain the order during deformation. Fourfold dislocations in the $D0_3$ structure have been observed in TEM (Crawford et al. 1973; Crawford and Ray 1977; Leamy et al. 1969b). However, as the composition, temperature, or degree of order changes,

(a) $\vec{b} = \frac{a_0}{2}<111>$ ⊥xxxxxxxxxxx⊥

(b) $\vec{b} = \frac{a_0}{2}<111>$ ⊥xxxxxxxxxxx

(c) $\vec{b} = a_0<111>$ ⊥

Figure 9-5. Schematic illustration of possible B2 superlattice dislocations and their imperfect variants. a_0 is the lattice parameter for the B2 cell (after Ehlers, S. K., and Mendiratta, M. G., 1982. *Tensile Behavior of Fe*-31 *at% Al alloy*, AFWAL-TR-82-4089. Wright Patterson Air Force Base, OH: Air Force Wright Aeronautical Laboratories; Leamy, H. J., and Kayser, F. X., 1969a. *Phys. Stat. Solidi* 34: 765–80; Leamy, H. J., Kayser, F. X., and Marcinkowski, M. J., 1969b. *Phil. Mag.* 20: 763–77; Marcinkowski, M. J. and Brown, N., 1961. *Acta Metall.* 9: 764–86).

the energy of the two types of APB change, leading to the formation of several different dislocation variants. Within the $D0_3$ structure, the possible superlattice dislocations and their imperfect variants are shown in Fig. 9-4b–e (Ehlers and Mendiratta 1982; Leamy and Kayser 1969; Marcinkowski and Brown 1961; Saburi, Yamauchi, and Nenno 1972). Associated with the imperfect variants will be deformation-induced APB trails (Fig. 9-4c–e).

At compositions of 24–26% Al, ordinary single dislocations of the $1/4a'_0\langle 111\rangle$ type, trailed by slip-induced $1/4a'_0\langle 111\rangle$-type APB's (similar to Fig. 9-4e), have been shown to be responsible for the deformation (Saburi, Yamauchi, and Nenno 1972). This has been attributed to the relatively low energy of the $1/4a'_0\langle 111\rangle$APB (enclosing the B2 domains) in this alloy (Marcinkowski and Brown 1961). NNNAPB trails have been observed (Ehlers and Mendiratta 1982; Mendiratta et al. 1987), suggesting that, with a low degree of order, NNNAPB energies in the $D0_3$ structure are low enough to allow for dissociation of the superdislocations into imperfect variants of the $1/4a'_0\langle 111\rangle$-type (Fig. 9-4c), leaving NNNABP trails in their wake (Ehlers and Mendiratta 1982, Leamy and Kayser 1969a; Marcinkowski and Brown 1961).

As the aluminum content increases within the $D0_3$ phase, the APB energy associated with the $D0_3$ order (NNNAPB) decreases while the APB energy associated with the B2 order (NNAPB) increases (Crawford and Ray 1977; Leamy and Kayser 1969a). Above about 30% Al, the second-nearest-neighbor interaction energy associated with the inner pair of

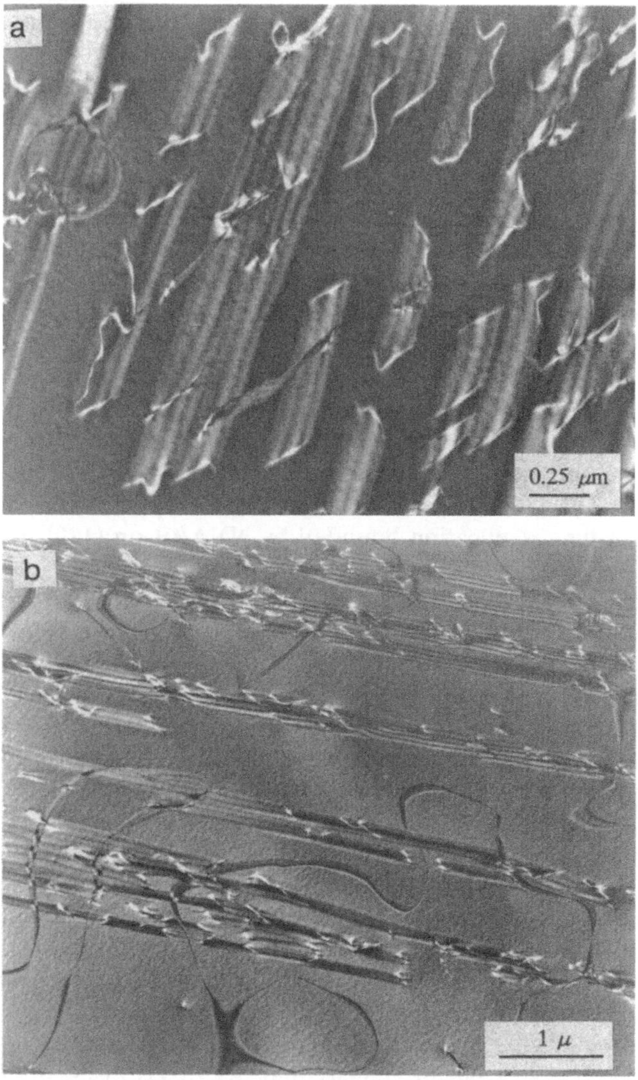

Figure 9-6. Transmission electron micrographs of Fe-26 at.% Al, (a)⟨200⟩ dark field image showing ⟨111⟩ faults between dislocations, (b)⟨111⟩ dark field image showing APB trails produced by the passage of dissociated superlattice dislocations (Courtesy of J. A. Horton, Oak Ridge National Laboratory).

dislocations becomes so low that the fourfold $D0_3$ dislocation configuration may not be able to hold together. Instead, only two $1/4a'_0\langle 111\rangle$ dislocations may move together, leaving a $1/2a'_0\langle 100\rangle$ APB behind them (Fig. 9-4c). Thus deformation by $1/4a'_0\langle 111\rangle$ dislocation pairs (forming $1/2a'_0\langle 111\rangle$ imperfect superlattice dislocations, Fig. 9-4b), sometimes with trailing NNNAPBs (as in Fig. 9-4d), would be expected and have been seen to occur (Crawford and Ray 1977; Mendiratta et al. 1987). Also, the number of superlattice dislocations with $\langle 111\rangle$ fault vectors are expected to decrease with an increase in aluminum level, while those with $\langle 100\rangle$ fault vectors should increase. Because APB energies in the {112} and {321} planes are not much less than APB energies in the {110} plane, cross slip to those planes in the $\langle 111\rangle$ zone can also occur, just as in b.c.c. metals. In the case of Fe_3Al which is quenched from above the $D0_3$-B2 transition temperature, resulting in a mixture of $D0_3$ and B2 order (Taylor and Jones 1958), or in the case of deformation at higher temperatures (> 300° C) (Mendiratta et al. 1987), deformation occurs by the movement of two $1/2a_0\langle 111\rangle$-type dislocation bounded by NNAPB on {110} planes, as was theorized by Marcinkowski and Brown (1961).

In the B2 phase, in which ordering involves only the first neighbors, the dislocation configuration should theoretically consist of two $1/2a_0$ $\langle 111\rangle$ dislocations connected by a $1/2a_0\langle 111\rangle$ antiphase boundary (APB) (Fig. 9-5a), where a_0 is the lattice parameter of the B2 unit cell (Crawford et al. 1973; Marcinkowski and Brown 1961, 1962). However, studies have indicated that room-temperature deformation of B2-ordered FeAl occurs by motion of $a_0\langle 111\rangle$-type dislocations on {110} planes (Fig. 9-5c) (Mendiratta, Kim, and Lipsitt, 1984; Munroe and Baker 1991). Especially in compositions above approximately 36 at% Al, the NNAPB energy becomes so high that the spacing between the two dislocations that make up the superdislocation cannot be resolved by TEM and only one dislocation is usually observed. In these high-aluminum compositions, a transition to the $\langle 100\rangle$ slip direction occurs at temperatures of around 0.4 to 0.5 of the melting temperature (T_m) (Baker and Gaydosh 1987; Baker and Monroe 1990; Crawford 1976; Munroe and Baker 1991; Yamagata and Yoshida 1973). TEM results suggest that the $\langle 100\rangle$ dislocations may form (even at room temperature) as a result of energetically favorable reactions between $\langle 111\rangle$ dislocations, resulting in transgranular cleavage on {100} planes (Baker and Gaydosh 1987; Munroe and Baker 1991). Since for FeAl, T_m increases with decreasing aluminum content, the temperature at which $\langle 100\rangle$ slip occurs also increases with decreasing aluminum content and a mixture of both types can be observed in TEM at intermediate temperatures [Mendiratta et al. 1984; Umakoshi and Yamaguchi 1981].

5. MECHANICAL PROPERTIES

5.1. Tensile Strength

The mechanical properties of iron aluminides are very sensitive to many factors, including aluminum content, order (type, amount, size of ordered domains), heat treatment, test temperatures, alloying additions, environment, microstructure, and defects. In general, the room-temperature tensile yield strength of binary alloys tends to rise with increasing aluminum addition, reaches a peak near the Fe_3Al stoichiometric composition and then decreases again (Crimp and Vedula 1986; Justusson et al. 1957; Leamy and Kayser 1969; Leamy et al. 1969a; Mendiratta et al. 1987; Morgand, Mouturat, and Sainfort 1968). Similar trends have been observed in the hardness (Nagpal and Baker 1990; Westbrook 1956).

For binary alloys, the highest yield strength is associated with the Fe_3Al stoichiometric composition, as shown in Fig. 9-7a. The reason for this higher strength appears to be twofold. First, a much higher degree of $D0_3$ order is attainable at this composition. The many APB's produced on ordering serve as barriers to movement of dislocations. Also, since the $D0_3$ superlattice dislocation is comprised of four individual dislocations, it has great difficulty in cross slipping from one plane to another. The interaction of dislocations with APB's, whether the APB's are produced thermally or by movement of superpartial dislocations (as described above), contributes significantly to strengthening (Leamy and Kayser 1969). For example, a maximum in hardness at a long-range order parameter of 0.4–0.5 in Fe-24.6 at% Al has been attributed to the interaction of imperfect superlattice dislocations with domain boundaries (Davies 1964; Davies and Stoloff 1963). Second, in the Fe_3Al stoichiometric composition, variations in the degree and type of order can be obtained by quenching from temperatures where the disordered or B2- or $D0_3$-ordered phases exist. Such heat treatments, especially those to produce a dispersion of disordered α phase (Inouye 1985), can result in increased room-temperature tensile yield strength.

As a function of test temperature, the ordered Fe_3Al compositions experience a yield peak near the $D0_3$-B2 transition temperature, as observed for the 28% and 35.8% Al binary alloys in Figure 9-8a. Because the nucleation, interaction, and pileup of superlattice dislocations on slip planes during deformation, combined with variations in temperatures and interactions with APB's, result in complicated relationships between yielding and work hardening, several theories to explain the observed deformation characteristics near the $D0_3$-B2 transition have been proposed in the literature (Leamy and Kayser 1969a; Leamy et al. 1969a; Marcinkowski

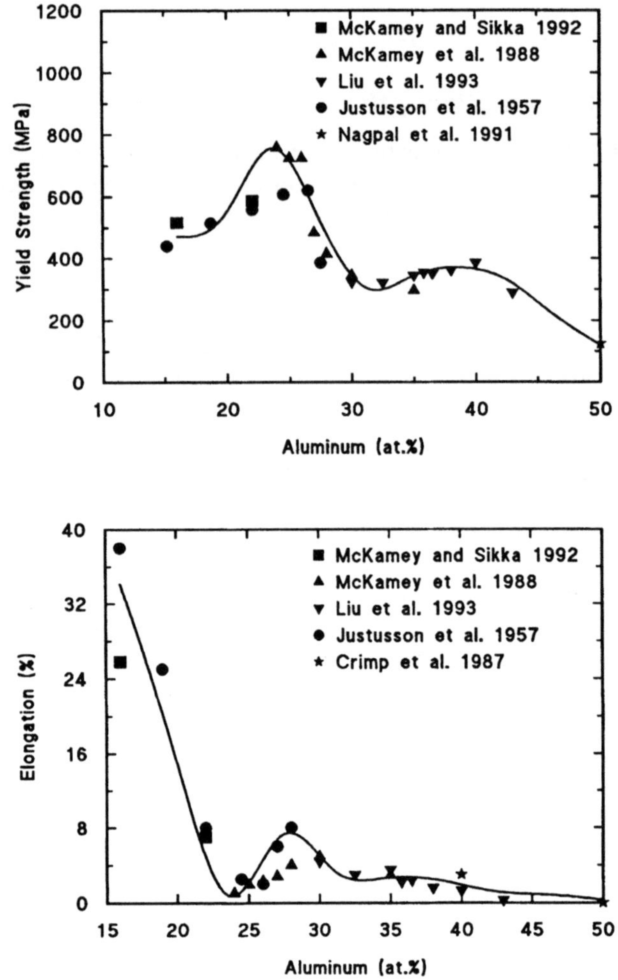

(b)

Figure 9-7. Room temperature tensile (a) yield strength and (b) elongation as a function of aluminum concentration in Fe-Al binary alloys (Crimp, M. A., Vedula K. M., and Gaydosh, D. J., 1987, MRS Symp. Proc. 81: 499–504. Justusson, W., Zackay, V. F., and Morgan, E. R., 1957. *Trans. ASM* 49: 905–21; Liu, C. T., Sikka, V. K., and McKamey, C. G., 1993. *Alloy Development of FeAl Aluminide Alloys for Structural Use in Corrosive Environments,* ORNL/TM-12199. Oak Ridge, TN: Oak Ridge National Laboratory; McKamey, C. G., and Sikka, V. K., 1992. Unpublished data; McKamey, C. G., Liu, C. T., David, S. A., Horton, J. A., Pierce, D. H., and Campbell, J. J., 1988. *Development of Iron Aluminides for Coal Conversion Systems,* ORNL/TM-10793. Oak Ridge, TN: Oak Ridge National Laboratory; Nagpal, P., and Baker, I., 1990. *Met Trans.* A21: 2281–82).

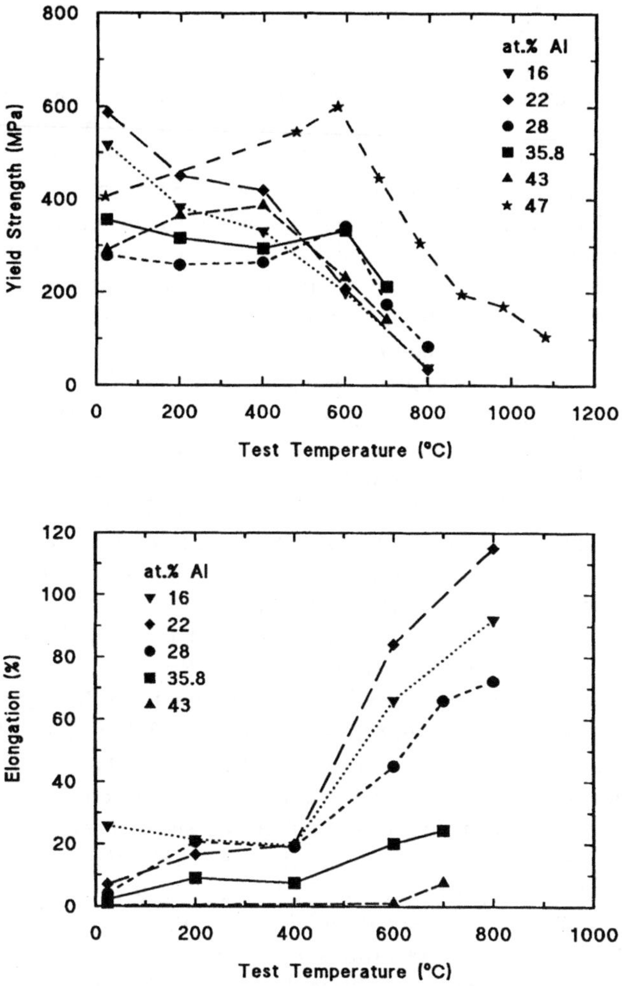

(b)

Figure 9-8. Tensile (a) yield strength and (b) elongation as a function of test temperature for several Fe-Al binary alloys (Liu, C. T., Sikka, V. K., and McKamey, C. G., 1993. *Alloy Development of FeAl Aluminide Alloys for Structural Use in Corrosive Environments,* ORNL/TM-12199. Oak Ridge, TN: Oak Ridge National Laboratory; McKamey, C. G., and Sikka, V. K., 1992. Unpublished data).

and Brown 1961; Stoloff and Davies 1964; Umakoshi and Yamaguchi 1980). Depending on the heat-treatment condition, a peak in flow stress can occur at intermediate temperatures of approximately 450° C to 600° C (Stoloff and Davies 1964). Stoloff and Davies (1964, 1966) postulated that the peak is a consequence of a transition from the motion of perfect superlattice dislocations in a highly ordered lattice at lower temperatures to the motion of individual dislocations in a partially ordered matrix at higher temperatures. The increased order at lower temperatures results in restraining forces on the individual dislocations because of APB creation leading to hardening of the matrix, while a lowering of order with increasing temperature results in lower APB energies and dissociation of the superdislocations into partial superlattice dislocations or unit dislocations. This theory suggests that the position of the peak with respect to T_c and the deformation behavior near the peak depend on the temperature dependence of the degree of long-range order (S). The temperature dependence of S can vary significantly with composition near Fe_3Al, resulting in the appearance of yield peaks at temperatures either below or above T_c, depending on aluminum composition (Hanada et al. 1981; Lawley and Cahn 1961). Recent studies have indicated that the precipitation of α phase significantly affects the anomalous yield behavior in hypostoichiometric alloys, where a first-order transition occurs from the α phase to the $D0_3$ structure (Inouye 1985).

With increasing aluminum content, the $D0_3$ APB energy decreases and the B2 energy increases (Crawford and Ray 1977; Leamy and Kayser 1969). Strengthening in the B2 FeAl compositions has been associated with retention of nonequilibrium vacancies (Baker and Munroe 1990; Nagpal and Baker 1990). As in other B2 compounds, deviations from the stoichiometric composition are accommodated by constitutional defects, that is, by either vacancies on the sublattice sites of the deficient element or antisite atoms of the element in excess (Bradley and Taylor 1937). Vacancies can be produced in FeAl-based alloys by quenching from elevated temperatures (Nagpal and Baker 1990; Paris et al. 1975; Riviere and Grilhe 1972). Concentrations of quenched-in vacancies as high as 1.85 at% have been reported (Paris, Lesbats, and Levy 1975). Studies have also suggested an increase in strength with grain refinement following a Hall–Petch relationship (Baker et al. 1991; D. Morris and M. Morris 1990).

With increasing temperatures above about 600° C, the strength of iron aluminides decreases drastically, while the elongation increases, as shown in Figure 9-8. Along with low room-temperature ductility, the decrease in strength above 600° C is a factor that has limited the use of these materials for structural applications at these high temperatures where their excellent corrosion resistance would be very beneficial. Efforts are continuing in a number of laboratories to develop iron aluminides with improved high-temperature tensile, creep, and fatigue properties (Diehm and Mikkola 1987; McKamey et al. 1991).

5.2. Ductility, Fracture and Environmental Embrittlement

At room temperature, binary compositions below approximately 18 at% Al fail in a ductile manner by void nucleation and coalescence and exhibit tensile ductilities of over 20% (see Fig. 9-7b). With the onset of the ordered $D0_3$ and B2 structures, however, ductility decreases rapidly, even though single crystals of both Fe_3Al and FeAl are reasonably ductile in compression (Crawford 1976; Leamy et al. 1969b; Yamagata and Yoshida, 1973). At temperatures below the ductile-brittle transition. FeAl and Fe_3Al characteristically exhibit limited tensile ductility and brittle fracture. Typical tensile elongation values are less than 10% for Fe_3Al-based binary compositions and less than 4% for aluminum compositions above approximately 35%, see Fig. 9-7b (Crimp et al. 1987; Justusson et al. 1957; McKamey et al. 1991; Vyas et al. 1992). The fracture mode of these alloys depends on aluminum content; those alloys containing $< 40\%$ Al fracture by transgranular cleavage (Kerr 1986; Mendiratta et al. 1987), while those with $> 40\%$ Al fracture intergranularly (Liu et al. 1991; Marcinkowski, Taylor, and Kayser 1975). Although the ordered iron aluminides possess low room-temperature ductility, they have all the intrinsic characteristics potentially required for good ductility. For example, they deform by $\{110\}\langle 111\rangle$ slip (like disordered b.c.c. metals), and therefore have more than five independent slip systems. They do not have the typical high-strength values at room temperature that would cause low ductility (recrystallized yield strengths are on the order of 300–500 MPa [McKamey et al. 1991]). Although FeAl fractures intergranularly due to weak grain boundaries (Liu et al. 1991), Fe_3Al does not suffer from weak grain boundaries. Until recently, the poor ductility of Fe_3Al-based compositions was assumed to reflect an intrinsically low cleavage strength (Mendiratta et al. 1987), in combination with a retardation of cross slip due to the ordered phase (Marcinkowski et al. 1975). However, Liu et al. (1989, 1990b, 1991; Liu and McKamey 1990) have now shown that the ductilities of both FeAl and Fe_3Al are significantly decreased through the reaction of water vapor in the test environment with aluminum in the alloy. Studies have indicated that this type of environmental embrittlement is a major cause of low ductility of many intermetallic alloy systems that contain large amounts of reactive elements such as aluminum, silicon, and titanium (Kuruvilla and Stoloff 1987; Liu and McKamey 1990; McKamey and Liu 1992, Stoloff 1988). Because iron-aluminide alloys contain relatively large concentrations of aluminum, it is postulated that the aluminum reacts with the water vapor molecule, producing alumina (or a precursor oxide) and chemisorbed hydrogen, which induces classic hydrogen embrittlement. (Environmental embrittlement is covered more extensively in another chapter of this book.)

Both Fe_3Al and FeAl are susceptible to environmental embrittlement in the presence of water vapor, although the degree of sensitivity of Fe_3Al appears to be less than FeAl (Liu et al. 1989, 1990, 1991; Liu and McKamey 1990). In a series of tests in which an environmental test

Table 9-2 Effect of Test Environment on Room-Temperature Tensile Properties of FeAl (36.5% Al) and Fe_3Al (28% Al)

Test Environment (Gas pressure)	Elongation (%)	Yield Strength (MPa)	Ultimate Tensile Strength (MPa)
ANNEALED 1H / 900° C + 2h / 700° C			
FeAl			
Air	2.2	360	412
Vacuum ($\sim 1 \times 10^{-4}$ Pa)	5.4	352	501
Oxygen (6.7×10^4 Pa)	17.6	360	805
Ar + 4% H_2 (6.7×10^4 Pa)	6.2	379	579
H_2O Vapor (67 Pa)	2.4	368	430
Fe_3Al			
Air	4.1	387	559
Vacuum (-1×10^-4 Pa)	12.8	387	851
Ar + 4% H_2(6.7×10^4 Pa)	8.4	385	731
Oxygen (6.7×10^4 Pa)	12.0	392	867
H_2O Vapor (1.3×10^3 Pa)	2.1	387	475
ANNEALED 1H / 850° C + 5D / 500° C			
Fe_3Al			
Air	3.7	279	514
Vacuum ($\sim 1 \times 10^{-4}$ Pa)	12.4	316	813
Oxygen (6.7×10^4 Pa)	11.7	298	888
H_2O Vapor*	2.1	322	439

*Air saturated with water vapor was leaked into the vacuum chamber.

Source: McKamey and Liu 1992

chamber was first evacuated to a pressure of $< 1 \times 10^{-4}$ Pa, various gaseous species (including oxygen, hydrogen, and water vapor) were leaked into the test chamber. By this method, the embrittling agent was identified as hydrogen produced by the dissociation of water vapor. For Fe_3Al the maximum ductility is attained in either oxygen or vacuum environments (see Table 9-2) (McKamey and Liu 1992), while the best ductility for FeAl containing 36.5 at% Al (17.6%) is obtained in dry oxygen (Table 9-2 and Fig. 9-9) (Gaydosh and Nathal 1990; Liu et al. 1991). In the presence of large amounts of oxygen, the oxidation reaction at the crack tip occurs between aluminum and oxygen, not between aluminum and water vapor, thereby reducing the amount of hydrogen that is generated.

The binary Fe_3Al alloy always shows cleavage fracture, independent of ductility and test environment. The FeAl alloy containing 36.5% Al, on the other hand, exhibits a distinct change in fracture mode from transgranular cleavage to intergranular fracture as environmental embrittlement is minimized and ductility increases (Fig. 9-9) (Liu et al. 1989, 1991

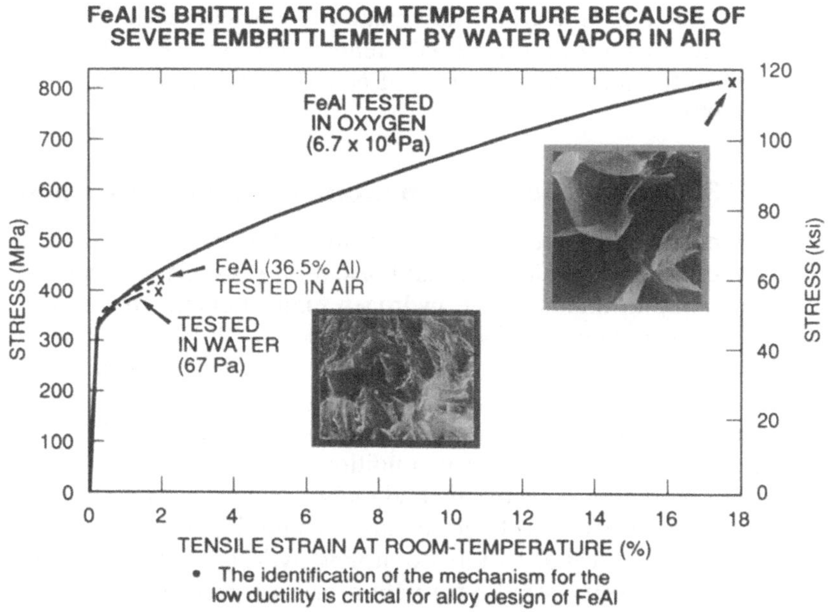

Figure 9-9. The effect of test environment on ductility and fracture behavior of FeAl(36.5% Al) tested at room temperature (Liu, C. T., and McKamey, C. G., 1990a. In *High Temperature Aluminides and Intermetallics* 133–51. Eds. S. H. Whang, C. T. Liu, D. P. Pope, and J. O. Stiegler. Warrendale PA: The Materials Society).

McKamey and Liu 1992). The comparison of the fracture modes indicates that the grain boundaries in Fe_3Al are basically strong enough to resist crack nucleation and propagation, even after extensive deformation that results in high-stress concentrations at these boundaries. In Fe-36.5% Al, the occurrence of cleavage fracture in air tests indicates that the cleavage planes are more susceptible to embrittlement than are the grain boundaries. When the embrittlement effect of moisture is removed by testing in dry oxygen, the fracture path follows the grain boundaries, which are the (intrinsic) weakest links in the aluminide. The stoichiometric FeAl composition has almost no ductility, fracturing on the grain boundaries before yielding (Crimp, Vedula, and Gaydosh, 1987). Reducing the aluminum composition within the B2 phase field reduces the yield stress, resulting in the observation of some room-temperature ductility (see Fig 9-7b).

Much more work needs to be done to unravel the details of environmental embrittlement in iron aluminides. What is clear from these recent results, however, is that Fe_3Al and FeAl are intrinsically much more ductile than previously realized, and that the low ductility commonly observed at room temperature is the result of an extrinsic factor, namely moisture-induced hydrogen embrittlement. At temperatures above the ductile transition, fracture of all compositions occurs in a ductile manner and tensile ductilities increase with temperature (see Fig. 9-8b), reaching over 100% for some compositions at 600° C (McKamey et al. 1991; Sikka et al. 1990).

5.3. Minimization of Environmental Embrittlement

Because of the great potential of iron aluminides for use in applications where superior sulfidation and oxidation resistance is needed, efforts are underway to minimize the hydrogen embrittlement that is a result of the reaction of aluminum with water vapor in the atmosphere. Results indicate that improved room-temperature tensile ductilities can be produced in iron aluminides by controlling the (a) composition, (b) microstructure (grain size and shape, dislocation character), (c) ordered phase structure, and (d) surface conditions.

As noted above, environmental embrittlement in iron aluminides has been associated with the large amount of aluminum in the alloys, which reacts with water vapor to release atomic hydrogen. However, that same reaction is necessary for the formation of a protective alumina film that gives iron aluminides their excellent corrosion resistance. For applications where the requirement for corrosion resistance can be relaxed somewhat, room-temperature ductilities of over 20% can be attained by lowering the aluminum content to less than approximately 18 at% (see Fig. 9-10) (Sikka

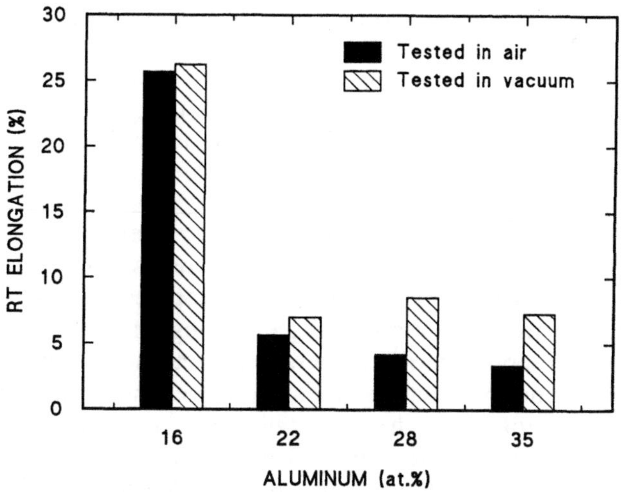

Figure 9-10. Effect of aluminum addition on the tensile ductility at room temperature. The difference between air and vacuum tests is an indication of the effect of environmentally-induced hydrogen embrittlement in the ordered compositions of ≥ 22% Al (Sikka, V. K., Viswanathan, S., and McKamey, C. G., 1993. *Structural Intermetallics*, pp. 483–491. Eds. R. Darolia, J. J. Lewandowski, C. T. Liu, P. L. Martin, D. B. Miracle, and M. V. Nathal. Warrendale, PA: The Materials Society).

et al. 1993). This results not only in an alloy with less reactive aluminum, but also moves the composition out of the ordered regions of the phase diagram. The ordered structures, especially the $D0_3$ structure, have been shown to be most susceptible to this type of hydrogen embrittlement (Castagna and Stoloff 1992; McKamey and Pierce 1993).

Studies have shown that the ductility of the ordered Fe_3Al-based aluminide can be substantially improved by increasing aluminum content from 25 to 28 at% (McKamey et al. 1987) and by adding chromium at a level of 2–6% (McKamey et al. 1989). The increase in aluminum concentration sharply decreases the yield strength of the aluminide by eliminating the hardening effect of precipitated α phase (Inouye 1985, McKamey et al. 1987). The beneficial effect of chromium on ductility has not been totally explained, but studies suggest several possible mechanisms. These include

1. Enhanced cleavage strength and partially suppressed cleavage fracture (McKamey et al. 1989).
2. Reduced APB energy, which increases the separation between the four superpartial dislocations, allowing easier dissociation of the superdislocations and easier cross slip (McKamey et al. 1989).

3. Stabilization of the B2 structure with concomitant reduction in $D0_3$ character (Munroe and Baker 1993).
4. Reduced environmental embrittlement due to some as-yet unexplained effect on the water vapor/aluminum reaction, e.g., composition of oxide scale, recombination of atomic hydrogen to molecular hydrogen, etc. (McKamey and Liu 1990, Morris et al. 1993b).

These alloys can be improved further by thermomechanical treatment and alloying with other elements to improve strength (Sikka 1991b). Table 9-3 shows room-temperature tensile properties for several Fe_3Al-based alloys versus heat treatment. Depending on which elements are added for strengthening, room-temperature ductilities of over 16% can be achieved by varying the heat treatment and quenching procedure. Stress-relief heat treatment below the recrystallization temperature, followed by oil quenching (heat treatment B in Table 9-3), has produced dramatic increases in ductility, which are believed to be due to three factors

1. Oil quenching after heat treatment leaves a residue of oil in any cracks present on the specimen surface and thus minimizes the entrance of hydrogen during stressing (Sikka 1991b).
2. Heat treating at 700–750° C relieves stress in sheet but maintains the elongated grain morphology of the rolled microstructure. This unrecrystallized microstructure provides a minimum of transverse cleavage planes, thereby disrupting the fracture path and making it more difficult for hydrogen to enter (McKamey and Pierce 1993; Viswanathan et al. 1993). This is further indicated in Figure 9-11, which shows a minimum ductility

Table 9-3 Effect of Heat Treatment and Composition on Room Temperature Tensile Properties of Fe_3Al-based Alloys.

	HEAT TREATMENT A[a]			HEAT TREATMENT B[b]		
Alloy Composition (at%)	Yield (MPa)	Ultimate (MPa)	Elong. (%)	Yield (MPa)	Ultimate (MPa)	Elong. (%)
Fe-28Al	279	514	3.7	434	716	8.0
Fe-28Al-4Cr	228	554	8.2	382	752	14.6
Fe-28Al-5Cr-.1Zr-.05B	312	546	7.2	480	973	16.4
Fe-28Al-5Cr-.5Nb-.2C	320	679	7.8	384	930	16.9
Fe-28Al-5Cr-.5Nb-.5Mo-.1Zr-.2B	379	630	5.0	589	965	10.2

[a]Heat treatment A = 1 h at 850° C plus 5-7 d at 500° C in air.

[b]Heat treatment B = 1 h at 750° C in air plus oil quench.

Source: McKamey and Liu 1992

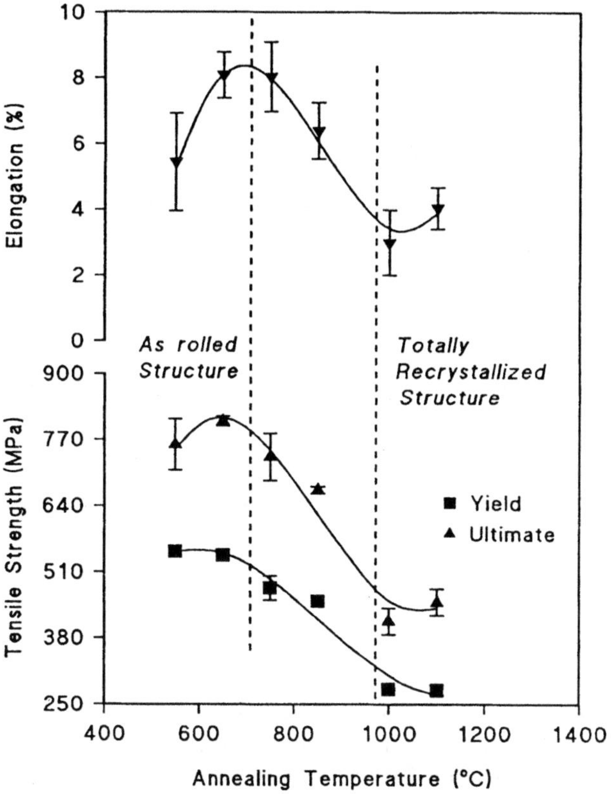

Figure 9-11. Room temperature tensile properties of a Fe_3Al-based alloy as a function of the pre-test annealing temperature (McKamey, C. G., and Pierce, D. H., 1993a; reprinted from *Scripta Metallurgica et Materialia*, Volume 28, pages 1173–1176, copyright 1993, with kind permission from Elsevier Science Ltd., The Boulevard, Langford Lane, Kidlington OX5 1GB, UK).

for totally recrystallized specimens and a maximum for specimens annealed at temperatures that produced stress-relief with minimum recrystallization.

3. Heat treatment at 700–750° C followed by rapid quenching maximizes the retention of B2-ordered phase and minimizes the amount of $D0_3$ structure (Sun et al. 1993), which has been shown, through tensile and fatigue tests, to be more environmentally sensitive than the B2 structure in the Fe_3Al-based compositions (Castagna and Stoloff 1992; McKamey and Pierce 1993; Mendiratta et al. 1987; Sikka 1991b).

Minimization of hydrogen in the testing atmosphere for FeAl alloys of 40% Al and higher has not resulted in the same significant increases in ductility observed in alloys with less aluminum. Table 9-4 shows the tensile

Table 9-4 Effect of Aluminum Concentration and Boron Addition on Ductility and Fracture Behavior of FeAl Alloys Tested in Air and Oxygen Atmospheres

Nominal Al Concentration (at%)	Test Environment	STRENGTH (MPA) Yield	Ultimate	Ductility (%)	Fracture mode[a]
36.5	Air	360	412	2.2	TF
36.5	Oxygen[b]	360	805	17.6	GBF
40.0	Air	390	405	1.2	GBF
40.0	Oxygen[b]	402	537	3.2	GBF
40.0 + B[c]	Air	391	577	4.3	TF
40.0 + B[c]	Oxygen[b]	392	923	16.8	TF

[a]GBF = grain boundary fracture; TF = mainly transgranular fracture (cleavage).

[b]Oxygen pressure = 6.7×10^4 Pa (500 torr).

[c]Doped with 300 wt ppm boron.

Source: Liu and George 1990; reprinted with permission from *Scripta Metallurgica et Materialia*, Volume 24, C. T. Liu and E. P. George, pages 1285–1290, 1990, Elsevier Science Ltd., Oxford, England.

results obtained from Fe-36.5 and 40 at% Al tested in air and in dry oxygen (oxygen pressure = 6.7×10^4 Pa)(Liu and George 1990, 1991). The 36.5% Al alloy showed a large increase in tensile ductility (2.2% to 17.6%) when the test environment was changed from air to oxygen. This result was not observed for the 40% Al alloy. Examination of the fracture modes in scanning electron microscopy (SEM) (Table 9-4 and Fig. 9-12a, b) indicated that the 36.5% Al alloy failed by transgranular cleavage in air, while the 40% Al alloy showed brittle grain-boundary fracture. This observation is consistent with the general trend observed in Fe_3Al-FeAl alloys: the propensity for grain-boundary fracture increases with increasing aluminum level (Baker and Gaydosh 1987, Crimp et al. 1987, Horton, Liu and Koch 1984; Kerr 1986; McKamey et al. 1987), and a change in fracture mode usually occurs around 40% Al. The 40% Al alloy showed only a small increase in ductility when tested in oxygen and its fracture mode was not affected by the test environment. These results suggest that the grain boundaries in the 40% Al alloy are relatively weak, and brittle intergranular fracture is the major cause for its low ductility at room temperature. Even in a controlled atmosphere, the alloy fractures with limited ductility because its weak grain boundaries fracture prematurely. Boron, which has been added to Ni_3Al and NiAl for grain boundary strengthening (Aoki and Izumi 1979; George and Liu 1990; Liu, White, and Horton 1985; Liu and Sikka 1986; Taub, Huang, and Chang 1984), has a strong tendency to segregate to grain boundaries and has been shown to

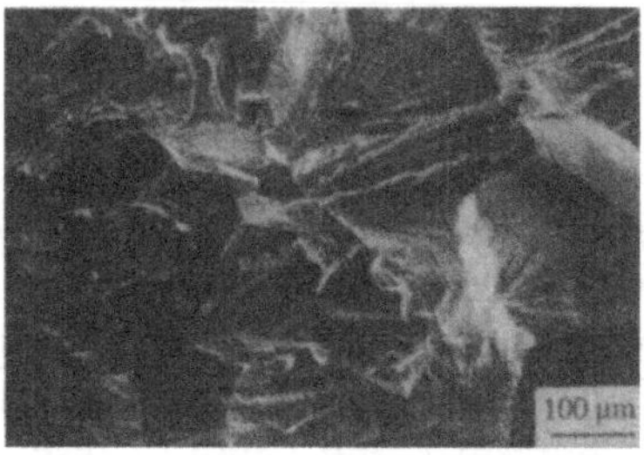

(*a*)

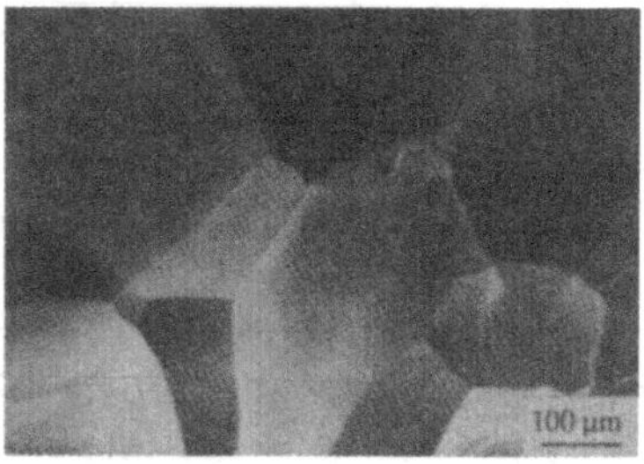

(*b*)

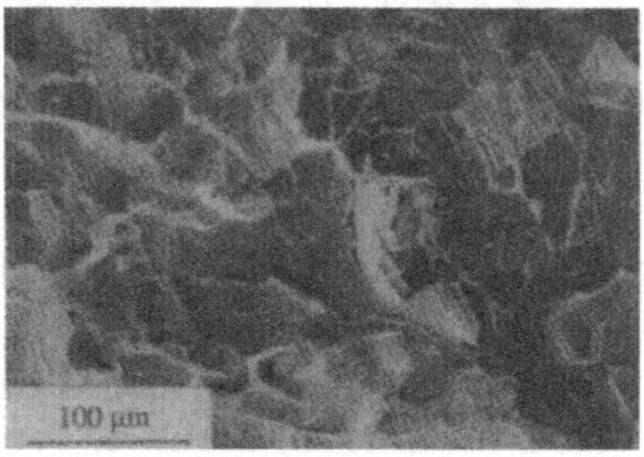

(*c*)

Figure 9-12. Fracture modes of FeAl alloys tested in air at room temperature: (a) 36.5% Al, (b) 40% Al, and (c) boron-doped 40% Al (Liu, C. T., and George, E. P., 1990c; reprinted from *Scripta Metallurgica et Materialia*, Volume 24, pages 1285–1990, copyright 1990, with kind permission from Elsevier Science Ltd., The Boulevard, Langford Lane, Kidlington OX5 1GB, UK).

suppress brittle intergranular fracture in FeAl (40% Al) produced by a powder-metallurgy process (Crimp and Vedula 1986; Crimp et al. 1987; Gaydosh, Draper, and Nathal 1989). Addition of only 300 wt ppm boron to melted and cast Fe-40% Al resulted in transgranular cleavage (Fig. 9-12c), with some grain-boundary facets visible. An increase in ductility in air of about 3% was seen, which is in agreement with the data reported for the powder -metallurgy product. The striking feature in Table 9-4 is that the boron-doped alloy shows a significant increase in ductility when tested in oxygen, indicating that boron-doped FeAl alloys ($\geq$ 40% Al), like the 36.5% Al alloy, exhibits environmental embrittlement in air which can be alleviated by testing in oxygen. Also, the boron-doped 40% Al alloy showed mainly transgranular fracture, even after extensive plastic deformation in the oxygen atmosphere, indicating that boron is effective in strengthening grain boundaries and suppressing grain-boundary fracture in Fe-Al-based aluminides. This boron segregation was confirmed by Auger analysis (Liu and George 1990, 1991).

5.4. Creep and Fatigue

High-temperature creep in ordered iron aluminides is assumed to be controlled by many of the same mechanisms that operate in ordinary metals and alloys (e.g., diffusional processes, dislocation glide and climb, and grain-boundary sliding) (Sherby and Burke 1968), but with some additional effects due to the ordered atomic structure (Stoloff and Davies 1966). The activation energy for creep approximates that for self-diffusion in pure iron and the rate-controlling mechanism is the climb of dislocations, as is observed for pure metals (Lawley, Coll, and Kahn 1960, McKamey, Masziasz, and Jones 1992). An increase in creep resistance upon ordering is associated with the increase in activation energy for diffusion necessary to overcome unfavorable bonding in an ordered lattice. At temperatures above T_c, binary iron aluminides follow a power law for creep (Lawley et al. 1960).

Although, in general, ordered alloys have creep-activation energies that are higher and secondary creep rates that are lower than those found in random alloys, theoretically resulting in comparatively better creep resistance in the ordered intermetallic alloys, iron aluminides have been known for some time to have poor high-temperature creep resistance, mainly due to their b.c.c. crystal structure (McKamey et al. 1991; Sikka and Baldwin 1992). The beneficial effects of long-range order may also be canceled by the effects of constitutional defects and thermal vacancies, especially in the B2 structure (Baker and Munroe 1990). The importance of additional effects, such as nonstoichiometry, antiphase domain boundaries, and the nature and structure of superdislocations and grain boundaries, and their role in deformation and fracture behavior, are still unknown.

Only a few studies of creep properties and processes in iron aluminides have been conducted (Jung and Sauthoff 1989; Lawley et al. 1960; McKamey et al. 1992; McKamey et al. 1993; Sikka and Baldwin 1992; Whittenberger 1983, 1986). The disordered low-aluminum binary-iron aluminides have creep resistance similar to other b.c.c. metals, but creep resistance increases slightly in the ordered compositions. Creep-rupture resistance in binary Fe_3Al at temperatures near 600° C is very poor, with failure occurring through gross separation of grain boundaries with limited plastic deformation of individual grains. The evidence indicates that the short creep life, high creep rate and brittle intergranular failure of the binary Fe_3Al alloy is probably due to a combination of weak high-angle grain boundaries, the reduced ability of dislocations to glide or multiply to produce matrix plasticity, and to formation of large creep cavities at triple points (McKamey et al 1992). Creep-rupture resistance can be improved by alloying additions, as will be discussed later in this chapter.

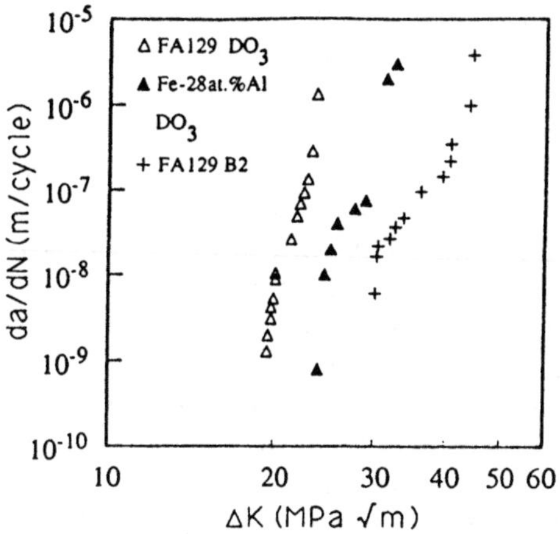

Figure 9-13. Comparison of fatigue crack growth in $D0_3$ and B2 conditions for Fe_3Al alloy FA-129 (Fe-28 at.% Al-5 at.% Cr-0.5 at.% Nb-0.2 at.% C) with binary Fe-28 at. % Al in the $D0_3$ condition (Castagna, A., and Stoloff, N. S.. 1992; reprinted from *Scripta Metallurgica et Materialia*, Volume 26, pages 673–678, copyright 1992, with kind permission from Elsevier Science Ltd., The Boulevard, Langford Lane, Kidlington OX5 1GB, UK).

Fatigue behavior of ordered iron aluminides is influenced by composition, temperature, and type and degree of order (Fuchs and Stoloff 1988). Although detailed data on the effect of these variables on low-cycle fatigue is lacking, recent studies of high-cycle fatigue at room temperature indicates that crack initiation in both B2 and $D0_3$-ordered Fe_3Al-based alloys occurs at grain boundaries at or near the surface (Stoloff et al. 1987). The crack then propagates transgranularly to failure. The low crack-growth resistance is a consequence of environmental hydrogen embrittlement, with the $D0_3$-ordered phase being more susceptible than the B2-ordered phase (see Fig. 9-13) (Castagna and Stoloff 1992). Fatigue resistance falls sharply with increasing temperature as a result of a decrease in long-range order and the onset of a creep-fatigue interaction. Fatigue resistance at elevated temperatures has been reported to be higher in the two-phase $\alpha + D0_3$ compositions than in the more highly ordered Fe_3Al compositions (Fuchs and Stoloff 1988).

5.5. Alloying Efforts

The mechanical, physical, welding, fabrication, and corrosion properties of iron aluminides are very sensitive to alloying additions. The results of past studies of the effect of alloying on the properties of Fe_3Al-based iron aluminides have been reviewed previously (Hardwick and Wallwork 1978; Holladay 1961; Lepkowsi and Holladay 1957; Mendiratta et al. 1987b; McKamey et al. 1991; Prakash et al. 1991).

Only chromium has been found to improve the room-temperature tensile ductility of Fe_3Al (Culbertson and Kortovich 1986; McKamey et al. 1989; McKamey and Liu 1990, Morris et al. 1993a, b; Sun et al. 1992). Additions of 2–6% Cr have resulted in a doubling of the room temperature ductility, while lowering the room-temperature yield strength (see Fig. 9-14) (Diehm et al. 1989; Knibloe et al. 1992; McKamey et al. 1989). As noted earlier, the exact mechanism whereby chromium increases the room-temperature ductility is unknown, but it may act to reduce environmental embrittlement in this system through some as-yet unexplained effect on the water vapor/aluminum reaction (McKamey and Liu 1990). Recent studies also suggest that chromium reduces the $D0_3$ character and stabilizes the B2 structure (Munroe and Baker 1993), which is also the phase believed to be less susceptible to environmental embrittlement (Castagna and Stoloff 1992). It may also mitigate the embrittling effects of

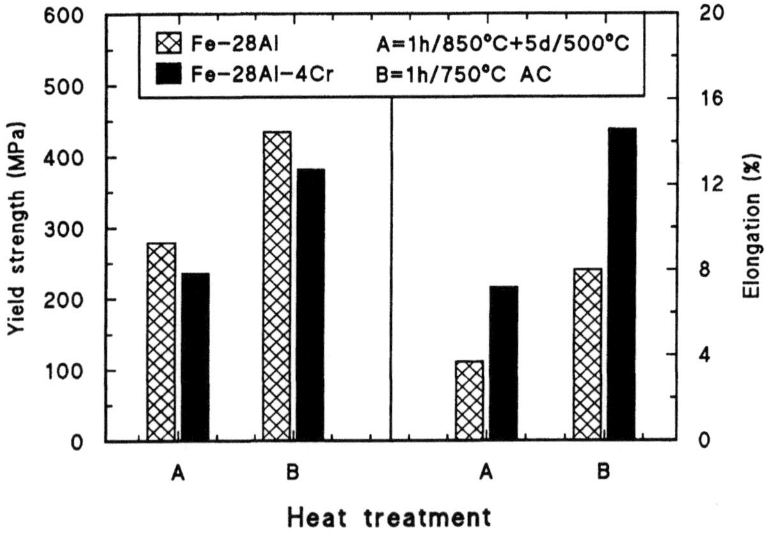

Figure 9-14. Effect of chromium addition on the room temperature tensile properties of Fe-28 at.% Al. Heat treatment A results in $D0_3$ order, while B produces some degree of B2 order (McKamey, C. G., and Sikka, V. K., 1992, unpublished data; McKamey, C. G., Horton, J. A., and Liu, C. T., 1989. *J. Mat. Res.* 4: 1156–63; reprinted with permission).

elements such as molybdenum and titanium, which are added for strength enhancement (Diehm et al. 1989). Recent work also suggests a change in slip plane from {110} and {112} to {123} with addition of chromium (Sun et al. 1992).

High temperature tensile and creep strengths in all iron-aluminide compositions have been shown to be improved with additions of elements such as Si, Ta, Ce, Ti, Mo, Zr, Hf, or Nb, but usually at the expense of room-temperature ductility (Bordeau 1987; Culbertson and Kortovich 1986; Diehm and Mikkola 1987; Dimiduk et al. 1988; Fortnum and Mikkola 1987; Longworth and Mikkola 1987; McKamey et al. 1991, 1992; Sun et al. 1992). Additions of molybdenum or niobium in combination with a small amount of zirconium result in a dramatic improvement in creep-rupture strength of binary Fe_3Al (from 2–5 h up to 100–300 h life at 593° C and 207 MPa) (Maziasz and McKamey 1992; McKamey et al. 1992). Silicon and titanium are potent strengtheners in both tension and stress-rupture (Justusson et al. 1957; Sikka et al. 1993), and silicon has also been reported to increase the oxidation resistance (Schmatz and Zackay 1959). All of these elements have low solubilities in iron aluminides and will readily form either intermetallic precipitates or carbides or borides. These precipitates, especially if they can be precipitated as fine-matrix particles, are potent strengtheners and grain-size refiners (Bordeau 1987; Culbertson and Kortovich 1986; Hardwick and Wallwork 1978; McKamey et al. 1993). By controlling the microstructures with heat treatments and by adding small amounts of the elements listed above (usually to a total of much less than 10 at%) along with microadditions of boron or carbon, strengthening in both tensile (Fig. 9-15) and creep (Fig. 9-16) conditions can be produced with a minimum reduction in room-temperature ductility (Diehm et al. 1987; McKamey et al. 1992, 1993). Among other theories, strengthening by these elemental additions has been explained in terms of precipitation effects (pinning of grain boundaries and dislocations) (Diehm et al. 1987, Maziasz and McKamey 1992b; McKamey et al. 1992), dislocation-APB interactions (Prakash, Buckley, and Jones 1991), or to an increase in T_c (Fortnum and Mikkola 1987).

It is generally agreed that carbide strengthening is most feasible for the iron aluminides. Additions of up to 0.2 wt% carbon to binary Fe_3Al have been reported to have no effect on room or elevated properties (Hardwick and Wallwork 1978). However, more than this amount in the binary alloy can result in the precipitation of Fe-Al-carbides along grain boundaries, resulting in brittle grain-boundary failure (Kerr 1986). Almost any other element added to Fe_3Al with carbon will have a higher affinity for carbon than the iron and aluminum, particularly those reactive elements listed above. The alloying elements to be added and the melting

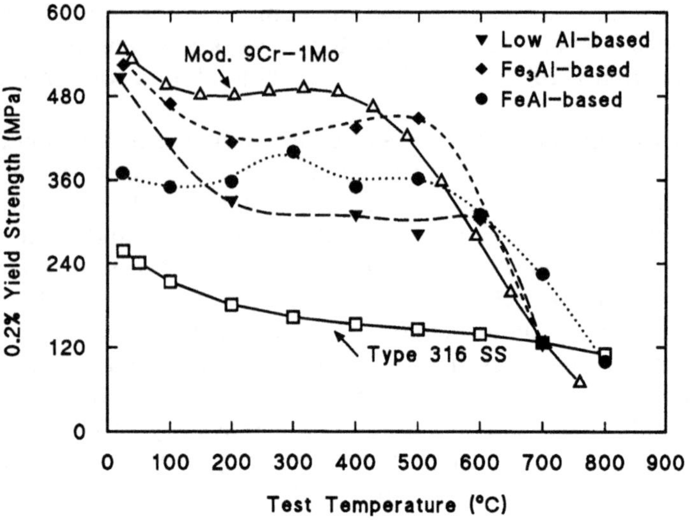

Figure 9-15. Comparison of iron aluminide tensile properties versus test temperature with type 316 stainless steel and modified 9Cr-1Mo steel (Liu, C. T., Sikka, V. K., and McKamey, C. G., 1993. *Alloy Development of FeAl Aluminide Alloys for Structural Use in Corrosive Environments,* ORNL/TM-12199. Oak Ridge, TN: Oak Ridge National Laboratory; McKamey, C. G., and Sikka, V. K., 1992; McKamey, C. G., DeVan, J. H., Tortorelli, P. F., and Sikka, V. K., 1991. *J. Mat. Res.* 6(8): 1779–1805; Sikka, V. K., Viswanathan, S., and McKamey, C. G., 1993. *Structural Intermetallics,* pp. 483–491. Eds. R. Darolia, J. J. Lewandowski, C. T. Liu, P. L. Martin, D. B. Miracle, and M. V. Nathal. Warrendale, PA: The Materials Society).

and fabrication processes should be chosen to ensure that the carbides will precipitate from the liquid, resulting in both refined cast structures on solidification and refined wrought structures after working operations.

Using alloy development and thermomechanical processing techniques to control grain and precipitate sizes and morphologies, Fe_3Al-based alloys with room temperatures ductilities of 10–20%, yield strengths of 500 MPa at temperatures up to 600° C (see Fig. 9-15), and improved creep-rupture strength (see Figs. 9-16 and 9-17) have been produced in experimental-size castings (0.5–7 kg) (Sikka et al. 1990; Sikka 1991b). These properties make Fe_3Al-based alloys more competitive with conventional austenitic and ferritic steels. However, further studies are certainly required to characterize the alloying effects on environmental embrittlement and to maximize their effectiveness in producing improved room-temperature ductility and high-temperature strength in these alloys.

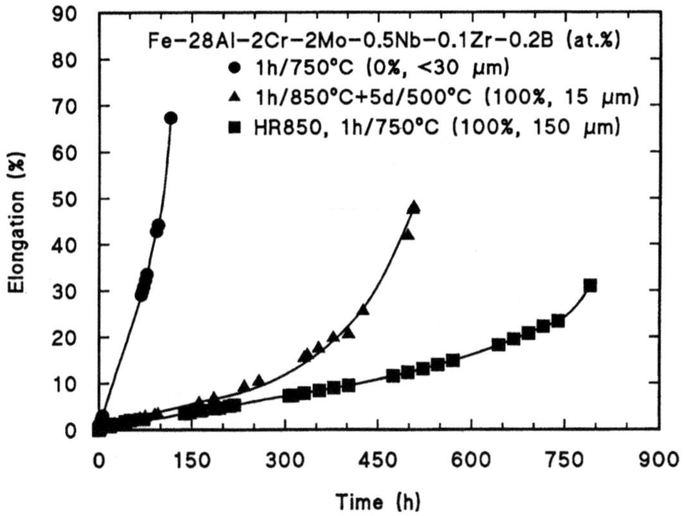

Figure 9-16. Effect of the microstructure of an Fe_3Al-based alloy on the stress-rupture life at a test temperature of 593° C (1100° F) and a test stress of 207 MPa (McKamey, C. G. 1993, Unpublished data).

In binary alloys of greater than approximately 40 at% Al, intrinsically weak grain boundaries, rather than environmental embrittlement, is the dominant mechanism contributing to their poor mechanical properties (Baker and Munroe 1990; Stephens 1985; Vedula and Stephens 1987), resulting in low ductility and brittle grain boundary fracture (see Fig. 9-12b and Table 9-4). As noted above, grain boundaries in the higher aluminum-containing alloys can be strengthened with the addition of small amounts of boron (Crimp and Vedula 1986; Crimp et al. 1987; Gaydosh and Nathal 1990; Liu and George 1990c). However, once the grain boundaries have been strengthened, then the effect of environmental embrittlement becomes evident, producing reduced room-temperature ductilities in the presence of water vapor (Gaydosh and Nathal 1990; Liu and George 1990). Tensile tests of the boron-doped 40% Al alloy produced a ductility of 4.3% in air, but 16.8% in oxygen (see Table 9-4) (Liu and George 1990).

In addition to boron, strengthening in FeAl has been noted for alloys with refined grain sizes of approximately 10 μm (Whittenberger 1983) and by alloying with many different elements to produce solid-solution or second phase strengthening (Crimp and Vedula 1986; Gaydosh et al 1989; Liu et al. 1993; Titran, Vedula, and Anderson). The addition of small

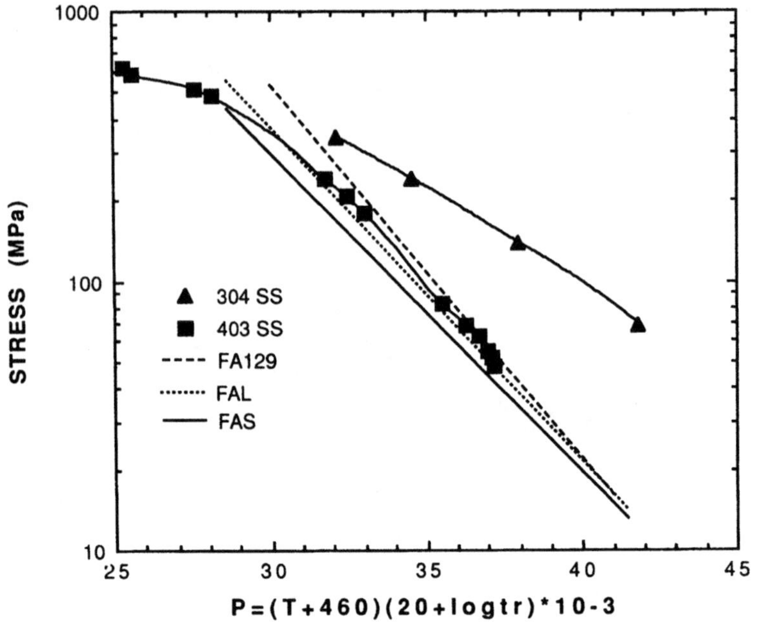

Figure 9-17. Comparison of Larson-Miller parameter plots of time-to-rupture for three iron aluminide alloys with types 403 and 304 stainless steel [Sikka, V. K., and Baldwin, R. H., 1992. SAMPE Qrt 24(1): 2–9; reprinted with permission of the Society for the Advancement of Material and Process Engineering].

amounts of zirconium or hafnium, in combination with carbon to promote formation of precipitates and boron for grain-boundary strengthening, has produced increased tensile strength in an Fe-40% Al alloy (Gaydosh et al. 1989; Liu et al. 1993].

6. WELDABILITY

Welding is a major area of concern for fully realizing the potential benefits of the unique properties of iron aluminides. The welding processes used must be capable of producing high-quality weldments under field and shop conditions that are free of defects and also have mechanical properties comparable to the base metal. Studies of the weldability of iron aluminides have identified a general propensity for cold cracking and a specific dependance of hot cracking on minor changes in alloying composition (David et al. 1989; David and Zacharia 1991; Maziasz et al. 1992; McKamey et al. 1993).

Cold cracking in this system is believed to be mainly related to the sensitivity of iron aluminides to environmentally-generated hydrogen (McKamey et al. 1993). For many iron-aluminide compositions, cold cracking, which can develop several hours or days after welding, can be minimized by using a minimum preheat of 200° C and a postweld heat treatment of 400° for one hour. Pre- and postweld-heat treating, in general, may lower residual stress due to differential thermal expansion and most certainly drives off residual hydrogen (Maziasz et al. 1992, McKamey et al. 1993).

Preliminary screening tests for hot-crack susceptibility have been performed using the Sigmajig test (Goodwin 1987). This test uses a preapplied transverse stress during autogenous GTA welding of a 50 mm by 50 mm sheet specimen to determine the threshold stress above which cracking in the fusion zone always occurs. The applied stress is sequentially increased until the specimen cracks, and this threshold cracking stress (TCS) is then recorded for comparison to other materials tested in the same way. Using this test, good weldability of iron aluminides in terms

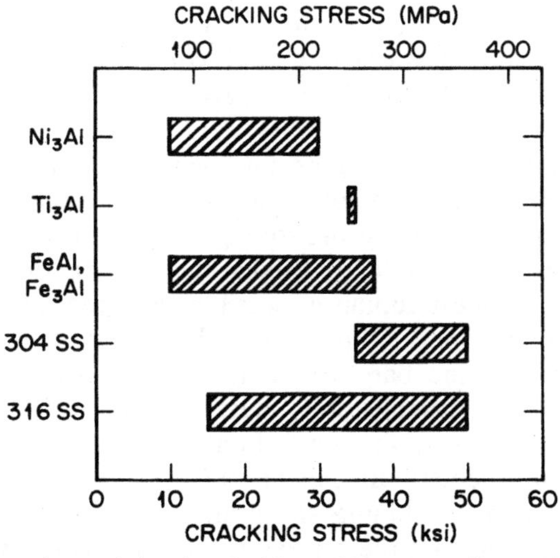

Figure 9-18. Comparison of the threshold cracking stress (from Sigmajig tests, the stress necessary to produce hot cracking during gas-tungsten-arc welding) of Fe_3Al- and FeAl-based alloys to other commonly used alloys (Goodwin, G. M., McKamey, C. G., Maziasz, P. J., and Sikka, V. K., 1993. In Proceedings of the *Seventh Annual Conference on Fossil Energy Materials*, ORNL/FMP-93/1, pp. 181–188. Compiled by N. C. Cole and R. R. Judkins. Oak Ridge, TN: Oak Ridge National Laboratory. July 1993).

of resistance to hot cracking has been found to depend strongly on alloying additions (McKamey et al. 1993). In general, niobium, chromium, and small amounts of carbon were found to enhance weldability, while zirconium, boron, and TiB_2 were found to be detrimental to weldability. Figure 9-18 shows that the TCS values for the Fe_3Al-based alloys are within the range of austenitic stainless steels and some other aluminides.

7. FABRICATION

Melting and fabrication of commercial-size heats will be necessary for the successful application of iron aluminides. Early studies of melting and fabrication have been reviewed (Hardwick and Wallwork 1978; Holladay 1961; Lepkowski and Holladay 1957; McKamey et al. 1991). More recent studies have shown that melting of iron aluminides can be accomplished by many different methods, including air induction melting (AIM), vacuum induction melting (VIM), vacuum are remelting (VAR), and electroslag remelting (ESR) (Sikka et al. 1990, 1993; Sikka 1991a). However, because of the reaction of aluminium with moisture, a large amount of hydrogen can be generated and dissolved in the molten metal during air melting. Therefore, to avoid gas porosity and control microstructure, vacuum melting techniques such as VIM and VAR are recommended (Sikka et al. 1990, 1993).

Conversion of Fe_3Al ingots into various product forms, such as billets, rod, wire, sheet, and tube, can be accomplished by hot or warm working of the as-cast ingot using rolling, forging, or extruding techniques at temperatures of 600–1100° C (Sikka et al. 1990, 1991). Production of iron-aluminide seamless tubing is possible by three fabrication methods: centrifugal casting from the liquid metal, hot rotary piercing of the bar, or by hot extrusion of the bar over a mandrel (Sikka et al. 1990). Warm working of the FeAl compositions must be done at higher temperatures (900–1100° C). However, the low aluminum compositions can be cold finished with intermediate anneals at 800° C (Sikka et al. 1993).

Techniques for machining and grinding of iron aluminides have not fully developed. However, these processes are extremely important for the production of finished products. Because iron aluminides tend to work harder, successful cutting without the generation of cracks requires sharp tools (e.g, high-speed tool steel) and slow machining speeds (on the order of 35 cm per sec or less) (Woodyard and Sikka 1991). Much of the cracking behavior during cutting or grinding has been attributed to environmental hydrogen embrittlement (Ferguson and Liu 1992; Woodyard and Sikka

1991). Therefore water-free environments (i.e., oil- or sulfur-based lubricants) should be used during these processes.

Fe_3Al-based powders can be produced by either nitrogen- or argon-gas atomization (Bordeau 1987; Culbertson and Kortovich 1986; Diehm, Kemppainen, and Mikkola 1989; Sikka et al. 1991). Subsequent consolidation to full density can be accomplished by extruding at 1000° C to an area reduction ratio of $\geq$ 9:1, resulting in fully dense alloys with a grain size of less than 100 μm. Tensile tests have indicated that alloys produced from the nitrogen-gas atomized powder have slightly enhanced properties over the argon-gas atomized powder (Sikka et al. 1991). However, without careful control of impurity levels, aluminides produced from powder can contain large volumes of oxides that can decrease mechanical properties, especially tensile ductility.

Rapid solidification processing (RSP) has been used extensively during recent years in attempts to enhance the strength and ductility of intermetallics. RSP fabrication produces grain refinement, increases chemical homogeneity, extends the solid solubility of elemental additions, and in some cases produces nonequilibrium structures (Baker, Schulson, and Stoloff, 1986; Koch 1985). Most RSP iron aluminides have been produced by powder processes, with consolidation by hot extrusion or pressing (Bordeau 1987; Culbertson and Kortovich 1986; Slaughter and Das 1980). However, recently melt spinning, which involves ejecting the molten alloy onto a rotating wheel, producing quench rates on the order of 10^6 °C/sec, has been used to study structure-property relationships in FeAl (Décamps et al. 1991; Wittig et al. 1987).

8. OTHER PROPERTIES

Although corrosion properties of intermetallics are covered more extensively in another chapter of this book, several comments on the corrosion resistance of iron-aluminide alloys are appropriate at this point. The ability of this system to form a protective Al_2O_3 coating that protects the underlying material from the damaging effects of oxidizing and/or sulfidizing environments at temperatures to 1400° C is well documented (DeVan 1989; Tomaszewicz and Wallwork 1978; Tortorelli and DeVan 1992). The minimum aluminum concentration needed for the formation of the protective Al_2O_3 coating is approximately 16–18 at% (Boggs 1971; Tomaszewicz and Wallwork 1978; Tortorelli and DeVan 1992), but this minimum aluminum requirement can be reduced slightly by the addition of a small amount of chromium (Hagel 1965; Tomaszewicz and Wallwork 1978; Tortorelli and DeVan 1992). Below this minimum amount of alu-

Table 9-5 Miscellaneous Properties of Fe_3Al-Based Alloys

Density, psi (g/cc)	0.237 (6.58)
Melting temperature, ° C	1510
Recrystallization temperature, ° C	750 (for 1 hr)[a]
Electrical resistivity, μohm-cm	160–200
Modulus of elasticity, 10^6 psi	23–29
Thermal conductivity, Btu/ft^2/hr/° F/ft	6.3
Coefficient of thermal expansion, in./in./° F	
68–212° F	7.4×10^{-6}
68–1300° F	11.0×10^{-6}

[a]Cold-worked sheet, 0.02–0.03 in. thick.

Source: Holladay 1961, Nachman and Buehler 1956.

minum, the formation of iron oxides, accompanied by internal oxidation, results in excessive degradation. With the exception of zirconium, which increases the pitting resistance (McKamey et al. 1991), the corrosion properties do not appear to be affected significantly by alloying additions.

Physical-, electrical-, and thermal-properties data on the iron-aluminide system are very limited. Table 9-5 lists some of the available information on Fe_3Al-based alloys (Holladay 1961; Köster et al. 1980–86; Nachman and Buehler 1954, 1956; Sykes and Bampfylde 1934). Several of these properties are worth noting; the thermal conductivity is low and the electrical resistivity is high compared to many other structural alloys. Thermal expansion is comparable to 300-series steels up to a temperature of about 800° C, above which the expansion of Fe_3Al is slightly higher. The coefficient of thermal expansion increases with increasing aluminum concentration (Clark and Whittenberger 1984; Stephens 1985).

The magnetic properties of iron aluminides have been studied extensively (Kuentzler 1983; McQueen and Kuczynski 1959; Nachman and Buehler 1954, 1956; Oki et al. 1981) and depend strongly on aluminum composition and heat treatment. Alloys with less than approximately 20 at% Al are all ferromagnetic and the magnetization decreases with increasing aluminum content (Kuentzler 1983). In the composition range between approximately 20 and 50 at% Al, the magnetic properties are more complex because of the variability of the amount and type of ordered phases that can be produced by changing the heat treatment and composition. For example, on cooling from 800 *K*, ordered alloys between 27 and 30 at% are first paramagnetic, then ferromagnetic below about 400 *K*, and finally mictomagnetic below about 100 K (Shukla and Wortis 1980; Shull, Okamoto, and Beck (1976). Alloys in this same composition range that have been disordered by heat treatment experience a decrease in magnetization with increasing aluminum similar to the low -aluminum alloys. A calculated magnetic-phase diagram is available for this system (Shukala and Wortis, 1980). Of compositions up to 50 at% Al, the best "soft" magnetic properties have been reported for alloys near 28 at% Al with 1.5–2 at % Mo (Nachman and Buehler 1956).

9. SUMMARY

The most unique and important attribute of iron aluminides containing greater than about 18 at% Al is their corrosion resistance in oxidizing and sulfidizing environments, particularly at elevated temperatures. This resistance derives from the thermodynamic stability and relatively slow growth of the alumina-surface product formed in the presence of relatively low partial pressures of oxygen. This unique advantage has been recognized for many years. However, these alloys have not been used extensively as structural materials because of their limited ductility at ambient temperatures and their lack of strength above approximately 600° C. Although a difficult task, if these two properties could be improved, iron-aluminide alloys have the potential to replace many steels and nickel-base alloys, especially in applications where corrosion resistance is desirable.

Recent advances in the understanding of environmental embrittlement in intermetallics, and particularly in iron aluminides, have indicated that iron aluminides are embrittled at room temperature by atomic hydrogen as a result of a reaction between moisture in the air and aluminum on the surface of the specimen. Although the exact mechanism has not been fully characterized, additions of chromium have been found to improve room-temperature tensile ductility in Fe_3Al-based alloys.

Controlled additions of Ti, Si, Nb, Mo, Zr and other elements can improve the hot strength through precipitation mechanisms. Other effects of alloying additions, such as increased ordering temperature, may also be a factor in the increased strength. Melting and fabrication techniques that result in controlled compositions and microstructures are being developed, and will allow for the production of commercial-size heats of Fe_3Al-based alloys. With control of alloying additions and welding techniques, the weldability of this system is comparable to some steels and many other intermetallic systems. These improved properties have resulted in increased interest in these alloys as structural material. Currently, Fe_3Al-based alloys with room-temperature ductilities of 10–15% and strengths comparable to many presently used structural alloys are being produced and tested in many different applications.

REFERENCES

ALLEN, S. M., and CAHN, J. W., 1975, *Acta Metall.* 23: 1017–1026.

AOKI, K., and IZUMI, O., 1979, *Nippon Kinzoku Gakkaishi* 43: 1190.

BAKER, I., and GAYDOSH, D. J., 1987, *Mat. Sci. Eng.*. 96: 147–158.

BAKER I., and MUNROE, P. R., 1990, in *High-Temperature Aluminides and Intermetallics* pp. 425–451. Warrendale, PA: The Materials Society. eds. S. H. Whang, C. T. Liu, D. P. Pope, and J. D. Stiegler.

BAKER, I., NAGPAL, P., LIU, F. and MUNROE, P. R., 1991, *Acta Metall. Mater.* 39(7): 1637–1644.

BAKER, I., SCHULSON, E. M., and STOLOFF, N. S., 1986, In *Mechanical Behavior of Rapidly Solidified Materials* pp. 257–277. Warrendale, PA: The Materials Society. eds. S. M. L. Sastry and B. A. MacDonald.

BEELER, J. R., 1967, In *Intermetallic Compounds* pp. 233–268. New York: John Wiley & Sons. Ed. J. H. Westbrook.

BOGGS, W. E., 1971, *J. Electrochem. Soc.* 118: 906.

BORDEAU, R. G., 1987, *Development of Iron Aluminides,* AFWAL-TR-4009. Wright-Patterson Air Force Base, OH: Air Force Wright Aeronautical Laboratories.

BRADLEY, A. J., and TAYLOR, A., 1937, *Proc. Roy. Soc.* A159: 56.

CASTAGNA, A., and STOLOFF, N. S., 1992, *Scripta Metall. Mater.* 26: 673–678.

CLARK, R. W., and WHITTENBERGER, J. D., 1984, *Proc. of the 8th International Thermal Expansion Symposium* pp. 189–196. New York: Plenum Press.

CRAWFORD, R. C. 1976, *Phil. Mag.* 33(3): 529–555.

CRAWFORD, R. C., and RAY, I. L. F., 1977, *Phil. Mag.* 35: 549–565.

CRAWFORD, R. C., RAY, I. L. F., and COCKAYNE, D. J. H., 1973, *Phil. Mag.* 27: 1–7.

CRIMP, M. A., and VEDULA, K., 1986, *Mat. Sci. Eng.* 78: 193–200.

CRIMP, M. A., VEDULA, K. M., and GAYDOSH, D. J., 1987, in *High-Temperature Ordered Intermetallics II*, MRS Symp. Proc. 81: 499–504.

CULBERTSON, G., and KORTOVICH, C. S., 1986, *Development of Iron Aluminides* AFWAL-TR-85-4155. Wright-Patterson Air Force Base, OH: Air Force Wright Aeronautical Laboratories.

DAVID, S. A., HORTON, J. A., MCKAMEY, C. G., ZACHARIA, T., and REED, R. W., 1989, *Welding J.* 68: 372–381*s*.

DAVID, S. A., and ZACHARIA, T., 1991, in *Heat Resistant Materials* pp. 169–73. Materials Park, OH: ASM Int. eds. K. Natesan and D. J. Tillack.

DAVIES, R. G., 1964, *Trans. Met. Soc. AIME* 230: 903–907.

DAVIES, R. G., and STOLOFF, N. S., 1963, *Acta Metall.* 11: 1187–1189.

DÉCAMPS, B., GIBSON, M. A., MORTON, A. J., and WOLFENDEN, A., 1991 In *High-Temperature Ordered Intermetallics IV*, MRS Symp. Proc. 213: 685–690.

DEVAN, J. H., 1989, In *Oxidation of High-Temperature Intermetallics* pp. 107–115. Warrendale, PA: The Materials Society. eds. T. Grobstein and J. Doychak.

DIEHM, R. S., KEMPPAINEN, M. P., and MIKKOLA, D. E., 1989, *Mater. Man. Proc.* 4: 61–72.

DIEHM, R. S., and MIKKOLA, D. E., 1987, In *High-Temperature Ordered Intermetallics II*, MRS Symp. Proc. 81: 329–334.

DIMIDUK, D. M., MENDIRATTA, M. G., BANERJEE, D., and LIPSITT, H. A., 1988, *Acta Metall.*. 36: 2947–2958.

EHLERS, S. K., and MENDIRATTA, M. G., 1982, *Tensile Behavior of Fe-31 at% Al Alloy,* AFWAL-TR-82-4089. Wright Patterson Air Force Base, OH: Air Force Wright Aeronautical Laboratories.

FERGUSON, P. A., and LIU, C. T., 1992, *Scripta Metall. Mater*. 26(11): 1669–1674.

FORTNUM, R. T., and MIKKOLA, D. E., 1987, *Mat. Sci. Eng*. 91: 223–231.

FUCHS, G. E., and STOLOFF, N. S., 1988, *Acta Metall.* 36(5): 1381–1387.

GAYDOSH, D. J., DRAPER, S. L., and NATHAL, M. V., 1989, *Metall. Trans*. A20: 1701–1714.

GAYDOSH, D. J., and NATHAL, M. V., 1990, *Scripta Metall. Mater*. 24: 1281–1284.

GEORGE, E. P., and LIU, C. T., 1990 *J. Mat. Res*. 5: 754.

GOODWIN, G. M., 1987, *Welding J*. 77(2): 33s.

GOODWIN, G. M., MCKAMEY, C. G., MAZIASZ, P. J., and Sikka, V. K., 1993, In *Proceedings of the Seventh Annual Conference on Fossil Energy Materials,* ORNL/FMP-93/1. Oak Ridge, TN: Oak Ridge National Laboratory. Compiled by N. C. Cole and R. R. Judkins, July 1993, pp. 181–188.

HAGEL, W. C., 1965, *Corrosion* 21: 316.

HANADA, S., WATANABE, S., SATO, T., and IZUMI, O., 1981, *Scripta Metall.* 15: 1345–1348.

HARDWICK, D., and WALLWORK, G., 1978, *Rev. High Temp. Mater*. 4: 47–74.

HOLLADAY, J. W., 1961, *Review of Development in Iron-Aluminum-Base Alloys*, DMIC Memo 82. Columbus, OH: Defense Metals Information Center, Battelle Memorial Institute.

HORTON, J. A., LIU, C. T., and KOCH, C. C. 1984, In *High-Temperature Alloys: Theory and Design* pp. 309–321. Warrendale, PA: The Materials Society. Ed. J. O. Stiegler

INOUYE, H., 1985, In *High-Temperature Ordered Intermetallics*, MRS Symp. Proc. 39: 225–261.

JUNG, I., and SAUTHOFF, G., 1989, *Z. Metall.* 80(7): 484–489.

JUSTUSSON, W, ZACKAY, V. F. and MORGAN, E. R., 1957, *Trans. ASM* 49: 905–921.

KERR, W. R., 1986, *Metall. Trans*.. A17: 2298–2300.

KNIBLOE, J. R., WRIGHT, R. N., SIKKA, V. K., BALDWIN, R. H., and Howell, C. R., 1992, *Mat. Sci. Eng*. A152: 382–386.

KOCH, C. C., 1985, In *High-Temperature Ordered Intermetallics*, MRS Symp. Proc. 39: 397–410.

KÖSTER, W., and GÖDECKE, T., 1980. *Z. Metall.* 71, 765–769.

KÖSTER, W., and GÖDECKE, T., 1981, *Z. Metall.* 72, 569–574

________, 1981, *Z. Metall.* 72, 707–711.

________, 1982, *Z. Metall.* 73, 111–114.

________, 1982, *Z. Metall.* 73, 502–510.

________, 1983, *Z. Metall.* 74, 705–708.

Köster, W., and Gödecke, T., 1984, *Z. Metall.* 75, 161–169.

_______, 1984, *Z. Metall.* 75, 432–435.

_______, 1986, *Z. Metall.* 77, 408–411.

Köster, W., Wachtel, E., and Gödecke, T., 1985, *Z. Metall.* 76, 382–387.

Köster, W., Wachtel, E., and Gödecke, T., 1985, *Z. Metall.* 76, 676–683.

Kuentzler, R., 1983, *J. Phys.* 44: 1167–1178.

Kuruvilla, A. K., and Stoloff, N. S., 1987, In *High-Temperature Ordered Intermetallics II*, MRS Symp. Proc. 81: 229–238.

Lawley, A., and Cahn, R. W. 1961, *J. Phys. Chem. Solids* 20: 204.

Lawley, A., Coll, J. A., and Cahn, R. W. 1960, *Trans. Met. Soc. AIME* 218: 166–176.

Leamy, H. J., and Kayser, F. X., 1969, *Phys. Stat. Solidi* 34: 765–780.

Leamy, H. J., Kayser, F. X., and Marcinkowski, M. J., 1969a, *Phil. Mag.* 20: 763–777.

Leamy, H. J., Kayser, F. X., and Marcinkowski, M. J., 1969b, *Phil. Mag.* 20: 779–798.

Lepkowski, W. J., and Holladay, J. W., 1957, *The Present State of Development of Iron-Aluminum-Base Alloys.* Titanium Metallurgical Laboratory, Battelle Memorial Institute.

Liu, C. T., Fu, C. L., George, E. P., and Painter, G. S., 1991, *ISIJ Intern.* 31(10): 1192–1200.

Liu, C. T., and George, E. P., 1990, *Scripta Metall. Mater.* 24: 1285–1290.

Liu, C. T., and George, E. P. 1991, In *High-Temperature Ordered Intermetallics IV*, MRS Symp. Proc. 213: 527–532.

Liu, C. T., Lee, E. H., and McKamey, C. G., 1989, *Scripta Metall.* 23: 875–880.

Liu, C. T., and McKamey, C. G., 1990, in *High Temperature Aluminides and Intermetallics* 133–151. Warrendale PA: The Materials Society. eds. S. H. Whang, C. T. Liu, D. P. Pope, and J. O. Stiegler.

Liu, C. T., McKamey, C. G., and Lee, E. H., 1990, *Scripta Metall. Mater.* 24: 385–389.

Liu, C. T., and Sikka, V. K., 1986, *J. Metals* 38: 19.

Liu, C. T., Sikka, V. K., and McKamey, C. G., 1993, *Alloy Development of FeAl Aluminide Alloys for Structural Use in Corrosive Environments.* ORNL/TM-12199. Oak Ridge, TN: Oak Ridge National Laboratory.

Liu, C. T., White, C. L., and Horton, J. A., 1985, *Acta Metall.* 33: 213.

Longworth, H. P., and Mikkola, D. E., 1987, *Mat. Sci. Eng.* 96: 213–229.

Marcinkowski, M. J., and Brown, N., 1961, *Acta Metall.* 9: 764–786.

_______, 1962, *J. Appl. Phys.* 33(2): 537–552.

Marcinkowski, M. J., Taylor, M. E., and Kayser, F. X., 1975, *J. Mat. Sci.* 10: 406–414.

Massalski, T. B., 1986, *Binary Alloy Phase Diagrams.* Materials Park, OH: ASM International.

Maziasz, P. J., Goodwin, G. M., Liu, C. T., and David, S. A. 1992, *Scripta Metall. Mater.* 27: 1835–1840.

MAZIASZ, P. J., and MCKAMEY, C. G., 1992, *Mat. Sci. Eng.*. A152: 322–334.
MCKAMEY, C. G., 1993, Unpublished data.
MCKAMEY, C. G., DEVAN, J. H., TORTORELLI, P. F., and SIKKA, V. K., 1991, *J. Mat. Res.*. 6(8): 1779–1805.
MCKAMEY, C. G., HORTON, J. A., and LIU, C. T., 1987, In *High-Temperature Ordered Intermetallics II*, MRS Symp. Proc. 81: 321–327.
MCKAMEY, C. G., HORTON, J. A., and LIU, C. T., 1989, *J. Mat. Res.* 4: 1156–1163.
MCKAMEY, C. G., and LIU, C. T., 1990. *Scripta Metall. Mater.* 24: 2119–2122.
_______, 1992, *In Proceedings of First International Symposium on Environmental Effects on Advanced Materials* paper #17. Houston, TX: NACE. ed R. D. Kane.
MCKAMEY, C. G., LIU, C. T., DAVID, S. A., HORTON, J. A., PIERCE, D. H., and CAMPBELL, J. J., 1988, *Development of Iron Aluminides for Coal Conversion Systems*. ORLN/TM-10793. Oak Ridge, TN: Oak Ridge National Laboratory.
MCKAMEY, C. G., MAZIASZ, P. J., GOODWIN, G. M., and ZACHARIA, T., 1994, *Mat. Sci. Eng. A*174: 59–70.
MCKAMEY, C. G., MAZIASZ, P. J., and JONES, P. W., 1992, *J. Mat. Res.* 7(8): 2089–2106.
MCKAMEY, C. G., AND PIERCE, D. H., 1993, *Scripta Metall. Mater.* 28: 1173–1176.
MCKAMEY, C. G., AND SIKKA, V. K., 1992, Unpublished Data.
MCQUEEN, H. J., AND KUCZYNSKI, 1959, *Trans. Met. Soc. AIME* 215: 619–622.
MENDIRATTA, M. G., EHLERS, S. K., CHATTERJEE, D. K., and LIPSITT, H. A., 1987, *Metall. Trans.* A18: 283–291.
MENDIRATTA, M. G., EHLERS, S. K., DIMIDUK, D. M., KERR, W. R., MAZDIYASNI, S., and LIPSITT, H. A., 1987, In *High-Temperature Ordered Intermetallics II*, MRS Symp. Proc. 81: 393–404.
MENDIRATTA, M. G., KIM, H.-M., and LIPSITT, H. A., 1984, *Metall. Trans.* A15: 395–399.
MORGAND, P., MOUTURAT, P., and SAINFORT, G., 1968, *Acta Metall.* 16: 867–875.
MORRIS, D. G., DADRAS, M. M. and MORRIS, M. A. 1993a, *Acta Metall. Mater.* 41: 97–111.
_______, 1993b, MRS Symp. Proc. 288: 623–628.
MORRIS, D. G., and MORRIS, M. A., 1990, *Mat. Sci. Eng. A*125: 97–106.
MUNROE, P. R., and BAKER, I., 1991, *Acta Metall. Mater.* 39: 1011–1017.
_______, 1993, *J. Mat. Sci.* 28: 2299–2303.
NACHMAN, J. F., and BUEHLER, W. J., 1954, *J. Appl. Phys.* 25(3): 307–313.
_______, 1956, *Application, Properties, and Fabrication of Thermenol Type Alloys*, NAVORD Report 4237. Silver Spring, MD: U.S. Naval Ordnance Laboratory. Available from Defense Technical Information Center. Alexandria, VA.
NAGPAL, P., and Baker, I, 1990, *Metall. Trans.* A21: 2281–2282.
NAGPAL, P., BAKER, I. LIU, F., and MUNROE, P. R., 1991, In *High-Temperature Ordered Intermetallics IV*, MRS Symp. Proc. 213: 533–536.
OKAMOTO, H., and BECK, P. A., 1971, *Metall. Trans.* 2: 569–574.

OKI, K., HASAKA, M., and EGUCHI, T., 1973, *Japan. J. Appl. Phys.* 12: 1522–1530.

OKI, K., TOWATA, S., TAMIYA, M., and EGUCHI, T., 1981, *Trans. Japan Inst. Metals* 22: 771.

PARIS, D., LESBATS, P., and LEVY, J., 1975, *Scripta Metall.* 9: 1373–1378.

PRAKASH, U., BUCKLEY, R. A., and JONES, H. 1991, MRS Symp. Proc. 213: 691–696.

PRAKASH, U., BUCKLEY, R. A., JONES, H., and SELLARS, C. M., 1991, *ISIJ I*NTER. 31(10): 1112–1125.

RAY, I. L. F., CRAWFORD, R. C., and COCKAYNE, D. J. H., 1970, *Phil. Mag.* 21: 1027–1032.

RIVIERE, J. P., and GRILHE, J., 1972, *Acta Metall.* 20: 1275–1280.

SABURI, T., YAMAUCHI, I., and Nenno, S., 1972, *J. Phys. Soc. Japan* 32(3): 694–701.

SCHMATZ, D. J., and ZACKAY, V. F., 1959, *Trans. ASM* 51: 299.

SHERBY, O. D., and BURKE, P. M., 1968, *Prog. Mat. Sci.* 13: 325.

SHUKLA, P., and Wortis, M., 1980, *Phys. Rev.* B21: 159.

SHULL, R. D., OKAMOTO, H., and BECK, P. A., 1976, *Solid State Comm.* 20: 863.

SIKKA, V. K., 1991a, MRS Symp. Proc. 213: 907–912.

________, 1991b, *SAMPE Qrt.* 22(4): 2–10.

SIKKA, V. K., and BALDWIN, R. H., 1992, *SAMPE Quarterly* 24(1): 2–9.

SIKKA, V. K., BALDWIN, R. H., REINSHAGEN, J. H., KNIBLOE, J. R., and WRIGHT, R. N., 1991, MRS Symp. Proc. 213: 901–906.

SIKKA, V. K., MCKAMEY, C. G., HOWELL, C. R., and BALDWIN, R. H., 1990, *Fabrication and Mechanical Properties of Fe_3Al-Based Aluminides*, ORNL/TM-11465. Oak Ridge, TN: Oak Ridge National Laboratory.

________, 1991, *Properties of Large Heats of Fe_3Al-Based Alloys*, ORNL/TM-11796. Oak Ridge, TN: Oak Ridge National Laboratory.

SIKKA, V. K., VISWANATHAN, S., and MCKAMEY, C. G., 1993, *Structural Intermetallics*, Warrendale, PA: The Materials Society. pp. 483–491. eds. R. Darolia, J. J. Lewandowski, C. T. Liu, P. L. Martin, D. B. Miracle, and M. V. Nathal.

SLAUGHTER, E. R., and DAS, S. K. 1980, In *Rapid Solidification Processing: Principles and Technologies, II*, pp 354–363. Baton Rouge, LA: Claitor's Pub. eds. R. Mehrabian, B. H. Kear, and M. Cohen.

STEPHENS, J. R., 1985, MRS Symp. Proc. 39: 381–395.

STOLOFF, N. S., 1988, *J. Metals* 40: 23.

STOLOFF, N. S., and DAVIES, R. G., 1964, *Acta Metall.* 12: 473–485.

________, 1966, *Prog. Mat. Sci.* 13: 1.

STOLOFF, N. S., FUCHS, G. E., KURUVILLA, A. K., and CHOE, S. J., 1987, MRS Symp. Proc. 81: 247–261.

SUN, Y., GUO, J., ZHANG, L., and ZHANG, J., 1992, *Acta Metall. Sin.* A5(1): 7–12.

SUN, Z. Q., HUANG, Y. D., YANG, W. Y., and CHEN, G. L., 1993, MRS Symp. Proc. 288: 885–890.

SWANN, P. R., DUFF, W. R., and FISHER, R. M., 1969, *Trans. AIME* 245: 851–853.

SYKES, C., and BAMPFYLDE, J., 1934, *J. Iron Steel Inst.* 130: 389.
TAUB, A. I., HUANG, S. C., and CHANG, K. M., 1984, Metall. Trans. 15*A*: 399.
TAYLOR, A., and JONES, M., 1958, *J. Phys. Chem. Solids* 6: 16.
TITRAN, R. H., VEDULA, K. M., and ANDERSON, G. G., 1985, MRS Symp. Proc. 39: 309–317.
TOMASZEWICZ, P., and Wallwork, G. R., 1978, *Rev. High Temp. Mater.* 4: 76–103.
TORTORELLI, P. F., and DeVan, J. H., 1992, *Mat. Sci. Eng.* A153: 573–577.
UMAKOSHI, Y., and YAMAGUCHI, M., 1980, *Phil. Mag.* A41: 573–588.
_______, 1981, *Phil. Mag.* A44: 711–715.
VEDULA, K., and Stephens, J. R., 1987, MRS Symp. Proc. 81: 381–391.
VISWANATHAN, S., SHELTON, B. R., WRIGHT, J. K., and, V. K., 1993, *Scripta Metall. Mater.* 29: 589–594.
VYAS, S., VISWANATHAN, S., and Sikka, V. K., 1992, *Scripta Metall. Mater.* 27: 185–190.
WESTBROOK, J. H., 1956, *J. Electrochem. Soc.* 103: 54.
WHITTENBERGER, J. D., 1983, *Mat. Sci. Eng.* 57: 77–85.
_______, 1986, *Mat. Sci. Eng.* 103–113.
WITTIG, J. E., VOGT, E., MÖLLER, R., and FROMMEYER, G., 1987, *Scripta Metall.* 21: 721–726.
WOODYARD, J. R., and SIKKA, V. K., 1991, Unpublished results.
YAMAGATA, T., and YOSHIDA, H., 1973, *Mat. Sci. Eng.* 12: 95–100.
ZIEGLER, N., 1932, *Trans. AIME* 100: 267.

Chapter 10

Advanced Intermetallics

K. S. Kumar

1. INTRODUCTION

The need for materials for very high-temperature applications with a simultaneous desire for low density, a balance of mechanical properties (e.g., creep resistance and low-temperature damage tolerance), and environmental resistance has directed research beyond the Fe-, Ni-, and Ti-aluminides. Thus, for example, there is considerable ongoing research in the area of silicides with interest residing in obtaining not only a fundamental understanding of the structure and deformation behavior but also in learning how to produce, fabricate, machine, join, and test components of such materials. For this reason, there is ample justification in dedicating a separate chapter in this book to the silicides and therefore, this area is not covered in this chapter. The scope of this chapter includes the trialuminides (Al_3X) with the DO_{22} and $L1_2$ structures, some aluminides that are frequently not included under other headings (e.g., CoAl, Nb_3Al, and RuAl), beryllides, and chromides based on the AB_2 stoichiometry (Laves Phase).

The quest for a lightweight, oxidation-resistant high-temperature material has led to extensive investigation of aluminum-rich intermetallics of the type Al_3X (commonly referred to as the trialuminides), where X is usually Ti, Zr, Hf, V, Nb, or Ta. These compounds crystallize with the DO_{22} (or DO_{23}) structure, and in an attempt to enhance their low-temperature ductility, theoretical and experimental efforts have been expended to transform the DO_{22} structure to the related, more symmetric, cubic $L1_2$ structure.

Some of the highest melting aluminides with densities less than or comparable to currently used superalloys include the B2 compounds NiAl, CoAl, and RuAl, and the A15 compound Nb_3Al. CoAl is somewhat a scientific curiosity because it is extremely brittle relative to its companion NiAl. RuAl is the latest to arrive on the stage and appears promising in terms of mechanical properties. The A15 compound, Nb_3Al, has always been an attraction as a superconducting material and currently, there is more attention focused on it as a potential high-temperature structural material in Japan than in the United States.

The refractory metal beryllides are an attractive class of materials and were extensively examined in the early 1960s but research was terminated because of the lack of any ductility in these materials up to fairly high temperatures. This was attributed to inadequate care in controlling interstitial levels in these materials and recently, with the advent of sophisticated processing techniques, there has been a surge in research activity to reexamine the potential of these low density, high-melting compounds. The compounds being examined primarily include $NbBe_{12}$, Nb_2Be_{17}, and $TiBe_{12}$. Another compound that has received research attention is NiBe with the B2 structure.

Finally, Laves phase (AB_2 stoichiometry)-containing two-phase systems, particularly Cr-based, are currently being examined with the intent of obtaining creep resistance from the Laves phase (Cr_2Nb, Cr_2Ti, Cr_2Hf, and Cr_2Ta) and low-temperature damage tolerance from the Cr phase via mechanisms such as crack blunting, crack bridging and crack branching.

In this chapter, physical and mechanical properties, deformation modes, and environmental sensitivity of these various compounds are reviewed. When available, data on phase equilibria and solubility of ternary and quaternary additions are included. A final discussion section examines the application potential of these advanced intermetallics and in the process identifies areas requiring further effort.

2. TRIALUMINIDES

2.1. Compounds with the DO_{22} (or DO_{23}) Structure

The group IV-A (Ti, Zr, Hf) and group V-A (V, Nb, Ta) elements form binary compounds with Al of the type Al_3X (X = Ti, Zr, Hf, V, Nb, or Ta), frequently referred to as *trialuminides*, that crystallize with the

DO_{22} or the related DO_{23} structures. These structures are long-period superlattices derived from the $L1_2$ structure by the periodic introduction of antiphase boundaries on the (001) plane with a $1/2\langle 110 \rangle$ displacement vector. A ball-and-stick model of each of these three crystal structures and the relative stabilities of these structures are shown in Figure 10-1. These Al_3X compounds form congruently (Al_3Nb) or incongruently (Al_3Ti) in the temperature range 1600 K–1900 K and are of technological interest due to their low density coupled with their reasonably high melting points and potentially good oxidation resistance. These intermetallics, however, frequently exist as line compounds, making it near impossible to produce them as single-phase material, and this coupled with their lack of any ductility at low and warm temperatures have deterred progress. In a recent review of Al-rich ternary intermetallics (Kumar 1990), it was concluded that while intersolubility between two Al_3X-type compounds was possible in the pseudobinary section of the ternary space, the line compound nature was essentially retained except in the Al-Ti-Mo system where the compositional range of existence of Al_3Ti was significantly enhanced by Mo addition.

The binary compound Al_3Ti has been examined in the single and polycrystalline forms (Yamaguchi, Umakoshi, and Yamane 1987) and its deformation behavior as a function of test temperature has been extensively characterized. Compression studies of the polycrystalline material revealed some ductility that was subsequently attributed to the presence of a small amount of aluminum second phase. Single crystals of single-phase Al_3Ti have been successfully grown by the floating zone method and tested in compression as a function of orientation. The major deformation mode at ambient temperature was determined to be ordered twinning that occurs without disturbing the DO_{22} structure. Such twinning occurs by the movement of $1/2\langle 112]$ partial dislocations on $\{111\}$ planes, yielding four deformation systems. At higher temperatures, slip of $\langle 100]$ and $\langle 110]$ dislocations have been observed in addition to ordered twinning (Yamaguchi et al. 1987; Hug, Douin, and Veyssiére 1989; Vasudevan, Wheeler, and Fraser 1989; Morris and Lerf 1991). The $\langle 110]$ dislocations are more widely dissociated on the (001) plane than on the $\{111\}$ planes, indicating highly anisotropic antiphase boundary (APB) energies. The measured APB energy on the (001) plane according to Vasudevan et al. (1989) is ~ 32 mJ/m^2 and ~ 25 mJ/m^2 according to Hug et al. (1989) and Morris and Lerf (1991). The corresponding values on the $\{111\}$ plane was 200 mJ/m^2 (Hug et al. 1989) and 190 mJ/m^2 (Morris and Lerf 1991). In addition to these observations, Vasudevan et al. (1989) also reported for the first time, the presence of $1/2\langle \bar{1}12]$ perfect dislocations that likely dissociate into $1/6[\bar{1}12] + 1/3[\bar{1}12]$. Subsequently, Morris and Lerf (1991)

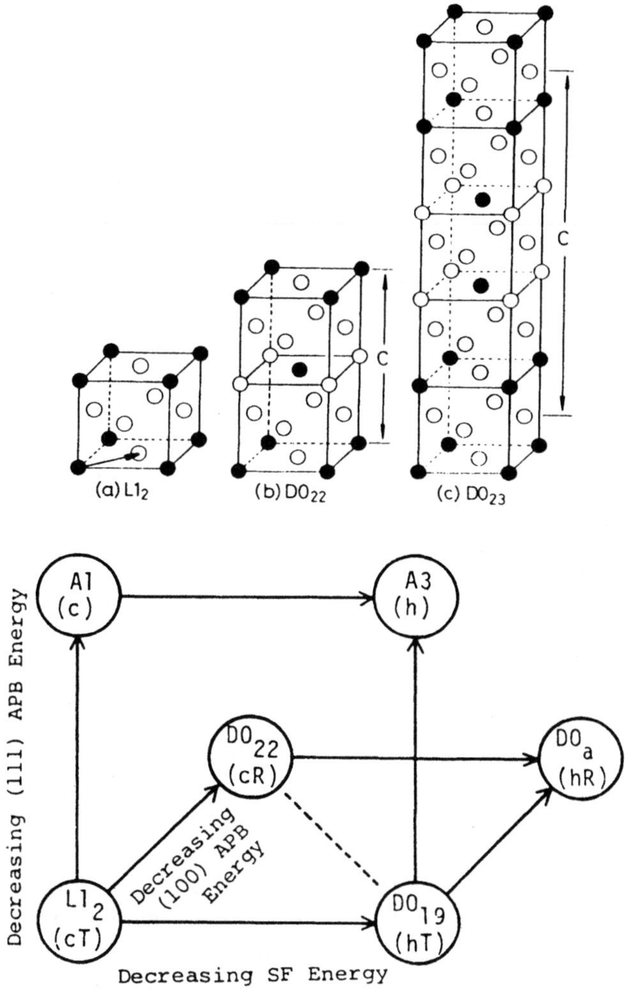

Figure 10-1. (a) The $L1_2$, DO_{22} and DO_{23} crystal structures (Yamaguchi, Shirai, and Umakoshi, 1988, with permission from The Minerals, Metals, and Materials Society); and (b) the influence of APB and SF energies on their relative stability. (Mishima, Oya and Suzuki, 1985, with permission from the Materials Research Society).

confirmed the previous observations of Vasudevan et al. (1989) on the occasional presence of dissociated $1/2\langle 112]$ dislocations that glide on $\{111\}$ planes; in addition, Morris and Lerf (1991) deduced an intrinsic stacking fault energy of $\sim 100\ \mathrm{mJ/m^2}$, which according to these investigators, is surprisingly high for a material that exhibits extensive twinning. Thus, they interpret the observation of extensive twinning as being due to difficulty in dislocation propagation rather than easy twinning.

To improve the low temperature ductility of Al_3Ti, it is necessary to enhance the twinning ($\{111\}\langle 11\bar{2}]$) and slip activities ($(001)\langle 110]$). This would require alloying, and an approach to identify the appropriate alloying element using crystal structure data for Al_3X and X_3Ti-type intermetallic compounds has been proposed (Yamaguchi and Umakoshi 1990). In essence, this approach identifies alloying elements that will lower the superlattice intrinsic stacking fault (SISF) energy to enhance twinning and APB energy on cube planes to enhance slip. The former may be achieved by identifying elements that stabilize the DO_a phase (e.g, Cu_3Ti) with respect to the DO_{22} phase, whereas the latter would require stabilizing the DO_{23} or $L1_2$ structure with respect to the DO_{22} structure (e.g., Hf, Zr, Li). Using these arguments it has been shown that the low temperature compressive ductility of Al_3Ti can be improved, although such improvements are not substantial.

The compound Al_3V is isostructural with Al_3Ti and exhibits similar melting point and a slightly higher density. Like Al_3Ti, it is brittle at room temperature and deformation is mainly controlled by $\langle 110\rangle$-type dislocations; however, unlike Al_3Ti, twinning is not observed at low temperatures in Al_3V (Umakoshi et al. 1988). At higher temperatures, dislocation motion occurs with $b = \langle 110]$ and $b = \langle 100]$. The partial substitution of Ti for V in Al_3V was shown to encourage twinning in addition to $\langle 110\rangle$ dislocations at low temperatures, thereby enhancing ductility. The variation in compressive yield strength and fracture strain in compression with test temperature for various levels of Ti substitution in Al_3V (Fig. 10-2a, b) reveals a minimum in strength and a maximum in fracture strain for the composition $Al_3(V_{0.95}Ti_{0.05})$. It is likely that the twinning activity reaches a maximum at this composition. An examination of the deformed microstructure of Al_3V and $Al_3(V_{0.95}Ti_{0.05})$ reveals only dislocation activity in the former whereas the latter exhibits extensive deformation twins and stacking faults (Fig. 10-3a, b). At higher Ti levels, stacking-fault energy is claimed to increase and this may discourage $\langle 110\rangle$-type dislocation motion.

Of the various binary DO_{22} trialuminides, Al_3Nb is perhaps the most attractive with a high melting point ($\sim$ 1880 K) and a fairly low density ($\sim$ 4.5 g/cm^3). The deformation behavior of polycrystalline Al_3Nb was examined (Schechtman and Jacobson 1975) and it was shown that ordered twinning, similar to that observed in Al_3Ti, is the dominant deformation mode (Note: The apparent difference in the reported twinning system in Al_3Nb by Schechtman and Jacobson (1975) relative to Al_3Ti (Yamaguchi

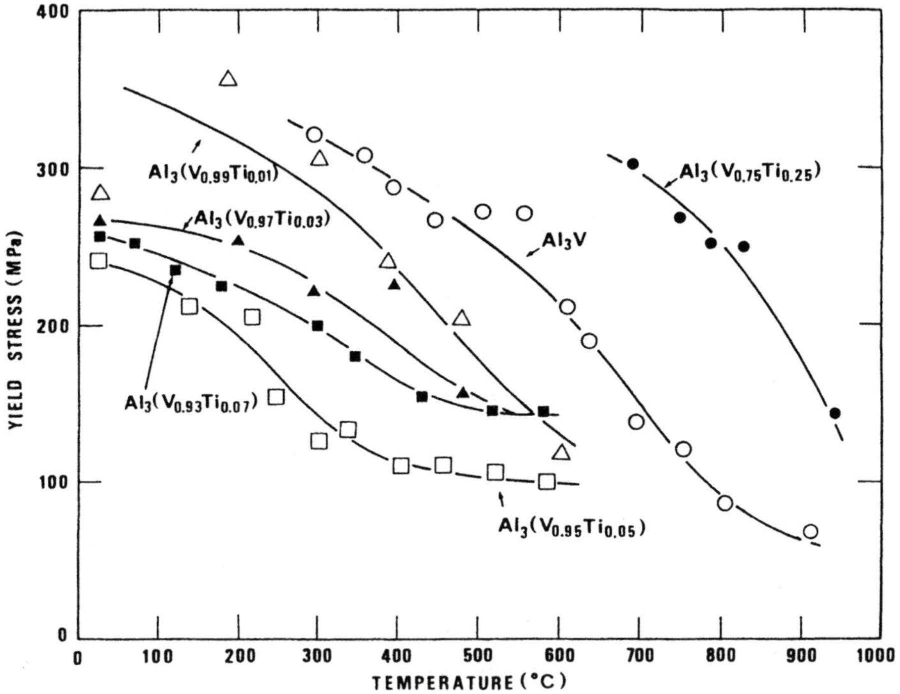

(a)

Figure 10-2. The effect of Ti substitution for V on the compression response of Al_3V: (a) yield strength versus temperature, and (b) fracture strain versus temperature. (Umakoshi et al. 1988).

et al. 1987) is a consequence of Miller indices in Al_3Ti being based on the underlying f.c.c. structure rather than the DO_{22} structure.)

Mechanical properties of polycrystalline Al_3Nb have been evaluated and Raisson and Vignes (1970) reported it being brittle in compression up to ~ 1100 K, although no strength values were included. Recently, Sauthoff (1990) has provided the variation in compressive yield strength with temperature of Al_3Nb, Al_3(Nb,Ti) and multiphase Al_3Nb-NiAl-NbNiAl. The substitution of Ti for Nb in single-phase Al_3Nb does not appear to influence the high-temperature compressive yield strength. The strain to failure in bending of Al_3Nb is less than 1% at 1423 K, although the substitution of Ti for Nb improves the deformability to some extent. The three-phase alloy showed considerable improvements in ductility relative to the single-phase material. Schneibel, Becher, and Horton (1988)

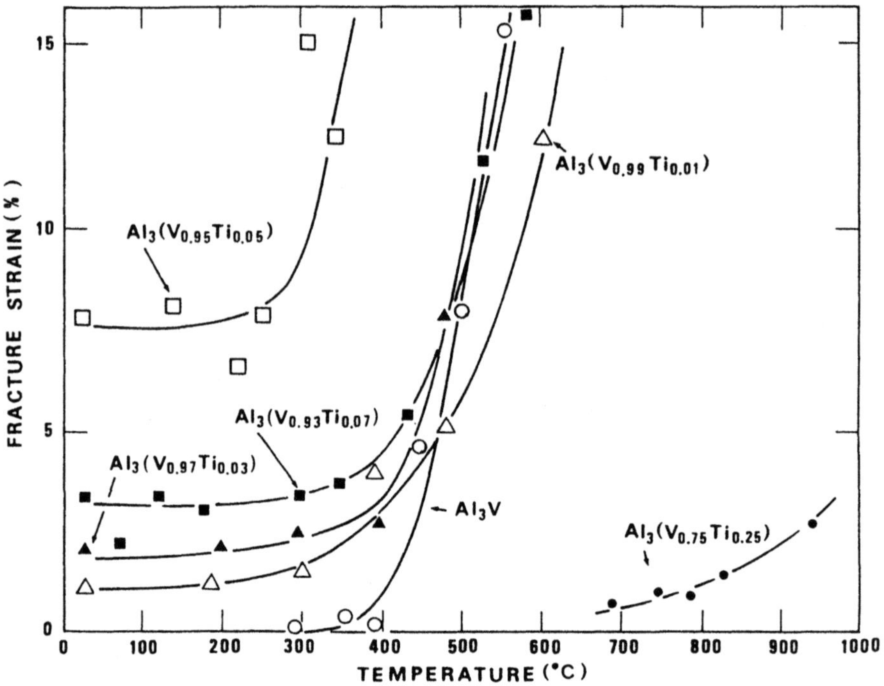

(b)

Figure 10-2. (Continued)

attempted to fabricate Al_3Nb via the powder route but found substantial porosity in the hot-pressed disks in spite of pressing at temperatures as high as $0.95T_m$. Nevertheless, they went ahead and measured fracture toughness of 2.5 ± 0.5 MPa$\sqrt{m}$ in an alloy of nominal composition Al-25.5 at%Nb. The oxidation behavior of Al_3Nb was examined (Perkins, Chiang, and Meier 1988) and it was shown that in the temperature range 1573 K–1623 K in air, binary Al_3Nb does not form a continuous protective alumina scale. This was attributed to the formation of an Al-depleted Nb_2Al layer at the scale-metal interface. Decreasing the Al content to 50 at%, however, allowed the formation of protective alumina scales, if Ti and Cr or V were also incorporated. Hebsur and co-workers (1992) have managed to obtain the desired oxidation resistance in Al_3Nb while retaining the high Al content, by adding 7 at%Cr and 0.5 at%Y. They finally came up with a multicomponent, multiphase alloy of the type Nb-67 at% Al-7 at% Cr-0.5 at%Y-0.25 at%W. This alloy was produced by rapid solidification and consolidated by a variety of techniques. They were then

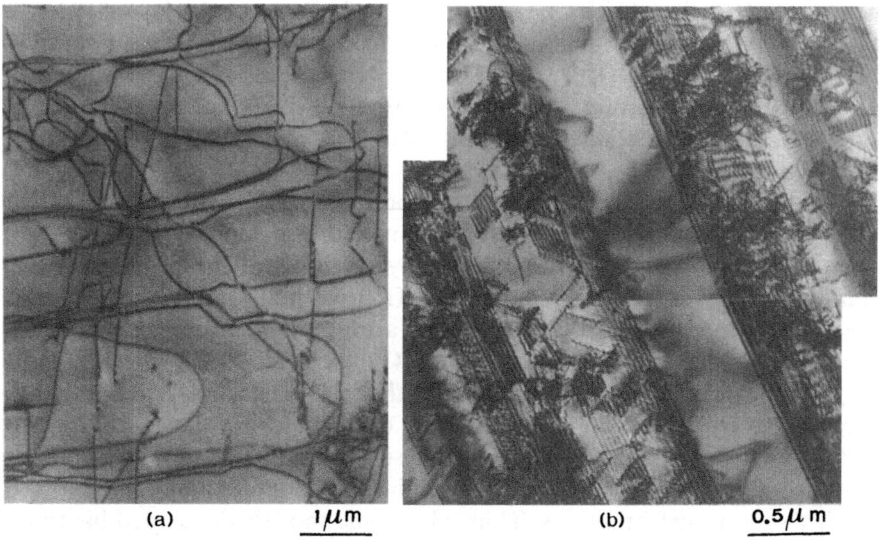

Figure 10-3. Deformed microstructures: (a) binary Al_3V at 528 K (255°C) and (b) $Al_3(V_{0.95}Ti_{0.05})$ at 298 K (25°C). (Reprinted with permission from *Progress in Material Science*, Volume 34, M. Yamaguchi and Y. Umakoshi, pages 1–148, 1990, Elsevier Science Ltd., Pergamon Imprint, Oxford, England).

evaluated in compression in the temperature range 300 K–1300 K, and in bending. The ductile-to-brittle transition temperature (DBTT) in bending was ~ 1000 K. Constant load-compressive creep tests in the 1200 K–1300 K range yielded a power-law dependency with a stress exponent of ~ 3.0 and an associated activation energy of ~ 250 ± 20 kJ/mol.

2.2. The DO_{22} to $L1_2$ Transformation

The relationship between the DO_{22} and the $L1_2$ structures was shown in Figure 10-1 and, at least in principle, by influencing the APB energy on the cube planes, it should be possible to transform one structure to another. Such a transformation would be beneficial to ductility, as the $L1_2$ structure has a higher symmetry and hence a higher number of slip systems. Minor alloying, retaining the DO_{22} structure did not significantly enhance the ductility, as was evident in the previous section.

Research efforts of various groups in the Soviet Union and Germany that had shown the existence of ternary aluminum-rich $L1_2$ compounds in the Al-Ti-Cu, Al-Ti-Ni, and Al-Ti-Fe systems with compositions of the form Al_5Ti_2Cu, $Al_{67}Ti_{25}Ni_8$, and $Al_{22}Ti_8Fe_3$ were recently reviewed (Kumar 1990). More recently, the occurrence of such $L1_2$ compounds has been noted in the Al-Ti-Cr, Al-Ti-Mn, Al-Ti-Pd, and Al-Ti-Ag systems. Similarly, Al-rich $L1_2$ compounds are also present in the Al-Zr-X (X = Cu, Ni, Fe, Co) and Al-Hf-X (X = Cu) systems, although they were not detected in the Al-group VA-X (X = Cu, Ni, Fe...) systems. These $L1_2$ compounds exhibit a small but finite range of compositions over which they are stable, and densities that are typically < 4.0 g/cm^3 (for the Ti-containing compounds) with melting points similar to Al_3Ti or Al_3Zr. It is tempting to suggest that these ternary cubic compounds are derived from their binary DO_{22} counterparts by selective alloying, although recently, there is an alternate suggestion (Durlu, Inal, and Yost 1991) that these compounds are Al_2Ti derivatives, a binary compound that has only recently drawn some research attention (Potez et al. 1992).

The currently more popular belief that these ternary $L1_2$ compounds are derived from the binary Al_3X DO_{22} compounds by alloying has led to a flurry of research activities, where physicists have teamed with materials scientists to perform first principle calculations to understand and perhaps predict such transformations. Thus, electronic models developed based on the existence of ternary $L1_2$ compounds in the Al-group IVA-X systems, attempted to predict potential alloying additions to transform Al_3Nb, Al_3Ta, and Al_3V from the DO_{22} to the $L1_2$ structure. Early in the theoretical efforts, a few alloying additions were recommended to transform Al_3Nb to an $L1_2$ structure, although experimental studies showed that such additions were not successful in producing a ternary $L1_2$ structure. Subsequently, theoretical research was aimed at explaining why such a transformation occurred in the Al-group IVA-X (X = Fe, Ni, Cu, Cr, Mn...) systems but not in the companion Al-group VA-X systems.

Attempts to understand the relative phase stability of various types of A_3B compounds include the generation of structural-stability maps based on extensive experimental data. Villars's maps (1983), however, used classical coordinates and therefore had drawbacks, including being unable to account for the existence of certain structures until Pettifor (1984) emphasized the necessity to use quantum mechanics. Subsequently, Pettifor proposed a chemical scale that ordered all the elements in the periodic table along a single axis, χ, while retaining their quantum characteristics. He then used χ_a and χ_b as the coordinates for constructing two-dimensional structural maps for compounds of various stoichiometries, partitioning the various structures of a particular stoichiometry into discrete domains. Such an AB_3 map (Pettifor 1991) has been used to rationalize the observed DO_{22}-to-$L1_2$ transformations in the trialuminides.

Other investigations to explain the DO_{22}-to-$L1_2$ transformation of Al_3Ti with alloying additions include the work of Nicholson et al. (1989), who emphasized the role of the tetragonal distortion in stabilizing the

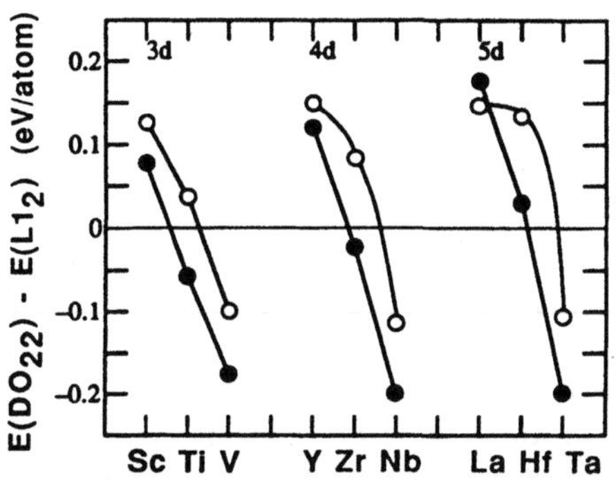

Figure 10-4. Calculated total-energy differences between the DO_{22} and $L1_2$ crystal structures for MAl_3 compounds, where M = group III, IV, or V transition metal. The open circles denote ideal c/a ratio and the solid circles the relaxed c/a ratio. (Carlsson and Meschter 1989, with permission from the Materials Research Society)

DO_{22} structure of Al_3Ti, and of Eberhart, Kumar, and MacLaren (1990), who used an electronic model based on cluster calculations and attributed the observed DO_{22}-to-$L1_2$ phase transition to the *d*-character provided in the Al *sp*-bonding region by the transition metal substitution that preferentially orders the Al second-neighbor atoms. In their calculations, Eberhart et al. (1990) used c/a = 2.0 rather than the actual tetragonal distortion and this is certain to influence the nature of first- and second-neighbor interactions. Carlsson and Meschter (1989) used total energy calculations, measuring the structural energy differences between the DO_{22} and $L1_2$ structures for Al_3X compounds where X = group III-A, IV-A or V-A element. Their findings are shown in Figure 10-4 and in essence can be summarized as:

1. The stability of the DO_{22} structure relative to the $L1_2$ increases rapidly as the transition metal *d*-electron count increases.
2. Atomic size effects do not appear to be important.
3. Energy differences are largest for group V trialuminides.

and

4. Calculated electronic density of states (DOS) show each structure to have a minimum in the DOS distribution at a characteristic *d*-electron count and the preferred structure for a compound is one in which the Fermi level lies in the minimum.

This study not only explained why Al_3Ti could be transformed from the DO_{22} to the $L1_2$ structure by alloying, but also why such an analogous behavior is not observed for Al_3Nb. The importance of the tetragonal distortion (i.e., relaxed c/a ratio versus ideal c/a ratio in Fig. 10-4) was emphasized in the Al_3Ti situation.

2.3. The $L1_2$ Trialuminides

In addition to the group IV-A and group V-A elements, Sc, Y, and the rare earth elements Er, Yb, Ho, Tm, Lu, and U form trialuminide compounds, and of these, Sc, Er, Yb, and U trialuminides crystallize in the $L1_2$ structure. From melting-point and density considerations, Al_3Sc is the most desirable of these compounds, although the exorbitant cost of Sc precludes widespread research. Nevertheless, there has been some theoretical and experimental work on the compound (Fu 1990; Schneibel and George 1990a; Schneibel and Hazzledine 1992). First principles calculations (Fu 1990) of the elastic moduli of Al_3Sc are in good agreement with the experimental data obtained via ultrasonic techniques (George et al. 1990). The compression behavior of cast and homogenized Al_3Sc containing ~ 500 ppm of oxygen was characterized in the range 77 K–1273 K (Schneibel and George 1990a). Strength at 77 K was marginally higher than at room temperature and a mild positive-temperature dependence above room temperature with a strength peak at ~ 500 K was noted. The strength of Al_3Sc is fairly low, not exceeding 150 MPa at any temperature; yet the specimens fail by brittle-transgranular cleavage, the cleavage planes being predominantly of the {011} type, although {100}-type planes were also identified (George et al. 1990). The fracture toughness at room temperature of Al_3Sc was estimated to be ~ 3 $MPa\sqrt{m}$ and the brittle fracture was attributed to a low K/G ratio (bulk to shear modulus), consistent with the Pugh criterion (Pugh 1954). The low value of K/G was attributed to a low K, implying low cleavage strength for such materials.

Recently, Schneibel and Hazzledine (1992) indicated that although not a primary cause, the fracture characteristics of Al_3Sc exhibit some similarities to the stress-corrosion cracking of f.c.c. materials and Cu_3Au. Further, they argue that brittle fracture of Al_3Sc cannot be rationalized exclusively on the basis of surface-energy considerations or models that classify fracture as either fully brittle or fully ductile and that detailed atomistic simulations would be required to obtain a full understanding of fracture. The dislocations in an Al_3Sc specimen compressed at 1350 K were shown to dissociate into $a/2\langle 110\rangle$ superpartials bounding APB's and gliding on the {111} planes (George et al. 1990). Superpartial separation of 3.7 nm yields an APB energy of 313 mJ/m^2 as compared to the

670 mJ/m^2 obtained by Fu (1990) from first-principles calculations. These calculations by Fu gave an APB energy of 450 mJ/m^2 on the (001) planes and a SISF energy of 265 mJ/m^2 on the (111) planes of Al_3Sc. It is interesting to note that in spite of a low SISF energy relative to the APB energy, the superpartials prefer to dissociate via APB separation.

The elastic constants of the ternary $L1_2$ trialuminides in the Al-Ti-X (X = Ni, Fe, Cr, Mn) and in the Al-Zr-Fe system have been measured (Table 10-1) and it is evident that the Poisson's ratio ν is low for all these compounds compared to metals, and further, some of the compounds exhibit significantly lower values than others (e.g., $Al_{67}Mn_8Ti_{25}$ has $\nu = 0.11$ whereas $Al_{67}Ti_{25}Ni_8$ has $\nu = 0.2$). The implications of a low Poisson's ratio are not fully understood, although they are suggestive of nonmetallic behavior. In a recent investigation (Nishitani et al. 1991), the electrical resistivity in the temperature range 4.2 K to 300 K for binary Al_3T_i was shown to increase with temperature as expected, whereas for $Al_{22}Ti_8Fe_3$, it remained essentially constant and decreased with increasing temperature for $Al_{67}Ti_{25}Mn_8$. This decrease for the last of these compounds continued up to 900 K. It is evident from these studies that the electronic structure of the various ternary $L1_2$ trialuminides exhibits significant differences and therefore, it is not unreasonable to expect a range of mechanical properties from these compounds including defect energies and dislocation separation schemes. Thus, in light of these findings, instead of generalizing the mechanical properties of these trialuminides as a group, they are treated individually in the following paragraphs. Of the various $L1_2$ ternary trialuminides, the Al-Ti-X (X = Cu, Ni, Fe, Cr, Mn) have received the most attention, although the Al-Zr-Cu, Al-Zr-Fe, and Al-Zr-Mn systems have also been examined (Varin, Virk, and Winnicka 1991).

Table 10-1 Elastic Properties of the $L1_2$ Trialuminides

Compound	E	G	K	ν	K/G	Reference
$Al_{67}Ti_{25}Ni_8$	200	82	116	0.22	1.41	Turner, Powers and Wert 1989.
	188	78	105	0.20	1.3	Mikkola et al. 1991.
$Al_{67}Ti_{25}Fe_8$	192	84	89	0.14	1.06	George et al. 1990.
	181	78	90	0.16	1.2	Mikkola et al. 1991.
$Al_{67}Zr_{25}Fe_8$	166	68	103	0.22	1.51	George et al. 1990.
$Al_{67}Ti_{25}Mn_8$	174	79	74	0.11	0.9	Mikkola et al. 1991.
$Al_{67}Ti_{25}Cr_8$	168	75	75	0.13	1.0	Mikkola et al. 1991.
Al_3Sc	166	68	99	0.22	1.46	George et al. 1990.

An early investigation (Kumar and Pickens 1988) on the variation of total crack length on the surface of a polished sample resulting from hardness indentations versus the applied load indicated the $L1_2$ compound Al_5Ti_2Cu to be more crack resistant than the binary DO_{22} Al_3Ti, although the $L1_2$ compound $Al_{22}Ti_8Fe_3$ was even better than its Cu-containing counterpart. A similar observation was made by Nic, Zhang, and Mikkola (1990) and by Varin et al. (1991). In addition, these latter two investigations also showed a decreasing hardness trend in these $L1_2$ trialuminides as the atomic number of the ternary addition decreased from Cu to Cr. The lattice parameter of the $L1_2$ phase exhibited an opposite trend and this was argued to be a consequence of weakening of interatomic bonds (Nic et al. 1990). The compressive yield strength-temperature profile for Al_5Ti_2Cu showed a mild positive temperature dependence, with a peak occurring at ~ 650 K and a strength of ~ 520 MPa. The addition of up to 500 ppm of boron to this compound increased the compressive yield strength at room temperature and did not beneficially influence ductility (Winnicka and Varin 1991).

The deformation behavior and fracture characteristics of $Al_{67}Ti_{25}Ni_8$ have been reported (Turner, Powers, and Wert 1989), and at room temperature, in compression, a yield strength of 355 MPa was obtained and substantial compressive plasticity was noted. Deformation occurred by $\{111\}\langle 110\rangle$ slip and the dislocations were found to dissociate with a 2 nm separation, indicative of high $\{111\}$ APB and SISF energies. The presence of sessile, intrinsic stacking faults on $\{001\}$ planes of annealed $Al_{67}Ti_{25}Ni_8$ resulting from the dissociation of $\langle 001\rangle\{001\}$ dislocations has been reported although these did not appear to participate directly in the deformation processes in this compound. Specimens deformed at 573 K revealed $\langle 110\rangle$ dislocations on $\{111\}$ and $\{100\}$ planes, those lying on the cube planes being dissociated into $1/2\langle 110\rangle$, whereas the dislocations on the octahedral planes were undissociated and frequently present as dipoles (Vasudevan et al. 1989). When the deformation temperature was increased to 873 K, the population of dissociated $\langle 110\rangle$ dislocations on the $\{001\}$ planes increased, although those on the $\{111\}$ planes remained undissociated and as dipoles. These investigators attributed the brittle behavior of this compound to the high dipole density they observed, following a previously proposed model that involves rapid work hardening due to dipoles.

Notched bend specimens failed by brittle-transgranular cleavage at room temperature and this was attributed to a high activation energy barrier for dislocation emission from crack tips (Turner et al. 1989) along the ideas originally proposed by Rice and Thomson (1974). Subsequently, it was shown that several of these $L1_2$ compounds exhibited dislocation dissociation and yet failed by brittle-transgranular cleavage, whereas the Rice–Thomson model could only be applied strictly for the case of undissociated dislocations. A recent investigation (Bartholomeusz and Wert 1992) has incorporated dislocation dissociation into the Rice–Thomson model and has shown that the results of the extended model are consistent with the observed fracture modes in these $L1_2$ compounds containing dissociated dislocations.

Although there have been numerous studies on these $L1_2$ trialuminides over the past five to seven years, there are several disagreements and inconsistencies in the published literature regarding mechanical properties and even dislocation-dissociation schemes for a particular compound. These differences stem from a variety of sources, including inadequate knowledge of the size and shape of the $L1_2$ phase field, its variation with temperature, and the sluggish kinetics of phase decomposition in these systems. Frequently, "nominally" single-phase materials as observed by optical microscope contain second phases that are resolvable only at the TEM level. Likewise, the role of stoichiometry within the single-phase field in influencing the mechanical properties and dislocation schemes is not well understood. Compounding these problems are difficulties involved in producing sufficient quantities of fully dense material for testing and characterization with a known composition and grain size. Thus, often material containing porosity and second phase are tested without paying significant attention to deviations from the intended composition or mi-

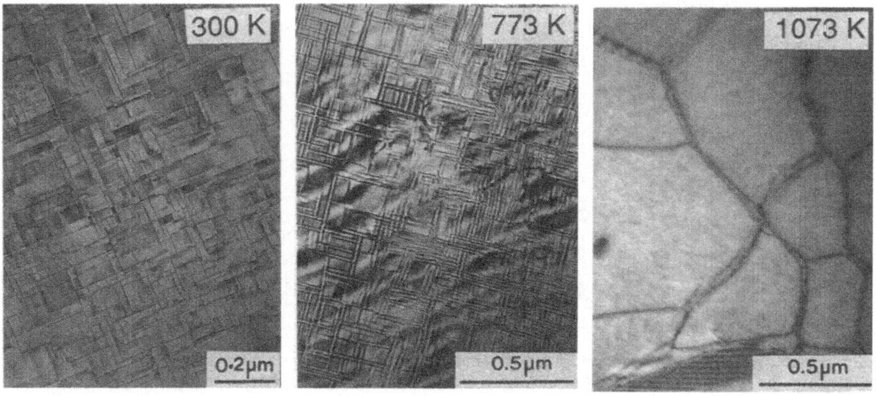

(a) (b) (c)

Figure 10-5. The presence of fine $Al_2Ti(Fe)$ precipitates in an Al-Ti-Fe based $L1_2$ matrix in (a) the as hot-pressed condition and (b) after deformation at 773 K; (c) a substantially clean single phase matrix after deformation at 1073 K.

crostructure. These problems are perhaps best illustrated in the case of investigations on the $L1_2$ trialuminide $Al_{22}Ti_8Fe_3$.

The size and shape of the $L1_2$ phase field in the Al-Ti-Fe system was recently evaluated at 1473 K and 1073 K (Mazdiyasni et al. 1989). With a decrease in temperature, not only does the single-phase field shrink in size but also the geometric center of the field shifts towards the Al-rich end. There is no information on the size and shape of the phase field below 1073 K. Attempts to measure the size and shape of the $L1_2$ phase field by microprobe techniques were futile (Kumar et al. 1991) because of the fine interwoven two-phase microstructures that result in the specimens. An example of an apparently single-phase material as observed in an optical microscope reveals fine precipitates with a lath morphology (Fig. 10-5a) that was originally erroneously identified (DiPietro, Kumar, and Whittenberger 1991) as $Fe_2Ti(Al)$ but was later reconciled to be $Al_2Ti(Fe)$. These will influence the mechanical properties measured and further, their precipitation in the matrix can be suppressed by air cooling and subsequent precipitation during high-temperature mechanical testing can occur (Lerf and Morris 1991).

The presence of these Al_2Ti-type precipitates after compression testing at 773 K (Fig. 10-5b) but not after testing at 1073 K (fig. 10-5c) confirms their dissolution between these temperatures, and their conspicu-

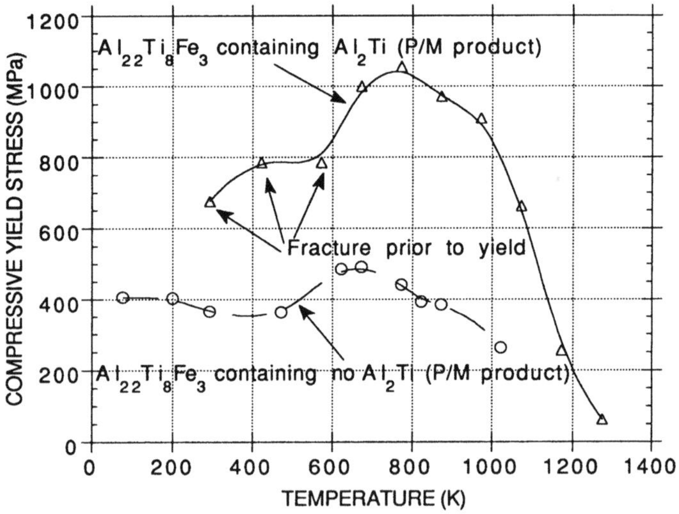

Figure 10-6. Compressive yield strength-temperature profiles for two Al-Ti-Fe alloys, one containing the Al_2Ti phase (the high-strength version) and the other with a substantially single-phase $L1_2$ structure.

ous absence even on a fine scale in the air-cooled 1073 K specimen endorses the previous proposal of the sluggish precipitation kinetics. These fine precipitates provide significant strengthening to the matrix up to ~ 1000 K, beyond which strength decreases precipitously and is coincident with the dissolution of these fine precipitates (Fig. 10-6). In this figure, data from an early study (Kumar and Pickens 1988) on $Al_{22}Ti_8Fe_3$ containing no Al_2Ti is included and it is evident that the overall strength is significantly lower at room and warm temperatures. Similar differences in strength-temperature profiles have been reported by Wu, Pope, and Vitek (1990) in single crystals and by Morris, Lerf, and Hollrigl (1991) for spray-deposited polycrystalline material. From these observations, it appears that this difference is more likely due to Al_2Ti rather than grain size, and further, the critical Ti content for Al_2Ti precipitation appears to be ≥ 28.0 at% with the Fe-level being maintained between 8 and 9 at%.

Bend tests on polycrystalline $Al_{22}Ti_8Fe_3$ have shown brittle failure up to 873 K (Kumar et al. 1991) with fracture propagating by transgranular cleavage. Serrated flow has been reported in these materials in the temperature range 573 K–900 K (Lerf and Morris 1991; Mabuchi et al. 1990; Potez, Lapasset and Kubin 1992) and has been attributed to repeated ⟨110⟩ dislocation dissociation by Lerf and Morris (1991) but disputed by Potez et al. (1992) as being due to the dynamic interaction of solute atoms and mobile dislocations (Portevin–Le Chatelier effect), although the nature of solute atoms involved was speculated as being either interstitial (oxygen) or substitutional (intrinsic). An interesting observation is that such serrations were not observed in binary Al_3Sc tested in compression in this temperature range (Schneibel and George 1990b), which may suggest that the ternary species plays an active role in the observed serrated flow.

Using single-crystal $Al_{22}Ti_8Fe_3$, it was shown that while the slip behavior was similar to that of Ni_3Al, the temperature dependence of the critical-resolved shear stress (CRSS) for {111} slip was characteristic of Pt_3Al (Wu et al. 1990). The nature of dislocation dissociation in $Al_{22}Ti_8Fe_3$ following deformation at various temperatures has been examined and the reported schemes are conflicting. In a specimen deformed at room temperature, Hu et al. (1991) indicated the presence of a/3⟨112⟩ superpartials separated by SISF on {111} planes, subsequently confirmed by Inui et al. (1992) who reported in addition a strong dipole-formation tendency. The interpretation of Inui et al. has however been questioned by Veyssiére and Morris (1992) and the latter prefer to interpret the observations of Inui et al. as being APB separation. Following deformation at 873 K, however, Inui et al. (1992) show the superdislocations to dissociate on {001} planes into a/2⟨110⟩ superpartials that bounded APB's. In an earlier investigation, George and co-workers (1990) reported an APB dissociation scheme in Al-23Ti-5V-6Fe alloy (essentially single-phase) deformed to 1.5% strain at room temperature with a/2⟨110⟩ superpartials bounding the APB and gliding on {111} planes. An associated APB energy of 274 mJ/m^2 was reported. These compression specimens were obtained from extruded material and whether the observed dissociated dislocations were a consequence of the compression test or prior deformation during hot extrusion is not evident.

Morris and his group at the University of Neuchatel in Switzerland have given significant research attention to dislocation-dissociation schemes in these $L1_2$ trialuminides (Lerf and Morris 1991; Morris 1991, 1992). Initially, they reported undissociated ⟨110⟩ dislocations in Al_5Ti_2Fe deformed 1% in compression at room temperature and also the presence of dipoles, both their densities increasing with increasing strain (Lerf and Morris 1991). Specimens deformed 1% at 773 K, however, showed the presence of 1/2⟨110⟩ superpartials bounding APB's on the {111} plane with an associated APB energy of 140 mJ/m^2, which is about half of that previously reported by George et al. (1990). In a specimen deformed 1% at 973 K, dissociated superdislocations were noted on {111} and {001} planes. In addition, fine precipitation was observed in the matrix, indicating a change in matrix composition. Measured APB energies on the {111} and {001} planes were 72 mJ/m^2 and 45 mJ/m^2, respectively, and the change in APB energy on the {111} plane with temperature was attributed to the change in matrix composition due to the observed precipitation. Such a large change in the {111} APB energy due to relatively small compositional changes implies the need for caution in using these values in future research. In a later report (Morris 1992), weak, localized dissociations were noted along the length of dislocations in a specimen deformed 1% at room temperature (Fig. 10-7a), whereas, when the specimen was deformed to 4% strain, complete dissociation occurred for all dislocations (Fig. 10-7b). These dissociated dislocations were of the a/2⟨110⟩ type on {111} planes bounding APB's and conflicting the earlier proposal (Hu et al. 1991; Inui et al. 1992) of a/3⟨112⟩ dislocations bounding SISF's on the {111} planes. The influence of annealing, for short times, at temperatures as low as 773 K, on dislocation configurations produced by prior deformation (4%; 300 K) has been studied (Morris 1991) and it was shown that dislocation dissociation is assisted and the separation enhanced by annealing (Figs. 10-8a, b). Based on this observation, it follows that previous estimates of APB energies and dislocation spacings following high-temperature deformation could be erroneous.

Two research groups independently reported the occurrence of ternary Al-rich $L1_2$ compounds in the Al-Ti-Cr and Al-Ti-Mn systems

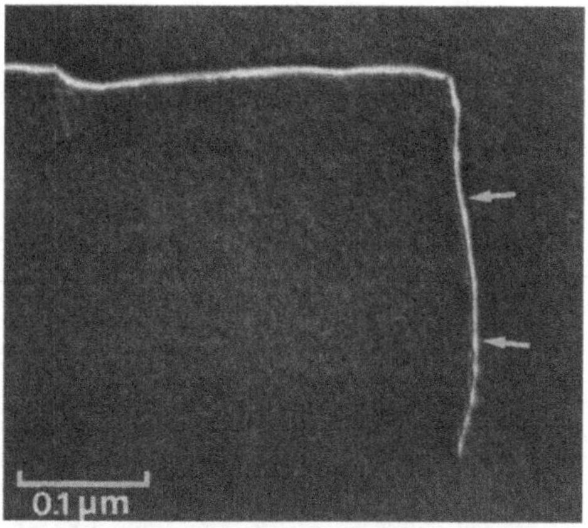

(a)

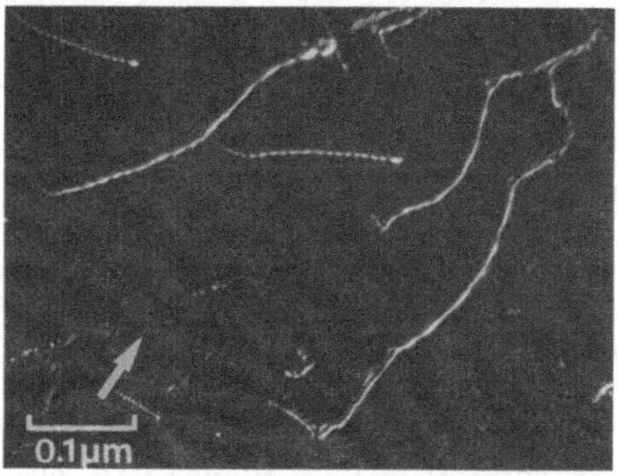

(b)

Figure 10-7. The effect of strain on the extent of superdislocation dissociation in Al_5Ti_2Fe deformed at room temperature (a) after 1% strain ($g = \bar{1}\bar{1}1$), and (b) after 4% strain ($g = 1\bar{1}1$). (Morris 1992a, with permission from the Materials Research Society)

(Mikkola et al. 1991; Mabuchi, Hirukawa, and Nakayama 1989; Mabuchi et al. 1990). These compounds, $Al_{66}Ti_{25}Cr_9$ and $Al_{67}Ti_{25}Mn_8$, were shown to exhibit significantly improved compressive ductility relative to $Al_{22}Ti_8Fe_3$ and even some bend ductility at room temperature. The variation in compressive yield strength with temperature for these compounds in the polycrystalline form, with various stoichiometries has been reported in several investigations (Mikkola et al. 1991; Brown, Kumar, and Whittenberger 1990; Mabuchi et al. 1990) and some of the data are shown

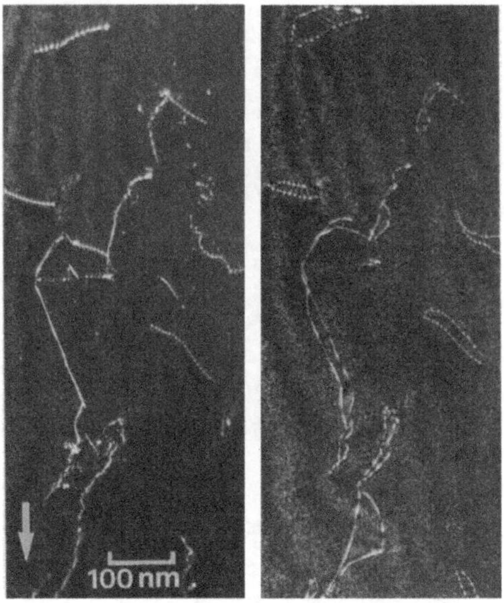

(a) (b)

Figure 10-8. Dislocations in Al_5Ti_2Fe deformed 4% in compression at room temperature (a) and after in situ annealing for 5 mins at 973 K; (b) Foil orientation is (011), diffraction vector is $[\bar{1}\bar{1}1]$. (Reprinted with permission from *Scripta Metallurgica et Materialia*, Volume 25, D. D. Morris, pages 713–716, 1991, Elsevier Science Ltd., Oxford, England).

in Figure 10-9. Compressive-creep studies have also been conducted on these materials (Whittenberger et al. 1991; Schneibel, Horton, and Porter 1992) in the 1000 K–1200 K range and the response has been shown to be strongly grain-size dependent.

Bend tests on forged $Al_{66}Ti_{25}Mn_9$ and $Al_{67}Ti_{25}Cr_8$ as a function of temperature revealed finite plastic strain in the Mn-based material at 298 K but not in the Cr-based counterpart up to 623 K (Kumar, Brown, and Whittenberger 1991) in contradiction[1] to reports of ambient temperature bend ductility in $Al_{67}Ti_{25}Cr_8$ (Mikkola et al. 1991; Schneibel et al. 1992).

[1]Extensive annealing of the forged material prior to bend testing lowered the bend ductile-to-brittle transition temperature of $Al_{67}Ti_{25}Cr_8$ to 473 K, where 0.5% plastic strain was consistently achievable for a range of grain sizes and over four orders of magnitude of strain rate, both in air and in oil. The material, however, still failed elastically at room temperature (Kumar and Brown 1992a).

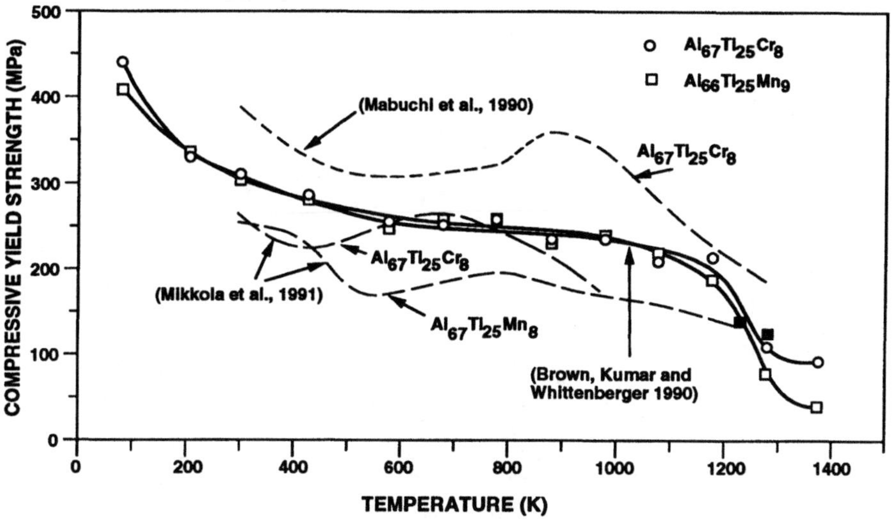

Figure 10-9. Compression yield strength-temperature profiles for $Al_{66}Ti_{25}Cr_9$ and $Al_{67}Ti_{25}Mn_8$ according to various investigations.

Dislocation dissociation schemes have been characterized in specimens deformed in compression (Morris 1992) at room temperature and the dissociations were shown to be larger than in $Al_{22}Ti_8Fe_3$, whereas the type of dissociation was similar.

Uniaxial tension tests on buttonhead specimens of forged $Al_{67}Ti_{25}Cr_8$ and $Al_{66}Ti_{25}Mn_9$ have been performed in air in the 298 K–1073 K temperature range (Kumar and Brown 1992b, 1992c). The variation in yield strength and ductility with test temperature are shown in Figure 10-10. The Mn-based compound exhibits a finite plastic strain at room temperature that increase up to 623 K before reaching a minimum at 773 K. At higher temperatures, ductility continues to increase. In contrast and in agreement with earlier results from bend tests, the Cr-containing compound exhibits ductility only at 623 K and higher. The tensile yield stress up to 623 K is higher than the compressive yield stress at any given temperature, although at higher temperatures, fairly good agreement is noted. Fracture transitions from transgranular cleavage at room temperature in $Al_{66}Ti_{25}Mn_9$ to intergranular failure at high temperatures, although there is not a direct correlation between the onset of intergranular failure and the ductility minimum. In $Al_{67}Ti_{25}Cr_8$, a similar behavior is noted except that the onset of intergranular failure occurs at higher

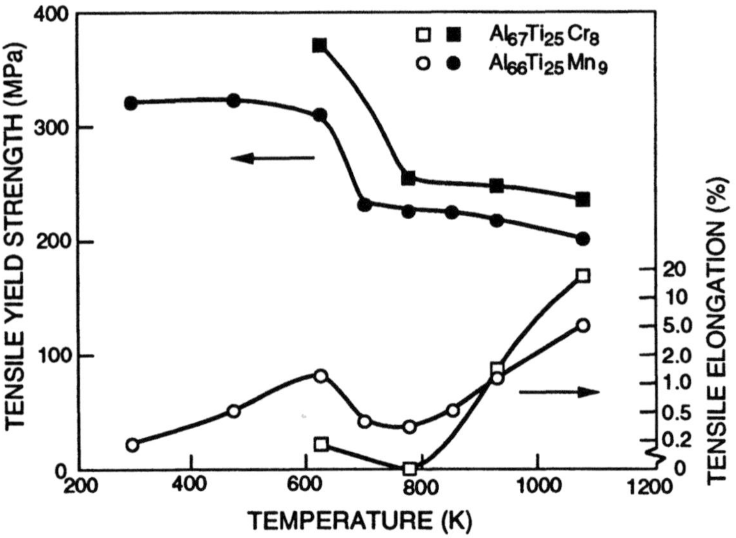

Figure 10-10. Variation in tensile yield strength and ductility with temperature for $Al_{66}Ti_{25}Cr_9$ and $Al_{67}Ti_{25}Mn_8$.

temperatures (Fig. 10-11a, b). In the 1073 K specimen of $Al_{67}Ti_{25}Cr_8$ that deformed to ~ 19% plastic strain, serrated grain boundaries were observed (Fig. 10-12a) and frequently, these serrated peaks were associated with sub-grain boundaries (Fig. 10-12b). Fracture in this specimen proceeds intergranularly and the presence of grain-boundary serrations produces a "delamination" effect along grain boundaries on the fracture surface. Mullendore and Grant (1961) reported that grain-boundary serrations occurred only in conjunction with grain-boundary sliding and migration, forming after an elongation sufficient to produce a substructure adjacent to the boundary. They contended that the peaks of the serrations represented the intersection of subgrains with the grain boundary and were thus similar to a grain-boundary triple point in acting as a crack-nucleation source. Lytton, Barrett, and Sherby (1965) rationalized the formation of serrated grain boundaries as a sequence of events that included the formation of shear faults along a grain boundary, thereby introducing lattice-bending stresses. Edge dislocations would then accumulate and polygonize at such shear faults so as to form tilt boundaries that relieved these bending stresses. The grain boundary migrates toward these tilt boundaries to equalize the interfacial tension, causing the boundary to assume a serrated morphology.

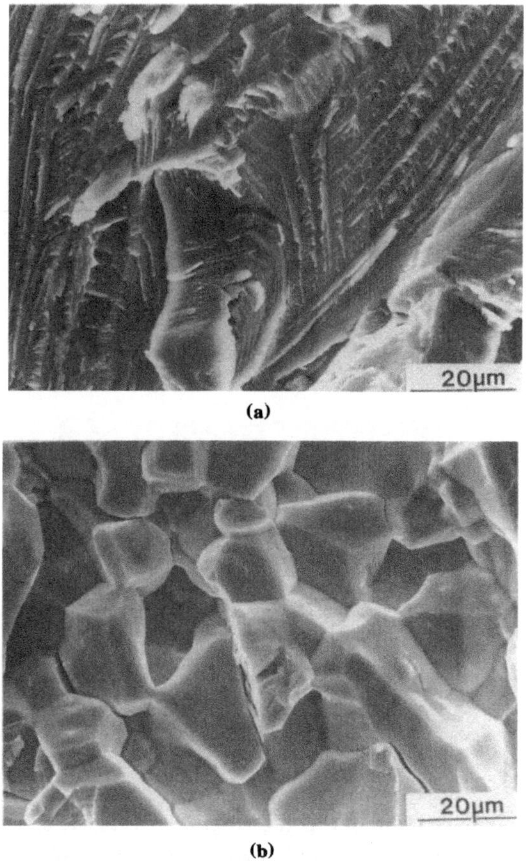

Figure 10-11. Fracture surfaces of tensile specimens of forged $Al_{67}Ti_{25}Cr_8$ tested at (a) 773 K and (b) 923 K.

Isothermal oxidation-resistance of three of these $L1_2$ compounds ($Al_{67}Ti_{25}Cr_8$, $Al_{67}Ti_{25}Ag_8$, and $Al_{66}Ti_{25}Mn_9$) was compared against binary Al_3Ti and TiAl (Hirukawa, Mabuchi, and Nakayama 1991), and at 1273 K, all of the trialuminides were superior to TiAl, although $Al_{67}Ti_{25}Ag_8$ was significantly worse than the other compounds in its category. The compound $Al_{67}Ti_{25}Cr_8$ appeared to be the only one that formed Al_2O_3 and no TiO_2 and was better than $Al_{66}Ti_{25}Mn_9$, although at a lower temperature of 1073 K, they exhibited comparable resistance. Cyclic oxidation-resistance (1473 K/air/200 cycles, 1h each) of the

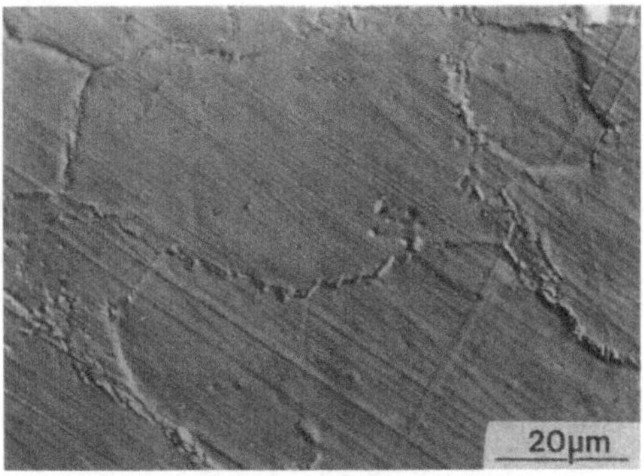

(a)

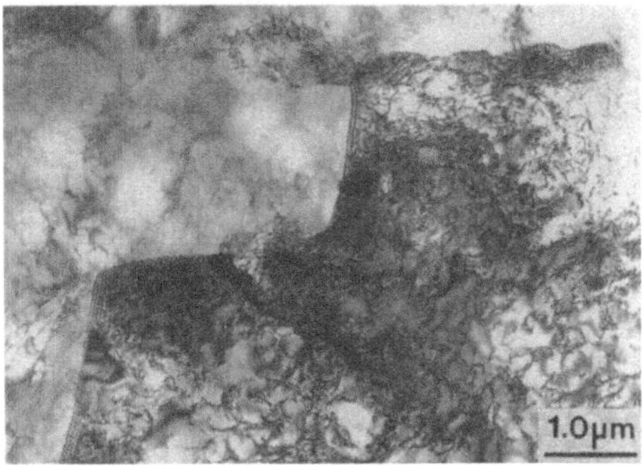

(b)

Figure 10-12. Cross-section of the gage region of a tensile specimen of $Al_{67}Ti_{25}Cr_8$ that was strained to ~ 19% at 1073 K showing (a) serrated grain boundaries, and (b) association of a subgrain boundary with a serrated peak.

Cr-containing $L1_2$ compound was shown to be superior to the Mn-based counterpart (Parfitt et al. 1991) although in a later study (Smialek 1991), it was shown that at lower temperatures, the two compounds exhibited comparable resistance.

Limited interdiffusion data and associated activation energies have been reported (Kumar and Whittenberger 1992) for the system $Al_{67}Ti_{25}Cr_8$-$Al_{66}Ti_{25}Mn_9$. The calculated activation energy, Q = 350 kJ/mol is in reasonable agreement with activation energy data for compressive creep of such materials (Whittenberger et al. 1991), indicating the dominance of volume diffusion. These diffusion studies also confirmed the existence of continuous solid solution between the ternary compounds

$Al_{22}Ti_8Fe_3$, $Al_{67}Ti_{25}Cr_8$, and $Al_{66}Ti_{25}Mn_9$, and quaternary and quinary compositions have been cast, isothermally forged, and evaluated in compression as a function of temperature (Kumar and Brown 1992d). It was found that progressive substitution of Fe for Mn in $Al_{66}Ti_{25}Mn_9$ produced significant strength improvements, particularly at high temperatures but was accompanied by a loss in low- and warm-temperature ductility.

2.4. Composites Using DO_{22} and $L1_2$ Trialuminides as Matrices

Anton (1988) and Alman, Stoloff, and Otsuki (1989) have examined Al_3Ta as a matrix for long-fiber and discontinuous reinforcement. Compressive yield strength of the monolithic material was measured at room temperature and at 1023 K in the latter study.

More recently, the role of TiB_2 particulates reinforcement in affecting the compressive strength of the $L1_2$ trialuminides (Fe-based, Cr-based, and Mn-based) as a function of temperature and strain rate at high temperature was examined (DiPietro et al. 1991; Whittenberger et al. 1991). These TiB_2 particles were typically 1–3 μm in size and up to 20 vol% of particulate reinforcement was considered. In general, the particulates significantly enhanced room- and warm-temperature strength but did not improve the high-temperature strength or extend the usefulness range of the monolithic material. In fact, above $\sim$ 1000 K, the composite containing 20 vol% TiB_2 in the Fe-based $L1_2$ trialuminide matrix was weaker than the monolithic counterpart (DiPietro et al. 1991).

The effects of strain rate on the flow stress of the monolithic trialuminide and the particulate composites in the temperature range 1000 K–1200 K have been examined (Whittenberger et al. 1991; DiPietro et al. 1991). A very interesting observation is that the composites are weaker than the monolithic material at fast strain rates at both 1100 K and 1200 K, but are stronger at the slow strain rates; the difference in strength at the two extreme strain rates considered progressively increased with increasing TiB_2 content. Stress exponents and activation energies for these materials were determined based on a power-law fit and both these parameters increase with increasing TiB_2 content. The observed "dispersion-weakening" effect has been explained using a previous proposal due to Edwards, McNelley, and Sherby (1975), namely that above a certain threshold stress, macroscopic flow stress is dominated by mobile dislocations, generated at particle-matrix interfaces by local stresses arising from incompatibilities between the matrix and the reinforcements. The reversal in behavior at the slower strain rates was argued on the basis that the strength levels are low and therefore the threshold stress was not reached;

thus, the composite was stronger. The effect of reinforcing $Al_{66}Ti_{25}Cr_9$ with up to 30 vol% of randomly oriented Al_2O_3 short fibers was explored using reactive hot-isostatic pressing methods (Alman et al. 1991) and it was found that while the compressive strength was improved with reinforcement, the compressive strain to failure decreased monotonically with increasing volume fraction of reinforcements leading the investigators to conclude that these materials were flaw intolerant.

3. OTHER ALUMINIDES

In this section, two aluminides with the B2 structure, CoAl and RuAl, and one with the A15 structure, Nb_3Al, are discussed. The compound CoAl exhibits similar density, melting point, and stoichiometric range of existence as its Ni-base counterpart, NiAl, but has received considerably less research attention. This is mainly due to the extremely brittle characteristics of CoAl relative to NiAl, reasons for which are not clear to date.

The defect structure in CoAl has been characterized (Neumann 1980) and a significant number of antistructure cobalt atoms exist even when Co vacancies are the predominant defect (i.e., Al concentration > 50 at%) and a large number of Co vacancies exist in compositions on the Co-rich side of stoichiometry even though Co-antistructure atoms are the dominant defect. At the stoichiometric composition, triple defects (i.e., two vacancies on Co sites and a cobalt atom on an Al site) were reported. A minimum in the diffusion coefficient was reported at the stoichiometric composition at 1523 K (Nix and Jaumot 1951).

The variation of the elastic modulus of CoAl as a function of stoichiometry (48.5 to 55.8 at%Co) and temperature (300 K–1300 K) is adequately described by the equation,

$$E(\text{CoAl}) = 327.88 - 0.169T - 0.0063X^2 + 0.0022XT \qquad (10.1)$$

where E is in GPa, X is at%Al and T is absolute temperature (Harmouche and Wolfenden 1986). Clark and Whittenberger (1984) measured the thermal expansion of binary CoAl as a function of temperature in the range 300 K–1273 K and found that composition did not significantly affect the thermal expansion behavior and further, the variation in thermal expansion with temperature was shown to closely resemble that for NiAl.

The brittle nature of CoAl has discouraged tensile studies on it and to date, no tensile data are available. Single- and polycrystalline CoAl have however been examined in compression as a function of temperature and at high temperature, as a function of strain rate. While the single-crystal study was restricted to stoichiometric CoAl (Hocking, Strutt, and Dodd 1971), the effect of stoichiometry on compressive creep response has been evaluated for the polycrystalline counterpart. In the single-crystal form, CoAl is extremely creep resistant at temperatures below 1223 K under stress conditions that cause high creep rates in NiAl. Correspondingly, dislocation climb leading to sub-boundary formation during steady-state creep is observed in NiAl above 873 K, whereas in CoAl, such sub-boundaries were uncommon even at temperatures up to 1323 K. In CoAl deformed at 1223 K, however, loops formed by double cross slip were frequently observed. Yaney and Nix (1988) observed sub-boundaries in CoAl (49.3 at%Al) that was extruded at 1505 K using a reduction ratio of 16 : 1. Based on their studies of this material, they argued that the superior high-temperature strength of CoAl over NiAl is neither related to differences in diffusivity nor stiffness at temperature between the two compounds; rather, they postulate that a greater lattice-friction-strengthening effect is responsible for the observed strength superiority of CoAl. The dislocation structure in the extruded material was analyzed (Yaney, Pelton, and Nix 1986) and both, $b = \langle 100 \rangle$ and $b = \langle 111 \rangle$ were observed, the former on both {110} and {100} planes and the latter on {110} planes. In spite of the high deformation temperature, $b = \langle 110 \rangle$ dislocations were not observed, suggesting that the difficulty in activating $\{110\}\langle 110 \rangle$ slip in CoAl may be greater than in NiAl.

Whittenberger (1985) has studied the compressive flow-stress response to strain rate in the 1100 K–1400 K range for fine-grained (10 μm) powder metallurgy processed CoAl for Co content varying from ~ 48.5 at% to 56.2 at%. The variation of flow stress with nominal strain rate was fitted to a temperature-compensated power-law equation of the form,

$$\mathring{\varepsilon} = B\sigma^n \exp(-Q/RT) \tag{10.2}$$

where B is a constant, Q is an activation energy, R is the universal gas constant, and T is the absolute temperature. The variation in stress exponent n and activation energy, Q, with composition is summarized in Table 10-2. The flow stress was observed to reach a maximum for an Al-to-Co ratio of 0.975 and it was suggested that the deformation mechanisms are likely influenced by stoichiometry. Ternary solid-solution alloys of Co-Fe-Al and Co-Ni-Al have also been similarly examined (Whittenberger 1987) in the 1100 K–1400 K range. Substituting Co with Fe or Ni enhanced the flow stress at slow strain rates (e.g., $\sim 2 \times 10^{-6}$ s^{-1}), Fe being more effective than Ni. The stress exponent and activation energy for compressive creep for two ternary alloys containing ~ 44 at%Al and a grain size of ~ 8–10 μm can be directly compared against

Table 10-2 Stress Exponent and Activation Energy for Compressive Creep of CoAl-Based Materials

Composition (at%)	Temperature Range (K)	n	Q (kJ/mol/K)
BINARY ALLOYS			
51.31 Al	1300–1400	4.74	238.6
49.89 Al	1300–1400	4.47	346.1
48.50 Al	1200–1400	4.47	337.7
43.65 Al	1100–1400	2.70	199.4
TERNARY ALLOYS			
43.8 Al-14.6 Ni	1100–1300	3.72	293.3
44.6 Al-15.4 Fe	1200–1400	3.79	330.4

Source: Whittenberger 1985; 1987

binary CoAl containing a similar Al level and comparable grain size (Table 10-2). The ternary alloys behave more like the binary alloys containing higher levels of Al (i.e., ~ 48 at%Al). Further, in the temperature regime under consideration, the ternary aluminides could be deformed up to 10% compressive strain before failure occurred by grain-boundary cracking; in contrast, the binary CoAl frequently failed by transgranular cleavage.

During the course of a search for high-melting intermetallics (i.e., > 1773 K) with reasonable densities (< 10.0 g/cm^3), Fleischer and Zabala (1990a) first identified the B2 compound RuSc as a potential candidate. Preliminary studies indicated that this compound exhibited some room-temperature damage tolerance. This prompted an examination of the B2 compound RuAl, which exhibits a slightly lower melting point (2473 K versus 2333 K) and similar density but considerably less expensive than RuSc. Binary, hypo- and hyper-stoichiometric RuAl have been characterized in terms of their physical and mechanical properties. In general, the Ru-rich compositions were tough at room temperature whereas the Al-rich compositions were brittle. This brittleness was associated with intergranular fracture and the addition of 0.5 at%B significantly enhanced the compressive ductility irrespective of the binary composition examined (Fleischer, Field, and Briant 1991). Dislocations with ⟨100⟩, ⟨110⟩ and ⟨111⟩ Burgers vectors were observed in $Ru_{53}Al_{47}$ deformed to 17% true strain. Neither stoichiometry nor the addition of B to the binary compositions significantly influences variations in microhardness with temperature, and hardness is retained up to ~ 1423 K. The substitution of Sc in the alloy for Ru appears beneficial for high-temperature strength without a concomitant loss in room-temperature toughness, and B addition to the Sc-containing alloys also improves ambient toughness. Several ternary B2 alloys in the systems Al-Ru-Co, Al-Ru-Fe, Al-Ru-Cr, and Al-Ru-Ti based on RuAl have been explored and show promise (Fleischer 1991a). Isothermal-oxidation studies revealed the formation of an alumina scale at temperatures below 1373 K, whereas at higher temperatures ($>$ 1473 K), owing to the high vapor pressure of the oxide, Ru is volatile and excessive weight losses (linear behavior) were observed (McKee and Fleischer 1991). The addition of 5% Cr to RuAl increases its oxidation resistance (Fleischer 1991a). One potential drawback of the RuAl compound is the cost and availability of Ru and this issue has been addressed by Fleischer (1991b).

Several compounds with the A15 structure (complex cubic; Cr_3Si-type) have been examined as potential superconducting material (Dew-Hughes 1975; Kammerdiner and Luo 1972) and examples include V_3Ga, Nb_3Sn, Nb_3Al, and $Nb_3(Al, Sn)$. In addition to its superconductivity, the compound Nb_3Al has been recently examined for its potential as a high-temperature structural material, both in the United States and in Japan. Microstructure and mechanical behavior of single-phase, binary Nb_3Al, two-phase Nb_3Al-Nb, and ternary Nb_3AlX (X = Mo, Ta, Cr, Ti, Si, Sn, and Ge) have been examined.

The A15 compound, Nb_3Al, forms incongruently as a consequence of a peritectic reaction at $\sim$ 2333 K and a Nb content of 78.5 at%. The compositional range of existence of single phase Nb_3Al extends from $\sim$ 17.5 at%Al to 21.5 at%Al and, in fact, does not include the peritectic composition. Likewise, the stoichiometric composition of 25 at%Al lies in the two-phase field of $Nb_3Al + Nb_2Al$. On the Nb-rich side, Nb_3Al is in equilibrium with Nb solid solution. The stability temperature of the A15 phase varies considerably with composition because of the nature of the phase diagram in the vicinity of the single-phase field. Metastable phases result from nonequilibrium solidification and even solid-state quenching in these compositional ranges and these have been characterized in several investigations. The ability to retain a metastable b.c.c. solid-solution phase has been verified by Kammerdiner and Luo (1972), and more recently by Yasuda et al. (1991). In these studies, a variety of techniques ranging from solid-state quenching to melt spinning and sputtering from the vapor state were used. In another recent study (Aindow et al. 1991), the presence of a metastable B2 phase in melt-spun ribbons of Nb-18 at%Al alloy is claimed. The ability to retain such metastable phases, particularly, the disordered b.c.c. structure, may provide an opportunity to work such materials at moderately low temperatures.

Ternary substitutions in Nb_3Al have been explored and two recent studies include the work of Shah and Anton (1991) and that by Yasuda and co-workers (1991). The influence of ternary alloying on the size, shape, and extension of the Nb_3Al phase field is shown in Figure 10-13. The phase field extends along Al-isoconcentration lines for Mo and Ta substitutions suggesting a direct replacement for Nb, whereas Ge, Sn, and Sb appear to substitute preferentially for Al. Based on the shape of the Nb_3Al phase-field extension into the ternary system, it appears that V, Ti, and Cr substitute for both Nb and Al. Of these various ternary substitutions, V and Cr are claimed to yield the most useful and tough microstructures. Yasuda et al. (1991), however, claim Cr and Ti substitutions on Nb sites only and Zr substitution on Al sites. It has to be recognized that the way a phase field lobes into ternary compositional space may suggest a preferential substitution for a species but does in no way imply a selective site occupancy and care must be exercised in differentiating between the two. Another point that is worth recognizing at this juncture is that neither of these investigations specifies how many specimens were examined, how stable the characterized microstructures were, and what the interstitial levels were in these alloys, and these features are certain to influence the observations.

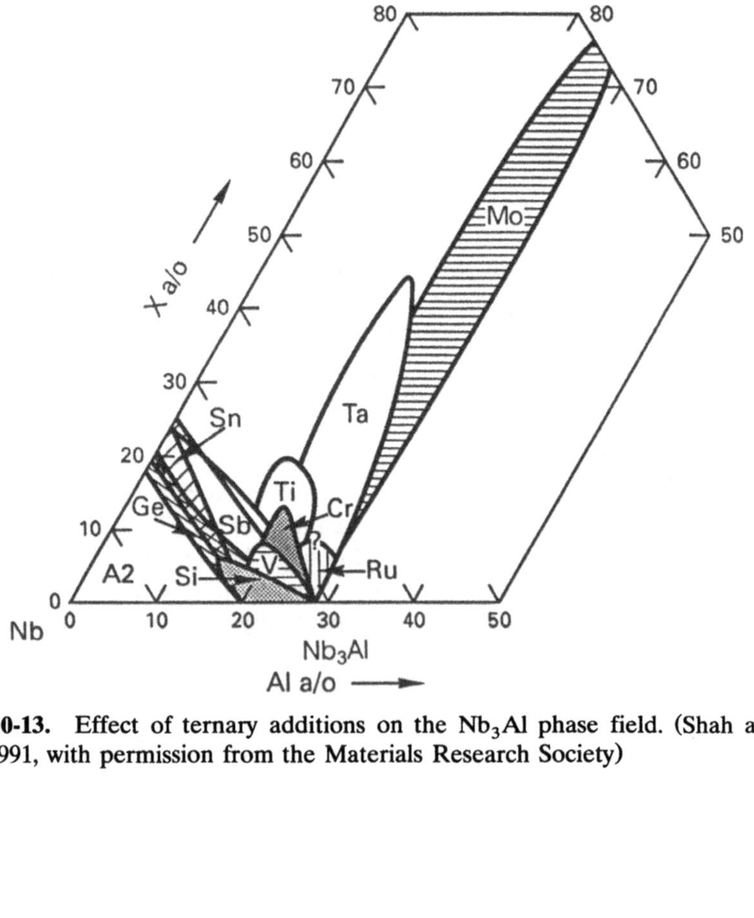

Figure 10-13. Effect of ternary additions on the Nb_3Al phase field. (Shah and Anton 1991, with permission from the Materials Research Society)

Mechanical properties of binary and ternary alloys and the deformation behavior of binary Nb_3Al are currently under investigation and preliminary results are reported here. The ductile-to-brittle transition temperature and the variation in elastic modulus with temperature for a binary Nb-18 at%Al alloy were obtained from four-point bend specimens (Anton and Shah 1989); fracture was found to transition from transgranular failure through intergranular cracking to ductile failure with increasing temperature. In an attempt to incorporate some damage tolerance in Nb_3Al, two phase alloys of Nb_3Al-Nb produced by a melt infiltration technique have been characterized (Kumagai, Hanada, and Saito 1991). As expected, the ductile-to-brittle transition temperature decreased with increasing volume fraction of Nb solid solution, although the high-temperature flow stress was shown to be very sensitive to strain rate and volume of the solid-solution phase.

The variation in compressive yield strength with temperature for binary Nb-25 at%Al and ternary Nb-20Al-10Ti is reported and the substitution of Ti leads to an overall decrease in strength at all temperatures examined; in fact, all the ternary additions examined (Fujiwara, Yasuda, and Kodama 1991), show a decrease in the 1773 K yield strength relative to the binary alloy. According to the recent binary-phase diagram (Kattner 1990), this stoichiometric composition would lie in a two-phase field and it is not clear what effects the ternary additions have on the two-phase microstructure in the binary alloy. Fujiwara et al. (1991) also showed that increasing the Al content in the binary alloy in the range 21–31 at%Al (i.e., progressively increasing amounts of Nb_2Al) causes a steady increase in the 1773 K compressive yield strength from $\sim$ 100 MPa to $\sim$ 275 MPa. Based on the recent report of Umakoshi (1991), however, it appears that single-phase Nb_3Al is stronger than the two-phase alloys of Nb_3Al + Nb_2Al at lower temperatures (1373 K–1573 K). There is not enough data available at present to decide whether these apparently conflicting results are internally consistent, although the superiority in compression creep resistance of Nb_2Al over Nb_3Al at 1473 K was verified by Anton and Shah (1991). In earlier compression creep studies conducted at 1273 K and 1473 K, a binary alloy of Nb-18 at%Al revealed a steady-state stress exponent $n = 2.86$ and an associated activation energy for creep of 83.1 kcal/mol which is in good agreement with the chemical diffusion coefficient, D for stoichiometric Nb_3Al of 87.5 kcal/mol (Anton and Shah 1989).

The oxidation resistance of Nb_3Al is, however, extremely poor at 1473 K and according to Anton and Shah (1989), in less than 24h at 1473 K, the Nb_3Al specimen "crumbled into yellowish oxide flakes."(p. 368). Likewise, the oxidation behavior of Nb_2Al was categorized as catastrophic at this temperature (Anton and Shah 1991a).

The nature of defects in deformed Nb_3Al has been characterized to some extent. Slip at room temperature occurs via the {100}⟨001⟩ system. Following deformation at 1473 K, the defects dissociate into partials, creating stacking faults on {001} planes with a/4⟨001⟩ displacement vector (Marieb et al. 1991). Murayama et al. (1991), however, contend that while {001}⟨100⟩ slip is operative at high temperatures also, the observed stacking faults exhibit an a/2⟨100⟩ displacement vector, extending in the ⟨120⟩, ⟨130⟩, and ⟨110⟩ directions. They postulated high-temperature deformation to proceed via dynamic recrystallization. Grown-in dislocations in melt-spun ribbons of Nb_3Al were shown to contain ⟨100⟩ dislocations, predominantly screw in character and dissociating into two 1/2⟨100⟩ partials that bound a complex stacking fault with a habit plane of {012} (Aindow et al. 1991).

4. BERYLLIDES

As a family of intermetallics, the beryllides have perhaps received the least attention in spite of their attractive specific properties. In the 1960s, refractory metal beryllides were evaluated for their potential as high-temperature structural materials (Lewis 1961a, 1961b) but research was terminated because of the brittle characteristics exhibited by these materials at room temperature. It was not evident then whether the lack of ductility was a consequence of impurity levels in the material or if it was intrinsic to these compounds. In recent years, sophisticated processing capabilities that allow careful control of impurity levels have encouraged researchers to revisit the ductility issues in these compounds. In spite of such advances, research on beryllides is still restricted to a few laboratories, mainly due to the stringent requirements and caution that must be exercised in processing and characterizing beryllium-based products.

Of the various refractory metal beryllides, the most extensively examined are the niobium beryllides, $NbBe_{12}$, and to a lesser extent, Nb_2Be_{17}. Zirconium- and titanium-beryllides have also been studied. The B2 beryllide, NiBe has recently been characterized in terms of microstructure and deformation response at low and high temperatures. The physical properties of some of these beryllides are summarized in Table 10-3.

The congruent-melting B2 compound, NiBe, with a melting point of 1878 K at the stoichiometric composition, exhibits an asymmetry in its homogeneity range, extending from ~ 25 at%Ni to 51.6 at%Ni at 1513 K and from 46 at%Ni to 51 at%Ni at 473 K (Okamoto and Tanner 1988). The solidus of the Be-rich compositions drops quickly from 1878 K to 1509 K with increasing Be content. The B2 compound is in equilibrium

Table 10-3 Crystal Structure, Density, Melting Point, and Elastic Properties of Some Beryllides

Compound	Crystal Str.	Density (g/cm^3)	E (GPa)	ν	G (GPa)	K (GPa)	T_m (K)
$Be_{12}Nb$*	$D2_b$ (I 26)	2.48	220	0.291	85.8	177	1945
$Be_{17}Nb_2$	(hR 19)	3.23	320	0.121	143	141	2073
$Be_{12}Ti$	$D2_b$ (I 26)	2.25	282	0.099	128	117	1823
$Be_{13}Zr$	$D2_3$ (cF 112)	2.72	289	0.033	131	122	1918
$Be_{13}Hf$	$D2_3$ (cF 112)	2.4	—	—	—	—	1893
BeNi (50 at%Ni)	B2	6.25	186	—	—	—	1878**

* highly anisotropic; see Fleischer, Gilmore, and Zabala (1989) for details.

** stoichiometric composition.

with Ni solid solution via a eutectic reaction occurring at 1423 K. The microstructure and mechanical behavior of NiBe have been recently evaluated and it was shown that a hardness minimum occurs at room temperature at the stoichiometric composition (Nieh, Wadsworth, and Liu 1989). The oxygen content of the alloys investigated revealed a profile similar to the hardness variation with stoichiometry and it was argued that the observed hardness trend was substantially a consequence of interstitial hardening with some contributions from antisite defect hardening. Based on the absence of oxides in optical micrographs, a maximum oxygen solubility of 0.10 at% ($\sim$ 240 ppm by weight) in NiBe is claimed. Whether this is true on an electron microscope scale was not verified. If, indeed, fine submicron oxides precipitate intragranularly at lower oxygen contents ($<$ 0.1 at%), then the reported hardening due to interstitial oxygen may incorporate a dispersion-hardening component.

Recently, Pharr et al. (1991) hot extruded (1373 K) induction melted, and chill-cast nominally stoichiometric NiBe using a 7:1 reduction ratio and performed bend and tension tests at room temperature, and compression tests in the temperature range 293 K–1093 K. A room-temperature tensile stress-true strain profile for NiBe (Fig. 10-14a) reveals a 0.2% offset-yield strength of 301 MPa, a failure stress of 370 MPa and total strain comprising the elastic and plastic components of 1.3%. Fracture occurs predominantly by grain-boundary failure (Fig. 10-14b). Slip in the extruded material is claimed to occur primarily in the $\langle 001 \rangle$ direction although the primary slip plane was not identified. Compression studies were performed at 873 K and 1023 K in the strain-rate regime $2 \times$

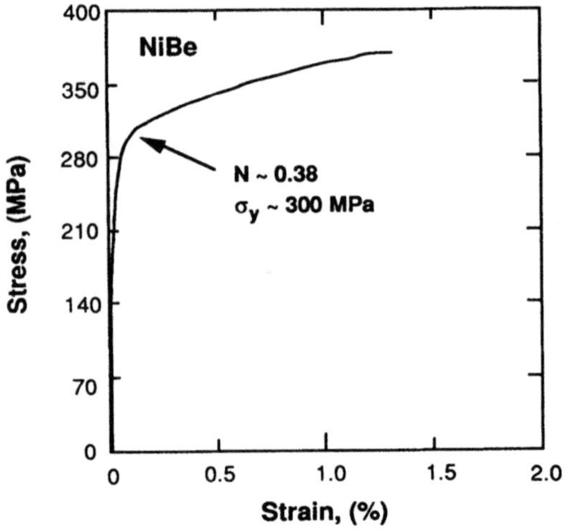

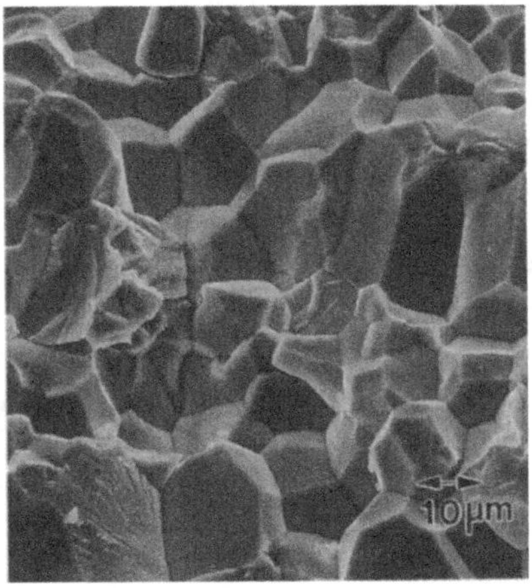

(b)

Figure 10-14. Tensile test at room temperature of extruded, stoichiometric NiBe: (a) true stress-true strain curve and (b) associated fracture surface. (Pharr et al. 1991)

10^{-5} s^{-1} to 5×10^{-4} s^{-1} and the variation in compressive-flow stress as a function of temperature and strain rate was reported. There are preliminary indications according to Pharr et al. (1991) that NiBe may be immune to hydrogen embrittlement.

Isothermal oxidation studies on arc-cast stoichiometric NiBe in the temperature range 1073 K–1473 K revealed superior oxidation resistance up to 1373 K and this was attributed to the formation of a thin, adherent, protective BeO scale (Lee and Nieh 1989). It was also shown that BeO adhered better to NiBe than it did to pure Be and this response was thought to arise as a consequence of a small lattice mismatch between the BeO and NiBe.

Thus, in comparison with the B2 aluminides such as NiAl, NiBe is competitive in terms of specific mechanical properties with a potential for tensile ductility in the polycrystalline material. The toxicity issue associated with handling Be-based materials will, however, act as a major deterrent in the advancement of this compound, particularly since it does not exhibit an overwhelming advantage in specific strength over its B2 aluminide counterpart.

As previously mentioned, the physical and mechanical properties of several refractory-metal beryllides were evaluated at room and elevated temperatures by notably two groups: one with The Brush Beryllium Company (Stonehouse, Paine, and Beaver 1960) and the other with The General Electric Research Company (Lewis 1961a, 1961b). It is not the intent in this section to review all of this early work and the interested reader is referred to the above-referenced articles. Wherever pertinent, information from the early work will be drawn for purposes of comparison with recent data. Fleischer and Zabala (1989) have reported on the microhardness variation with temperature of $Be_{12}Ti$, $Be_{12}Nb$, $Be_{13}Zr$, and $Be_{17}Nb_2$ produced by sintering powder compacts in vacuum. In all cases, hardness was substantially retained to a little over half the melting point and the temperature above which cracks were not formed at indentations was in excess of 1273 K for all compounds examined. The isothermal oxidation behavior of some of these beryllides in the temperature range 1366 K–1589 K was examined by Grensing (1989). The study showed the weight-gain response to follow a logarithmic growth behavior with the primary oxidation product being BeO. It was found that the BeO formed by the breakdown of the beryllide to a lower beryllide (e.g., $NbBe_{12}$ forming some Nb_2Be_{17} and liberating Be atoms; likewise, Nb_2Be_{17} formed $NbBe_2$).

Recently, the niobium beryllides have been the focus of attention at the Battelle Pacific Northwest Laboratories, where systematic studies are in progress to evaluate $Be_{12}Nb$ for its elevated-temperature capabilities, as well as to understand its deformation behavior at low and high temperatures. The binary Be-Nb phase diagram according to Okamoto and Tanner (1990) consists of six stable intermediate phases which include $Be_{12}Nb$, $Be_{17}Nb_2$ (stable only above 1688 K), Be_5Nb, Be_3Nb, Be_2Nb, and Be_2Nb_3, all of which are line compounds. Subsequently, a small but finite stoichiometric range of existence was claimed for $Be_{12}Nb$ (Brimhall, Charlot, and

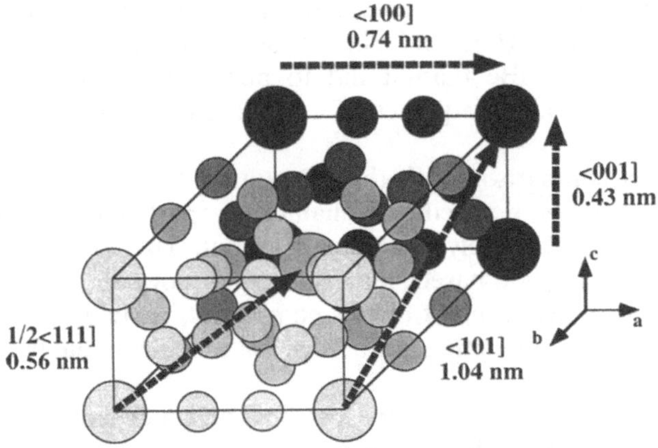

Figure 10-15. The body-centered tetragonal cell of $Be_{12}Nb$ with 26 atoms per unit cell. Large circles at the corners and body center are Nb atoms and smaller circles are Be atoms. Change in the level of shade from grey to black indicates depth along *b* axis. (Bruemmer et al. 1992)

Bruemmer 1991) and the deviation from stoichiometry was speculated to be accommodated by vacancies on the Nb sublattice; further, the stability of $Be_{17}Nb_2$ was extended to room temperature and the Be_5Nb-phase was found not to be present. In the same study (Brimhall et al. 1991), $Be_{12}Nb$ was claimed to be preceded by a $Be_{12}Nb'$ metastable precursor with a b.c.c. structure, and in addition, a second metastable b.c.c. phase was also identified. The $Be_{12}Nb'$-b.c.c. phase was argued to be a highly faulted $Be_{12}Nb$-phase and was claimed to be a precursor also to the $Be_{17}Nb_2$ compound.

The crystal structure and the potential slip vectors for $Be_{12}Nb$ are shown in Figure 10-15. It is essentially a body-centered tetragonal structure consisting of 26 atoms per unit cell with a $c/a = 0.58$. The Nb atoms occupy the body center and corner positions and a tight cage of Be atoms surrounds each Nb atom. Vacuum hot-pressed $Be_{12}Nb$ has been shown to contain annealing twins, planar faults (initiating from grain and twin interfaces and terminating at partial dislocations in the matrix) and short range faults (Fig. 10-16a–c). The twins and extended faults exhibited $\{101\}$ habit planes and $\langle 011 \rangle$ twinning directions, while the short-range faults exhibited (001) habit planes. A schematic of the twinning orientation as proposed by Charlott et al. (1991) is reproduced in Figure 10-16d.

The compression strength of vacuum hot-pressed $Be_{12}Nb$ has been measured as a function of temperature in the range 300 K–1473 K

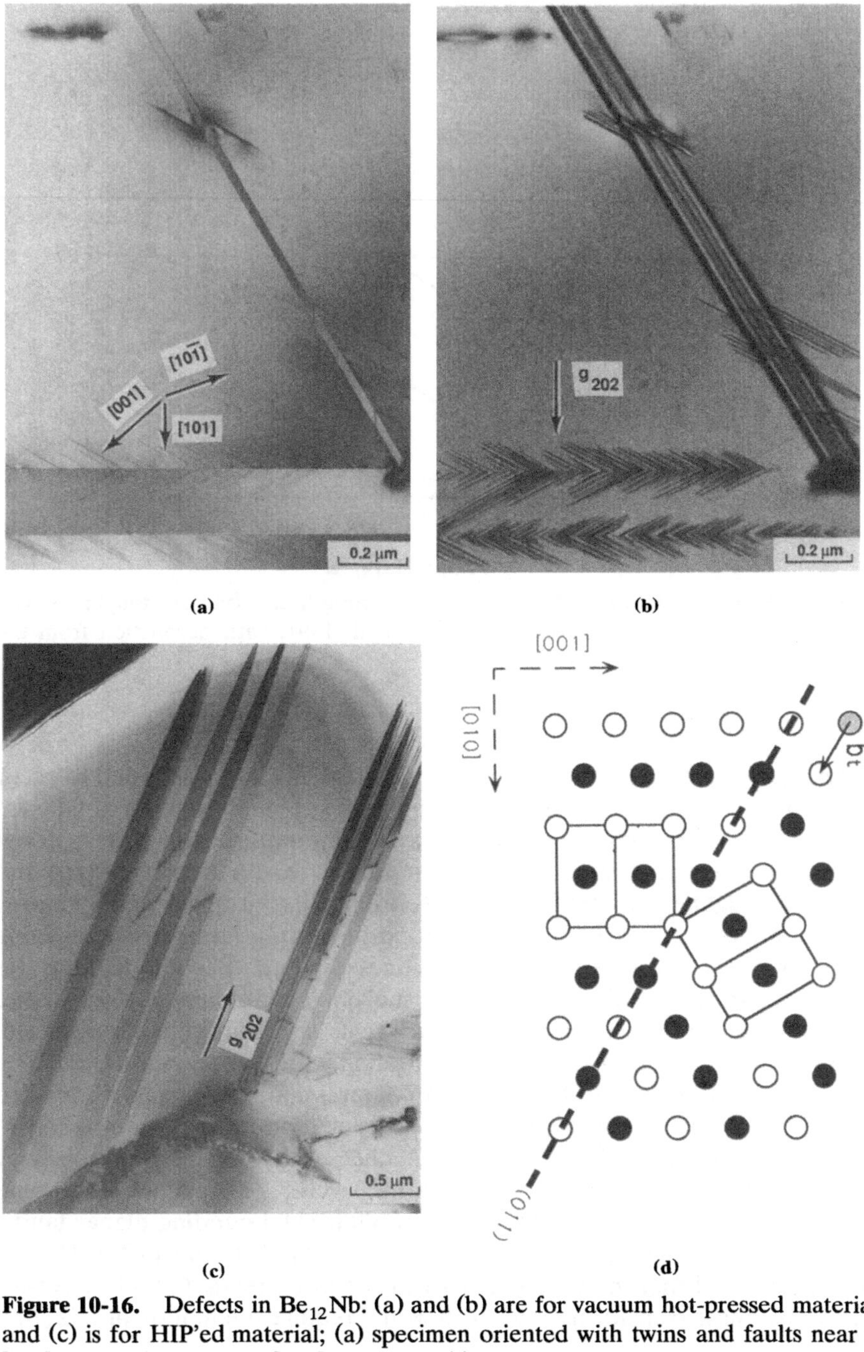

Figure 10-16. Defects in $Be_{12}Nb$: (a) and (b) are for vacuum hot-pressed material and (c) is for HIP'ed material; (a) specimen oriented with twins and faults near a [010] zone; (b) tilted to a [131] zone; and (c) extended faults terminating at partial dislocations; (d) a schematic of the twinning orientation showing only the Nb atoms along [100] projection. Filled atoms are below the plane of the paper. (Reprinted with permission from *Scripta Metallurgica et Materialia*, Volume 25, L. A. Charlot, J. L. Brimhall, L. E. Thomas, S. M. Bruemmer, and J. P. Hirth, pages 99–103, 1991, Elsevier Science Ltd., Oxford, England)

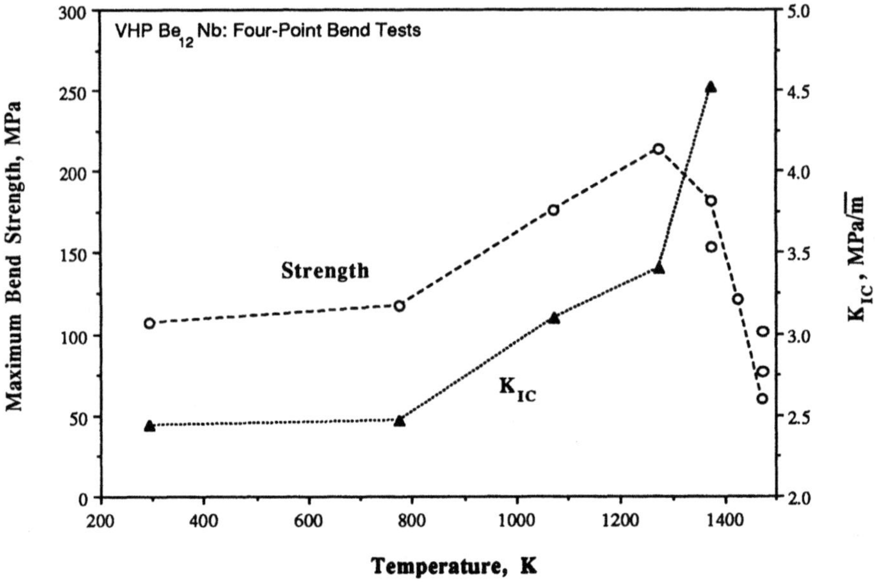

Figure 10-17. Variation in four-point bend strength and fracture toughness with test temperature for $Be_{12}Nb$. (Bruemmer et al. 1991, with permission from the Materials Research Society)

(Henager, Jacobson, and Bruemmer 1992) and yielding was found to occur only above 973 K. Variations in bend strength and fracture toughness (based on four-point bend tests) with temperature for this material (Bruemmer et al. 1991) are shown in Figure 10-17. Bend strength and fracture toughness at 300 K and 773 K are similar; at higher temperatures in the range 773 K to 1273 K, both properties exhibit an increase implying that low-temperature strength is toughness-limited. Beyond 1273 K, the strength drops precipitously. Fracture morphology remains essentially the same, being a mixture of intergranular facets and transgranular cleavage, the extent of intergranular failure increasing with test temperature.

Dislocations were observed in a compression specimen deformed to ~ 5% plastic strain at 1473 K. These dislocation networks were subsequently characterized (Bruemmer et al. 1992) and consisted of perfect and partial dislocations. Most dislocations were partials of the type $1/2\langle 101]\{121)$, $1/2\langle 101]\{101)$, and $1/2\langle 100]\{011)$, bounding planar faults. The perfect dislocations were of the type $1/2\langle 111]\{101)$ and $1/2\langle 111]\{121)$. It is postulated that the partial dislocations may be related to a Be_{12}Nb-to-$Be_{17}Nb_2$ phase transformation which would require the motion of $1/2\langle 100]$

transformation dislocations on every third {002) plane along with diffusional composition adjustment. It has been suggested that the mobility of the observed 1/2⟨101] and 1/2⟨100] partials may control the onset of ductility in this compound and thus relies on the stacking fault energy. Thus, alloying additions that lower the fault energy have the potential to lower the brittle-to-ductile transition temperature. Research is currently underway to examine phase stability between $Be_{13}X$, $Be_{12}X$, and $Be_{17}X_2$ phases. Examples of such ternary compounds include $Be_{12}(Nb, Zr)$, $Be_{17}(Nb, Zr)_2$, and $Be_{13}(Nb, Zr)$.

Other research on beryllides include the study of two-phase alloys such as $Be_{12}Ti$ + Be and $Be_{13}Zr$ + Be (Fleischer and Field 1990) that showed no enhancement in toughness but a rapid loss in high-temperature hardness relative to the single-phase beryllide. Recently, Moshier et al. (1991) reported on the room-temperature tensile-fracture strength of $TiBe_{12}$ produced as a foil by DC Magnetron sputtering, thereby maintaining a low interstitial content and a submicron grain size. They found strength to be dependent on the Be concentration, varying from 96 MPa at 90.5 at%Be to 205 MPa at 94.7 at%Be. Except for some early tensile-test data on cold-pressed and sintered Nb_2Be_{17} (Marder and Stonehouse 1988), this appears to be the only tensile testing performed on such materials, particularly when the interstitial levels are controlled. Finally, in recent years, there has been significant research emphasis to develop refractory fibers with a high coefficient of thermal expansion (CTE) for reinforcing brittle materials such as polycrystalline NiAl, which also exhibits a high CTE and has minimal tolerance to cracking during thermal cycling as a consequence of CTE mismatch with currently available ceramic fibers. The refractory beryllides (e.g., Nb_2Be_{17} or $TiBe_{12}$) exhibit an excellent CTE match with NiAl, making them potential reinforcements. Several efforts to produce beryllide fibers are underway.

5. CHROMIDES AND OTHER ADVANCED INTERMETALLICS

A large number of intermetallics compounds with high melting points crystallize with the AB_2 stoichiometry and the C14, C15, or C36 crystal structures. These compounds are often referred to as the Laves phases. In binary refractory metal—chromium systems such as Cr-Ti, Cr-Zr, Cr-Hf, Cr-Nb, and Cr-Ta, these Laves phases are in equilibrium with Cr, and frequently with the refractory metal as well, via eutectic reactions. Alloy compositions in the two-phase regions where the intermetallic is the predominant phase, can in principle be damage tolerant at room tempera ture and yet be sufficiently creep resistant at elevated temperatures. It is evident that in the Cr-Hf, Cr-Nb, and Cr-Ta systems, alloy compositions in the Cr-Cr_2X side will be less dense and likely more oxidation resistant than on the X-Cr_2X region of the phase diagram. Research has proceeded mainly in two directions: the first has involved a fundamental understanding of the deformation behavior, defects and polymorphic transformations in the Laves phase; and the second has examined the effect of microstructure on mechanical properties of two-phase alloys with primary emphasis on toughening mechanisms that may be operative in such alloys. Studies on single-phase Laves compounds as well as two-phase alloys of the type discussed above are by no means restricted to the Cr-containing systems and results of investigations on systems containing Laves compounds such as Co_2Nb and HfV_2-V solid solution alloys are briefly reviewed for completeness.

The Laves phase in several of the Cr-based systems (e.g., Cr-Hf, Cr-Nb, Cr-Ta) undergoes a polymorphic transformation from a C14 (hexagonal 2H) structure at high temperatures to a C15 (cubic) structure at low temperatures, the transformation temperature being system specific but typically above 1273 K, and the kinetics and complexity of transformation within one system depending on the composition of the compound (Allen 1985) within the small but finite single-phase range. The C14, C15, and C36 structures essentially only differ in the stacking of similarly built layers, and such transformations suggest that the stacking fault energy should be low in such compounds; further these transformations are extremely sluggish, reflecting the small energy differences between these structures. Allen, Delavignette, and Amelinckx (1972) have shown extensive faulting in the basal plane of C14 Cr_2Ti and indicate the occurrence of a C36 (4H) intermediate structure during the C14 to C15 transformation. The transformation from C14 to C15 itself is believed to be a martensitic shear transformation of the 2H- to 3R-type resulting from structural faulting accompanying the glide of shockley partials, the difference being that these are "synchroshear" partials rather than ordinary shockley partials. Dislocation glide in complex crystal structures requires the coordinated reshuffling of atoms in multiple layers, the process being referred to as "zonal glide" of zonal dislocations; synchroshear is the simplest form of zonal glide and involves only one intermediate atomic layer (Kronberg 1957).

Mechanical behavior of single-phase Laves compounds has been characterized in single-crystal and polycrystalline form for $MgCu_2$ with the C15 structure (Moran 1965), $MgZn_2$ with the C14 structure (Paufler and Schulze 1967) and for $MgNi_2$ and $MgCu_{0.4}Zn_{1.6}$ with the C36 structure (Livingston and Hall 1991). Sauthoff (1989) reported on the compression

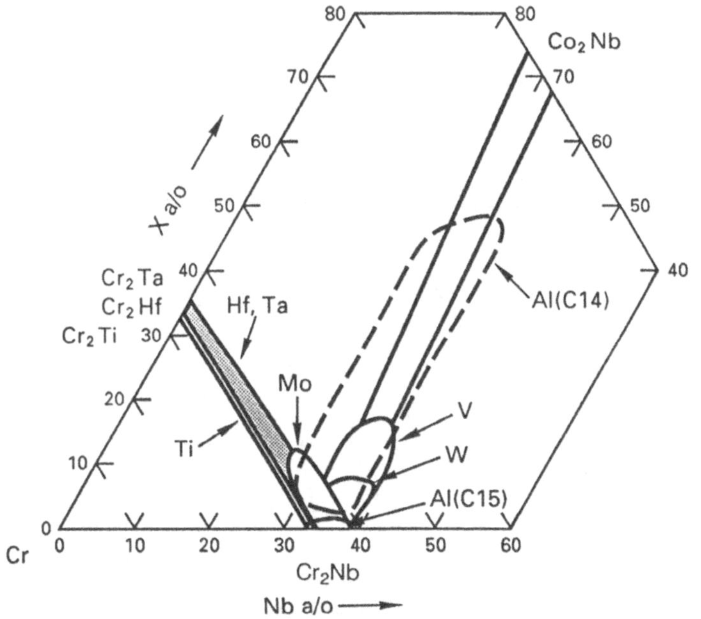

Figure 10-18. Effect of ternary alloying on the Cr_2Nb phase field. (Shah and Anton 1991, with permission from the Materials Research Society)

behavior at high temperatures of binary, single-phase Cr_2Ti and showed that the substitution of Si for Cr in Cr_2Ti to yield a composition $TiCr_{1.5}Si_{0.5}$, significantly enhanced the high-temperature strength. Recently, the effect of ternary alloying additions on the size and shape of the Cr_2Nb phase field (Shah and Anton 1991) (Fig. 10-18) and the properties of polycrystalline, single-phase binary Cr_2Nb as well as ternary compositions such as $Cr_2Nb(Al)$, $(Cr, Co)_2Nb$, and $Cr_2(Nb,Ta)$ have been reported (Anton and Shah 1991a). In the last of these ternary compositions, a two-phase microstructure was evident, the second phase constituting approximately 50 vol% of the disordered (Nb,Ta) solid solution. Compressive creep studies at 1473 K revealed that the Al-containing ternary compound was inferior to binary Cr_2Nb, whereas the other two ternary compounds showed improvement over their binary counterpart. For the binary compound, a minimum compressive creep rate of 7.2×10^{-7} s^{-1} was noted at 1473 K and a stress level of 69 MPa with an associated activation energy of 31 kcal/mol and a stress exponent of $n = 2$. In this study, a similar compound, Co_2Nb, was also characterized, which under similar test conditions exhibited a minimum creep rate of 5.6×10^{-6} s^{-1}, $Q = 85$ kcal/mol

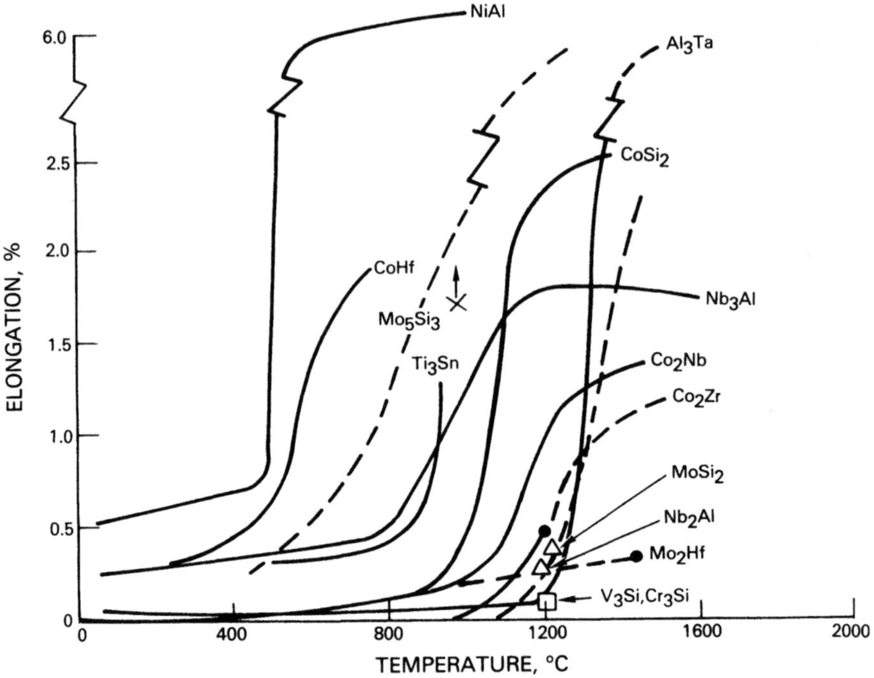

Figure 10-19. The strain to failure in bending versus temperature for several binary intermetallics providing a ductile to brittle transition temperature in many cases. (Anton and Shah 1991b)

and $n = 3.8$. The DBTT in bending for Cr_2Nb, Co_2Nb and other advanced binary intermetallics are shown in Figure 10-19. The two Laves compounds, Cr_2Nb and Co_2Nb, exhibit a similar DBTT of ~ 1273 K–1373 K. The cyclic oxidation behavior for these two compounds is claimed to be comparable to that of Ni-base superalloys at this temperature.

Various two-phase alloys on the Cr-Cr_2Nb, as well as the Nb-Cr_2Nb side of this binary system are under investigation. The Cr-Cr_2Nb eutectic composition, the solvus curves on the Cr-rich side and Nb-rich side, and the range of existence of Cr_2Nb have all been redefined recently (Thoma and Perepezko 1992), and in two hypoeutectic alloys that were examined by Takeyama and Liu (1991), a fine dispersion of Cr_2Nb was observed in the primary Cr matrix and was attributed to the significant Nb solubility in Cr at high temperatures. These dispersions were extremely resistant to coarsening and are responsible for high compressive-strength retention at elevated temperatures. Fleischer and Zabala (1990b) examined a series of binary two-phase alloys in the Cr-Ti system and found that substantial amounts of Ti were necessary (composition corresponding to 50Cr-50Ti) to provide any measurable resistance to cracking of the two-phase material at room temperature and resulted in a significant loss in high-temperature strength relative to single-phase Cr_2Ti. Ternary substitutions for Ti in the Cr-Ti alloys were explored (Al, Zr, and Nb) and subsequently, Ti-Cr-Nb alloys were evaluated in compression and in bending (Fleischer and Zabala 1990c). A significant difference in yield strength was noted between these two types of tests at 296 K and 1273 K. Plastic strain was not measurable at 296 K and fracture toughness measured using notched-bend specimens was $\sim 5\text{–}6\ \mathrm{MPa}\sqrt{m}$ at room temperature for all alloys examined.

A final note on the subject of the Laves phase-containing alloys concerns the two-phase alloy (HfV_2-V solid solution) in the V-Hf-Nb system which can be rolled without cracking at room temperature (Inoue, Kuroda, and Tachikawa 1979). The HfV_2 phase exhibits the C15 structure and a recent study of the rolled two-phase material (Livingston and Hall 1990) concluded that $\{111\}\langle 112\rangle$ twinning was a major deformation mode. The presence of concentrated shear bands was reported and an examination of the as-cast microstructure verified the observed features in the Laves phase to be a consequence of the deformation induced by rolling.

6. DISCUSSION

It is perhaps a misnomer to title this chapter "Advanced Intermetallics" because the more "advanced" the intermetallic is, the less is known about its physical and mechanical characteristics. Several of these compounds have been characterized, but only in a cursory manner. In many instances, essential research tools such as reliable phase-diagram data and physical properties are simply not available, leading to erroneous experimentation and interpretation of observations. At present, mechanical properties for several of these compounds are sparse and those available are not entirely dependable due to either contamination from interstitials or incomplete densification. Frequently, the extent of characterization has not gone past a few hardness indentations on one or two specimens and, rightly or wrongly, mechanical characteristics of the material have been inferred from such indentations. Several of these compounds are difficult to produce via casting technology, either due to their high melting points or their inability to sustain thermal shock during cooling. This had led to these compounds being produced frequently via powder-metallurgy techniques (particularly processes taking advantage of the exothermic reaction that accompanies the formation of these compounds) and thus the inevitable contamination problem that accompanies these processes has to be acknowledged. In spite of these drawbacks, research efforts are in progress at an aggressive pace to produce and characterize these materials using a host of innovative processing routes to keep contamination to a minimum. Likewise, there is a strong need to generate reliable phase-diagram information that can provide the basis not only for alloy selection and optimized heat treatment but also to understand, rationalize, and eliminate undesirable reactions, be they in the bulk or only at surfaces. Ironically, the least understanding exists in the two areas that have been the basis for selecting intermetallics—low-temperature ductility and high-temperature creep resistance. Although tremendous volumes of information comprising experimental evidence, theories, and models have been and are being generated regarding the ductility issue in these materials, there appears to be little in the way of any definite approach that promises success, and to date, trial-and-error techniques are still dominant. Thus, for example, some Laves phase-containing alloys can be cold rolled, whereas others are brittle up to very high temperatures for reasons that are not altogether clear. Likewise, some $L1_2$ compounds exhibit extensive ductility while others fail by brittle-transgranular cleavage. While several of these advanced intermetallic compounds have extremely enviable properties such as low density, high melting point, oxidation resistance, and high stiffness at elevated temperatures, it appears at present, that they will have very specific application niches (e.g., laminated foils of beryllides, for small parts in a missile that has a short life span) and are unlikely in the near future to replace currently used high-temperature structural materials (e.g., Ni-base superalloys in jet engine applications) in any large volume. Pertinent associated issues that must be addressed before any of these compounds can be considered technologically useful include economical machining, joining, thermal and mechanical shock resistance, and above all, the ability to produce these materials in commercial quantities with reproducible properties.

All this, however, does not mean that these materials should be ignored. On the contrary, intense fundamental research must continue on these compounds at universities and laboratories around the world, preferably in collaboration with the national and industrial laboratories that provide the universities with good quality material and guidance on application requirements and goals. It is with such joint research efforts that such attractive but difficult-to-process materials can one day be brought to technological maturity.

REFERENCES

AINDOW, M., SHYUE, J., GASPAR, T. A., and FRASER, H. L., 1991, *Phil. Mag. Lett.* 64(2): 59–65.

ALLEN, C. W., 1985, In *High Temperature Ordered Intermetallic Alloys*, MRS Symp. Proc. 39: 141–146.

ALLEN, C. W., DELAVIGNETTE, P., and AMELINCKX, S., 1972, *Phys. Stat. Sol.* (*a*) 9: 237–246.

ALMAN, D., DIBBLE, A., BOSE, A., GERMAN, R. M., STOLOFF, N. S., and MOORE, B., 1989, In *Processing of Ceramic and Metal Matrix Composites: Vol.* 17, ed. H. Mostaghaci, pp. 217–227. Pergamon Press, New York.

ALMAN, D. E., STOLOFF, N. S. and OTSUKI, M., 1991, In *Intermetallic Compounds—Structure and Mechanical Properties*, ed. Osamu Izumi, pp. 891–899. Sendai, Japan: The Japan Institute of Metals.

ANTON, D. L., 1988, In *High Temperature/High Performance Composites*, Vol. 120, ed. F. D. Lemkey, S. G. Fishman, A. G. Evans, and J. R. Strife, pp. 57–64. Pittsburgh, PA: Materials Research Society.

ANTON, D. L., and SHAH, D. M., 1989, In *High Temperature Ordered Intermetallic Alloys III*, MRS Symp. Proc. 133: 361–371.

________, 1991a, In *Intermetallic Compounds—Structure and Mechanical Properties*, ed. O. Izumi, pp. 379–386. Sendai, Japan: The Japan Institute of Metals.

________, 1991b, Development potential of advanced intermetallic materials. Final report for the period January 1987–April 1990, submitted to the Wright Research and Development Center, Wright Patterson AFB, OH. (Report # WRDC-TR-90-4122).

BARTHOLOMEUSZ, M. F., and WERT, J. A., 1992, *Acta. Metall. et Mater.* 40(4): 673–682.

BRIMHALL, J. L., CHARLOT, L. A., and BRUEMMER, S. M., 1991, *Scr. Metall. et Mater.* 25(3): 553–558.

BROWN, S. A., KUMAR, K. S., and WHITTENBERGER, J. D., 1990, *Scr. Metall. et Mater.* 24(10): 2001–2006.

BRUEMMER, S. M., AREY, B. W., JACOBSON, R. E., and HENAGER, C. H., JR., 1991, In *High Temperature Ordered Intermetallics Alloys IV*, MRS Symp. Proc. 213: 475–480.

BRUEMMER, S. M., CHARLOT, L. A., BRIMHALL, J. L., HENAGER, C. H., JR. and HIRTH, J. P. 1992, *Phil. Mag. A* 65(5): 1083–1094.

CARLSSON, A. E. and MESCHTER, P. J., 1989, *J. Mat. Res.*, 4(5): 1060–1063.

CHARLOT, L. A., BRIMHALL, J. L., THOMAS, L. E., BRUEMMER, S. M., and HIRTH, J. P., 1991, *Scr. Metall. et Mater.* 25(1): 99–103.

CLARK, R. W. and WHITTENBERGER, D. J., 1984, In *Thermal Expansion 8*, ed. T. A. Hahn, pp. 189–196. New York and London: Plenum Press.

DEW-HUGHES, D., 1975, *Cryogenics.* 15(8): 435–454.

DIPIETRO, M. S., KUMAR, K. S., and WHITTENBERGER, J. D., 1991, *J. Mat. Res.*, 6(3): 530–538.

DURLU, N., INAL, O. T., and YOST, F. G., 1991, *Scr. Metall. et Mater.* 25(11): 2475–2479.

EBERHART, M. E., KUMAR, K. S., and MACLAREN, J. M., 1990, *Phil. Mag. B* 61(6): 943–956.

EDWARDS, G. R., MCNELLEY, T. R., and SHERBY, O. D., 1975, *Phil. Mag.* 32(6): 1245–1264.

FLEISCHER, R. L., GILMORE, R. S., and ZABALA, R. J., 1989, *Acta Metall.* 37(10): 2801–2803.

FLEISCHER, R. L., and ZABALA, R. J., 1989, *Metall. Trans.* 20A(7): 1279–1282.

________, 1990a, *Metall. Trans.* 21A(10): 2709–2715.

________, 1990b, *Metall. Trans.* 21A(7): 1951–1957.

________, 1990c, *Metall. Trans.* 21A(8): 2149–2154.

FLEISCHER ROBERT L., 1991a, *ISIJ Inter.* 31(10): 1186–1191.

________, 1991b, In *Intermetallic Compounds—Structure and Mechanical Properties*, ed. O. Izumi, pp. 157–164. Sendai, Japan: The Japan Institute of Metals.

FLEISCHER, R. L. and FIELD, R. D. 1990. Development potential of advanced intermetallic materials. Final report for period September 1985 to January 1990. Submitted to the Materials Laboratory, Wright Research and Development Center. WRDC-TR-90-4046.

FLEISCHER, R. L., FIELD, R. D., and BRIANT, C. L., 1991, *Metall. Trans.* 22A(2): 403–413.

FU, C. L., 1990, *J. Mat. Res.* 5(5): 971–979.

FUJIWARA, T., YASUDA, K., and KODAMA, H., 1991, In *Intermetallics Compounds—Structure and Mechanical Properties*, ed. O. Izumi, pp. 663–637. Sendai, Japan: The Japan Institute of Metals.

GEORGE, E. P., HORTON, J. A., PORTER, W. D., and SCHNEIBEL, J. H., 1990, *J. Mater. Res.* 5(8): 1639–1648.

GRENSING, FRITZ C., 1989, In *Oxidation of High Temperature Intermetallics*, ed. T. Grobstein and J. Doychak, pp. 279–293. Warrendale, PA: The Minerals, Metals and Materials Society.

HARMOUCHE, M. R., and WOLFENDEN, A., 1986, *Mat. Sci. Eng.* 84(1): 35–42.

HEBSUR, M. G., LOCCI, I. E., RAJ, S. V., and NATHAL, M. V., 1992, *J. Mat. Res.* 7(7): 1–11.

HENAGER, C. H. JR., JACOBSON, R. E., and BRUEMMER, S. M., 1992, *Mat. Sci. Eng.*, A153(1–2): 416–421.

HIRUKAWA, K., MABUCHI, H., and NAKAYAMA, Y., 1991, *Scr. Metall. et Mater.* 25(5): 1211–1216.

HOCKING, L. A., STRUTT, P. R., and DODD, R. A., 1971, *J. Inst. of Metals* 99: 98–101.

HU, G., CHEN, S., WU, X., and CHEN, X. 1991, *J. Mat. Res.* 6(5): 957–963.

HUG, G., DOUIN, J., and VEYSSIÉRE, P., 1989, In *High Temperature Ordered Intermetallic Alloys III*, MRS Symp. Proc. 133: 125–130.

INOUE, K., KURODA, T., and TACHIKAWA, K., 1979, *IEEE Trans. on Magnetics.* Mag-15(1): 635–638.

INUI, H., LUZZI, D. E., PORTER, W. D., POPE, D. P., VITEK, V., and YAMAGUCHI, M., 1992, *Phil. Mag.* A65(1): 245–259.

KAMMERDINER, L., and LUO, H. L., 1972, *J. Appl. Phys.* 43(11): 4728–4731.

KATTNER, U. R., 1990, In *Binary Alloy Phase Diagrams, Second Edition, Vol. 1* ed. T. B. Massalski, H. Okamoto, P. R. Subramanian, and L. Kacprzak, pp. 179–181. Materials Park, OH: ASM International.

KRONBERG, M. L., 1957, *Acta Metall.* 5(9): 507–524.

KUMAGAI, T., HANADA, S., and SAITO, S., 1991, In *Intermetallic Compounds—Structure and Mechanical Properties.*, ed. O. Izumi, pp. 1039–1044. Sendai, Japan: The Japan Institute of Metals.

KUMAR, K. S., 1990, *Inter. Mater. Rev.* 35(6): 293–327.

KUMAR, K. S., and BROWN, S. A. 1992a, Unpublished results.

_______, 1992b, *Phil. Mag. A* 65(1): 91–109.

_______, 1992c, *Acta. Metall. et Mater.* 40(8): 1923–1932.

_______, 1992d, *Scr. Metall. et Mater.* 26(2): 197–202.

KUMAR, K. S., BROWN, S. A., MONTOYA, K., WHITTENBERGER, J. D., and DIPIETRO, M. S., 1991, In *Intermetallic Compounds—Structure and Mechanical Properties*, ed. O. Izumi, pp. 667–671. Sendai, Japan: The Japan Institute of Metals.

KUMAR, K. S., BROWN, S. A., and WHITTENBERGER, J. D., 1991, In *High Temperature Ordered Intermetallic Alloys IV*, MRS Symp. Proc. 213: 481–486.

KUMAR, K. S., and PICKENS, J. R., 1988, In *Dispersion Strengthened Aluminum Alloys*, ed. Y.-W. Kim, and W. M. Griffith, pp. 763–786. Warrendale, PA: The Minerals, Metals and Materials Society.

KUMAR, K. S., AND WHITTENBERGER, J. D., 1992, *J. Mat. Res.* 7(5): 1043–1045.

LEE, J. S., and NIEH, T. G., 1989, In *Oxidation of High Temperature Intermetallics*. ed. T. Grobstein, and J. Doychak, pp. 271–277. Warrendale, PA: The Minerals, Metals and Materials Society.

LERF, R., and MORRIS, D. G., 1991, *Acta. Metall. et Mat.* 39(10): 2419–2430.

LEWIS, J. R., 1961a, *Journal of Metals* 13(5): 357–362.

_______, 1961b, *Journal of Metals* 13(11): 829–832.

LIVINGSTON, J. D., and HALL, E. L., 1990, *J. Mat. Res.* 5(1): 5–8.

_______, 1991, In *High Temperature Ordered Intermetallic Alloys IV*, MRS Symp. Proc. 213: 443–448.

LYTTON, J. L., BARRETT, C. R., and SHERBY, O. D., 1965, *Trans. of TMS-AIME*, 233(7): 1399–1408.

MABUCHI, H., HIRUKAWA, K., and NAKAYAMA, Y., 1989, *Scr. Metall.* 23(10): 1761–1766.

MABUCHI, H., HIRUKAWA, K., TSUDA, H. and NAKAYAMA, Y., 1990, *Scr. Metall. et Mater.* 24(3): 505–508.

MARDER, J. M. and STONEHOUSE, A. J., 1988, In *The Proceedings of the 2nd International SAMPE Metals Conference*, August 2–4, pp. 357–367.

MARIEB, T. N., KAISER, A. D., NUTT, S. R., ANTON, D. L., and SHAH, D. M., 1991, In *High Temperature Ordered Intermetallic Alloys IV*, MRS Symp. Proc. 213: 329–336.

MAZDIYASNI, S., MIRACLE, D. B., DIMIDUK, D. M., MENDIRATTA, M. G., and SUBRAMANIAN, P. R., 1989, *Scr. Metall.* 23(3): 327–331.

MCKEE, D. W., and FLEISCHER, R. L., 1991, In *High Temperature Ordered Intermetallic Alloys IV*, MRS Symp. Proc. 213: 969–974.

MIKKOLA, D. E., NIC, J. P., ZHANG, S., and MILLIGAN, W. W., 1991, *ISIJ Inter.* 31(10): 1076–1079.

MISHIMA, Y., OYA, Y., and SUZUKI, T., 1985, In *High Temperature Ordered Intermetallic Alloys*, MRS Symp. Proc. 39: 263–277.

MORAN, J. B., 1965, *Transactions of TMS-AIME* 233(8): 1473–1482.

MORRIS, D. G., 1991, *Scr. Metall. et Mat.* 25(3): 713–716.

_______, 1992, *J. Mat. Res.* 7(2): 1–10.

MORRIS, D. G. and LERF, R., 1991, *Phil. Mag. A* 63(6): 1195–1206.

MORRIS, D. G., LERF, R., and HOLLRIGL, G., 1991, Paper presented at The International Conference on PM Aerospace Materials 1991. Lausanne, Switzerland.

MOSHIER, W. C., CARPENTER, B. F., MISRA, M. S., BERRY, A., and NORDIN, C., 1991, In *Microcomposites and Nanophase Materials*, ed. D. C. Van Aken, G. S. Was, and A. K. Ghosh, pp. 115–123. Warrendale, PA: The Minerals, Metals and Materials Society.

MULLENDORE, A. W., and GRANT, N. J., 1961, In *Structural Processes in Creep*, special report No. 70. pp. 44–55. London, The Iron and Steel Institute.

MURAYAMA, Y., HANADA, S., OBARA, K., HIRAGA, K., and ISHIDA, S., 1991, In *Intermetallic Compounds—Structure and Mechanical Properties*, ed. O. Izumi, pp. 1033–1037, Sandai, Japan: The Japan Institute of Metals.

NEUMANN, J. P., 1980, *Acta Metall.* 28(8): 1165–1170.

NIC, J. P., ZHANG, S., and MIKKOLA, D. E., 1990, *Scr. Metall. et Mat.* 24(6): 1099–1104.

NICHOLSON, D. M., STOCKS, G. M., TEMMERMAN, W. M., STERNE, P., and PETTIFOR, D. G., 1989, In *High Temperature Ordered Intermetallic Alloys III*, MRS Symp. Proc. 133: 17–22.

NIEH, T. G., WADSWORTH, J., and LIU, C. T., 1989, In *High Temperature Ordered Intermetallic Alloys III*, MRS Symp. Proc. 133: 743–748.

NISHITANI, S. R., MASAKI, K., SHIRAI, Y., and YAMAGUCHI, M., 1991, *Scr. Metall. et Mat.* 25(8): 1921–1925.

NIX, F. C., and JAUMOT, F. E., JR., 1951, *Phys. Rev.* 83: 1275–1276.

OKAMOTO, H., and TANNER, L. E., 1988, *Bulletin of Alloy Phase Diagrams* 9(5): 563–573.

_______, 1990, In *Binary Alloy Phase Diagrams* Vol. 1, ed. T. B. Massalski, H. Okamoto, P. R. subramanian, and L. Kacprzak, pp. 670–673. Materials Park, Ohio: ASM International.

PARFITT, L. J., SMIALEK, J. L., NIC, J. P., and MIKKOLA, D. E., 1991, *Scr. Metall. et Mat.* 25(3): 727–731.

PAUFLER, P., and SCHULZE, G. E. R., 1967, *Phys. Stat. Sol. (a)* 24: 77–87.

PERKINS, R. A., CHIANG, K. T., and MEIER, G. H., 1988, *Scr. Metall.* 22(3): 419–424.

PETTIFOR, D. G., 1984, *Solid State Comm.* 51(1): 31–34.

________, 1991, In *Intermetallic Compounds—Structure and Mechanical Properties*, ed. O. Izumi, pp. 149–156. Sendai, Japan: The Japan Institute of Metals.

PHARR, G. M., COURINGTON, S. V., WADSWORTH, J., and NIEH, T. G., 1991, *J. Mat. Res.* 6(12): 2653–2659.

POTEZ, L., LAPASSET, G., and KUBIN, L. P., 1992, *Scr. Metall. et Mat.* 26(5): 841–846.

POTEZ, L., LOISEAU, A., NAKA, S. and LAPASSET, G., 1992, *J. Mat. Res.* 7(4): 876–882.

PUGH, S. F., 1954, *Phil. Mag.* 45(7): 823–843.

RAISSON, G., and VIGNES, A., 1970, *Rev. Phys. Appl.* 5(3): 535–544.

RICE, J. R., and THOMSON, R., 1974, *Phil. Mag.* 29(1): 73–97.

SAUTHOFF, G., 1989, *Zeit. Metallkde* 80(5): 337–344.

________, 1990, *Zeit. Metallkde* 81(12): 855–861.

SCHNEIBEL, J. H., BECHER, P. F., and HORTON, J. A., 1988, *J. Mat. Res.* 3(6): 1272–1276.

SCHNEIBEL, J. H., and GEORGE, E. P., 1990a, *Scr. Metall. et Mat.* 24(6): 1069–1074.

________, 1990b, Private communication.

SCHNEIBEL, J. H., and HAZZLEDINE, P. M., 1992, *J. Mat. Res.* 7(4): 868–875.

SCHNEIBEL, J. H., HORTON, J. A., and PORTER, W. D., 1992, *Mat. Sci. Eng.*, Special issue: Proceedings of the International Conference on High-Temperature Aluminides and Intermetallics in San Diego, CA. Sept. 1991, A152(1–2): 126–131.

SHAH, D. M., and ANTON, D. L., 1991, In *High Temperature Ordered Intermetallic Alloys IV*, MRS Symp. Proc. 213: 63–68.

SHECHTMAN, D., and JACOBSON, L. A., 1975, *Metall. Trans.* 6A(7): 1325–1328.

SMIALEK, J. L., 1991, Private communication (also presented at the ASM-sponsored AEROMAT 1991 meeting in Long Beach, CA in May 1991).

STONEHOUSE, A. J., PAINE, R. M., and BEAVER, W. W., 1960, In *Mechanical Properties of Intermetallic Compounds.*, ed. J. H. Westbrook, pp. 297–319. London: John Wiley and Sons.

TAKEYAMA, M., and LIU, C. T., 1991, *Mat. Sci. Eng.* A132: 61–66.

THOMA, D. J., and PEREPEZKO, J. H., 1992, *Mat. Sci. Eng.* A156(1): 97–108.

TURNER, C. D., POWERS, W. O., and WERT, J. R., 1989, *Acta Metall.* 37(10): 2635–2643.

UMAKOSHI, Y., 1991, *Bulletin of Japan Inst. of Metals* 30(1): 72–79.

UMAKOSHI, Y., YAMAGUCHI, M., YAMANE, T., and HIRANO, T., 1988, *Phil. Mag. A*, 58(4): 651–666.

VARIN, R. A., VIRK, I., and WINNICKA, M. B., 1991, In *Mechanical Behavior of Materials-VI*, Vol. 3, ed. M. Jono and T. Inoue, pp. 179–186. Tokyo: Pergamon Press.

VASUDEVAN, V. K., WHEELER, R., and FRASER, H. L., 1989, In *High Temperature Ordered Intermetallic Alloys III*, MRS Symp. Proc. 133: 705–710.

VEYSSIÉRE, P., and MORRIS, D. G., 1993, *Phil. Mag. A.* 67: 491–495.

VILLARS, P., 1983, *J. Less-Comm. Met.* 92: 215–238.

WHITTENBERGER, D. J., 1985, *Mat. Sci. Eng.* 73: 87–96.

________, 1987, *Mat. Sci. Eng.* 85: 91–99.

WHITTENBERGER, D. J., KUMAR, K. S., BROWN, S., DIPIETRO, M. S., and FARMER, S. C., 1991, In *Lightweight Alloys for Aerospace Applications II.*, ed. E. W. Lee and Nack J. Kim, pp. 327–349. Warrendale, PA: The Minerals, Metals and Materials Society.

WINNICKA, M. B., and VARIN, R. A., 1991, *Scr. Metall. et Mat.* 25(6): 1289–1294.

WU, Z. L., POPE, D. P., and VITEK, V., 1990, *Scr. Metall. et Mat.* 24(11): 2191–2196.

YAMAGUCHI, M., SHIRAI, Y., and UMAKOSHI, Y., 1988, In *Dispersion Strengthened Aluminum Alloys*, ed. Y.-W. Kim and W. M. Griffith, pp. 721–740. Warrendale, PA: The Minerals, Metals and Materials Society.

YAMAGUCHI, M., and UMAKOSHI, Y., 1990, *Progress in Mat. Sci.* 34(1): 1–148.

YAMAGUCHI, M., UMAKOSHI, Y., and YAMANE, T., 1987, *Phil. Mag. A* 55(3): 301–513.

YANEY, D. L., and NIX, W. D., 1988, *J. Mat. Sci.* 23: 3088–3098.

YANEY, D. L., PELTON, A. R., and NIX, W. D., 1986, *J. Mat. Sci.* 21: 2083–2087.

YASUDA, K., FUJIWARA, T., KODAMA, H., and SUWA, M., 1991, In *High Temperature Ordered Intermetallic Alloys, IV*, MRS Symp. Proc. 213: 865–870.

Chapter 11

Silicides

Michael J. Maloney and Dilip Shah

1. INTRODUCTION

The formation of silicides is reviewed with the focus on the disilicides, 5-3 silicides and monosilicides, as the three principal useful groups. For the refractory metal-based disilicides, the relationship between the crystal structures $C11_b$, C40, C49, and C54 is examined in terms of stacking sequences and contrasted in relation to the structures of aluminides. The role of interstitial elements and the in situ composite approaches are emphasized for the complex 5-3 silicides and the monosilicides as being the most adjacent phases to refractory metal-solid solutions. For most silicides with a noncubic crystal structure, the effect of associated anisotropy of the coefficient of thermal expansion (CTE) on the mechanical integrity through processing and application of the material is brought forth as a potentially critical issue. Molybdenum disilicide has received much attention for use as a high-temperature structural material. Recent efforts to produce $MoSi_2$-base composites are reviewed.

2. MONOLITHIC SILICIDES

2.1. Structure of Silicides

To understand the formation of silicides in a convenient manner, it is best to consider the projection of a pseudoquarternary-phase diagram

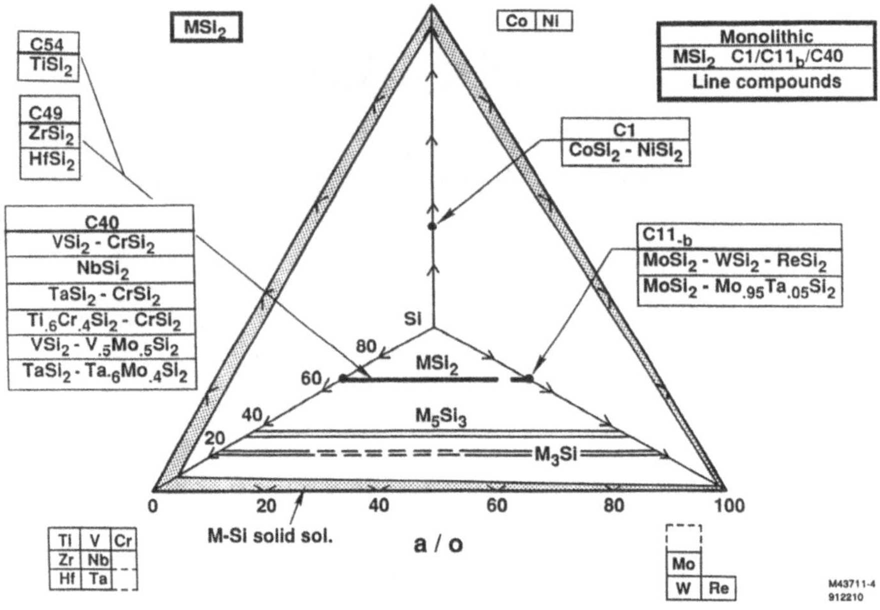

(a)

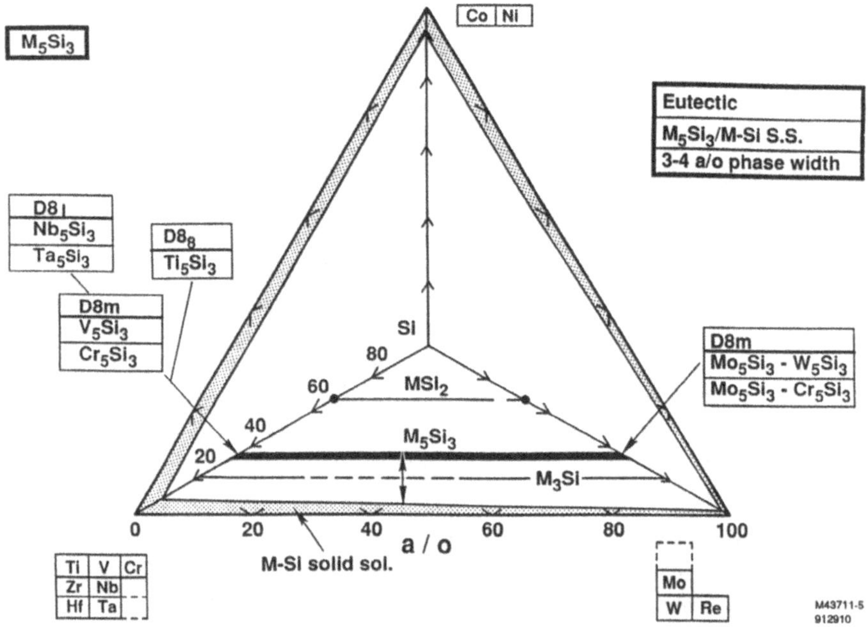

(b)

Figure 11-1. Projections of a pseudoquarternary phase diagram summarizing the formation of useful silicides: (a) disilicides, (b) 5-3 silicides, and (c) monosilicides. (Shah et al. 1992)

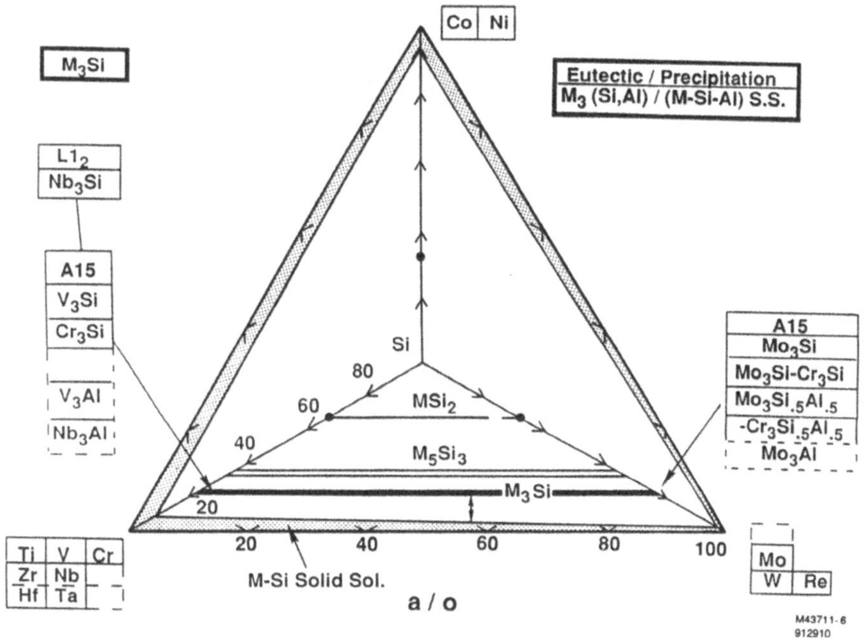

Figure 11-1. (Continued)

(Anton et al. 1989) with Si at the apex, onto the base ternary formed by three groups of transition elements, as shown in Figure 11-1. The specific groupings of the transition elements in Figure 11-1 is not based on a precise criterion but each group loosely consists of transition elements forming isostructural silicides. The objective is to graphically organize the available information from standard reference sources (Massalski 1986; Villars and Calvert 1985) on the formation of silicides and their alloys.

The pseudoquarternary projections presented in Figure 11-1 are not intended to represent all silicides or Si-containing phases. Si is known to be soluble up to 20 a/o in topologically closed packed (TCP) phases (Laves and Sigma) formed between transition elements (Wernick 1967). This is the phase region along the plane of projection in Figure 11-1. However, rather arbitrarily these phases are generally not considered silicides and are also ignored here. In many other binary systems, such as Mn-Si (Massalski 1986), several intermediate silicides with various stoichiometries are found, but many of these line compounds have been ignored here to retain the focus on the major systems listed in Table 11-1.

Table 11-1 Review of Recent Evaluation of Silicides

Crystal Structure	Silicide System	Melting Point, °C	Density, gr/cm^3	PROCESSING: D.S. Single Crystal	PROCESSING: Cast	PROCESSING: Powder Processed	Physical Properties	MECHANICAL PROPERTIES: Ductility/Toughness s	MECHANICAL PROPERTIES: Strength	MECHANICAL PROPERTIES: Creep	ENVIRONMENTAL: Oxidation	ENVIRONMENTAL: Pesting Behavior	Reference
C1	$CoSi_2$	1326	4.98		X			X	X	X	X	X	Anton 1991
				X					X				Ito et al. 1991
	$NiSi_2$	994	4.83		X				X	X			Takeuchi et al. 1991
	(Ni, Co)Si_2				X				X	X			Takeuchi et al. 1991
$C11_b$	$MoSi_2$	2030	6.24			X	X		X				Lugscheider et al. 1991
	$MoSi_2$				X	X					X	X	Berztiss et al. 1992
	$MoSi_2$					X				X			Weiderhorn et al. 1992
	$MoSi_2$					X				X			Bose 1992
	$MoSi_2$ + C					X		X					Maloy et al. 1992
	(Mo, W) Si_2					X				X			Sadannanda et al. 1992
C54	$TiSi_2$	1542	4.04			X	X		X			X	Lugsheider et al. 1991
C40	VSi_2	1673	4.63	X			X						Kakamura 1991
	$CrSi_2$	1477	4.63	X			X						Kakamura 1991
				X									Paine et al. 1960
	$NbSi_2$	1920	5.66		X	X					X	X	Alexander 1988
	NbSi2 + Al, Ti				X						X	X	Alexander 1988

C40	$TaSi_2$	2204	9.08		X	X					X	X	Alexander 1988
	TaSi2 + Al				X						X	X	Alexander 1988
$D8_x$	Ti_5Si_3	2130	4.32										Anton 1991
	Ti_5Si_3 + V, Nb			X	X	X			X				Frommeyer et al. 1990
	Ti_5Si3/Ti 3Al + Al, Nb Eutectic				X			X	X	X	X		Murata et al. 1991
	Ti5Si3 + Al					X				X			Frommeyer et al. 1990
	Ti5Si3/Ti Eutectic			X			X	X	X	X			Frommeyer et al. 1990
	Nb_5Si_3	2484	7.16		X						X	X	Alexander 1988
	Nb_5Si_3/Nb Eutectic				X	X					X	X	Alexander 1988
	Nb_5Si_3 + Ti, Al				X						X	X	Alexander 1988
	Nb5Si_3 + Ta, Mo, Zr				X						X	X	Alexander 1988
	Ta_5Si_3	2505	13.4		X						X	X	Alexander 1988
	Ta_5Si_3 + Ti, Al				X						X	X	Alexander 1988
	Mo_5Si_3	2180	8.24		X				X	X	X	X	Anton 1991
Al5	V_3Si	1973	5.71		X	X			X	X	X	X	Anton 1991
	Cr_3Si	1773	6.46		X	X			X	X	X	X	Chang and Pope 1991
				X					X				

Source: Shah et al. 1992

As summarized in Figure 11-1a, there are primarily three groups of disilicides. These are Ni- and Co-disilicides with the cubic CaF_2(C1) structure; Mo-, Re-, and W-disilicides with the tetragonal $C11_b$ structure, and Cr and Ti groups of disilicides with the hexagonal C40 and C54 structures, respectively. Among these, the disilicides because of the C1 and the $C11_b$ structures have limited alloying options with low solubility of other elements. Additionally, the Ni- and Co-based cubic silicides with the C1 structure bear no simple crystal-structure relationship to other disilicides. The disilicides with $C11_b$, C40, and C54 structures are closely related by differing stacking sequence, as shall be discussed later. The disilicides with the C40 structure present the broadest latitude in alloying, by far. Note that disilicides present no known opportunities for in situ ductile phase-toughened composites with a terminal metal-solid solution. Brittle-brittle in situ eutectic composites with neighboring 5-3 silicides, however, may be considered.

With the 5-3 silicides, as shown in Figure 11-1b, the alloying opportunities are broadened and this may be even further extended if the interrelationships between various complex $D8_x$ structures were known. The 5-3 silicides are some of the highest melting intermetallics. The ternary intermetallic compound $Zr_3Ti_2Si_3$ with the $D8_8$ structure has been recently suggested as an ultra high-temperature structural material (Celis, Kagawa, and Ishizaki 1991). In many systems, with the suppression of other silicides or TCP phases, a eutectic is formed between the 5-3 silicide and the refractory metal-solid solution. As listed in Table 11-1, Nb/Nb_5Si_3 and Ti/Ti_5Si_3 are two such systems currently being explored (Frommeyer, Rosenkranz, and Ludecke 1990; Mendiratta et al. 1991).

Lastly, the monosilicides with the A15(Cr_3Si) structure form the third largest group of silicides occurring next to the refractory metal-solid solution, as shown in Figure 11-1c. For this group of cubic silicides, alloying with other refractory metals is limited, though complete solid solubility exists between Cr_3Si and Mo_3Si. The alloying possibilities are expanded, however, when considered in conjunction with isostructural aluminides such as Nb_3Al (Anton and Shah 1991).

2.2. Alloying of Silicides with Aluminum

Among the nontransition elements such as Al, Ga, Ge, Sn, As, and Sb, with which silicides may be alloyed, Al is apparently most attractive from the standpoint of enhancing the environmental resistance. As shown with a pseudoquarternary diagram in Figure 11-2, Al is soluble to a great degree in conjunction with other refractory elements. Note that Figure 11-2 is not an exhaustive summary of alloying with Al. Addition of Al is

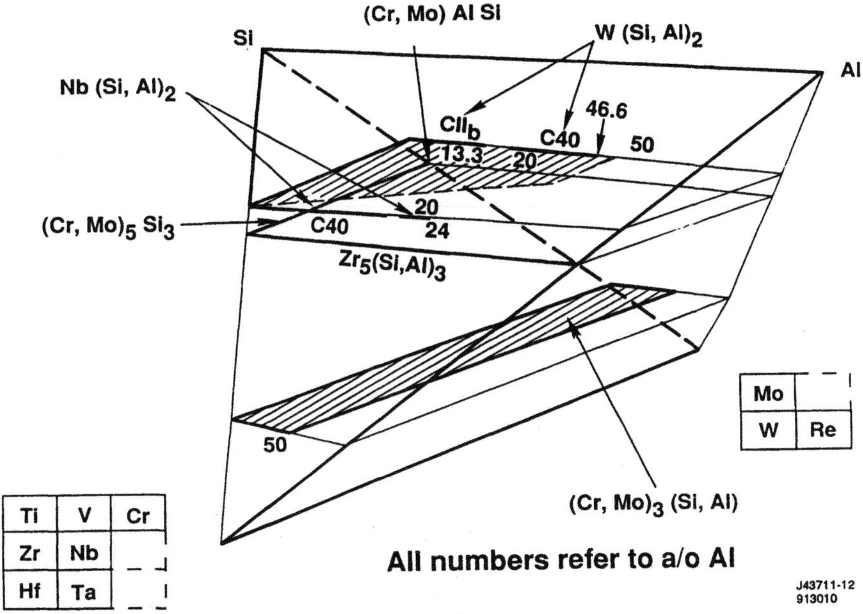

Figure 11-2. Schematic pseudo ternary showing the effect of alloying with aluminum on silicide phase fields. (Shah et al. 1992)

expected to impart more metallic character to the silicides, but the benefits of alloying with Al are not clearly defined. With the historical-experience base available with aluminides, understanding a link between the aluminides and silicides poses an interesting fundamental problem.

2.3. Atomic Stacking in Disilicides

Given the mature state of understanding of the deformation behavior of aluminides with the $L1_2$ and related structures, in terms of atomic stackings and fault energies, it is natural to extrapolate the same reasoning to disilicides. As summarized in Figure 11-3, the relationships between DO_{19}, $L1_2$, DO_{24}, DO_{22}, and DO_{23} crystal structures are well understood in terms of variation in the geometry of ordering and the stacking sequence of the close-packed planes. A similar pattern of stacking relationships exists for disilicides with the $C11_b$, C40, and C54 structures as summarized in Figure 11-4. This relationship, though known for a long time (Mass et al. 1984), has been only recently considered with renewed interest for understanding the deformation behavior of silicides (Umakoshi et al. 1990). Beyond the generic analogy, however, there is a clear

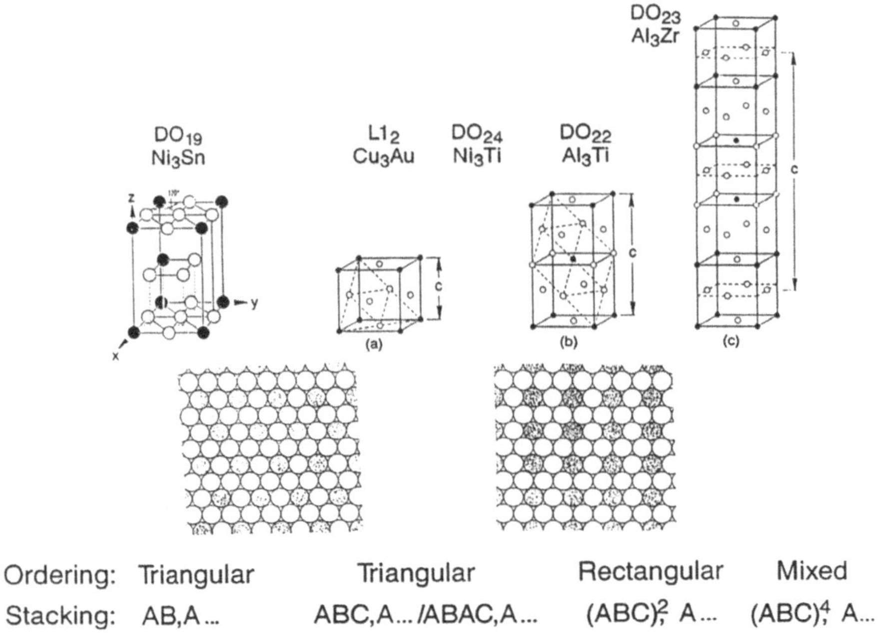

Figure 11-3. Interrelationships between geometrically closed packed structures related to $L1_2$ in terms of the nature of ordering and the stacking sequence of the closed packed planes. (Shah et al. 1992)

geometrical difference in the stacking of silicides versus aluminides as compared in Figure 11-5. While the atoms in the next layer in the disilicides occupy the saddle positions between two Si atoms (Fig. 11-5a), in the case of aluminides the atoms fit into the natural pockets between three contiguous atoms (Fig. 11-5b). This results in four equivalent stacking positions for the silicides but only three for the aluminides. Interestingly, for the simpler stacking sequences in silicides with the $C11_b$ or C40 structures, Si atoms remain closely bonded in sheets normal to the stacking plane (Mass et al. 1984). Only for the C54 ($TiSi_2$) structure, with all equivalent stacking positions used, that Si-Si chain is interrupted with refractory metal atoms as shown in Figure 11-5a. The resultant increased disturbance of diffusion path along the Si-Si bonds is considered responsible for the decreased texture sharpness in $TiSi_2$ coatings (Mass et al. 1984). From another viewpoint, the breaking of the Si-Si chains qualitatively implies increased metallic character and may be given as a simplistic reason for increased solubility of Al in the disilicides with C40 and C54 structures.

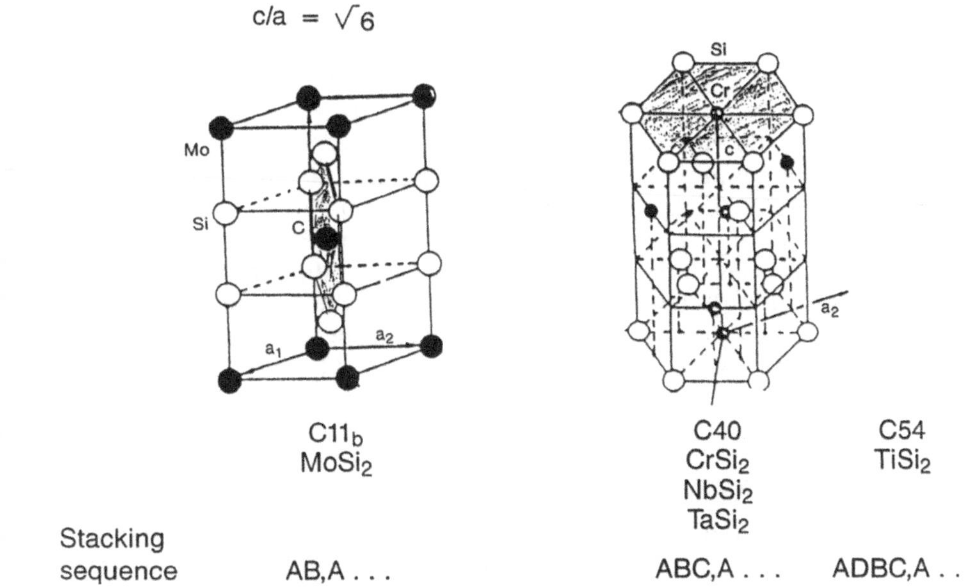

Figure 11-4. Interrelationships between $C11_b$, C40, and C54 structures in terms of the stacking sequence of the closed packed planes. (Shah et al. 1992)

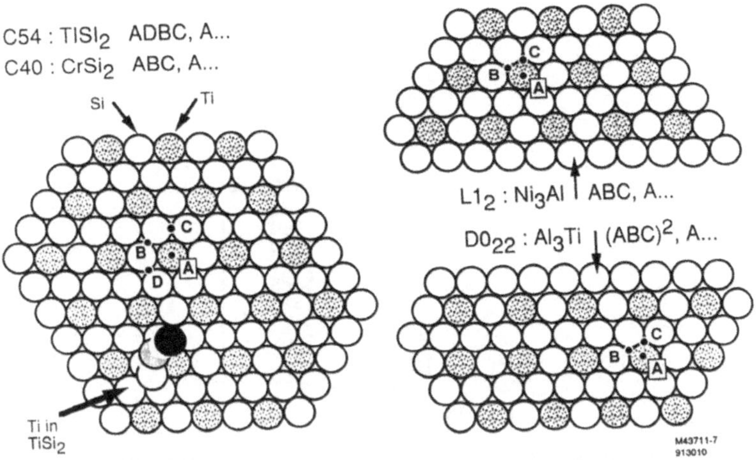

Figure 11-5. Comparison of Geometry of Stacking in Disilicides versus Trialuminides. (Shah et al. 1992)

One missing link in understanding the relationship between silicides and aluminides is the $ZrSi_2$(C49) structure. It is claimed that the orthorhombic C49 structure belongs to the tetragonal $Al_3Ti(DO_{22})$ family and becomes quasi-tetragonal in $TiAl_{0.3}Si_{1.7}$ (a = c = .358 nm, b = 1.349 nm) (Schubert 1967). Strictly the C49(Cmcm) structure belongs to the point group mmm, which under special conditions may be promoted to the point group 4/mmm, to which the DO_{22}(I4/mmm) structure belongs. This can occur if one of the axes acquires a fourfold symmetry, as it approximately does in the quasi-tetragonal $ZrSi_2$(C49) structure based on lattice parameters. Similarly, some correlations in the atomic arrangement between the C49 and the DO_{22} structures can be discerned at certain special positions of the Zr and Si atoms in $ZrSi_2$. However, this aspect is not entirely clear.

2.4. Structure of A15 and $D8_8$ Silicides

Closely related to the disilicide structures are the defect disilicides based on $MnSi_{2-x}$. These are pseudosolid solution of the type $(Mn, T)_n Si_{2n-m}$; where T may be Cr, Fe or Co; n may vary between 4 and 31 and $2n - m$ may vary between 7 and 54; with the c/a ratio ranging from 6 to 24 (Nowotny, Boller, and Zwilling 1972). Atomic structural

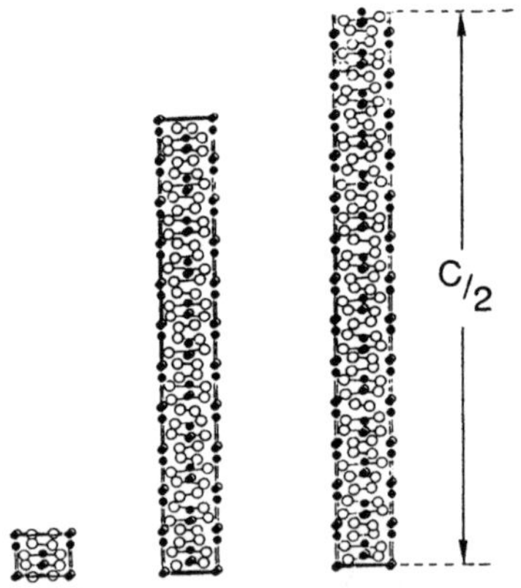

Figure 11-6. Structure of defect disilicide $MnSi_{2-x}$ in terms of its basic building block $TiSi_2$(C54). (Shah et al. 1992)

arrangements for some typical defect silicides are shown in Figure 11-6 along with the basic building block based on $TiSi_2$(C54).

The crystal structure of the A15(Cr_3Si) and $D8_8$(Ti_5Si_3, prototype Mn_5Si_3) are schematically shown in Figure 11-7a and b, respectively. Note that in the case of A15 structure the atoms of the refractory element form three orthogonal chains, which criss-cross the body-centered cubic arrangement of nontransition elements. In the $D8_8$ structure only part of the atoms (Ti(I) in Fig. 11-7b) of the refractory element form chains along the hexagonal axis, octahedrally surrounded by Si atoms. The A15 structure is considered the simplest of the topologically closed-packed (TCP) structures, which by definition should have no octahedral interstitialcies (Khantha, Vitek, and Pope 1989). By contrast, in the $D8_8$ structure, the second group of atoms of refractory elements (Ti(II) in Fig. 11-7a) form columns of octahedrons connected along the c-axis, with the octahedral interstices partially or completely filled by interstitials. It is known that carbon addition to non-$D8_8$ 5-3 silicide stabilizes the $D8_8$ structure and is usually referred to as a Nowotny phase (Taylor 1961). The preceding discussion was held with the recognition that as high-temperature structural materials, it is important to know the solubility of interstitial elements and their effect on the strength, stability, and environmental resistance of the silicides.

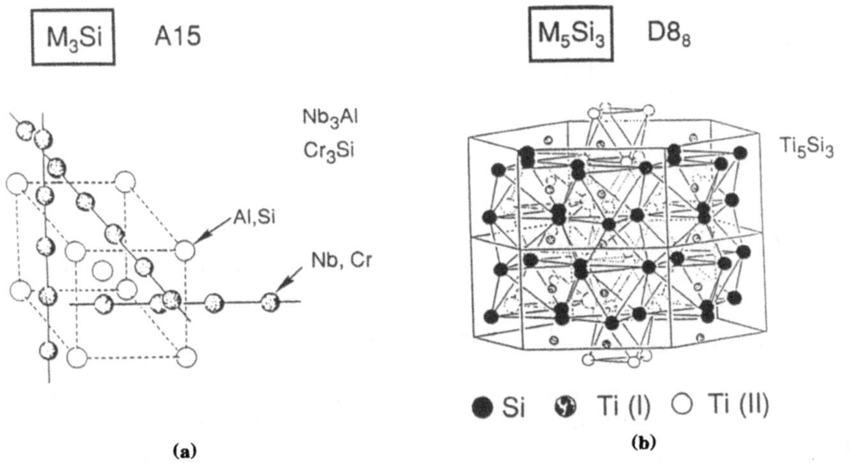

Figure 11-7. Crystal structure of (a) Cr_3Si (A15) and (b) Ti_5Si_3 ($D8_8$). (Shah et al. 1992)

3. PHYSICAL AND MECHANICAL PROPERTIES OF SILICIDES

Owing to their semimetallic bonding characteristics, silicides present an interesting combination of physical and mechanical properties. The following discussion is limited to the properties of immediate interest for high-temperature structural materials.

3.1. Thermal Expansion and Elastic Modulus

In both processing as well as application of high-temperature materials, thermal stresses are generated either due to imposed thermal gradients or thermal cycling. The magnitude of the stresses affect the processibility and the durability of the material. It is well known that microcracks are induced in ceramics due to anisotropy of the coefficient of thermal expansion (CTE) as well as elastic anisotropy (Tvergaard and Hutchinson 1988).

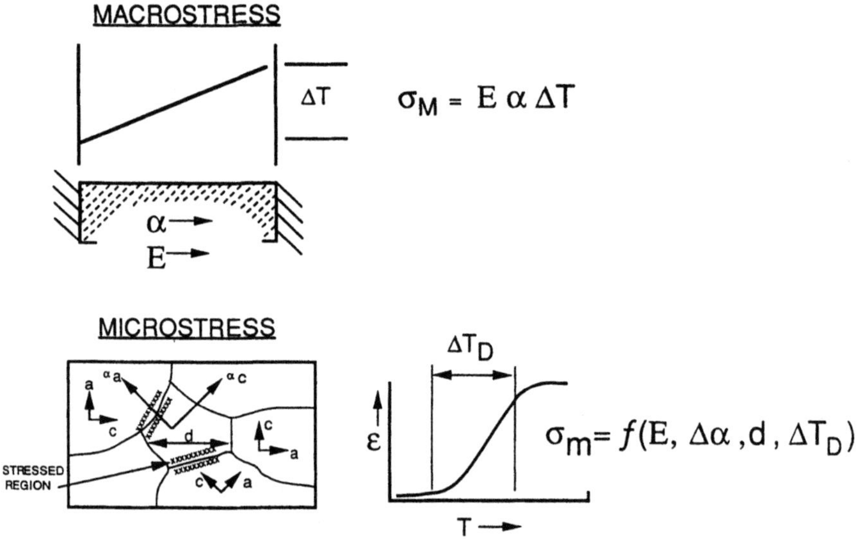

Figure 11-8. Schematic representation and parametric dependence of macrostress caused by thermal gradient and residual microstress caused by anisotropy of CTE. In the mathematical expressions σ is thermal stress, E is elastic modulus, α is CTE, and d is average grain size. $\Delta\alpha$, ΔT, and ΔT_D represent difference in CTE in two directions, temperature gradient, and the difference between the DBTT and room temperature, respectively. (Shah et al. 1992)

As schematically shown in Figure 11-8, in constrained bodies macrostress is caused by a thermal gradient and residual microstress results in materials with an anisotropic CTE. In practice, such a situation commonly occurs for complex components subjected to thermal transients or hot spots. The magnitude of the macrostress is proportional to the product of the elastic modulus, CTE, and temperature gradient. The residual microstress is confined to grain boundaries below the ductile-to-brittle transition temperature (DBTT) owing to differences in the CTE between neighboring grains along the same direction. Such a stress at room temperature is a function of the elastic modulus, anisotropy in CTE, grain size, and the DBTT. The macrostress may be minimized by more uniform heating or cooling. However, in principle there is no way to eliminate the residual microstress in brittle polycrystalline materials with a highly anisotropic CTE. In all situations, low modulus and low value of CTE are beneficial. On the basis of the fundamental postulate of crystal physics known as Neumann's principle, the CTE, being a second-rank tensor, must be isotropic for cubic materials. For most of the silicides, which are noncubic, the CTE is expected to be anisotropic. The magnitude of the anisotropy, however, is not predictable a priori. No approximate correlations have been developed between the extent of anisotropy and the nature of atomic ordering for specific crystal-structure type. To appreciate the magnitude of anisotropy, an attempt has been made to collect the CTE for several silicides from various sources (Frommeyer et al. 1990; Samsonov 1964; Berkowitz-Mattuck 1966) in Table 11-2. Note that no data has been found for any orthorhombic disilicide, for which three CTE's must be specified. The data suggest near isotropy of the CTE for the noncubic disilicides, but high anisotropy for the hexagonal 5-3 silicides. This is consistent with the experimental observations of transgranular cracking in Ti_5Si_3, as shown in Figure 11-9, with no reported evidence of such cracking in any disilicides. The restoration of isotropy for the CTE of Nowotny Ta_5Si_3 as quoted in Table 11-2 is attractive, but it will be more useful to understand the physical basis for the significant change. The average elastic modulus for several silicides are plotted as a function of temperature in Figure 11-10. For purposes of comparison, the extent of elastic anisotropy at room temperature for nickel-base alloys, Nb and Al, is also shown. The data have been collected from various sources (Anton and Shah 1991; Kakamura 1991; Lugsheider et al. 1991 and in most cases, having been measured by acoustic methods, are considered very reliable. As indicated, several results are based on four-point bend tests (Anton and Shah 1991). With the notable exception of compounds with the $C11_b$ structure ($MoSi_2$, WSi_2), all silicides have moduli comparable to nickel-base alloys. For VSi_2, $CrSi_2$, $MoSi_2$, and WSi_2, the complete elastic-con-

Table 11-2 Anisotropy of Coefficient of Thermal Expansion in Silicides

Compound	Crystal System	Coefficient of Thermal Expansion X 10^{-6}, °C^{-1}			Temperature Interval, °C	Reference
		α_1	α_2	α_3		
$CoSi_2$	Cubic					
Cr_3Si	Cubic	10.5	10.5	10.5	20–1000	Samsonov 1964
V_3Si	Cubic	12.0	12.0	12.0	620–820	Samsonov 1964
$MoSi_2$	Tetragonal	8.25	8.25	8.25 + 0.8	20–1070	Berkowitz-Mattuck 1966
$ReSi_2$	Tetragonal	6.6	6.6	6.6	20–1070	Samsonov 1964
Ti	Hexagonal	9.6	9.6	10.7	20	Frommeyer et al. 1990
$TiSi_3$	Hexagonal	3.05	3.05	10.7	20	Frommeyer et al. 1990
Ta_5Si_3	Hexagonal	5.5	5.5	8.0	20–1000	Samsonov 1964
Ta_5Si_3 (Nowotny)	Hexagonal	6.3	6.3	6.6	20–1000	Samsonov 1964
$TaSi_2$	Hexagonal	8.9	8.9	8.8	20–1000	Samsonov 1964
$TiSi_2$	Orthorhombic					

Source: Shah et al. 1992

Figure 11-9. Transgranular microcracking in as-cast Ti_5Si_3. (Shah et al. 1992)

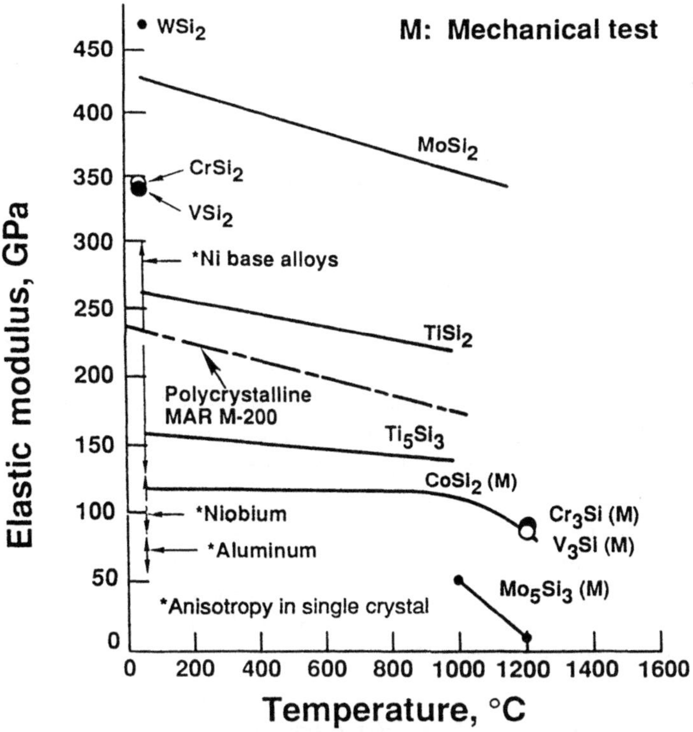

Figure 11-10. Comparison of average elastic modulus as a function of temperature. (Shah et al. 1992)

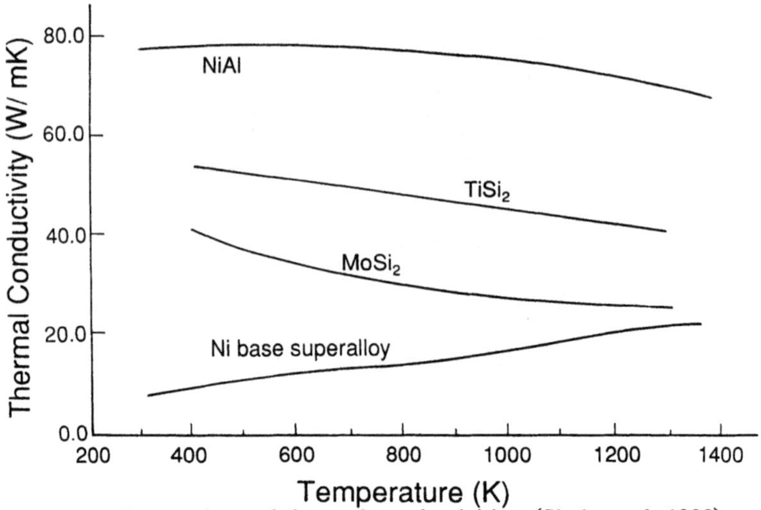

Figure 11-11. Comparison of thermal conductivities. (Shah et al. 1992)

stant matrices at room temperature are known (Kakamura 1991). While as expected, both VSi_2 and $CrSi_2$, with hexagonal structures, are transversely isotropic in the basal plane, the latter is elastically more isotropic than the former. Such information may prove more useful at a later stage in the development of silicides.

3.2. Thermal Conductivity

A material with high thermal conductivity can dissipate thermal transients more rapidly, thereby improving its performance at high temperature under thermal gradients. This has been one of the motivations behind the development of single-crystal NiAl as a high-temperature structural material (Darolia 1991), in spite of only a moderate advantage in melting temperature over nickel-base superalloys. The thermal conductivity data for NiAl and nickel-base superalloys, from reference (Darolia 1991), are compared with data for $TiSi_2$ and $MoSi_2$ (Lugsheider et al. 1991) in Figure 11-11. Note that if extrapolated to temperatures above 1000°C, the thermal conductivities of both silicides are comparable to those of nickel-base superalloys.

3.3. Strength, Ductility and Toughness

The strength of silicides may be evaluated directly using bend tests (Anton and Shah 1991, Lugsheider et al. 1991) or in compression

(Frommeyer et al. 1990; Ito et al. 1992; Takeuchi, Hashimoto, and Shibuya 1991), or indirectly compared using hardness measurements (Es-Souni et al. 1991, Lugsheider et al. 1991, Murata et al. 1991). In many silicides the deformation behavior of single crystals has been studied and operative slip systems as well as the values of critical resolved shear stresses have been determined (Mass et al. 1984; Ito et al. 1992; Chang and Pope 1991; Umakoshi et al. 1991). Bend- and compression-test results are compared in Table 11-3 at both room and high temperature (1000°C to 1200°C). For brittle materials it is not surprising to find a large discrepancy between the strength measured in bending versus that in compression, as appears to be the case with $CoSi_2$. Note that for $CoSi_2$ and $TiSi_2$, the maximum bend strength shows a moderate increase in strength with temperature. It is, however, premature to conclude this to be anomalous behavior (Lugsheider et al. 1991), since with increased fracture strain with increasing temperature, higher fracture strength is achieved.

The results of attempts to define the ductile-to-brittle transition temperature (DBTT) using four-point bend testing are summarized in Figure 11-12 (Anton and Shah 1991). As depicted in Figure 11-12a, the DBTT for $CoSi_2$ was defined using repeated bend testing to successively lower temperatures until the sample fractured below a strain of 0.5%. Following this, additional intermetallic samples were used to define the elongation as a function of temperature using four-point bend test, as plotted in Figure 11-12b. For purposes of comparison, a typical plot for polycrystalline NiA1 is included. A single data point for Mo_5Si_3 with an arrow pointing to higher elongation represents unexplained serrated failure for the as-cast material with no clear fracture.

There is no well-defined approach proposed for lowering the DBTT of silicides. This is not surprising, given the infant state of the fundamental understanding of the deformation behavior of this class of material. However, the prospect of improving the room-temperature toughness by using the ductile-phase-toughened in situ composite approach appears promising. This has been demonstrated for the unidirectionally solidified Ti/Ti_5Si_3 eutectic where the fracture toughness has been shown to improve from 2 $MPam^{1/2}$ for the monolithic material to 12 $MPam^{1/2}$ for the two-phase material at room temperature. At 400°C, the two-phase material has an even higher fracture toughness of 25 $MPam^{1/2}$. Considerable work has been carried out investigating the toughness of the $Nb/Nb5Si_3$ system as both a discontinuously reinforced structure (Lewandowski et al. 1988; Mendiratta, and Dimiduk 1989; Nekanti and Dimiduk 1990; Rigney and Lewandowski 1991; Mendiratta et al. 1991) and a laminate (Kajuch, Rigney, and Lewandowski 1992). An interesting observation was that with thermomechanical processing, ductility of the ductile phase could be enhanced and room-temperature fracture-toughness values

Table 11-3 Comparison of Strengths of Various Silicides

		ROOM TEMPERATURE		HIGH TEMPERATURE			
Silicide System	Test	Y.S. MPa	Max. Stress MPa	Temp. °C	Y.S. MPa	Max. Stress MPa	Reference
$CoSi_2$, HIP	Bend	—	51.7	1000	60.7	66.9	Anton and Shah 1990
$CoSi_2$, Cast	Comp.	—	812	1000	50	—	Ito et al. 1991
$(Co, Ni)Si_2$	Comp.	> 500	—	1100	20	—	Takeuchi et al. 1991
$MoSi_2$, HIP	Bend		325	1000	290	—	Lugsheider et al. 1991
$MoSi_2$, HIP	Bend	—	—	1200	165	—	Lugsheider et al. 1991
$TiSi_2$, HIP	Bend	—	75	1000	—	200	Lugsheider et al. 1991
Mo_5Si_3, Cast	Bend	—	210	1200	—	12.4	Anton and Shah 1990
Ti_5Si_3, HIP	Comp.	—		1200	—	550	Frommeyer et al. 1990
Ti_5Si_3/Ti, DS	Comp.	1,100	—	1000	100	—	Frommeyer et al. 1990
V_3Si, HP	Bend	—	—	1200	—	101.4	Anton and Shah 1990
Cr_3Si, HP	Bend	—	—	1200	—	154.4	Anton and Shah 1990

Source: Shah et al. 1992

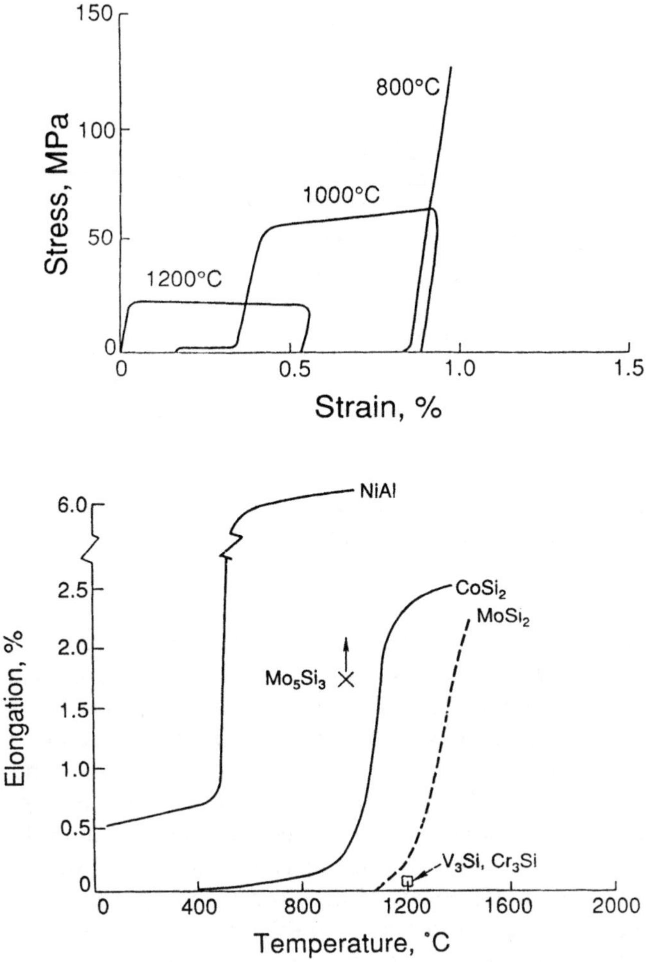

Figure 11-12. (a) Depiction of stress-strain response for $CoSi_2$ in evaluation of DBTT; (b) Strain to failure versus temperature illustrating the DBTT. (Shah et al. 1992)

of up to 21 $MPam^{1/2}$ could be obtained (Mendiratta et al. 1991).

3.4. High-Temperature Creep Resistance

For several binary and alloyed silicides, the minimum creep rate in compression versus stress is plotted in Figure 11-13. The data for Ti_5Si_3

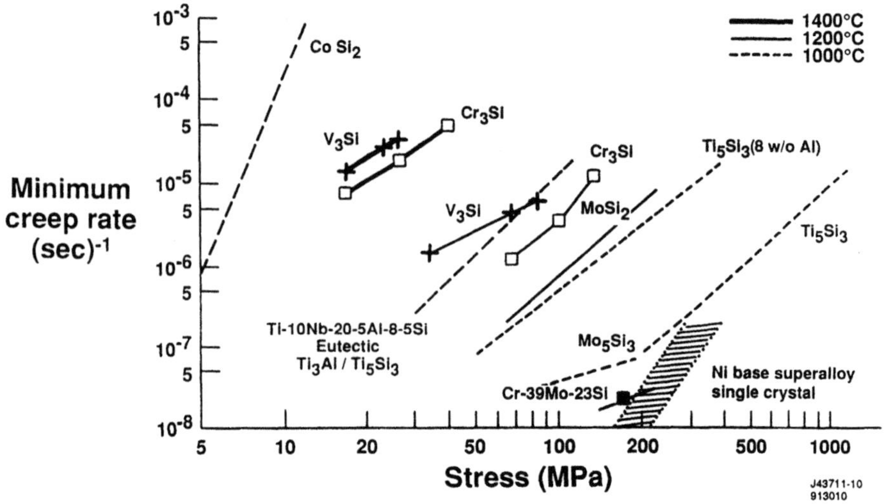

Figure 11-13. Comparison of minimum creep rate in compression versus stress for several silicides and their alloys. (Shah et al. 1992)

and the Ti_3Al/Ti_5Si_3 eutectic are from Frommeyer et al. (1990) and Es-Souni et al. (1991), respectively. A range of minimum creep rates for the current generation of nickel-base superalloy single crystals is shown at 1000°C (Cetel and Duhl 1988), for the purpose of comparison. Based on a direct comparison at 1000°C, the creep resistance of Ti_5Si_3 and Mo_5Si_3 is comparable to those of superalloys, whereas the low-melting $CoSi_2$ is several orders of magnitude weaker than superalloys. Similarly, a reasonable extrapolation of the creep rates at 1200°C to 1000°C show that V_3Si is considerably weaker. However, $MoSi_2$ and Cr_3Si appear to be promising with extrapolated creep rates comparable to superalloys. This is further confirmed with the creep data for the Mo-modified Cr_3Si(Cr-39 a/o Mo-23 a/o Si). In this alloy, as plotted in Figure 11-13, the minimum creep rate at 1200°C is comparable to those of superalloys at 1000°C at 172 MPa.

The addition of tungsten to $MoSi_2$ has been shown to significantly improve creep resistance over that of the binary silicide (Wiederhorn et al. 1992; Bose 1992; Sadanada, Feng, and Jones 1992). Examination of the available data indicates a fairly large scatter in the creep-rate results. Silica is a common impurity in $MoSi_2$ and the inconsistent results may be due to the varying levels of amorphous silica present in the materials (W. Maxwell 1952; Cotton, Kim and Kaufman 1991; Maloy et al. 1991; Maloy, Lewandowski, and Heuer 1992).

4. ENVIRONMENTAL RESISTANCE

4.1. High-Temperature Oxidation Resistance

Poor environmental resistance of the refractory metals is one of the strongest underlying motivators for pursuing intermetallics, and in particular, oxidation resistant silicides, as high-temperature structural materials. A prime requirement for structural materials is achievement of some degree of noncatastrophic environmental resistance, so that the integrity of the component is not solely dependent on the oxidation-resistant coating (prime-reliant coating). Typically for high-temperature oxidation resistance, this may be expressed in terms of a metal recession rate of 2.5 μm per hour or less. This may include external metal loss, a loss due to internal oxidation, as well as the depth of interstitial embrittlement.

Based on earlier work (Paine, Stonehouse, and Beaver 1960; Alexander 1989; Anton and Shah 1991), it may be concluded that disilicides generally possess good oxidation resistance at high temperatures due to the formation of a protective, adherent silica scale. At lower temperatures a catastrophic type of oxidation phenomenon has been observed that may completely reduce the material to powder. This phenomenon was first observed for $MoSi_2$ and termed "pesting" due to similarity to the tin-pest reaction (Fitzer 1956). Many disilicides, including WSi_2, $NbSi_2$, $FeSi_2$, and $MnSi_2$, also have been observed to undergo pesting oxidation (Westbrook and Wood 1964). Due to the recent interest in $MoSi_2$ as a high-temperature structural material, a great deal of research has been carried out to understand pesting phenomena (Berztiss et al. 1992; McKamey et al. 1992). It is generally believed that two conditions must be met for pesting of $MoSi_2$ to occur. First, the temperature must be low enough so that condensed molybdenum oxide forms during oxidation; and second, cracks or pores must be present at the intermetallic surface. It is believed that formation of the voluminous molybdenum oxide in the preexisting cracks or pores continuously wedges the material apart. The addition of Ge to $MoSi_2$ has been shown to suppress pesting by helping to stabilize a protective SiO_2 layer (Mueller et al. 1992).

While $MoSi_2$ is known to form essentially pure SiO_2 at high temperatures above approximately 600°C, most other disilicides form mixed oxides, although the formation of mixed oxides in some cases is not necessarily detrimental. Table 11-4 gives the oxidation resistance of selected silicides. The most oxidation resistant of these materials are $MoSi_2$, WSi_2 and Ti_5Si_3 (Vasudévan and Petrovic 1992). Figure 11-14 shows the isothermal-oxidation kinetics of $MoSi_2$, selected intermetallic aluminides, and a single-crystal nickel-base superalloy, PWA 1484 (Vasudévan and

Table 11-4 Oxidation Resistance of Selected Silicides

Silicide	Oxidation Temperature, °C	Oxidation Time, HR	Weight Change, mg/cm²
$MoSi_2$	1200	4	0.3
$MoSi_2$	1500	4	1.3
WSi_2	1200	4	−17
WSi_2	1500	4	−23
$NbSi_2$	1200	6	40
$TaSi_2$	1200	4	60
$TiSi_2$	1200	4	22
Mo_5Si_3	1500	4	−67
W_5Si_3	1500	4	−205
Ta_5Si_3	1500	1	125
Ti_5Si_3	1500	2	32

Source: Vaudévan and Petrovic 1992

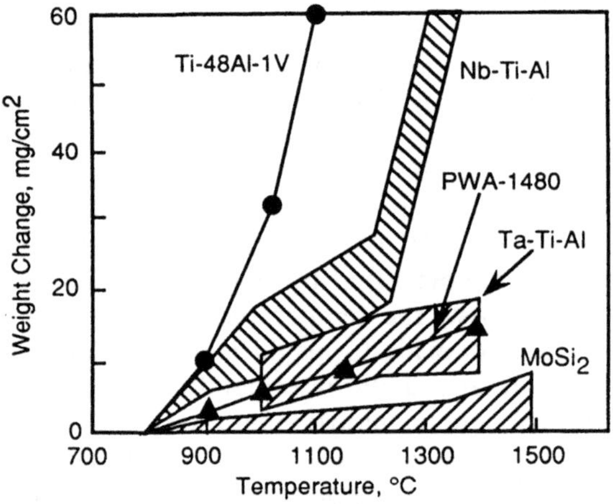

Figure 11-14. Comparison of isothermal oxidation behavior of $MoSi_2$ with selected intermetalics and PWA 1484. (Vasudévan and Petrovic 1992)

Petrovic 1992). The figure illustrates that $MoSi_2$ has significantly improved resistance to oxidation compared to aluminides and a nickel-base superalloy, especially at elevated temperatures. In cyclic oxidation testing (1149°C/55 minutes/5 minutes), both $MoSi_2$ and $CoSi_2$ show excellent oxidation resistance, even though the latter forms SiO_2 mixed with 16 v/o of Co_2SiO_4 cubic spinel (Anton and Shah 1991). The effect of alloying with A1 is complex. With the addition of aluminum at a level below the

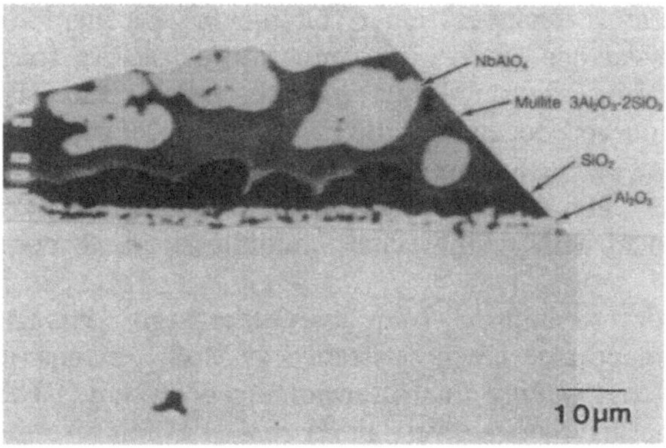

Figure 11-15. Typical mixed oxide scale formed on Al-Containing $NbSi_2$ (Nb-5A1-62Si) at 1370°C. (Shah et al. 1992)

solubility limit in $NbSi_2$ (and $TaSi_2$) mixed oxide scales are formed during static exposure at 1370°C, as shown in Figure 11-15. This is accompanied by a significant aluminum depletion just below the oxide layer. If the aluminum concentration is increased, a relatively pure protective alumina scale is formed on $NbSi_2$ (to a limited extent on $TaSi_2$) and no A1 depletion occurs below the surface. Apparently the A1 saturation is maintained with scattered grains of free A1 in the substrate (Alexander 1989).

Among 5-3 silicides, only Ti_5Si_3 is known to have reasonable oxidation resistance (Anton and Shah 1991; Paine et al. 1960). Mo_5Si_3 and Nb_5Si_3 are known to oxidize rapidly at high temperature (Alexander 1989). Consistent with these observations, alloying of Nb_5Si_3 with Ti shows significant improvement in oxidation resistance. Alloying with Ta, Mo, Zr, and Hf is not beneficial and alloying with A1 is only marginally better in conjunction with Ti. Nb_5Si_3 also undergoes pesting at intermediate temperatures (Westbrook and Wood 1964). Among a limited number of monosilicides, only Cr_3Si is known to possess good oxidation resistance, though somewhat poorer than $MoSi_2$ (Anton and Shah 1991).

5. MOLYBDENUM DISILICIDE COMPOSITE DEVELOPMENT

The use of molybdenum disilicide as a high-temperature structural material for turbine engine applications has been investigated for over four decades (M. Maxwell 1949). Properties of $MoSi_2$ that are desirable for a high-temperature structural material include excellent high-temperature oxidation resistance, good high-temperature fracture strength, high thermal conductivity, and ductility at elevated temperature. The use of $MoSi_2$ has been hindered because of the brittle nature of the material at ambient temperatures and a high creep rate at temperatures above 1200°C. Also, the relatively high thermal expansion of $MoSi_2$, compared to SiC and Si_3N_4, coupled with a high elastic modulus, leads to poor thermal shock resistance.

In order to improve both low-temperature damage tolerance and elevated-temperature creep resistance of $MoSi_2$, extensive work has been carried on compositing this intermetallic compound. The two basic approaches that have been investigated consist of reinforcement with discontinuous phases and reinforcement with continuous phases.

5.1. Discontinuously Reinforced $MoSi_2$ Composites

Most of the work carried out on development of $MoSi_2$ composites has involved discontinuous reinforcement and this work has recently been reviewed (Vasudévan and Petrovic 1992). The addition of both brittle and ductile phases to $MoSi_2$ has been investigated. The brittle-phase reinforcements include SiC (Gac and Petrovic 1985; Carter, Gibbs, and Petrovic 1989; Vasudévan et al. 1991; Bose 1992), ZrO_2 (Petrovic et al. 1992), ZrB_2, HfB_2, and TiB_2 (Aikin 1992), and the ductile phase reinforcements include Nb (Xiao and Abbaschian 1992; Lu et al. 1991; Alman et al. 1992) and Ta (Castro et al. 1992). Work with discontinuous ductile-phase reinforcement has shown some improvement in fracture toughness; however, all the work involved the use of uncoated refractory metals. Due to the interdiffusion that occurs between $MoSi_2$ and all of the refractory metals, these composites are of little use at elevated temperatures for any reasonable length of time.

Of the work carried out with brittle-phase reinforcement, the most encouraging results have been obtained with SiC additions. Significant improvements in creep rate have been demonstrated as shown in Figure 11-16, which plots the creep rate at 1100°C of $MoSi_2$, $MoSi_2$ + 20 vol/0 SiC whiskers, $(Mo, W)Si_2$, and $(Mo, W)Si_2$ + 20 vol/0 SiC whiskers (Sadananda et al. 1992). It is shown that the addition of SiC whiskers is effective in decreasing the creep rate of both $MoSi_2$ and $(Mo, W)Si_2$ by approximately two orders of magnitude.

Improvements also have been demonstrated in fracture toughness by discontinuous reinforcement. Figure 11-17 shows the fracture toughness of $MoSi_2$ compared to $MoSi_2$ containing increasing levels of ZrO_2 and SiC (Vasudévan et al. 1991). Pure $MoSi_2$ has a fracture toughness that is less

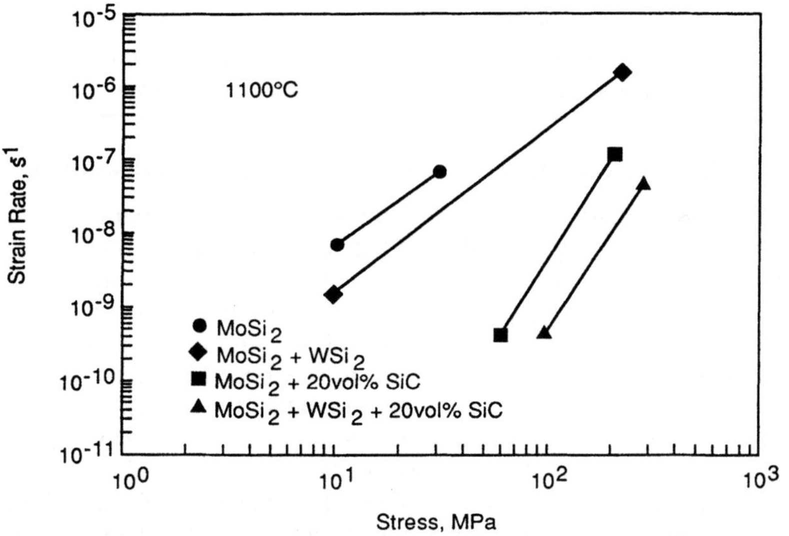

Figure 11-16. Effect of SiC whisker reinforcement on the creep of $MoSi_2$ and (Mo, W)Si_2. (Sadananda et al. 1992)

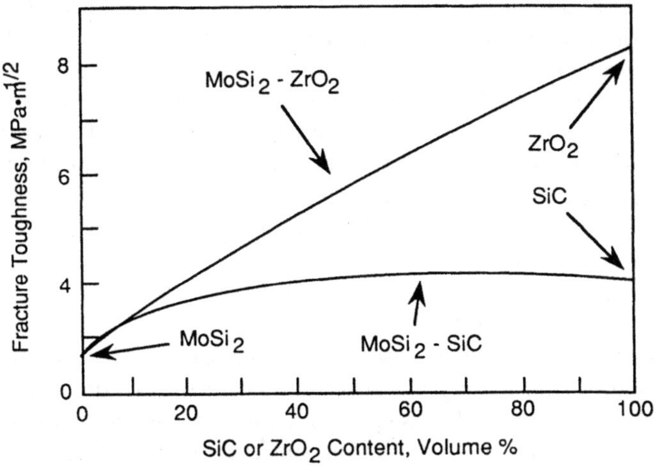

Figure 11-17. Improvement in room-temperature toughness due to additions of SiC and ZrO_2 to $MoSi_2$. (Vasudévan et al. 1991)

than 3 $MPam^{1/2}$. The addition of SiC whiskers increased the fracture toughness only moderately to a value slightly greater than 4 $MPam^{1/2}$. The mechanism for this improvement is believed to be a crack deflection/crack branching phenomenon. The addition of ZrO_2 is much more effective in improving toughness and is the result of transformation toughening caused by the transformation of metastable tetragonal ZrO_2 to monoclinic.

5.2. Continuously Reinforced $MoSi_2$ Composites

The advantages of using continuous-fiber reinforcement include more effective load-carrying capability of the reinforcement and, especially with the use of a brittle matrix, the ability to obtain significant improvements in toughness. Disadvantages to the use of continuous-fiber reinforcement include use of complex fabrication techniques and anisotropy of mechanical properties. Fibers that can be used to reinforce $MoSi_2$ can be classified as brittle ceramic fibers or ductile metallic fibers. The development philosophy with the use of ceramic fibers parallels that carried out with ceramic matrix composites and depends largely upon the development of a suitable fiber/matrix interface. The fiber/matrix interface should be strong enough to transfer load from the matrix to the fiber, however the interface must be weak enough to allow pull-out of the fiber from the matrix during fracture and yield a suitable level of toughness (Budiansky, Hutchinson, and Evans 1986; Evans 1989). The development philosophy with the use of ductile metallic fibers is different. Depending upon the strength of the metallic fiber, the fiber may or may not serve to strengthen the composite over that of monolithic $MoSi_2$. Toughness of a ductile-fiber-reinforced $MoSi_2$ composite is not derived directly from fiber/matrix interfacial properties, but from the toughness or work of fracture of the fibers themselves. This effect, termed "ductile phase toughening," has been theoretically modeled to demonstrate the potential improvement in fracture toughness (Sigl et al. 1988; Kunz-Douglass, Beaumont, and Ashby 1980).

Fibers investigated for reinforcing $MoSi_2$ include SiC, sapphire, polycrystalline alumina, and the refractory metals, Nb, W, Mo. The advantages and disadvantages of the use of these various reinforcements as they relate to compositing $MoSi_2$ has recently been summarized (Maloney and Hecht 1992). Significant disadvantages with the use of refractory metal fibers are poor resistance to oxidation and the need for a fiber coating to prevent the interdiffusion between fiber and matrix (Fitzer and Schmidt 1971). SiC and sapphire are chemically compatible with $MoSi_2$ (Nowotny et al. 1954; Yang and Jeng 1991); however, this is of little consequence because a fiber coating capable of debonding is required with these fibers.

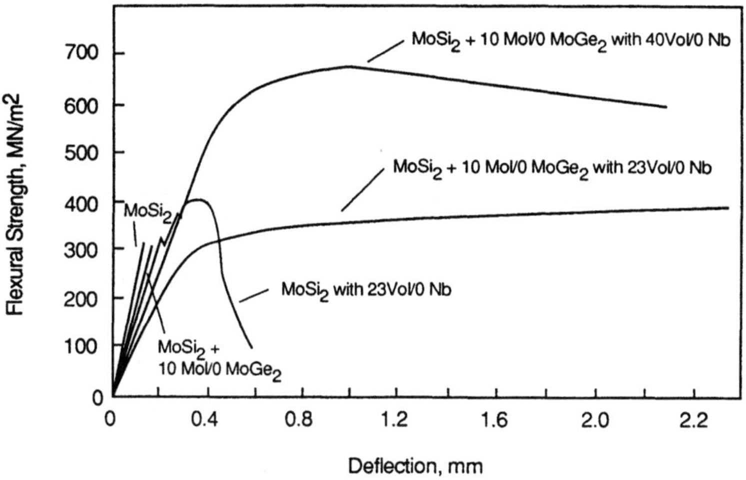

Figure 11-18. Stress strain diagram for $MoSi_2$, $MoSi_2$ + $MoGe_2$, $MoSi_2$ + Nb, and $MoSi_2$ + $MoGe_2$ + Nb. (Fitzer and Schmidt 1971)

Sapphire, polycrystalline alumina, and Nb have a thermal expansion coefficient that matches that of $MoSi_2$ very closely. The other fibers listed have a considerably lower thermal expansion coefficient, which results in residual stresses being generated in the composite during thermal cycling.

Significantly more work has been carried out with ductile fiber reinforcement than brittle fiber reinforcement to $MoSi_2$. It has been shown that the maximum toughening effect is achieved from a ductile reinforcement that is continuous and aligned as opposed to a randomly oriented fiber or a particulate (Alman et al. 1992). The first reported effort involving ductile-phase toughening of $MoSi_2$ involved the addition of pure Nb fibers to $MoSi_2$ and $MoSi_2$ containing additions of $MoGe_2$ (Fitzer, Gadow, and Remmele 1983; Fitzer and Remmele 1985). Figure 11-18 shows room-temperature flexure strength as a function of displacement for monolithic $MoSi_2$, monolithic $MoSi_2$ + 10 mole% $MoGe_2$, Nb fiber-reinforced $MoSi_2$ and Nb fiber-reinforced $MoSi_2$ + 10 mole % $MoGe_2$. The addition of $MoGe_2$ had little effect on the elastic modulus or flexure strength of $MoSi_2$. The Nb fiber-reinforced $MoSi_2$ + 10 mole % $MoGe_2$ matrix composite containing 40 vol% Nb was shown to have twice the ultimate flexure strength of the monolithic materials. A significant finding of this work was that the addition of $MoGe_2$ led to separation of the Nb fiber from the matrix along the fiber/matrix interface during fracture. This enabled the fibers to neck and undergo extensive plastic deformation. Conversely, the Nb fibers in the pure $MoSi_2$ matrix did not separate from the matrix during crack propagation and fractured in an intergranular, brittle manner. The lack of separation of fiber and matrix during fracture was attributed to a strong fiber/matrix interfacial bond in the Nb/pure $MoSi_2$ composite system.

Due to the interdiffusion between $MoSi_2$ and the Nb that occurs at high temperatures, a diffusion-barrier fiber coating is required for this system to be used at elevated temperatures for reasonable lengths times. Work has been carried out to identify fiber-coating materials that are chemically compatible with both the matrix and Nb reinforcing fiber and serve as an effective diffusion barrier (Xiao and Abbaschian 1992). Coatings that have been evaluated include $A1_2O_3$ and ZrO_2. It was found that both materials are chemically stable with both Nb and $MoSi_2$, however a limited amount of diffusion of Si occurred through the coating and reacted with the fiber.

Of the refractory metal fibers investigated, the molybdenum alloy, NS-Mo, has been shown to be one of the most promising as a reinforcement for an $MoSi_2$ composite system (Déve, Weber, and Maloney 1992; Maloney and Hecht 1992). NS-Mo is doped with approximately 1060 ppm Si, 150 ppm K and 36 ppm A1. Unlike niobium, this alloy has the potential to both toughen and strengthen $MoSi_2$. Tungsten and most molybdenum alloys undergo recrystallization and become embrittled at temperatures typically required to consolidate $MoSi_2$, thereby rendering the alloys useless as a ductile-phase reinforcement. NS-Mo is processed in such a manner that it remains ductile following recrystallization. Figure 11-19

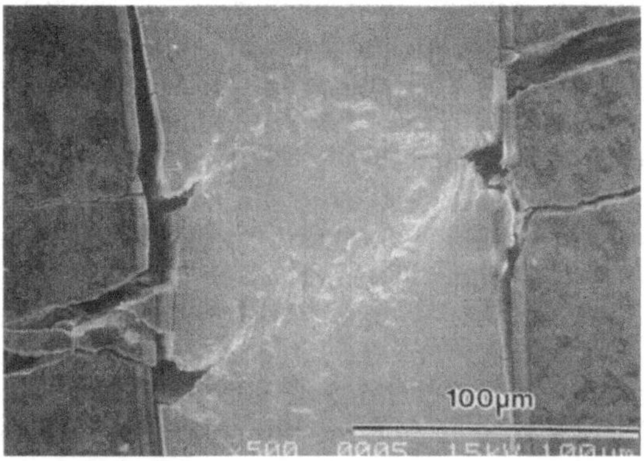

Figure 11-19. NS-Mo reinforced $MoSi_2$ + SiC composite undergoing fracture. (Reprinted with permission from *Materials Science and Engineering*, Volume A153, H. E. Déve, C. H. Weber, and M. Maloney, pages 668–675, 1992, Elsevier Science Ltd., Oxford, England)

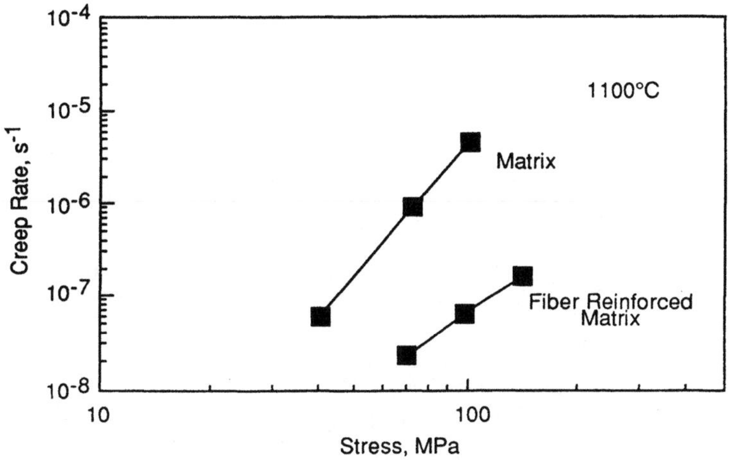

Figure 11-20. Minimum creep rate of NS-Mo reinforced composite versus load. (Reprinted with permission from *Materials Science and Engineering*, Volume A153, H. E. Déve, C. H. Weber, and M. Maloney, pages 668–675, 1992, Elsevier Science Ltd., Oxford, England)

shows a micrograph of the NS-Mo-reinforced $MoSi_2$ matrix composite. This composite was fabricated using an alumina fiber coating to limit matrix/fiber interdiffusion and using SiC particulate additions to the $MoSi_2$ matrix to match reinforcement/matrix thermal expansion (Maloney and Hecht 1992). Toughness, flexure strength, and creep resistance have been measured for the composite and compared the matrix values (Déve et al. 1992). Toughness values for the composite were measured at 37.5 kJm^2 versus 0.047 kJm^2 for the matrix. Room-temperature ultimate strength for the composite was 340 MPa versus 190 MPa for the matrix. Figure 11-20 shows the minimum creep rate of the NS-Mo-reinforced composite compared to that of the matrix. It is shown that well over an order of magnitude improvement was obtained (Déve et al. 1992).

Limited work has been carried out on brittle-fiber-reinforced $MoSi_2$ composites. The use of SiC fibers resulted in matrix cracking, even when low thermal expansion phases were added to the $MoSi_2$ matrix to minimize thermal expansion mismatch (Maloney and Hecht 1992). Fine-diameter polycrystalline alumina fibers have been incorporated into $MoSi_2$ by an injection molding process; however the composite properties have not been measured (Alman et al. 1992). Sapphire fiber has been used to reinforce $MoSi_2$ and has been shown to significantly improve creep resistance. Figure 11-21 shows a micrograph of a sapphire-reinforced $MoSi_2$ composite and strain-versus-time curve generated at 1371°C and a

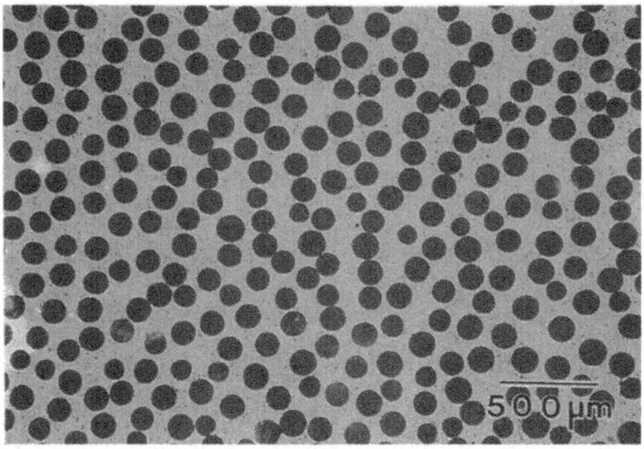

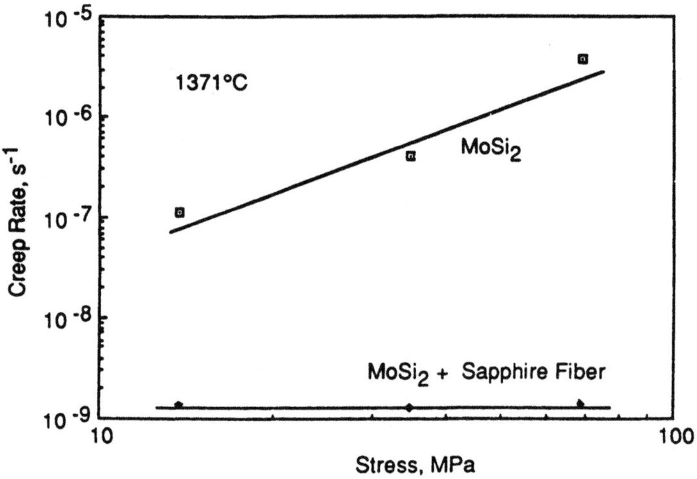

Figure 11-21. (a) Sapphire reinforced $MoSi_2$ composite microstructure and (b) Minimum creep rate versus load at 1371°C for composite and monolithic material.

load of 69 MPa. Although the improvement in creep rate is significant for this composite compared to monolithic $MoSi_2$, the fibers were so strongly bonded to the matrix such that a debonding fiber coating would be required to obtain a degree of damage tolerance. (Maloney and Hecht 1992).

6. SUMMARY

Silicides with almost metallic-like physical properties offer interesting opportunities as high-temperature structural materials. Among closely related disilicides based on refractory metals, compounds with the hexagonal C40($CrSi_2$) structure and the orthorhombic C54($TiSi_2$) structure present broader alloying opportunities than compounds with the tetragonal $C11_b$($MoSi_2$) structure. A better fundamental understanding of the relationships among the crystal structures of these disilicides in terms of atomic stacking and ordering arrangements may provide useful insight for future development. Most disilicides show excellent high-temperature oxidation resistance, but suffer from low-temperature disintegration due to the pest effect. This may be rooted in their noncubic crystal structure and associated anisotropic CTE. Disilicides such as $CoSi_2$ and $NiSi_2$ with the cubic C1(CaF_2) structure are exceptional, but and being low melting, are of marginal interest. With the exception of Ti_5Si_3, 5-3 silicides in general show poor high-temperature oxidation resistance. With wider phase fields, 5-3 silicides offer a greater latitude for alloying, not only with the transition metals, but also with the nontransition metals, including Nowotny phases stabilized with soluble interstitial elements. This class of silicides with high strength and creep resistance has potential for enhanced room-temperature toughness as in situ composites with a ductile refractory metal phase. The limited number of monosilicides with a cubic A15(Cr_3Si) structure, are most promising in terms of environmental resistance, strength, and creep resistance.

Of the various silicides, molybdenum disilicide is perhaps the most promising in terms of materials for high-temperature structural application. Advantages to the use of this material are excellent high-temperature oxidation resistance, ductility at temperatures above approximately 1000°C, and a high thermal conductivity. Significant problems with the use of this material are low fracture toughness at ambient temperatures and poor creep resistance at elevated temperatures. Various compositing approaches have been investigated to improve the properties. The use of SiC-whiskers has been shown to significantly improve creep resistance, especially in combination with tungsten alloying. The use of SiC, however, had minimal effect on improving toughness. The use of continuous, ductile molybdenum-alloy fibers has shown improvements in both toughness and creep strength.

REFERENCES

AIKIN, R. M., JR., 1992, *Mat. Sci. Eng.* A155: 121–133.

ALEXANDER, D., 1989, *Development of High Temperature Metallics for Structural Aerospace Applications*, Interim Technical Report for Period August, 1987 to September, 1988; under DARPA Order No. 6154, Contract No. N00014-87-C-0862.

ALMAN, D. E., SHAW, K. G., STOLOFF, N. S., and RAJAN, K., 1992, *Mat. Sci. Eng.* A155: 85–93.

ANTON, D. L., and SHAH, D. M., 1991, *Development Potential of Intermetallic Compounds*, U.S. Air Force Report WRDC-TR-90-4122.

________, 1991, In *MRS Symp. Proc.* 213: 733–738.

________, 1991, In *MRS Symp. Proc.* 213: 63–68.

ANTON, D. L., SHAH, D. M., DUHL, D. N., and GIAMEI, A. F., 1989, *J. Met.* 41(9): 12–17.

BERKOWITZ-MATTUCK, J. B., 1966, Final Report, NASA Contract No. NASW-1167.

BERZTISS, D., CERCHIARA, R. R., GULBRANSEN, E. A., PETTIT, F. S., and MEIER, G. H., 1992, *Mat. Sci. Eng.* A155: 165–181.

BOSE, S., 1992, *Mat. Sci. Eng.* A155: 217–225.

BUDIANSKY, B., HUTCHINSON, J. W., and EVANS, A. G., 1986, *J. Mech. Phys. Solids* 2: 167–189.

CARTER, D. H., GIBBS, W. S., and PETROVIC, J. J., 1989, *Proceedings for the Third International Symposium on Ceramic Materials and Components for Engines*. Westerville, OH: American Ceramic Society. 977–986.

CASTRO, R. G., SMITH, R. W., ROLLET, A. D., and STANEK, P. D., 1992, *Mat. Sci. Eng.* A155: 101–107.

CELIS, P. B., KAGAWA, E., and ISHIZAKI, K., 1991, *J. Mat. Res.* 6(10): 2077–2083.

CETEL, A. D., and DUHL, D. N., 1988, In *Superalloys 1988*, ed. D. N. Duhl, G. Maurer, S. Antolovich, C. Lund, and S. Reichman, pp. 235–245, Warrendale, PA: The Minerals, Metals and Materials Society.

CHANG, C. S., and POPE, D. P., 1991, In *High Temperature Ordered Intermetallic Alloys IV*, *MRS Symp. Proc.* 213: 745–750.

COPLEY, S. M., and KEAR, B. H., 1967, *Trans. TMS-AIME.* 239: 977–983.

COTTON, J. D., KIM, Y. S., and KAUFMAN, M. J., 1991, *Mat. Sci. Eng.* A144: 287–291.

DAROLIA, R., 1991, *J. Met.*, 43(3): 44–49.

DÉVE, H. E., WEBER, C. H., and MALONEY, M., 1992, *Mat. Sci. Eng.* A153: 668–675.

Es-Souni, M., Chen, D., Dogan, B., Wagner, R., Beaven, P. A., and Bartels, A., 1991, In *Procd. of Int. Symp. on Intermetallic Compounds (JIMIS-6)*, ed. O. Izumi, pp. 525–529. Sendai, Japan: The Japan Institute of Metals.

Evans, A. G., 1989, *Mat. Sci. Eng.* A107: 227–239.

Fitzer, E., Gadow, G., and Remmele, W., 1983, In *Ceramic Components for Engines, Proceedings of the First International Symposium*, ed. S. Somiya, E. Kanai, and K. Ando, pp. 799–807, New York: Elsevier Science Publishers.

Fitzer, E., and Remmele, W., 1985, In *Proceedings of the Fifth International Conference on Composite Materials, ICCM-V*, ed. W. C. Harrigan, Jr., J. Strife, and A. K. Dhingra, pp. 515–530. Warrendale, PA: Metallurgical Society of AIME.

Fitzer, E., and Schmidt, F. K., 1971, *High Temp. High Press.* 3: 443–460.

Fitzer, E., 1956, *Proceedings of the Second Plansee Seminar*, ed. F. Benesovsky, 56, London: Pergamon Press.

Frommeyer, G., Rosenkranz, R., and Ludecke, C., 1990, *Z. Metal..* 81: 307–313.

Gac, F. D., Petrovic, J. J., 1985, *J. Am. Ceram. Soc.* 68: c200–c201.

Ito, K., Inui, H., Hirano, T., and Yamaguchi, M., 1992, *Mat. Sci. Eng. A* 152: 153–159.

Kajuch, J., Rigney, J. D., and Lewandowski, J. J., 1992, *Mat. Sci. Eng. A* 155(1-2): 59–65.

Kakamura, M., 1991, *Proceedings of International Symposium on Intermetallic Compounds (JIMIS-6)*, ed. O. Izumi, pp. 655–659. Sendai, Japan: The Japan Institute of Metals.

Kampe, S. L., Swope, G. H., and Christodoulou, L., 1990, *MRS Symp. Proc.* 194: 97–103.

Khantha, M., Vitek, V., and Pope, D. P., 1989, *MRS Symp. Proc.* 133: 99–104.

Kunz-Douglass, S., Beaumont, P. W. R., and Ashby, M. F., 1980, *J. Mat. Sci.* 15: 1109–1123.

Law, C. C., Shah, D. M., and Lin, J., 1988, *Fundamental Understanding of the Intrinsic Ductility in Nickel-base $L1_2$ Type Alloys,* Final Report (FR-20424, Bolling AFB, DC) prepared for the Air Force Office of Scientific Research.

Lewandowski, J. J., Dimiduk, D., Kerr, W., and Mendiratta, M. G., 1988, In *MRS Symp. Proc.* 120: 103–109.

Lu, T. C., Evans, A. G., Hecht, R. J., and Mehhrabian, R., 1991, *Acta Metall. Mater..* 39. 1853–1862.

Lugsheider, E., Westermann, U., Wonka, J., Meinhardt, H., Neisius, E., and Arnold, E., 1991, *Proceedings of International Symposium on Intermetallic Compounds (JIMIS-6)*, ed. O. Izumi, pp. 621–625. Sendai, Japan: The Japan Institute of Metals.

Maloney, M. J., and Hecht, R. J., 1992, *Mat. Sci. Eng.* A155: 19–31.

Maloy, S. A., Heuer, A. H., Lewandowski, J., and Petrovic, J., 1991, *J. Am. Ceram. Soc.* 74: 2704–2706.

Maloy, S. A., Lewandowski, J. J., and Heuer, A. H., 1992, *Mat. Sci. Eng.* A155: 159–163.

MASS, J., BASTIN, G., VAN LOU, F., and METSELAAR, R., 1984, *Z. Metal.* 75: 140–145.

MASSALSKI, T. B., 1986, *Binary Phase Diagrams*. Metals Park, OH: American Society For Metals.

MAXWELL, M. G., 1949, *Properties of Certain Intermetallics as Related to Elevated-Temperature Applications 1-Molybdenum Disilicide*. Report E9G01. Washington: National Advisory Committee for Aeronautics.

MAXWELL, W. A., 1952, *Some Factors Effecting Fabrication and High-Temperature Strength of Molybdenum Disiliside*. Report RME52B06. Washington, DC: National Advisory Committee for Aeronautics.

MCKAMEY, C. G., TORTORELLI, P. F., DEVAN, J. H., and CARMICHAEL, C. A., 1992, *J. Mat. Res.* 7: 2747–2755.

MENDIRATTA, M. G., and DIMIDUK, D. M., 1989, *MRS Symp. Proc.* 133: 441–446.

MENDIRATTA, M. G., LEVANDOWSKI, J. J., and DIMIDUK, D. M., 1991, *Met. Trans.* 22A: 1573–1583.

MUELLER, A., WANG, G., RAPP, R. A., and COURTRIGHT, E. L., 1992, *J. Elect. Soc.* 139: 1266–1275.

MURATA, Y., HIGUCHI, T., TAKEDA, Y., MORINAGA, M., and YUKAWA, N., 1991, *Proceedings of International Symposium on Intermetallic Compounds (JIMIS-6)*, ed. O. Izumi, pp. 627–631. Sendai, Japan: The Japan Institute of Metals.

NEKKANTI, R. M., and DIMIDUK, D. M., 1990, *Intermetallic Matrix Composites, MRS Symp. Proc.* 194: 175–182.

NOWOTNY, H., BOLLER, H., and ZWILLING, G., 1972, *Proceedings of the Fifth Materials Research Symposium*, National Bureau of Standards Publication 364: 487–504.

NOWOTNY, H., PARTHE, E., KIEFFER, R., and BENESOVSKY, F., 1954, *Monat. Chem.* 85: 255–256.

PAINE, R. M., STONEHOUSE, A. J., and BEAVER, W. W., 1960, *WADC Technical Report 59-29-Part I*, Dayton: OH, United States Air Force.

PETROVIC, J. J., BHATTACHARYA, A. K., HONNELL, R. E., MITCHELL, T. E., WADE, R. K., and MCCLELLAN, K. J., 1992, *Mat. Sci. Eng.* A155: 259–266.

RIGNEY, J. D., and LEWANDOWSKI, J. J., 1991, In *Ceramic Transactions*, vol. 19, pp. 519–525. ed. M. D. Sacks. Westerville, OH: The American Ceramic Society.

SADANANDA, K., FENG, C. R., and JONES, H., 1992, *Mat. Sci. Eng.* A155: 227–239.

SAMSONOV, G. V., 1964, *Handbook of High Temperature Materials*, No. 2, p. 130. New York: Plenum Press.

SCHUBERT, K., 1967, *Intermetallic Compounds*, pp. 118–119. New York: John Wiley & Sons.

SHAH, D. M., BERCZIK, D., ANTON, D. L., and HECHT, R., 1992, *Mat. Sci. Eng.* A155: 45–57.

SIGL, L. S., MATAGA, P. A., DALGLEISH, B. J., MCMEEKING, R. M., and EVANS, A. G., 1988, *Acta Metall. Mater.* 36: 945–953.

SMARSLY, W., ROSENKRANZ, R., FROMMEYER, G., 1992, *Mat. Sci. Eng.* A152: 288–294.

TAKEUCHI, S., HASHIMOTO, T., and SHIBUYA, T., 1991, *Proceedings of International Symposium on Intermetallic Compounds (JIMIS-6)*, ed. O. Izumi, pp. 645–649. Sendai, Japan: The Japan Institute of Metals.

TAYLOR, A., 1961, *X-ray Metallography* p. 432. New York: John Wiley & Sons.

TVERGAARD, V., and HUTCHINSON, J. W., 1988, *J. Am. Ceram. Soc.* 71: 157–166.

UMAKOSHI, Y., HIRANO, T., SAKAGAMI, T., and YAMANE, T., 1990, *High Temperature Aluminides and Intermetallics*, ed. S. H. Wang, C. T. Liu, D. P. Pope, and J. D. Stiegler, pp. 111–129, Warrendale, PA: The Minerals, Metals and Materials Society.

UMAKOSHI, Y., NKASHIMA, T., YAMANE, T., and SENBA, R. E., 1991, In *Proceedings of International Symposium on Intermetallic Compounds (JIMIS-6)*, ed. O. Izumi, pp. 639–643. Sendai, Japan: The Japan Institute of Metals.

VASUDÉVAN, A. K., and PETROVIC, J. J., 1992, *Mat. Sci. Eng.* A155: 1–17.

VASUDÉVAN, A. K., PETROVIC, J. J., and SADANANDA, K., 1991, In *Metal Matrix Composites, 12th Riso Proceedings*, eds. N. Hanson, D. J. Jensen, P. Leffers, H. Lihlt, T. Loren, A. S. Peterson, O. B. Peterson, and B. Ralph. Roskilde, Denmark: Riso National Library.

VILLARS, P., and CALVERT, L. D., 1985, *Pearson's Handbook of Crystallographic Data for Intermetallic Phases*. ed. P. Villars and L. D. Calvert. Metals Park, OH: American Society for Metals.

WERNICK, J. H., 1967, *Intermetallic Compounds: 203*. New York: John Wiley & Sons.

WESTBROOK, J. H., and WOOD, D. L., 1964, *J. Nuc. Mat.* 12: 208–215.

WIEDERHORN, S. M., GETTINGS, R. J., ROBERTS, D. E., OSTERTAG, C., and PETROVIC, J. J., 1992, *Mat. Sci. Eng.* A155: 209–215.

XIAO, L., and ABBASCHIAN, R., 1992, *Mat. Sci. Eng.* A155: 135–145.

YANG, J. M., and JENG, S. M., 1991, *J. Mat. Res.* 6: 505–513.

Part III

Environmental Effects

Chapter 12

Environmental Embrittlement

N. S. Stoloff

1. INTRODUCTION

Recent studies have shown that low ductility and brittle fracture in intermetallics are caused not only by intrinsic factors (such as lack of sufficient deformation modes, poor cleavage strength, etc.) but also by extrinsic factors. Environmental degradation, an extrinsic factor, is found to be a major cause of brittle fracture in many ordered intermetallics, particularly those with high crystal symmetries (i.e., cubic Ll_2, and B2 and hexagonal DO_{19}).

There are two types of environmental degradation observed in intermetallics. One is hydrogen-induced embrittlement occurring at ambient temperatures. Many intermetallic alloys show a substantial decrease in room-temperature tensile ductility due to moisture-induced hydrogen embrittlement in air, or as a result of direct exposure to hydrogen by precharging or testing in hydrogen gas. The other is oxygen-induced embrittlement in air at elevated temperatures. In most cases, the embrittlement is due to a dynamic effect involving generation and penetration of an embrittling agent (i.e., hydrogen or oxygen) during testing. In this chapter, the two types of environment-sensitive degradation are treated separately in sections 2 and 3. In section 3.3, accelerated oxidation, resulting in the "pest" phenomenon, is discussed. The last section discusses metallurgical means to alleviate environmental degradation.

2. EMBRITTLEMENT AT AMBIENT TEMPERATURES

Hydrogen, introduced either by exposure to moist environments, by cathodic charging, or by testing in hydrogen gas, has been shown to embrittle a large number of single-phase intermetallics. Table 12-1 lists those alloys that have been tested for susceptibility to environmental embrittlement. Most notable is the susceptibility to moisture or hydrogen of all Ll_2 intermetallics that have been examined to date, for example, Ni_3Al, Ni_3Al + B, Ni_3Al + Cr, Ni_3Fe, Ni_3Si, Co_3Ti, $(Fe,Ni)_3V$, $(Co,Fe)_3V$, and $Ni_3(Al,Mn)$. Therefore, it is highly likely that all Ll_2 intermetallics containing substantial amounts of the transition metals Fe, Ni or Co will prove to be susceptible to such embrittlement. Data for other superlattice structures remain scarce. Two B2 alloys are embrittled by hydrogen: FeCo-2%V and FeAl. The DO_{19} superlattice Ti_3Al (α_2) and its alloys with Nb are reported to form hydrides when exposed to hydrogen gas and are embrittled by hydrogen. TiAl also is susceptible to embrittlement in air or in hydrogen gas, perhaps by hydrides, but the effect is much smaller than in

Table 12-1 Alloys Embrittled by Moisture or Hydrogen at Ambient Temperature

Alloy	Crystal Structure	Environmental Embrittlement	Reference
$Ni_3(Al,Ti)$ (single crystal)	Ll_2	o	Takasugi 1991
Ni_3Al + B	Ll_2	o	
Ni_3Al + Be	Ll_2	o	Takasugi et al. 1986
$Ni_3(Al,Mn)$	Ll_2	o	Masahashi et al. 1988
Ni_3Si	Ll_2	o	Liu and Oliver 1991
$Ni_3(Si,Ti)$	Ll_2	o	Takasugi et al. 1991
$Ni_3(Si,Ti)$ + B	Ll_2	x	Masahashi et al. 1988
Co_3Ti	Ll_2	o	Takasugi and Izumi 1986
Co_3Ti + Fe, Al	Ll_2		Liu et al. 1989
$(Co,Fe)_3V$	Ll_2	o	Nishimura and Liu 1991
Ni_3Fe	Ll_2	x	Liu and Takeyama 1990
TiAl	Ll_0	o	Nakamura 1991
FeAl	B2	o	Liu et al. 1989
Fe_3Al	DO_3	o	Liu et al. 1990

o Environmental embrittlement observed in oxidizing or moist environment Observed but reduced by alloying

x Not observed

Source: (Takeyama and Liu, 1992; reprinted from *Materials Science and Engineering*, Volume A153, pages 538–547, copyright 1992, with kind permission from Elsevier Science Ltd., The Boulevard, Langford Lane, Kidlington OX5 1GB, UK).

α_2 alloys. Ni_2Cr is highly susceptible to hydrogen embrittlement in both the ordered and disordered conditions.

2.1. $L1_2$ Alloys

In terms of environmental degradation, ordered intermetallics can be grouped into two categories: (a) alloys containing no reactive elements, and (b) alloys containing reactive elements (e.g., Al or Si). For the first category, the alloys are severely embrittled only when being charged with hydrogen, such as by cathodically charging. For the second category, the alloys themselves are capable of generating hydrogen from hydrogen-containing environments at ambient temperatures. The most striking case is severe embrittlement of iron aluminides in moist air at room temperature.

2.1.1. $L1_2$ Alloys Containing No Reactive Elements. Ni_3Fe is a model material for study of hydrogen embrittlement in alloys containing no reactive elements. Ni_3Fe can be disordered by quenching from above its critical ordering temperature, T_c. This has permitted a direct comparison to be made between the ordered and disordered conditions of Ni_3Fe (see Fig. 12-1; Camus, Stoloff, and Duquette 1989). In each case, hydrogen was cathodically charged for one hour prior to tensile test at room temperature. Note that Ni_3Fe is slightly embrittled by precharging in both the ordered and disordered conditions; simultaneous charging and testing cause greater loss of ductility than precharging, with an especially severe effect for the ordered condition, Figure 12-1b. Fractographic studies confirm a change in fracture mode from microvoid coalescence to intergranular fracture in hydrogen accompanying the reduction in elongation. However, the intergranular embrittlement zone is only about one-third as deep in precharged samples of ordered Ni_3Fe as in disordered samples.

Embrittlement of the ordered condition is much more severe when charging occurs during testing, and unlike precharged samples, the entire fracture surface of the simultaneously charged samples is intergranular. The fracture surfaces of disordered samples display a mixed mode. The diffusivity of hydrogen in ordered Ni_3Fe should be lower than that in the disordered condition, although no data are available. Therefore, only a shallow layer can be embrittled by hydrogen. However, when plastic deformation accompanies hydrogen charging, dislocations can carry hydrogen inward to a much greater depth than by diffusion alone; hence embrittlement is more severe, and the fracture path is intergranular to a greater depth than in precharged material.

As shown in Figure 12-1, both ordered and disordered Ni_3Fe display extensive ductility and ductile fracture modes in air, indicating that moisture does not embrittle the alloys. This is because Ni_3Fe contains no

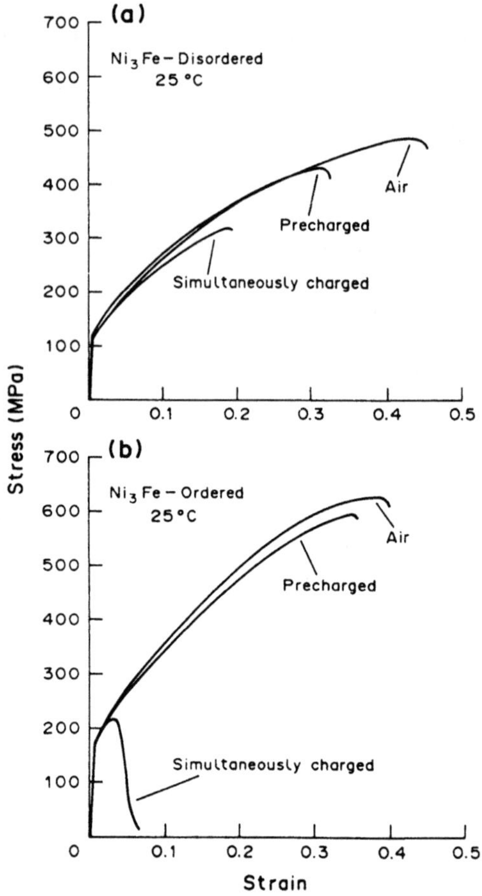

Figure 12-1. Embrittlement of Ni_3Fe by electrolytically charged hydrogen (Reprinted with permission from *Acta Metallurgica*, Volume 37, G. M. Camus, N. S. Stoloff, and D. J. Doquette, pages 1497–1501, 1989, Elsevier Science, Pergamon Imprint, Oxford, England).

reactive elements (such as Al, Ti, V, Si, etc.) and is not able to release hydrogen from moisture in air. Similarly, Ni_3Mn exhibits transgranular fracture and good ductility in air at room temperature.

*2.1.2. **Intermetallics Containing Reactive Elements.*** Most intermetallics of interest for structural applications, including Ll_2 compounds, contain reactive elements such as Al (for aluminides), Si (for silicides), Ti,

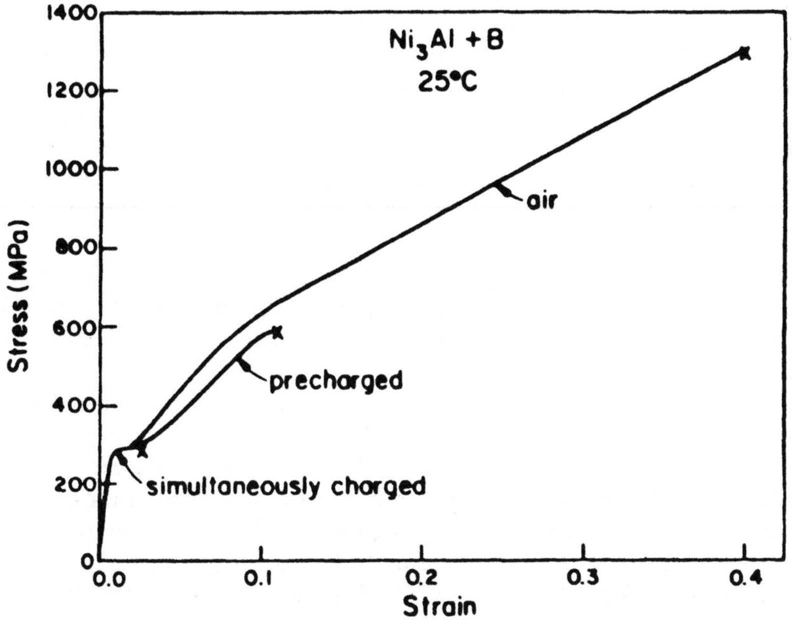

Figure 12-2. Effect of hydrogen on ductility of Ni_3Al + B (Kuruvilla and Stoloff 1985b).

and V. These intermetallics show severe degradation not only in the hydrogen-charged condition, as seen e.g., in Figure 12-2 for Ni_3Al + B (Kuruvilla and Stoloff 1985b), but also in hydrogen-releasing atmospheres such as moist air. Severe embrittlement has been observed in many $L1_2$ intermetallics, including $Ni_3(Al,Mn)$, Be-doped Ni_3Al, B-doped Ni_3Al, Co_3Ti, $(Co,Fe)_3V$, Ni_3Si, and $Ni_3(Si,Ti)$, tested in moist air at room temperature. In these cases, embrittlement involves the reaction of reactive elements with moisture in air and generation of atomic hydrogen that penetrates into crack tips and facilitates crack propagation.

Ni_3Al Alloys. Although the embrittling effects of hydrogen on fracture behavior of Ni_3Al alloys have been well documented (e.g., Kuruvilla and Stoloff 1985b; Masahashi, Takasugi, and Izumi 1988), only recently has evidence been provided for moisture-induced cracking in these alloys.

Liu (1992) has reported that the tensile ductility of Ni-24at%Al is increased from 2.6% to 7.2% by testing in oxygen rather than in air. A similar result was noted with Ni-23.5at%Al (2.5% vs. 8.2% elongation in air and oxygen, respectively).

Even more striking effects of the environment have been reported by George, Liu, and Pope (1992) for a Ni-22.65Al-0.26Zr alloy recrystallized from a ⟨110⟩-oriented single crystal and tested at room temperature in two orientations, see Table 12-2. This boron-free alloy displayed 8.7% elongation in water, 13.2% in air, and 50.6% in oxygen. Fracture was predominantly intergranular in all environments. Although the results could have been influenced by Zr content, by texture (unknown in these alloys), or by a preponderance of fracture-resistant grain boundaries in this material, the authors suggest that the principal role of boron in ductilizing Ni_3Al may be to suppress environmental embrittlement. Grain-boundary strengthening, permitting transgranular failure, would be a secondary effect.

Wan, Zhu, and Jing 1992 have reported that the tensile ductility of Ni-23at%Al-120wppm B is only 0.6% in 0.1 MPa hydrogen gas and rises to 10.1% in air. However, when tested in 1.3×10^{-3} Pa vacuum or in 0.1 MPa oxygen, ductilities of 20% or more are achieved. Although preoxidation can suppress embrittlement by cathodically charged hydrogen (Kurvilla and Stoloff 1985b) embrittlement by hydrogen gas was little affected by preoxidizing for 24 h. at 900°C (Wan et al. 1992). Previous work by Takasugi and Izumi (1986) carried out on Ni_3Al with 500wppm B, had shown no effect of the test environment on ductility. Wan et al. (1992) conclude that higher boron contents may suppress grain-boundary decohesion in the presence of hydrogen released from water vapor. Clearly, however, boron does not prevent hydrogen embrittlement under cathodic charging conditions (Kurvilla and Stoloff 1985b).

Figure 12-3 shows tensile data for $Ni_3(Al_{0.4}Mn_{0.6})$ tested in various environments and at several strain rates at room temperature (Masahashi et al. 1988). Manganese was added to Ni_3Al in order to enhance ductility. At a low strain rate of 10^{-5} s^{-1}, the alloy showed about 40% elongation in vacuum but only 5% in air at room temperature. The ductility was further reduced by hydrogen charging, followed by testing in air. The loss in ductility in air was attributed to moisture-induced hydrogen embrittlement. Consistent with the environmental effect, the ductility increases with increasing strain rate, and the yield strength is essentially independent of both test environment and strain rate (see Fig. 12-3). The decrease in ductility due to environmental embrittlement is accompanied by changing fracture mode from ductile-transgranular, through mixed mode, to brittle-intergranular. This observation suggests that hydrogen, generated either by the moisture/aluminum reaction or by hydrogen charging, diffuses mainly along grain boundaries and causes intergranular fracture. Note that a similar reduction in ductility and change in fracture mode has been observed in other Ll_2 intermetallic alloys (see Table 12-1).

Table 12-2 Effects of Environment and Specimen Orientation on Tensile Properties of Ni-22.65%Al-0.26%Zr, 25°C, $\dot{\epsilon} = 5.3 \times 10^{-3}\ s^{-1}$

Specimen Orientation	Test Environment	Elongation to Fracture (%)	Yield Strength (MPa)	Ultimate Tensile Strength (MPa)	$\sigma_{10} - \sigma_5$[a] (MPa)	$\sigma_{15} - \sigma_5$[a] (MPa)
0°	Water	8.7	322	528	—	—
	Air	13.2	324	661	133	—
	Oxygen	50.6	326	1451	127	250
45°	Water	6.3	331	473	—	—
	Air	10.7	341	603	127	—
	Oxygen	47.8	327	1438	143	271

[a] Difference in flow stress at 10 and 5% (or 15 and 5%) strain.

Source: (Reprinted with permission from *Scripta Metallurgica*, Volume 27, E. P. George, C. T. Liu, and D. P. Pope, pages 365–370, 1992, Elsevier Science Ltd., Oxford, England).

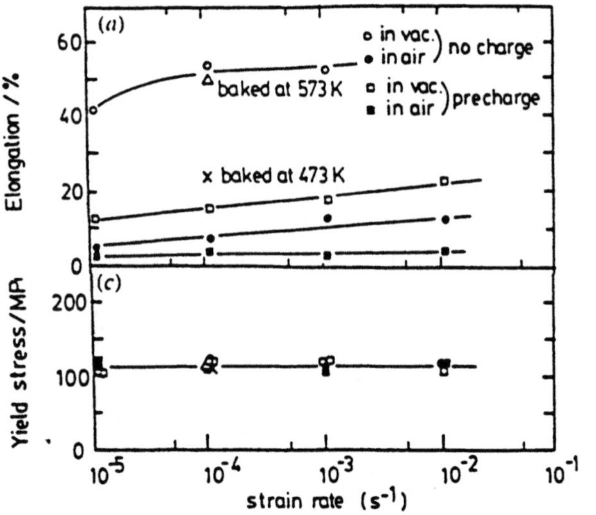

Figure 12-3. Environmental embrittlement in $Ni_3(Al_{0.4}Mn_{0.6})$ (Masahashi et al. 1988).

Ricker et al. (1991) have shown that stress-corrosion cracking of Ni_3Al alloys such as IC-50 (Ni-22at%Al, 0.01Zr, 0.08B) and IC-218 (17at%Al, 8.18Cr, 0.12Zr, 0.098B) in acidic solutions (HNO_3, H_2SO_4) also arises from the liberation of hydrogen. Slow strain-rate tests in low pH solutions show a large reduction in ductility and a change in fracture mode from transgranular to intergranular compared to tests in neutral or alkaline solutions. Preexposure to sulfuric acid also reduces ductility in a subsequent slow strain-rate test conducted in air, again due to release of hydrogen at the specimen surface (Ricker et al. 1991).

Ni_3Si ALLOYS. Recently, intergranular fracture and environmental embrittlement have been studied in alloys based on Ni_3Si (Liu and Oliver 1991; Takasugi, Suenaga, and Izumi 1991). Ni_3Si showed no appreciable plastic deformation when tested in moist air but displayed an elongation of 7.5% when tested in dry oxygen (Table 12-3), demonstrating that Ni_3Si is prone to environmental embrittlement. Since the elimination of the environmental effect by testing in dry oxygen does not lead to extensive ductility (e.g., 30% or more) and complete suppression of intergranular fracture in Ni_3Si, moisture-induced hydrogen embrittlement appears not to be the sole source of grain-boundary brittleness in the silicide. This is further indicated by comparison of the tensile data for Ni_3Si and $Ni_3(Si,Ti)$.

Table 12-3 Environmental Embrittlement of Ni_3Si and $Ni_3(Si,Ti)$ Alloys[a]

Alloy	Test Environment	Tensile Ductility (%)	Yield Strength MPa	Ultimate Strength MPa
Ni_3Si	Air	0[b]	b	627
Ni_3Si	Vacuum	4.7	677	853
Ni_3Si	Oxygen	7.5	685	1040
$Ni_3(Si,Ti)$	Air	7	606	813
$Ni_3(Si,Ti)$	Vacuum	29	586	1323
B-doped $Ni_3(Si,Ti)$[c]	Air	36	599	1323
B-doped $Ni_3(Si,Ti)$[c]	Vacuum	34	599	1323

[a]Alloys containing 9.5 at.% Ti (Takasugi et al. 1991).

[b]Fracture prior to macroscopic yielding.

[c]Doped with 50 ppm boron (Takasugi et al. 1991).

As shown in Table 12-3, $Ni_3(Si,Ti)$ also is susceptible to environmental embrittlement (Liu and Oliver 1991). The $Ni_3(Si,Ti)$ alloy, however, exhibited considerable ductility (7%) when tested in air at room temperature. It displayed excellent ductility (29%) when tested in vacuum, indicating that titanium enhances the intrinsic grain-boundary cohesion of Ni_3Si. The effects of boron and carbon additions on alleviation of environmental degradation will be discussed in detail in section 4.4.

$(Co,Fe,Ni)_3V$ ALLOYS. Environmental embrittlement also is severe in the $(Co,Fe)_3V$ $L1_2$ system, as shown in Figure 12-4 for $(Co_{78}Fe_{22})_3V$ samples tested in various environments at 25°C (Nishimura and Liu 1991). As is typical, yield and flow stresses were unaffected by the test environment, while ductility and tensile strength decreased as the environment changed from dry oxygen to vacuum to air and distilled water. A change from transgranular to intergranular failure was noted with increasing severity of embrittlement. Other $L1_2$ alloys belonging to the same group of $(Fe,Co,Ni)_3V$ and $(Co,Fe)_3V$ alloys have previously been shown to be embrittled by exposure to high-pressure hydrogen or to hydrogen produced by electrolytic charging (Kuruvilla and Stoloff 1985c). Therefore, there is little reason to doubt that hydrogen is the prime agent of embrittlement when such alloys are tested in moist air.

Hydrogen gas at 1 atm pressure has no effect on fatigue-crack growth in a disordered $(Fe,Ni)_3V$ alloy at room temperature, but there is a

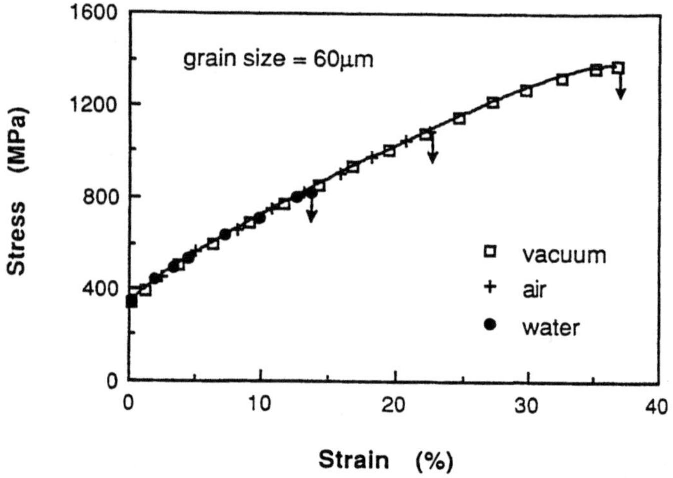

Figure 12-4. Stress-strain behavior of $(Co_{78}Fe_{22})_3V$ at 25°C (Reprinted with permission from *Materials Science and Engineerring*, Volume A152, C. Nishimura and C. T. Liu, pages 146–151, Elsevier Science Ltd., Oxford, England).

marked effect on crack-growth rate in the ordered condition, see Figure 12-5. These results are consistent with those observed in tension under static loading in demonstrating the heightened susceptibility to hydrogen embrittlement when the lattice is ordered. The influence of hydrogen was severe at all cyclic stress-intensity levels studied, and the fracture surface revealed a primarily intergranular path, compared to mostly faceted, stage I growth in argon. Embrittlement under cyclic loading also has been studied in Fe_3Al alloys (Castagna and Stoloff 1992a, 1992b); see section 2.2.2.

Co_3Ti ALLOYS. Both polycrystals and single crystals of Co_3Ti alloys are embrittled by testing in air or in hydrogen gas. Typical tensile data for $Co_{74}Ni_3Ti_{23}$ single crystals are shown in Figure 12-6. Fracture in all Co_3Ti binary and ternary single crystals was by cleavage along {001}, demonstrating that even in Ll_2 alloys grain boundaries are not necessary for embrittlement to occur (Y. Liu et al, 1989).

2.2. Iron Aluminides

The two iron aluminides, FeAl and Fe_3Al, form b.c.c.-ordered crystal structures, B2 and DO_3, respectively. Despite the difference in crystal structure, these aluminides show somewhat similar embrittlement in hydrogen-containing atmospheres at room temperature. Both aluminides exhibit only a few percent ductility (1% to 5%) when tested at ambient

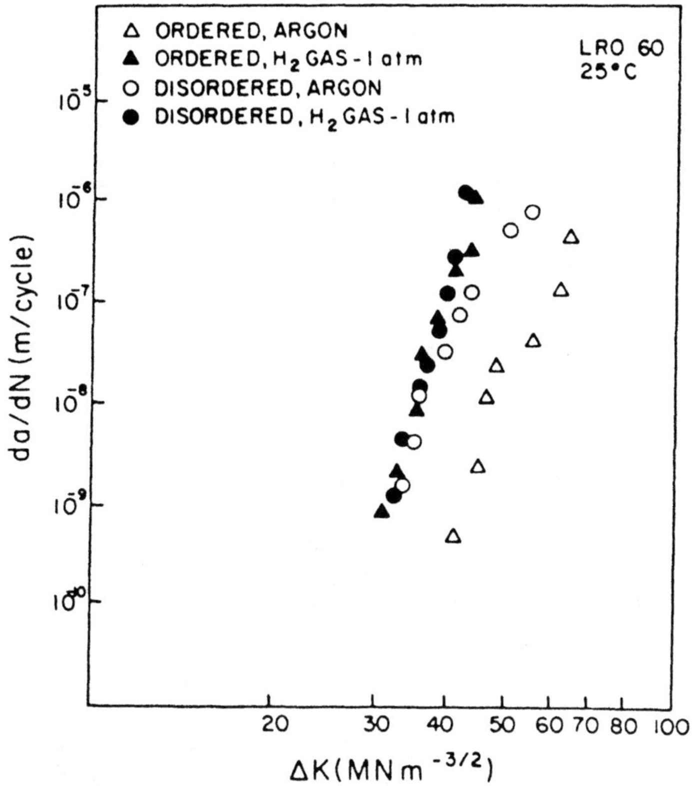

Figure 12-5. Influence of hydrogen gas on crack growth in $(FeNi)_3V$ (Kuruvilla and Stoloff 1985a).

temperature in air. This led many researchers to conclude that these alloys are inherently brittle. However, it has been demonstrated conclusively that when water vapor and hydrogen are eliminated from the external environment, both alloys exhibit considerable ductility.

2.2.1. FeAl. For an FeAl (36.5 at%Al) alloy, tensile elongations up to 18% have been achieved in "dry" environments such as oxygen, see Figure 12-7 (C. T. Liu et al. 1989). The increase in ductility from 2% to 18% was accompanied by a change in fracture mode from transgranular cleavage in air to mainly grain-boundary separation in dry oxygen. These observations suggest that cleavage planes are more susceptible to embrittlement than are grain boundaries. The maximum degree of moisture-induced embrittlement occurs on either side of room temperature. At higher temperatures, in situ protective oxide films can form readily on specimen surfaces, while at low temperatures the aluminum-moisture

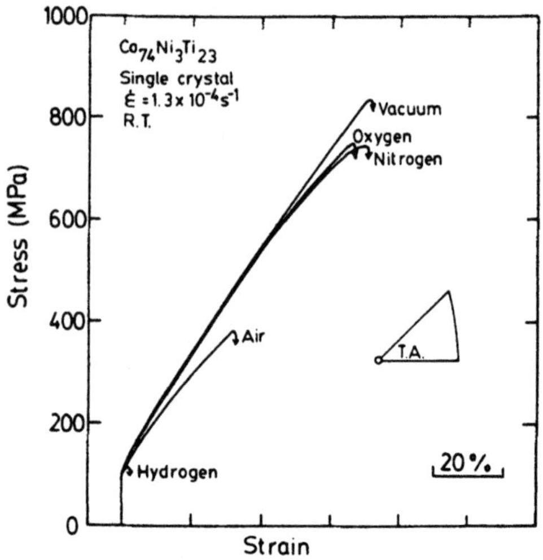

Figure 12-6. Effect of environment on tensile stress-strain behavior of $Co_{74}Ni_3Ti_{23}$ single crystals (Reprinted with permission from *Acta Metallurgica*, Volume 37, Y. Liu, T. Takasugi, O. Izuma, and T. Yamada, pages 507–518, 1989, Elsevier Science Ltd., Pergamon Imprint, Oxford, England).

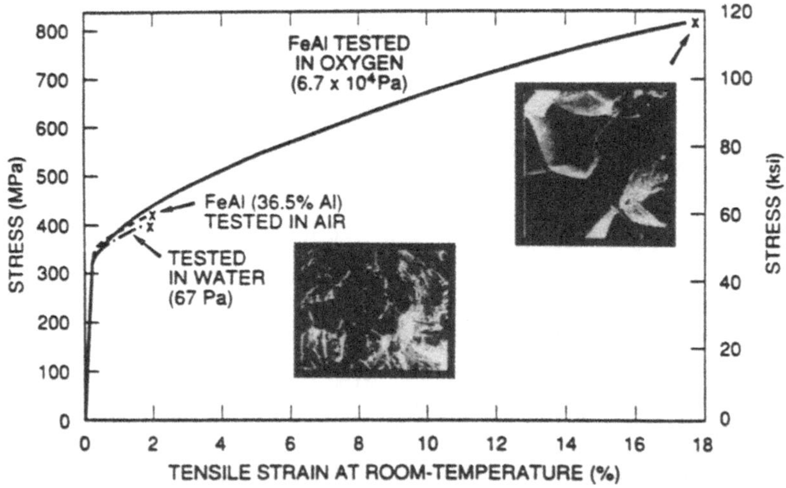

Figure 12-7. Influence of environment on room-temperature ductility of Fe-36at%Al (Reprinted with permission from *Scripta Metallurgica*, Volume 23, C. T. Liu, E. H. Lee, and C. G. McKamey, pages 875–880, 1989, Elsevier Science Ltd., Oxford, England).

reaction is slowed and the equilibrium moisture content in air also is lowered.

The environmental sensitivity of FeAl is markedly reduced when the aluminum content is higher than 38 at%. For Fe-40%Al ductility is about 4% in air, in 4×10^{-7} torr vacuum, or after hydrogen charging (Shea et al. 1991). Gaydosh and Nathal (1990) reported that the ductility of Fe-40%Al is sensitive to microstructure (annealed and furnace-cooled material displayed 9% elongation in vacuum compared to 5% for an as-extruded sample). However, B increased ductility of as-extruded Fe-40%Al to 9% in vacuum. Fe-50%Al, on the other hand, was brittle in vacuum for both the as-extruded and furnace-cooled conditions. For Fe-43 at%Al, the ductility is nil in air as well as in dry oxygen; all specimens fail intergranularly. This difference in behavior with aluminum content suggests that grain boundaries in FeAl alloys with Al > 38% are intrinsically brittle. Therefore, environmental embrittlement and intrinsic effects must be distinguished in order to establish strategies for reducing brittleness. It has been demonstrated that the intrinsic grain-boundary brittleness in FeAl and other intermetallics can be alleviated by microalloying with boron (Gaydosh and Nathal 1990; Liu and George 1990).

2.2.2. Fe_3Al. Fe_3Al alloys form in the range of about 18–32a%Al. When near-stoichiometric alloys are quenched from above the critical ordering temperature, *Tc*, they display a partially ordered B2 structure. Slow cooling through *Tc* or extended annealing just below *Tc* results in a highly ordered DO_3 structure at room temperature. There is considerable

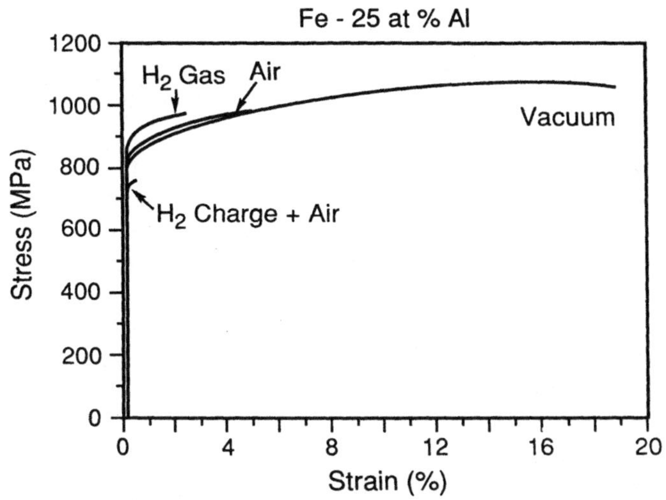

Figure 12-8. Stress-strain behavior of Fe_3Al in various environments (Shea et al. 1991).

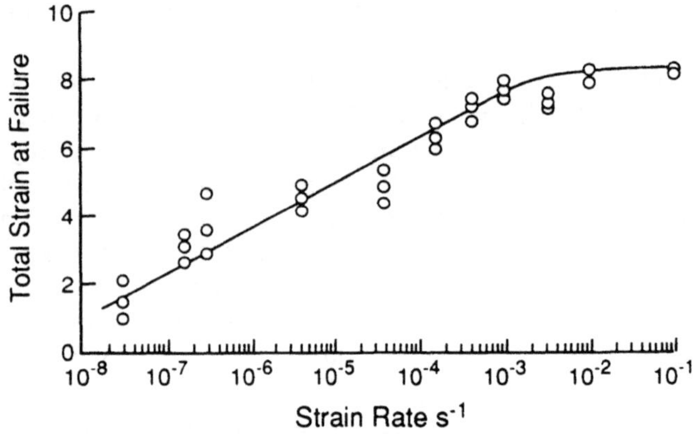

Figure 12-9. Influence of strain rate on ductility of Fe-24at%Al tested in air (Scott et al. 1992).

evidence that environmental embrittlement, resulting from moisture in air, is severe in binary Fe_3Al alloys. For example, tensile elongation in air is 4%, but 19% in vacuum, while less than 1% strain is noted after precharging with hydrogen; see Figure 12-8 (Shea, Castagna, and Stoloff 1991). Ductility of Fe-28%Al alloys in air is significantly increased by the addition of chromium (McKamey and Liu 1990; Scott, Duquette and Stoloff 1992). While the mechanism of improvement in ductility is unknown, it may result from modification of the Al_2O_3 coating that naturally forms on Fe_3Al in such a way that hydrogen liberation from water vapor is reduced. It is unlikely that Cr_2O_3 replaces Al_2O_3 as the protective oxide at these Cr levels. Unfortunately, chromium additions do not suppress embrittlement by hydrogen or by hydrogen introduced by electrolytic charging (Shea et al. 1991; Scott et al. 1992).

Reducing strain rate sharply reduces the ductility of Fe-24at%Al tested in air at room temperature, see Figure 12-9 (Scott et al. 1992). Such behavior further supports the concept of environmental embrittlement, since lower strain rates provide more time for release of hydrogen from the water vapor in air, and for penetration into the alloy. The influence of strain rate on ductility of Fe-Al-Cr alloys in air has been confirmed by Shau and Lin (1992). Further, Shea et al. (1991), have shown the same trend of increasing embrittlement with lower strain rate in hydrogen-charged Fe-23.5at% Al.

Considerable work on cyclic loading of Fe_3Al alloys in various environments has been carried out. Typical results for an alloy containing 6a%Cr are shown in Figures 12-10a and b for the DO_3 and B2 conditions, respectively (Castagna and Stoloff 1992b). Note the poor crack-growth

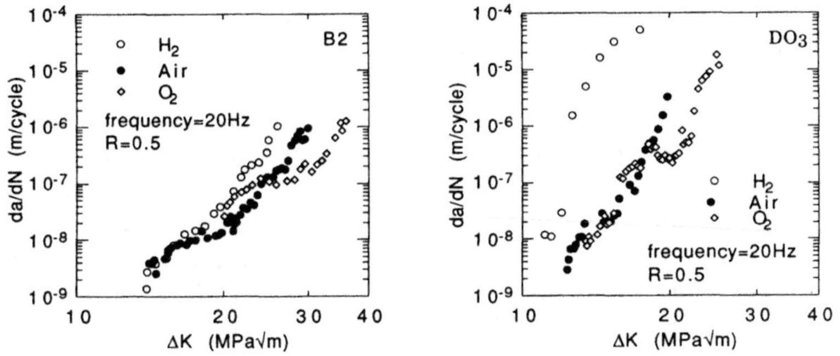

(a) (b)

Figure 12-10. Fatigue-crack growth of Fe_3Al alloy FA-129 at room temperature: a) B2 condition; b) DO_3 condition (Castagna and Stoloff 1994).

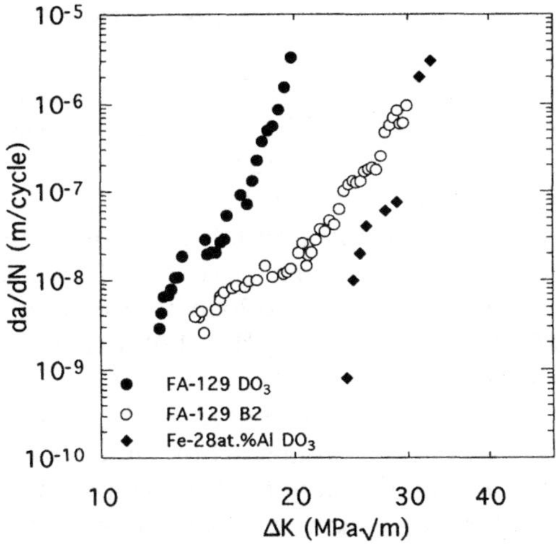

Figure 12-11. Comparison of fatigue-crack growth in DO_3 and B2 conditions for Fe_3Al alloy FA-129 with DO_3 Fe-28%Al (Castagna and Stoloff 1994).

resistance of the alloy in either ordered condition when tested in 1 atm. hydrogen gas, the intermediate behavior noted in air, and the beneficial effects of testing in vacuum (10^{-6} torr) or in dry oxygen. The environmental effect is greater in the DO_3 condition, as may be seen in Figure 12-10b. Note also that crack-growth rates in either air or oxygen are lower in the B2 condition than in the DO_3 condition, see Figure 12-11. Additional evidence for the detrimental influence of DO_3-type order on fracture behavior in hydrogen-containing environments is the lowered critical-stress intensity for fracture calculated from the critical crack length, see Table

5-4. The values of ΔK_c range from 35.9 MPa$\sqrt{m}$ for the B2 condition in oxygen to 17.4 MPa$\sqrt{m}$ for the DO_3 condition tested in 1 atm. hydrogen gas. (These values are for comparison purposes only, as ASTM E-399 conditions for a valid fracture toughness test were not met; such values would probably be considerably lower than the ΔK_c values recorded in Table 5-4).

A fundamental question with respect to environmental embrittlement of Fe_3Al alloys relates to the role of composition and, to lesser extent, the type and degree of long-range order. Early work on Fe-Al alloys showed that alloys with less than 8.5w%Al (16.3at%Al) exhibited substantial ductility at room temperature (Justusson, Zackay, and Morgan 1957; Marcinkowski, Taylor, and Kayser 1975). Recently, Vyos, Viswanathan, and Sikka (1992) have shown that binary Fe-16.3at%Al, which is a single-phase disordered alloy, is not susceptible to environmental embrittlement at 25°C. Microstructural changes seemed to have no effect on ductility, which was near 25% for the 16.3at%Al alloy in the fully recrystallized condition. In contrast, an alloy containing 22at%Al (displaying a two-phase ordered + disordered microstructure) showed tensile elongation of about 5% in air, with slightly higher elongations in vacuum.

With respect to chromium content, beneficial effects of chromium on environmental embrittlement in tension of Fe-28%Al alloys have been documented (McKamey and Liu 1990). Highest ductilities are produced in a partially recrystallized, quenched condition which produces partial B2-type order. In contrast, under cyclic loading a marked susceptibility of a Fe-28at%Al-4.8Cr-0.47Nb-0.2C alloy (FA-129) to environmental embrittlement has been noted, as shown in Figures 12-10 and 12-11 (Castagna and Stoloff 1992b). The effect is exacerbated in the slow-cooled DO_3 condition, but does not seem to be influenced by the microstructure, since both fully recrystallized and partly recrystallized samples showed similar crack-growth rates. The differing behavior between monotonic and cyclic loading may be due to the repeated rupturing of oxide films that occurs under cyclic loading. Therefore, any beneficial effect that chromium confers on the protectiveness of Al_2O_3 under monotonic conditions is lost under cyclic loading.

Alloy FA-129 also has been shown to be susceptible to intergranular stress-corrosion cracking in neutral pH chloride-containing solutions (Bavarian, Harutouni, and Zamanzadeh 1992). However, little effect of cathodic potentials on SCC of Fe_3Al was noted, in marked contrast to the high susceptibility of a Ti-Al-Nb-V-Mo alloy under similar charging conditions. These results conflict with the earlier findings of Shea et al. (1991) and Kasul and Heldt (1991), who showed that binary Fe-24.6a%Al is very susceptible to hydrogen charging, as well as work that showed that the maximum stress intensity at fracture of FA-129 is very sensitive to atmosphere (Castagna and Stoloff 1992b). These differences may be due in part to pH effects, since Ricker et al. (1991) have reported that hydrogen-induced SCC of Ni_3Al occurs only in low pH solutions. A more detailed description of SCC appears in Chapter 15.

2.3. Titanium Aluminides

2.3.1. ***Ti_3Al.*** Considerable effort has been devoted to determining the hydrogen susceptibility of γ TiAl (Ll_0 structure) and α_2 Ti_3Al alloys (DO_{19} structure). Ti_3Al takes up hydrogen much more readily than TiAl, but not as rapidly as titanium, see Figure 12-12 (Ronald 1988). Hydride formation occurs in both binary Ti_3Al and Ti-Al-Nb or Ti-Al-Nb-Mo alloys, as well as in binary TiAl alloys containing small amounts of α_2.

The structure of hydrides found in Ti_3Al alloys depends upon alloy composition and hydrogen content. Thompson (1991) reported that a γ hydride forms as thin platelets in Ti-24Al-11Nb. Although this hydride is sometimes called TiH_2, there is sufficient solubility range that TiH may be a more appropriate description. For low hydrogen contents, the hydride is f.c.c., but at higher hydrogen contents the structure is f.c.t. Gao et al. (1991), on the other hand, have reported that the hydrides that form in Ti-24Al-11Nb at 500°C under 1 atm. pressure of H_2 are b.c.t., TiH_2-type, with $c/a = 1.32$. This apparent discrepancy disappears when it is taken into account that the b.c.t. structure is equivalent to a f.c.t. structure with

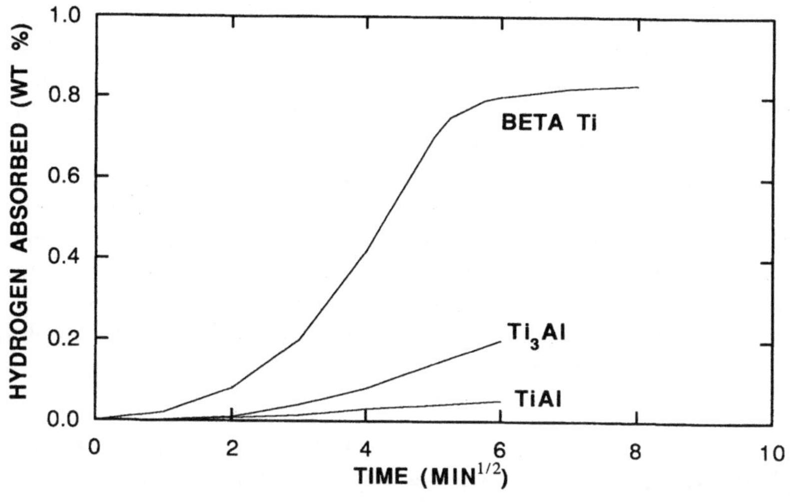

Figure 12-12. Solubility of hydrogen in various titanium alloys at 600°C (Ronald, 1988).

c/a = 0.94. The hydrides have the following orientation relationship with the α_2 matrix:

$$(0001)\,\alpha_2//(110)\mathrm{TiH_2} \text{ and } \{1120]\,\alpha_2//[001]\mathrm{TiH_2}$$

Transmission microscopy reveals that hydrides lie on the prism planes in Ti-24Al-11Nb and on the basal plane in stoichiometric Ti_3Al (Shih, Scarr, and Wasielewski 1989; Gao et al. 1991). It appears that hydride formation is reversible in these two alloys, since outgassing at 700°C removed the hydrides. However, when present the hydrides crack and lead to much lower ductility in tension. Similarly, both fracture stress and fracture toughness are sharply reduced with increasing hydrogen content, as shown in Figure 12-13 (Chu and Thompson 1992). Delayed failure occurs in Ti-24-Al-11Nb in hydrogen gas (13.8 MPa pressure). Super α_2 (Ti-25Al-10Nb-3V-1Mo) displays significant decreases in ductility at both 25°C and 204°C (Sullivan 1988). Hydride-induced cracking in super α_2 occurred primarily in the α_2 phase and at the α_2/β interface, while in the same alloy tested in helium, fracture occurs in the β phase (Jewett and Matejczyk 1988). Zero elongation is observed in super α_2 for hydrogen contents in excess of 320 wt.ppm H_2.

Recently, it has been shown that microstructural changes produced by heat treatment can have a significant effect on the susceptibility of Ti-24Al-11Nb to hydrogen embrittlement under elevated-temperature sustained-load creep conditions in hydrogen gas and by slow strain-rate tensile tests on previously exposed creep specimens. A fine basketweave structure displayed no change in total elongation with increasing hydrogen content at levels to 1300 wppm, while two other microstructures showed lower elongation with increasing hydrogen content. This hydrogen-resistant fine basketweave microstructure consisted of fine α_2 platelets and discontinuous β phase. Lower hydrogen intake rates were observed with this microstructure than with the coarse basketweave structure or the equiaxed structure. The discontinuous β was assumed to be responsible for improved hydrogen tolerance because of the resulting limit on the size of hydride particles. However, it must be noted that continuous β is required for higher ductility and toughness in air. Therefore, a proper balance of properties in air and in hydrogen needs to be established through further, detailed microstructural observations and mechanical tests.

2.3.2. TiAl. It has been reported that hydrogen does not react with single-phase TiAl to form hydrides (Hjelm 1988). However, Ti-48Al and various ternary and quaternary TiAl alloys contain small amounts of α_2,

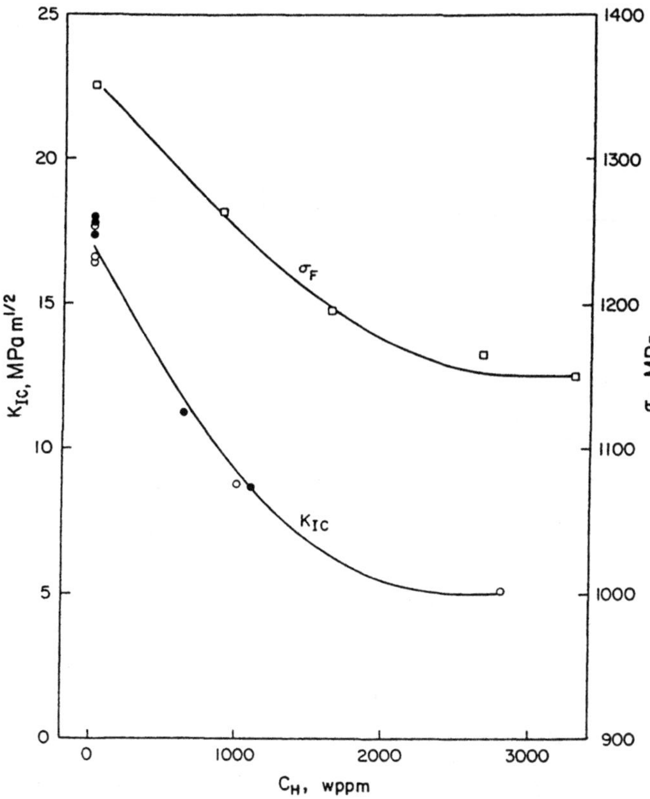

Figure 12-13. Effect of hydrogen on mechanical properties of Ti-24Al-11Nb (Chu and Thompson 1992).

and, therefore, may be embrittled by hydrides. For example, Kerr (1988) reported that a TiAl alloy containing less than 5% α_2 particles exhibits reduced ductility at 815°C in 13.8 MPa hydrogen gas. Other work also has indicated that two-phase TiAl alloys are susceptible to embrittlement (Sastry, Soboyejo, and Lederick 1990). Thompson (1991) has shown that the level of hydrogen solubility in Ti-50at%Al (single-phase γ) is lower than for Ti-24Al-11Nb (α_2) over a range of temperatures. Hydrides with the structure of f.c.t. TiH have been found in Ti-50Al for hydrogen contents below about 1000 wppm. One or two additional hydride structures may form at higher hydrogen concentrations. Although no fracture data were reported, Thompson (1991) noted that at hydrogen levels above 2000 ppm, cracking can occur on cooling of charged alloys. Other work (Kane and Chakachery 1991) has shown that Ti-48Al-2.5Nb-0.3Ta, which is brittle even in air at 23°C, displays substantially lower tensile strength and elongation in high-pressure (13.8 MPa) hydrogen gas than in air at 650°C. Further reductions in tensile properties of γ TiAl were noted at 23°C after prolonged exposure to hydrogen gas at 815°C. Interestingly, these workers found no evidence of increased hydrogen content after exposure to hydrogen gas at 650°C, but hydrogen content in excess of 2000 wppm was noted after exposure at 815°C to high-pressure hydrogen. It appeared from this work that hydrides might be responsible for embrittlement after exposure to high-pressure hydrogen gas at 815°C, but that at 650°C the reduction in tensile properties resulted from accelerated crack growth due to hydrogen on the surface of the alloy. No embrittlement resulted from exposure to low-pressure hydrogen, either at 650°C or 23°C.

As in the case of Ti_3Al, the design of hydrogen-tolerant microstructures in TiAl has been a subject of considerable interest. Christodoulou and Clarke (1990) have reported that Ti-45Al-3V, reinforced with 7.5vol%TiB_2 by the XD process, when tested in tension at room temperature, is insensitive to exposure to 1 atm. gaseous H_2 at 815°C for periods up to 82 hours, independent of prior microstructural state. This is in marked contrast to the behavior of Ti-48Al-2.5Nb-0.3Ta when exposed to H_2 at 815°C, as described above. No hydrides were observed in the XD material, which may account for the difference in susceptibility of the two alloys to hydrogen exposure. Finally, Fritzmeyer and Jacinto (1990) reported that the γ alloy Ti-33w%Al-5%Nb-1%Ta was not embrittled by testing in 13.8 MPa hydrogen gas at temperatures from -130°C to 204°C. It should be noted that the amount of hydrogen that can enter titanium aluminides from the gas phase is very dependent upon surface condition. Surface oxides inhibit hydrogen entry and formation of hydrides, so that careful analyses of experimental conditions is necessary when comparing data from different sources.

Environmental embrittlement of TiAl (50-50) in air and in hydrogen gas has been reported recently, see Figure 12-14 (Nakamura 1991). Note that bend ductility is considerably higher in vacuum than either in air or in hydrogen gas at 500 mm Hg pressure. Liu and Kim (1992) have shown that a two-phase γ TiAl alloy containing Nb, Cr, Mn and Si displays only 0.2% plastic strain to fracture in air (55% humidity) versus 1.2% strain in pure oxygen. No change in fracture mode was observed. Yamaguchi (1991) also has reported that single crystals of polysynthetically twinned (PST) TiAl are embrittled by moisture at room temperature. The degree of embrittlement of these crystals was strongly dependent upon crystal orientation. It is particularly important that studies of environmental embrittlement in brittle alloys such as TiAl be carried out in tension and that oxygen rather

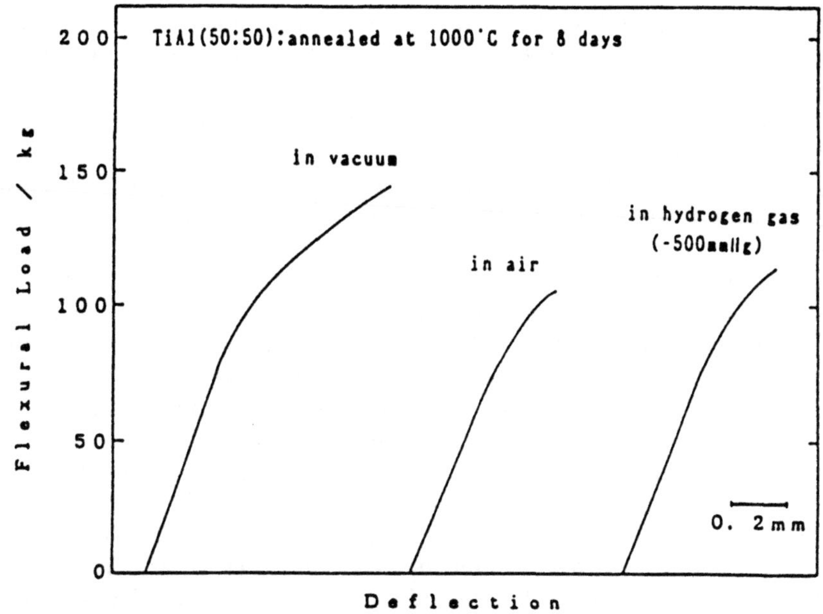

Figure 12-14. Environmental embrittlement of Ti-50at%Al (Nakamura 1991).

than vacuum be used as the reference environment in order to maximize the effect (Liu and Kim 1992).

The environmental embrittlement of TiAl is somewhat surprising, since neither hydride formation nor embrittlement was detected in TiAl exposed to high-pressure hydrogen (13.8 MPa) at temperatures to 204°C by Fritzmeyer and Jacinto (1990).

2.4. Other Intermetallics

There have been few reports of embrittlement of ordered intermetallics by hydrogen or moisture, apart from those systems described above. The influence of hydrogen charging on the tensile ductility of Ni_2Cr was found to be a maximum for disordered and nearly ordered Ni_2Cr (Berkowitz and Miller 1980). Recent fatigue-crack growth studies have shown that nearly ordered Ni_2Cr displays the lowest crack-growth rate and highest threshold in air and in 3.5%NaCl, while disordered Ni_2Cr had the highest growth rates and lowest threshold (Pao et al. 1992). Better fatigue-crack growth resistance in partially ordered and nearly ordered Ni_2Cr was attributed to the occurrence of planar slip and the resultant cyclic-slip reversibility. However, cracks grow much more rapidly and thresholds are significantly lower in 3.5%NaCl than in air for all conditions of order, (S = 0, 0.5 or 0.9).

Limited embrittlement studies have been carried out on the B2 alloy FeCo-2%V (Kuruvilla and Stoloff 1985b). This alloy can be disordered by quenching from above the critical temperature, *Tc*, of 720°C. Embrittlement occurs in the fully ordered (S = 1), partially disordered (S = 0.4) and disordered (S = 0) conditions. Brittle transgranular cleavage is observed in ordered material in both air and in hydrogen. Dimpled fracture is observed in partially ordered and disordered samples tested in air, while cleavage is again observed in hydrogen. These results support the conclusion that transgranular cleavage can be induced by hydrogen in B2 polycrystals (FeAl and FeCo-2V) as well as in Ll_2 single crystals (Co_3Ti). The influence of hydrogen on cleavage energy of these cubic alloys needs to be established by both theoretical and experimental methods.

Theoretical calculations already have been provided for the influence of hydrogen on cleavage energy of FeAl (Fu 1991). This work suggested that hydrogen embrittlement in FeAl is associated with a decrease in directional d-bonding charge on Fe sites, coupled with a large increase in charge on the H sites. These calculations implied that hydrogen embrittlement will be more severe in hypostoichiometric (Fe-rich) FeAl.

2.5. Embrittling Mechanisms

As mentioned in the foregoing sections, many ordered intermetallics show environmental embrittlement in hydrogen (charged or uncharged) environments or in moist air at ambient temperatures. It has been demonstrated that some intermetallics containing reactive elements, such as FeAl and Fe_3Al, exhibit more severe embrittlement in moist air than in dry hydrogen. The proposed chemical reaction for the moisture-induced embrittlement involves (Liu, Lee and McKamey 1989):

$$xM + yH_2O \rightarrow M_xO_y + 2yH \qquad (12.1)$$

where *M* is a reactive element in intermetallics. It is the high-fugacity atomic hydrogen that rapidly penetrates into crack tips and causes severe embrittlement. For the B2 structure the {200} planes offer the maximum amount of aluminum to react with water vapor. Fracture in FeAl is by transgranular cleavage in moisture or hydrogen, mixed mode in vacuum, and intergranular in dry oxygen. Thus, the underlying mechanism of moisture-induced embrittlement in FeAl, Fe_3Al, Ni_3Al, Co_3V and other intermetallics is similar to embrittlement by hydrogen with the principal difference being the manner in which atomic hydrogen is generated and absorbed at crack tips. The yield strength of the intermetallics is found to be insensitive to test environment as is typical with conventional alloys. Strain-rate effects are very significant (Scott et al. 1992), due to the time dependence of hydrogen diffusion. The highest ductility is generally obtained in dry-oxygen environments because oxygen suppresses the reaction in equation 12.1 and allows more rapid formation of oxides than in the presence of moisture (Liu 1991).

Hydrogen embrittlement is known to be a very complex phenomenon even in conventional metals and alloys. The underlying mechanisms suggested for hydrogen embrittlement in ordered intermetallics resemble those for structural alloys, and can be grouped into four categories:

1. Reduction of atomic bonding across cleavage planes
2. Reduction of cohesive strength across grain boundaries
3. Reduction of dislocation mobility and crack tip plasticity
4. Formation of brittle hydrides.

Environmental and hydrogen embrittlement in the b.c.c.-ordered iron aluminides and in FeCo-V occurs along cleavage planes rather than grain boundaries, suggesting the reduction of cleavage strength by absorbed or adsorbed hydrogen. Experimental results are supported by recent first-principles quantum-mechanical calculations, which indicate that absorbed hydrogen significantly reduces the cleavage strength and energy of FeAl (by as much as 20–70%, depending on the hydrogen concentration; Fu 1991). Superdislocations have been suggested to be the carriers for enhanced diffusion of hydrogen at crack tips (Kuruvilla and Stoloff 1985a). In the case of f.c.c.-ordered intermetallics, hydrogen embrittlement usually takes place along grain boundaries, suggesting that reduction of cohesive strength along the boundaries is responsible. However, single crystals of Co_3Ti and ternary Co_3Ti alloys also are embrittled (Takasugi and Izumi 1986), so that grain boundaries clearly are not necessary to observe reduced ductility. Bond, Robertson, and Birnbaum (1989) have shown by in-situ TEM that hydrogen enhances dislocation mobility and crack growth in Ni_3Al, but these authors also suggested that decohesion occurs in Ni_3Al.

Limited studies by TEM indicate that hydrogen accumulated at crack tips may introduce a large number of stacking faults in Co_3Ti that hinder the emission and motion of dislocations and thereby decrease the associated plastic work at crack tips (Liu et al. 1989). Hydrides have been detected only in titanium aluminides and are probably responsible for lowering tensile ductilities in hydrogen environments. However, embrittlement in the absence of hydrides cannot be excluded.

3. EMBRITTLEMENT AT ELEVATED TEMPERATURES

Environmental degradation in ordered intermetallics occurs not only at ambient temperatures but also at elevated temperatures, see Table 12-4 (Takeyama and Liu 1992). The embrittling mechanism and agent are, nevertheless, quite different in these two cases. Hydrogen is the major embrittling agent and oxygen is beneficial at room temperature, whereas oxygen is the major embrittling agent at elevated temperatures (typically above 300°C). At present, only a few intermetallic systems have been studied for environmental degradation at elevated temperatures, and data are available mainly for Ni_3Al and Ni_3Si alloys.

Table 12-4 Elevated Temperature Embrittlement of Ordered Alloys

Alloy	Crystal Structure	Environmental Embrittlement	Reference
Ni_3Al + Hf + B	$L1_2$	o	Liu and Sikka, 1986
Ni_3Al + Cr	$L1_2$		Liu and Sikka, 1986
$(Ni,Co)_3Al$	$L1_2$	o	Taub et al. 1986
Ni_3Si	$L1_2$	o	Oliver, 1989
Ni_3Si + Cr	$L1_2$		Oliver, 1989
$Ni_3(Si,Ti)$	$L1_2$	o*	Takasugi et al. 1991
$(Fe_{22}Co_{78})_3V$	$L1_2$	o	Miura and Liu, 1992
$Ni_3(Si,Ti)$ + B	$L1_2$	o	Takasugi et al. 1991
Co_3Ti	$L1_2$	x	Takasugi and Izumi, 1986
FeAl	B2	x	Baker and Gaydosh, 1987
Fe_3Al	DO_3	x	Baker and Gaydosh, 1987
Ti_3Al	DO_{19}	o	Balsone, 1988

o Environmental embrittlement is observed when tested in oxidizing environments.

*No difference in elevated-temperature tensile ductility between air and vacuum, but environmental embrittlement is possibly masked because of a poor vacuum. Observed but reduced by alloying with Cr.

x Not observed.

Source: (Takeyama and Liu, 1992).

3.1. $L1_2$ Alloys

Ni_3Al ALLOYS. Tensile properties of Ni_3Al are sensitive to test temperature and environment. Figure 12-15 compares the tensile elongation of a Ni_3Al alloy (Ni-21.5Al-0.5Hf-0.1%) (IC-145), (Liu and Sikka 1986) tested in air and vacuum (10^{-3} Pa) as a function of test temperature. The alloy tested in air showed distinctly lower ductility than that tested in vacuum at temperatures above 300°C, and the severest embrittlement occurred around 750°C, despite the fact that Ni_3Al alloys exhibited good oxidation resistance in air. The loss in ductility generally is accompanied by a change in fracture mode from ductile transgranular to brittle intergranular. Similar embrittlement has been observed in other Ni_3Al alloys, such as B-doped Ni_3Al containing Fe or Hf (Liu 1987) and B-doped $(Ni,Co)_3Al$ alloys. In these cases, oxygen has been identified as the embrittling agent. The similarity in the profile of the two curves in Figure 12-15 further indicates that embrittlement cannot be completely suppressed by a conventional vacuum of 10^{-3} Pa. The role of air pressure in embrittlement of Ni_3Al alloys has been shown clearly for IC-136 (Ni-23at%Al-0.5%Hf-0.07%B) tested at 760°C. There is a rapid increase in elongation as air pressure is reduced from 1 to 10^{-3} torr, and continued increase in ductility to a pressure of 10^{-7} torr (Liu 1987).

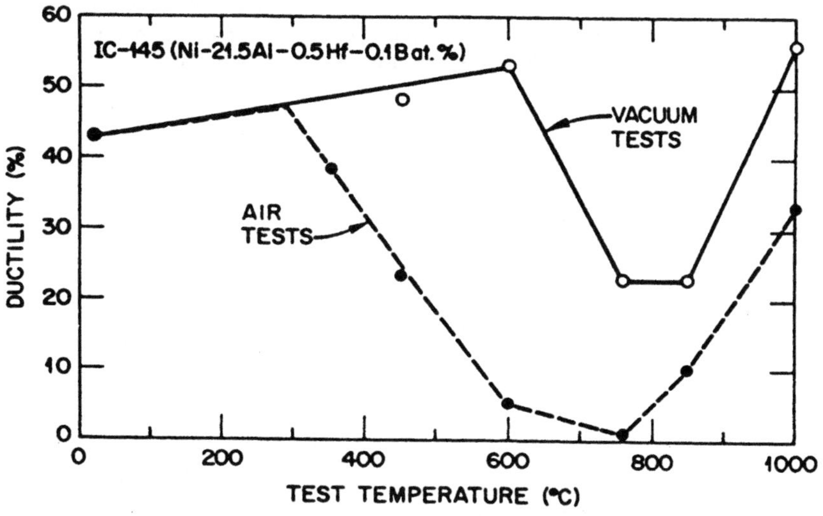

Figure 12-15. Influence of air vs. vacuum on ductility of IC-145 (Ni-21.5at%Al-0.5Hf-0.1B [Liu and Sikka, 1986; reprinted with permission from *JOM* (formerly *Journal of Metals*) Volume 38, No. 5, pages 19–21, a publication of The Materials, Metals and Minerals Society, Warrendale, PA 15086].

The fracture toughness of advanced Ni_3Al alloys such as wrought IC-218 LZr and cast IC-396 also has been shown to decrease sharply at 650°C in air, (Alexander 1991). Low toughness is accompanied by a shift in fracture mode from transgranular at room temperature to intergranular at 650°C, as in the case of tensile experiments, suggesting that embrittlement is caused by oxygen at the crack tip.

Test environments also affect the fatigue life of boron-doped Ni_3Al (24 at%Al) at elevated temperatures (Stoloff et al. 1987). The alloy showed a sharp drop in fatigue life at temperatures above 500°C when tested in a conventional vacuum (10^{-3} Pa). The drop in the fatigue life was accompanied by a change in fracture mode from transgranular to intergranular.

Ni_3Si ALLOYS. Like Ni_3Al, Ni_3Si alloys also exhibit severe environmental embrittlement in oxidizing environments at elevated temperatures. For Ni_3Si and $Ni_3(Si,Ti)$ alloys doped with and without B, tensile ductility decreases sharply at temperatures above 300°C in moist air. The ductility of low Si alloys (e.g., 19 at%) reaches a minimum at 600°C, as seen in Figure 12-16 (Oliver 1989), and above that temperature the ductility increases sharply. Tests in vacuum result in an increase in ductility at 600°C by a factor as high as twenty. Embrittlement in air is effectively reduced by 2-6at%Cr (Oliver 1989). For high (Si + Ti) alloys (e.g., 21%) ductility decreases continuously with increasing temperature and approaches zero above 600°C. The alloys showed no improvement in ductility at temperatures above 600°C in a conventional vacuum.

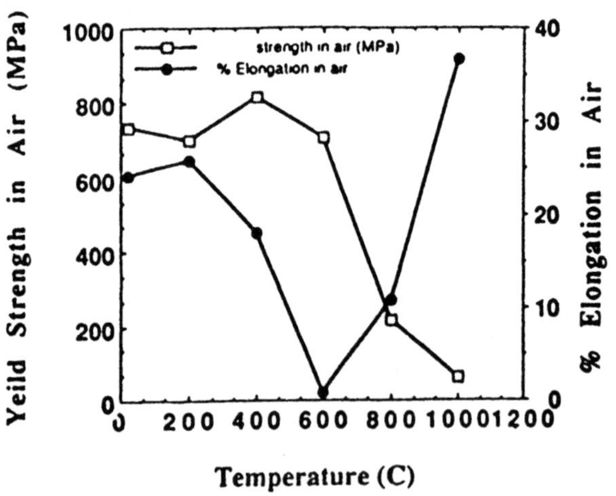

Figure 12-16. Yield strength and tensile elongation of Ni-18.9at%Si in air vs. temperature (Oliver 1989).

3.2. Other Intermetallics

$(CoFe)_3V$ alloys, previously described as susceptible to embrittlement by water or moist air (Nishimura and Liu 1991), also display a sharp loss in ductility at temperatures above 500°C; minimum ductility in both air and vacuum ($\sim 5 \times 10^{-4}$ Pa) is noted near Tc ($\simeq$ 910°C) (Miura and Liu 1992). Ductility is lower in air than in vacuum between 500°C and 910°C. Embrittlement below Tc was attributed to oxygen-induced penetration of grain boundaries. A sharp increase in ductility and environmental insensitivity above Tc was attributed to the lack of long-range order.

The titanium aluminide Ti_3Al, is prone to oxygen-induced embrittlement at elevated temperatures (Perkins 1991). Elevated-temperature exposure of a Nb-modified α_2 alloy for 24 hours in various atmospheres at 900°C showed some striking differences in room temperature ductility; see Figure 12-17 (Meier and Pettit 1992). Exposure to vacuum for 24 hours resulted in 12% elongation, while annealing in various oxidizing atmo-

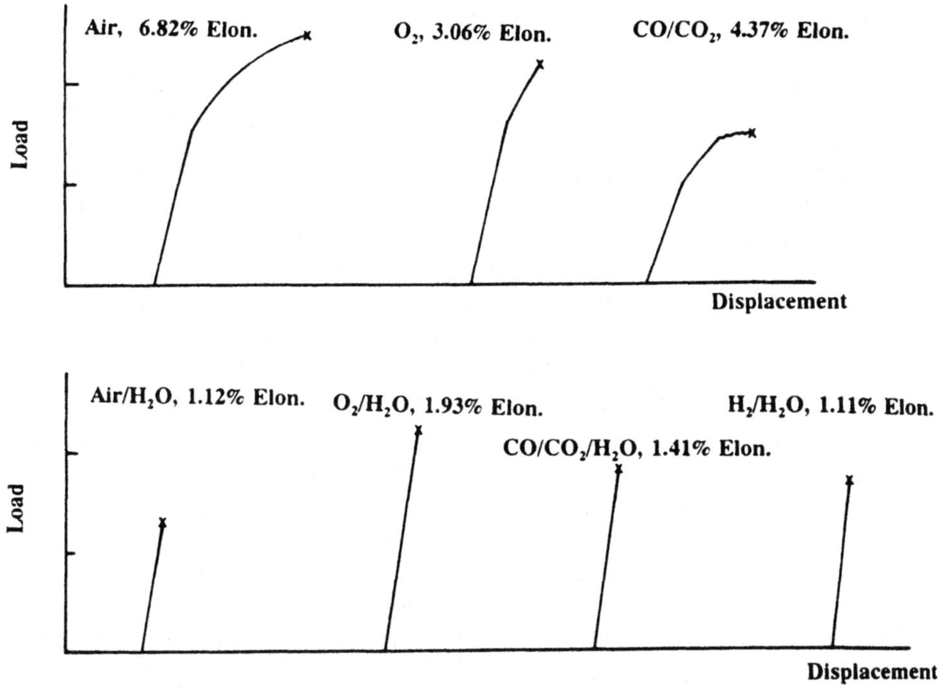

Figure 12-17. Bend load vs. displacement for Nb-modified Ti_3Al after exposure to various environments for 24 hours at 900°C (Reprinted with permission from *Materials Science and Engineering*, Volume A153, G. H. Meier and F. S. Pettit, pages 548–560, 1992, Elsevier Science Ltd., Oxford, England).

spheres produced ductilities ranging from 3–6.8%. These results indicate that oxygen diffusion into α_2 results in severe embrittlement. When water vapor (30–40vol%) was present in the atmosphere at 900°C, ductilities were in the 1–2% range. Therefore, embrittlement of Ti-Al-Nb alloys seems to be exacerbated by the release of hydrogen from water vapor, as in the case of FeAl, Fe_3Al, and Ni_3Al. Iron aluminide alloys based on Fe_3Al or FeAl, on the other hand, exhibit no indication of elevated-temperature embrittlement in oxidizing environments, even though they show severe ambient-temperature embrittlement in moist air. The reason for the absence of elevated-temperature embrittlement in these aluminides is not well understood, possibly related to the lack of a substantial yield anomaly, together with rapid formation of protective oxide films due to rapid diffusion in these b.c.c. materials. Further studies are required to clarify these points.

3.3. Pest Reaction

The "pest" reaction occurs in many intermetallics during oxidation at relatively low temperatures (400–600°C). Alloys affected include many high-melting silicides, berylides, and aluminides. In extreme cases, samples will disintegrate into powder.

Detailed studies of the pest reaction have been carried out at 500°C on single- and polycrystalline $MoSi_2$ (Chou and Nieh 1992). Pested polycrystalline samples completely disintegrated into powdery products including MoO_3 whiskers, SiO_2 clusters, and $MoSi_2$ crystals coated with a Si-Mo-O oxide. Pesting was much more localized in single crystals, although similar reaction products were found. Substantial volume expansion caused by the formation of oxides, particularly MoO_3, is considered to be the most likely cause of disintegration of polycrystalline samples.

3.4. Embrittling Mechanisms

Embrittlement (other than via the pest reaction) has been suggested to be caused by a dynamic effect simultaneously involving high localized-stress concentration, elevated temperature, and gaseous oxygen (Takeyama and Liu 1992). Such a dynamic effect involves repeated weakening and cracking of grain boundaries as a result of oxygen absorption and penetration at crack tips. Based on a detailed study of crack growth in Ni_3Al alloys tested in oxidizing environments, a fracture mechanism of stress-assisted grain-boundary oxygen penetration has been suggested to explain the elevated temperature embrittlement. This model, shown schematically in Figure 12-18, (Hippsley and Devan 1989) consists of four sequential steps: (i) occurrence of surface cracks at the initial stage of deformation,

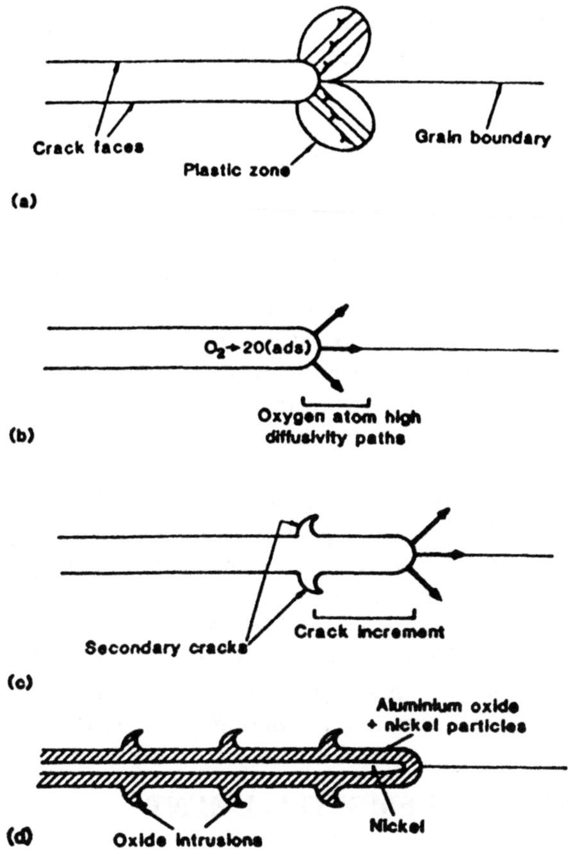

Figure 12-18. Schematic of proposed mechanism of dynamic embrittlement at elevated temperatures (Reprinted with permission from *Acta Metallurgica*, Volume 37, C. A. Hippsley and J. H. Devan, pages 1485–1496, 1989, Elsevier Science Ltd., Pergamon Imprint, Oxford, England).

(ii) chemisorption of gaseous oxygen to the crack tips where a high localized stress field is involved, (iii) oxygen penetration in its atomistic form to the stress field ahead of tips, and (iv) inward development of surface cracks preferentially along the grain boundaries, leaving some secondary cracks (Hippsley and Devan 1989). Steps (ii) and (iv) proceed continuously and repeatedly during deformation, leading to premature fracture and severe loss in ductility at elevated temperatures in oxidizing environments.

Liu and Oliver (1989) have shown that grain shape can drastically change the susceptibility of Ni_3Al to embrittlement by oxygen. This is discussed in section 4.2.

The fatigue-crack growth rate of Ni_3Al and the Ni_3Al-base alloy IC-221 increases with increasing temperature, between 25°C and 600°C, even when testing is carried out in moderate vacuum (Matuszyk et al. 1990). This behavior indicates that environmental embrittlement due to oxygen is occurring, especially since the flow stress increases with temperature over the same range. These observations were consistent with those of Hippsley and Devan (1989) for static crack growth. By contrast, when Fe_3Al is tested at elevated temperatures in air the fatigue crack-growth rate decreases, perhaps because moisture-induced embrittlement is maximized near room temperature, as in the case of structural steels.

Alternative mechanisms for high-temperature embrittlement also have been suggested by observations of low elevated-temperature ductility in Ni_3Al-base single crystals tested in air (Aoki and Izumi 1979; Jishan et al. 1989; Nazmy 1991). It is possible that the increase in flow stress to a maximum in the temperature regime of minimum ductility plays a significant role in elevated-temperature embrittlement. Further, Yizhang et al. (1989) suggest that in cast polycrystalline Ni_3Al alloys, Cr may enhance ductility through its effect on γ-γ' eutectic size and distribution and the replacement of brittle γ'-γ' grain boundaries by tough γ'-γ-γ' boundaries. Clearly, the role of Cr in improving ductility in air requires further study.

4. ALLEVIATION OF EMBRITTLEMENT

Environmental degradation has been identified as a main cause of the low ductility and brittle fracture in many ordered intermetallics. This problem has to be solved satisfactorily in order to use intermetallic alloys as engineering materials. Despite their different embrittling agents, ambient-temperature and elevated-temperature embrittlement can be treated together because both involve surface reactions and are sensitive to localized stress concentrations. Results generated to date indicate that embrittlement can be alleviated or reduced by, (a) control of surface conditions, (b) control of grain shape, and (c) alloy additions.

4.1. Surface Conditions

Control of surface conditions is a simple way to alleviate environment degradation involving surface reactions. In several cases, preoxidation and formation of protective oxide scales were proven to be beneficial in reducing environmental embrittlement at ambient and elevated temperatures. Preoxidation at 1000°C effectively reduced ambient-temperature embrittlement in B-doped Ni_3Al charged with hydrogen (Kuruvilla and Stoloff 1985b). Formation of protective oxide scales increases the tensile ductility of FeAl and Fe_3Al alloys in air at ambient temperatures and the ductility of boron-doped Ni_3Al alloys at elevated temperatures. Unfortunately, the oxide films crack after stretching a few percent, and their protective effect disappears. Surface coatings also should be useful in protecting underlying alloys from hydrogen or oxygen penetration along grain boundaries or bulk material; however this effect has not yet been well demonstrated.

Takeyama and Liu (1992) have discussed the complex interactions among grain size, preoxidation, and elevated-temperature ductility. The ductility of preoxidized samples decreases with increasing grain size at 600°C and 760°C in vacuum. However, the ductility of bare specimens is nearly unaffected by grain size. Penetration of oxygen along grain boundaries occurs during preoxidation of coarse-grained samples, but not for fine-grained samples. Therefore, only coarse-grained samples are embrittled. A nickel-rich, nonprotective oxide formed on coarse-grained samples, while a protective Al-rich oxide formed on fine-grained samples. These results suggest that the nature of the surface oxide, rather than the grain size, is the key factor influencing ductility of preoxidized specimens.

4.2. Grain Size and Shape

Columnar-grained structures have proven to be effective in reducing environmental embrittlement in Ni_3Al alloys tested in moist air at room- and elevated temperatures. For instance, formation of a columnar grain structure in boron-doped Ni_3Al produced by directional levitation-zone remelting increases the ductility in air from 0.2% to 33% at temperatures in the range 600–760°C, (Liu and Oliver 1989). The loss of ductility in air is accompanied by a change in fracture mode from microvoid coalescence to intergranular in equiaxed material. However, the columnar structure displayed mainly transgranular failure in both air and vacuum. The beneficial effect of the columnar-grained structure with grain boundaries oriented parallel to the stress axis is attributed to minimizing the normal stress across grain boundaries, thereby suppressing nucleation and propagation of cracks along boundaries even when those boundaries are weakened by oxygen penetration. Hirano (1991) has reported high ductility in stoichiometric Ni_3Al tested both parallel to and perpendicular to the growth direction. A detailed analysis of grain-boundary chemistry and misorientation is required to fully understand these results. Further studies should be extended to other intermetallic systems.

4.3. Other Processing Techniques

Boron-doped Ni_3Al alloys prepared by rapid solidification are more prone to environmental embrittlement than alloys produced by conventional casting and thermomechanical processing (Taub Chang, and Liu 1986). The latter exhibit much higher ductility at 760°C in vacuum than does the rapidly solidified material, although at 600°C there is little effect of microstructure. The same workers reported that spray-formed Ni-10at%Co-24%Al with boron-exhibited zero ductility in vacuum at 760°C, but ductility could be substantially improved by subsequent thermomechanical processing. A number of factors could be influencing the results obtained with differing processing techniques, including:

1. grain size and shape
2. grain boundary energy
3. oxygen and other impurities
4. mobile dislocation density
5. thermal history
6. texture

Of these, the only factor definitely linked with low ductility in many alloys is impurity content.

The role of grain size seems to be the same whether alloys fracture intergranularly (e.g., Ni_3Al with H_2 or O_2) or transgranularly (e.g., Fe_3Al and FeAl). As in the case of structural steels, refining grain size tends to reduce susceptibility to embrittlement.

4.4. Alloy Additions

Considerable effect has been devoted to alleviating environmental embrittlement by alloying additions in intermetallic alloys. The most prominent case involves Co_3Ti containing 23 at%Ti (see Figure 12-6, Liu et al. 1989). The elements yttrium, tantalum, chromium, molybdenum, tungsten, and germanium at a level of 3 at% have no beneficial effects on moisture-induced embrittlement in air at room temperature, whereas 3% of iron or aluminum completely eliminate the moisture-induced embrittlement in air. It has been suggested that iron may create a more homogeneous electron distribution at grain boundaries; however, there is no direct evidence to support this hypothesis.

Boron has been found to be effective in reducing moisture-induced embrittlement of grain boundaries in certain Ll_2 intermetallics. As shown in Table 12-3 (Liu and Oliver 1991), B-free Ni_3Si and $Ni_3(Si,Ti)$ are prone to environmental embrittlement, whereas B-doped $Ni_3(Si,Ti)$ is insensitive to test environment at room temperature (Takasugi et al. 1991). This result clearly indicates that boron is very effective in alleviating ambient-temperature embrittlement in $Ni_3(Si,Ti)$. Carbon-doped $Ni_3(Si,Ti)$ also exhibited high ductility, independent of test environment. Boron and carbon are known to segregate strongly to grain boundaries in $Ni_3(Si,Ti)$ and their beneficial effect has been suggested to arise from slower hydrogen diffusion through reduction in site occupation by hydrogen at the boundaries. Boron, on the other hand, does not alleviate embrittlement in Co_3Ti, possibly because of no strong segregation of boron to Co_3Ti grain boundaries. Note that boron eliminates environmental embrittlement in moist air, but it does not prevent embrittlement induced by hydrogen charging in Ni_3Al alloys.

Fe_3Al and FeAl exhibit severe embrittlement when tested in moist air at ambient temperatures. Recent efforts on alloy design showed that the ductility of Fe_3Al in air can be substantially improved by increasing aluminum concentration from 25% to 28% and by addition of chromium at a level of 2% to 6% (McKamey and Liu 1990). The increase in aluminum concentration sharply lowers the yield strength of the aluminide. The chromium additions double the tensile ductility when there are oxide scales formed on specimen surfaces during hot rolling or subsequent heat treatment in air. Refinement of grain structure and control of degree of recrystallization have been proven to be effective in improving the ductility of Fe_3Al and FeAl alloys tested at room temperature in moist air. Also, avoidance of water lubrication during machining of Fe-40at%Al and Fe-40at%Al + 500 wppm B has been shown to prevent surface cracking during fabrication (Ferguson and Liu 1992).

A systems study of alloying additions on the tensile ductility of Ni_3Al showed that no elements, except chromium, significantly alleviate elevated-temperature embrittlement in air. Alloying Ni_3Al + B with 8at%Cr is effective in alleviating embrittlement at elevated temperatures, as shown in Figure 12-19 (Liu and Sikka 1986). The chromium addition increases the minimum ductility of polycrystalline Ni_3Al and Ni_3Al + Fe alloys at 700°C from 6% to above 20%. Chromium also is effective in reducing elevated-temperature embrittlement in Ni_3Si alloys (Oliver 1989). Cyclic loading tests indicate that chromium additions improve the fatigue resistance of Ni_3Al alloys tested in air at elevated temperatures (Matuszyk et al. 1990). The beneficial effect of chromium has been suggested to arise from rapid formation of protective chromium oxide films that reduce the rate of penetration of oxygen into Ni_3Al grain boundaries (Liu and Sikka 1986). However, Ni_3Al alloys containing chromium display higher ductility in vacuum than in air, indicating that chromium does not entirely elimi-

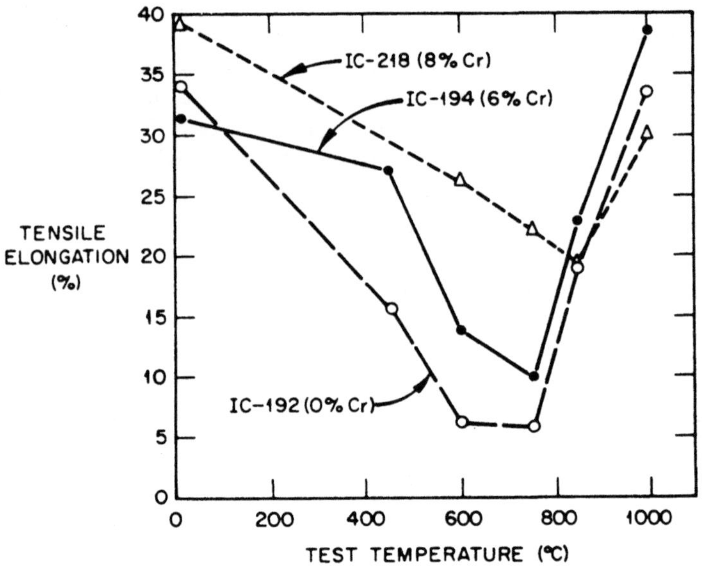

Figure 12-19. Effect of Cr on elevated temperature ductility of Ni_3Al [Liu and Sikka 1986; reprinted with permission from *JOM* (formerly *Journal of Metals*) Volume 38, No. 5, pages 19–21, a publication of the Minerals, Metals and materials Society, Warrendale, PA 15086].

nate the environmental effect. Another problem with the hypothesis that oxygen is responsible for grain-boundary embrittlement is the observation that even single crystals of binary Ni_3Al (Aoki and Izumi 1979) and alloyed Ni_3Al (Nazmy 1991) display sharply reduced ductility at elevated temperatures. Therefore, an inherent loss of ductility in the temperature region of rapid rise in flow stress cannot be ruled out as a major factor in the low ductility of polycrystalline Ni_3Al alloys.

5. SUMMARY

Environmentally induced embrittlement is both pernicious and complex. The low ductility and toughness of many ordered alloys, especially aluminides, at room temperature is now known to be an extrinsic effect. Exclusion of moisture or other sources of hydrogen from the environment causes Ni_3Al, Ni_3Si, Fe_3Al, and FeAl to display high elongations. However, even when hydrogen is excluded, other aluminides (e.g., NiAl, CoAl, TiAl, and Ti_3Al) exhibit little or no room-temperature ductility. The reasons for this difference in behavior are complex; the number of independent slip systems, bonding, and differences in environmental susceptibility are the most obvious factors affecting ductility. Elevated-temperature embrittlement seems to be confined to fewer alloys, notably $L1_2$ alloys such as Ni_3Al and Ni_3Si, as well as Ti_3Al. The nature of, and rates of formation of surface oxides, especially as the latter compare to diffusivity of oxygen along grain boundaries seem to govern the occurrence and severity of embrittlement. Fortunately, alloying of the $L1_2$ alloys Ni_3Al and Ni_3Si with Cr sharply reduces the severity of oxygen-induced embrittlement. However, in the case of Ti_3Al, development of an effective coating seems to offer the best chance of minimizing such embrittlement.

Grain size and shape, alloy texture, and impurity content are among the factors that influence environmental embrittlement. Considerably more work is needed to establish the importance of each of these factors for most intermetallics of interest for possible structural applications.

REFERENCES

ALEXANDER, D. J., 1991, Technology Transfer Conference and Workshop on Nickel and Iron Aluminides, Oak Ridge, TN.

AOKI, K., and IZUMI, O., 1979, *J. Mat. Sci.* 14:1800–1806.

BAKER, I., and GAYDOSH, D. J., 1987, In *High Temperature Ordered Intermetallic Alloys II*, MRS Symp. Proc. 81:315–320.

BALSONE, S. J., 1988, In *Oxidation of High Temperature Intermetallics*, eds. T. Grobstein and J. Doychak, p. 219, The Minerals, Metals and Materials Society, Warrendale, PA.

BAVARIAN, B., HARUTOUNI, S., and ZAMANZADEH, M., 1992, *Mat. Sci. Eng. A* A153:613–618.

BERKOWITZ, B. J. and MILLER, C., 1980, *Metall. Trans.* 11:1877–1881.

BOND, G. M., ROBERTSON, I. M., and BIRNBAUM, H. K., 1989, *Acta Metall.* 37:1407–1413.

CAMUS, G. M., STOLOFF, N. S., and DUQUETTE, D. J., 1989, *Acta Metall.* 37:1497–1501.

CASTAGNA, A., and STOLOFF, N. S., 1992, *Scripta Metall.* 27:673–678.

CASTAGNA, A., and STOLOFF, N. S., 1994, In *Proc. Ninth Annual Conference on Fossil Energy Materials* May 12–14, 1994, Oak Ridge, Tennessee, in press.

CHAN, KWAI, S., 1992, *Metall. Trans. A* 23A:497–507.

CHOU, T. C., and NIEH, T. G., 1992, *Scripta Metall.* 27:19–24.

CHRISTODOULOU, L., and CLARKE, J. A., 1990, In *Hydrogen Effects on Material Behavior*, eds. N. R. Moody and A. W. Thompson, pp. 515–521, Warrendale, PA: The Minerals, Metals and Materials Society.

CHU, W.-Y., and THOMPSON, A. W., 1992, *Metall. Trans. A* 23A:1299–1312.

CHU, W.-Y., THOMPSON, A. W., and WILLIAMS, J. C., 1990, In *Hydrogen Effects on Material Behavior*, eds. N. R. Moody and A. W. Thompson, pp. 543–554, Warrendale, PA: The Minerals, Metals and Materials Society.

FERGUSON, P. A., and LIU, C. T., 1992, *Scripta Metall.*, to be published.

FRITZMEYER, L. G., and JACINTO, M. A., 1990, In *Hydrogen Effects on Material Behavior*, eds. N. R. Moody, and A. W. Thompson, pp. 533–542, Warrendale, PA: The Minerals, Metals and Materials Society.

FU, C. L., 1991, *Intermetallic Compounds—Structure and Mechanical Properties* 6, ed. O. Izumi, pp. 387–396, Sendai, Japan.

GAO, M., BOODEY, J. BART, and WEI, ROBERT P., 1991, In *Environmental Effects on Advanced Materials*, ed. R. H. Jones and R. E. Ricker, pp. 47–55, Warrendale, PA: The Minerals, Metals and Materials Society.

GAYDOSH, D. J., and NATHAL, M. V., 1990, *Scripta Metall.* 24:1281–1284.

GEORGE, E. P., LIU, C. T., and POPE, D. P., 1992, *Scripta Metall.* 27:365–370.

HIPPSLEY, C. A., and DEVAN, J. H., 1989, *Acta Metall.* 37:1485–1496.

HIRANO, T., 1991, *Scripta Metall.* 25:1747–1752.

HJELM, L., 1988, In *Proc. Second Workshop on Hydrogen-Materials Interactions*, ed. H. G. Nelson, pp. 18–27, National Aerospace Plane (NASP) Workshop Publ. 1004.

IZUMI, O., and TAKASUGI, T., 1988, *J. Mat. Res.* 3:426–440.

JEWETT, R., and MATEJCZYK, D., 1988, In *Proc. Second Workshop on Hydrogen-Materials Interactions*, ed. H. G. Nelson, pp. 137–139, National Aerospace Plane (NASP) Workshop Publ. 1004.

JISHAN, Z., ZHIYA, Z., YAJUN, T., JINGHUA, Z., YANG, Y., YINGAO, L., DONGSHENG, D., and ZHUANGQI, H., 1989, In *High Temperature Ordered Intermetallic Alloys III*, MRS Symp. Proc. 133:549–54.

JUSTUSSON, W., ZACKAY, V. F., and MORGAN, E. R., 1957, Trans. ASM 49:905–921.

KANE, R. H., and CHAKACHERY, E. A., 1991, In *Environmental Effects on Advanced Materials*, eds. R. H. Jones, and R. E. Ricker, pp. 35–46, Warrendale, PA: The Minerals, Metals and Materials Society.

KASUL, D. B., and HELDT, L. A., 1991, In *Environmental Effects on Advanced Materials*, eds. R. H. Jones and R. E. Ricker, pp. 67–75, Warrendale, PA: The Minerals, Metals and Materials Society.

KERR, W., 1988, Presented at *Second Workshop on Hydrogen-Materials Interactions*, Scottsdale, AZ, June 1–3, 1988.

KURUVILLA, A. K., and STOLOFF, N. S., 1985a, *Metall. Trans. A* 16A:815–820.

________, 1985b, *Scripta Metall.* 19:83–87.

KURUVILLA, A. K., and STOLOFF, N. S., 1985c in *High Temperature Ordered Intermetallic Alloys*, MRS Symp. Proc. 39:229–238.

LIU, C. T., 1987, In *High Temperature Ordered Intermetallic Alloys II*, MRS Symp. Proc. 81:355–367.

________, 1992, *Scripta Metall.* 27:25–28.

LIU, C. T., and GEORGE, E. P., 1990, *Scripta Metall.* 24:1285–1290.

LIU, C. T., and KIM, Y. W., 1992, *Scripta Metall.* 27:599–603.

LIU, C. T., LEE, E. H., and MCKAMEY, C. G., 1989, *Scripta Metall.* 23:875–880.

LIU, C. T., MCKAMEY, C. G., and LEE, E. H., 1990, *Scripta Metall. et Mater.* 24:385–390.

LIU, C. T., and OLIVER, W. C., 1989, *J. Mat. Res.* 4, 294–299.

Liu, C. T., and Oliver, W. C., 1991, *Scripta Metall. et Mater.* 25:1933–1937.

Liu, C. T., and Sikka, V. K., 1986, *J. Metals* 38(5):19–21.

Liu, C. T., and Takeyama, M., 1990, *Scripta Metall.* 24:1583–1586.

Liu, C. T., and White, C. L., 1987, *Acta Metall.* 35:643–649.

Liu, Y., Takasugi, T., Izumi, O., and Yamada, T., 1989, *Acta Metall.* 37:507–518.

Marcinkowski, M. J., Taylor, M. E., and Kayser, F. X., 1975, *J. Mat. Sci.* 10:406–414.

Masahashi, N., Takasugi, T., and Izumi, O., 1988, *Metall. Trans. A* 19A:353–358.

Matuszyk, W., Camus, G., Duquette, D. J., and Stoloff, N. S., 1990, *Metall. Trans. A* 21A:2967–2976.

McKamey, C. G., and Liu, C. T., 1990, *Scripta Metall.* 24:219–222.

Meier, G. H., and Pettit, F. S., 1992, *Mat. Sci. Eng. A* 153:548–560.

Miura, S., and Liu, C. T., 1992, *Scripta Metall.* 26:1753–1758.

Nakamura, M., 1991, National Research Institute for Metals, Tokyo, Japan, unpublished work.

Nazmy, M., 1991, ABB Power Generation, Ltd., Baden, Switzerland, unpublished research.

Nishimura, C., and Liu, C. T., 1991, *Scripta Metall.* 25:791–794.

________, 1992, *Mat. Sci. Eng.*, A152:146–151.

Oliver, W. C., 1989, In *High Temperature Ordered Intermetallic Alloys III*, MRS Symp. Proc. 133:397–402.

Pao, P. S., Gill, S. J., Feng, C. R., and Michel, D. J., 1992, *Mat. Sci. Eng.* A153:532–537.

Perkins, R., 1991, Reported by Meier, G. H. and Pettit, F. S. 1992, *Mat. Sci. Eng.*, A153:548–560.

Ricker, R. E., 1991, Technology Transfer Conf. and Workshop on Nickel and Iron Aluminides, Oak Ridge, TN.

Ricker, R. E., Bertocci, U., Fink, J. L., and Stoudt, M. R., 1991, In *Environmental Effects on Advanced Materials*, 213–225, Warrendale, PA: The Minerals, Metals and Materials Society.

Ronald, T., 1988, In *Proc. Second Workshop on Hydrogen-Materials Interactions* ed. H. G. Nelson, pp. 137–139. National Aerospace Plane (NASP) Workshop Publ. 1004.

Sastry, S. M., Soboyejo, W. O., and Lederick, R. J., 1990, In *Proc. Third Workshop on Hydrogen-Materials Interactions*, ed. H. G. Nelson, pp. 191–200, National Aerospace Plane (NASP) Workshop Publ. 1007.

Scott, J., Duquette, D. J., and Stoloff, N. S., 1992, Rensselaer Polytechnic Institute, unpublished research.

Shau, A., and Lin, D., 1992, *Scripta Metall.* 27:95–99.

Shea, M., Castagna, A., and Stoloff, N. S., 1991, In High Temperature Ordered Intermetallics IV, MRS Symp. Proc. 213:609–16.

Shih, D. S., Scarr, G. K., and Wasielewski, G. E., 1989, *Scripta Metall.* 23:973–978.

Stoloff, N. S., 1990, In *Hydrogen Effects on Materials Behavior*, eds. N. R. Moody, and A. W. Thompson, pp. 483–97, TMS-AIME, Pittsburgh, PA.

STOLOFF, N. S., FUCHS, G. E., KURUVILLA, A. K., and CHOE, S. J., 1987, In *High Temperature Ordered Intermetallics II*, MRS Symp. Proc. 81:247–261.
SULLIVAN, S., 1988, In *Proc. Second Workshop on Hydrogen Materials Interactions*, ed. H. G. Nelson, pp. 142–148, National Aerospace Plane (NASP) Workshop Publ. 1004.
TAKASUGI, T., 1991, *Acta Metall.* 39:2157–2167.
TAKASUGI, T., and IZUMI, O., 1986, *Acta Metall.* 34:607–618.
TAKASUGI, T., MASAHASHI, N., and IZUMI, O., 1986, *Scripta Metall.* 20:1317–1321.
TAKASUGI, T., SUENAGA, H., and IZUMI, O., 1991, *J. Mat. Sci.* 26:1179–1186.
________, 1991, *J. Mat.* in press.
TAKEYAMA, M., and LIU, C. T., 1992, *Mat. Sci. Eng.* A153:538–547.
TAUB, A. C., CHANG, K. M., and LIU, C. T., 1986, *Scripta Metall.* 20:1613–1618.
THOMPSON, A. W., 1991, In *Environmental Effects on Advanced Materials*, eds. R. H. Jones, and R. E. Ricker, pp. 21–33, Warrendale, PA: The Minerals, Metals, and Materials Society.
VYOS, S., VISWANATHAN, S., and SIKKA, V. K., 1992, *Scripta Metall.* 27:185–190.
WAN, X. J., ZHU, J. H., and JING, K. L., 1992, *Scripta Metall.* 26:473–477.
YAMAGUCHI, M., 1991, Presented at Workshop "Scientific Issues on Intermetallic Compounds-Status and Forecast for Fundamental Research," La Jolla, CA, September 11–14.
YIZHANG, Z., TIANXIANG, Z., YINGJIE, T., BINGDA, Z., YAOXIAO, Z., and ZHUANGQI, H., 1989, In *High Temperature Ordered Intermetallic Alloys III*, MRS Symp. Proc. 133:555–559.

Chapter 13

Aqueous Corrosion of Intermetallic Alloys

R. A. Buchanan, J. G. Kim, R. E. Ricker and L. A. Heldt

1. INTRODUCTION

New engineering alloys based on intermetallic compounds continue to undergo extensive development primarily for high-temperature applications. Accordingly, their strengths and oxidation/sulfidation resistances at elevated temperatures are of paramount importance. However, these alloys will not be continuously exposed to elevated-temperature conditions, and if any detrimental aqueous corrosion occurs near room temperature, their overall performances could be compromised. Consequently, the aqueous-corrosion properties are also of major concern. Moreover, the aqueous-corrosion properties of certain intermetallics in certain aqueous solutions at and near room temperature are sufficiently promising that principal applications may be warranted in these areas.

Currently, the major alloy systems based on intermetallic compounds are the nickel aluminides, iron aluminides, and titanium aluminides. Of these, the nickel and iron aluminides have received considerable attention relative to aqueous-corrosion properties, whereas the titanium aluminides have received little attention. Consequently, this chapter will concentrate on the aqueous-corrosion properties of the nickel and iron aluminides. Before doing so, however, a brief section will be devoted to reviewing the more common methods utilized in evaluating aqueous-corrosion behavior.

2. REVIEW OF EXPERIMENTAL CORROSION METHODS

Many of the more rapid aqueous-corrosion characterization techniques involve electrochemical methods. Two of the more common methods are polarization resistance and cyclic anodic polarization. Both methods start by immersing the corrosion specimen in the electrolyte of concern and monitoring the specimen potential, E, until a steady-state value, E_{corr}, is established and maintained. E_{corr} is known as the free-corrosion potential, the corrosion potential, the open-cell potential, or the open-circuit potential, and is specifically the potential at which the specimen exists under natural, freely corroding conditions with respect to a reference electrode with a known potential compared to the standard hydrogen electrode.

The polarization-resistance method allows determination of the average corrosion rate for the specimen at E_{corr}. Several steps are involved. With a potentiostatic or potentiodynamic system, and an electrochemical cell arrangement (specimen, reference electrode, and auxiliary electrode), the specimen potential is slowly stepped or scanned through E_{corr} while the external-circuit current, I_{ex}, is measured. A small potential range is employed (e.g., $E_{corr} \pm 10$ mV). The I_{ex} values are converted to external current density values, i_{ex}, by dividing by specimen area A, (i.e., $i_{ex} = I_{ex}/A$). the parameter of concern is the slope of the E vs. i_{ex} curve at E_{corr}, $(dE/di_{ex})_{E_{corr}}$, which in low-resistivity electrolytes is the polarization resistance, R_P, of the specimen. From R_P, the corrosion current density. i_{corr}, is determined at E_{corr} by application of the Stern–Geary equation:

$$i_{corr} = [\beta_a \beta_c/\{2.3(\beta_a + \beta_c)\}]/(dE/di_{ex})_{E_{corr}} = k/R_P \quad (13.1)$$

where the β values are anodic and cathodic Tafel constants and must be experimentally determined (or reasonable values assumed). The corrosion current density relates to the corrosion reaction $M \rightarrow M^{m+} + me$ (metal corroding to M^{m+} ions in solution with release of m electrons). The rate of material loss per unit area per unit time (corrosion intensity, CI) is calculated from i_{corr} by application of Faraday's law:

$$CI = (Mi_{corr})/(mF) \quad (13.2)$$

where M is the atomic weight and F the Faraday constant (charge of Avogadro's number of electrons). It is often more appropriate to express the corrosion rate as an average corrosion penetration rate (CPR), which follows when Equation 2 is divided by density, ρ:

$$CPR = (Mi_{corr})/(mF\rho) \tag{13.3}$$

The common unit for CPR is μm/day (or, more traditionally, mils (0.001 inch) per year (mpy), where 1 mpy = 0.0696 μm/day). Upon inserting the value for F, and recognizing that alloys (rather than pure elements) are often of concern, Equation 3 becomes:

$$CPR(\mu m/day) = (0.895\ M_{avg} i_{corr})/(m_{avg}\,\rho) \tag{13.4}$$

where M_{avg} is the atom-fraction-average atomic weight for the alloy (g/mol), m_{avg} the atom-fraction-average ion valence, ρ the density (kg/m^3) and i_{corr} the corrosion current density (mA/m^2).

The cyclic-anodic polarization method allows determination of corrosion behavior over a large range of anodic potentials, and also allows determination of susceptibility to localized corrosion. Descriptions of the method and the types of information it provides will be given relative to the schematic experimental curves in Figure 13.1. The test starts at the free-corrosion potential, E_{corr}. Then the potential of the specimen, E, is slowly increased while the external current density, i_{ex}, is measured. At potentials greater than E_{corr} by approximately 50 mV, the i_{ex} values are approximately equal to the corrosion current densities at these potentials, i'_{corr}. To determine the corrosion rates, the $i_{ex} \approx i'_{corr}$ values are substituted for i_{corr} in the Faraday law expressions. Figure 13-1a represents a material that only undergoes active corrosion, that is, i_{ex} (and, therefore, the corrosion rate) continuously increases with increasing E because a protective, passive film never forms at the material surface. Figure 13-1b shows a material that initially undergoes active corrosion, but then at a higher potential and at a critical anodic current density, i_{crit}, a passive film begins to form. Upon development of the passive film, i_{ex} is significantly reduced to the passive current density, i_p. With increasing potential, i_p remains low until a breakdown potential, E_B, is reached at which the passive film begins to undergo dissolution and i_{ex} begins to increase. During a cyclic test, the potential scan direction is reversed when i_{ex} reaches a relatively high predetermined value, often 5×10^4 mA/m^2(5×10^3 μA/cm^2). In Figure 13-1b, no hysteresis loop is generated on scan reversal, which generally means that the material is not susceptible to localized corrosion (pitting or crevice corrosion) in the test electrolyte. Figure 13-1c illustrates ideal behavior: passive corrosion at a low rate and no susceptibility to localized corrosion. In Figure 13-1d, the material exhibits passive corrosion, but then on breakdown and scan reversal,

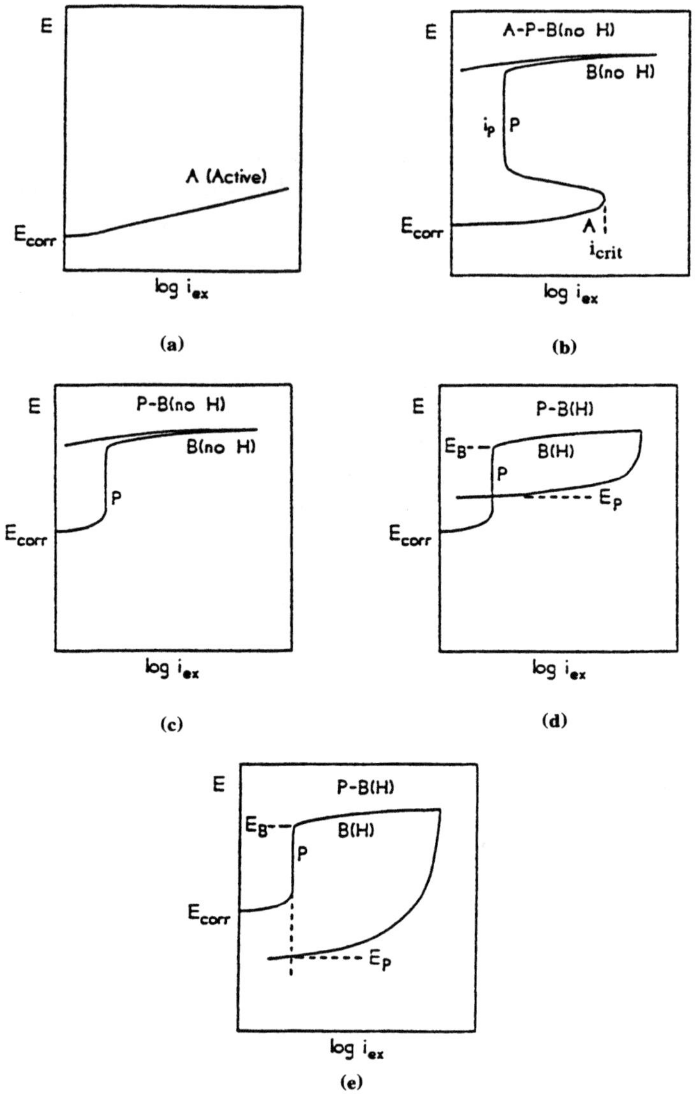

Figure 13-1. Schematic illustrations of experimental cyclic-anodic-polarization curves.

exhibits a hysteresis loop which indicates susceptibility to localized corrosion. During the reverse-scan path, the potential at which i_{ex} again reaches i_p is called the protection potential or the repassivation potential, E_P. Given that a material is susceptible to localized corrosion (Figures 13-1d and 13-1e), its relative resistance is proportional to $(E_P - E_{corr})$, (i.e., one would like to have an E_P value as far above E_{corr} as possible). In summary then, the cyclic anodic polarization behavior of a material in a given electrolyte can indicate the potential ranges of active corrosion and/or passive corrosion, whether or not the material is susceptible to localized corrosion, and if susceptible, the material's relative resistance to localized corrosion.

The more rapid electrochemical corrosion-evaluation methods are often supplemented with longer term mass-loss immersion tests. The immersion tests are employed to confirm electrochemical-based predictions and to reveal any time dependent processes not fully characterized by the short-time electrochemical tests, (e.g., times associated with initiation of localized corrosion). Specimens are weighed, exposed to an electrolyte for a set time, cleaned, and reweighed. After exposure, the surfaces are examined for the specific type of corrosion, uniform or localized. The amount of corrosion, based on mass loss, is often expressed as an average corrosion-penetration rate (CPR), given by:

$$\text{CPR}\ (\mu\text{m/day}) = (10^6 W)/(\rho A t) \qquad (13.5)$$

where W is the mass loss (kg), ρ the density (kg/m^3), A the specimen area (m^2), and t the exposure time (days). Care should be exercised in using the average CPR. If the corrosion is uniform, it is a physically meaningful parameter; if the corrosion is localized, for example, pitting corrosion, it is a less meaningful parameter since the corrosion penetration at the pit sites is much more than the calculated average CPR.

When evaluating corrosion-sensitive embrittlement phenomena, for example, stress-corrosion cracking and/or hydrogen embrittlement, the slow-strain-rate test (SSRT) is often employed. The SSRT is essentially a tensile test conducted at a very slow strain rate, (e.g., 10^{-6}/s), in an aqueous environment of concern. The reason for the slow strain rate is to allow ample time for operation of the corrosion-related component of the embrittlement mechanism (including slip-step dissolution as related to stress-corrosion cracking and/or cathodic-hydrogen absorption as related to hydrogen embrittlement). A measure of the embrittlement is the reduction in ductility and/or strength in the corrosive environment relative to that in an inert environment. SSRT's are often conducted at several electrochemical potentials, including the free-corrosion potential, E_{corr}, and applied anodic and/or cathodic potentials, E_{app}. Tests at E_{corr} will indicate the susceptibility to embrittlement in the corrosive environment under naturally corroding conditions; tests at anodic E_{app} values provide additional information on stress-corrosion-cracking mechanisms as related to passive-film stability, pitting corrosion, etc.; and tests at cathodic E_{app} values provide additional information on hydrogen-embrittlement mechanisms, since the amount of atomic hydrogen produced at the material

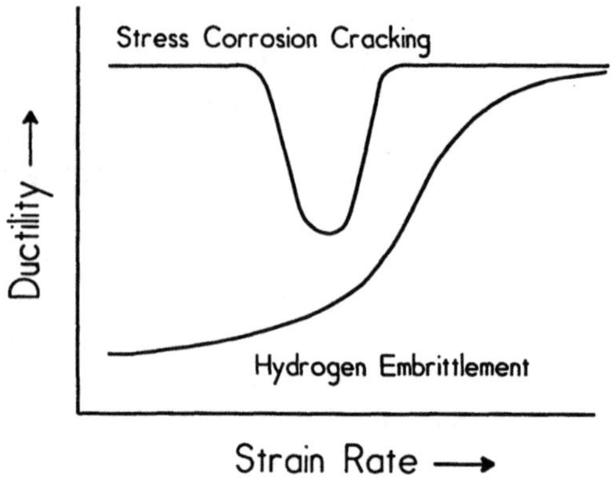

Figure 13-2. Schematic illustrations of typical strain-rate responses for stress-corrosion cracking and hydrogen embrittlement.

surface increases as the magnitude of the applied cathodic potential increases.

SSRT's are often carried out over a range of strain rates to provide additional information on corrosion-sensitive embrittlement mechanisms. The stress-corrosion-cracking and hydrogen-embrittlement mechanisms generally have different strain-rate responses, as indicated in Figure 13-2. Stress-corrosion cracking tends to occur over a limited range of strain rates, whereas hydrogen embrittlement tends to increase continuously with decreasing strain rate.

When hydrogen embrittlement is a possible corrosion-related embrittlement mechanism, an important consideration is the relationship between the potential (either E_{corr} or an applied potential E_{app}) and the equilibrium concentration of hydrogen at the material surface, with the hydrogen being produced by the reduction component of the corrosion process. This relationship is given by the Nernst equation for the $2H^+ + 2e = H_2$ and $2H_2O + 2e = 2OH^- + H_2$ equilibrium reactions at 25° C:

$$E'_H[\text{mV(SHE)}] = 59\,\text{pH} - 29.5\log P_{H_2} \quad (13.6)$$

$$E'_H[\text{mV(SCE)}] = -241 - 59\,\text{pH} - 29.5\log P_{H_2} \quad (13.7)$$

where E'_H[mV(SHE)] and E'_H[mV(SCE)] are the equilibrium potentials of both reactions in millivolts relative to the standard hydrogen reference electrode (SHE) and the saturated calomel reference electrode (SCE), respectively, and P_{H_2} is the equilibrium hydrogen fugacity, which is taken to be equal to the hydrogen partial pressure. The relationship is given in terms of two reference-electrode scales, SHE and SCE, because both are found commonly in the literature.

During a given corrosion process, assuming that equilibrium is established for the above hydrogen reactions, then at the E_{corr} value, $E_{corr} = E'_H$, and at an E_{app} value, $E_{app} = E'_H$. Therefore, Equations 6 and 7 become:

$$E_{corr} \text{ or } E_{app}[\text{mV(SHE)}] = -59\,\text{pH} - 29.5 \log P_{H_2} \tag{13.8}$$

$$E_{corr} \text{ or } E_{app}[\text{mV(SCE)}] = -241 - 59\,\text{pH} - 29.5 \log P_{H_2} \tag{13.9}$$

Thus, at a given pH, as the potential decreases, P_{H_2} increases, that is, more hydrogen is being produced by the cathodic reactions $2H^+ + 2e \rightarrow H_2$ and/or $2H_2O + 2e \rightarrow 2OH^- + H_2$. The hydrogen initially produced is atomic hydrogen adsorbed to the specimen surface, that is, $H^+ + e \rightarrow$ H(ads) or $H_2O + e \rightarrow OH^- +$ H(ads). The adsorbed atomic hydrogen is removed from the surface either by formation of hydrogen molecules according to the recombination reaction H(ads) + H(ads) $\rightarrow H_2$, by electrochemical desorption according to the reaction $H^+ +$ H(ads) + e $\rightarrow H_2$, or by being absorbed into the metal according to the reaction H(ads) $\rightarrow$ H(abs). The absorbed atomic hydrogen [H(abs)], or the adsorbed atomic hydrogen [H(ads)], are the species that can cause hydrogen embrittlement. Thus, at a given pH, for E_{corr} or E_{app} values corresponding to $P_{H_2} < 10^5$ Pa (1 atm.), the hydrogen produced partly remains on the surface at some coverage [H(ads)], partly dissolves into the solution (H_2) and partly diffuses into the material [H(abs)]. For more negative E_{corr} or E_{app} values corresponding to $P_{H_2} \geq 10^5$ (1 atm.), the greater amount of hydrogen produced partly remains on the surface at some coverage [H(ads)], partly forms vapor bubbles (H_2) and again partly diffuses into the material [H(abs)]. A significant question is at what potential will sufficient atomic hydrogen be produced to embrittle a material that is susceptible to hydrogen embrittlement. Generally, work to be described on the nickel and iron aluminides has shown that this will occur at potentials approximately equal to or less than that corresponding to $P_{H_2} = 10^5$ (1 atm.). This critical hydrogen potential will be designated as $E_{crit,H}$ and from Equations 6 and 7:

$$E_{crit,H}[\text{mV(SHE)}] = -59\,\text{pH} \tag{13.10}$$

$$E_{crit,H}[\text{mV(SCE)}] = -241 - 59\,\text{pH} \tag{13.11}$$

Thus, at a given pH, if E_{corr} or E_{app} is equal to or less than $E_{crit,H}$, significant quantities of atomic hydrogen are being produced at the specimen surface which may or may not produce hydrogen embrittlement, depending on the inherent resistance of the material to hydrogen embrittlement.

3. NICKEL ALUMINIDES

The aqueous-corrosion and corrosion-sensitive embrittlement properties of nickel aluminides based on the Ni_3Al stoichiometry have received considerable study over the past several years. Designations for these alloys by the Oak Ridge National Laboratory (IC-50, IC-218, and IC-221) and their respective chemical compositions are given in Table 13-1. It is noted that the alloys contain a small amount of boron (0.02 wt.% = 200 ppm) which is primarily responsible for the significantly enhanced ductility of these new "ductile" polycrystalline nickel aluminides (Liu and Stiegler 1984; Taub and Fleischer 1989).

3.1. Aqueous Corrosion Behavior

Corrosion-rate data based on 200-hour continuous-exposure mass-loss immersion tests at 25° C and 95° C are given in Table 13-2 for the nickel aluminides in a wide range of electrolytes (Buchanan and Kim 1988; Kim et al. 1989). In arriving at this data, the mass-loss values were experimentally determined, then Equation 13.5. was used to calculate the average corrosion-penetration rates (CPR's).

It is seen in Table 13-2 that at 25° C all three nickel-aluminide alloys exhibit quite low average CPR's (generally less than 0.07 μm/day or 1 mpy) in the oxalic and acetic acids, the sodium chloride solution, and the sodium and ammonium hydroxide basic solutions. The CPR's are moderately low in the lower concentration sulfuric and hydrochloric acids, and in the hydrofluoric and phosphoric acids. The average CPR's increase slightly in the higher concentration sulfuric acid and increase greatly in the higher

Table 13-1 Designations and Chemical Compositions of Nickel Aluminides

	ELEMENT: WEIGHT % (ATOMIC %)				
Alloy	Al	Cr	B	Zr	Ni
IC-50	11.3 (21.7)	—	0.02 (0.10)	0.6 (0.3)	bal
IC-218	8.5 (16.7)	7.8 (7.9)	0.02 (0.10)	0.8 (0.46)	bal
IC-221	8.5 (16.7)	7.8 (8.0)	0.02 (0.10)	1.7 (0.99)	bal

Table 13-2 Average corrosion penetration rates for nickel aluminides as calculated from 200 hour mass-loss data

	Concentration		Average Penetration Rate (μM / day) from Mass-Loss Data, 200 Hour Continuous Exposure					
			25° C			95° C		
Solution	mol/L	Weight Percent	IC-50	IC-218	IC-221	IC-50	IC-218	IC-218
Sulfuric Acid	1	9	0.14	0.14	0.20	> 34 (e)	21	> 60 (e)
	18	96	0.49	0.42	0.42	4.7	1.3	1.4
Hydrochloric Acid	1	4	0.35	0.20	0.20	35	> 37 (e)	77
	12	37	8.4	13	26	> 35 (e)	> 35 (e)	93 (e)
Nitric Acid	1	6	30	> 111 (a)	37	> 36 (e)	> 37 (e)	21
	8	40	> 850 (b)	0.07	0.20	> 33 (e)	3.1	3.9
	16	70	4.4	0.07	0.20	> 36 (e)	4.1	14
Hydrofluoric Acid	1	2	0.20	0.20	0.20	> 37 (e)	> 34 (e)	28
Phosphoric Acid	1	10	0.14	0.14	0.14	> 33 (e)	> 36 (e)	52
Oxalic Acid	1	10	0.07	0.07	0.07	(f)	0.20	0.28
Acetic Acid	1	6	0.20	0.014	0	0.28	0.28	0.28
Sodium Chloride	0.6	3.5	0.014	0.014	0.007	0.014	(f)	(f)
Sodium Hydroxide	10	30	0.028	0.014	0.014	(f)	(f)	(f)
Ammonium Hydroxide	10	18	0.42	0.020	0.014	(f)	(f)	(f)
Ferric Chloride	0.2	20	> 410 (c)	> 320 (d)	62	> 36 (e)	> 34 (e)	> 79 (e)

(a) Total dissolution in < 65 hours.
(b) Total dissolution in < 8 hours.
(c) Total dissolution in < 17 hours.
(d) Total dissolution in < 24 hours.
(e) Total dissolution in < 200 hours.
(f) Very small mass gain.

Source: Buchanan and Kim 1988; Kim et al. 1989.

concentration hydrochloric acid. The ferric-chloride solution is extremely damaging to the nickel aluminides, indicating that a highly oxidizing environment (as produced by ferric ions) in combination with chlorides should be avoided. In the nitric acid, an oxidizing acid, the CPR's are high at all concentrations for the IC-50 alloy. However, for the IC-218 and IC-221 alloys, which contain chromium (7.8 wt.%, the CPR's decrease significantly at the higher concentrations, indicating the development of passivation.

At the higher test temperature, 95° C, it is seen in Table 13-2 that the CPR's are generally increased. However, the average CPR's still remain quite low in the organic acids (oxalic and acetic), in the basic solutions (sodium and ammonium hydroxides), and in the sodium-chloride solution.

For the 0.6 mol/L sodium-chloride solution, even though the average CPR's are quite low at both 25° C and 95° C, evidence of pitting corrosion was reported. Thus, the *average* CPR values in Table 13-2 may have little practical significance.

Cyclic anodic-polarization studies also have been conducted on the IC-50, IC-218, and IC-221 nickel aluminides in electrolytes similar to those in Table 13-2 and at temperatures of 25° C and 95° C (Buchanan and Kim 1988; Kim et al. 14989). The results will be summarized relative to the schematic polarization curves of Figure 13-1. In 0.5 mol/L sulfuric acid at 25° C, active/passive behavior is demonstrated by all three alloys, as schematically shown in Figure 13-1b, with E_{corr} in the active range, indicating that a protective passive film does not form under freely corroding conditions. At 95° C, only active corrosion (no passive film) is produced at all potentials (Figure 13-1a). In 1 mol/L nitric acid at 25° C, all three alloys undergo active/passive corrosion (Figure 13-1b), but the critical anodic current density, i_{crit}, is considerably lower for the chromium-containing IC-218 and IC-221 alloys, indicating greater ease of passivation. With reference to Table 13-2 at 25° C, it is apparent that the IC-218 and IC-221 alloys passivate in nitric acid at the higher concentrations. In 1 mol/L nitric acid at 95° C, the polarization behavior is active only (Figure 13-1a) for the IC-50 alloy, indicating no chance of passivation, and active/passive for the IC-218 and IC-221 alloys (Figure 13-1b) but with high passive current densities, i_p. These polarization results are consistent with the mass-loss results at 95° C in Table 13-2. In 1 mol/L hydrofluoric acid, at both 25° C and 95° C, the polarization results for all three alloys show only active corrosion (Figure 13-1a), indicating the role of halide ions in preventing (or retarding) the formation of passive films. The polarization behaviors in 1 mol/L acetic acid and 1 mol/L sodium hydroxide at 25° C indicate passive corrosion, that is, E_{corr} is in the passive range (Fig. 13-1c); thus, the formation of protective passive films is responsible for the excellent behavior shown by the mass-loss data of Table 13-2. At 95° C, active/passive behavior is produced in 1 mol/L acetic acid (Fig. 13-1b) and passive behavior in 1 mol/L sodium hydroxide (Fig. 13-1c). In 0.6 mol/L sodium chloride at both 25° C and 95° C, the cyclic-anodic polarization behaviors of all three nickel aluminides show passive corrosion with generation of hysteresis loops (Fig. 13-1d). This behavior indicates the formation of passive films but also susceptibility to chloride-induced localized corrosion. (i.e., pitting or crevice corrosion). After generation of the polarization curves, the specimen surfaces show pitting corrosion. These results are consistent with the observations previously described with regard to the mass-loss results of Table 13-2. Pitting corrosion also is observed for the IC-50 alloy in 0.5 mol/L sulfuric acid at 25° C, not at E_{corr}, but over the potential range corresponding to the active-to-passive transition (Bertocci et al. 1990).

Studies comparing the aqueous-corrosion properties of the IC-50 nickel aluminide to those of pure nickel and pure aluminum in a range of electrolytes have been reported (Bertocci et al. 1990, Gram, Bertocci, and Ricker 1991; Ricker, Hall, and Fink 1990a). Review of thermodynamic data contained in Pourbaix (potential-pH) diagrams indicates that neither nickel nor aluminum should form protective passive films in the low pH range. In the near-neutral pH range, passive films may form; and in the high pH range, nickel is usually passive but aluminum usually dissolves rapidly as aluminate ions (AlO_2^-) (Bertocci et al. 1990). It was of interest to determine if the nickel aluminide behaved more like nickel, more like aluminum, or exhibited unique corrosion behavior.

Free-corrosion potentials (E_{corr} values) for the IC-50 nickel aluminide were measured in a series of electrolytes over a wide range of pH (Bertocci et al. 1990; Ricker et al. 1990). The electrolytes included 0.5 mol/L H_2SO_4 (pH ≈ 0.7), 0.5 mol/L HNO_3 (pH ≈ 0.6), 0.5 mol/L Na_2SO_4 (pH ≈ 6.5), 0.5 mol/L NaCl (pH ≈ 6.3) and 0.5 mol/L NaOH (pH ≈ 13.1). In all cases, the E_{corr} values for the nickel aluminide are found to be closer to those of nickel than those of aluminum. These results indicate that the behavior of the nickel aluminide is closer to that of nickel, rather than aluminum, in all of the solutions. Moreover, polarization behaviors of the IC-50 nickel aluminide and pure nickel were determined and compared in 0.5 mol/L H_2SO_4, 1.0 mol/L NaCl, and 0.5 mol/L NaOH (Bertocci et al. 1990; Ricker et al. 1990). In all cases, the behavior of the nickel aluminide is very similar to that of pure nickel. In the H_2SO_4 and HNO_3 acids, both the nickel aluminide and nickel are active (no passive films) at the free-corrosion potentials, but undergo active/passive transitions at higher anodic potentials. In the near-neutral NaCl solution, both the nickel aluminide and nickel passivate, but pitting corrosion is observed to occur during anodic polarization. The breakdown potential for pitting corrosion $E_{B,pit}$ (Figure 13-1d), is about the same for the nickel aluminide and nickel, and about 700 mV higher than that usually observed for aluminum. In the high-pH NaOH solution, both the nickel aluminide and nickel passivate over a wide potential range. These results suggest that while oxides or hydroxides of aluminum may be present on the surface or incorporated in the passive film, the protective passive film behaves essentially identically to that which causes passivity of pure nickel. The overall conclusions of these comparative studies are that (a) the electrochemical behavior of the nickel aluminide studied is essentially identical to that of pure nickel, and (b) in terms of aqueous corrosion, the nickel aluminide may be considered as a substitute for

Table 13-3 Cavitation Erosion Rates for Tests at 41 Mpa (6000 psi)

Material	Erosion Rate (mg/hour)	Relative Erosion Rate (Ref. to 308 SS)
IC-50 Nickel Aluminide (cold worked)	1.4	0.04
IC-221 Nickel Aluminide	2.5	0.08
IC-218 Nickel Aluminide	2.9	0.09
IC-50 Nickel Aluminide	4.1	0.1
IRECA from Hydro-Quebec (stick)	5.6	0.2
Stoody Deloro HQI (5/32″ stick)	6.1	0.2
IC-50 Nickel Aluminide (plasma weld overlay)	7.0	0.2
Stellite 21 (stick)	7.3	0.2
Stoody Deloro HQI (1/16″ wire, 100% Ar shielding gas)	8.8	0.3
Stoody Deloro HQI (1/16″ wire, 75% Ar-25% CO_2 shielding gas)	9.1	0.3
Iron Aluminide	11.6	0.4
Nitronic 60 bar stock	12.8	0.4
McKay Tubealloy APO (1/16″ wire)	14.0	0.4
Wilbanks Intl. aluminum oxide ceramic plate	18.4	0.6
308 SS overlay (stick)	31.6	1.0
Carbon steel (cold-rolled bar)	32.3	1.0
304 SS plate	33.2	1.1
Aluminum Bronze	84.4	2.7

Source: March 1988, Johnson et al. 1990.

nickel, if other properties, such as higher strength or lower density, make it attractive.

3.2. Erosion Corrosion Behavior

Studies have been conducted to determine the cavitation-erosion resistance of intermetallic alloys relative to other materials (March 1988; Johnson et al. 1990). The cavitation-erosion phenomenon is a major problem for the operation of hydraulic equipment, (e.g., valves and fittings, pump impellers, hydroturbines and ship propellers). A cavitating jet erosion test facility was used to impinge a water jet at 41 MPa (6000 psi) onto test specimens of various materials. The average erosive wear rate (mg/hour) was measured and also was expressed as a relative erosion rate normalized to that of 308 stainless steel. The results for IC-50, IC-218, and

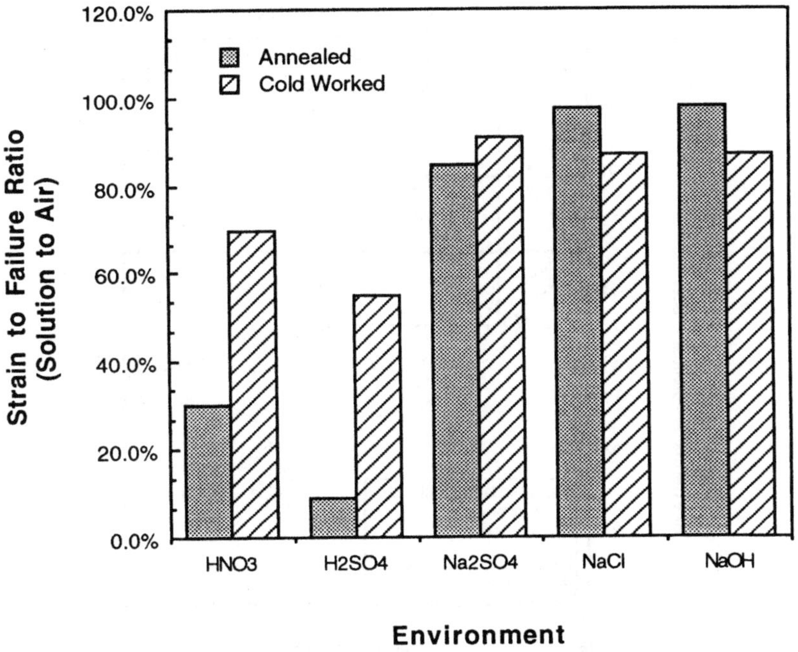

Figure 13-3. Strain-to-failure ratios for annealed and cold-worked Ni_3Al (IC-50) in different aqueous environments (Ricker et al. 1990; reprinted with permission from *Scripta Metallurgica*, Volume 24, pages 291–296, copyright 1990, with kind permission from Elsevier Science Ltd., The Boulevard, Langford Lane, Kidlington, OX5 1GB, UK).

IC-221 nickel aluminides, an iron aluminide, and a variety of other erosion-resistant materials are given in Table 13-3. It is seen that the nickel aluminides display the lowest cavitation-erosion rates. The iron aluminide also produces rates lower than several of the other materials. The conclusion is that the intermetallic alloys, especially the nickel aluminides, have remarkable resistance to cavitation-erosion damage, and therefore have a high potential for applications in areas requiring this excellent resistance.

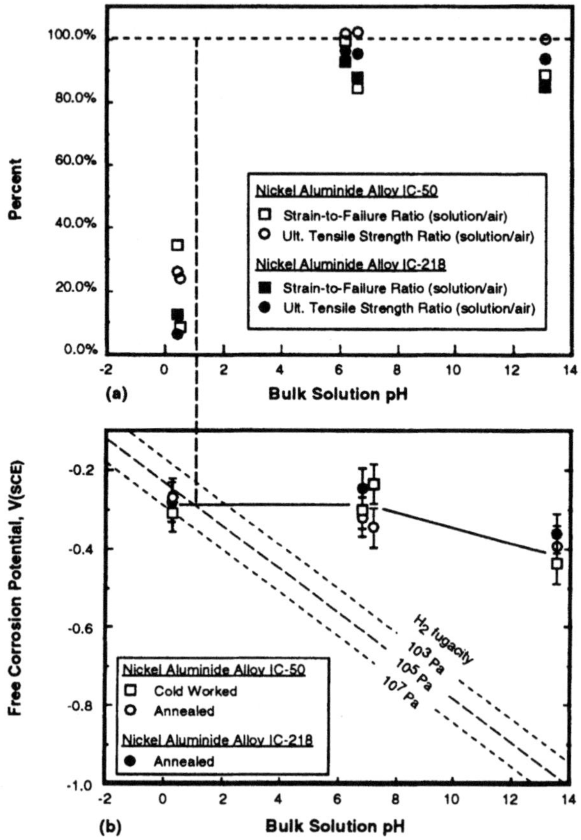

Figure 13-4. Results of slow-strain-rate tensile tests in solutions of varying pH on two different Ni_3Al alloys: (a) strain-to-failure or ultimate tensile strength as a percentage of the value observed in air, (b) measured free-corrosion potentials in the same solutions with the equilibrium potentials for hydrogen fugacities of 10^3, 10^5 and 10^7 Pa shown,

3.3. Corrosion-Sensitive Embrittlement

To study the corrosion-sensitive embrittlement characteristics of the IC-50 and IC-218 nickel aluminides, slow-strain-rate tests (SSRT's) at 1.0×10^{-6}/s were conducted in aqueous solutions over a wide range of pH (Ricker et al. 1990; Ricker, Stoudt, and Fink 1990; Ricker et al. 1991). Annealed IC-50 and IC-218 (1100° C, 1 h) and 40% cold-worked IC-50 were evaluated in 0.5 mol/L HNO_3 (pH ≈ 0.6), 0.5 mol/L H_2SO_4 (pH ≈ 0.7), 0.5 mol/L Na_2SO_4 (pH ≈ 6.5), 0.5 mol/L NaCl (pH ≈ 6.3) and 0.5 mol/L NaOH (pH ≈ 13.1). The strain-to-failure ratios, relative to that in air, for IC-50 in the different aqueous environments are shown in Figure 13-3 (Ricker et al. 1990). The strain-to-failure ratios and the ultimate-tensile-strength ratios for annealed IC-50 and IC-218 relative to the

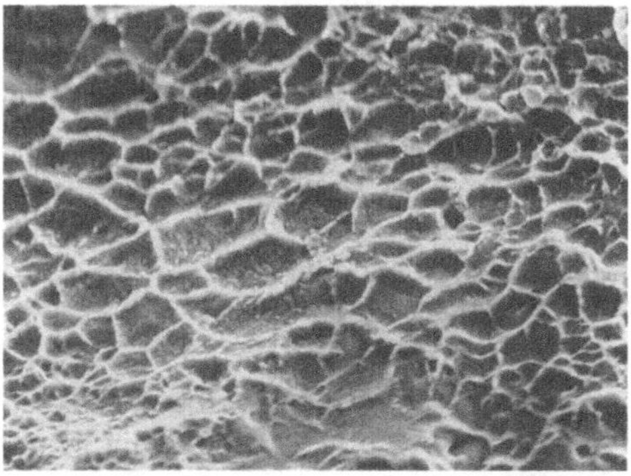

Figure 13-4. (Continued) (c) fractograph of a sample tested in 0.5 mol/L Na_2SO_4, and (d) fractograph of a sample tested in 0.5 mol/L H_2SO_4 (Ricker et al. 1991).

solution pH values are given in Figure 13-4a (Ricker et al. 1991). It is seen that significant embrittlement occurs in the low-pH solutions (HNO_3 and H_2SO_4 acids), but not in the near-neutral solutions (Na_2SO_4 and NaCl) or in the high-pH solution (NaOH). The fracture mode is intergranular in the low-pH solutions (Figure 13-4d), but ductile transgranular in the near-neutral and high-pH solutions (Figure 13-4c). To investigate the reason for embrittlement in low-pH solutions, the free-corrosion potentials (E_{corr} values) of the nickel aluminides in the various solutions were plotted against the solution pH, as shown in Figure 13-4b. Also in this figure, the equilibrium hydrogen potentials (E'_H, Equation 13.7) corresponding to hydrogen partial pressures (fugacities) of 10^3 Pa (10^{-2} atm.), 10^5 Pa (1.0 atm.), and 10^7 Pa (10^2 atm.) are plotted as a function of solution pH. It is seen that the free-corrosion potentials in the acidic solutions create large quantities of hydrogen by virtue of the cathodic reduction process, with hydrogen partial pressures of 10^5 Pa (1 atm.) or greater, whereas the free-corrosion potentials in the near-neutral and high-pH solutions produce vanishingly small quantities of hydrogen. These results strongly suggest hydrogen embrittlement as the mechanism responsible for embrittlement in the sulfuric and nitric acids. To verify the role of hydrogen, SSRT's at 1.0×10^{-6}/s were conducted on annealed IC-50 under the following conditions (Fig. 13-5): (1) tested in air, (2) tested in the 0.5 mol/L H_2SO_4, (3) preexposed to 0.5 mol/L H_2SO_4 for 10h, then tested in air, and (4) preexposed to 0.5 mol/L H_2SO_4 for 10 h, then vacuum "baked" at room temperature for 100 h, then tested in air (Ricker et al. 1990, 1991). The results shown in Figure 13-5 are clearly consistent with hydrogen embrittlement.

With regard to hydrogen embrittlement of nickel aluminides in neutral solutions, work has shown that embrittlement occurs during SSRT's if hydrogen is artificially produced by cathodic charging (Kuruvilla and Stoloff 1985). Additional, similar results are shown in Figure 13-6, where SSRT's (1.0×10^{-6}/s) were conducted on IC-50 in near-neutral 0.5 mol/L NaCl (pH $\approx$ 6.3) and 0.5 mol/L Na_2SO_4 (pH $\approx$ 6.5) solutions, and highly-basic 0.5 mol/L NaOH solution (pH $\approx$ 13.1) (Ricker et al. 1990). In each case, the SSRT's were conducted over a range of applied potentials. Also shown for each electrolyte are the potentials corresponding to hydrogen partial pressures of 10^3 Pa (0.01 atm.), 10^5 Pa (1.0 atm.), and 10^7 Pa (100 atm.), based on Equation 13-7 and the electrolyte pH. In general, it is seen that when the applied potential is made more negative that the potential corresponding to $P_{H_2} = 10^5$ Pa (1 atm.), embrittlement and intergranular fracture occurs. These results, as well as those in Figures 13-4a and 13-4b, indicate that a good estimate of the critical hydrogen-embrittlement potential, $E_{crit,H}$, is that corresponding to $P_{H_2} = 10^5$ Pa (1 atm.) at a given pH (Equation 13.11). It is noted in Figure 13.4

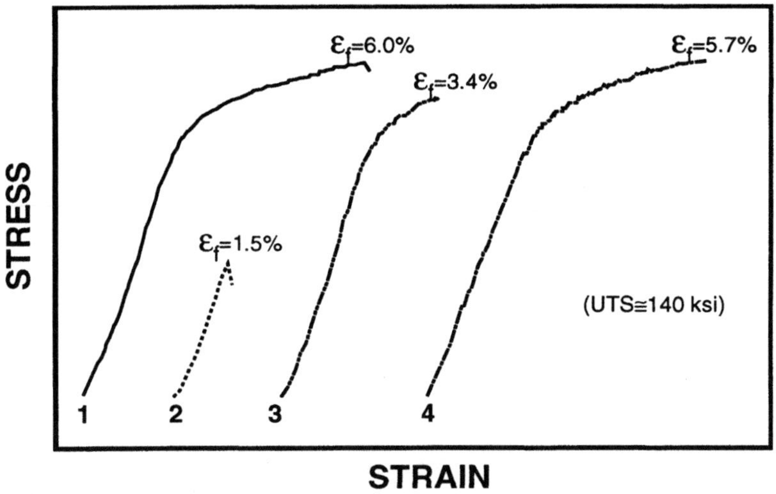

Figure 13-5. Diagram representing the stress vs. strain curves obtained (1) in air, (2) in sulfuric acid, (3) in air after pre exposure to sulfuric acid for 10 hours, and (4) in air after pre exposure to sulfuric acid for 10 hours followed by a vacuum "bake-out" at room temperature for 100 hours (Ricker et al. 1990, 1991).

that the free-corrosion potentials (E_{corr} values) in the neutral and basic electrolytes are much higher than $E_{crit,H}$ and that under normal, freely corroding conditions hydrogen embrittlement does not occur.

However, there are several situations that could possibly cause the E_{corr} in neutral solutions to drop below $E_{crit,H}$, and therefore possibly result in hydrogen embrittlement. These situations include: (a) galvanic coupling of the nickel aluminide to a more active alloy, (b) local acidification at pitting-corrosion or crevice-corrosion sites, and (c) passive-film-rupture/repassivation events associated with dynamic loading, for example, crack propagation during corrosion fatigue. Each of these situations has been considered (Ricker et al. 1990, 1991). An example of the galvanic coupling situation would be if the nickel aluminide were coupled to an aluminum alloy such as 7075, which has an E_{corr} of approximately −740 mV(SCE) in 0.5 mol/L NaCl. If the area of the aluminum alloy is large relative to the nickel aluminide, the coupled corrosion potential will be close to that of the aluminum alloy, −740 mV(SCE), which is less than $E_{crit,H}$ for the nickel aluminide (Fig. 13-6a). Therefore, in this coupled condition, the nickel aluminide would be expected to undergo hydrogen embrittlement (Ricker et al. 1990).

In occluded geometries associated with pitting or crevice corrosion, high metal-ion concentrations result which lead to local acidification by

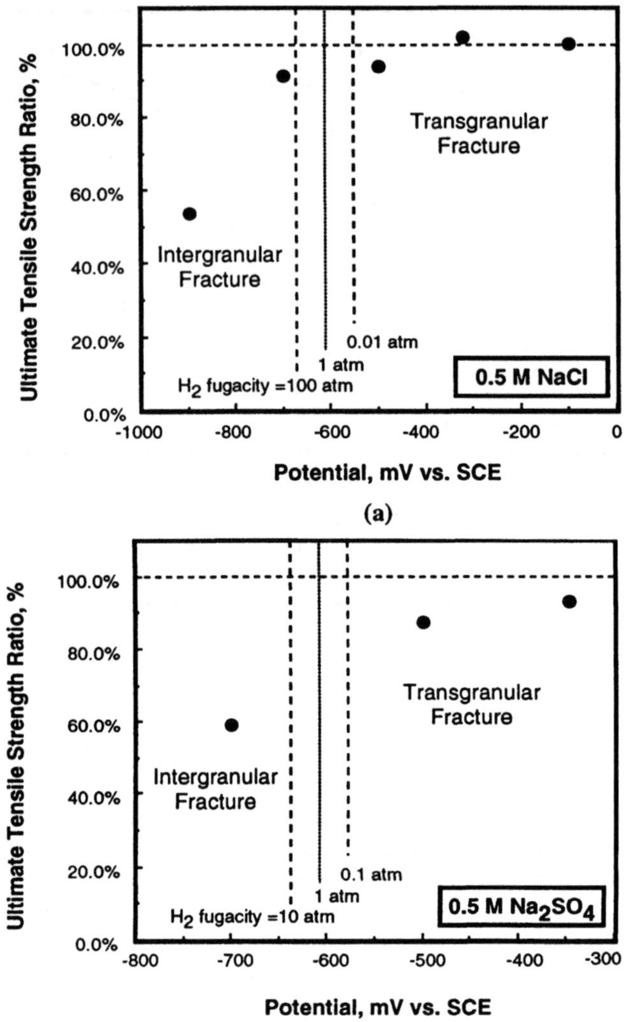

Figure 13-6. Ratio of the ultimate tensile strength of Ni_3Al (IC-50) observed in (a) 0.5 mol/L NaCl, (b) 0.5 mol/L Na_2SO_4.

hydrolytic reactions of the form:

$$M^{m+} + H_2O \rightarrow M(OH)^{(m-1)+} + H^+ \tag{13.12}$$

where M^{m+} represents metal ions and the end product may be ions in the

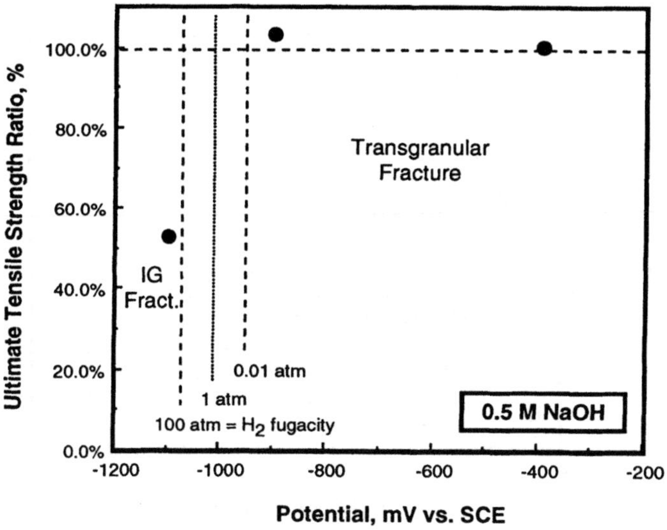

(c)

Figure 13-6. (Continued) (c) 0.5 mol/L NaOH at different potentials to the value measured in air (Ricker et al. 1990)

solution or insoluble hydroxides or oxides (Ricker et al. 1991). Thus, in these local sites, as the metal-ion concentration increases, the pH decreases. At a given potential, as the pH decreases, the quantity of hydrogen produced increases (Equations 13.8 and 13.9), and therefore the chance of hydrogen embrittlement increases. For nickel aluminide, analyses have shown that a greater reduction in pH is associated with aluminum ions than nickel ions. Using the worst case aluminum-ion condition to simulate this effect, SSRT's (1.0×10^{-6}/s) were conducted in 0.17 mol/L (pH = 2.7) and 1.7 moL/L (pH = 1.4) $AlCl_3$ solutions; the results are shown in Figure 13-7 along with results previously shown in Figure 13-4a. It is seen that the 0.17 mol/L concentration of $AlCl_3$ produces essentially no embrittlement with ductile transgranular fracture, whereas the 1.7 mol/L concentration produces some embrittlement with mixed transgranular and intergranular fracture. It is concluded that since the aluminum-chloride concentrations employed were quite high and would not be expected, except perhaps under wet/dry immersion cycles, the nickel aluminide will not undergo hydrogen embrittlement at pit or crevice sites due to local acidification in bulk neutral chloride solutions (Ricker et al. 1991).

Under dynamic loading conditions, such as corrosion fatigue, dislocation movement to the surface causes passive-film rupture, thus exposing

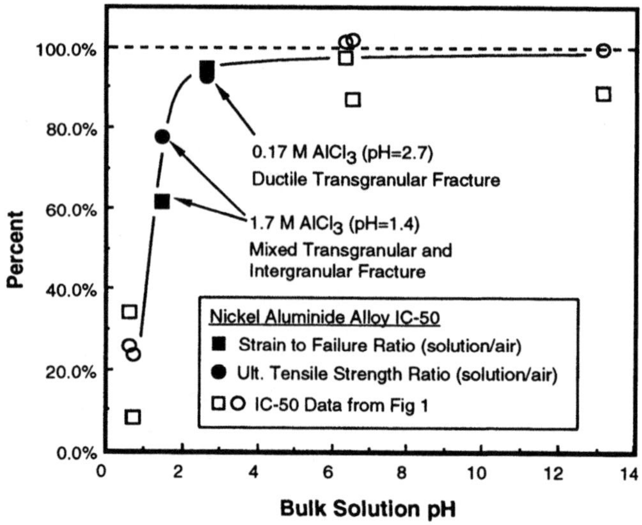

Figure 13-7. Results of slow-strain-rate tensile tests on Ni_3Al (IC-50) in $ALCl_3$ solutions plotted against pH, along with results previously shown in Fig. 4a (Ricker et al. 1991).

the bare metal to the electrolyte. This event causes a change in the local E_{corr} value, a change in the negative direction. On repassivation at the local site, E_{corr} moves back toward its original value. The question here is whether the minimum in the potential transient will reach $E_{crit,\,H}$ such that local hydrogen embrittlement will occur. To address this question, scratch tests were performed in 0.5 mol/L Na_2SO_4, NaCl, and $NiCl_2$ solutions, whereby the passive films were suddenly removed by a sapphire scribe (Ricker et al. 1991). In each case, the potential transient during the film-rupture/repassivation event was measured. The average minimum potential for the IC-50 nickel aluminide and the critical hydrogen potential ($E_{crit,\,H}$) are given for each solution in Table 13-4. In all cases, the minimum potential is greater than the critical hydrogen-embrittlement potential, and the conclusion is that the nickel aluminide would not undergo local hydrogen embrittlement due to dynamic loading in these types of solutions.

4. IRON ALUMINIDES

The aqueous-corrosion properties of iron aluminides have received considerable study over recent years. This section describes the corrosion behav-

Table 13-4 Average Minimum Potentials (mV(SCE)) Measured during Scratch Repassivation Transients for Pure Nickel, Pure Aluminum and Nickel Aluminide (IC-50) in Sodium Sulfate, Sodium Chloride and Nickel Chloride

Solution	Pure Ni	Pure Al	Ni_3Al	$E(H^+/H_2) = 10^5$ Pa (1 atm.)*
0.5 mol/L Na_2SO_4	−450	−1,518	−391	−626
0.5 mol/L NaCl	−561	−1,568	−493	−614
0.5 mol/L $NiCl_2$	−588**	−1,296**	−430**	−537

* The potential for an equilibrium hydrogen fugacity of 1 atmosphere as calculated for the measured pH of the solution.

**Average of two measurements. All others are the average of three measurements.

Source: Ricker et al. 1991.

iors of Fe_3Al-based iron aluminides and low-aluminum Fe-Al alloys in a range of electrolytes, and also describes the corrosion sensitive embrittlement behaviors of these alloys under various test conditions.

Most of the Fe_3Al-based iron aluminides described in the following sections contained 28 at.%Al and were normally given a heat treatment prior to corrosion testing to produce the B2 superlattice structure; with reference to the Fe-Al phase diagram (Swann and Duff 1969; Kattner and Burton 1993), the B2 heat treatment was 1 h at 750° C in air followed by oil quenching. Some of the alloys were heat treated to produce the DO_3 superlattice structure; the DO_3 heat treatment was 1 h at 850° C in air, 2 d at 500° C in air, then air cooled. Limited corrosion tests indicate that the two heat treatments do not produce major differences in aqueous corrosion properties (Buchanan and Kim 1990, 1991a). The low aluminum alloys contained 16 at.%Al (disordered α structure) and were given a heat treatment prior to corrosion testing of 1 h at 800° C followed by air cooling. Most of the alloys are indexed according to their Oak Ridge National Laboratory designations; these designations and the alloy compositions are given in Table 13-5.

Corrosion results are described relative to acidic, basic, chloride, and sulfur-compound solutions. Some explanation is in order for the sulfur-compound solutions, $Na_2S_2O_3$ (sodium thiosulfate) and $Na_2S_4O_6$ (sodium tetrahionate). A potential application for the iron aluminides is in the fossil energy production area. In this area the primary source of sulfur compounds is the fuel itself, and more directly, the fuel-combustion products. Any structural material exposed to the combustion products, as in flue-gas clean-up systems, may be susceptible to accelerated corrosion

Table 13-5 Designations and Chemical Compositions of Iron Aluminides (at.%).

Alloy	Al	Cr	Mo	Nb	Zr	Y	C	B	Fe
FA-84	28	2	—	—	—	—	—	0.05	bal
FA-129	28	5	—	0.5	—	—	0.2	—	bal
FAL-Mo	28	5	1	—	0.08	—	—	0.04	bal
FAP-Y	16.1	5.4	1.1	—	0.1	0.06	0.1	—	bal
CM-Mo	15.8	4.9	1.6	—	0.15	0.06	0.05	—	bal

and stress-corrosion cracking (SCC) due to the sulfur compounds. Such structural components normally operating at elevated temperature may develop scales containing sulfides. During shutdown, for example, for inspections, and upon exposure to the atmosphere, the sulfides can interact with moisture and oxygen to form aggressive sulfur-bearing acids. Of particular importance are the polythionic acids ($H_2S_xO_6$) (Ahmad et al. 1983a, 1983b; Scarberry, Pearman, and Crum 1976). A number of investigators have indicated that the most aggressive polythionic-acid component is tetrahionic acid ($H_2S_4O_6$) (Berge and Donati 1981; Newman and Sieradzki 1983). This acid can be formed by adding $Na_2S_4O_6$ to water, which undergoes hydrolysis to form the acid. An ion not formed in polythionic-acid solution, but closely related to the tetrahionate ion, is the thiosulfate ion ($S_2O_3^{2-}$) (Horowitz 1983). It also has been linked to SCC problems in petroleum and electric-power industries (Newman and Sieradzki 1983; Horowitz 1983; Newman, Roberge and Bandy 1983; Bandy, Roberge, and Newman 1983).

4.1. Aqueous Corrosion Behavior

Fe_3Al-Based Alloys in Acidic, Basic and Sulfur-Compound Solutions. Corrosion rates at the free-corrosion potentials (E_{corr} values) for FA-84, FA-129, and FAL-Mo iron aluminides as determined by the electrochemical polarization-resistance method in acidic, basic, and sulfur-compound solutions are shown in Table 13-6 (Buchanan and Kim 1991a, 1992a, 1992b). In 1 mol/L HCI acid, the iron aluminides and the annealed 304L stainless steel (SS) reference material exhibit very high corrosion-penetration rates. The iron aluminides also produce high corrosion rates in 0.5 mol/L H_2SO_4 and 1 mol/L HNO_3 acids, and in 0.1 mol/L $Na_2S_2O_3$ and 0.1 mol/L $Na_2S_4O_6$ sulfur-compound solutions; whereas, the 304L SS passivates in these electrolytes and produces very low corrosion rates. It is noted that in the 1 mol/L HNO_3 acid, FA-129 (5Cr) and FAL-Mo (5Cr + 1Mo) produce significantly lower corrosion rates than FA-84 (2Cr).

Table 13-6 Corrosion Rates by the Polarization-Resistance Method for Fe_3Al-Based Iron Aluminides in Acidic, Basic and Sulfur-Compound Solutions

	AVERAGE PENETRATION RATES (μm / DAY)			
	MATERIALS			
Solutions	FA-84	FA-129	FAL-Mo	304L SS
1 mol/L HCl	430	52	14	9.7
0.5 mol/L H_2SO_4	400	72	120	0.014
1 mol/L HNO_3	160	3.5	1.4	0.007
1 mol/L NaOH	0.042	0.021	0.056	0.007
0.1 mol/L $Na_2S_2O_3$	7.8	9.9	2.6	0.004
0.1 mol/L $Na_2S_4O_6$	21	6.5	6.5	0.020

Source: Buchanan and Kim 1991a, 1992a, 1992b.

In the 1 mol/L NaOH basic solution, the iron aluminides and the 304L SS produce very low corrosion rates consistent with passivation for all of these materials. Corrosion rates in the acids and base also were determined by three-week mass-loss immersion tests (Equation 13.5) (Buchanan and Kim 1992a, 1992b). The corrosion-rate trends are very similar to those presented in Table 13-6.

The cyclic anodic polarization behaviors of FA-84, FA-129, and FAL-Mo (B2 heat treatment), and 304L SS as a reference material, also have been studied (Buchanan and Kim 1991a, 1992b). The results are described relative to the schematic cyclic-anodic polarization curves of Figure 13-1. In 1 mol/L HCl acid, a very strong acid-chloride solution, only active corrosion is exhibited by all three aluminides and 304L SS (Fig. 13-1a), indicating that protective passive films are not formed. In 0.5 mol/L H_2SO_4 acid, 304L SS is passivated at the free-corrosion potential, E_{corr}, and remains passivated until a relatively high breakdown potential (Figure 13-1c). However, in 0.5 mol/L H_2SO_4, the three iron aluminides undergo active corrosion at E_{corr}, then passivate at higher potentials, and finally undergo breakdown, which is the active-passive type behavior illustrated in Figure 13-1b. In mol/L HNO_3 acid, again the 304L SS is passivated at E_{corr} and remains passivated until breakdown at a higher potential. However, the three iron aluminides all exhibit active corrosion at E_{corr}, then passivate at higher potentials, and finally undergo breakdown of the passive films. The critical anodic-current densities, i_{crit}, vary among the iron aluminides. The i_{crit} values for FA-129 and FAL-Mo, both of which contain 5% Cr, are much lower than the i_{crit} value for FA-84, which contains 2% Cr.

In 1 mol/L NaOH, a highly basic solution, all three iron aluminides and the 304L SS are passivated at E_{corr} and remain passivated until a relatively high breakdown potential. This ideal polarization behavior, as schematically illustrated in Figure 13-1c, indicates very good overall corrosion resistance.

In the 0.1 mol/L $Na_2S_2O_3$ and 0.1 mol/L $Na_2S_4O_6$ solutions, the iron aluminides exhibit only active corrosion (Figure 1a) indicating that protective passive films are not formed. On the contrary, the 304L SS exhibits passivation at E_{corr} and remains passive to a relatively high breakdown potential. It should be noted that these sulfur-compound solutions are at high concentration levels, almost at saturation. Work recently has been performed at lower concentrations, 0.01 and 0.001 mol/L $Na_2S_2O_3$ and $Na_2S_4O_6$.[27] The polarization results show that FA-84 and FA-129 continue to undergo active corrosion at the lower concentrations; however, the Cr-and-Mo-containing FAL-Mo passivates with significantly lower corrosion rates in $Na_2S_2O_3$ and $Na_2S_4O_6$ at both 0.01 and 0.001 mol/L concentrations, indicating much better corrosion behavior.

Fe_3Al-Based Alloys in Chloride Solutions. The average corrosion-penetration rates for the FA-84, FA-129, and FAL-Mo alloys, as determined by three-week mass-loss immersion tests, are given in Table 13-7 for three chloride solutions: (1) an atmospheric-corrosion acid-rain-type mild acid-chloride solution (pH = 4 (H_2SO_4), 200 ppm Cl^-), (2) a 3.5 wt.% NaCl solution, and (3) synthetic sea water (Buchanan and Kim 1992a, 1992b). Generally, low *average* corrosion penetration rates are produced. However, localized corrosion proves to be a problem. As seen in Table 13-7, localized corrosion is observed to initiate on the non-

Table 13-7 Results of Three-Week Mass-Loss Immersion Tests on Fe_3Al-Based Iron Aluminides in Chloride Solutions

	AVERAGE PENETRATION RATES (µm / DAY)			
	MATERIALS			
Cloride Solutions	FA-84	FA-129	FAL-Mo	304L SS
Mild Acid-Chloride	0.16*	0.11*	0.06	< 0.007
3.5 wt% NaCl	0.19*	0.13*	0.11	< 0.007
Synthetic Seawater	0.07*	0.07*	0.06	< 0.007

* Localized corrosion initiated within 24h.

Source: Buchanan and Kim 1992a, 1992b.

Mo-containing aluminides (FA-84 and FA-129) within 24 hours in all three chloride solutions. Only FAL-Mo and the reference material, 304L SS, remain totally passivated and therefore free of localized corrosion for the three-week time period.

Immersion tests were conducted on a number of iron aluminides in the mild acid-chloride solution to determine the relative times for initiation of localized corrosion (Buchanan and Kim 1991a). The results, given in Table 13-8, indicate that both Cr and Mo are essential for improved localized-corrosion resistance in chloride solutions, and that the necessary levels are approximately 4–6% Cr and 1–2% Mo.

In order to further elucidate the corrosion behavior of iron aluminides in chloride environments, and the effects of Cr and Mo alloying additions, cyclic-anodic polarization studies were performed. In the mild acid-chloride solution, the cyclic-anodic polarization behaviors of a binary Fe-28Al alloy and three ternary alloys, Fe-28Al-2Cr, Fe-28Al-4Cr, and Fe-28Al-6Cr (at.%), all with the DO_3 heat treatment, are shown in Figure 13-8 (Buchanan and Kim 1990, 1991a). It is apparent that the Cr additions

Table 13-8 Localized-Corrosion Initiation Times for Fe_3Al-Based Iron Aluminides in a Mild Acid-Chloride Solution

Iron Alumindes		Localized-Corrosion Initiation Times (days)
Fe-28Al*		1
Fe-28Al plus:	2Cr*	2
	2Cr-0.05B (FA-84)†	1
	4Cr*	2
	5Cr-0.2C-0.5Nb (FA-129)	1
	6Cr*	27
	1Mo*	2
	2Cr-1Mo*	41
	4Cr-0.5Mo*	59
	4Cr-1Mo*	> 122**
	4Cr-2Mo*	> 122**
	5Cr-1Mo-0.04B-0.08Zr (FAL-Mo)+	> 259**
304L Stainless Steel		> 259**

* DO3 heat treatment.

† B2 heat treatment.

** Tests terminated.

Source: Buchanan and Kim 1991a

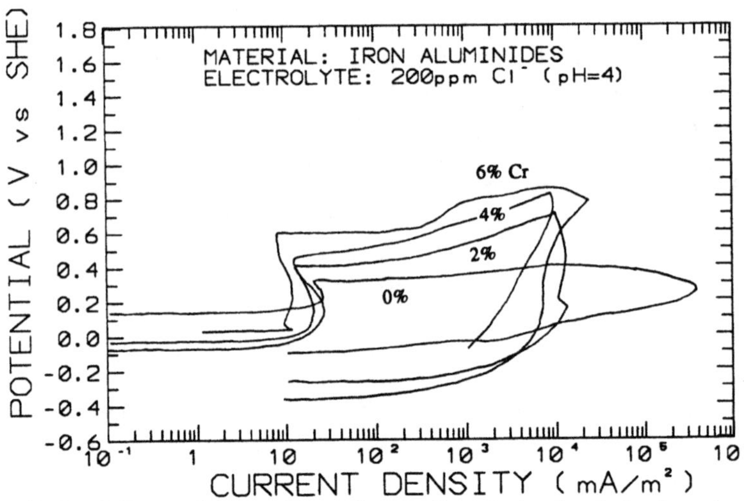

Figure 13-8. Effects of Cr on the cyclic anodic polarization behavior of annealed iron aluminide in aqueous solution containing 200 ppm Cl^- at pH = 4 (Buchanan and Kim 1990, 1991a).

have major effects on the polarization behavior. The behaviors are analogous to the schematic curve in Figure 13-1e. All of the aluminides are passivated at the free-corrosion potential, E_{corr}, and remain passivated until reaching a breakdown potential E_B. At E_B, pitting corrosion initiates. Upon scan reversal, it is seen that hysteresis loops are generated and that the protection potentials, E_p, are always below the corrosion potentials, E_{corr}. In accordance with the previous discussion on the forms of cyclic polarization curves (relative to Fig. 13-1), the results indicate that these iron aluminides are susceptible to localized corrosion in chloride environments. However, the results also indicate that increasing Cr levels increase the breakdown potential (critical pitting potential), and therefore aid in the resistance to initiation of localized corrosion.

Based on the cyclic anodic polarization behaviors in Figure 13-8, it was recognized that the localized corrosion resistance of the iron aluminides in chloride solutions could be improved if the protection potential, E_p, could be increased by alloy additions to a point above the free-corrosion potential, E_{corr}. In this regard, Cr-containing iron aluminides with Mo additions were evaluated (Buchanan and Kim 1990, 1991a, 1991b). Figure 13-9 shows the polarization curves for Fe-28Al-4Cr (at.%) alloys containing 0.5, 1.0, and 2.0 at.% Mo (DO_3 heat treatment) in the mild acid-chloride solution. It is seen that additions of Mo significantly

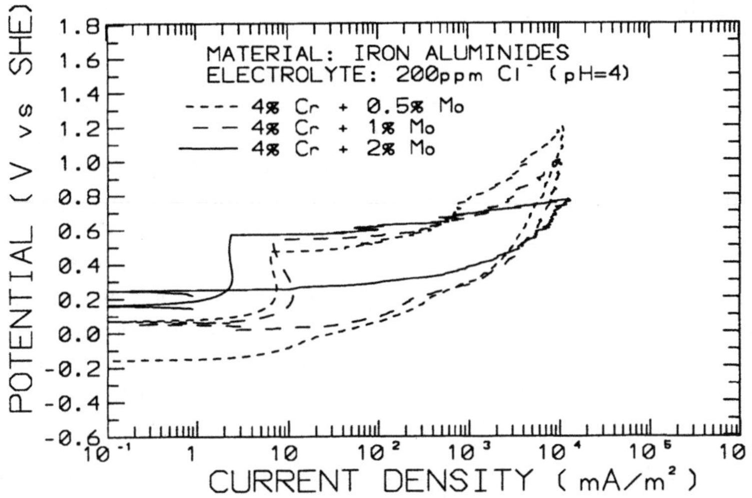

Figure 13-9. Effects of Mo on the cyclic anodic polarization behavior of 4% Cr iron aluminide in aqueous solution containing 200 ppm Cl^- at pH = 4 (Buchanan and Kim 1990, 1991a, 1991b).

increase E_P (relative to that in Fig. 13-8), and that Mo levels of 1% and 2% result in E_P values that are approximately equal to or greater than E_{corr}. These results indicate that for improved resistance to localized attack in chloride solutions, the iron aluminides should contain both Cr and Mo.

Polarization behaviors of FA-84 (2Cr), FA-129 (5Cr), and the Mo-containing FAL-Mo (5Cr-1Mo), all with the B2 heat treatment, were determined in the mild acid-chloride solution, 3.5 wt.% NaCl solution and synthetic sea water (Buchanan and Kim 1992b). For comparison, the behavior of annealed 304L SS also was determined. In all three chloride solutions, only FAL-Mo and 304L SS exhibit protection potentials approximately equal to or higher than the free-corrosion potentials. These results suggest that in chloride solutions the corrosion behavior of the Cr and Mo containing FAL-Mo is generally comparable to that of 304L SS.

Crevice Corrosion of Fe_3Al-Based Alloys. In view of the improved localized corrosion resistance of the Cr-and-Mo-containing iron aluminides, FAL-Mo was further tested under more severe conditions (along with FA-84 and 304L SS as reference materials). Three-week crevice-corrosion tests were conducted in the mild acid-chloride solution and in the 3.5 wt% NaCl solution using crevice-forming acetal-resin serrated washers that provided 60 crevice sites per specimen. (Buchanan and Kim

1992a, 1992b; Kim and Buchanan 1993a). In the mild acid-chloride solution, the FA-84, containing only 2% Cr and no Mo, undergoes severe crevice corrosion (52 of the 60 sites were attacked at a maximum depth of 6 μm). The 304L SS is quite resistant (only 1 of 60 sites was attacked at a depth of 4 μm). However, the FAL-Mo alloy (5Cr + Mo) is highly resistant (0 sites were attacked). These results again indicate the beneficial effects of Cr and Mo additions to the iron aluminides with respect to chloride-induced localized corrosion. In the much-higher-chloride solution, 3.5 wt.%NaCl, the FAL-Mo alloy does not perform as well as 304L SS. The number of sites that undergo crevice corrosion is higher for FAL-Mo (5 vs. 2) and the maximum depth of attack is significantly greater (50μm vs. 4μm). These results indicate that at high chloride concentrations (e.g., marine environments), crevice corrosion problems could develop even with the Cr-and-Mo-containing iron aluminides.

Passive Film Compositions of Fe_3Al-Based Alloys. As previously discussed, the iron aluminides initially passivate in chloride solutions, but the stabilities of the passive films are very dependent on the Cr and Mo concentrations of the alloys. It was of interest to determine how the bulk-alloy composition influenced the passive-film characteristics of the iron aluminides. In this regard, Fe-28Al (at.%) alloys containing 4 at.% Cr and 0, 0.5, 1, and 2 at.% Mo were immersed in the mild acid-chloride solution for 18 hours, cleaned with distilled water and methanol, then submitted to surface chemicals analyses by x-ray photoelectron spectroscopy (XPS) (Buchanan and Kim 1992a). The XPS method is ideal for analysis of extremely thin passive films since the depth from which compositional data are obtained is only on the order of 1–2 nm.

The XPS results on passive-film compositions are given in Table 13-9 and compared to the bulk-alloy compositions. The passive films are

Table 13-9 X-ray Photolectron Spectroscopy Results for Fe_3Al-Based Iron Aluminides

	Compositions: Surface (at.%) / Bulk (at.%)			
Iron Aluminides (at.%)	Fe	Al	Cr	Mo
Fe-28Al-4Cr	45.0/68	43.7/28	11.3/4	—
Fe-28Al-4Cr-0.5Mo	14.9/67.5	65.5/28	16.3/4	3.2/0.5
Fe-28Al-4Cr-1Mo	15.7/67	65.4/28	13.0/4	6.0/1
Fe-28Al-4Cr-2Mo	13.5/66	65.9/28	12.5/4	8.2/2

Source: Buchanan and Kim 1992a, 1992b, Kim and Buchanan 1993a.

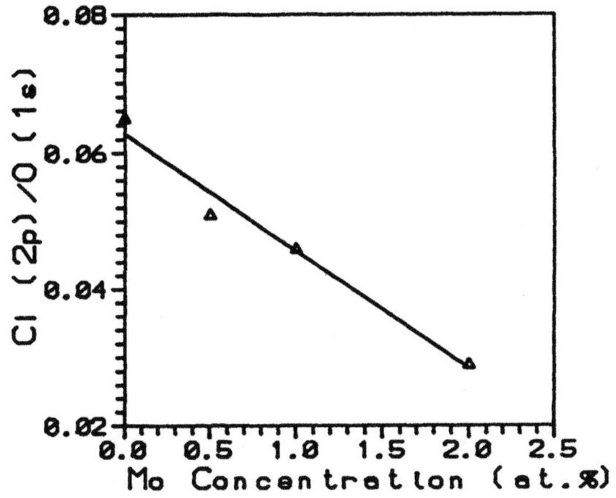

Figure 13-10. Variation of passive-film chloride concentration with bulk-alloy Mo concentration of iron aluminides (Buchanan and Kim 1992a).

depleted in Fe, due to preferential dissolution, and enriched in Al, Cr, and Mo relative to the bulk alloys. The presence of Mo appears to promote a greater degree of Al enrichment in the passive films. Chlorides were also detected. Figure 13-10 shows the ratio of chloride-to-oxygen in the passive films as a function of Mo concentration in the bulk alloys. It is seen that increasing Mo concentration reduces the quantity of chloride ions either within the passive films or adsorbed to the passive-film surfaces, and therefore would be expected to reduce the localized corrosion damage associated with chloride ions.

Low-Aluminum Fe-Al Alloys. Corrosion behaviors for two of the new low-aluminum (16 at.% Al) alloys, FAP-Y and CM-Mo, were determined by polarization-resistance and cyclic anodic-polarization methods in three aerated solutions: the mild acid-chloride solution, 3.5 wt.%NaCl and 1 mol/L NaOH (Kim and Buchanan 1993a, 1993b). FAL-Mo (28 at.%Al), one of the best Fe_3Al-based alloys in terms of aqueous-corrosion resistance, also was evaluated as a reference. It is noted that FAP-Y, CM-Mo, and FAL-Mo all contain Cr and Mo (Table 13-5), which in combination have been shown to be very beneficial in terms of aqueous-corrosion behavior. As determined by the polarization-resistance method, all three alloys produce low passive-corrosion rates in these chloride and basic solutions. Furthermore, the cyclic anodic-polarization behaviors are found to be very similar, reflecting comparable resistances to localized corrosion in the chloride solutions, and complete passivation in the basic solution.

4.2. Corrosion-Sensitive Embrittlement

Fe_3Al-Based Alloys. As background information, it is noted that the tensile ductilities of Fe_3Al-based iron aluminides are found to be sensitive to the moisture content of the testing environment, for example, laboratory air (Liu, Lee, and McKamey 1989, 1990a; Liu and McKamey 1990; Liu and George 1990; Gaydosh and Nathal 1990; Shea, Castagna, and Stoloff 1991; Kasul and Heldt 1991a, 1991b, 1994; Lynch, Heldt, and Milligan 1991, 1992; Lynch et al. 1994). For example, when a binary Fe-33Al (at.%) alloy is subjected to tensile tests in air at a strain rate of 3.3×10^{-4}/s, it is found that only a few percent relative humidity leads to significant embrittlement (Lynch et al. 1992). The mechanism responsible is believed to be hydrogen embrittlement, with the atomic hydrogen being produced by the following reaction at the iron-aluminide surface (Liu et al. 1989, 1990a; Ricker and Duquette 1988):

$$2Al + 3H_2O \rightarrow Al_2O_3 + 6H \tag{13.13}$$

This overall reaction may be viewed as the sum of the following anodic and cathodic reactions, respectively:

$$2Al + 3H_2O \rightarrow Al_2O_3 + 6H^+ + 6e \tag{13.14}$$

$$6H^+ + 6e \rightarrow 6H. \tag{13.15}$$

Considerable evidence points to hydrogen embrittlement as the responsible mechanism. For example, whereas the ductility of an Fe-35Al alloy is not sensitive to strain rate when tested in vacuum, the ductility decreases with decreasing strain rate when tested in air (Kasul and Heldt 1994). As previously discussed (Review of Experimental Corrosion Methods), this strain-rate dependence in air is clearly consistent with hydrogen embrittlement. Furthermore, when an Fe-24.6Al (at.%) alloy is cathodically charged with hydrogen in H_2SO_4 acid, then tensile tested in air, the ductility is found to decrease continuously with increasing precharging time, that is, increasing hydrogen content (Kasul and Heldt 1991b). On the other hand, if the precharged specimens are subjected to an adequate "bake-out" treatment before tensile testing, ductility is completely restored (Kasul and Heldt 1991a). This reversibility is characteristic of the hydrogen-embrittlement phenomenon.

Of principal concern in this chapter is the relative severity of the hydrogen-embrittlement effect in aqueous corrosion environments. According to previous discussion (Review of Experimental Corrosion Methods), the quantity of hydrogen produced at a metal surface during aqueous corrosion is related to the solution pH and the potential (E) of the metal. For the nickel aluminides, it was found that at a given pH, a sufficient quantity of absorbed hydrogen is produced to cause embrittlement when the potential is equal to or less than a critical hydrogen potential, $E_{crit,H}$, where $E_{crit,H}$ is the equilibrium hydrogen potential corresponding to $P_{H_2} = 10^5$ Pa (1 atm.) (Equations 13.10 and 13.11).

The aqueous-corrosion effects on the tensile ductility (strain rate = 3.3×10^{-4}/s) of the binary Fe-24.6Al alloy have been evaluated in 0.5 mol/L H_2SO_4 (pH = 0.5), 1 mol/L Na_2SO_4 (pH = 6.2), and 1 mol /L NaOH (pH = 13.1) (Kasul and Heldt 1991b). In each solution, the ductilities were measured at several different potentials, as illustrated in Figure 13-11. $E_{crit,H}$ is shown as a function of pH by the solid line corresponding to $P_{H_2} = 10^5$ Pa (1 atm.) Thus, if hydrogen embrittlement is to occur, one might expect it to occur in the H_2SO_4 acid at −1000 mV(SCE) and at the free-corrosion potential (open-circuit potential, E_{corr}) but not at +400 nor +2500 mV(SCE), in the Na_2SO_4 solution at −800 but not at +100 mV(SCE), and in the NaOH solution at

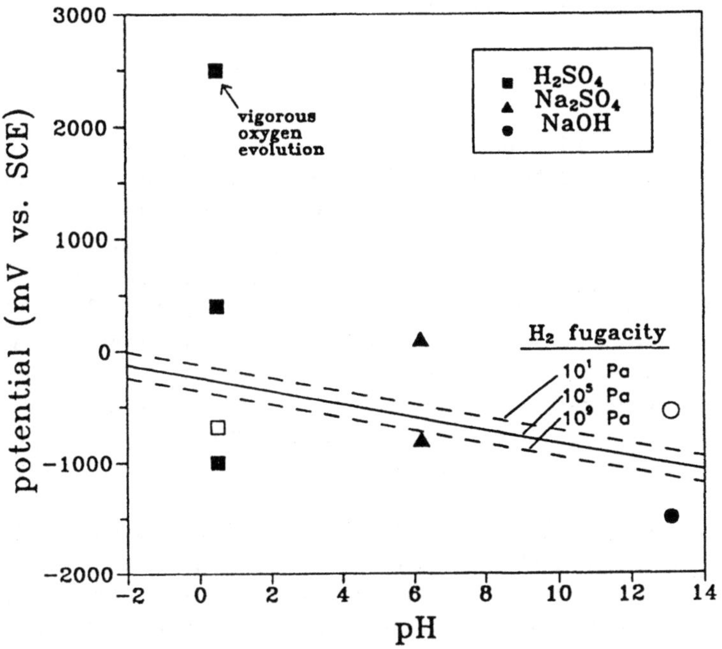

Figure 13-11. Electrochemical potential-pH conditions for in situ tensile tests on Fe-24.6% Al alloy; open symbols represent free-corrosion conditions (Kasul and Heldt 1991b).

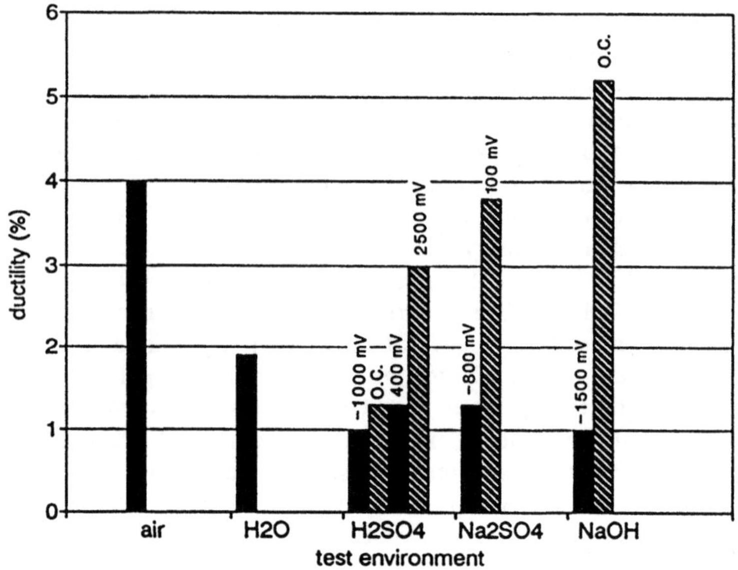

Figure 13-12. Ductility of an Fe-24.6% Al alloy under conditions of controlled potential and pH (Electrochemical Potentials are vs. SCE) (Kasul and Heldt 1991b).

−1500 mV(SCE) but not at the free-corrosion potential. The measured tensile ductilities in H_2SO_4, Na_2SO_4, and NaOH at the various potentials are shown in Figure 13-12, along with those measured in air and distilled water. Generally, the predictions cited above are consistent with the results and therefore consistent with hydrogen embrittlement. The only exceptions are in the H_2SO_4 at +400 and +2500 mV(SCE), where embrittlement is not expected, but did occur. The investigators explain this apparent anomaly as possibly due to an experimental problem wherein the specimens were exposed to the H_2SO_4 for a short time at the free-corrosion potential (E_{corr}) before the anodic potentials could be applied; the observed embrittlement could have occurred during the short time at E_{corr}.

It is noted in Figure 13-12 that for the Fe-24.6Al alloy, distilled water causes significant embrittlement relative to that in air, a result to be expected based on the relative-humidity effect. It is pointed out that all of these results are from tensile tests conducted within the water or water-vapor environments. On the other hand, if specimens (Fe-33Al) are preexposed to water or humid air, and then tested in vacuum, no embrittlement occurs (Lynch et al. 1992). These results suggest that oxide/hydroxide films (passive films) on the surfaces may be effective barriers to hydrogen absorption if they are chemically stable in the environment and are not mechanically broken by dislocation motion associated with plastic strain (thereby exposing bare metal to the environment) (Liu et al. 1989; Lynch et al. 1992).

For the Cr-containing Fe_3Al-based iron aluminides, U-bend corrosion tests have been conducted at room temperature on the FA-84 and FA-129 alloys (B2 heat treatment) (Buchanan and Kim 1991a). The tests were conducted for a maximum time of 200 hours (when failure did not occur) under freely corroding conditions (E_{corr}) in three electrolytes: the mild acid-chloride solution, 0.1 mol/L sodium thiosulfate, and 0.1 mol/L sodium tetrahionate. The E_{corr} values were continuously measured. Within the 200-hour time period, the FA-84 and FA-129 iron aluminides underwent cracking failures in the thiosulfate and tetrathionate solutions but not in the mild acid-chloride solution. Recall that these iron aluminides passivate in the mild acid-chloride solution, but do not passivate in the sulfur-compound solutions. Whereas the E_{corr} values in the acid-chloride solution (pH = 4) are found to be near or above $E_{crit,H} = -236$ mV(SHE), based on Equation 13.10, the E_{corr} values in the thiosulfate and tetrahionate solutions always are found to be below the $E_{crit,H}$ values (thiosulfate, pH = 5, $E_{crit,H} = -295$ mVHE); tetrahionate, pH = 4, $E_{crit,H} = -236$ mV(SHE)). These results suggest hydrogen embrittlement as the cause of U-bend cracking failures in the 0.1 M thiosulfate and tetrahionate solutions.

Additional 200-hour U-bend corrosion tests were conducted in the mild acid-chloride solution over a range of potentials: an applied anodic potential of +300 mV(SHE), E_{corr}, and applied cathodic potentials of −900, −1000 and −1500 mV(SHE) (Buchanan and Kim 1991a). It is again noted that in this solution (pH = 4), $E_{crit,H} = -236$ mV(SHE). Thus, at the anodic potential of +300 mV(SHE), hydrogen essentially is not produced; but at the cathodic potentials of −900, −1000 and −1500 mV(SHE), hydrogen is produced in large quantities. Within the 200-hour tests, FA-84 (2%Cr) does not undergo cracking failure at +300 mV(SHE), E_{corr} or −900 mV(SHE), but does undergo cracking failure in 45 hours at −1000 mV(SHE) and in 17 hours at −1500 mV(SHE). However, FA-129 (5%Cr) is resistant to cracking failure at all potentials over the 200-hour duration of the test. These results again suggest that hydrogen embrittlement is the cause of the U-bend cracking failures, and that the higher Cr FA-129 is more resistant to hydrogen embrittlement than FA-84 under U-bend corrosion test conditions.

Slow-strain-rate tests (SSRT's) at 1.2×10^{-6}/s also were conducted on the FA-84 and FA-129 iron aluminides (B2 heat treatment) in the mild acid-chloride solution (Buchanan and Kim 1992a, 1992b, 1991c). The tests were performed under applied anodic potentials (+300 and

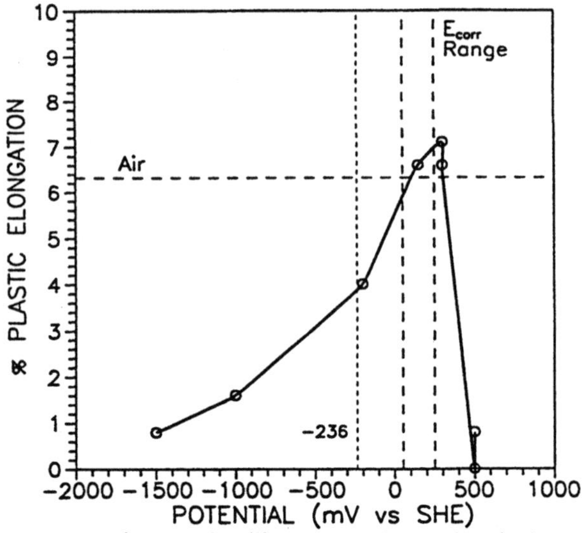

Figure 13-13. Slow-strain-rate ductility versus electrochemicals potential for FA-129 iron aluminide (Buchanan and Kim 1992a, 1992b, 1991c).

+500 mV(SHE)), at the free-corrosion potentials, and under applied cathodic potentials (−200, −1000 and −1500 mV(SHE)). For comparison, tests also were conducted in laboratory air at the same slow-strain-rate. Although the U-bend tests show different results for the FA-84 and FA-129 alloys, the SSRT's show very similar results. As typified by the behavior of FA-129 in Figure 13-13, the maximum ductilities (and maximum fracture stresses) occur at, or close to, the free-corrosion conditions (E_{corr} ranges), and these values are generally comparable to, or greater than, those measured in laboratory air. At the highest anodic potential +500 mV(SHE), and at the applied cathodic potentials, −200, −1000 and −1500 mV(SHE), losses in ductility (and strength) occur.

It is recalled that in this mild acid-chloride solution at pH = 4, $E_{crit,\,H} = -236$ mV(SHE), which is indicated in Figure 13-13. The iron-aluminide ductility continuously decreases at potentials less than −236 mV(SHE). These results again suggest hydrogen embrittlement as the responsible mechanism. Above −236 mV(SHE), the iron-aluminide ductility increases toward a maximum value, suggesting decreasing levels of hydrogen embrittlement due to decreasing amounts of hydrogen production. Beyond the maximum ductility value, and in particular at the +500 mV(SHE) anodic potential, the ductility again decreases. In this high-potential range, pitting corrosion initiates at the specimen surfaces.

For two reasons, it is possible that hydrogen production is again accelerated within the corrosion pits: (1) local acidification due to hydrolytic reactions, and (2) IR drops within the pits which could cause the actual potential at the bases of the pits to be less than $E_{crit,\,H}$. These effects could again cause severe hydrogen embrittlement at the highest potentials. On the other hand, the corrosion pits themselves act as stress-concentration sites, and this effect could be associated with the loss of ductility at the highest potential. Within the free-corrosion-potential range in the mild acid-chloride solution, the ductilities and strengths are generally comparable to, or higher than, the values measured in air. This result indicates that in terms of mechanical embrittlement phenomena at the slow-strain rate employed, the aqueous-corrosion-medium effect was no worse than that of laboratory air. This statement does not imply the absence of hydrogen embrittlement in the corrosive medium under naturally corroding conditions or in laboratory air (due to interaction with water vapor), but rather that the levels of hydrogen embrittlement in the two cases are generally comparable.

Unlike the U-bend corrosion test results, which indicate better hydrogen-embrittlement resistance for the higher-Cr FA-129, the SSRT results indicate, for the most part, similar behavior for FA-84 and FA-129. However, it must be recognized that the respective test conditions are very different. Whereas in the U-bend corrosion test, the passive film is not mechanically disturbed, in the SSRT, the passive film is continuously disrupted by dislocation movement to the surface. In the U-bend tests where mechanical passive-film disruptions are not occurring, at the applied hydrogen-charging potentials, the results indicate that the higher Cr FA-129 passive film serves as a better barrier to hydrogen absorption than the FA-84 passive film. On the other hand, in the SSRT's where mechanical passive-film disruptions are continuously occurring, one might expect that the passive-film characteristics will have less influence on hydrogen embrittlement, that is, that FA-84 and FA-129 might exhibit more similar behavior, as the results indicate.

Low-Aluminum Fe-Al Alloys. Limited work has been performed to characterize the hydrogen-embrittlement susceptibilities of two of the new low-aluminum alloys, FAP-Y and CM-Mo (16 at.% Al, disordered α structure), relative to that of the Fe_3Al-based FA-129 alloy (28 at.% Al, ordered B2 structure) (Kim and Buchanan 1993a, 1993b). Slow strain-rate tests (SSRT's) at 1.2×10^{-6}/s were conducted at the free-corrosion potentials and at a potentiostatically controlled cathodic potential of -1500 mV(SHE) in the mild acid-chloride solution. The cathodic potential was substantially below $E_{crit,\,H} = -236$ mV(SHE) for the pH = 4

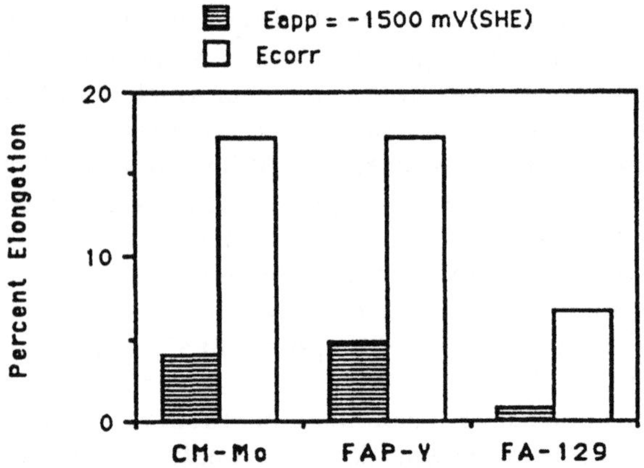

Figure 13-14. Comparison of hydrogen embrittlement susceptibilities for low-aluminum CM-Mo and FAP-Y, and Fe_3Al-based FA-129 (Kim and Buchanan 1993a, 1993b).

solution and therefore generated significant amounts of hydrogen at the specimen surfaces. The results are shown in Figure 13-14 as percent elongation with hydrogen charging (potential = −1500 mV[SHE]) and without hydrogen charging (potential = free corrosion potential, E_{corr}). The ductilities of the low aluminum alloys are much higher than those of the higher aluminum alloy both at the hydrogen-charging cathodic potential and at the free-corrosion potential. The severity of hydrogen embrittlement is obviously influenced by alloying content, specifically by the aluminum content, and possible by structure (disordered vs. ordered). The fracture surface of FAP-Y with hydrogen charging exhibited mostly brittle-transgranular cleavage cracking with little evidence of plastic deformation. In contrast, the fracture surface at the free-corrosion potential exhibited a ductile dimpled structure.

5. TITANIUM ALUMINIDES

Little information is currently available on the aqueous-corrosion properties and corrosion-sensitive embrittlement characteristics of titanium aluminides. Much more information is needed in these areas. Relative to aqueous corrosion, the titanium aluminides involve an interesting combination of elements that perform in entirely different ways in chloride environments. At room temperature, pure titanium passivates and is immune to localized pitting/crevice corrosion. On the other hand, pure

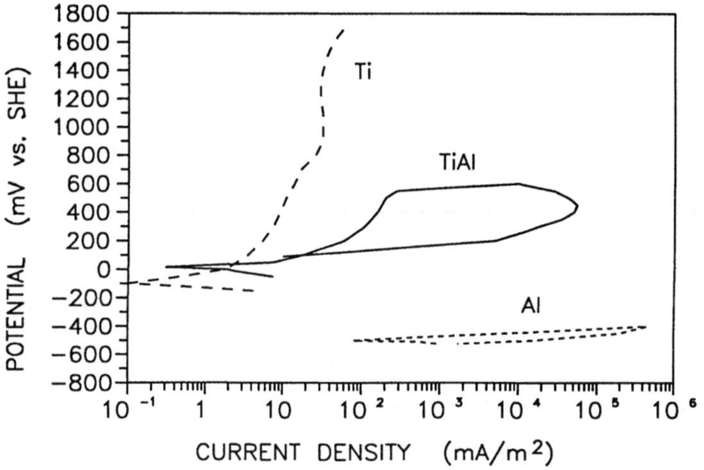

Figure 13-15. Anodic polarization behaviors of pure Ti, pure Al and single-phase TiAl (46Ti-54Al, at.%) in aerated 3.5 wt.% NaCl (Buchanan and Corallo 1993).

aluminum passivates, but immediately undergoes pitting corrosion. The question, of course, is how the titanium aluminides will perform. Shown in Figure 13-15 are anodic polarization behaviors of pure titanium, pure aluminum, and single-phase TiAl (46Ti-54Al, at.%) evaluated at room temperature is aerated 3.5 wt.% NaCl solution (Buchanan and Corallo 1993). Hysteresis is demonstrated by the TiAl, as schematically indicated in Figure 13-1d, which indicates that the TiAl may be susceptible to chloride-induced localized corrosion. However, the protection potential, E_P, is above the free-corrosion potential, E_{corr}, which suggests reasonable resistance to localized corrosion under freely corroding conditions. For a Ti_3Al-based alloy (Ti-25Al-10Nb-3V-1Mo, at.%) recent work has shown that in 3.5 wt.%NaCl solutions at pH = 3, 7 and 10, pitting corrosion occurs at potentials greater than the breakdown potentials (pitting potentials) (Bavarian, Harutouni, and Zamanzadeh 1992). However, whether or not the alloy will undergo pitting corrosion at E_{corr} is yet to be determined.

With regard to environmentally sensitive embrittlement, more work probably has been devoted to the hydrogen-embrittlement susceptibilities of titanium aluminides than to all of the other intermetallic alloys combined (Isserow 1993). Hydrogen embrittlement is associated with hydride formation in the titanium aluminides, and the degree of embrittlement is strongly dependent on the chemical composition and microstructure. Most of the studies have involved exposures to hydrogen-containing gaseous environments or to electrolytes under potentiostatically controlled hydrogen-charging conditions, with subsequent analysis of the embrittling effects due to absorbed hydrogen. An excellent, recently published guide to the literature in this area is available (Isserow 1993).

With regard to embrittling effects during aqueous corrosion, very little information is available. A recent study conducted slow strain-rate tests on a Ti_3Al-based alloy (Ti-25Al-10Nb-3V-1Mo, at.%) in an acidic (pH = 3) 3.5 wt.% NaCl solution over a range of applied potentials (Bavarian et al. 1992). The results are shown in Figure 13-16, where the strain ratio is plastic strain observed in the solution over that observed in air. For this alloy, hydrogen embrittlement occurs at hydrogen-producing cathodic potentials with a transgranular fracture mode (identified in the figure as transgranular stress-corrosion cracking, TGSCC) and embrittlement occurs at anodic potentials greater than the pitting potential with an intergranular fracture mode (identified in the figure as intergranular stress corrosion cracking, IGSCC). However, at the intermediate potential range,

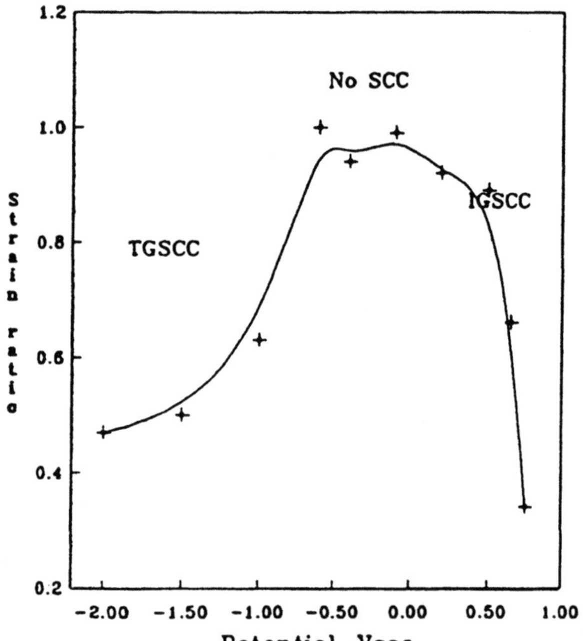

Figure 13-16. Strain ratio as a function of applied potential during slow-strain-rate testing of a Ti_3Al-based alloy in 3.5% NaCl solution at pH = 3 (E_{corr} = −0.62, E_{pit} = +0.60 V(SCE)) (Reprinted with permission from *Materials Science and Engineering*, Volume A153, B. Bavarian, S. Harutouni, and M. Zamanzadeh, pages 613–618, 1992, Elsevier Science Ltd., Oxford, England).

which includes the free corrosion potential, E_{corr}, no embrittlement occurs. More corrosion-sensitive embrittlement studies are needed for other titanium aluminides and in other solutions.

6. SUMMARY

Nickel aluminides based on Ni_3Al are found to exhibit excellent resistance to uniform corrosion at room temperature in organic acids (oxalic and acetic acids), bases (sodium and ammonium hydroxides), and a neutral sodium-chloride solution. However, there is some susceptibility to localized pitting/crevice corrosion in sodium-chloride solutions; this area needs additional study. In 1 mol/L concentrations of nonoxidizing inorganic acids (sulfuric, hydrochloric, hydrofluoric, and phosphoric acids), the nickel aluminides show moderate resistance to uniform corrosion; however, caution should be exercised due to hydrogen-embrittlement susceptibility. The corrosion rate increases with increasing concentration for these acids, and is unacceptably high for highly oxidizing nitric acid. The nickel aluminides corrode at extremely high rates in ferric chloride solutions, which not only contain chloride ions but also are highly oxidizing. At a higher test temperature of 95° C, the nickel aluminides still maintain good uniform-corrosion resistance in the organic acids, bases and sodium-chloride solutions. Comparative studies among nickel aluminides, pure nickel, and pure aluminum show that the nickel-aluminide behavior is essentially the same as that of pure nickel. A remarkable characteristic of the Ni_3Al based nickel aluminides is their exceptional resistance to cavitation-erosion damage. The nickel aluminides are susceptible to hydrogen embrittlement during aqueous corrosion and are shown to undergo hydrogen embrittlement in sulfuric and nitric acids. However, tests indicate that hydrogen embrittlement will not occur under freely corroding conditions in neutral and basic solutions.

The Fe_3Al-based iron aluminides show excellent resistance to uniform corrosion in basic (sodium hydroxide) and neutral-chloride solutions. There is a susceptibility to localized pitting/crevice corrosion in chloride solutions, but alloys containing both Cr and Mo show relatively high resistance to localized attack (generally on the order of that of 304L stainless steel). The iron aluminides corrode rapidly in acid environments (hydrochloric, sulfuric, and nitric acids). In sulfur-bearing thiosulfate and tetrahionate solutions, the iron aluminides can corrode rapidly or at very low rates depending on the alloy composition and the solution concentration. Alloys containing both Cr and Mo are significantly more resistant to the sulfur-bearing solutions. The iron aluminides are sensitive to hydrogen embrittlement through interactions with water vapor in air or water in aqueous solutions. However, slow-strain-rate tests for the Cr-containing iron aluminides in a mild acid-chloride solution indicate that the embrittlement is not enhanced by the aqueous solution over that observed in air of normal humidity. The new low-aluminum alloys (16 at% Al), which contain Cr and Mo, display aqueous-corrosion properties comparable to that of the Cr- and Mo-containing Fe_3Al-based iron aluminides. Moreover, their ductilities during slow-strain-rate testing in an aqueous solution are significantly higher than that of the Fe_3Al-based alloys.

Little work has been performed relative to the aqueous-corrosion and corrosion-sensitive embrittlement characteristics of titanium aluminides. TiAl and Ti_3Al alloys undergo pitting corrosion in chloride solutions during anodic polarization, but may not undergo this form of attack under freely corroding conditions. The titanium aluminides are susceptible to hydrogen embrittlement during exposure to hydrogen-bearing atmospheres or after cathodically precharging with hydrogen; the degree of embrittlement strongly depends on alloy composition and microstructure. In an aqueous acid-chloride solution, a Ti_3Al-based alloy is found to undergo hydrogen embrittlement at applied hydrogen-producing cathodic potentials, and embrittlement at applied anodic potentials greater that the pitting potential, but not to undergo embrittlement at the free-corrosion potential. More investigations of the corrosion related behaviors of titanium aluminides are needed.

REFERENCES

AHMAD, S., MEHTA, M. L., SARAF, S. K. and SARASWAT, I. P., 1983a, *Corrosion* 39(8): 333.

________, 1983b, *Corrosion* 39(8): 330.

BANDY, R., ROBERGE, R., and NEWMAN, R. C., 1983, *Corrosion* 39(10): 391.

BAVARIAN, B., HARUTOUNI, S., and ZAMANZADEH, M., 1992, *Mat. Sci. Eng.* A153: 613–618.

BERGE, P. H., and DONATI, J. L., 1981, *Nuclear Technology* 55: 88.

BERTOCCI, U. J., FINK, L., HALL, D. E., MADSEN, P. V., and RICKER, R. E., 1990, *Corro. Sci.* 31: 471–478.

BUCHANAN, R. A., and CORALLO, C. S., 1993, Dept. of Materials Science and Engineering, Univ. of Tennessee, Knoxville, TN (unpublished research).

BUCHANAN, R. A., and KIM, J. G., 1988, *Aqueous Corrosion Characteristics of Nickel Aluminides*. Final Report. Toronto, Ontario: The Nickel Development Institute.

________, 1990, In *Proceedings of the Fourth Annual Conference on Fossil Energy Materials*, pp. 383–390. Oak Ridge, TN: Oak Ridge National Laboratory.

________, 1991a, *Aqueous Corrosion Characteristics and Corrosion-Related Cracking Susceptibilities of Fe_3Al-Type Iron Aluminides*. Report ORNL/Sub/88-07685CT92/01. Springfield, VA: National Technical Information Service.

________, 1991b, In *High-Temperature Ordered Intermetallic Alloys IV*, MRS Symp. Proc. 213: 945–950, L. A. Johnson D. P. Pope, and J. O. Stiegler, eds. Pittsburgh: Materials Research Society.

________, 1991c, In *Proceedings of the Fifth Annual Conference on Fossil Energy Materials*, pp. 271–279. Oak Ridge, TN: Oak Ridge National Laboratory.

________, 1992a, In *Proceedings of the Sixth Annual Conference on Fossil Energy Materials*, pp. 315–323. Oak Ridge, TN: Oak Ridge National Laboratory.

________, 1992b, *Fe_3Al-Type Iron Aluminides: Aqueous Corrosion Properties in a Range of Electrolytes and Slow-Strain-Rate Ductilities During Aqueous Corrosion*. Report ORNL/Sub/88-07685CT92/02. Springfield, VA: National Technical Information Service.

GAYDOSH, D. J., and NATHAL, M. V., 1990, *Scripta Metall.* 24: 1281.

GRAM, D. M., BERTOCCI, U., and RICKER, R. E., 1991, Technical Report No. 3 (for the period April–June 1991). Doc. No. 855.05:91-44. Gaithersburg, MD: National Institute of Standards and Technology.

HOROWITZ, H. H., 1983, *Corro. Sci.* 23: 353–362.

ISSEROW, SAUL, 1993, *Environmental Effects on Intermetallic Compounds: A Guide to the Literature*. ARL-MR-112. Watertown, MA: U.S. Army Research Laboratory.

JOHNSON, M., MIKKOLA, D. E., MARCH, P. A., and WRIGHT, R. N., 1990, *Wear* 140: 279–289.

KASUL, D. B., and HELDT, L. A., 1991a, *Scripta Metall.* 25: 1047.

________, 1991b, In *Environmental Effects on Advance Materials*. R. H. Jones and R. E. Ricker eds. Warrendale, PA: The Minerals, Metals & Materials Society.

KASUL, D. B., and HELDT, L. A., 1994, *Metall. Trans.* 25A: 1285–1290.

KATTNER, U. R., and BUTRON, B. P, 1993, In *Phase Diagrams of Binary Iron Alloys*. H. Okamoto, ed. pp. 12–28. Materials Park, OH: ASM Intl.

KIM, J. G., and BUCHANAN, R. A., 1993a, *Aqueous Corrosion Properties and Slow-Strain-Rate Ductilities of Fe_3Al-Based and Lean-Aluminum Iron Aluminides*. Report ORNL/Sub/88-07685CT92/03. Springfield, VA: National Technical Information Service.

________, 1993b, In *Proceedings of the Seventh Annual Conference on Fossil Energy Materials*. Oak Ridge, TN: Oak Ridge National Laboratory.

KIM, J. G., BUCHANAN, R. A., SIKKA, V. K., and WEIR, J. R., 1989, In NACE *Corrosion Research Syrmposium*, p. 3. Houston: National Asso. of Corrosion Engineers.

KURUVILLA, A. K., and STOLOFF, N. S., *Scripta Metall.* 19: 83–87.

LIU, C. T., and GEORGE, E. P., 1990, *Scripta Metall.* 24: 1285.

LIU, C., T., LEE, E. H., and MCKAMEY, C. G., 1989, *Scripta Metall.* 23: 875.

LIU, C. T. and MCKAMEY, C. G., 1990, In *High Temperature Aluminides and Intermetallics*. S. H. Whang, C. T. Liu, D. P. Pope, and J. O. Stiegler, eds. pp.

133–151. Warrendale, PA: The Minerals, Metals, and Materials Society.
LIU, C. T., MCKAMEY, C. G., and LEE, E. H., 1990a, *Scripta Metall.* 24: 385.
LIU, C. T., and STIEGLER, J. O., 1984, *Science* 226: 636–642.
LYNCH, R. J., HARBURN, M., MAUCIONE, L., and HELDT, L. A., 1994, *Scripta Metallurgica et Materialia* 30: 1157–1160.
LYNCH, R. J., HELDT, L. A., and MILLIGAN, W. W.,1991, *Scripta Metall.* 25: 2147.
LYNCH, R. J., MILLIGAN, W. W., and HELDT, L. A., 1991, In *Parkins Symposium on Fundamental Aspects of Stress Corrosion Cracking*. S. M. Bruemmer, E. I. Meletis, R. H. Jones, W. W. Gerberich, F. P. Ford, and R. W. Staehle, eds. Warrendale, PA: The Minerals, Metals & Materials Society pp. 117–130.
MARCH, PATRICK A., 1988, Report No. WR28-4-900-213, TVA Engineering Laboratory. Norris, Tennessee: Tennessee Valley Authority.
NEWMAN, R. C, ROBERGE, R., and BANDY, R., 1983, *Corrosion* 39(10): 386.
NEWMAN, R. C., and SIERADZKI, K., *Corro. Sci.* 23: 363–378.
RICKER, R. E., BERTOCCI, U., FINK, J. L., and STOUDT, M. R., 1991, In *Environmental Effects on Advanced Materials*: R. H. Jones and R. E. Ricker, eds. pp. 213–225. Warrendale, PA: TMS.
RICKER, R. E., and DUQUETTE, D. J., 1988, *Metall. Trans. A* 19A: 1775.
RICKER, R. E., HALL, D. E., and FINK, J. L, 1990, *Scripta. Metall.* 24: 291–296.
RICKER, R. E., STOUDT, M. R., and FINK, J. L., 1990, In *Hydrogen Effects on Materials Behavior IV*. N. R. Moody and A. W. Thompson, eds. pp. 499–506. Warrendale, PA: TMS.
SCARBERRY, R. C., PEARMAN S. C., and CRUM, J. R., 1976, *Corrosion* 32: 401.
SHEA, M., CASTAGNA, A., and STOLOFF, N. S., 1991, In *High-Temperature Ordered Intermetallic Alloys IV*. MRS Symp. Proc. 213: 609–616, L. A. Johnson, D. P. Pope, J. O. Stiegler, eds.. Pittsburgh; Materials Research Society.
SWANN, P. R., and DUFF, W. R., 1969, *Transactions TMS-AIME* 245: 851–853.
TAUB, A. I., and FLEISCHER, R. L., 1989, *Science* 243: 616–621.

Part IV

Processing

Chapter 14

Processing of Aluminides

V. K. Sikka

1. INTRODUCTION

Intermetallic compounds of nickel, iron, and titanium with aluminum have been studied extensively. Research on these materials has been published in the proceedings of several symposia and in journal articles. Detailed articles describing the development of nickel aluminides (Liu 1994; Noebe, Bowman, and Nathal 1994), iron aluminides (McKamey 1994), and titanium aluminides (Froes and Suryanarayana 1994) are the subject of other chapters in this book. In order to take advantage of their unique properties in various applications, methods for their processing need to be identified. In addition, the selected processing methods need to be implemented in the real-world manufacturing sector for the economical production of these materials. The processing of these materials is possible by both casting and powder metallurgy methods. The powder metallurgy processing is described in detail in a separate chapter (German and Iacocca 1994) of this book. This chapter deals with the processing techniques involving melting and casting into net or near-net shapes or ingots for secondary breakdown.

---

2. MELTING OF ALLOYS

There are several factors that need consideration in the melting of aluminides. These include:

1. Large differences between the melting point of aluminum and the aluminide forming elements (nickel, iron, and titanium)
2. Large amounts of aluminum (16 to 50 at%) as opposed to < 1 at% used in most commercial alloys
3. The exothermic nature of the intermetallic compound-formation reaction
4. The melting point of the resultant intermetallic, which can be higher than either of the elements, as is the case with NiAl
5. General reactivity of the elements involved in forming the aluminide.

In addition to the above factors, consideration is also required of the melting equipment and process parameters. The melting method can vary from the simplest air-induction melting to a method such as vacuum-arc double-electrode remelt (VADER). The process parameters require considerations of furnace linings (crucibles); melting atmosphere (air, vacuum, and inert gas); furnace loading (low- and high-melting-point metals); purification techniques (control of carbon, nitrogen, sulfur, etc.); and use of recycle material.

The extent of exothermic reaction in aluminum binary systems with various elements is shown in Figure 14-1 (German 1992). It is obvious from this figure that the formation of NiAl is the most exothermic reaction. The proper use of the exotherm during melting can result in substantial savings of energy and time needed for melting. The proper use of the exotherm can also result in good chemistry control of the melt, especially the oxidation products (slag formation). The improper use or lack of considering the exotherm can have serious consequences in the melting of aluminides. For example, the temperature of the molten metal can overshoot by several hundred degrees, which causes delays in cooling the metal to the desired pouring temperature; this may oxidize some of the alloying elements.

An optimized furnace loading arrangement for Ni_3Al-based alloys is shown in Figure 14-2. As the furnace is heated, the molten aluminum reacts exothermally with nickel to form NiAl. The superheated NiAl liquid of temperature exceeding its melting point of 1640°C flows downward, causing heating and melting of the charge underneath. Within a very short time, after the melting of the aluminum, the entire bath becomes molten, dissolving the various alloying elements. At this point, the furnace power can be adjusted to obtain the desired tapping temperature of the molten

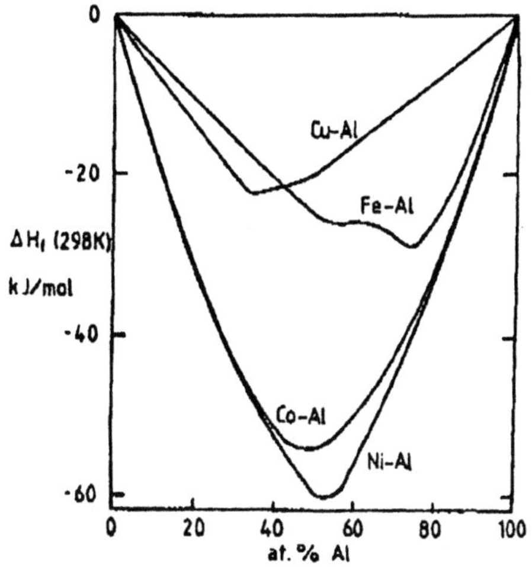

Figure 14-1. Extent of exothermic reaction in aluminum binary systems with transition metals: iron, cobalt, nickel, and copper (German 1992).

metal. The furnace loading shown in Figure 14-2 has been successfully used (Sikka 1993) during the commercial melting of 2500 kg heats of Ni_3Al-based alloys. The furnace loading sequence shown in Figure 14-2 is also applicable for melting iron and titanium aluminides. Instead of nickel, the iron or titanium for such melts will be divided between the top and bottom of the furnace to take advantage of the exothermic reactions for the formation of FeAl and TiAl compounds. The same furnace loading will result in the use of the exothermic reaction during melting in air or vacuum.

The Ni_3Al- and Fe_3Al-based aluminides can be melted in air (Sikka 1991). The high levels of aluminum in these alloys promote the rapid formation of a continuous Al_2O_3 film on the top of the molten metal. The oxide film is impervious to any transfer of oxygen or nitrogen from air into the molten metal, which is evidenced by the very low values of these elements in air-melted heats. Typical oxygen and nitrogen values in nickel and iron aluminides are < 40 ppm by weight (Sikka 1989). These values are similar to oxygen levels of less than 20 ppm by weight possible in aluminum deoxidized steels (Fruehan 1985). Compared to nickel aluminide, iron aluminides are very sensitive to pickup of hydrogen during air melting (Rishel and Brody 1991). The hydrogen results from the reduction of water vapor by aluminum. The water vapor can be present in the melt crucible or associated with the moist charge. The Fe_3Al-based iron alu-

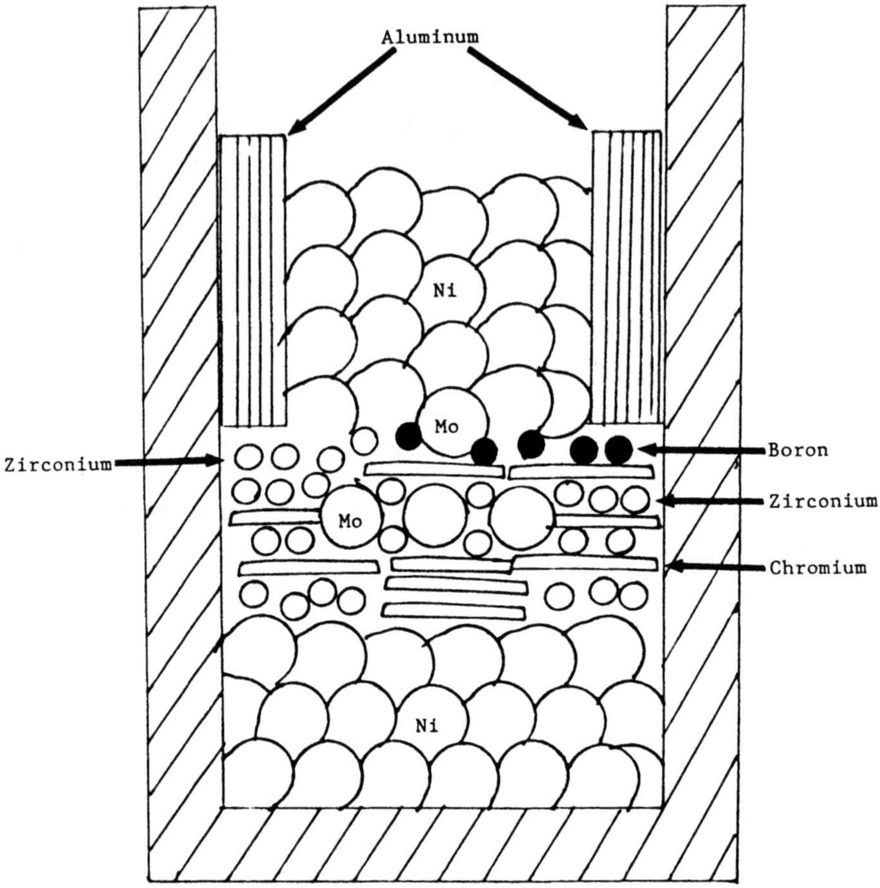

Figure 14-2. Furnace-loading sequence to take advantage of heat of formation of Ni_3Al during the melting of nickel-aluminide alloys. Similar loading sequence, replacing nickel by iron and by titanium, will work for the melting of iron and titanium aluminides (Sikka, unpublished research, 1993).

minide can dissolve 20 cc H_2 per 15 cc of alloy at the melt temperature (Rishel and Brody 1991). The dissolved hydrogen in molten metal and its rejection during solidification can cause gross porosity in the cast product. An example of gross porosity observed in a 230 kg air-melted iron aluminide is shown in Figure 14-3. The hydrogen-related gas porosity in iron aluminides can be eliminated (Rishel and Brody 1991) by: (a) using dry charge of both iron and aluminum; (b) purging the molten metal with

Figure 14-3. Gross porosity observed in a 230 kg air-melted ingot of iron aluminide (Sikka, unpublished research, 1990).

argon gas, which lowers the hydrogen level significantly; or (c) vacuum melting of the alloy. The typical value of 4 ppm by weight of hydrogen in air-melted material can be reduced to 2 ppm by argon blowing, and hydrogen levels of 1 ppm can be obtained by vacuum-induction melting (VIM).

The aluminum-oxide film formation on the top of the melt not only results in low levels of oxygen and nitrogen but also provides very high recovery of various elements. The recovery of various alloying elements during air melting of iron-aluminide alloys is shown in Table 14-1 (Sikka 1991). The retention of various elements as a function of holding time of the melt is important to meet the chemistry requirements during the casting of intermetallic alloys. Samples taken at the start and end of 8 min. of casting time of a nickel-aluminide alloy (see Table 14-2) show that the composition remains very stable. This aspect is again attributed to the highly protective nature of the Al_2O_3 film on top of the molten metal of these alloys.

Table 14-1 Recovery of Various Elements during Air-Induction Melting of 7-kg Iron-Aluminide Heats

	ALLOY 1		ALLOY 2		ALLOY 3		ALLOY 4	
Element	Target (wt%)	Recovery (%)	Target (wt%)	Recovery (%)	Target (wt%)	Recovery (%)	Target (wt%)	Recovery (%)
C	—	—	—	—	0.04	250.00[a]	—	—
Cr	2.157	98.75	5.46	91.03	5.46	95.24	2.194	95.72
Mo	0.995	100.00	—	—	—	—	—	—
Nb	1.542	100.00	—	—	0.21	100.00	—	—
Al	15.672	98.86	15.88	96.73	15.88	97.54	15.932	96.67
B	0.011	81.82	0.01	80.00	—	—	0.011	27.27
Zr	0.189	100.00	0.19	100.00	—	—	—	—
Fe	*b*	—	*b*	—	*b*	—	*b*	—

[a]Recovery of greater than 100% indicates pickup of carbon from external sources. For example, a graphite rod was used for stirring the liquid metal, and a small fraction of the graphite was probably dissolved in the metal.

[b]Balance (100 minus total of all other elements).

Source: Sikka et al. 1991

Table 14-2 Effect of Holding Time on the Chemical Analysis of a Nickel-Aluminide Alloy, IC-221M, Air Melted with an Argon Gas Cover

	WEIGHT PERCENT	
	Holding time (min)	
Element	0	8
Ni	81.2	81.2
Al	7.61	7.59
Cr	7.66	7.63
Mo	1.42	1.43
Zr	1.46	1.47
B	0.009	0.009
Nb	0.02	0.02
Ti	0.14	0.14
C	0.009	0.010
S	0.001	0.001
N	0.006	0.006
O	0.011	0.010
Fe	0.44	0.44
Si	0.04	0.04

Source: Sikka, unpublished research, 1993

Although air melting meets the recovery and chemistry requirements for nickel and iron aluminides, it is recommended that the melting be carried out under an argon-gas cover. One of the reasons for this recommendation is that during commercial melting, the Al_2O_3 film is frequently disturbed in order to take several chemical-analysis samples. Each time the oxide film is disturbed, it exposes fresh molten metal to oxidation. Using an argon-gas cover minimizes such oxidation possibilities.

The high aluminum content of the aluminides raises an issue regarding the suitability of various crucible materials. The magnesia crucible (74% MgO + 24% Al_2O_3) is most commonly used in commercial melting. A comparative study (Sikka 1991) on melting of iron aluminides in magnesia versus alumina crucible showed that the aluminum in the aluminide reduces part of the MgO, with resultant pickup of magnesium in the alloy. In fact, for an identical melt, the magnesium content of the alloy melted in a magnesia crucible was 20 ppm versus 10 ppm by weight in the Al_2O_3 crucible. The homogenously distributed magnesium does not seem to cause any problem up to levels of 20 ppm. However, a segregated magnesium-rich area was found to cause alligatoring during the hot rolling of iron aluminide (Sikka, unpublished research, 1991). A magnesium addition of 50 to 200 ppm by weight was found to increase the high-temperature ductility and hot workability of a Ni_3Al-Cr-based alloy (Peili et al. 1992). The beneficial effects of alloying with small additions of magnesium were suggested (Peili et al. 1992) to result from the enhancement of grain-boundary cohesion due to the segregation of magnesium in the grain boundaries, which inhibits the initiation and propagation of grain-boundary cracks.

Air melting is the simplest and least expensive process for melting nickel and iron aluminides, and for many applications, such a process is more than adequate. However, for improved fatigue life through the reduction of inclusions and improved yield through superior surface quality of ingots, other processes can also be used for the aluminides which include: VIM, vacuum-arc remelting (VAR), VADER, and electroslag remelting (ESR). The results of VIM, VAR, and ESR processes for the iron aluminides are described in the literature (Sikka 1991). The ESR (Sikka 1992) and VADER (Sikka 1989) processes for nickel aluminides are also discussed in the literature. The major benefits of the vacuum processes are to reduce the gas content (oxygen, nitrogen, and hydrogen). The VAR process also reduces the amount of sulfur and magnesium. This process also provides a microstructure with a grain orientation that is suitable for ease of hot processing. The VADER process is a further refinement of the VAR process and produces similar impurity control but produces an even finer solidification structure (Sikka 1989). The proper selection of slag during the ESR process can reduce undesirable impurities such as sulfur, magnesium, and even silicon. The ESR process also produces a highly desirable microstructure for hot deformation (see Figure 14-4). The surface quality of the ESR-processed ingots is such that they can be used for subsequent processing without additional surface conditioning. The surface quality of a 403-mm-diam by 2.7-m-long ESR ingot of iron-aluminide alloy FA-129 is shown in Figure 14-5.

The induction-skull melting (ISM) is one of the methods used for the preparation of titanium aluminides (Chronister et al. 1986; Breig and Scott 1989). The ISM process combines some features of consumable-arc-skull melting and conventional ceramic crucible-induction melting. The primary difference between ISM and conventional crucible-induction melting is in the nature of the crucible. The ISM process uses a multislotted copper crucible containing internal passageways for cooling water. The copper crucible is necessary to avoid the reaction and contamination associated with the ceramic crucible, and the slots are necessary to allow the induction fields to couple with the charge. Even with the slots, however, much more power is required for ISM than for ceramic crucible-induction

Figure 14-4. Favorable grain structure for hot deformation produced in nickel-aluminide alloy IC-218 by electroslag remelting (ESR). The photograph shows the top section of a 100 mm-diam ESR ingot [Reprinted from Sikka (1989), pages 1–24 by courtesy of Marcel Dekker, Inc.].

Figure 14-5. Photograph showing excellent surface quality obtained for an electroslag-remelted (ESR) ingot of iron-aluminide alloy FA-129 measuring 403 mm-diam by 2.7 m long. The ESR process was carried out at Precision Rolled Products, Inc. (Florham Park, New Jersey) [Sikka, unpublished research, 1992].

melting for the same size charge. The extra power provides an important side benefit in that the stirring action from the fields is even more intense than in conventional induction melting. The intense stirring produces a molten bath that is both chemically and thermally homogeneous, and when poured into a mold, produces a casting with a uniform chemistry.

Compared to consumable-arc-skull melting, ISM does not require the fabrication of an electrode. In addition to the elimination of the cost of pressing or welding scrap together, the potential for oxygen and nitrogen contamination during welding of scrap pieces is also eliminated. The ISM process is very flexible with regard to charge stock. A wide variety of shapes and sizes of charge materials may be melted. This flexibility in charge stock translates to an ability to easily blend the proper elements for alloying. In addition, the stirring action from induction melting further ensures that the alloying elements are mixed effectively. In the ISM process, the induction fields and constant stirring of metal maintain a high level of superheat throughout the melt, so that the temperature of the initial metal poured is virtually identical to the temperature of the final metal poured from a heat. The temperature uniformity and high degree of superheat allow for lap-free filling of sections as thin as 0.127 cm in statically cast cold-investment molds.

The ISM process has been successfully used (Breig and Scott 1989) in melting titanium-aluminide compositions. In the current practice, titanium scrap and master alloys are charged to the crucible and melted under argon. An argon melt is employed to minimize rapid outgassing and splashing of the charge material as it melts. Once melted, the molten bath is held to ensure all alloying additions are completely melted. While still under argon melt, aluminum is added to the molten bath by a vibratory feeder ramp. Once the alloying additions have melted and mixed uniformly, power is shut off and the chamber is evacuated. The power is again increased to melt the solidified metal. The molten metal can be poured into graphite molds for ingots or investment molds for shapes.

The ISM process can produce titanium aluminides with the maximum impurity levels of 350 ppm carbon, 200 ppm nitrogen, 135 ppm hydrogen, 800 ppm oxygen in γ alloys, and 1000 ppm oxygen in α_2 alloys. The lower oxygen levels are obtained for γ alloys because of the large amounts of aluminum present in the alloy. The aluminum charge material has very low base levels of oxygen and nitrogen.

The melting and investment-casting methods for conventional titanium alloys such as Ti-6 Al-4 V with special alterations have been used (London et al. 1993) to process γ-titanium aluminides. The VAR process is used to produce a γ-titanium-aluminide ingot from the master alloying elements. The ingots are double- or triple-VAR processed to ensure chemical homogeneity throughout the typical 70–160 kg ingots.

The plasma-arc cold-hearth melting (PACHM) is another method being tried (Johnson et al. 1993) for the melting of titanium aluminides. The melting is carried out in a water-cooled copper hearth to separate the three stages of the process (i.e., melting, refining, and casting of the alloy). The hearth also eliminates the possibility of contamination by a ceramic containment material and provides an opportunity for the removal of unwanted inclusions. The PACHM method takes place under a positive pressure (approximate 1.1 bar) of argon, which virtually eliminates the loss by evaporation of high vapor pressure-alloying elements such as aluminum, magnesium, and chromium. The plasma furnace described by Johnson et al. (1993) is capable of producing ingots of 100 and 150 mm nominal diam and up to 1.4 m in length. The furnace is powered by two 150 kW servohydraulic, computer-controlled, transferred-arc plasma torches that use helium as the plasma gas. One torch melts the feed into a water-cooled copper hearth with the initial formation of a skull. The liquid metal then flows over the skull from the hearth into the water-cooled copper mold with a retractable base to form a continuous ingot. Before melting starts, the furnace is evacuated to a pressure of $< 40\text{-}\mu$ bar and backfilled with argon to a pressure of 1.1 bar, which is maintained throughout the melting process. Details of the process parameters on the quality of the ingots are available in the paper by Johnson et al. (1993).

3. CASTING OF ALLOYS

Many applications of nickel, iron, and titanium aluminides require cast components. The casting methods of interest for these materials include: sand, investment, centrifugal, directional solidification, and innovative near-net shaping. The castability responses of the aluminides are related to the process as well as product. The process-related castability concerns deal with the steps involved in the casting process, whereas the product-related issues deal with the porosity and microstructural features of the cast components.

3.1. Casting-Process Issues for Aluminides

The availability of molten metal of controlled chemistry is the primary requirement for any of the casting processes. It is obvious from the previous section that the nickel- and iron-aluminide molten metal can be prepared by several conventional methods, and with some special care, inclusions and gas content can be minimized. Because of their reactivity, titanium aluminides can only be prepared by a few selected methods.

Oxidation or slag-forming tendency of the molten metal is also an important concern with the casting of components from the aluminides. The high aluminum content of the aluminides causes them to form an alumina skin (crust) even after a very short exposure of liquid to oxygen. For the casting of nickel and iron aluminides in air, the alumina formation can be minimized by a liquid-metal pouring process known as *tea kettle*. In this method, the clean metal under the alumina slag is poured into the mold without exposing to air. This method has been successfully applied to both nickel and iron aluminides. The other method to avoid such oxidation is to melt and cast in vacuum, which has been used for the casting of turbochargers of nickel (Sikka et al., unpublished research, 1993) and titanium aluminides (London et al. 1993; Breig and Scott 1989).

Molten-metal superheat retention during the time it takes to cast a batch of components is important to obtain reproducible quality of cast components. The molten-metal heat losses can be minimized during air casting of nickel and iron aluminides by using an insulating cover over the ladle holding the metal. This practice was found to be extremely useful during sand casting of nickel aluminide, where molten-metal temperature in the holding ladle had dropped several hundred degrees without a cover (Sikka, Reinholz, and Patel, unpublished research, 1993). When melting and casting are carried out in vacuum, as is the case with the investment-cast components of nickel and titanium aluminides, the heat loss becomes less of a concern. During vacuum casting, the molten metal is directly poured from the furnace into the preheated investment mold. The vacuum in the melting and casting chamber also eliminates the possibility of heat losses by convection.

Handling and recovery of excess metal during air melting and casting of nickel and iron aluminides are important. During a typical casting run, the excess metal is poured into sand molds in what is known as "pigs." The nickel and iron aluminides react with the sand to form a crust around the pig castings. These pigs are commonly used in the subsequent melting charges, and if not cleaned, they can increase the silicon content of the molten bath. Thus, it is recommended that the crust be removed from the pigs by some mechanical device to minimize the silicon pickup and make an efficient recovery of excess metal from a melting run. During the vacuum melting and casting, the melt charges are carefully monitored to minimize any excess material.

The chemical analysis of aluminides is also considered an important issue. The aluminum content of the nickel, iron, and titanium aluminides is beyond the testing capability of commercial equipment commonly used for chemical analysis. Experience to date has indicated that the standard equipment shows approximately 1 wt% lower aluminum in Ni_3Al-based alloys than actual amount. As a result, the commercial tendency is to add more aluminum to the melt, which pushes the aluminum level into the range where the boron addition is no longer effective in improving the ductility of Ni_3Al (Liu, White, and Horton 1985). Similarly, the control of aluminum content is important in obtaining an optimum combination of strength and ductility properties in iron and titanium aluminides. In order to ensure the needed control of properties, the aluminum analysis-standard development is critical for the range of aluminum levels encountered in the aluminides of interest.

3.2. Casting-Product-Related Issues for Aluminides

The reaction of aluminides with mold materials is an important consideration in the quality and properties of their castings. For example, when nickel-aluminide alloys containing zirconium are cast in conventional investment molds, the zirconium in the alloys reduces the SiO_2 in the zircon coat or Al_2O_3 in the cobalt aluminate which is used as a grain refiner. The result of such a reaction is the formation of ZrO_2 intrusions; see Figure 14-6a. These intrusions can act as crack initiators during cyclic loading and result in lower fatigue life. Special mold materials and washes are needed to minimize the reaction. For nickel aluminide, a carbon-containing shell can completely eliminate the formation of ZrO_2; see Figure 14-6b [Sikka and Reinholz, unpublished research, 1992]. Howmet Turbine Components Corporation (Whitehall, Michigan) has developed special mold washes for titanium-aluminide castings (London et al. 1993).

The extent of reaction in nickel aluminide was found to depend on its zirconium content (see Table 14-3). The ZrO_2 intrusions were totally absent for a zirconium content of 0.85 wt% (Sikka, unpublished research, 1993). The alloy composition can also be modified to eliminate the reaction of zirconium with the conventional mold material. One such example is the addition of hafnium, which preferentially forms a very thin continuous layer of HfO_2 on the casting surface and prevents the possibility of a zirconium reaction with the mold; see Figure 14-6c (Sikka and Reinholz, unpublished research, 1993). Similar alloy modifications should also work for titanium aluminides; however, no such data exist at present. The iron aluminides do not show any reaction with the mold material.

In addition to near-surface reaction, the properties of castings are also limited by casting defects and microstructural features. The casting

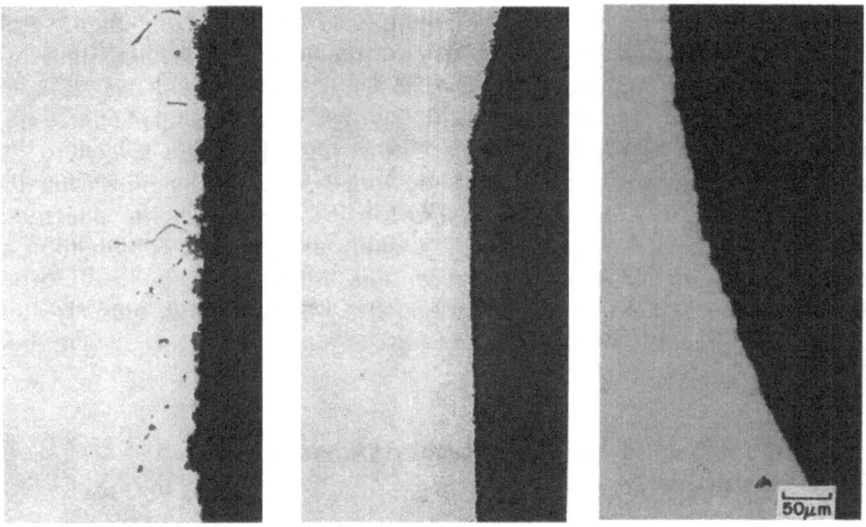

(a) (b) (c)

Figure 14-6. Surface reaction of nickel-aluminide alloy with mold material and methods of eliminating: (*a*) nickel aluminide of 1.7 wt% Zr cast in an alumina-silicate mold, (*b*) same composition cast in a clay-graphite crucible, and (*c*) same composition modified by hafnium and cast in zircon shell with a cobalt-aluminate wash (Sikka, unpublished research, 1992; Sikka and Reinholz, unpublished research, 1993).

Table 14-3 Effect of Zirconium Content in Nickel-Aluminide Alloy on Surface Reaction with the Investment Mold Shell Material

Alloy Zirconium content (wt%)	Reaction depth (mm)
0.85	0
1.28	0.051
1.70	0.076

Source: Sikka, unpublished research, 1993

defects are macroporosity, microporosity, gas porosity, and inclusions. The microstructural features include the secondary dendrite arm spacing, grain size, and the size and distribution of various phases in the alloy. The cooling rate during casting has been reported (Ho, Cheng, and Sekhar

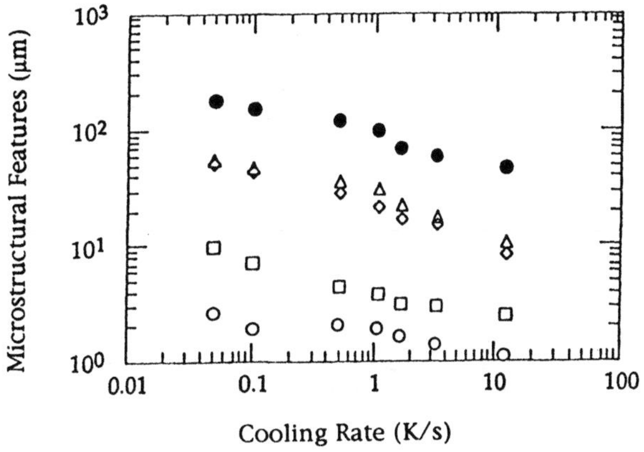

Figure 14-7. Size of various microstructural features as a function of cooling rate for nickel-aluminide alloy IC-396M containing 0.85 wt% Zr (Ho et al. 1991).

1991) to be the most important factor in controlling both the casting and microstructural features in nickel-aluminide alloys. Figure 14-7 shows a summary plot of cooling-rate effect on the radius of macroporosity, microporosity, the γ' particles, γ' cells, and the secondary dendrite arm spacing, $\lambda 2$ for a nickel-aluminide alloy containing 0.85 wt%Zr. It is obvious from this figure that the size of all features can be reduced by a factor of approximately three by increasing the cooling rate by two orders of magnitude.

Besides cooling rate, composition also affects the volume fraction of microporosity in nickel aluminide (Cheng 1992; Sekhar, Liu, and Cheng 1991); see Figure 14-8. The IC-221M alloy, which contains 1.7 wt%Zr, shows a significantly lower microporosity volume fraction than alloy IC-396M, which contains 0.85 wt%Zr. The size of the microporosity is also somewhat lower in the IC-221M alloy as opposed to IC-396M. The IC-221M alloy contains a greater amount of Ni_5Zr_3 eutectic (*mp* = 1172°C). It is the presence of the low-melting eutectic that reduces the volume fraction of the microporosity by being the last liquid to freeze. The

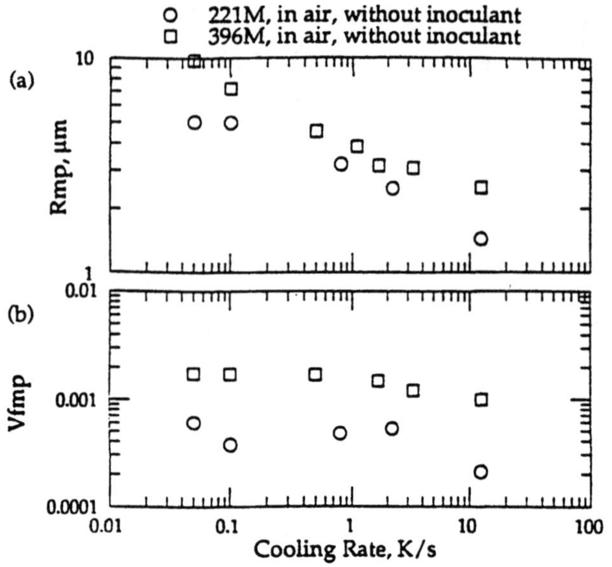

Figure 14-8. Effect of cooling rate on volume fraction and size of microporosity of nickel-aluminide alloys IC-396M containing 0.85 wt% Zr and IC-221M containing 1.7 wt% Zr (Cheng 1992).

same eutectic is also considered responsible for slightly reducing the microporosity size. The volume fraction of the microporosity is essentially independent of the cooling rate. However, the volume fraction of the macroporosity is significantly lowered by increasing the cooling rate; see Figure 14-9.

Increasing the zirconium content in nickel aluminide reduces the volume fraction and size of the macroporosity. However, as shown in Table 14-3, it also increases the depth of reaction product from reaction with the mold-shell material. The competing effects are shown schematically in Figure 14-10. This figure suggests that the zirconium level in nickel aluminides needs to be chosen to minimize both the surface-reaction depth and the volume fraction of microporosity.

The knowledge of cooling-rate effects on the casting defects and microstructural features helps in the mold design to control the porosity in various sections of a complex casting. No such effects of cooling rate on

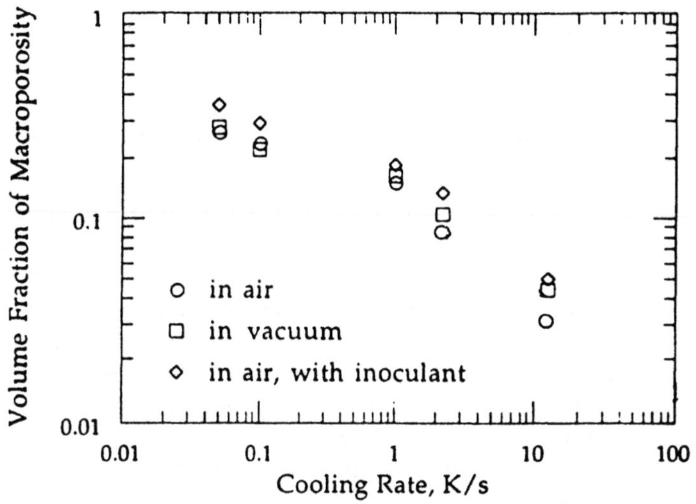

Figure 14-9. Effect of cooling rate on the macroporosity volume fraction of a nickel-aluminide alloy IC-221M containing 1.7 wt% Zr (Cheng 1992).

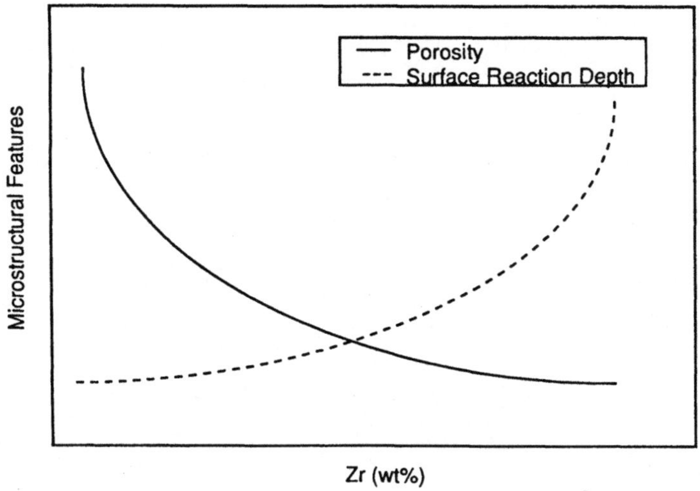

Figure 14-10. Schematic representation of effects of zirconium content on microporosity and surface reaction depth for nickel-aluminide alloys (Sikka, unpublished research, 1992).

the casting defects and microstructural feature for iron and titanium aluminides are known at present.

3.3. Castings of Aluminides by Near-Net-Shape Techniques

Casting of aluminides into sheet and bar by near-net-shape methods are of great interest for their low-cost manufacturing. The sheet is directly cast from liquid metal by bringing the liquid in contact with a rotating drum. The description of the process variables that control the final dimensions and quality of the sheet are beyond the scope of this chapter. Suffice it to say that this method has been successfully used for the casting of 1–2 mm-thick by 300 mm-wide continuous sheet coils of nickel aluminides; see Figure 14-11. The sheet coils cast from the air-melted alloy had excellent surface quality and microstructure (Sikka 1989). The direct casting of iron-aluminide sheet coils has also been successful. However, iron aluminide, which has very low ductility in the cast condition (Sikka

Figure 14-11. Direct-cast sheet of 1–2 mm-thick by 300 mm-wide nickel-aluminide alloys IC-50 and IC-218. Sheet casting was carried out at Allegheny Ludlum Corporation (Brackenridge, PA) [Reprinted from Sikka (1989), pages 1–24 by courtesy of Marcel Dekker, Inc.].

et al. 1991), becomes difficult to uncoil and roll from the direct cast sheet at room temperature. Since the direct sheet-casting process is available only for conventional melting crucibles and is carried out in air, it has not been applied to the sheet casting of titanium aluminides.

Direct bar casting from liquid metal is carried out by directly connecting the mold to the induction furnace. Initially, the hole in the mold is plugged by a dummy bar stock, and the process is started by pulling this bar horizontally through a set of pinch rolls. The nickel aluminides have been successfully cast (Sikka 1989) into 8 mm and 12 mm-diam rods; see Figure 14-12. The iron aluminides have been successfully cast in 75 mm-diam bar stock (Sikka, unpublished research, 1992). In principle, titanium aluminides can also be cast into bar by this process. The limiting factors are the capability of melting titanium aluminides and avoiding interactions with the crucibles and the mold material.

Figure 14-12. Direct-cast bar of 8 mm-diam nickel-aluminide alloy IC-221M. Casting was carried out at Harrison Alloys (Harrison, NJ) [Reprinted from Sikka (1989), pages 1–24 by courtesy of Marcel Dekker, Inc.]).

4. PROCESSING OF INTERMETALLICS

Many applications of intermetallic alloys use castings. However, many components also require their use in the fine-grain wrought condition. The commercial utilization of the intermetallic alloys at competitive cost requires their fabrication by conventional processes including: hot extrusion; hot forging; hot-flat and bar rolling; hot swaging; cold-flat and bar rolling; and cold drawing into tube, rod, and wire. The hot-working operations are desirable because of low required-flow stresses, which put less power requirements on the processing equipment. The low flow stresses at high temperatures also promote a more homogeneous deformation across the section thickness.

The general criteria that determine the processing response of an alloy include: ductility (30% or greater) of the cast structure at desirable hot-processing temperature and at industrial strain rates approaching 10^{-1} s^{-1}; the flow stress of the cast structure at processing temperatures and strain rates should be one-fifth to one-tenth of the flow stress at room temperature; a broad temperature range with high ductility; absence of the low melting-point liquid formation at the processing temperature; and absence of environmental effect and good intermediate-temperature ductility ($\geq$ 10%) to prevent cracking during cooling from the hot-processing temperature.

Table 14-4 The Ni_3Al-Based Alloy Compositions Selected for Commercial Applications

	Alloy (Atomic Percent)			
Element	IC-50[a]	IC-218LZr[b]	IC-396M[c]	IC-221M[d]
Al	21.73	17.01	15.94	15.93
Cr	—	8.22	8.00	7.96
Mo	—	—	1.70	0.80
Zr	0.34	0.12	0.50	1.00
B	0.01	0.01	0.025	0.04
Ni	77.92	74.64	73.84	74.27

[a] Cold workable.

[b] Hot and cold workable.

[c] Castable alloy for static applications (some microporosity).

[d] Castable alloy for dynamic applications (no harmful microporosity).

Source: Reprinted with permission from *Materials Science and Engineering*, Volume A153, V. K. Sikka, S. Viswanathan, and C. G. McKamay, pages 712–721, 1992, Elsevier Science Ltd., Oxford, England

Table 14-5 The Fe_3Al-Based Alloy Compositions Selected for Commercial Applications

	ALLOY (ATOMIC PERCENT)		
Element	FAS[a]	FAL[b]	FA-129[c]
Al	28.08	28.03	28.08
Cr	2.02	5.03	5.04
B	0.04	0.04	—
Zr	—	0.08	—
Nb	—	—	0.51
C	—	—	0.20
Mo	—	—	—
Y	—	—	—
Fe	69.86	66.81	66.17

[a] Sulfidation-resistant alloy.

[b] High room-temperature (RT) tensile ductility and cyclic oxidation resistant.

[c] High-temperature strength with good RT ductility.

Source: Sikka et al., 1991

There are four nickel-aluminide (Ni_3Al-based) compositions (see Table 14-4) and three iron-aluminide (Fe_3Al-based) compositions (see Table 14-5) that are currently being commercialized for a broad range of applications. There are numerous titanium-aluminide (TiAl-based) compositions that have been investigated in the literature. The one that has received the most attention for commercial application is the one based on the General Electric composition: Ti-48Al-2Cr-2Nb (at%) (Huang 1993). The other compositions of titanium aluminide will be mentioned as needed to bring out the data.

Ductility and its variation with test temperature, composition, strain rate, grain size, microstructure, and test environment are the general issues in the processing of intermetallics by conventional methods mentioned above. Effects of various variables on the ductility of nickel, iron, and titanium aluminides and how they influence their processing are described below.

4.1. Nickel-Aluminide Alloys

The tensile ductility of IC-50 nickel aluminide (see Table 14-4) in the cast and wrought conditions is shown as a function of temperature in Figure 14-13 (Sikka 1989). The nickel aluminide in the cast condition has over 10% ductility at room temperature (see Figure 14-13a), which goes to less than 5% at 1000°C. Furthermore, the ductility becomes less than 1% at 800°C due to dynamic oxygen embrittlement (Liu and Sikka 1986; Sikka 1989). The combination of low high-temperature ductility and dynamic oxygen embrittlement prevent the hot working of IC-50-type alloy in the cast condition by conventional methods in the bare condition. The cast ingots can be encapsulated in a stainless steel can and extruded at temperatures in the range of 1100 to 1150°C. Extrusion-can design by the finite-element method has been described by Semiatin et al. (1991). The cast or extruded stock can be cold rolled with intermediate anneals in the range of 1050 to 1100°C. The refined grain structure improves the ductility of IC-50 both at room temperature and high temperature; (see Figure 14-13a). However, the low ductility at 800°C due to dynamic oxygen embrittlement still exists in the fine-grain material.

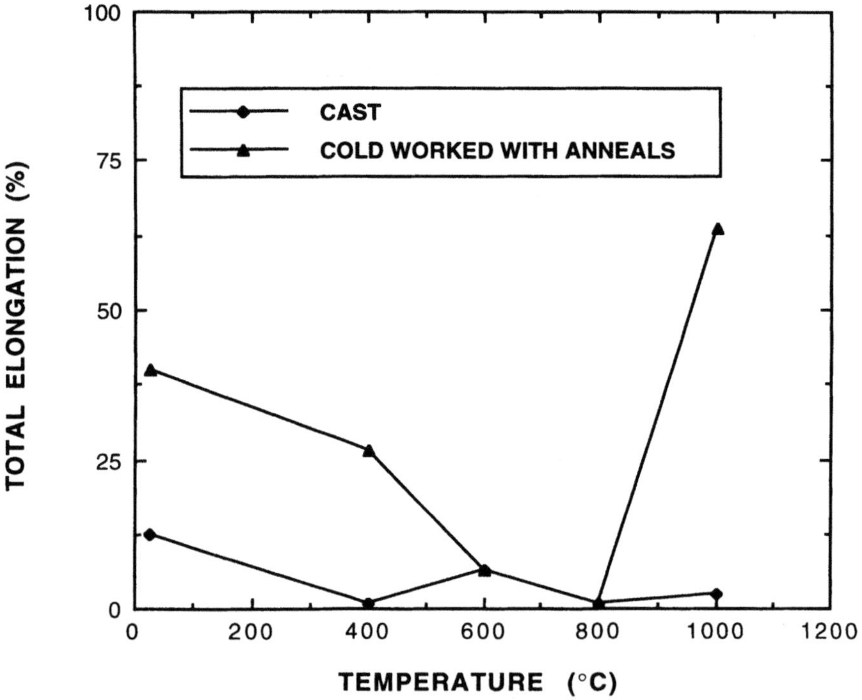

(a)

Figure 14-13a. Effect of grain refinement by cold working with intermediate anneals on ductility of Ni_3Al-based alloy IC-50. Tests were conducted in air at a strain rate of 3×10^{-3} s^{-1} [Reprinted from Sikka (1989), pages 1–24 by courtesy of Marcel Dekker, Inc.].

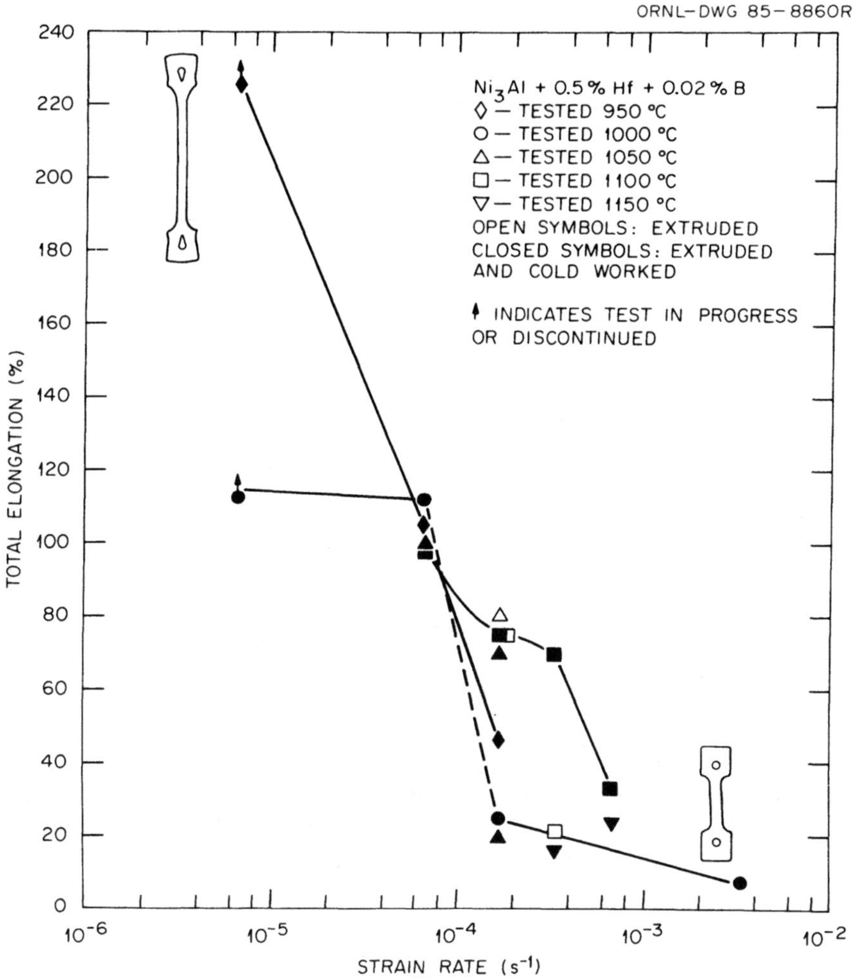

Figure 14-13b. Effect of strain rate and test temperature on tensile ductility of Ni_3Al alloy containing 0.5 wt% Hf and 0.02% B (Sikka 1987; reprinted with permission of the American Society of Mechanical Engineers).

The ductility of the fine-grained nickel aluminide (IC-50) is also affected (Sikka 1987) by test temperature and strain rate at temperatures ≥ 950°C; see Figure 14-13b. At strain rates $\leq 10^{-4}$ s^{-1}, the ductility values increased sharply for all test temperatures. Indications of super-plastic behavior are visible for a test temperature of 950°C and a strain rate of 6×10^{-5} s^{-1}. At strain rates $\geq 10^{-4}$ s^{-1}, higher ductility values are obtainable by increasing the test temperature. The lower ductility

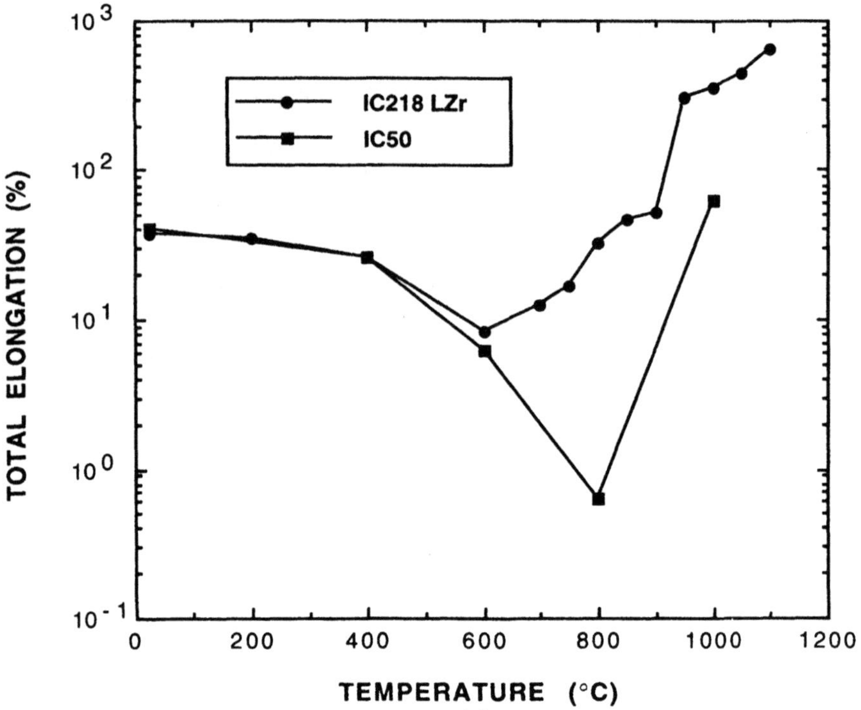

(a)

Figure 14-14a. Effect of chromium in reducing the dynamic-oxygen-embrittlement effect in Ni_3Al-based alloys. Both alloys were tested in air at a strain rate of $6.7 \times 10^{-4}\ s^{-1}$ in the wrought condition [Reprinted from Sikka (1989), pages 1–24 by courtesy of Marcel Dekker, Inc.].

values for a test temperature of 1150°C are probably related to the grain growth.

The dynamic oxygen embrittlement in Ni_3Al-based alloy (IC-50) is reduced by an addition of chromium; see Figure 14-14a. The chromium-containing alloy is designated as IC-218; see Table 14-4. The ductility of this alloy in the cast and wrought conditions is shown in Figure 14-14b. This figure also reinforces the significant effect of ductility improvement through grain refinement. The superplastic flow is also evident in fine-grain material at test temperatures of $\geq$ 950°C.

The hot-compression response for nickel-aluminide alloy IC-218LZr for a range of temperatures and percent reductions is shown in Figure 14-15. The curve showing the temperature above which the cast structure can be hot forged without significant surface defects is also included in

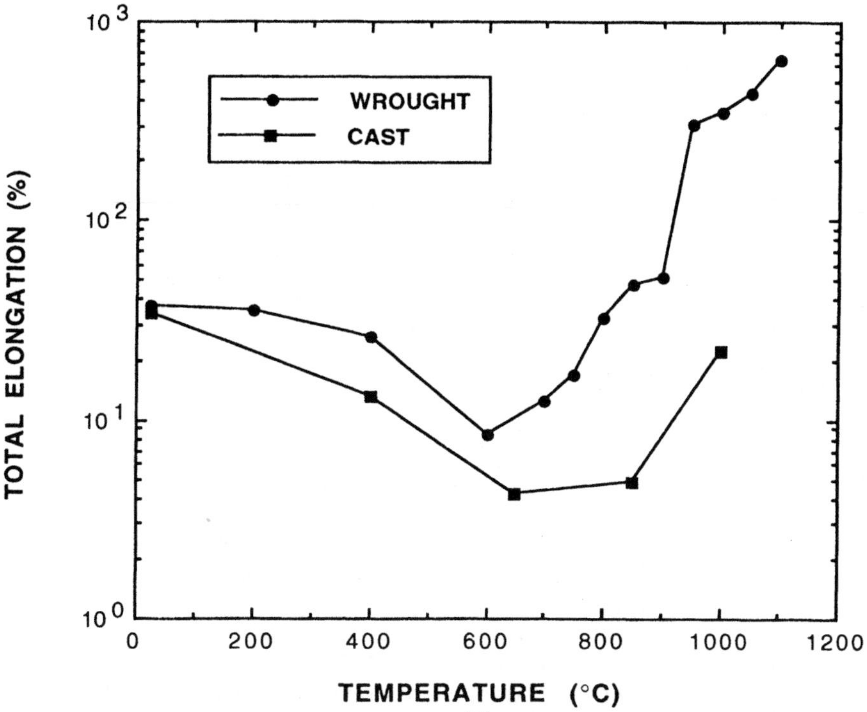

(b)

Figure 14-14b. Effect of grain-structure refinement through hot working followed by cold working on ductility of Ni_3Al-based alloy IC-218LZr. All of the specimens were tested in air at a strain rate of 6.7×10^{-4} s^{-1} [Reprinted from Sikka (1989), pages 1–24 by courtesy of Marcel Dekker, Inc.].

Figure 14-15. The temperature range of 1150°C to 1200°C is considered safe for hot forging of the nickel-aluminide alloy IC-218LZr in the cast condition. The safe processing temperature was not influenced by changing from the diameter of the experimental to commercial-size heats (approximately 75–403 mm) (Sikka et al. 1992). The cast structure, which can be hot forged as discussed above, still lacks tensile ductility for direct extrusion or rolling. In order for the cast structure to be extruded successfully, it needs to be encapsulated in a mild steel can. The cast billets encapsulated in a mild steel can are extrudable at 1100°C to 1150°C. The extrusion process produces an equiaxed fine-grain microstructure that can be further refined by cold working and annealing; see Figure 14-16 (Sikka et al. 1992). Once the large-grain cast structure of alloy IC-218LZr is refined, its grain size remains stable up to 1200°C; see Figure 14-17 (Sikka, unpublished research, 1992).

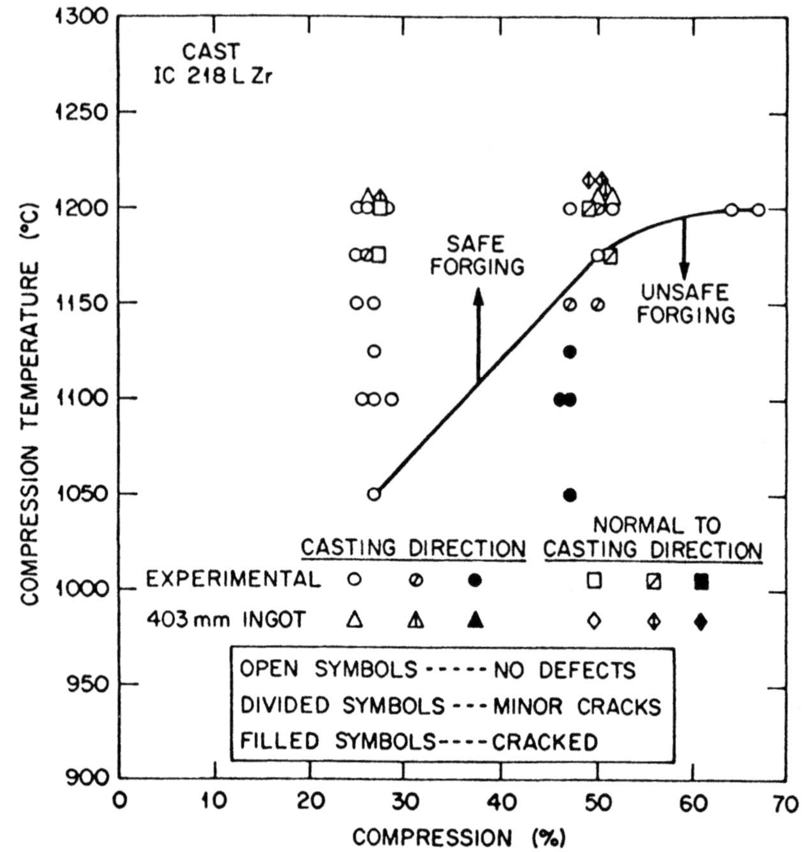

Figure 14-15. Hot-compression response of samples taken from the experimental (75 mm-diam) and commercial (403 mm-diam) ingots of Ni_3Al alloy containing chromium (IC-218LZr) (Sikka et al. 1992; reprinted from *Materials Science and Engineering*, Volume A153, pages 712–721, copyright 1993, with kind permission from Elsevier Science Ltd., The Boulevard, Langford Lane, Kidlington OX5 1GB, UK).

To avoid canning of billets, the best practice for conventional processing of nickel-aluminide alloy IC-218LZr is to: (a) break the coarse-grain cast structure by hot forging (80% in reduction of area), preferably rotary forging at 1150°C to 1200°C, and (b) homogenize the cast structure at 1200°C for times exceeding 8 h to improve the hot-deformation response of the alloy. However, care must be taken to prevent near-surface oxidation (Sikka et al. 1992). If near-surface oxidation is observed, it should be removed by machining to avoid surface cracking during hot forging.

The nickel-aluminide alloys IC-396M and IC-221M (see Table 14-4) are alloyed by the addition of molybdenum and high zirconium content.

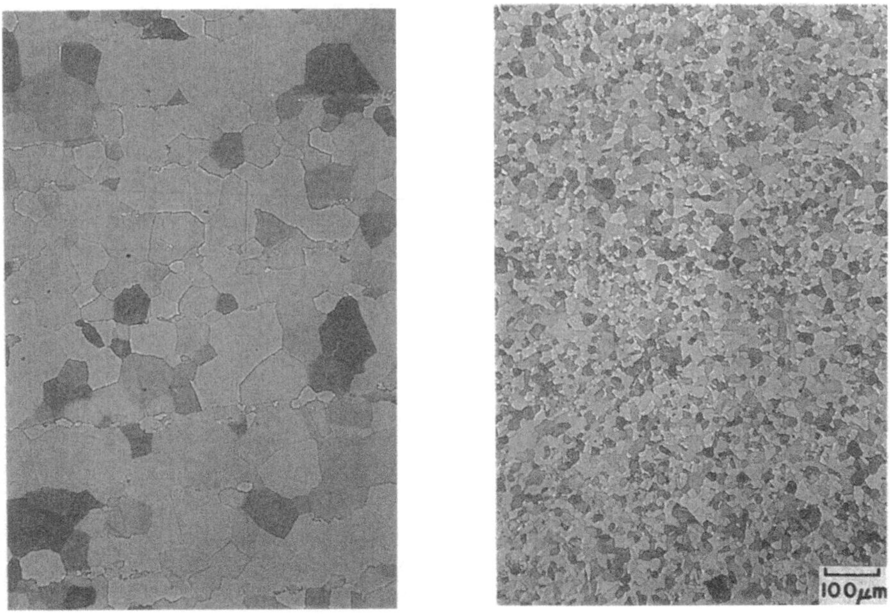

(a) (b)

Figure 14-16. Grain refinement produced in Ni_3Al-based alloy IC-218LZr by: (a) hot extrusion of cast ingot and (b) further refinement by 35% cold work followed by anneal at 1100°C for 1 h (Sikka et al. 1992; reprinted from *Materials Science and Engineering*, Volume A153, pages 712–721, copyright 1992, with kind permission from Elsevier Science Ltd., The Boulevard, Langford Lane, Kidlington OX5 1GB, UK).

Alloying with molybdenum increases the room-temperature strength of the castings through solid-solution strengthening. The zirconium addition reduces the micro- and macroporosity of the castings through the formation of low melting-point eutectic (Ni_5Zr_3), which freezes last during solidification. The high alloy content of IC-396M and IC-221M results in castings of acceptable tensile and fatigue strengths, but these alloys are not readily hot workable. The low melting-point (1172°C) eutectic (Ni_5Zr_3), which reduces casting porosity, becomes the major issue in the hot-processing temperature range of 1150°C to 1200°C identified for nickel-aluminide alloy IC-218LZr; see Figure 14-15.

4.2. Iron Aluminides

The ductility of iron-aluminide alloy FA-129 (see Table 14-5 for composition) in the cast and wrought conditions is shown in Figure 14-18 (Sikka et al. 1991). The ductility of this alloy in the cast condition is not

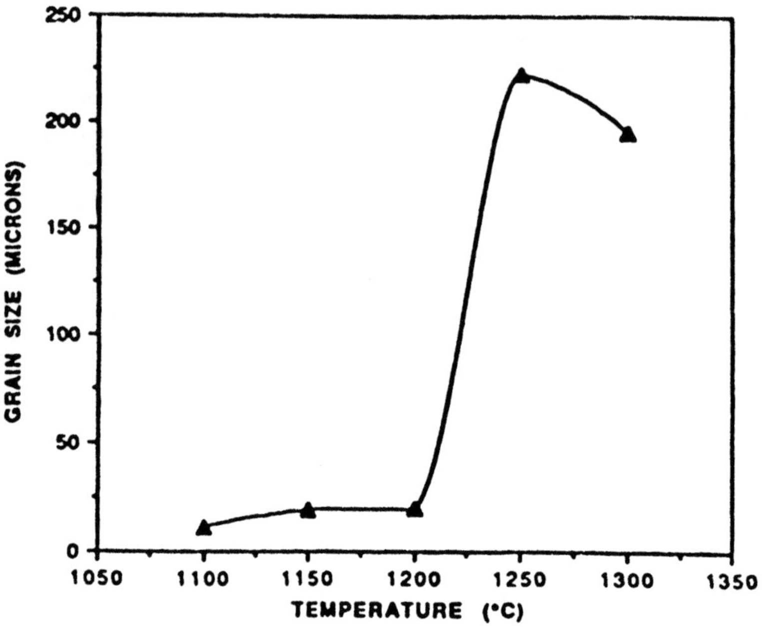

Figure 14-17. Grain size as a function of temperature for Ni_3-Al-based alloy IC-218LZr. The 57% cold-worked samples were annealed for 1 h at the indicated temperatures for this study (Sikka, unpublished research, 1992).

even measurable up to 650°C. However, the ductility values increase to over 40% at 800°C, which remains constant up to 1000°C. The high ductility values over a broad range of temperatures allow the processing of iron-aluminide ingots by conventional methods. For example, the ingots can be extruded without a can. They can be conventionally hot forged and hot rolled. The bars fabricated by hot-bar rolling of 102 mm and 152 mm-diam ingots of alloy FA-129 are shown in Figure 14-19.

The working of the cast structure improves it ductility for the entire test-temperature range; see Figure 14-18. However, it is the combination of processing and heat treatment that produces the maximum ductility. The processing steps for maximizing the ductility consist of hot forging 50% at 1000°C to 1100°C, hot rolling 50% at 800°C, and final rolling of 50–70% at 650°C. The last rolling step produces a highly elongated microstructure with minimal transverse-grain boundaries; see Figure 14-20. Such a microstructure, when oil quenched from 700–750°C, results in a partially recrystallized B2 structure. The partially recrystallized B2 structure has been reported to produce the maximum ductility at room temperature (Sikka 1991; Sikka et al. 1991). The iron aluminides are also susceptible to moisture-related hydrogen embrittlement at temperatures

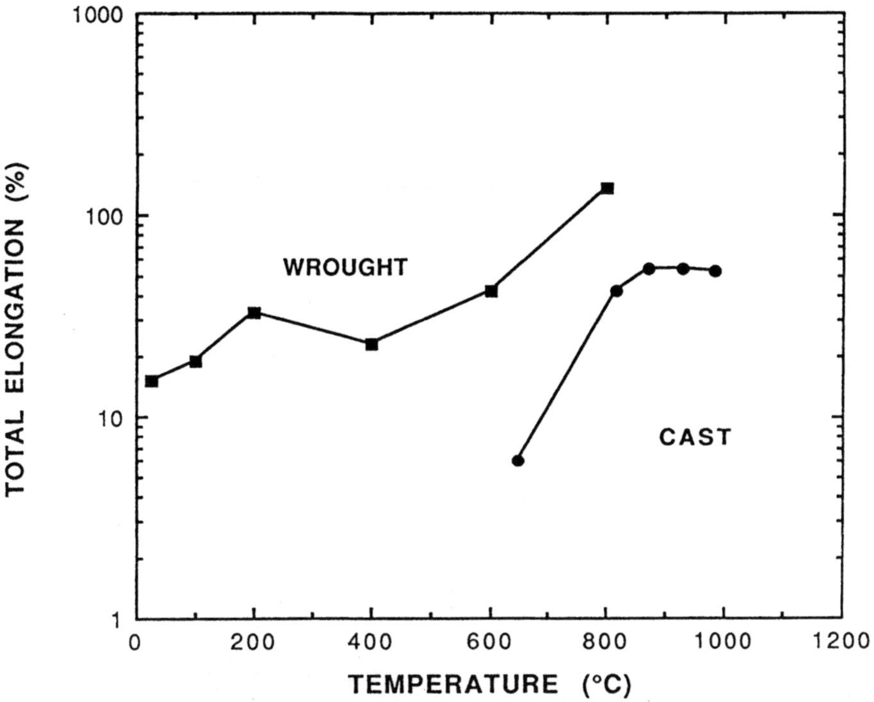

Figure 14-18. Comparison of cast and wrought tensile elongation of iron-aluminide alloy FA-129 (Sikka 1991).

≤ 200°C (Liu et al. 1989; McKamey et al. 1991; Sikka, Reinholz, and Patel 1993). The partially recrystallized B2 structure results in maximum ductility by minimizing the environmental effect at ≤ 200°C (Sikka et al. 1993).

The iron aluminides are very ductile in both as-cast and wrought conditions at temperatures of ≥ 650°C; see Figure 14-18. The ductility of iron aluminides is significantly improved at low temperature through thermomechanical processing described above. However, at the intermediate processing steps, for example, after 800°C rolling, the tensile ductility is still too low to permit the drawing operations to make rod and wire. Warm drawing is possible where the drawing die and the work piece are maintained at 650°C. In practice, even such a drawing process becomes difficult to manage because a drop of a few degrees in temperature of the work piece causes breakage due to brittle-cleavage fracture, and a few degrees rise in temperature causes ductile failure due to extensive necking. Thus, the only way to draw wire from iron aluminides is to produce high ductility, elongated B2 microstructure through thermomechanical

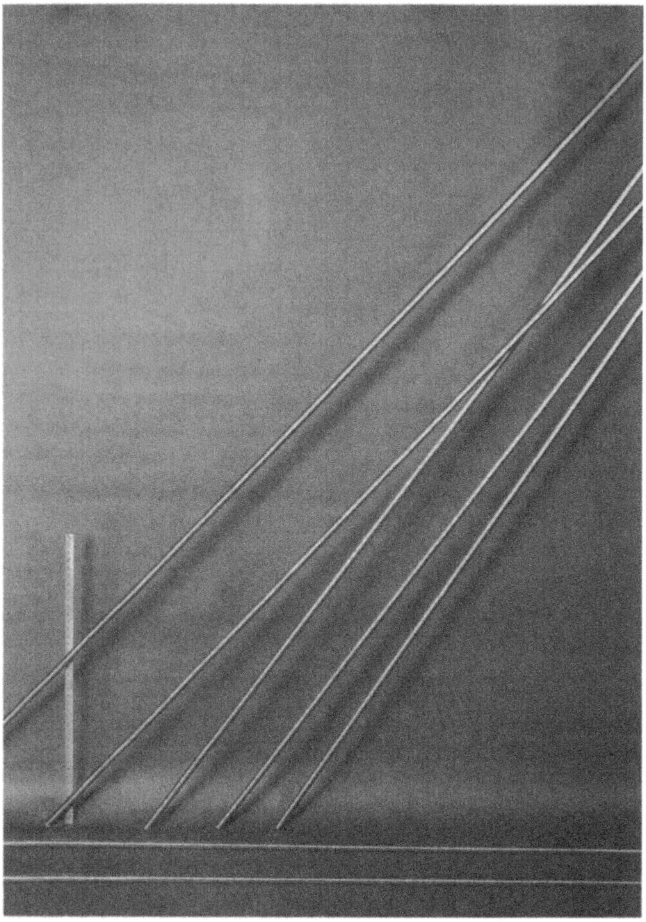

Figure 14-19. Photograph of 25 mm-diam bars fabricated by hot-bar rolling of 102 and 152 mm-diam ingots of iron-aluminide alloy FA-129. Melting and fabrication were carried out by special metals corporation (New Hartford, NY) (Sikka 1991).

processing steps described above and make no more than total reductions of approximately 20% in between heat treatments. The reduction between the heat treatments increases as the wire size decreases through the progressive drawing steps. The frequently required heat treatments are possible through resistance heating of the wire in air followed by oil quenching. Similar steps will likely have to be followed to produce foil from the sheet of iron aluminides by cold rolling.

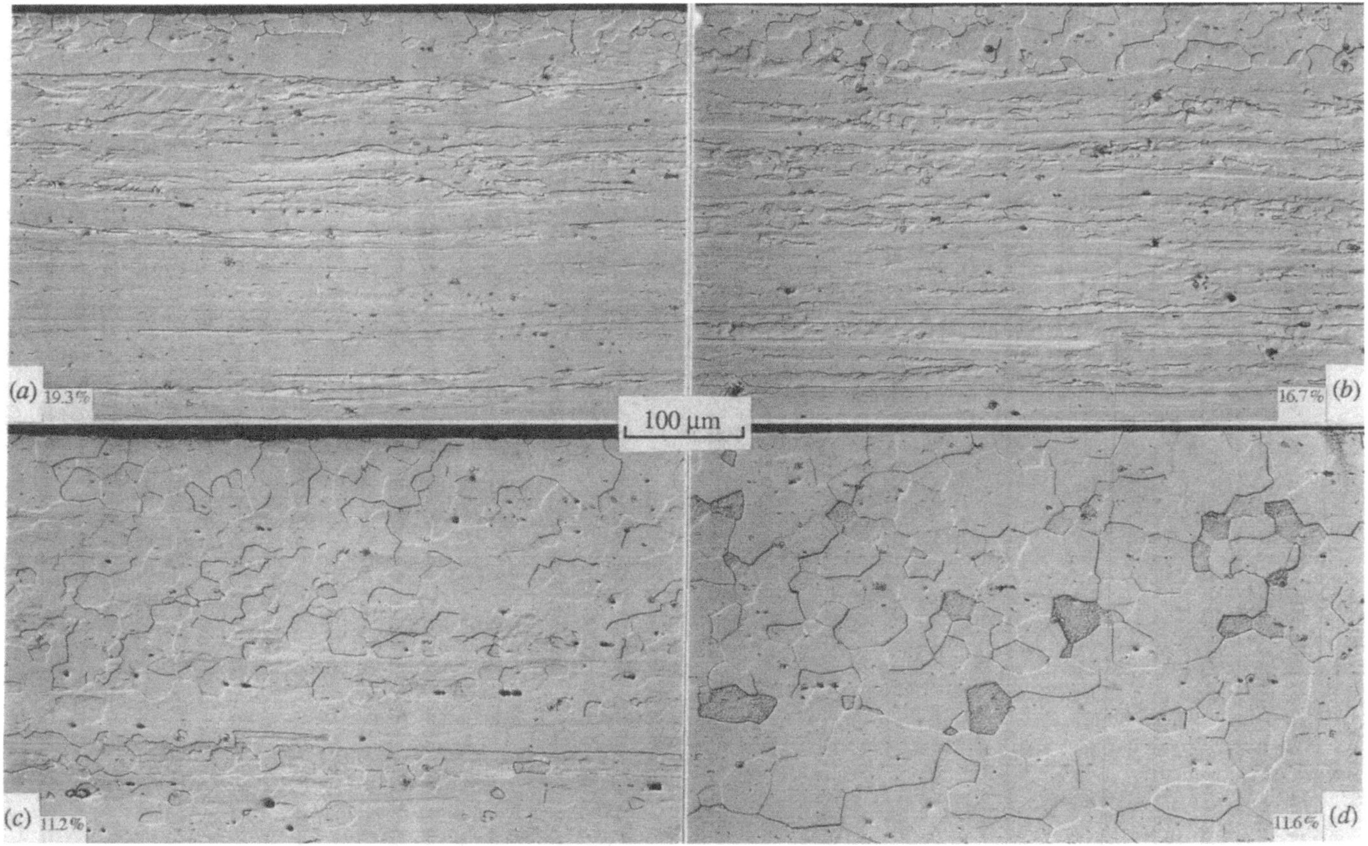

Figure 14-20. Effect of thermomechanical processing on room-temperature tensile ductility of iron-aluminide alloy FAL. Total elongation values are marked on each micrograph for microstructures produced after heat treating the warm-rolled sheet for 1 h at: (a) 700°C, (b) 750°C, (c) 800°C, and (d) 900°C followed by oil quenching (Sikka 1993).

Table 14-6 Tensile Elongation Data on Cast Specimens at Two Different Strain Rates of γ-TiAl Alloy Ti-49.5Al-2.5Nb-1.1Mn (at%)

Temperature (°C)	Nominal strain rate (s^{-1})	Elongation (%)	Reduction of area (%)
1050	10^{-2}	< 0.2	0
1150	10^{-2}	50.7	44.6
1150	1	0	0
1250	1	0	0

Source: Seetharaman et al., 1991a

The processing of iron-aluminide alloy FA-129 has been used as an example. Other alloys (FAS and FAL), see Table 14-5, also respond very similarly in their response to processing.

4.3. Titanium-Aluminide Alloys

Unlike nickel and iron aluminides, essentially no tensile-ductility data are available for γ-TiAl alloys in the as-cast condition over a broad range of temperatures. Limited tensile-ductility data at room temperature and 700°C are available for six experimental alloys from Anton et al. (1993). One of the best alloys showed the tensile elongation values of 1.24% and 1.94%, respectively, at room temperature and 700°C.

Seetharaman et al. (1991a) conducted tensile tests on induction-skull-melted cast ingots of a Ti-49.5Al-2.5Nb-1.1Mn (at%) alloy. The ingots were hot-isostatically pressed at 1175°C and 105 MPa for 4 h followed by slow cooling prior to testing. Tensile elongation data on the cast specimens at two different strain rates are shown in Table 14-6. These data indicate nil ductility for the cast Ti-49.5Al-2.5Nb-1.1Mn (at%) alloy for test temperatures up to 1050°C for a strain rate of 10^{-2} s^{-1} and nil ductility up to 1250°C for a strain rate of 1 s^{-1}. The poor ductility of γ-TiAl alloys in the as-cast condition is the cause of their limited hot workability by conventional processing methods.

Isothermal compression tests to 80% true strain at a wide range of temperatures (975–1200°C) and strain rates (1×10^{-1} to 3×10^{-3} s^{-1}) have been carried out by Shih and Scarr (1991) on cast cylinders of Ti-48Al-2Cr-2Nb (at%) alloy. The true stress-true strain curves from Shih and Scarr (1991) are reproduced in Figure 14-21 for the highest (1×10^{-1} s^{-1}) and lowest (3×10^{-3} s^{-1}) strain rates. Typically, the flow stress peaks at ≈ 2.5% true strain with the exception of 1200°C tests in which flow stress decreases monotonically with strain. As much as 40% softening

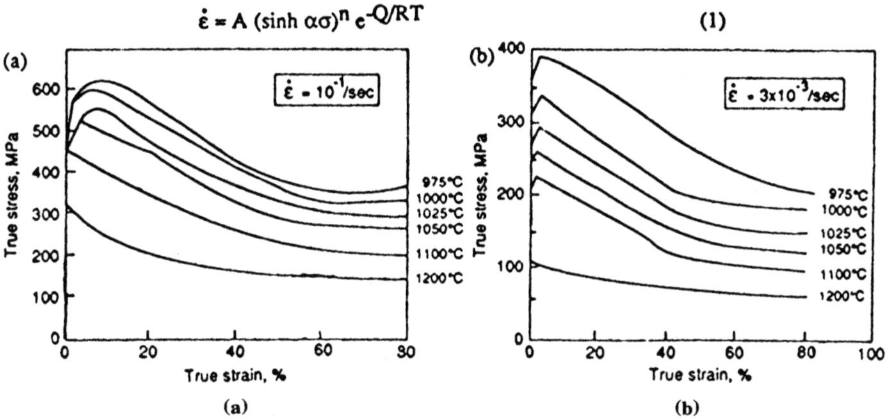

Figure 14-21. Stress-strain curves for (a) high- and (b) low-strain-rate compression tests on samples taken from arc-melted Ti-48Al-2Cr-2Nb (at%) alloy ingots (Shih and Scarr 1991).

is observed and is thought to primarily result from dynamic recovery and recrystallization. The adiabatic heating from deformation processing is also suggested as the cause of flow stress softening by Semiatin et al. (1990). The flow stress normally increases with increasing strain rate at constant temperature and strain; see Figure 14-21.

The hot-compression test data on γ-TiAl alloy have been plotted as a deformation map by Shih and Scarr (1991). The map (see Figure 14-22) illustrates the soundness of deformation and isostress contours in terms of deformation temperature and strain rate. The dashed line in Figure 14-22, occurring between 450 MPa and 500 MPa in peak flow stress, divides the region of good and poor workability on the deformation map. The map indicates that sound hot working of γ-TiAl can be carried out over a broad range of temperatures and strain rates. The use of high strain rates have been observed by Shih and Scarr (1991) to produce inhomogeneous microstructures. Thus, Shih and Scarr recommend low strain rate (approximately 10^{-3} s^{-1}) for good hot working.

The isothermal forging of cylindrical billets has become the common practice for hot working of γ-TiAl alloys. It is carried out in a typical vacuum of approximately 4×10^{-3} Pa. The temperature varies somewhat with the alloy composition but generally falls in the range of 1050°C to 1200°C. The low strain rates of 10^{-3} to 10^{-2} s^{-1} are preferred for homogeneous deformation (Semiatin et al. 1991b; Masahashi et al. 1991; Kim 1992). During isothermal forging, the cylinders are reduced to reduc-

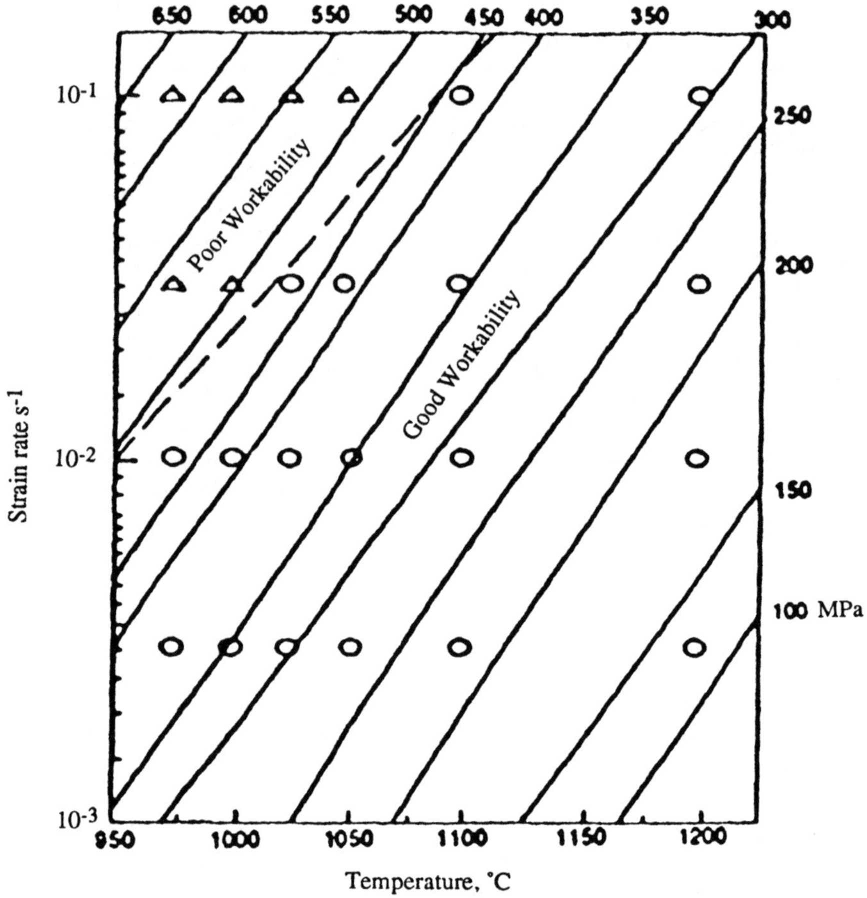

Figure 14-22. Deformation map illustrating soundness of deformation isostress contours for a range of temperatures and strain rates for Ti-48Al-2Cr-2Nb (at%) alloy (Shih and Scarr 1991).

tions of 4 : 1 and 6 : 1. Such reductions impart sufficient work in the material to cause recrystallization either in the hot-worked condition or during subsequent heat treatment (Semiatin et al. 1991b). The only major problem reported (Semiatin et al. 1991a) during isothermal pancake forging is the development of cracks at the bulge-free surfaces, the occurrence of which can affect the overall yield if they propagate extensively to the interior of the workpiece. Seetharaman et al. (1991a) have analyzed the free-surface cracking problem for a single-phase γ alloy,

Ti-49.5Al-2.5Nb-1.1Mn (at%), and established that the source of these defects is the development of secondary stresses on the bulge-free surfaces. The degree of bulge and, thus, the level of tensile stresses are a function of the aspect ratio (height to diameter) of the workpiece and die-workpiece interface friction conditions. Seetharaman et al. (1991) used the maximum tensile work criterion to explain the bulge-free surface-cracking problem. Semiatin et al. (1991a) used the advanced finite element code ALPID (analysis of large plastic incremental deformation) to explain the cracking problem. Their analysis and experimental verification showed that reductions of 6 : 1 at 1150°C and a strain rate of 10^{-3} s^{-1} resulted in no free-surface fractures. The analysis also predicted that a strain rate of 3×10^{-2} s^{-1} at 1150°C will result in tensile type of free-surface cracks for the same alloy even for a reduction ratio of 4 : 1. Thus, it becomes critical to control the strain rate for the chosen processing temperature during isothermal forging to produce crack-free products from γ-TiAl alloys.

The breakdown of the cast structure by isothermal forging results in significant improvement in the ductility of a range of γ-TiAl alloys. Tensile elongation data on isothermally forged alloys reported by Hashimoto and Kimura (1993), Yang, Lee, and Mukherjee (1993), and Clemens et al. (1993) are plotted in Figure 14-23. The high-temperature ($\geq$ 800°C) tests by Hashimoto and Kimura (1993) and Yang et al. (1993) were conducted in vacuum. Data in Figure 14-23 show that tensile ductility values of over 10% can be reached in a range of γ-TiAl alloys at temperatures $\geq$ 600°C. In fact, the alloys show superplasticity at temperatures $\geq$ 1200°C.

Isothermal forging in the temperature range of 1050°C to 1200°C at a strain rate of 10^{-3} to 10^{-2} s^{-1} is the best method of breaking down the coarse-grain cast structure. However, the lowest deformation rate achieved on the industrial equipment used by Clemens et al. (1993) was 1 s^{-1}, which is at least two to three orders of magnitude higher than the desirable rate listed above. In order to deform the Ti-48Al-2 Cr ingots on industrial equipment, they were canned, which minimized the loss of heat and protected the specimen against oxidation. The canning of ingots provided "quasi-isothermal" deformation conditions. The pancakes with diameters up to 570 mm were produced (Clemens et al., 1993) by forging the ingot encapsulated in a can at 1125°C and a strain rate of 1 s^{-1}.

In addition to isothermal forging, conventional extrusion processing is also used for the breakdown of the as-cast structure of γ-TiAl alloys. Because of the narrow temperature window, the extrusion process for γ-TiAl must rely on some form of can to minimize the effects of excessive heat loss to the extrusion container and die. These heat losses, coupled with secondary tensile stresses whose magnitude is a function of die design

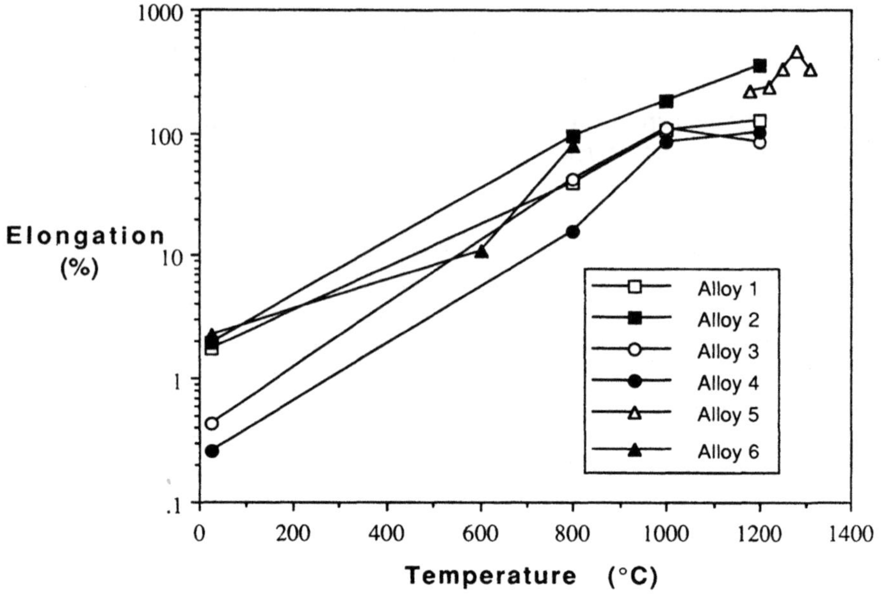

Alloy number	Composition (at. %)
1	Ti-46.8Al-2.0V
2	Ti-47.2Al-2.1Nb
3	Ti-45.6Al-2.0Mo
4	Ti-46.5Al-2.5Ta
5	Ti-47.3Al-1.9Nb
6	Ti-48Al-2Cr

Figure 14-23. Tensile elongation of TiAl alloys after isothermal forging and heat treatment. Alloys 1 through 4 tested at a strain rate of $5 \times 10^{-4}\ s^{-1}$, alloy 5 tested at $8 \times 10^{-5}\ s^{-1}$, and alloy 6 tested at $1\ s^{-1}$ (Hashimoto and Kimura 1993; Yang et al. 1993; Clemens et al. 1993).

(conical vs. streamline), can greatly affect microstructure development as well as give rise to fracture ranging from mild to very severe. Seetharaman et al. (1991b) described the effects of extrusion temperature and speed on the final grain size and grain-size variations through the cross-section for the single-phase gamma alloy Ti-49.5Al-2.5Nb-1.1Mn (at%). Semiatin et al. (1991a) reported the application of finite-element method to the design of extrusion cans and the failure modes associated with the improper can design. Commercial-purity titanium and stainless steel cans were investigated in the study by Semiatin et al. (1991a). The flow-stress difference between commercial-purity titanium and γ-TiAl is 10 : 1. Such a large difference in flow stresses causes tearing of the can, which leads to excessive chilling and gross failure of the billet. The flow-stress difference between stainless steel and γ-TiAl is 1 : 1.7. Such a flow-stress difference permits some deformation of the γ-TiAl billet before can thinning becomes significant. Selection of a can material with nearly matching flow stress will produce a defect-free extrusion of γ-TiAl alloys.

Secondary processing of γ-TiAl alloys is possible by isothermal closed-die forging (Semiatin et al. 1991a), pack sheet rolling (Semiatin et al. 1991a; Semiatin et al. 1990; and Clemens et al. 1993), and superplastic sheet forming (Graves and Ghosh 1989). The closed-die isothermal forging process allows the fabrication of near-net-shape components. Isothermal forging relies on the development of a uniform and fine two-phase microstructure (Semiatin and Altan 1983). Because of the tendency of ingot metallurgy near γ-TiAl alloys to form banded microstructures (Semiatin et al. 1991b; Semiatin and McQuay 1992; Nekkanti, Quay, and Semiatin 1992), some form of thermomechanical processing is needed for homogenization and subsequent breakdown of the lamellar $(\alpha_2 + \gamma)$ microstructure produced during homogenization. Details of various homogenization and thermomechanical treatments have been reported by Semiatin et al. (1991), Semiatin et al. (1991b), and Semiatin and McQuay (1992).

Sheet fabrication of γ-TiAl alloys is possible only by pack rolling. However, even the pack rolling of the γ-TiAl alloys is particularly difficult because of the narrow working-temperature range and the development of secondary tensile stresses in the rolling process. The secondary tensile stresses arise when a relatively high flow-stress material (γ-TiAl) is packed and rolled within a material with a lower flow stress (Semiatin and Piehler 1979; Semiatin et al. 1990). The tensile stresses generated in the rolling direction of the aluminide may result in tensile fracture. In this case, the failure usually occurs across the entire width of the work piece. The tendency for such failures can be exacerbated by heat losses from the work piece during transfer from the furnace to the rolling equipment, as well as due to roll chilling. The heat losses, coupled with the sharp dependence of ductility on temperature and strain rate, are important in the design of the rolling process for γ-TiAl alloys.

Clemens et al. (1993) have reported the successful fabrication of 700 × 220 × 1.8-mm sheet of a Ti-48Al-2Cr (at%) alloy. A modified, pack-rolling process (no details are available) that enabled the rolling in the ($\alpha + \gamma$) temperature region (1125°C to 1350°C) at rolling speeds of < 10 m/min was used.

5. APPLICATIONS

Intermetallic aluminides are clearly a new class of materials that offer significant advantages over conventional materials for a broad range of applications. The specific advantages and applications for each of the Ni_3Al-, Fe_3Al-, and TiAl-based alloys are briefly described here.

5.1. Ni_3Al-Based Alloys

These materials compete with the nickel-base alloys and stainless steels. Specific advantages include: (a) excellent oxidation resistance, (b) carburization resistance, (c) nitridation resistance, (d) high-temperature strength, (e) high-temperature wear resistance, (f) cavitation erosion resistance, and (g) fatigue resistance. Based on these advantages, several applications have been identified for the Ni_3Al-based alloys, which include:

1. Furnace Fixtures: This application deals with the furnace components for highly carburizing, nitriding, and oxidizing environments. For this application, nickel aluminides will replace highly alloyed stainless steels, and typical components are used in the cast condition.
2. Turbochargers for Diesel Engines: In this application, the Ni_3Al-based alloys will replace the nickel-base superalloy IN-713C and take advantage of the significantly superior fatigue strength. The turbochargers are manufactured by investment casting, and to maximize the fatigue strength, the melting and casting are carried out in vacuum.
3. Hydroturbines: This application takes advantage of the superior cavitation-erosion resistance of nickel aluminides. The application replaces 300-series stainless steel. The turbine components can be fabricated by the casting process. However, the greatest improvement in cavitation-erosion resistance has been observed for fine-grained material and, preferably, in the cold-worked condition. Thus, the actual application may use both cast and wrought components. For instance, the hub section might be cast, and the blades may be in the worked condition.
4. Catalytic Converter Substrates: There is a serious consideration under way to replace the ceramic substrate in the automotive catalytic converters with a metallic substrate. The key advantages of the metallic substrate are the possibility for significantly increasing the surface area and improved resistance to both thermal and mechanical shock. The Ni_3Al-based alloys are

serious contenders because they offer the needed attributes for this application. The material needed for this application is a 0.050 mm-thick foil. The 0.050 mm-thick foil will be prepared by cold-rolling with intermediate anneals of direct-cast sheet from liquid metal.

5. Manufacturing Tooling: This application uses nickel aluminide for forging dies and molds for permanent mold casting. This application takes advantage of the high-temperature strength and oxidation resistance of nickel aluminide. The manufacturing dies and permanent-casting molds are prepared by ingot-casting and sand-casting processes followed by some minor machining.
6. Replacement of Cobalt in WC Cutting Tools: Nickel aluminides as a whole or partial replacement of cobalt in WC cutting tools work effectively. Not only is the expensive element cobalt replaced, but the performance of the final cutting tool is also significantly enhanced. Nickel-aluminide powder is used for this application.

5.2. Fe_3Al-Based Alloys

These materials compete with the 300- and 400-series stainless steels and some nickel-base alloys. Specific advantages include: (a) excellent sulfidation resistance, (b) oxidation resistance, (c) lower in density, (d) good wear resistance, (e) cavitation resistance, and (f) potentially lower cost. Based on these advantages, several applications have been identified for the Fe_3Al-based alloys, which include:

1. Sintered Porous Gas-Metal Filters: In this application, the filters are used to clean gas in the coal-gasification process and other processes where the gas contains a high sulfur content. Advantage is taken of the superior sulfidation resistance of iron-aluminide alloys as opposed to any other currently available material. The filters are prepared by sintering of powders to a desired level of porosity.
2. Heating Elements: This application uses the iron-aluminide alloys wire for heating elements in toasters, ovens, and dryers. Advantage is taken of the high resistivity, which remains constant up to 1000°C, and the excellent oxidation resistance. The fabrication of wire has been hampered by limited ductility at room temperature.
3. Furnace Fixtures: This application takes advantage of the excellent oxidation resistance of iron-aluminide alloys for components such as retorts, rollers, beams, etc. The furnace fixtures are expected to be manufactured by various casting processes.
4. Catalytic Converter Substrate: This application takes advantage of the excellent oxidation resistance of iron aluminides. The 0.050 mm foil for this application is prepared primarily by warm rolling followed by cold rolling in the last few passes.

5. Regenerator Disks: In this application, iron aluminide is used as a heat exchanger in a gas turbine engine. Advantage is taken of the excellent oxidation and sulfidation resistance. The foil requirements for this application are similar to those for the catalytic converter substrates.
6. Components for Molten Salt Applications: Iron-aluminide alloys have excellent compatibility with oxidizing and carbonate salts. Processes dealing with molten salts have the potential of using iron aluminides for their containment, transfer, or even rotating components. The components for this application will be manufactured primarily by the casting process.
7. Shielding: This application uses iron aluminide as a shielding to prevent the excessive oxidation of tubes in power plants and incinerators. The shields are typically 3 to 4 mm thick and are shaped like tubes. These shields are typically manufactured by bending the warm-rolled sheet of the desired thickness.

5.3. γ-TiAl-Based Alloys

The low density of γ-TiAl, combined with good strength properties up to 900°C, makes them attractive for several applications. These materials compete with nickel-base superalloys and ceramics. The applications identified for these materials include:

1. Turbocharger Rotors: For this application, γ-TiAl alloys offer lower density than nickel-base alloys and better toughness than ceramics. The turbochargers are manufactured by the investment-casting process.
2. Automotive Exhaust Valve: This application takes advantage of the lower density, high-temperature wear resistance, and strength. Prototype valves of γ-TiAl alloys for automotive engine tests have been produced by investment casting, isothermal forging, and powder metallurgy processes. The γ-TiAl valves have performed extremely well in engine tests. The production in the type of quantities needed for automotive applications and machining at competitive prices is a major issue for their commercial applications.
3. Components for Aerospace: Many other components of γ-TiAl alloys have been identified for aerospace applications such as blades, vanes, and hangers. Several of these applications require γ-TiAl sheet and foil.

6. SUMMARY

Intermetallic aluminides, especially Ni_3Al, Fe_3Al, and γ-TiAl, have progressed the most in many aspects of basic understanding and their commercial applications. Several compositions for Ni_3Al- and Fe_3Al-based alloys have been identified to take advantage of their special properties. The γ-TiAl-based compositions are still being optimized along with heat treatments to maximize the low-temperature toughness, ductility, and high-temperature strength. The Ni_3Al- and Fe_3Al-based intermetallics are readily adaptable to the normal melting equipment that is currently used for stainless steels and nickel-based superalloys. However, because of their higher aluminum content and the exothermic nature of the compound formation, somewhat different procedures are needed to load the melting furnaces and protect the molten metal from excessive slagging. The γ-TiAl-based alloys require melting in nonceramic crucibles by processes such as ISM, PACHM, and VAR. All of these processes needed for γ-TiAl alloys are more costly to operate than those needed for Ni_3Al- and Fe_3Al-based alloys.

The Ni_3Al- and Fe_3Al-based alloys are castable by conventional processes such as sand, investment, and centrifugal casting. However, the room-temperature ductility of Fe_3Al-based alloys in the as-cast condition is extremely low, and this limits the use of such castings somewhat. The interaction of zirconium in the castable Ni_3Al-based composition with the investment-shell material is an issue. Special investment shell materials or alloy-composition modification can be used to eliminate the zirconium reaction. The investment casting of γ-TiAl also has the problem of reaction with the shell material. Special mold wash coatings have been developed to eliminate such reactions.

The processing of cast ingots of all three aluminide-based alloys is restricted. The Fe_3Al-based alloys are the easiest to hot work but have limited ductility for subsequent cold-working operations. The Ni_3Al-based alloys can be hot worked by conventional methods but with special care, such as a carbon steel can. Compared to the Fe_3Al-based alloys, the Ni_3Al-based alloys are cold workable by several conventional methods. The γ-TiAl-based alloys have limited hot and cold ductility and, thus, require isothermal slow-strain-rate type of processes for the breakdown of cast structure. Even the hot-worked, fine-grain material cannot be cold worked. As opposed to Ni_3Al- and Fe_3Al-based alloys, the processing response of the γ-TiAl alloys is very limited.

Many applications of all three aluminide-based alloys have been listed. Those requiring cast components are finding near-term applications. Much work is needed toward better understanding of methods to improve the hot and cold workability of aluminide-based alloys.

REFERENCES

ANTON, D. L., SNOW, D. B., NAZMY, M., and STAUBLI, M., 1993, in *ISSI Structural Intermetallics*, eds. R. Darolia, J. J. Lewandowski, C. T. Liu, P. L. Martin, D. B. Miracle, and M. V. Nathal, pp. 265–273. Warrendale, PA: The Minerals, Metals and Materials Society.

BREIG, P. G., and SCOTT, S. W., 1989, *Mater. Manuf. Proc.* 4(1): 72–83.

CHENG, C. T., 1992, *Microstructure and Porosity Evolution in Equiaxed Solidified Multicomponent Nickel Aluminide*, Ph.D. dissertation, University of Cincinnati, Cincinnati, OH.

CHRONISTER, D. J., SCOTT, S. W., STICKLE, D. R., and FROES, F. H., 1986, *JOM* 38(5): 51–54.

CLEMENS, H., SCHRETTER, P., WURZWALLNER, K., BARTELS, A., and KOEPPE, C., 1993, in *ISSI Structural Intermetallics*, eds. R. Darolia, J. J. Lewandowski, C. T. Liu, P. L. Martin, D. B. Miracle, and M. V. Nathal, pp. 205–214. Warrendale, PA: The Minerals, Metals and Materials Society.

FROES, F. H., and SURYANARAYANA, C., 1994, in *Physical Metallurgy and Processing of Intermetallic Compounds*, eds. N. S. Stoloff and V. K. Sikka, pp. 297–350. New York, NY: Chapman and Hall.

FRUEHAN, R. F., ed., 1985, in *Ladle Metallurgy Principles and Practices*, pp. 3–12. Chelsia, MI: Bookcrafter Inc.

GERMAN, R. M., 1992, in *Thermal Analysis in Metallurgy*, eds. R. D. Shull and A. Joshi, pp. 205–231. Warrendale, PA: The Minerals, Metals and Materials Society.

GERMAN, R. M., and IACOCCA, R. G., 1994, in *Physical Metallurgy and Processing of Intermetallic Compounds*, eds. N. S. Stoloff and V. K. Sikka, New York, NY: Chapman and Hall.

GRAVES, J. A., and GHOSH, A. K., 1989, in *High Temperature Ordered Intermetallic Alloys IV*, MRS Symp. Proc., 133: 317–322.

HASHIMOTO, K., and KIMURA, M., 1993, in *ISSI Structural Intermetallics*, eds. R. Darolia, J. J. Lewandowski, C. T. Liu, P. L. Martin, D. B. Miracle, and M. V. Nathal, pp. 309–318, Warrendale, PA: The Minerals, Metals and Materials Society.

HO, C. T., CHENG, C. T., and SEKHAR, J. A., 1991, *Metall. Trans.* 22A: 225–234.

HUANG, S. C., 1993, in *ISSI Structural Intermetallics*, eds. R. Darolia, J. J. Lewandowski, C. T. Liu, P. L. Martin, D. B. Miracle, and M. V. Nathal, pp. 299–307. Warrendale, PA: The Minerals, Metals and Materials Society.

JOHNSON, T. P., YOUNG, J. M., WARD, R. M., and JACOBS, M. H., 1993, in *ISSI Structural Intermetallics*, eds. R. Darolia, J. J. Lewandowski, C. T. Liu, P. L. Martin, D. B. Miracle, and M. V. Nathal, pp. 159–165. Warrendale, PA: The Minerals, Metals and Materials Society.

KIM, Y-W., 1992, *Acta Metall. Mater.* 40(6): 1121–1134.

LIU, C. T., LEE, E. H., and MCKAMEY, C. G., 1989, *Scripta. Metall. Mater.* 23: 875–880.

LIU, C. T., and SIKKA, V. K., 1986, *J. Metals* 38(5): 19–21.

LIU, C. T., WHITE, C. L., and HORTON, J. A., 1985, *Acta Metall.* 33: 213–229.

LONDON, B., LARSEN, D. E., JR., WHEELER, D. A., and AIMONE, P. R., 1993, in *ISSI Structural Intermetallics*, eds. R. Darolia, J. J. Lewandowski, C. T. Liu, P. L. Martin, D. B. Miracle, and M. V. Nathal, pp. 151–157. Warrendale, PA: The Minerals, Metals and Materials Society.

MASAHASHI, N., MIZUHARA, Y., MASTSUO, M., HASHIMOTO, K., HANAMURA, T., and FUJII, H., 1991, in *High Temperature Ordered Intermetallic Alloys IV*, *MRS Symp. Proc.*, 213: 795–800.

MCKAMEY, C. G., 1994, in *Physical Metallurgy and Processing of Intermetallic Compounds*, eds. N. S. Stoloff and V. K. Sikka, pp. 351–391. New York, NY: Chapman and Hall.

MCKAMEY, C. G., DEVAN, J. H., TORTORELLI, P. F., and SIKKA, V. K., 1991, *J. Mat. Res.* 6(8): 1779–1805.

NEKKANTI, R. M., MCQUAY, P. A., and SEMIATIN, S. L., 1992, *Scripta. Metall. Mater.* 26: 1089–1094.

NOEBE, R. D., BOWMAN, R. R., and NATHAL, M. V., 1994, in *Physical Metallurgy and Processing of Intermetallic Compounds*, eds. N. S. Stoloff and V. K. Sikka, pp. 212–296. New York, NY: Chapman and Hall.

PEILI, M., YING, Y., LINGUANG, H., YULIANG, L., and ZENGYONG, Z., 1992, *Mat. Sci. Eng. A* A153: 377–381.

RISHEL, L. L., and BRODY, H. D., 1991, in *Heat-Resistant Materials*, eds. K. Natesan and D. J. Tillack, pp. 149–157, Materials Park, OH: ASM International.

SEETHARAMAN, V., GOETZ, R. L., and SEMIATIN, S. L., 1991a, in *High Temperature Ordered Intermetallic Alloys IV*, *MRS Symp. Proc.* 213: 895–900.

SEETHARAMAN, V., MALAS, J. C., and LOMBARD, C. M., 1991b, in *High Temperature Ordered Intermetallic Alloys IV*, *MRS Symp. Proc.* 213: 889–894.

SEKHAR, J. A., LIN, C. S., and CHENG, C. J., 1991, in *Nature and Properties of Semi-Solids Materials*, eds. J. A. Sekhar and J. Dantzig, pp. 267–290, Warrendale, PA: The Minerals, Metals and Materials Society.

SEMIATIN, S. L., and ALTAN, T., 1983, *Isothermal and Hot-Die Forging of High Temperature Alloys*, Metals and Ceramics Information Center, Battelle Memorial Institute, Columbus, Ohio.

SEMIATIN, S. L., and PIEHLER, H. R., 1979, *Metall. Trans. A* 10A: 97–107.

SEMIATIN, S. L., VOLLMER, D. C., EL-SOUDANI, S. M., and SU, C., 1990, *Scripta. Metall. Mater.* 24: 1409–1413.

SEMIATIN, S. L., OH, S. I., FIORENTINO, R. J., SEETHARAMAN, and MALAS, J. C., 1991a in *Heat-Resistant Materials*, eds. K. Natesan and D. J. Tillack, pp. 175–186. Materials Park, OH: ASM International.

SEMIATIN, S. L., and MCQUAY, P. A., 1992, *Metall. Trans. A* 23A: 149–161.

SEMIATIN, S. L., MCQUAY, P., STUCKE, M., KIM, Y-W., and EL-SOUDANI, S. M., 1991b, in *High Temperature Ordered Intermetallic Alloys IV*, *MRS Symp. Proc.* 213: 883–888.

SHIH, D. S., and SCARR, G. K., 1991, in *High Temperature Ordered Intermetallic Alloys IV*, *MRS Symp. Proc.* 213: 727–732.

SIKKA, V. K., 1987, In *Interdisciplinary Issues in Materials Processing and Manufacturing Vol. II*, eds. S. K. Samanta, R. Kowanduri, R. McMeeking, M. M. Chen, and A. Tseng, pp. 451–467. New York, NY: The American Society of Mechanical Engineers.

________, 1989, *Mater. Manuf. Proc.* 4(1): 1–24, Marcel Dekker, Inc., New York.

________, 1990, Oak Ridge National Laboratory, Oak Ridge, Tennessee, unpublished research.

________, 1991a, in *Heat-Resistant Materials*, eds. K. Natesan and D. J. Tillack, 141–147. Materials Park, OH: ASM International.

________, 1991b, Oak Ridge National Laboratory, Oak Ridge, Tennessee, unpublished research.

________, 1991c, *SAMPE O.* 22(4): 2–10.

________, 1993, Oak Ridge National Laboratory, Oak Ridge, Tennessee, unpublished research.

________, 1992, Oak Ridge National Laboratory, Oak Ridge, Tennessee, unpublished research.

SIKKA, V. K., GIESEKE, B. G., and BALDWIN, R. H., 1991, in *Heat-Resistant Materials*, eds. K. Natesan and D. J. Tillack, pp. 363–371, Materials Park, OH: ASM International.

SIKKA, V. K., MAVITY, J. T., and ANDERSON, K., 1992, *Mat. Sci. Eng. A* A153: 712–721.

SIKKA, V. K., and REINHOLZ, E. D., 1993, Oak Ridge National Laboratory, Oak Ridge, Tennessee, unpublished research.

SIKKA, V. K., REINHOLZ, E. D., and PATEL, M. J., 1993, Oak Ridge National Laboratory, Oak Ridge, Tennessee, unpublished research.

SIKKA, V. K., VISWANATHAN, S., and MCKAMEY, C. G., 1993, in *ISSI Structural Intermetallics*, eds. R. Darolia, J. J. Lewandowski, C. T. Liu, P. L. Martin, D. B. Miracle, and M. V. Nathal, pp. 483–491. Warrendale, PA: The Minerals, Metals and Materials Society.

STOLOFF, N. S. and LIU, C. T., 1994, in *Physical Metallurgy and Processing of Intermetallic Compounds*, eds. N. S. Stoloff and V. K. Sikka, pp. 159–211. New York, NY: Chapman and Hall.

YANG, H. S., LEE, W. B., and MUKHERJEE, A. K., 1993, in *ISSI Structural Intermetallics*, eds. R. Darolia, J. J. Lewandowski, C. T. Liu, P. L. Martin, D. B. Miracle, and M. V. Nathal, pp. 69–76, Warrendale, PA: The Minerals, Metals and Materials Society.

Chapter 15

Powder Metallurgy Processing

R. M. German and R. G. Iacocca

1. INTRODUCTION

The processing of intermetallic compounds has similarities to ceramic processing, as both ceramics and intermetallics are stoichiometric, with limited compositional ranges and brittle behavior. Generally, the limited ductility forces a reliance on powder techniques for shaping and consolidation. The high melting points of many intermetallics make them attractive for high-temperature service, but this same attribute contributes to difficulty in fabricating and consolidating intermetallic powders.

Some of the intermetallics are ductile with good fabricability. The lower melting-temperature compounds can be processed by fusion techniques. The net shaping capabilities of powder metallurgy (P/M) are attractive because of material utilization and elimination of machining or deformation steps. This is particularly useful for the brittle compounds. Additionally, many of the intermetallics have a high melt viscosity that inhibits casting. Cast intermetallics also suffer problems with grain-size control and segregation. Thus, P/M techniques are suited for the processing of intermetallics because of the combination of net shaping, avoidance of deformation, and insensitivity to casting characteristics (Shulson 1987; Stoloff 1985; Vedula and Stephens 1987). The microstructure of the consolidated material can be controlled via the powder characteristics and the processing cycle. Indeed, P/M offers flexibility through a wide array of powder-fabrication and consolidation techniques. Intermetallic powders can be formed by gas atomization, reactive synthesis, centrifugal atomization, mechanical alloying, and melt spinning. Subsequently, the powders can be consolidated by sintering, hot pressing, hot isostatic compaction, explosive compaction, hot extrusion, or spray-forming techniques. This latitude offers considerable control over the microstructure, shape, composition, and processing expense.

Most of the interesting intermetallics for high-temperature service lack the desired combination of oxidation resistance, ductility, strength, and toughness. Hence, considerable effort has been put into intermetallic matrix composites (IMC's), which are also processed by powder techniques. The titanium- and nickel-aluminide systems have received much attention along these lines, where both strength and toughness have been improved by incorporation of various reinforcing phases. These typically are brittle systems and consequently there is major attention to microstructure, interface quality, and possible interfacial reactions between the matrix and reinforcement. The reinforcing phases are selected for stiffness, strength, or chemical compatibility from candidates that include continuous fibers (SiC, Nb, Al_2O_3, or W), chopped fibers (Al_2O_3 being typical), and particles (TiC, TiB_2, Al_2O_3, HfC, AlN, and Y_2O_3).

The P/M approach to the fabrication of intermetallics and intermetallic-matrix composites is recognized for its net-shaping advantage. Although the field is still under development, a few systems are reaching sufficient maturity to allow analysis and projection of future trends.

2. INTERMETALLIC POWDERS

Intermetallic powders are synthesized by a wide variety of techniques. Elemental powders are used to fabricate intermetallic compounds by reactive-synthesis techniques, with some purification due to self-heating (Oddone and German 1989; Martirosyan et al. 1976). Alternatively, many of the traditional powder fabrication techniques have been applied to the intermetallics.

2.1. Chemical Synthesis

Direct chemical synthesis of intermetallics is not common as yet. Two reports are available describing such routes for NiAl and Ni_3Al (Withers et al. 1991; Abe and Tsuge 1991). Very small powders are available through chemical-precipitation routes. However, high oxygen contents are a major problem and it may be some time before these powders compete with the purity available through atomization.

2.2. Milling and Mechanical Alloying

Milling techniques are generally useful for producing powders from brittle-cast alloys and for generation of alloyed powders using mechanical alloying. This is a slow process that can be accelerated by hydriding the powder (Suryanarayana, Sundaresan, and Froes 1990; Majima et al. 1992). Mechanical alloying employs the attritioning motion between agitated balls to create an alloyed composite powder. The process starts with a stoichiometric ratio of elemental powders. On a microscopic scale, the repeated milling, cold welding, and fracture of the particles produces the desired intermetallic powder. The homogenization that takes place during milling converts the initial elemental particles into an alloyed form over a long period of time.

Like other mechanical powder fabrication techniques, mechanical alloying suffers from contamination problems that can be minimized by using equipment made from the same material as the powder. The selection of a gas or fluid phase to shield the powder is important to controlling contamination. The mechanical-alloying approach has been demonstrated for the fabrication of many intermetallic compounds. Mechanical alloying produces an intermetallic powder that can be amorphous, nanocrystalline, or fully ordered. Also, mechanical alloying proves useful in forming oxide-dispersion-strengthened intermetallics (Vedula, Michel, and Figueredo 1988; Wang et al. 1988; Benn, Mirchandani, and Watwe 1988).

One novel mechanical-alloying approach uses Ti and Si_3N_4 powders to synthesize a $TiSi_2$ intermetallic with TiN reinforcement [Iwase, Koga, and Naguno 1992]. A reaction occurs during alloying, giving the desired intermetallic powder at the final composition. In cases where the reaction is suppressed by cooling during milling, it is possible to form an amorphous powder. The stored energy due to mechanical alloying is estimated to be less than 10 kJ/mol.

2.3. Inert Gas Atomization

Atomization involves the formation of powder from molten metal using dispersed fine droplets that solidify during free flight. Both elemental and prealloyed powders can be formed by such processes. The flexibility of the approach, coupled to its applicability to several alloys, make it an attractive route to the production of the lower melting-temperature intermetallics.

Inert gas atomization (IGA) uses helium or argon as a fluid for breaking up the molten metal (Lawley 1992). The liquid metal stream is

Figure 15-1. A scanning electron micrograph of a typical intermetallic alloy particle formed by inert-gas atomization of the melt.

disintegrated by rapid gas expansion out of a nozzle. The high velocity gas emerging from the nozzle crates a siphon, pulling molten metal into the gas-expansion zone where it forms a spray of molten droplets. During flight through the collection chamber, the droplets solidify. The extensive use of inert gas atomization for the production of superalloys makes it a familiar route to the production of high quality, alloyed powders. Components made from rapidly solidified powders exhibit improved microstructures over cast intermetallics. Powders provide a situation where small dimensions are present for rapid quenching. The particle cooling rate determines the homogeneity, crystal size, and degree of crystallinity. Figure 15-1 shows a typical powder fabricated by inert gas atomization, with a spherical shape and slight agglomeration. The internal microstructure of the powder reflects the solidification events during atomization (Maurer et al. 1989, Hsiung, Kuntz, and Wadley 1991).

Gas atomization can be performed totally under inert conditions, thereby maintaining the integrity of the intermetallic powder. The particle shape is spherical with a fairly wide size distribution generally ranging between 10 and 200 μm, but with adherent "satellites." The atomization parameters can be adjusted to tailor the powder characteristics. The main advantage of gas atomization is in the product homogeneity and the good packing properties available with the resulting spherical powder. It is favored for forming intermetallic powders with melting temperatures below approximately 1700°C from the less reactive metals, especially when subsequent consolidation is by hot isostatic compaction. IGA has been applied to the production of powders from Fe_3Al, NiAl, Ni_3Al, $NiAl_3$, and TiAl. Generally the powder microstructures are similar to rapidly solidified ribbons of similar compositions, consisting of the ordered primary-intermetallic phase and minor quantities of secondary-intermetallic phases. The quantity of each phase varies with cooling rate as reflected by a dependence on the particle size. Small voids may be observed in atomized powders due to trapped gas or shrinkage during solidification. Cooling rates for inert gas-atomized powder are as high as 10^5°C/s, with estimated solidification times of 0.1 s. Kaysser et al. (1991) reported success in forming a 1.5 μm powder size from NiAl by milling an inert-gas-atomized powder. Such a small powder gives excellent sintering without the application of external pressure.

The melt-explosion technique uses a hydrogen-saturated liquid and rapid desaturation in a vacuum to form a powder spray; this process is also termed vacuum atomization (VA). The melt is pressurized with 1–3 MPa of hydrogen. A siphon tube then exhausts the saturated melt into a large vacuum chamber where the high velocity and hydrogen desaturation cause the melt to explode into droplets that form powder. It has been applied to the fabrication of 50 μm Ni_3Al with a cellular microstructure (Wright, Williamson, and Knibloe 1990; Wright, Rabin, and Knibloe 1989).

2.4. Centrifugal Atomization

The desire to control particle size and the difficulties in fabricating powders from reactive metals have led to the development of centrifugal atomization. The theme is to use a fusion process coupled to a centrifugal force. The centrifugal force throws off the molten metal as a fine spray which solidifies into a powder. The plasma rotating electrode process (PREP) is the most prevalent application of this concept. Here a consumable electrode made from the desired material is melted at one end by a plasma arc. The consumable electrode rotates at velocities up to 50,000 revolutions per minute under an inert gas. The liquid forms a sheet that extends beyond the lip of the rotating anode. During free-fall after leaving the lip, the ligaments pinch into droplets. This process produces a spherical powder, having an average particle size of 250 μm. Rotating electrode powders have a lower oxygen level (approximately 250 ppm) as compared to gas-atomized powder, but a larger microstructural scale (Porter et al. 1990). This approach has been applied to Fe_3Al (McKamey et al. 1991), TiAl (Apgar and Eylon 1991, Tokizane, Ishonishi, and Kido 1990, Yolton et al. 1986, Porter et al. 1990, Graves et al. 1987; Schaefer 1992), and Ti_3Al (Osborne, Porter, and Eylon 1991; Kohmoto and Fraser 1989; Hsiung et al. 1991; Porter et al. 1990). A variant centrifugal atomizer pours molten metal onto a rapidly spinning disk. The resulting powder is quenched in helium at 10^5°C/s, which gives a mixture of dendritic and lamellar microstructures for Ni_3Al (Baker et al. 1984).

The benefits of centrifugal atomization are the powder cleanliness, spherical shape (high packing density and easy flow), uniform particle size, and lack of crucible contamination. There are several other forms of centrifugal atomization that rely on external melting. In these cases, the liquid metal is directed onto a chilled rotating disk, cup, mesh or wheel. The centrifugal force disintegrates the melt into a thin sheet to form particles. Melt extraction using a spinning wheel is applicable to form ribbon or coarse powder from higher melting-temperature alloys. The product initially solidifies onto the spinning wheel, but high centrifugal forces cause the material to detach. Coarse particle sizes can be obtained by texturing the wheel to control the solidified product size. Rapid cooling rates occur by conduction from the wheel, leading to amorphous or microcrystalline powders. A composite powder of NiAl-HfC with a grain size below 50 nm has been generated by melt spinning. This extremely small grain size provided good microstructure control in subsequent densification (Ray 1990).

2.5. Plasma Atomization

Another atomization technique useful for producing high-temperature intermetallic powders is plasma atomization (Cheney and Seydel 1990; Shaw, Misiolek and German 1991). An agglomerated powder is fed into a plasma torch where melting and rapid acceleration occur. A powder smaller than 30 μm can be generated via secondary impaction against a substrate. The high velocity and high temperature of the plasma arc result in a high superheat, making this approach useful for the higher temperature intermetallics. It has been applied to NiAl with success (Shaw and German 1991).

3. CONSOLIDATION TECHNIQUES

The fabrication of an intermetallic powder is the first major step in the P/M cycle. Next the power must be shaped and consolidated. Often these steps are combined to give a net shape, full-density component. There is a large body of research on forming and consolidation of intermetallics. Table 15-1 provides a summary of the processing cycles applied to several intermetallics and Table 15-2 gives similar information for intermetallic matrix composites. Only the higher performance systems have been included; the lower temperature compounds (for example those based on gold, copper, and tin) are not considered here.

3.1. Fundamentals

Fundamental to powder processing is densification and the elimination of residual pores. The intermetallics generally are brittle and not responsive to cold compaction. Gutmanas and Lawley (1990) report success in ultra high pressure compaction of a few compounds, but in general thermal processing is needed for densification. Sintering is effective in densification of small powders via high-temperature exposures that induce atomic diffusion. However, for composites, sintering proves ineffective since the reinforcing phase provides more strengthening than can be overcome by the sintering force. The properties of many intermetallics can exceed those possible through fusion techniques by achieving full density using powders. The ability to control microstructure, segregation, grain size, inclusion population, and material texture has motivated considerable exploration of full-density P/M processing (Coble 1978; James 1985; Kulkarni 1987).

To obtain high densities, three material characteristics are noteworthy. First, for most intermetallics the rate of strain hardening decreases with an increasing temperature. Second, the ductility increases with temperature. Third, the yield strength decreases as temperature increases for most materials, although some compounds such as Ni_3Al exhibit an intermediate temperature strengthening. As a consequence, elevated temperatures aid the densification of intermetallic powders because of increased workability. Therefore, consolidation techniques chosen for intermetallic powders must include the application of stress and temperature, with attention being given to particle size. Other factors, such as the powder type and strain rate are less significant.

The microstructural control possible with powder processing provides additional benefits. In a polycrystalline material, strength increases with the inverse square root of the grain size d, as quantified by the

Table 15-1 Powder Metallurgy Processing Cycles for Densification of Intermetallic Compounds

Compound	Process and Parameters	Reference
Co_2Ti	cold compaction 3 GPa, sinter 1000°C, 1 h	Gutmanas and Lawley 1990
FeAl	melt spun ribbon, 975°C hot extrude 16:1	Baker and Gaydosh 1986
Fe_3Al	IGA, hot extrude (9:1) 1000°C	Knibloe and Wright 1990
Fe_3Al	RS, HIP 1100°C, 1 h, 207 MPa	Rabin and Wright 1992
Fe_3Al	RS, VHP 1000°C, 70 MPa	Rabin and Wright 1991
Fe_3Al	RHIP 1200°C, 1 h, 172 MPa	Bose et al. 1991
Fe_3Al	RHIP 1000°C, 140 MPa	Rabin and Wright 1992
Fe_2Ti	cold compaction 3 GPa, sinter 1000°C, 1 h	Gutmanas and Lawley 1990
$MoSi_2$	IGHP 1700°C, 35 MPa	Yang and Jeng 1991
$MoSi_2$	VHP 1650°C, 1 h, 48 MPa	Vedula and Lisy 1990
$NbAl_3$	RS 1200°C, 1 h, vacuum	Murray and German 1992
$NbAl_3$	RHIP 1200°C, 173 MPa	Murray and German 1990
$NbAl_3$	RS, VHP 1350°C, 2.5 h, 70 MPa	Lu et al. 1992
NiAl	MA, HIP 1300°C, 3 h, 103 MPa	Wang et al. 1988
NiAl	IGA, MA, HIP 1200°C, 2 h, 100 MPa	Kaysser et al. 1991
NiAl	RS 800°C	German 1990
NiAl	RHIP 1200°C, 1 h, 172 MPa	Alman and Stoloff 1991b
NiAl	RHIP 1400°C, 0.5 h, 60 MPa	Nishiyama et al. 1991
NiAl	RS, VHP 1250°C, 5 min, 64 MPa	Ochiai et al. 1987
NiAl	IGA, VHP 1400°C, 2 h, 48 MPa	Rigney et al. 1988
NiAl	VA, hot extrude 1400°C, 16:1	Bowman et al. 1992
NiAl	VHP 1400°C, 2 h, 48 MPa	Rigney et al. 1988
Ni_3Al	IGA, HIP 1000°C, 1 h, 35 MPa	Wright et al. 1990
Ni_3Al	IGA, HIP 1000°C, 1 h, 150 MPa	Maurer et al. 1989
Ni_3Al	IGA, HIP 1127°C, 1 h, 80 MPa	Laag et al. 1989
Ni_3Al	IGA, VHP 1050°C, 2 h, 48 MPa	Rigney et al. 1988
Ni_3Al	IGA, HIP 1120°C, 103 MPa	Sikka 1988
Ni_3Al	IGA, HIP 1150°C, 2 h, 100 MPa	Chang et al. 1987
Ni_3Al	IGA, pseudo HIP 1150°C, 5 s, 827 MPa	Sikka 1988
Ni_3Al	MA, VHP 700°C, 0.5 h, 150 MPa	Shiga 1991a
Ni_3Al	MA, hot extrude 1150°C	Benn 1988
Ni_3Al	VA, hot extrude 1100°C, 8:1	Wright and Knibloe 1990
Ni_3Al	VA, HIP 1100°C, 2.5 h, 200 MPa	Wright et al. 1989
Ni_3Al	RS 750°C, 10 min, argon	Hwang and Lu 1990
Ni_3Al	RS 700°C, 15 min, vacuum	German 1987
Ni_3Al	RHIP 1100°C, 1 h, 170 MPa	German et al. 1990
Ni_3Al	VHP 1050°C, 1 h, 48 MPa	Rigney et al. 1988
Ni_3Al	plasma spray, HIP 1165°C, 2 h, 103 MPa	Taub et al. 1984
Ni_3Al	MA, HIP	Wang et al. 1988
$NiAl_3$	MA, VHP 700°C, 0.5 h, 150 MPa	Shiga et al. 1991b
Ni_3Fe	HIP 1200°C, 0.5 h, 103 MPa	Lograsso et al. 1988

Table 15-1 (Continued)

Compound	Process and Parameters	Reference
Ni_3Ti	cold compaction 3 GPa, sinter 1000°C, 1 h	Gutmanas and Lawley 1990
$TaAl_3$	RHIP 1200°C	Misiolek and German 1991a
TiAl	RS, HIP 900°C, 1 h, 35 MPa	Oddone and German 1989
TiAl	RHIP 1370°C, 4 h, 170 MPa	Oddone and German 1989
TiAl	RHIP 1200°C, 196 MPa	Kakitsuji et al. 1990
TiAl	IGA, HIP 1200°C, 2 h, 200 MPa	Beddoes et al. 1992
TiAl	RS, HIP 1250°C, 1 h, 200 MPa	Smarsly and Dahms 1990
TiAl	RS, HIP 1350°C, 1 h, 145 MPa	German t al. 1990
TiAl	MA, VHP 900°C, 1 h, 100 MPa	Tokizane et al. 1990a
TiAl	IGA, HIP 1000°C	Schaefer and Janowski 1992
TiAl	cold extrude 9:1, RHIP 800°C, 1 h, 200 MPa	Dahms 1990
TiAl	PREP, MA, VHP 950°C, 3 h, 200 MPa	Tokizane et al. 1990b
TiAl	PREP, HIP 875°C, 4 h, 275 MPa	Steele et al. 1989
TiAl	PREP, pseudo HIP 1500°C, 45 s, 1.2 GPa	Levi et al. 1988
Ti_3Al	RPEP, pseudo HIP	Porter et al. 1990
Ti_3Al	PREP, HIP 925°C, 2 h, 200 MPa	Moxson and Friedman 1986
Ti_3Al	plasma spray, HIP 1050°C, 1 h, 105 MPa	Hsiung et al. 1991
Ti_3Al	explosive compaction, HIP 1000°C, 1 h, 200 MPa	Shang and Meyers 1991
Ti_2CuAl_5	RS, HIP 950°C, 1 h, 204 MPa	Maeland and Narasimhan 1989
Ti_5Si_3	MA, VHP	Iwase et al. 1992

notations
RS = reactive sinter or reactive synthesis
IGA = inert gas atomized
MA = mechanically alloyed
PREP = plasma rotating electrode process
IGHP = inert gas hot press
VHP = vacuum hot press
HIP = hot isostatic press

Hall–Petch relation,

$$\sigma_y = \sigma_0 + \kappa d^{-1/2} \tag{15.1}$$

where κ is a stress concentration factor, σ_y is the yield strength, and σ_0 is a material constant. Many of the IMC's show a yield-strength increase over the monolithic intermetallic because of grain-size refinement. A form

Table 15-2 Examples of IMC's and Nominal P/M Fabrication Processes

Material System	Process and Parameters	Reference
FeAl-Y_2O_3	hot extrude 1175°C	Strothers and Vedula 1987
Fe_2Al-SiC	RS, hot forge	Munar et al. 1990
$MoSi_2$-Al_2O_3	HIP 1600°C, 2 h, 172 MPa	Alman and Stoloff 1991a
$MoSi_2$-SiC	HIP 1700°C, 1 h, 200 MPa	Yang and Jeng 1991
$MoSi_2$-TiB_2	VHP 1650°C, 1 h, 48 MPa	Vedula and Lisy 1990
$MoSi_2$-WSi_2-SiC	—	Sheppard 1990
$NbAl_3$	RS 1200°C, 1 h	Misiolek and German 1991b
$NbAl_3$	pseudo HIP	Anon. 1990
$NbAl_3$-Al_2O_3	RHIP 1200°C, 4 h, 170MPa	German et al. 1990
$NbAl_3$-Nb	RHP 1350°C, 1 h, 48 MPa	Lu et al. 1992
$NbAl_3$-NbB	RS	Sekhar 1992
NiAl	RS 800°C	Misiolek and German 1990
NiAl	VHP 1400°C, 2 h, 48 MPa	Rigney et al. 1988
NiAl-AlN	MA, extrude, HIP 1350°C, 5 h, 207 MPa	Whittenberger et al. 1990
NiAl-Al_2O_3	HIP 1200°C, 1 h, 172 MPa	Alman and Stoloff 1991a
NiAl-HfC	hot forge	Ray 1990
NiAl-TiB_2	VHP 1400°C, 2 h, 48 MPa	Rigney et al. 1988
NiAl-TiB_2	RS	German et al. 1988
NiAl-TiB_2	RHIP 1200°C, 1 h, 170 MPa	German et al. 1990
NiAl-TiB_2	HIP 1200°C, 0.5 h, 100 MPa	Misiolek and German 1991a
Ni_3Al	RS 700°C, 0.25 h, vacuum	Misiolek and German 1990
Ni_3Al	VHP 1050°C, 1 h, 48 MPa	Rigney et al. 1988
Ni_3Al	HIP 1165°C, 2 h, 103 MPa	Taub et al. 1984
Ni_3Al	MA, HIP	Wang et al. 1988
Ni_3Al-Al_2O_3	RHIP 800°C, 1 h, 172 MPa	German et al. 1988
Ni_3-Al-Al_2O_3	HIP 1150°C, 1 h, 172 MPa	German et al. 1988
Ni_3Al-Al_2O_3	VHP 1200°C, 1 h, 21 MPa	McKamey et al. 1988
Ni_3Al-Al_2O_3	VHP 900°C, 1 h, 35 MPa	Brennan et al. 1991
Ni_3Al-Al_2O_3	hot extrude 1100°C, 9:1 reduction	McKamey et al. 1991
Ni_3Al-Al_2O_3	RS	Fuchs 1990
Ni_3Al-SiC	RS	Lebrat et al. 1992
Ni_3Al-SiC	VHP 980°C, 1 h, 48 MPa	Yang et al. 1988
Ni_3Al-TiB_2	RS	Fuchs 1990
Ni_3Al-TiB_2	VHP 1050°C, 4 h, 48 MPa	Rigney et al. 1988
Ni_3Al-TiC	RS	Fuchs 1990
Ni_3Al-TiC	RS	Dummead et al. 1990
Ni_3Al-Y_2O_3	RHIP 800°C	German et al. 1988
Ni_3Fe-Y_2O_3	HIP 1200°C, 0.5 h, 103 MPa	Bose et al. 1992
Ni_3Si-Al_2O_3	RS 800°C	Lebrat et al. 1992
$TaAl_3$	RHIP 1200°C	Misiolek and German 1991b
$TaAl_3$-Al_2O_3	RS, VHP 1200°C	Anton 1988
$TaAl_3$-Fe	RHIP 1200°C, 4 h, 170 MPa	German et al. 1990
TiTaAl-W	VHP 1200°C, 1 h, 48 MPa	Déve and Maloney 1991
TiAl	RHIP 1370°C, 4 h, 170 MPa	Oddone and German 1989
TiAl-Al_2O_3	RS	Wrzesinski and Rawers 1990
TiAl-SiC	RS	Wrzesinski and Rawers 1990

Table 15-2 (Continued)

Material System	Process and Parameters	Reference
TiAl-$TiAl_3$	pressurized high energy forging	Persad et al. 1988
TiAl-ZrO_2	plasma sinter	Kimura and Kobayashi 1992
Ti_3Al-Nb	plasma spray + HIP 1050°C, 4 h, 105 MPa	Hsiung et al. 1991
Ti_3Al-SiC	foil-fiber VHP	Larsen et al. 1990
Ti_3Al-SiC	VHP	Brindley 1987
$Ti_3Fe_3Al_{22}$-TiB_2	VHP 1200°C, 4 h, HIP	DiPietro et al. 1991
$Ti_3Fe_3Al_{22}$-TiB_2	RS, VHP, hot forge	Kumar et al. 1991
Ti_5Si_3	MA, VHP	Iwase et al. 1992

similar to Equation 15.1 holds for the fracture strength as well as the yield strength. In addition to fracture toughness and strength, grain size also affects hardness, creep, toughness, and ductility.

3.2. Sintering

In traditional P/M, most products are formed using a two step process that involves first shaping the powder and subsequently sintering the compact. Sintering is the least expensive means of densifying powders. It relies on heating the powder compact to a temperature where diffusion rates are adequate to allow elimination of the surface energy associated with a powder. Diffusion bonds form between contacting particles and the particles densify as pores are eliminated. Small particles can be sintered to full density, especially at high sintering temperatures close to the melting temperature. For example 14 μm Ni_3Al has been sintered to 98% density using 1325°C for 1 h. The initial compact shape can be generated by die compaction, cold isostatic compaction, injection molding, slip casting, tape casting, powder rolling, and so on [Diehl and Stover 1990].

A simple quantitative index for sintering to full density is based on a dimensionless cluster typically encountered in diffusion,

$$\Psi = D_c t / D^2 \tag{15.2}$$

where D_c is the chemical diffusivity at the sintering temperature, t is the isothermal sintering time, and D is the particle size. The sintering index Ψ varies exponentially with temperature and increases with small powders, high temperatures, and long sintering times. As a first guess, sintering index values between 20 and 1000 are required to obtain full density by solid-state sintering. Most intermetallics and their composites prove difficult to densify using solid-state sintering because of large particle sizes and low diffusivities in ordered compounds. Only with very small powders can high sintered densities be attained without an external force. Formation of a liquid phase during sintering is useful since the liquid provides a higher diffusivity than encountered in the solid state, in addition to strong capillary forces that induce particle compression at point contacts (German 1985). Both factors contribute to rapid densification during sintering at temperatures between the liquidus and solidus temperatures for an off-stoichiometry compound.

3.3. Reactive Synthesis

Many studies have emphasized reactive synthesis (RS) as a process for forming intermetallics. The initial compact is composed of mixed elemental powders that are heated to a temperature where they react to form an intermetallic. Often the reaction occurs on the first formation of a liquid, which is typically a eutectic liquid at the interface between contacting particles (Munir 1992; German 1985). Figure 15-2 is a schematic binary-phase diagram for a reactive sintering system, where a stoichiometric mixture of A and B powders is used to form an intermetallic compound AB (German et al. 1990; German 1990). A transient liquid forms at the lowest eutectic temperature and spreads through the compact. Heat is liberated because of the thermodynamic stability of the high melting-temperature compound. Consequently, the reaction can be nearly spontaneous once the liquid forms. By appropriate selection of temperature, particle size, green density and composition, the liquid becomes self-propagating throughout the compact and persists for only a few seconds; thus the name self-propagating high-temperature synthesis. The liquid provides a capillary force on the structure which can lead to densification, although

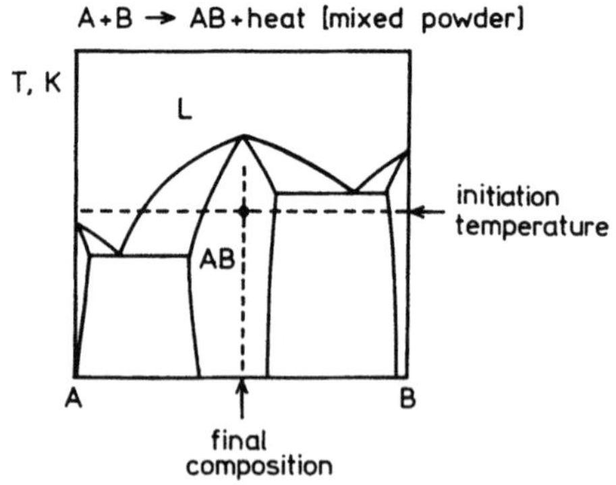

Figure 15-2. Reactive sintering involves heating a stoichiometric particle mixture to form a melt leading to generation of a stable compound, as evident on the phase diagram.

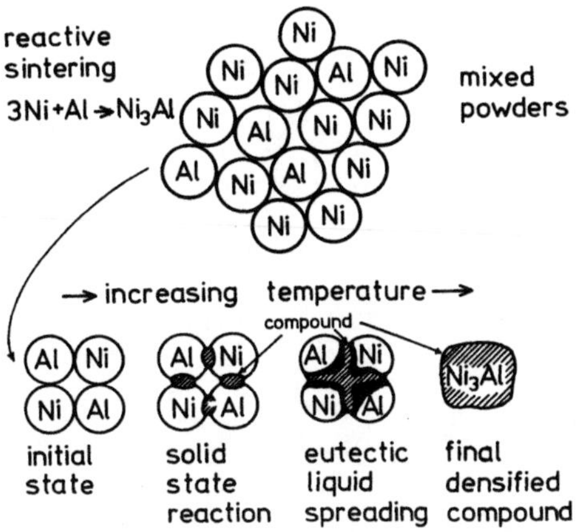

Figure 15-3. A sketch of the reactive sintering process showing the case for generation of Ni_3Al from mixed elemental powders.

swelling is typical due to imbalanced solubilities and improved atomic packing in the compound (Savitskii and Romanov 1990). Reactive sintering is applied to Ni_3Al where there is a large body of knowledge (German 1992). Other systems include TiAl, NiAl, Ti_2CuAl_5, $NbAl_3$, Fe_3Al, Nb_2Al, $TaAl_3$ and their composites.

The powders must be outgassed prior to reaction to avoid incorporation of impurities into the product. Usually, smaller powders give better reaction rates, but also contain higher contamination levels. The reaction velocities depend on the green density of the compact and the materials involved in the reaction, ranging between 0.1 and 100 mm/s. The temperature rise can be as high as 10^{7}°C/s over reaction regions of 50 to 400 μm width. Accordingly, the rapid heat rise and narrow reaction zone induce considerable stress that causes defects in the product phase. Stoichiometry control is important and is often achieved using an excess of the more volatile ingredient or intermediate chemical leaching to remove unreacted constituents.

Figure 15-3 is a sketch of the reactive sintering process. Powders are randomly mixed in a stoichiometric ratio and compressed to create particle-particle contact. On heating to the first eutectic temperature, liquid forms and rapidly spreads throughout the structure. The liquid consumes the elemental powders and forms a precipitated intermetallic solid behind the advancing liquid interface. Interdiffusion is quite rapid in the liquid phase and the compound generates heat which further accelerates the reaction. If the reaction is controlled, the compound will be nearly fully

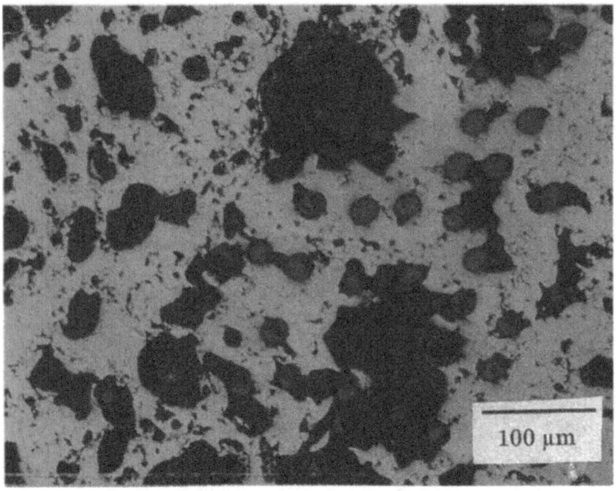

Figure 15-4. The porous microstructure of Ni_3Al-Al_2O_3 fabricated by reactive sintering. The alumina fibers inhibited densification of the matrix phase and for this reason pressure-assisted densification is required for IMC's.

densified and suitable for containerless hot isostatic compaction to full density. However, in the case of intermetallic matrix composites, reinforcing phases impede densification by lowering the reaction temperature (dilution) and slowing the reaction velocity, giving a porous structure as shown in Figure 15-4. The large residual pores represent the restraining effect of the ceramic-fiber network. Also, there is a reaction with the ceramic that can form embrittling phases (Lebrat, Varma and Miller 1992). Consequently, pressure-assisted densification is usually required to form IMC's via reactive synthesis when the reinforcing phase is not formed in situ. Alternatively, reactive sintering using mixed constituents can generate composites such as Nb_2Al-NbB-Nb, where ductile-phase toughening occurs (Bhattacharya 1992).

3.4. Hot Pressing

Diffusional creep is the dominant process in vacuum hot pressing (VHP) and hot isostatic pressing (HIP) to full density (German 1984). To aid in understanding the processes involved, densification maps are used to show the interactions between variables. Figures 15-5 and 15-6 give such maps for Ni_3Al and Ti_3Al consolidated via hot isostatic compaction. The density can be predicted versus various processing factors, including green density, pressure, temperature, particle size, and time (Artz, Ashby,

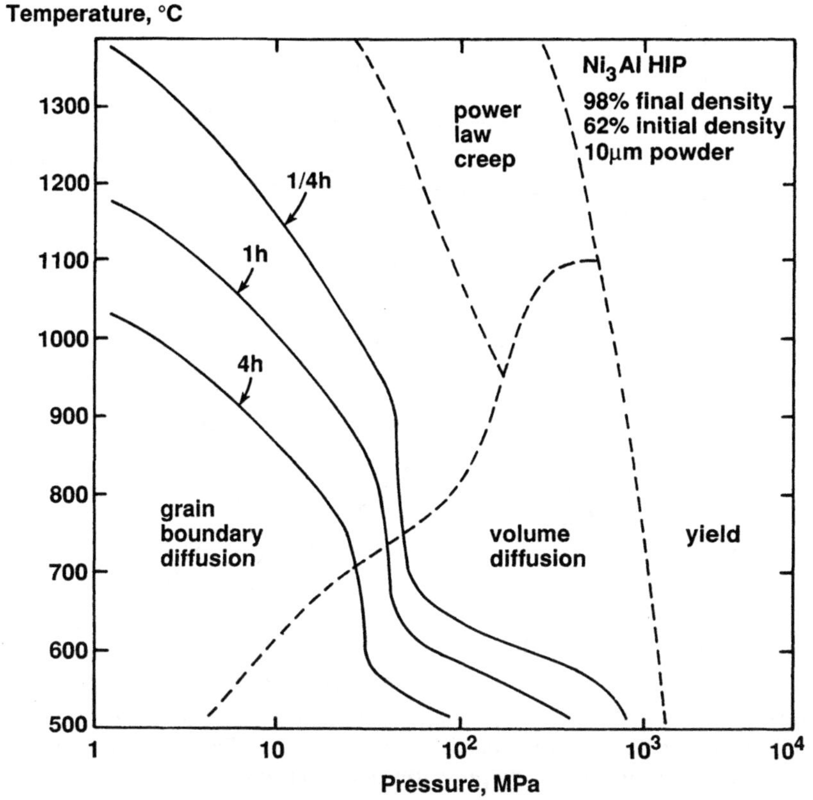

Figure 15-5. The temperature, pressure, and time combinations for attaining 98% density in Ni_3Al from a 10 μm powder. The regions of various densification mechanisms are indicated along with time for 0.25, 1, and 4 h HIP time.

and Easterling 1983; Coble 1978; Notis, Smook, and Krishnamachari 1975; Schaefer and Linzer 1991).

Vacuum hot pressing is a stress-enhanced densification process that is performed in a rigid die using uniaxial pressurization. For intermetallics the die is usually made from graphite to allow external induction heating. Typical maximum temperatures are 2200°C and maximum pressures are 50 MPa. Faster heating can be achieved using plasma techniques (Kimura and Kobayashi 1992). Typically, hot pressing occurs in vacuum (VHP) or inert gas (IGHP) to minimize possible contamination. The powders can be elemental, leading to reactive hot pressing, or prealloyed. As noted in Tables 15-1 and 15-2, the approach is widely used in forming intermetallics and their composites, including Fe_3Al, $MoSi_2$, $NbAl_3$, TiAl,

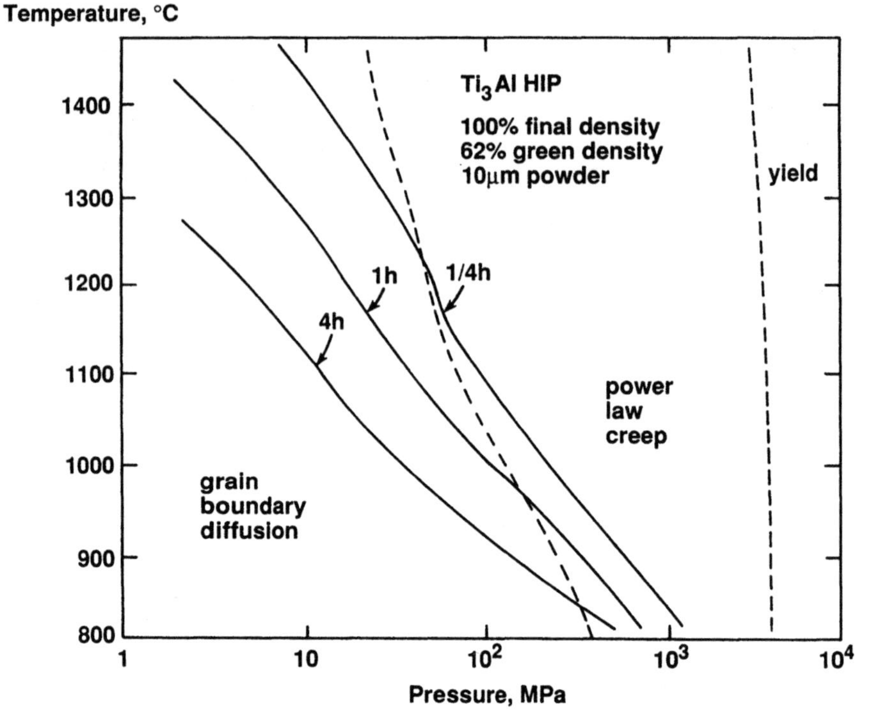

Figure 15-6. The HIP map of pressure, temperature, and time needed for 100% final density for Ti_3Al with a particle size of 10 μm. Densification is largely by grain boundary diffusion at low pressures and long times.

Ti_5Si_3, Ni_3Al-SiC, Ni_3Al-Al_2O_3, Ni_3Al-TiB_2, $MoSi_2$-TiB_2, $TaAl_3$-Al_2O_3, $NbAl_3$-Nb, and NiAl-TiB_2.

Flexible dies and hydrostatic pressurization are used in HIP. Figure 15-7 is a schematic of a HIP sequence. A gas-tight container is used to form the loose powder. The container is selected to deform at the HIP temperature, and is usually made of glass, steel, stainless steel, titanium, or tantalum, depending on the maximum temperature. Prior to HIP consolidation, the filled container is heated and vacuum degassed to remove volatile contaminants. The consolidation of the powder occurs in an internally heated, cold-wall pressure vessel. High pressure gas, such as argon or nitrogen, is used to transfer heat and pressure to the compact, giving densification. Temperatures up to 2200°C and pressures up to 200 MPa are possible using HIP. After consolidation the compact is removed and the container stripped from the densified compact. Temperatures higher than the minimum required for full densification often

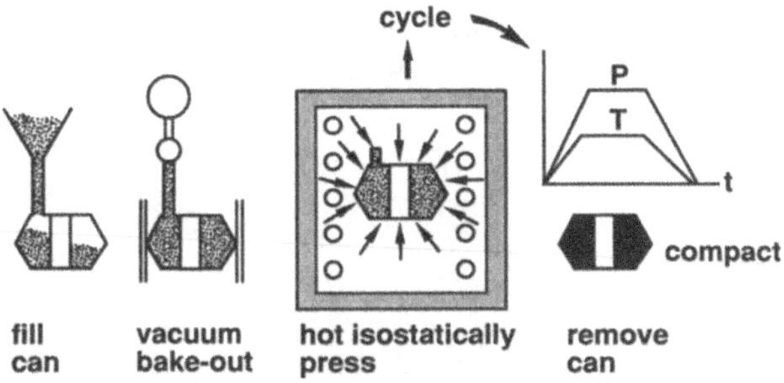

Figure 15-7. A Schematic of the hot isostatic pressing sequence. The pressure (P) and temperature (T) variations versus time (t) are shown in the accompanying schematic plot.

contribute to improved ductility (Wright et al. 1990). As noted in Tables 15-1 and 15-2, several intermetallics and IMC's have been successfully compacted via HIP and reactive HIP (RHIP). In some cases, off-stoichiometry or partially homogenized powders prove easier to densify because the excess energy in the powder promotes better diffusion during heating.

Typically the external surfaces of the compact are contaminated by the container and need to be removed by chemical dissolution, machining, or abrasion. Because consolidation is by hydrostatic stress, there is little shear on the particle surfaces; hence, prior particle boundaries may be evident after HIP. Figure 15-8 shows a micrograph of an oxide formed along the prior particle boundaries of a hot-pressed $NbAl_3$ fabricated using a 1500°C cycle for 1 h at 20 MPa in vacuum. Such oxides are invariably weak aspects of the microstructure and limit mechanical properties. Accordingly, for some applications it is necessary to subject the HIP compact to postconsolidation deformation such as extrusion. When properly executed, HIP provides a near-net shaped compact with full density and minimal microstructural defects or gradients. It has been used to fabricate monolithic and composite structures based on Ti_5Si_3, Ni_3Al, $NbAl_3$, NiAl, $MoSi_2$, TiAl, Til_2Mn, Ti_3Al, $TaAl_3$, and Fe_3Al. Initial calculations of target cycles can be performed using established models for pressure-assisted densification and emerging data collections for intermetallics (Ashby 1991; Maurer et al. 1989; Laag et al. 1990).

3.5. Pseudoisostatic Techniques

The generation of a hydrostatic pressure requires expensive pressurization and containment equipment. Several alternative approaches are in development that use uniaxial pressures acting on a viscous medium

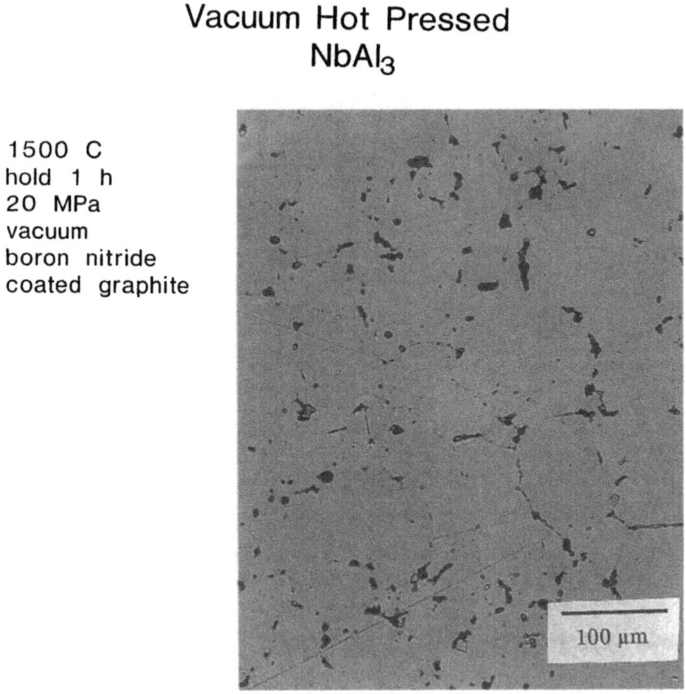

Figure 15-8. Micrograph of $NbAl_3$ processed by vacuum hot pressing showing oxide contamination of prior particle boundaries by oxides during densification.

containing a shaped component or a powder in a container (James 1985). Force transmission is through either a low melting-point liquid, soft metal, or granular solid (alumina or graphite). The medium transforms the uniaxial force into a pseudohydrostatic force, where the hydrostatic stress is about one-third of the applied stress. Figure 15-9 illustrates one such process using a uniaxial press and heated pellets that contain the powder compact. The pellets can be preheated prior to the consolidation cycle, or if they are made of graphite, they can be heated by passing electrical current through the granules. Such pseudoisostatic compaction processes are faster than HIP and can be performed directly on porous compacts because the granules will not penetrate the pores. Further, the consolidation is typically more like forging than HIP since the stresses exceed the yield strength. However, because the pressure transmission is not ideal the dimensional change in densification is not uniform. The approach has been successfully applied to $NbAl_3$, Ni_3Al, TiAl, and Ti_3Al (Sikka 1988; Levi et al. 1988; German et al. 1990; Porter et al. 1990).

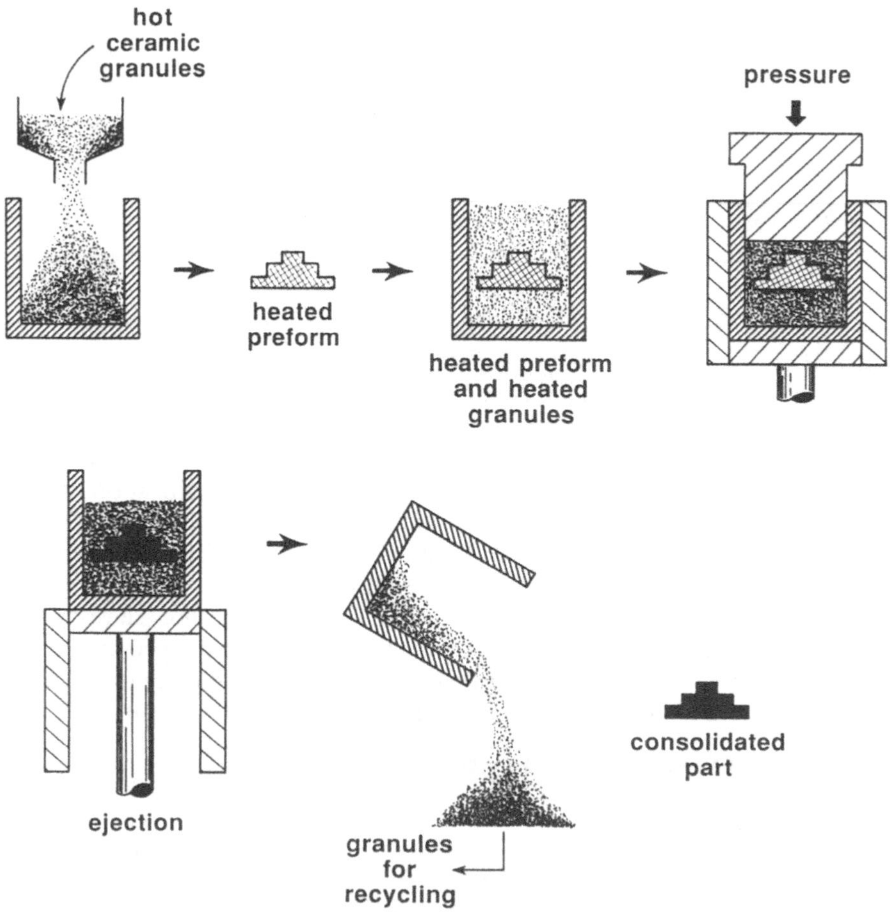

Figure 15-9. Pseudoisostatic conditions are used to consolidate powder by immersion of a shaped compact in heated graphite grains which are pressurized by uniaxially loading.

3.6. Powder Extrusion

The extrusion of an intermetallic powder at an elevated temperature is another approach to achieving full density. There are several variants to the basic process (Roberts and Ferguson 1991). One of the most common applications for hot extrusion is in consolidating mechanically alloyed powders. Both densification and shaping are achieved during one operation. Further, microstructure textures and substructures can be generated in extrusion that give superior mechanical properties (Ray 1990; Adams

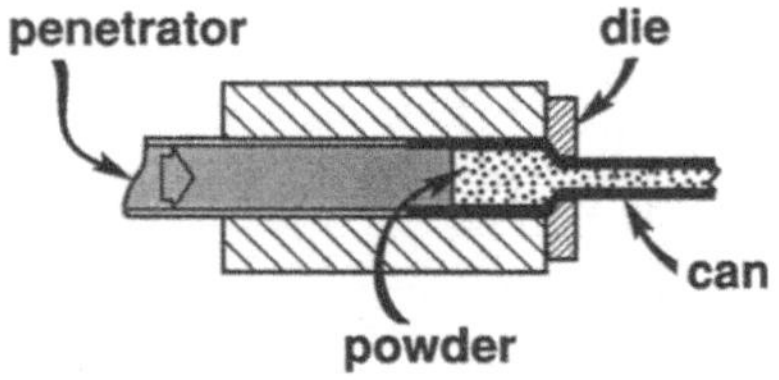

Figure 15-10. Powder extrusion is accomplished using canned powder and a penetrator ram.

et al. 1989; Bowman et al. 1992; Wright and Knibloe 1990; Wright et al. 1990).

A schematic drawing of powder extrusion is shown in Figure 15-10. A small penetrator in the extrusion ram directly stresses the powder and thereby avoids buckling of the can. The area reduction ratio usually exceeds 10 for adequate densification. Successful densification in the extrusion of prealloyed titanium aluminide powders (TiAl and Ti_3Al) require temperatures of 1000°C (Moll, Yolton, and McTiernan 1990). As an alternative Dahms (Dahms 1990) employed cold extrusion of mixed aluminum and titanium powders with a 9:1 reduction ratio. The powders were 95% dense and were then reactively hot isostatically pressed to full density. When IMC's are hot extruded the process temperature has a strong effect on reinforcement distribution (Adams et al. 1989).

3.7. Dynamic and Explosive Compaction

The high strain rates and instantaneous heating involved in explosive compaction have proven somewhat effective in consolidating intermetallics (Myers, Gupta, and Muir 1981; Prummer 1983; Wright, Korth, and Flinn 1987; Ferreira et al. 1991; Arkens et al. 1989; Persad et al. 1988; Balankin, Sokolov, and Troitzikiy 1990). These specialized techniques may eventually be successful in consolidating IMC's. Explosive compaction uses a shaped explosive charge surrounding a green powder mass to generate a shock wave for compaction. Dynamic compaction involves using an accelerated mass which strikes the powder at a high velocity to supply the shock wave. The self-heating can be used to initiate reactive synthesis of some elemental powder mixtures (Thadhani et al. 1990; Shang and Meyers 1991; Krueger, Mutz, and Vreeland 1992; Song and Thadhani 1992; Eliezer et al. 1990). However, the final microstructure is inhomogeneous and requires considerable further thermal processing.

Another variation of high strain-rate forming is high energy consolidation via electrical discharge through the powder. This technique is similar to dynamic and explosive compaction in that the process is quite rapid, with high pressure pulses. The consolidation energy increases with the yield strength of the material and small powders prove difficult to compact. For brittle IMC's, it is necessary to heat the compact to minimize fracture. Typical compaction wave velocities are in the 6 km/s range with pressure pulses of 30 GPa (300,000 atmospheres), but the pulse duration is only microseconds.

The dynamic techniques prove difficult to manage (Thadhani et al. 1990; Ferreira et al. 1991). Compacts made via these methods often experience cracking or pores due to trapped gas and secondary shock waves being reflected back through the component. Hot isostatic compaction has been used to cure these defects (Arkens et al. 1989). High energy consolidation techniques have been applied to titanium aluminides, nickel silicides, and nickel aluminides with densities generally reaching 95% of theoretical.

3.8. Spray Forming

Spray forming uses an inert gas atomizer and substrate closely positioned below the atomization nozzle (Annavarapu, Apelian, and Lawley 1990; Gutierrex-Miravete et al. 1989; Singer 1985). The spray is collected onto a substrate where it rapidly solidifies after undergoing splat deformation. Usually the spraying conditions are adjusted to have the droplets arrive in a semisolid condition. This allows rapid heat extraction and shear deformation to ensure near-full densification. The rapid heat extraction delivers microstructural homogeneity. In some instances the product is sufficiently dense for direct use while in other instances it is subjected to subsequent rolling, hot isostatic compaction, or hot forging. Because only a small volume of liquid is present at a given time, segregation is limited. One of the major drawbacks of this technique is that only simple shapes can be fabricated. IMC's are produced with this technique by co-spraying ceramic reinforcing phases into the semisolid molten stream. Typical deposition rates are 0.5 to 2 kg/s.

In a similar manner a plasma torch provides a means of depositing dense coatings (Apelian et al. 1983; Herman 1988). Powder is fed into a plasma arc and rapidly heated and accelerated. The intermetallic powder melts and is projected onto an external substrate where it splats to form a layered structure. Reinforcing phases can be trapped in the splat layers to form composites. Deposit density usually is about 85% of theoretical when deposition takes place at atmospheric pressure. If the spraying is performed in vacuum, then deposit densities of 95% to 99% of theoretical are possible. These can be fully densified by postspray HIP (Taub, Huang, and Chang 1984; Chang, Taub, and Huang 1984).

4. COMPOSITES SHAPING BY POWDER METALLURGY

The initial concern is with placement of the constituents and attaining the desired dispersion, homogeneity, and alignment using the fabrication approaches outlined earlier. For the lower temperature alloys it is possible to make a composite by infiltration or pressure-assisted casting into a ceramic fiber bundle. However, because of the high melting temperatures of IMC's, mixed powders are more widely employed with consolidation by hot compaction. Table 15-2 summarizes the several IMC's studied in the past and gives a nominal view of the processing cycles used in forming the composites, and more detailed processing information with mechanical properties on some of the systems are presented in Tables 15-4 and 15-5. Obviously, the references must be consulted to determine the details, including alloying or doping additions, raw material condition, reinforcement shape, size, and quantity.

4.1. Powder Injection Molding

Because of the high strength and low ductility, IMC's are difficult to machine. Therefore, the shaping technique must be selected to minimize postprocessing operations. One technique that provides a net-shape component and fiber alignment for discontinuous fibers is powder injection molding (PIM) (Bose and German 1988a). Figure 15-11 shows a schematic of the steps. A particulate or fiber-reinforcement phase is mixed with the matrix powder and a binder to form a thermoplastic feedstock. The binder is selected on the basis of powder wetting, ease of removal from the molded shape, and viscosity. Injection molding allows fiber alignment in the molding operation. Figure 15-12 is a demonstration of fiber alignment induced by the viscous flow of molding. Several of the problems associated with injection molding of IMC's are binder removal, nonuniform shrinkage, distortion, and detection of molding defects, and fiber breakage. Mixing and molding involve high shear rates that can damage the reinforcing phase. Intermetallic matrix composites reinforced with alumina have been injection molded with matrix phases of Ni_3Al (Moore et al. 1988; Bose and German 1988c), NiAl and $MoSi_2$ (Alman and Stoloff 1991a, 1991b), and $TaAl_3$ (Alman et al. 1989). These composites require separate consolidation steps such as HIP to achieve densification.

Table 15-3 P/M Processing and Properties for Intermetallic Compounds

Intermetallic	Processing Route	Powder Type	Test °C	Yield MPa	UTS MPa	Elong. %	Comments
Ni_3Al (Hf, B)	MA in air, HIP	prealloyed	RT		1014	1	1 wt% stearic acid added; A
Ni_3Al (Hf, B)	MA in Ar, HIP	prealloyed	RT	690	793	3	No stearic acid; A
Ni_3Al	VHP, hot extruded	IGA	RT	875	1525	34	IC-221; B
FeAl	extrusion	melt spun	823	100		20	Fe rich; C
FeAl	hot extrusion, warm worked	RSR, prealloyed	RT	115	138	17	D
FeAl	hot extrusion, recrystallized	RSR, prealloyed	RT	101	128	8	D
FeAl	extrusion	melt spun	RT	500	600	3	Fe rich; C
Fe_3Al	CIP HIP at 1100°C, 207 MPa	elemental	RT	884	1101	5	SHS processing, 2% Cr; E
Fe_3Al	CIP, HIP	elemental	RT	995	1060	3	SHS processing, 5% Cr; E
Fe_3Al	hot extruded	prealloyed	RT	358	726	12	1 h at 1100°C, 2 h at 750°C, oil quench; E

Source: A = Wang et al. 1988; B = Fuchs and Kao 1988; C = Baker and Gaydosh 1986; D = McKamey et al. 1991; E = Wright and Rabin 1992

Table 15-4 Mechanical Properties of P/M Ni_3Al Based IMCs

Matrix	Reinforcement	Fabrication	Test	Environ.	Yield	UTS	Elong.	Comments	Ref.
Ni_3Al	3% Al_2O_3	RHIP 800°C, 1 h, 170 MPa	RT	air		548		dense	A
Ni_3Al	20% Y_2O_3	RHIP 800°C, 1 h, 170 MPa	RT	air		463		dense	A
IC-15	20% Al_2O_3	VHP 1350°C, 1 h, 20 MPa	RT	air	227	388	10		B
IC-15	20% Al_2O_3	VHP 1350°C, 1 h, 20 MPa	1000	vacuum		114	*	brittle	B
IC-15	20% Al_2O_3	VHP 1335°C, 1 h, 20 MPa	1000	vacuum	102	102	0.4		B
IC-218	20% Al_2O_3	VHP 1170°C, 1.5 h, 20 MPa	1000	vacuum		134	0.4		B
IC-218	20% Al_2O_3	VHP 1200°C, 1.5 h, 20 MPa	RT	air	329	414	4		B
IC-218	20% Al_2O_3	VHP 1200°C, 1.5 h, 20 MPa	1000	vacuum	148	148	0.6		B
IC-218	20% Al_2O_3	VHP 1300°C, 1.5 h, 20 MPa	RT	air	250	348	6		B
IC-218	20% Al_2O_3	VHP 1300°C, 1.5 h, 20 MPa	1000	vacuum	144	144	0.4		B
IC-50	10% Al_2O_3	extrude 1100°C	25	air	660	870	9		C
IC-50	10% Al_2O_3	extrude 1100°C	1000	air	86	125	13		C
IC-50	2 wt% TiC		25	air	673	755	4		C
IC-50	2 wt% TiC		1000	air	225	270	2		C
IC-50	7% Al_2O_3		25	air	552	803	9	fiber	D
IC-50	9% Al_2O_3		25	air	488	985	13	fiber	D
IC-50	10% Al_2O_3		25	air	485	1027	24	fiber	D
IC-50	14% Al_2O_3		25	air	523	763	11	fiber	D
IC-50	22% Al_2O_3		25	air	515	748	19	fiber	D
Ni_2Al + Hf, B	3% Al_2O_3	HIP 1300°C, 3 h, 103 MPa	RT	air	—	924	—	MA 6 h	E
Ni_2Al + Hf, B	4% Al_2O_3	HIP 1300°C, 3 h, 103 MPa	RT	air	—	1241	—	MA 6 h	E
Ni_3Al + Hf, B	3% Y_2O_3	HIP 1300°C, 3 h, 103 MPa	RT	air	—	1172	—	MA 6 h	E
Ni_3Al + Hf, B	3% RhO_2	HIP 1300°C, 3 h, 103 MPa	RT	air	—	1172	—	MA 6 h	E
Ni_3Al + B	3% Al_2O_3	RHIP 800°C, 1 h, 104 MPa	25	air	474	548	1		F
Ni_3Al + B	20% Y_2O_3	RHIP 800°C, 1 h, 104 MPa	25	air	391	464	2		F
IC-218	5% Al_2O_3	HIP 1150°C, 1 h, 172 MPa	600	vacuum	662	889	4	whiskers	F
IC-221	5% Al_2O_3	HIP 1150°C, 1 h, 172 MPa	25	air	813	868	1	whiskers	F
IC-218	3% Al_2O_3	HIP 1150°C, 1 h, 172 MPa	25	air	499	756	6	whiskers	F

(yield = yield strength, UTS = ultimate tensile strength, * = fracture occurred before yield, MA = mechanical alloying)

Source: A = German 1987; B = McKamey et al. 1988; C = Sikka et al. 1991; D = Schneibel et al. 1991; E = Wang et al. 1988; F = German et al. 1988.

Table 15-5 Example Properties Available from Various P/M Processing Routes

Matrix	Additive	Form	Vol. %	Fabrication	Strength, MPa	Behavior	Ref.
FeAl	Y_2O_3	particle	1	hot extrude 975°C	930	ductile	A
Fe_3Al	—	—	—	RHIP 1100°C/2 h/207 MPa	1100	ductile	B
Fe_3Al	Al_2O_3	particle	10	RHIP 1100°C/2 h/207 MPa	—	small grain	B
$NbAl_3$	—	—	0	RS 1200°C	—	95% dense	C
$NbAl_3$	Al_2O_3	particle	30	RHIP 1200°C/4 h/170 MPa	—	98% dense	C
NiAl	—	—	0	RS 700°C	890	98% dense	C
NiAl	Al_2O_3	fiber	15	HIP 1200°C/2 h/172 MPa	163, 800°C	brittle	D
NiAl	TiB_2	particle	15	RHIP 1200°C/1 h/170 MPa	1060	dense	C
NiAl	TiB_2	particle	10	VHP 1400°C/2 h/48 MPa	673	grain size	E
NiAl	TiB_2	particle	20	RHIP 1200°C/1 h/170 MPa	1350	dense	C
Ni_3Fe	—	—	—	HIP 1200°C/0.5 h/103 MPa	627	37% elong	F
Ni_3Fe	Y_2O_3	particle	5	HIP 1200°C/0.5 h/103 MPa	649	10%	F
$TaAl_3$	—	—	0	RHIP 1200°C	531	98% dense	C
$TaAl_3$	Al_2O_3	fiber	35	RS VHP 1000°C	174	brittle	G
TiAl	—	—	0	RHIP 1370°C/4 h/170 MPa	533	brittle	C
TiAl	—	—	0	HIP	480	brittle	H
TiAl	Al_2O_3	particle	30	RHIP 1370°C/3 h/186 MPa	—	brittle	C
TiAl	SiC	particle	10	VHP 900°C/0.25 h/4 Mpa	95	porous	I
TiAl	TiB_2	particle	7	hot extrusion	875	brittle	J
Ti_3Al	Nb alloy	—	0	HIP 925°C/4 h/210 MPa	700	ductile	K
Ti_3Al	SiC	fibers	40	VHP powder-fiber laminate	700	reactions	L
Ti_3Al	SiC	fibers	35	VHP foil-fiber laminate	1480/270	anisotropic	H

(RHIP) = reactive hot isostatic pressing, VHP = vacuum hot pressing, HIP = hot isostatic pressing, RS = reactive synthesis, MA = mechanically alloyed)

Source: A = Strothers and Vedula 1987; B = Wright and Rabin 1992; C = Misiolek and German 1991b; D = Alman and Stoloff 1991a; E = Rigney et al. 1988; F = Bose et al. 1992; G = Anton 1988; H = Larsen et al. 1990; I = Wrzesinski and Rawers 1990; J = Adams et al. 1989; K = Moxson and Friedman 1986; and L = Brindley 1987.

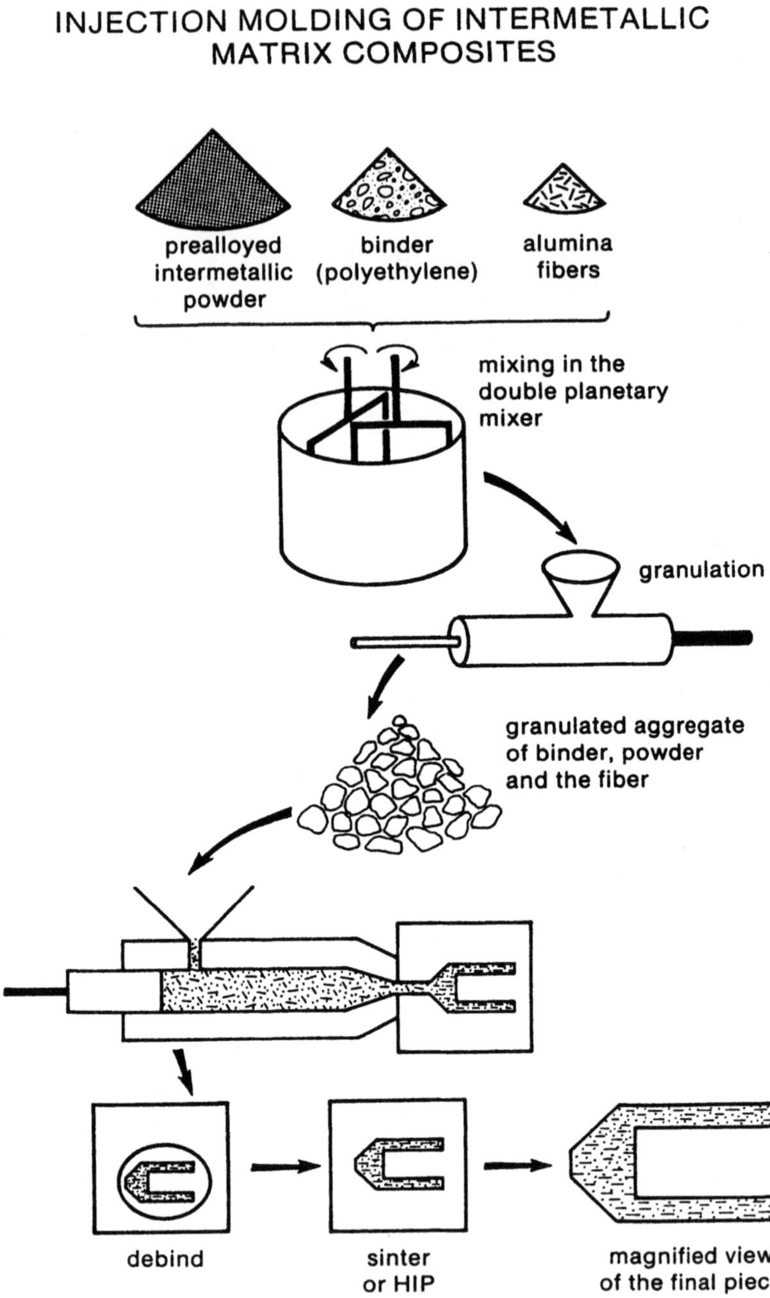

Figure 15-11. A sketch of the powder-injection molding process as used for form-aligned IMC's from powder-fiber mixtures.

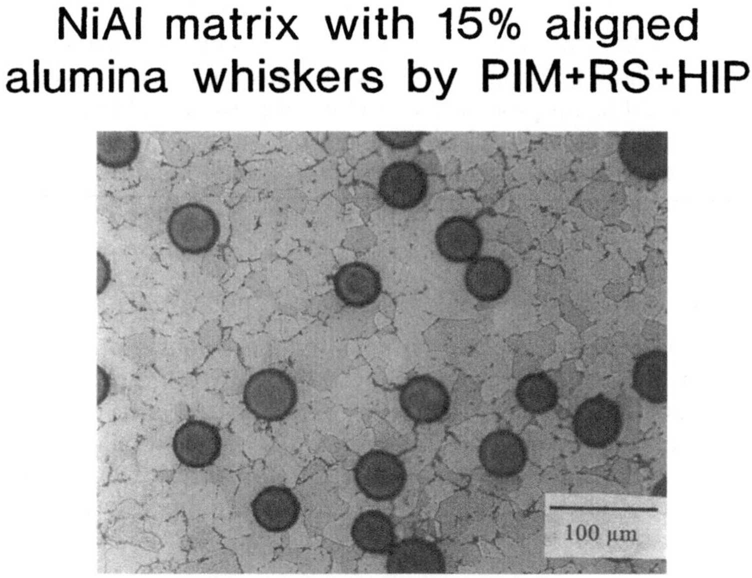

Figure 15-12. A micrograph showing aligned alumina fibers in a NiAl matrix formed by a combination of powder-injection molding, reactive sintering, and hot isostatic compaction (photo courtesy of D. E. Alman).

4.2. Spray Deposition Techniques

Spray deposition techniques adapt traditional gas atomization so that ceramic phases are injected into depositing intermetallic droplets to form an IMC. Plasma spraying uses a similar principle. It is possible to layup laminated IMC's with alternating layers of matrix and reinforcement. The reinforcement phase can also be fed directly into the plasma torch, thereby being co-deposited with the rapidly solidified intermetallic matrix. A random structure is generated by injecting the reinforcing phase into the deposit. Results by Tiwari et al. (1990) show that TiB_2 particles co-deposited with Ni_3Al in a plasma torch produce a higher ultimate tensile strength than Ni_3Al-TiB_2 composites formed by other methods. They cite difficulties in attaining target deposit compositions.

4.3. Binder-Assisted Shaping

Binder-assisted shaping is a loose hybridization of powder injection molding and traditional compaction methods aimed at forming oriented microstructures. As with injection molding, the powders and reinforcing phase are mixed with a binder; however, rather than being fed into an injection molding machine, the powders are compacted using more conventional forming techniques such as die pressing or cold isostatic pressing. One useful layup process uses a cloth of powder and binder as a basis for layering intermetallic phases, or encapsulation of fibers. The binder gives the newly formed component a higher initial handling strength, and prevents segregation of the powders (Bose and German 1988b; Alman and Stoloff 1991b; German et al. 1990).

4.4. Extrusion

This is a widely applied methods for the manufacture of IMC's. In addition to consolidating the material, the hot extrusion process will align any fibrous reinforcement along the extrusion direction. If the matrix powders are added in elemental form rather than prealloyed, the heat created by the exothermic reaction in the formation of the intermetallic compound can substantially lower the extrusion temperature required for prealloyed material.

4.5. Other Techniques

Several novel techniques have been demonstrated for IMC's. Typically, these have been applied to limited systems and are mentioned here as alternatives to the widely recognized processes covered previously.

Continuous fiber reinforced IMC's have been created using laminates of intermetallic powders and layers of ceramic fibers. The multiple-layer layup is densified by hot pressing or HIP. Systems processed by this approach are built around the Ni_3Al-Al_2O_3 or Ti_3Al-SiC compositions (Schneibel et al. 1991; Larsen et al. 1990). The aligned fibers give highly anisotropic properties and in the Ni_3Al-Al_2O_3 case gave little property benefit, but proved very beneficial in the Ti_3Al-SiC case (alloyed with 11% Nb with 35 vol.% fibers). The longitudinal strength (approximately 1460 MPa) was over five times higher than the transverse strength. For the longitudinal strength, there was 75% strength retention to 780°C.

TiAl-ZrO_2 composites with a gradient in the microstructure have been fabricated using a unique combination of mechanical alloying, hot pressing, and plasma sintering (Kimura and Kobayashi 1992). The functionally gradient composite is targeted to creating a transition from pure TiAl to pure ZrO_2 (with Y_2O_3 stabilization) over a distance of approximately 6 mm.

One variant is to create pseudofibers, where the fibers are actually regions of two phase structure (Misiolek and German 1991b; Nardone and Strife 1991). At low magnifications the two phase regions appear as fibers, but at high magnifications they are particulate reinforced portions of the structure. This approach solves two key problems in IMC's. First, new combinations can be explored in spite of a limited availability of fibers. Second, interfacial coupling between the fiber and matrix is ensured by selection of the binder phase for the pseudofiber regions. This approach has been applied to NiAl-TiB_2 and NiAl-B_4C.

5. MATERIAL EFFECTS ON CONSOLIDATION

Contamination of the powder is a major factor in P/M processing of intermetallics. This problem can be minimized by proper selection of processing aids and use of inert handling (Diehl and Stover 1990). The high-stability oxide-forming species such as titanium, aluminum, and silicon provide major problems with atmosphere control. Another concern is stoichiometry in powder fabrication and consolidation (Laag et al. 1991). High-temperature atomization and plasma processing cause preferential evaporation, leading to mixed-phase structures where a low melting eutectic might form unexpectedly. Other aspects relate to the actual material properties. The intermetallics exhibit property combinations that make processing difficult. The low ductility contributes to fracture, possibly induced just by cooling (Ferreira et al. 1991, Bowman et al. 1992). Therefore, thermal expansion matches are needed between the intermetallic and the processing dies or tools. Alloying also affects strength, strain hardening, and fracture toughness and these in turn dictate densification, microstructure, and properties.

The principal objective with IMC's is to improve the mechanical properties. This is especially important at higher temperatures, where the intermetallic matrix phase shows softening. Improved toughness and environmental resistance are key to high-temperature applications, and both depend on the material combination and interfacial quality. There are relatively few reinforcing phases that exhibit the properties required for high-performance IMC's. The typical choices are alumina, silicon carbide, titanium diboride, titanium carbide, and yttria. Fibers are not available in all of these, so often the reinforcement is in the form of particulates (Keraus et al. 1989).

The choice of a reinforcement phase in an IMC depends on material availability, oxidation resistance, thermal expansion coefficient, melting temperature (or other measures of thermodynamic stability), elastic modulus, and strength. Chemical compatibility between the intermetallic matrix and the reinforcement is required for long-term stability. A mismatch in thermal expansion coefficients leads to thermal fatigue during heating and cooling cycles (Mark et al. 1989). Thus, coupled thermal expansion is desired, but infrequently realized. Coatings are one route available to alleviate the incompatibility between phases. The interfacial coating must be tailored for the property goals—chemical stability, oxidation resistance, shear strength, or fracture resistance. Indeed, the needed interface chemistry varies with the mechanical properties of the matrix. An experimental and calculation protocol involving high-temperature exposures is emerging to aid in selection of viable combinations (Chou and Nieh 1990; Draper et al. 1990). This starts with thermochemical stability calculations, but necessarily involves experimental verification. The suitability of a material combination and selection for service involves thermal, mechanical, and environmental concerns coupled with assessment of interfacial stability using microstructural observations, damage tests, and environmental exposure typically to at least 1000°C.

6. PROPERTIES

The properties of intermetallics formed from powders largely depend on the composition, microstructure, and impurity level (Porter et al. 1990; Yang and Jeng 1992). A great deal of work has been conducted on determining factors that control the mechanical properties of the aluminides. The purpose of this section is to briefly introduce these results by noting the material and processing parameters. Table 15-3 lists the mechanical properties of representative P/M aluminides. These data are a representation of the mechanical-property data that have been assembled on intermetallic compounds; several recent review articles provide a more comprehensive description (McKamey et al. 1991; Lipsitt 1985; Vedula and Stephens 1986; Stoloff 1984). Likewise, much effort has been applied to the IMC's to correct problems with toughness, high-temperature strength, or microstructure control.

Tables 15-3, 15-4 and 15-5 provide an overview of the properties of several systems. Table 15-3 covers representative aluminides, Table 15-4 is restricted to Ni_3Al matrix composites, and Table 15-5 covers other IMC's. In each case there is brief information on the nominal composition, fabrication route, properties, and reference for further information. In the IMC's, the fibers limit the interfacial strength between the matrix and reinforcement phase. These are arranged by the nominal matrix chemistry. In many instances the matrix was not stoichiometric and was often alloyed for increased strength or environmental stability. The reinforcing phases were largely particles, although a few continuous fiber or whisker composites are known.

Although some of the entries are outstanding, many are not competitive with available commercial alternatives. The presence of brittle reaction layers at the interface prove detrimental to the fracture toughness. Accordingly, most intermetallic compounds exhibit minimal property improvements when used in composites. Furthermore, as of 1992 there are no known IMC's with maximum-use temperatures better than existing nickel-base superalloys (Doychak 1992). On a density basis, there is a strength advantage to the titanium aluminides. For many of the IMC's, problems originate with the brittle character of the matrix phase. The low-temperature brittleness causes failure during thermal cycling due to a mismatch between thermal expansion coefficients of the phases. There is a strong grain-size effect on the mechanical properties, with smaller grain sizes giving higher strengths at room temperature (Wang et al. 1988). For this reason, processing techniques that produce smaller grain sizes give higher strengths. Thus, mechanical alloying, dynamic compaction, spray forming, extrusion, and forging give higher apparent strengths.

6.1. Ni_3Al

The nickel aluminides are probably the most widely explored P/M intermetallics. The Ni_3Al composition needs extensive alloying to generate competitive properties, but even when alloyed the high-temperature properties are low; thus, the extensive interest in P/M composites based on this compound. The presence of aluminum in both the reinforcement and matrix phases aids the chemical stability in the Ni_3Al-Al_2O_3 system. A detailed comparison of tensile results for Ni_3Al composites is given in Table 15-4. The IC designations are shown where an alloyed form of Ni_3Al has been used for the matrix (for example IC-218 consists of Ni, 7.67% Cr, 8.18% Al, 0.8% Zr, and 0.02% B). Because of its relatively low melting temperature (about 1340°C), the Ni_3Al intermetallic can be formed from elemental powders, or through atomization from the melt. Although the system does not qualify for high-temperature applications, the ductility of Ni_3Al has sparked considerable interest. Various ceramic particles and fibers have been incorporated into the powder mixture, including Y_2O_3, Al_2O_3, TiB_2, and SiC. The composites include up to 30 vol.% ceramic phase.

The large body of processing research on the Ni_3Al-Al_2O_3 system provides a basis for understanding several other systems processed using powder techniques. Composites can be fabricated from mixed elemental powders using temperatures below 1100°C. Mechanical properties for such IMC's are summarized in Table 15-5. A classic RHIP cycle involves mixing alumina fibers with nickel and aluminum powder, with reaction and densification by heating to approximately 800°C, with a 0.5 h hold under a pressure of 104 MPa.

The Ni_3Al-Al_2O_3 system usually has inferior properties when compared to monolithic Ni_3Al. For example, in the HIP processing of prealloyed powders the yield strength, tensile strength, and ductility improved with higher HIP temperatures (Moore et al. 1988). The addition of 5 vol.% alumina fiber decreased all three properties, especially in the fully heat-treated condition. This is probably due to low inherent strength in the alumina and a poor interfacial bond. As indicated in Table 15-5, some of the composites showed increased strength, but the ductility decreased with alumina fiber additions. At room temperature HIP alloyed Ni_3Al has a yield strength in the 200 to 600 MPa range, tensile strength in the 1250 to 1350 MPa range and 40% to 60% elongation. For comparison, the HIP composites with 20 vol.% Al_2O_3 have similar yield strengths (225 to 480 MPa), lower tensile strengths (350 to 900 MPa), and ductilities generally below 10%. The property difference is even more pronounced at 1000°C. This degradation of properties is a disappointing characteristic of most IMC's.

6.2. NiAl

The NiAl compound is a low density (5.86 g/cm^3), higher melting temperature intermetallic (about 1649°C for the stoichiometric composition) with excellent oxidation resistance. The high melting temperature and high aluminum content are favorable for high-temperature use. However, the current estimate is a maximum-use temperature below 1000°C (Doychak 1992). High-temperature strength is improved by incorporation of various ceramic phases (TiC, TiB_2, Al_2O_3, AlN, or HfC), with most attention to TiB_2 (Rigney et al. 1988; Misiolek and German 1990; Alman and Stoloff 1991b; Wang and Arsenault 1991; Whittenberger et al. 1989). The strengthening is largely due to grain-size reduction during processing; the strength for 0% TiB_2 is between 310 and 550 MPa while 20% addition gives a strength of 1350 MPa (Stoloff and Alman 1990). Sapphire fibers resulted in a weak fiber-matrix interface and poor strength (Doychak 1992).

During reactive synthesis the exothermic heat release is sufficient to melt the compound. Consequently, prealloyed NiAl or a ceramic reinforcing phase are needed to dilute the reaction. One novel IMC based on NiAl is shown in Figure 15-12. This composite was formed by a combination of powder injection molding, reactive synthesis, and HIP (Alman and Stoloff 1991a, 1991b). The injection molding was performed using polyethylene as a binder to align the alumina whiskers in the flow field. The polyethylene was extracted and the compact reacted to synthesize the NiAl. Subsequently, the structure was encapsulated in borosilicate glass and subjected to hot isostatic compaction to attain full density. The final tensile properties at 800°C were not improved by the alumina fibers (135 MPa yield strength with 14% ductility for 0% fibers versus 142 MPa yield strength with 3% ductility with 15 vol.% fibers). Also, the fibers were damaged by the consolidation process.

The use of TiB_2 as the reinforcing phase gives higher strength because of grain-size reduction, but lowers oxidation resistance. In experiments with reactively synthesized NiAl, VHP was used to fabricate both NiAl-TiB_2 and NiAl-Al_2O_3 composites (Kumar, Mannan, and Visnandham 1992). The matrix toughness after hot pressing is typically near 9 MPa$(m)^{1/2}$, but this is not significantly improved by the addition of the ceramic phases. Part of the difficulty was that failure preferentially occurred along the interphase boundary. Rigney et al. (1988) were more successful by hot pressing at 1400°C, giving an improved toughness of 20 MPa$(m)^{1/2}$ with 10 vol.% TiB_2.

6.3. Ti_3Al and TiAl

Titanium aluminides provide possible improvements over classic titanium alloys in terms of specific strength, melting temperature, and oxidation resistance. The titanium aluminides have a low fracture toughness and require alloying with V, Nb, Mo, and Ta for ductility, and composites are used to improve toughness and strength (Larsen et al. 1990; Kumar et al. 1991). The light weight coupled with the recognized oxidation resistance make TiAl-SiC a widely favored system. Several approaches have been explored for consolidation, including VHP, hot extrusion, reactive synthesis of TiAl followed by VHP, and RHIP, generally with Al_2O_3, TiB_2, or SiC particles and Al_2O_3 or SiC fibers. Early data on Ti_3Al-SiC shows good properties and reasonable interfacial stability. Limited strength improvements were attributed to the brittle matrix. Subsequent evaluation determined poor environmental resistance and susceptibility to thermal fatigue (Larsen et al. 1990). Generally, the particle reinforced TiAl composites prove brittle (Adams et al. 1989; Whittenberger et al. 1989) and the properties are unattractive, especially creep, fatigue, and fracture toughness. Porter et al. (1990) compared two powder production techniques consolidated by a pseudoisostatic compaction technique. They reported an effect due to the original powder purity. Coatings are probably needed to impeded reactions between the phases at elevated temperatures. Otherwise, the ductility, formability, and toughness are unacceptable for use in high temperature systems

(Parameswaran 1992). Both TiAl and Ti_3Al prove susceptible to hydrogen embrittlement. Titanium aluminide composites display extreme property degradation in oxidizing environments with cyclic loading or cyclic heating. The TiAl-Ti_3Al composites have not improved on the inherent brittle behavior (Persad et al. 1988). However, some progress has occurred using ternary compounds, such as Ti_2NbAl. This compound has a higher specific strength than the binary titanium aluminides.

Since the titanium aluminides are low in ductility, one variant at a composite is to use a ductile reinforcing phase to provide toughening. Titanium alloys can be used for these applications, but use temperatures are limited.

6.4. $MoSi_2$

The $MoSi_2$ intermetallic formed by reactive synthesis from elemental powders finds common use as a heating element for high-temperature furnaces. The resistance to high-temperature oxidation is a major attraction. However, $MoSi_2$ has several limitations, including room-temperature brittleness and low strength at high temperatures. Porosity induces oxygen attack at temperatures over 500°C that further deteriorates the compound. The addition of SiC increases the high-temperature creep resistance of the otherwise soft $MoSi_2$ (Yang and Jeng 1992). Continuous SiC fibers induce a stress from a thermal expansion mismatch that causes matrix cracking. In comparison with other high-temperature ceramic matrix composites, $MoSi_2$-SiC proves inferior and limits the near-term applications to electrical components such as heating elements. There is some hope that the $MoSi_2$-WSi_2-SiC composite will have the desired high-temperature strength (Sheppard 1990).

7. MICROSTRUCTURES

Perhaps the most important factor controlling the properties of intermetallics and their composites is the microstructure. Indeed, to understand the properties of these materials, it is essential to have a complete microstructural evaluation; for without it, any information obtained from any mechanical, thermomechanical, and oxidation testing will be meaningless. Microstructure examination using optical, electron beam, and other methods provides details needed to explain the properties of the composite. For example, optical microstructural evaluation can often reveal if the matrix has reacted with the reinforcement phase during fabrication, or high-temperature exposure. Such degradation will decrease strength, ductility, and toughness.

7.1. Reinforcements

Much of the published literature on IMC's mention a nominal fiber-volume fraction to characterize the material. Such numbers are deceiving. During processing, it is common for fibers to clump together in a localized region, producing a very high volume fraction of fibers in one area, with the majority of the component being completely devoid of any fibers.

7.2. Microstructural Scale and Grain Size

Nonequilibrium microstructures have been developed in titanium aluminides using high-pressure shock waves to consolidate RSP powder (Vassiliou et al. 1990). As particle melting occurs on a very localized level, the minimal heat generated does not permit phase transformations to occur, thereby maintaining the rapidly solified microstructure.

In terms of powder production, RSP methods are desirable because of the refined grain size and the metastable structures. Such methods include melt spinning, the plasma rotating electrode process (PREP), and inert gas atomization (IGA). Novel initial microstructures have been obtained with high-energy mechanical alloying techniques (Koch and Cho 1992; Benn et al. 1988; H. Kimura and M. Kimura 1990; Wang et al. 1988); however, only limited quantities can be produced by such methods. Microstructures of the powder can vary from amorphous powder (H. Kimura and M. Kimura 1990) to nanocrystalline structures.

The addition of small particles or fibers is considered desirable for intermetallic properties, but this has not been borne out with experimental data. In many instances, reinforcement additions actually decrease strength. In those instances where there is an improvement in strength over the monolithic matrix, the increase is a result of a smaller grain size in the matrix created by grain-boundary pinning during consolidation.

On a practical level, there are distinct disadvantages to adding small reinforcements. As the reinforcement size decreases, so does the cross-sectional area. However, at the elevated temperatures encountered during processing, a certain perimeter of the reinforcement may be consumed. Additionally, fibers are likely to incur damage during processing. Larger fibers can be handled more easily, especially when laying up composites.

7.3. Homogeneity and Contamination

The issues of homogeneity and contamination by trace elements are two closely linked topics when dealing with intermetallics and are crucial to the future success. Since intermetallics form at a discrete stoichiometric ratio, alloy composition must be controlled to insure a single-phase compound. Contamination in the matrix, or the presence of trace impurity elements, can also produce undesirable results. This can be harmful if an unwanted phase nucleates at the grain boundaries, especially in those instances when the material is to be used at elevated temperatures. Contamination also can prove harmful if it occurs at the matrix-fiber interface. If one of the reaction products exhibits properties (mechanical, chemical, or thermal) that are lower than desired, the integrity of the entire structure may be jeopardized. Mendiratta and Dimiduk (1988) have taken advantage of this concept of off-stoichiometric compositions to develop in situ Nb_5Si_3-Nb composites.

7.4. Segregation and Interfacial Reactions

For high temperature IMC's, the reinforcement and matrix must be thermally compatible. However, as most of these composites are artificially produced, the fibers and the matrix will ultimately react with each other. The extent and the time required for these reactions is determined by the kinetics of the particular system. The reactions can occur in processing or use. Thermodynamic data exist that can predict the possibility of a reaction (Misra 1990; Chou and Nieh 1990); however, chemical reactivity is very sensitive to many factors; these predictions are not fully accurate. Verification of the chemical compatibility is obtained through experimentation. Typical tests involve 1000°C for 50 to 100 h in air, with interfacial reaction barriers often proving necessary to minimize reactions (Deve and Maloney 1991; Brennan et al. 1991; Chou and Nieh 1990; Tien and Kopp 1990).

Several studies have been performed on the interfacial stability in IMCs (Fuchs 1990; Yang and Jeng 1991; Brennan, Kao, and Katzman 1991; Yang, Kao, and Liu 1989; Draper et al. 1990; Nourbakhsh, Liang, and Margolin 1990; Deve and Maloney 1991; Wang and Arsenault 1991; Emiliani and Deve 1992). One such study was performed on SiC whiskers in a $MoSi_2$ matrix. The composite was hot pressed at 1700°C. Though the interface was intact with no gap between whiskers and the matrix, there was a reaction attributed to impurities present in the $MoSi_2$ powder, specifically iron and aluminum. This illustrates the effects of trace elements on interfacial integrity.

Reactions in the P/M Ni_3Al-SiC system are now well understood (Larkin, Interrante, and Bose 1990; Chou and Nieh 1990; Yang et al. 1988) and future efforts need to bring a similar level of understanding to other systems. Unfortunately, even with full density most of the IMC's show minimal property improvements, and more typically a decrement in properties due to impurities (Kim and Froes 1989). An example is given by Ni_3Al powder which was mixed with 15 vol.% TiC and consolidated to 100% density via HIP and extrusion (Fuchs 1989). The yield strength increased slightly from 700 to 780 MPa, but the ultimate strength decreased from 1050 MPa to 800 MPa and the elongation severely decreased from 10% to 0.5%. Similar trends persisted in tensile tests up to 1000°C.

7.5. Fiber Damage in Composites

Fiber damage is common in P/M processing of IMC's. It is caused by the shearing encountered in mixing, shaping, and densification. Fiber damage can also occur in hot densification when the deformation strain exceeds the strain to failure for the fiber. Such damage is less likely to occur during consolidation in applications that apply isostatic pressure to the green component.

8. PROBLEMS

Intermetallics are at a relatively early stage of development and P/M has been applied to several systems during this development stage. The initial enthusiasm for possible improved high-temperature structural materials has been tempered by the observed instabilities and environmental attacks. There are few examples of improved high-temperature properties for either the basic intermetallics or their composites fabricated by P/M. The compound powders are expensive when properly alloyed for high-temperature stability in an oxidizing atmosphere. Further, intermetallics have questionable high-temperature properties and are often unacceptably brittle at low temperature. Several attempts have been made to adjust the fundamental crystal structure and ductility. Even the systems that exhibit room-temperature ductility become embrittled at intermediate temperatures due to oxygen penetration of grain boundaries (Taub et al. 1984; Wang et al. 1988; Wright et al. 1990). This leads to a ductile-brittle transformation that is dependent on prior history. The lack of pure powders with good compositional control has impeded process development.

The method of powder fabrication can contribute to surface contamination. Oxygen contamination is a pervasive problem that requires attention through degassing treatments and inert handling. Auger analysis of Ni_3Al powders produces evidence of C and S contamination, with a 2 nm alumina film. The titanium aluminides represent a particularly sensitive system where 600 to 1000 ppm of oxygen is typical. Higher levels of oxygen are observed as the particle size decreases.

One consequence of powder surface contamination is the formation of inhomogeneities at interfaces in the consolidated microstructure. If the surface contamination is an oxide or similar inert film (usually a nitride or carbide), then particle bonding will be weak. Therefore, full density is not a direct assurance of competitive properties. There are three steps available to counteract prior particle boundary contamination. First, the powder must be processed in a manner that avoids detrimental segregation. Second, the powder must be maintained in a high purity, inert atmosphere to avoid the formation of surface films before consolidation. Finally, extensive plastic deformation during consolidation (e.g., by hot extrusion) ensures disruption of any continuous surface films which can degrade properties (Vedla and Stephens 1987).

Each of the P/M techniques for consolidation of the intermetallic compounds suffers from processing difficulties. For plasma spray there is difficulty in maintaining stoichiometry. Hot pressing and hot isostatic compaction give poor compact degassing and low precision final components. The mechanical properties are not necessarily optimal even when 100% density is achieved and generally higher temperatures are required for good particle bonding (Beddoes et al. 1992). Powder injection molding followed by sintering has been successfully applied with minimal contamination, but final densities are only 94% to 98% of theoretical. Forging techniques involving granular pressure-transmission media give surface contamination of the intermetallics. Explosive compaction often cracks the compact during consolidation. Thus, each approach has limitations in product quality and microstructure control. There is a clear need to control the grain size and impurities. A smaller powder gives a finer grain size with improved ductility and strength, but may limit creep behavior and introduce more impurities (Feest and Tweed 1992).

Intermetallic matrix composites fabricated by P/M suffer these same fundamental problems and have further complications involving stability between phases, fabrication, and long-term integrity. In many cases, the IMC's have inferior properties when compared to the monolithic intermetallics (Kumar et al. 1991; Wrzesinski and Rawers 1990). Although many potential reinforcing phases are discussed, in reality there are a relatively small number in use. Further, the discontinuous particle or whisker phases have been emphasized because of low cost, but with poor composite properties (Erich 1987). Consequently, only a few IMC systems fabricated by P/M are under serious consideration, and these have recognized limitations. Consider that TiAl-SiC is a popular system with attractive properties, but the combination is unstable since Ti_5Si_3 forms during high-temperature exposure (Wrzesinski and Rawers 1990).

Most of the research on candidate fabrication processes for IMC's have brought to light several processing problems. For example, with plasma spraying of two materials, matrix and reinforcement, there are now recognized compositional problems with loss of intermetallic stoichiometry and differential sticking of the two phases. Further, the entrapped spraying gas and lack of full densification lead to reduced properties. Another drawback is the lack of component precision. Likewise, HIP has been explored sufficiently for IMC's that key problems are now identified. These include contamination of the powders or fibers with resultant microstructural decorations. The lack of outgassing and shear during consolidation allow the impurities to severely limit properties. Most of the composite-forming techniques damage the fibers and in many systems this combines with the brittle character of the intermetallic phase to produce a brittle, low-performance composite.

The mechanical properties of the P/M IMC's are not overly convincing of improved properties beyond available materials. Toughness is often inferior and in cases where microstructural toughening has been successful, other problems become fatal. Fatigue is a complex problem in these systems that involves consideration of orientation, temperature and stress cycles, frequency, environment, and damage. The complex testing needed to fully understand these issues delays applications in high-performance areas. Further, thermal fatigue is a problem because of the differing thermal expansion coefficients between matrix and fiber. For example, in the Ti_3Al-SiC system the fibers prove very effective in promoting toughness. With 500 cycles from 150–815°C in air, the room-temperature tensile strength of the composite decreased from 1260 MPa to 120 MPa (Larsen et al. 1990).

Fundamentally, the interfaces are crucial to high-temperature service and are the major microstructural weakness (Draper et al. 1990; Chou and Nieh 1990; Tien and Koch 1990; Das 1990). The interface serves as a fast diffusion pipeline during P/M fabrication or service, leading to substantial degradation of the material. The widespread sensitivity to oxygen attack is a fundamental problem that may only be remedied by use of expensive coatings. Cracking and microcracking are common defects that appear to largely originate in fabrication. The low-cost fabrication techniques, such as powder injection molding, are still experimental and suffer problems with microstructural inhomogeneities and defects introduced during processing (Bose and German 1988c).

Another fundamental problem is the low production volumes currently anticipated. There is no clear leading P/M system, although many are under active evaluation. Accordingly, consumption in the United States is quite small. Because this is a relatively small market, there is no attention to dedicated production operations, leaing to poor economies of scale and long lead times in fabricating demonstration materials. The use of intermetallics produced by P/M is inhibited by the classic material development cycle time of 10 to 20 years.

9. APPLICATIONS

In the near term, most of the applications for intermetallics and intermetallic matrix composites formed using P/M techniques will be at low temperatures. Examples include dies for casting, bonds for carbides or diamonds, bearings, oxidation coatings, memory alloys, and corrosion barriers. The best developed application is in the spray coating field, where nickel-aluminum alloys are applied as semidense coatings to inhibit oxidation or corrosion. In the near term, it is expected that aluminides will find use in furnace hardware, especially for heat-treating operations where ferrous alloys adsorb carbon. Here the moderate strength at temperature is less important than the insolubility of carbon from the heat treating atmosphere. Another possible application for Fe_3Al is in cutting tools and passivation coatings for ferrous alloys. One of the first is use of Ni_3Al as a replacement for cobalt in bonding carbides and diamonds for abrasive and cutting tools. Other applications that appear to be maturing include furnace hardware, mining tools, diesel engine components, abradable seals, chemical filters, cutting tools, and structural alloys for low-temperature applications.

10. CONCLUSIONS, OPPORTUNITIES, NEEDS

Intermetallic compounds represent a novel hybrid of stoichiometric (ceramiclike) materials and metals. They have recognized uses as reinforcing phases in many metallurgical systems, especially superalloys. Some of the intermetallic compounds are stable to high temperatures and offer possible uses because of inherent high-temperature corrosion and oxidation resistance. Accordingly, developments in this field focus on the intermetallic compounds as well as possible composites. These materials are new and will require extensive effort to bring to maturity for aerospace components. However, as knowledge on composition, processing, structure, properties, and behavior emerge there are several near term applications using P/M.

Will these P/M materials ever achieve the envisioned uses in high-temperature jet engines? Based on the preliminary data there is little basis for anticipation of such an application. Probably the near-term applications will be in the 600°C to 1000°C temperature range (Larsen et al. 1990). Special intermetallics such as the nickel and iron aluminides may find applications in wear components, oxidation barriers, and uses involving certain corrosion fluids. Here the high-temperature strength is not important, but a composite might be useful for wear resistance or other tailored properties. Before acceptance as aerospace materials there is a need for more demonstration of mechanical reliability, especially over the anticipated use range.

The P/M-based processes generally offer a major advantage in their shaping capabilities. Accordingly, many of the IMC's will be developed using powder consolidation techniques. Not only are complex shapes possible, but the productivity, material utilization, energy expenditure, precision, and microstructural uniformity are attractions. The generic, widespread interest in net-shape forming has focused considerable research and development interest on powder approaches.

REFERENCES

ABE, O., and TSUGE, A., 1991, *J. Mat. Res*. 6: 928–934.

ADAMS, M. L., KAMPE, S. L., HARMON, A. R., and CHRISTODOULOU, L., 1989, *Advances in Powder Metallurgy* 3: 439–448.

ALMAN, D. E., and STOLOFF, N. S., 1991a, In *Low Density, High Temperature Powder Metallurgy Alloys*, 109–125. Warrendale, PA: The Minerals, Metals and Materials Society. W. E. Frazier, M. J. Koczak, and P. W. Lee, Eds.

________, 1991b, *International Journal of Powder Metallurgy* 27: 29–41.

ALMAN, D. E., DIBBLE, A., MOORE, B., BOSE, A., GERMAN, R. M. and STOLOFF, N. S. 1989, In *Processing of Ceramic and Metal Matrix Composites* 217–227. New York: Pergamon.

ANNAVARAPU, S., APELIAN, D., and LAWLEY, A., 1990, *Metall. Trans*. 21A: 3237–3256.

ANTON, D. L., 1988, In *High Temperature/High Performance Composites, MRS Symp. Proc*. 120: 57–64.

APELIAN, D., PALIWAL, M., SMITH, R. W., and SCHILLING, W. F., 1983, *Inter. Met. Rev*. 28: 271–294.

APGAR, L. S., and EYLON, D., 1991, *ISIJ Inter*. 31(8): 915–921.

ARKENS, O., DELAEY, L., DE TAVERNIER, J., HUYBRECHTS, B., BUEKENHOUT, L. and LIBOUTON, J. C. 1989. In *High Temperature Ordered Intermetallic Alloys III*, *MRS Symp. Proc.* 133: 493–498.

ARZT, E., ASHBY, M. F., and EASTERLING, K. E., 1983, *Metall. Trans.* 14A: 211–221.

ASHBY, M. F., 1991, *HIP 6.0 User Manual*, Cambridge, England: Cambridge University.

BAESLACK III, W. A., CIESLAK, M. J., and HEADLEY, T. J., 1988, *Scripta Metall.* 22(7): 1155–1160.

BAKER, I., and GAYDOSH, D. J., 1986, In *High Temperature Ordered Intermetallic Alloys II*, *MRS Symp. Proc.* 81: 315–320.

BAKER, I., ICHISHITA, F. S., SURPRENANT, V. A., and SCHULSON, E. M., 1984, *Metallog.* 17: 299–314.

BALANKIN, S. A., SOKOLOV, V. S., and TROITZIKIY, A. O., 1990, In *Proceedings PM-90*, *International Conference on Powder Metallurgy* 3: 37–39. London: The Institute of Metals.

BEDDOES, J. C., WALLACE, W., and DE MALHERBE, M. C., 1992, *Inter. J. Powder Met.* 28: 313–325.

BENN, R. C., MIRCHANDANI, P. K., and WATWE, A. S., 1988, In *Modern Developments in Powder Metallurgy*, 21: 479–493. Princeton, NJ: Metal Powder Industries Federation.

BHATTACHARYA, A. K., 1992, *J. Amer. Ceramic Soc.* 75: 1678–1681.

BOSE, A., CAMUS, G., GERMAN, R. M., DUQUETTE, D. J., and STOLOFF, N. S., 1993, *J. Mat. Res.*, 8: 430–437.

BOSE, A., PAGE, R. A., MISIOLEK, W. Z., and GERMAN, R. M., 1991, *Advances in Powder Metallurgy* 6: 131–145.

BOSE, A., and GERMAN, R. M., 1988a, In *Modern Developments in Powder Metallurgy* 18: 299–314. Princeton, NJ: Metal Powder Industries Federation.

________, 1988b, *Advanced Materials and Manufacturing Processes* 3: 37–56.

________, 1988c, *Industrial Heating* 55(5): 38–41.

BOSE, A., MOORE, B., GERMAN, R. M., and STOLOFF, N. S., 1988a, *J. Met.* 40(9): 14–17.

BOSE, A., RABIN, B. H., and GERMAN, R. M., 1988b, *Powder Metall. Inter.* 20(3): 25–30.

BOWMAN, R. R., NOEBE, R. D., RAJ, S. V., and LOCCI, I. E., 1992, *Metall. Trans.* 23A: 1493–1508.

BRENNAN, P. C., KAO, W. H., and KATZMAN, H. A., 1991, *J. Mat. Res.* 6(2): 355–360.

BRINDLEY, P. K., 1987, In *High-Temperature Ordered Alloys II*, *MRS Symp. Proc.* 81: 419–424.

BULJAN, S. T., PASTO, A. E., and KIM, H. J., 1989, *Ceramic Bull.* 68: 387–394.

CHANG, K. M., HUANG, S. C., TAUB, A. I., CHANG, G.-M., and MORRIS, JR., J. W., 1987, *Metall. Trans.* 18A: 1819–1820.

CHANG, K. M., TAUB, A. J., and HUANG, S. C., 1984, In *High-Temperature Ordered Intermetallic Alloys*, *MRS Symp. Proc.* 39: 335–42.

CHEN, J., GENG, Z., and CHIN, B. A., 1988, In *High-Temperature Ordered Intermetallic Alloys III*, *MRS Symp. Proc.* 133: 447–452.

CHENEY, R. F., and SEYDEL, E. R., 1990, *Metal Powder Report* 45(1): 43–46.

CHENG, S. C., WOLFENSTINE, J., and SHERBY, O. D., 1992, *Metall. Trans.* 23A: 1509–1513.

CHIANG, Y. M., HAGGERTY, J. S., MESSNER, R. P., and DEMETRY, C., 1989, *Ceramic Bull.* 68(2): 420–428.

CHOU, T. C., and NIEH, T. G., 1990, *J. Mat. Res.* 5(9): 1985–1994.

CHRISTODOULOU, L., PARRISH, P. A., and CROWE, C. R., 1988, In *High Temperature/High Performance Composites*, *MRS Symp. Proc.* 120: 29–34.

CLYNE, T. W., and MASON, J. F., 1987, *Metall. Trans.* 18A: 1519–1530.

COBLE, R. L., 1978, *Powder Met. Inter.* 10: 128–130.

CRANMER, D. C., 1989, *Ceramic Bull.* 68(2): 415–442.

DAHMS, M., 1990, In *Solid State Powder Processing* pp. 77–82, Warrendale, PA: The Minerals, Metals and Materials Society. A. H. Claver and J. J. Debarbadillo (Eds.)

DAS, G., 1990, In *Metal and Ceramic Matrix Composites: Processing, Modeling and Mechanical Behavior*, pp. 413–442. Warrendale, PA: The Minerals, Metals and Materials Society. R. B. Bhagat, A. H. Claver, P. Kumar and A. M. Ritter, Eds.

DÉVE, H. E., and MALONEY, M. J., 1991, *Acta Metall. Mater.* 39(10): 2275–2284.

DIEHL, W., and STOVER, D., 1990, *Metal Powder Report* 45(5): 333–338.

DIPIETRO, M. S., KUMAR, K. S., and WHITTENBERGER, J. D., 1991, *J. Mat. Res.* 6(3): 530–538.

DOYCHAK, J., 1992, *J. Met.* 44: 46–51.

DRAPER, S. L., GAYDOSH, D. J., NATHAL, M. V., and MISRA, A. K., 1990, *J. Mat. Res.* 5(9): 1976–1984.

DRIVER, D., 1990, In *High Temperature Materials for Power Engineering Conference*, pp. 883–902. Dordrecht: Kluwer Academic Publishers.

DUNMEAD, S. D., MUNIR, Z. A., HOLT, J. B., and KINGMAN, D. D., 1990, In *Combustion and Plasma Synthesis of High Temperature Materials*, pp. 229–237. New York: VCH Publishers.

EL-ESKANDARANY, M. S., SUZUKI, H., AOKI, K., and SUZUKI, K., 1990, *J. Japan Soc. Powder Powder Metall.* 38: 934–940.

ELIEZER, Z., LEE, B. H., HOU, C. J., PERSAD, C., and MARCUS, H., 1990, In *Metal and Ceramic Matrix Composites: Processing, Modeling and Mechanical Behavior* 401–412. Warrendale, PA: The Minerals, Metals and Materials Society. R. B. Bhagat, A. H. Claver, P. Kumar and A. M. Ritter, Eds.

ERICH, D. L., 1987, *Inter. J. Powder Met.* 23(1): 45–54.

FEEST, E. A., and J. H. TWEED, 1992, *Mater. Sci. Tech.* 8: 308–316.

FERREIRA, A., MEYERS, M. A., THADHANI, N. N., CHANG, S. N., and KOUGH, J. R., 1991, *Metall. Trans.* 22A: 685–695.

FISHMAN, S. G., 1986, *ASTM Standardization News* 14(6): 46–49.

FUCHS, G. E., 1989, In *High-Temperature Ordered Intermetallic Alloys III*, MRS Symp. Proc. 133: 615–620.

________, 1990, In *Metal and Ceramic Matrix Composites: Processing, Modeling and Mechanical Behavior Conference* pp. 391–400. Warrendale, PA: The Minerals, Metals and Materials Society.

FUCHS, G. E., and KAO, W. H., 1988, In *Modern Developments in Powder Metallurgy*, 20: 531–557. Princeton, NJ: Metal Powder Industries Federation.

GERMAN, R. M., 1984, *Powder Metallurgy Science*. Princeton, NJ: Metal Powder Industries Federation.

________, 1985, *Liquid Phase Sintering*. New York: Plenum Press.

________, 1987, In *Science of Sintering: New Directions for Materials Processing and Microstructural Control*, pp. 439–451. New York: Plenum Press. D. P. Uskokovic, H. Palmour and R. M. Spriggs (Eds.)

________, 1989, *Powder Injection Molding*. Princeton: Metal Powder Industries Federation.

________, 1990, *Advances in Powder Metallurgy* 2: 115–132.

________, 1992, In *Thermal Analysis in Metallurgy* pp. 205–231. Warrendale: The Minerals, Metals and Materials Society. R. D. Schull and A. Joshi (Eds.)

GERMAN, R. M., and BOSE, A., 1989, *Mat. Sci. Eng.*, A107: 107–116.

GERMAN, R. M., BOSE, A., ALMAN, D. E., 1990, *Proceedings PM-90, International Conference on Powder Metallurgy* 1: 310–323. London: The Institute of Metals.

GERMAN, R. M., BOSE, A., and STOLOFF, N. S., 1988, In *High Temperature Ordered Intermetallic Alloys III* MRS Symp. Proc. 133: 403–414.

GRAVES, J. A., PEREPEZKO, J. H., WARD, C. H., and FROES, F. H., 1987, *Scripta Met*. 21: 567–572.

GUTIERREX-MIRAVETE, E., LAVERNIA, E. J., TRAPAGA, G. M., SZEKELY, J., and GRANT, N. J., 1989, *Metall. Trans*. 20A: 71–85.

GUTMANAS, E., AND LAWLEY, A., 1990, *Advances in Powder Metallurgy* 2: 1–14.

HERMAN, H., 1988, *MRS Bulletin* 13(Dec): 60–67.

HSIUNG, L. M., KUNTZ, T. A., and WADLEY, H. N. G., 1991, In *Low Density, High Temperature Powder Metallurgy Alloys* pp. 21–34. Warrendale, PA: The Minerals, Metals and Materials Society.

HWANG, K. S., and LU, Y. C., 1990, *Advances in Powder Metallurgy* 2: 133–144.

IWASE, Y., KOGA, T., and NAGUMO, M., 1992, *J. Japan Soc. Powder Powder Metall.* 39: 174–178.

JAMES, W. B., 1985, *Inter. J. Powder Met*. 21: 163–182.

JANG, J. S. C., and KOCH, C. C., 1990, *J. Mat. Res*. 5(3): 498–510.

KAKITSUJI, A., TANIHARA, M., MIYAMOTO, H., TAKAHASHI, Y., MORI, M., MIYAMOTO, K., and INAMURA, S., 1990, *J. Japan Soc. Powder Powder Met*. 37: 665–670.

KAYSSER, W. A., LAAG, R., MURRAY, J. C., and PETZOW, G. E., 1991, *Inter. J. Powder Metallurgy* 27(1): 43–49.

KEAR, B. H., SIMS, C. T., STOLOFF, N. S., and WESTBROOK, J. H., eds., 1970, *Ordered Alloys: Structural Applications and Physical Metallurgy*. Baton Rouge: Claitor's Publishing Division.

KERANS, R. J., HAY, R. S., PAGANO, N. J., and PARTHASARATHY, T. A., 1989, *Ceramic Bull.* 68(2): 429–442.

KIM, Y. W., and FROES, F. H., 1989, *Advances in Powder Metallurgy* 3: 251–267.

KIMURA, H., and KIMURA, M., 1990, In *Solid State Powder Processing* pp. 365–377. Warrendale, PA: The Minerals, Metals and Materials Society.

KIMURA, H., and KOBAYASHI, S., 1992, *J. Japan Soc. Powder Powder Met.* 39: 287–290.

KNIBLOE, J. R., and WRIGHT, R. N., 1990, In *Proceedings PM '90 International Conference on Powder Metallurgy* 2: 310–323. London: The Institute of Metals.

KOCH, C. C., 1986, In *High-Temperature Ordered Intermetallic Alloys II, MRS Symp. Proc.* 81: 369–380.

KOCH, C. C., and CHO, Y. S., 1992, *Nanostructural Materials* 1: 207–212.

KOCH, C. C., LIU, C. T., and STOLOFF, N. S., eds., 1985, *High-Temperature Ordered Intermetallic Alloys*, MRS Symp. Proc 39.

KOHMOTO, H., and FRASER, H. L., 1989, *Advances in Powder Metallurgy* 3: 203–211.

KRUEGER, B. R., MUTZ, A. H., and VREELAND, JR., T., 1992, *Metall. Trans.* 23A: 55–58.

KULKARNI, K. L., ED., 1987, *Powder Metallurgy for Full Density Products*. Princeton, NJ: Metal Powder Industries Federation.

KUMAR, K. S., DIPETRO, M. S., BROWN, S. A., and WHITTENBERGER, J. D., 1991, *NASA Technical Memorandum* #NASA TM-103724, Lewis Research Center, Cleveland, OH.

KUMAR, K. S., MANNAN, S. K., and VISWANDHAM, R. K., 1992, *Acta Metall. Mater.* 40: 1201–1222.

LAAG, R., KAYSSER, W. A., MAURER, R., and PETZOW, G., 1989, In *Science of Sintering*, pp. 295–309. New York: Plenum Press. D. P. Uskokovic, H. Palmour and R. M. Sprigg.

________, 1991, *Hot Isostatic Pressing: Theory and Applications*, pp. 101–113. Materials Park, OH: ASM. OH. R. J. Schaefer and M. Linzer (Eds.)

LAAG, R., KAYSSER, W. A., GALINSKI, G., and MAURER, R., 1990, In *Proceedings PM-90, International Powder Metallurgy Conference* 2: 278–288. London: The Institute of Metals.

LARKIN, D. J., INTERRANTE, L. V., and BOSE, A., 1990, *J. Mat. Res.* 5: 2706–2717.

LARSEN, J. M., WILLIAMS, K. A., BALSONE, S. J., and STUCKE, M. A., 1990, In *High Temperature Aluminides and Intermetallics*, pp. 521–556. Warrendale, PA: The Minerals, Metals and Materials Society. S. H. Whang, C. T. Liu, D. P. Pope, and J. O. Stiegler (Eds.)

LASDAY, S. B., 1988, *Industrial Heating* 55: 14–15.

LAWLEY, A., *Atomization*, Metal Powder Industries Federation, Princeton, NJ, 1992.

LEBRAT, J. P., VARMA, A., and MILLER, A. E., 1992, *Metall. Trans.* 23A: 69–76.

LEVI, C. G., MEHRABIAN, R., OSLIN, B., ANDERSON, R. L., and SASTRY, S. M. L., 1988, *J. Mater. Shaping Tech.* 6: 125–132.

LIPSITT, H. A., 1985, In *High-Temperature Ordered Intermetallic Alloys*, *MRS Symp. Proc.* 39(1): 352–364.

LOGRASSO, B., BOSE, A., CAMUS, G., STOLOFF, N. S., DUQUETTE, D. J., and GERMAN, R. M., 1988, In *Modern Developments in Powder Metallurgy* 20: 559–572. Princeton, NJ: Metal Powder Industries Federation.

LU, L., KIM, Y. S., GOKHALE, A. B., and ABBASCHIAN, R., 1992, Unpublished results, University of Florida.

MAELAND, A. J., and NARASIMHAN, D., 1989, In *High-Temperature Ordered Intermetallic Alloys III*, MRS Symp. Proc., 133: 723–727.

MAJIMA, K., NIIMI, N., KATSUYAMA, S., TOMIZAWA, J., and NAGAI, H., 1992, *J. Japan Soc. Powder Powder Met.* 39: 197–202.

MARK, H. F., OTHMER, D. F., OVERBERGER, C. G., and SEABORG, G. T., 1989, *Encyclopedia of Chemical Technology, 3rd edition* 12: 459–481. New York: Wiley and Sons.

MARTIROSYAN, N. A., DOLUKHANYAN, S. K., MKRTCHYAN, G. M., BOROVINSKAYA, I. P., and MERZHANOV, A. G., 1976, *Soviet Powder Met. Metal Ceramic* 15: 522–525.

MAURER, R., GALINSKI, G., LAAG, R., and KAYSSER, W. A., 1989, In *High-Temperature Ordered Intermetallic Alloys III*, *MRS Symp. Proc.* 133: 421–426.

MCKAMEY, C. G., DEVAN, J. H., TORTORELLI, P. F., and SIKKA, V. K., 1991, *J. Mat. Res.* 6(8): 1779–1805.

MCKAMEY, C. G., POVIRK, G. L., HORTON, J. A., TIEGS, T. N., and OHRINER, E. K., 1988, In *High-Temperature Ordered Intermetallic Alloys III*, MRS Symp. Proc. 133: 609–614.

MEHRABIAN, R., 1988, In *High Temperature/High Performance Composites*, MRS Symp. Proc. 120: 2–21.

MENDIRATTA, M. G., and DIMIDUK, D. M., 1989, In *High-Temperature Ordered Intermetallic Alloys III*, MRS Symp. Proc. 133: 441–446.

MISIOLEK, W. Z., and GERMAN, R. M., 1990, *Advances in Powder Metallurgy* 2: 161–172.

________, 1991a, *Advances in Powder Metallurgy* 6: 167–175.

________, 1991b, *Mat. Sci. Eng.* A144: 1–10.

MISRA, A. K., 1990, *J. Mat. Res.* 5(7): 1561–1566.

MOLL, J. H., YOLTON, C. F., and MCTIERNAN, B. J., 1990, *Inter. J. Powder Met.* 26(2): 149–155.

MOORE, B., BOSE, A., GERMAN, R. M., and STOLOFF, N. S., 1988, In *High Temperature/High Performance Composites*, *MRS Symp. Proc.* 120: 51–56.

MORTENSEN, A., CORNIE, J. A., and FLEMINGS, M. C., 1988, *J. Met.* 40(2): 12–19.

MOXSON, V. S., and FRIEDMAN, G. I., 1986, *Progress in Powder Metallurgy* 42: 489–500.

MUNAR, B., MACLEAN, M., and DOWER, R., 1990, In *Proceedings PM-90, International Conference on Powder Metallurgy* 3: 89–93. London: The Institute of Metals.

MUNIR, Z. A., 1992, *Metall. Trans.* 23A: 7–13.

MURRAY, J. C., and GERMAN, R. M., 1990, *Advances in Powder Metallurgy* 2: 145–160.

_______, 1992, *Metall. Trans.*, 23A: 2357–2364.

MURRAY, J. C., LAAG, R., KAYSSER, W. A., and PETZOW, G., 1990, *Advances in Powder Metallurgy* 2: 233–242.

MYERS, M. A., GUPTA, B. B., and MURR, L. E., 1981, *J. Met.* 33(10): 21–26.

NARDONE, V. C., 1992, *Metall. Trans.* 23A: 563–572.

NARDONE, V. C., and STRIFE, J. R., 1991, *Metall. Trans.* 22A: 183–189.

NISHIYAMA, K., TOSHIDA, Y., MOHRI, M., EGUCHI, N., UMEKAWA, S., and SHITE, S., 1991, *J. Japan Soc. Powder Powder Met.* 38: 10–14.

NOTIS, M. R., SMOAK, R. H., and KRISHNAMACHARI, V., 1975, In *Sintering and Catalysis*, pp. 493–507. New York: Plenum Press. G. C. Kuczynski (ed.)

NOURBAKHSH, S., LIANG, F. L., and MARGOLIN, H., 1990, *Metall. Trans.* 21A: 213–219.

OCHIAI, S., NOGUCHI, O., and UENO, M., 1987, *J. Japan Inst. Met.* 51: 686–693.

ODDONE, R. R., and GERMAN, R. M., 1989, *Advances in Powder Metallurgy* 3: 475–489.

OSBORNE, N. R., PORTER, W. J., and EYLON, D., 1991, *SAMPE Quarterly* 22(4): 21–28.

PARAMESWARAN, V. R., 1992, *J. Met.* 44(6): 41–43.

PERSAD, C., LEE, B. H., HOU, C. J., ELIEZER, Z., and MARCUS, H. L., 1988, In *High-Temperature Ordered Intermetallic Alloys III, MRS Symp. Proc.* 133: 717–722.

PORTER, W. J., OSBORNE, N. R., EYLON, D., and CLIFFORD, J. P., 1990, *Advances in Powder Metallurgy* 2: 243–257.

PRUMMER, R., 1983, Powder compaction. In *Explosive Welding, Forming and Compaction*, pp. 369–395. New York: Applied Science. T. Z. Balazynski (ed.)

RABIN, B. H., and WRIGHT, R. N., 1991, *Metall. Trans.* 22A: 277–286.

_______, 1992, *Metall. Trans.* 23A: 35–40.

RABIN, B. H., BOSE, A., and GERMAN, R. M., 1988, In *Modern Developments in Powder Metallurgy* 21: 511–529. Princeton, NJ: Metal Powder Industries Federation.

RAY, R., 1990, *Metal Powder Report* 45(1): 56–59.

RIGNEY, J. D., KHADKIKAR, P. S., LEWANDOWSKI, J. J., and VEDULA, K., 1988, In *High-Temperature Ordered Intermetallic Alloys III, MRS Symp. Proc.* 133: 603–608.

ROBERTS, P. R., and FERGUSON, B. L., 1991, *Inter. Mat. Rev.* 36(2): 62–79.

SAVITSKII, A. P., and ROMANOV, G. N., 1990, *Sci. Sintering* 22(2): 73–77.

SCHAEFER, R. J., 1992, *Inter. J. Powder Met.* 28: 161–173.

SCHAEFER, R. J., and JANOWSKI, G. M., 1992, *Acta Metall. Mater.* 40: 1645–1651.

SCHAEFER, R. J., and LINZER, M., eds., 1991, *Hot Isostatic Pressing: Theory and Applications*, Materials Park: ASM International.

SCHNEIBEL, J. H., GEORGE, E. P., and MCKAMEY, C. G., 1991, *J. Mat. Res.* 6(8): 1673–1679.

SCHULSON, E. M., 1987, *Inter. J. Powder Met.* 23: 25–32.

SEKHAR, J. A., 1992, Presented at the Innovative Processing of Intermetallic Materials and Metal-Matrix Composites Workshop, 18–20 May 1992, Cornell University.

SHANG, S-S., and MEYERS, M. A., 1991, *Metall. Trans.* 22A: 2667–2676.

SHAW, K. G., and GERMAN, R. M., 1991, *Advances in Powder Metallurgy* 5: 263–275.

SHAW, K. G., MISIOLEK, W. Z., and GERMAN, R. M., 1991, *Advances in Powder Metallurgy* 6: 159–166.

SHEPPARD, L. M., 1990, *Ceramic Bull.* 69: 666–673.

SHIGA, S., ITSUKAICHI, T., UMEMOTO, M., AND OKANE, I., 1991a, *J. Japan Soc. Powder Powder Met.* 38: 963–966.

———, 1991b, *J. Japan Soc. Powder Powder Met.* 38: 971–975.

SHIH, D. S., SCARR, G. K., and CHESNUTT, J. C., 1988, In *High-Temperature Ordered Intermetallic Alloys III, MRS Symp. Proc.* 133: 167–172.

SIKKA, V. K., 1988, In *Modern Developments in Powder Metallurgy* 20: 543–557. Princeton, NJ: Metal Powder Industries Federation.

SIKKA, V. K., OHRINER, and ALLARD, L. F., 1991, In *P/M in Aerospace and Defense Technologies Symposium Proceedings* 137–145. Princeton, NJ: Metal Powder Industries Federation.

SIMS, D. M., BOSE, A., and GERMAN, R. M., 1987, *Progress in Powder Metallurgy* 43: 575–596.

SINGER, A. R. E., 1985, *Inter. J. Powder Met. Powder Tech.* 21(3): 219–234.

SMARSLY, W. G., and DAHMS, M., 1990, In *Proceedings PM-90, International Powder Metallurgy Conference* 2: 86–92. London: The Institute of Metals.

SONG, I., and THADHANI, N. N., 1992, *Metall. Trans.* 23A: 41–48.

STOLOFF, N. S., 1984, *Inter. Met. Rev.* 29: 123–135.

———, 1985, In *High-Temperature Ordered Intermetallic Alloys* MRS Symp. Proc. 39(1): 3–27.

STOLOFF, N. S., and ALMAN, D. E.,s 1990, *MRS Bull.* 15(12): 47–53.

STROTHERS, S., and VEDULA, K., 1987, *Progress in Powder Metallurgy* 43: 597–610.

SURYANARAYANA, C., SUNDARESAN, R., and FROES, F. H., 1990, In *Solid State Powder Processing*, pp. 55–64. Warrendale, PA: The Minerals, Metals and Materials Society.

TAUB, A. I., HUANG, S. C., and CHANG, K. M., 1984, In *High-Temperature Ordered Intermetallic Alloys, MRS Symp. Proc.* 39: 221–228.

THADHANI, N. N., COSTELLO, M. J., SONG, I., WORK, S., and GRAHAM, R. A., 1990, In *Solid State Powder Processing*, 97–109. Warrendale, PA: The Minerals, Metals and Materials Society.

TIEN, J. K., and KOPP, M. W., 1990, In *Metal and Ceramic Matrix Composites: Processing, Modeling and Mechanical Behavior*, pp. 443–456. Warrendale, PA: The Minerals, Metals and Materials Society.

TIWARI, R., HERMAN, H., SAMPATH, S., and GUDMUNDSSON, B., 1990, Presented at ASM Materials Congress, Detroit, to be published.

TOKIZANE, M., AMEYAMA, K., and SUGIMOTO, H., 1990, In *Solid State Powder Processing*, pp. 67–75. Warrendale, PA: The Minerals, Metals and Materials Society.

TOKIZANE, M., ISONISHI, K., and KIDO, S., 1990, In *Proceedings PM-90, International Powder Metallurgy Conference* 2: 93–96. London: The Institute of Metals.

VEDULA, K., and LISY, F., 1990, In *Proceedings PM-90: International Conference on Powder Metallurgy* 1: 548–556. London: The Institute of Metals.

VEDULA, K., MICHAL, G. M., and FIGUEREDO, A. M., 1988, In *Modern Developments in Powder Metallurgy* 20: 491–502. Princeton, NJ: Metal Powder Industries Federation.

VEDULA, K., and STEPHENS, J. R., 1986, In *Powder Metallurgy 1986: State of the Art*, pp. 205–214. Freiburg: Verlag Schmid GMBH.

________, 1987a, *Metal Powder Report* 42: 84–89.

________, 1987b, *Progress in Powder Metal* 43: 561–573.

WANG, L., and ARSENAULT, R. J., 1991, *Metall. Trans.* 22A: 3013–3018.

WANG, J. S. C., DONNELLY, S. G., GODAVARTI, P., and KOCH, C. C., 1988, *Inter. J. Powder Met.* 24: 315–325.

WESTBROOK, J. H., ed., 1967, *Intermetallic Compounds*. New York: John Wiley & Sons.

WHITTENBERGER, J. D., ARZT, E., and LUTON, M. J., 1990, *J. Mat. Res.* 5(12): 2819–2827.

WHITTENBERGER, J. D., VISWANADHAM, R. K., MANNAN, S. K., and KUMAR, K. S., 1989, *J. Mat. Res.* 4(5): 1164–1171.

WITHERS, J. C., SHIAO, H.-C., LOUTFY, R. O., and WANG, P., 1991, *J. Met.* 43: 36–39.

WRIGHT, R. N., and KNIBLOE, J. R., 1990, *Acta Met. Mater.* 38: 1993–2001.

WRIGHT, R. N., KORTH, G. E., and FLINN, J. E., 1987, *Adv. Mater. Processes* 137(4): 56–59.

WRIGHT, R. N., and RABIN, B. H., 1992, *Advances in Powder Metallurgy*, 9: 283–294.

WRIGHT, R. N., RABIN, B. H., and KNIBLOE, J. R., 1989, *Mater. Manuf. Proc.* 4(1): 25–37.

WRIGHT, R. N., WILLIAMSON, R. L., and KNIBLOE, J. R., 1990, *Powder Metall.* 33: 253–259.

WRZESINSKI, W. R., and RAWERS, J. C., 1990, *J. Mat. Sci. Let.* 9(4): 432–435.

YANG, J. M., and JENG, S. M., 1991, *J. Mat. Res.* 6(3): 505–513.

________, 1992, *J. Met.* 44: 52–57.

YANG, J. M., KAO, W. H., and LIU, C. T., 1988, In *High-Temperature Ordered Intermetallic Alloys III, MRS Symp. Proc.* 133: 453–458.

________, 1989, *Metall. Trans.* 20A: 2459–2469.

YOLTON, C. F., LIZZI, T., CHANDHOK, V. K., and MOLL, J. H., 1986, *Progress in Powder Metallurgy* 42: 479–487.

Chapter 16

Joining

S. A. David and M. L. Santella

1. INTRODUCTION

In recent years, ordered intermetallic alloys have attracted considerable interest because of the unique properties that make them attractive candidates for high-temperature structural applications (Stoloff and Davies 1966; Kear et al. 1970; Westbrook 1959, 1967; Lipsitt 1985; Liu 1984; Stoloff 1989). Some of the alloy systems that are currently under development or in use include those based on the following compounds: Ni_3Al, Fe_3Al, Ti_3Al, Co_3V, Ni_3V, TiAl, NiAl, and FeAl. Many of these compounds have low density and high modulus coupled with excellent high-temperature strength, which derives from the unique dislocation dynamics in ordered lattices and slow atomic mobility. Often they have unique oxidation and corrosion properties (Liu and Stiegler 1984). Intermetallic compounds, however, tend to have limited ductility, and this characteristic has deterred their widespread use for structural applications. Recent work has shown that the ductility and fabricability of several ordered intermetallic compounds can be substantially improved using physical metallurgy principles (Liu 1984; Stoloff 1989; Rhodes, Hamilton, and Paton 1987; Liu, White, and Horton 1985). Because of the unique properties of these intermetallic alloys, a wide range of applications is anticipated, ranging from aerospace to land-based transportation systems, energy systems, chemical systems, heating elements, and so on. The applications under consideration for intermetallic alloys will require that they be available in a variety of product forms, including sheet, plate, bar, tubing, and castings.

Although, to date, significant efforts have gone into understanding the behavior of these alloys, basic knowledge of the processing, fabrication, mechanical behavior, and environmental resistance is still very limited. Fusion welding is a key issue in the development of intermetallic alloys. The ability to join these alloys successfully will increase their use in engineering applications. A limited amount of work has been carried out on the weldability of intermetallic alloys, where weldability is defined as the ease with which defect-free welds can be made in a material. The weldability of many of these alloys has been carried out on thin sheets of materials and has been found to be marginal, and is a strong function of composition, process, and process variables. In some of these alloys (in particular, iron aluminides), cold-cracking related to hydrogen effects is a severe problem. This chapter will briefly review the behavior of selected intermetallic alloys under a variety of conditions relevant to fusion welding processes.

2. WELDING AND WELDABILITY

2.1. Nickel Aluminides

The nickel aluminide alloys of interest for engineering applications are based on Ni_3Al and NiAl. Although advancements in alloy design have led to renewed interest in NiAl-type alloys, most welding studies have been carried out on Ni_3Al-type alloys. These particular alloys, also referred to as γ' alloys, are intrinsically weak in polycrystalline form exhibiting brittle intergranular failure with no ductility. Polycrystalline Ni_3Al has been ductilized by the addition of small amounts (a few 100 ppm or less) of boron, and other important properties such as strength and environmental resistance have been improved by alloying as well (Aoki and Izumi 1979; Liu et al. 1985). The nominal compositions of some of these alloys are given in Table 16-1.

Weldability of several Ni_3Al-type alloys has been investigated using both gas-tungsten arc (GTA) and electron beam (EB) welding processes. The preliminary weldability studies were often performed on thin annealed sheets (typically 1 mm-thick) of these alloys by making autogenous welds and examining the specimens for heat-affected zone (HAZ) or fusion zone (FZ) cracking. On a quantitative basis, the sensitivity for FZ cracking has been established by the sigmajig weldability test (Goodwin 1987). This test ranks materials by determining the threshold stress (σ_0)

Table 16-1 Chemical Compositions of Ni_3Al Alloys (wt%) Investigated

Alloy	Al	Fe	Mn	Ti	B	Hf	Zr	Cr	Ni
IC-14	10.8	10.7	1.1	0.5	0.05	—	—	—	Balance
IC-15	12.7	—	—	—	0.05	—	—	—	Balance
IC-18	12.7	—	—	—	0.25	—	—	—	Balance
IC-19	12.7	—	—	—	0.01	—	—	—	Balance
IC-25	9.8	10.7	0.5	0.5	0.05	—	—	—	Balance
IC-50	12.5	—	—	—	0.02	1.7	—	—	Balance
IC-103	12.5	10.7	0.5	0.5	0.02	—	—	—	Balance
IC-218	8.8	—	—	—	0.02	—	0.9	7.0	Balance
IC-221B	8.0	—	—	—	0.03	—	—	7.7	Balance
IC-221M	8.0	—	—	—	0.03	—	1.7	7.7	Balance
IC-221W	8.0	—	—	—	0.03	—	3.0	7.7	Balance
IC-306	9.0	13.0	—	—	0.02	—	0.3	7.0	Balance

above which cracking occurs in the FZ during welding. Threshold stresses based on sigmajig testing range from 10–30 ksi (70-210 MPa) for these alloys as shown in Figure 16-1.

The first study of the weldability of Ni_3Al-type alloys established that their cracking resistance was low and was a strong function of both welding parameters and composition (David et al. 1985). Figure 16-2 shows the influence of boron on the cracking tendency of Ni_3Al, where cracking severity is the total number of cracks in the weld per unit length (number/cm). Cracking was observed in both the FZ and the HAZ. The addition of 10 at% iron to Ni_3Al dramatically improved the weldability of the alloys. Whereas all of the welds made on the boron-doped Ni_3Al contained cracks, crack-free welds were obtained only on an alloy containing 10 at%Fe + 500 ppm B (IC-25). However, the cracking behavior of the Fe-containing alloys was a function of the EB welding parameters, with relatively high heat input conditions (produced at low welding speeds) being preferred. Annealed sheet of the IC-25 alloy was single phase, but the weld fusion zones consisted of a $\gamma' + \beta'$ duplex microstructure as shown in Figure 16.3. The β' was distributed interdendritically. Also, the interdendritic regions were enriched in Al, Ti, and Mn, and they were depleted in Ni. Precipitation of β' on grain boundaries in the weld HAZs was also observed. It was shown that the β' phase, shown in Figure 16-3, could be dissolved by heat treatment at 1000° C.

Subsequent work (Santella, David and White 1985) on the same alloy systems confirmed that the addition of 10 at%Fe to Ni_3Al suppressed cracking, and showed that the boron concentration was an important

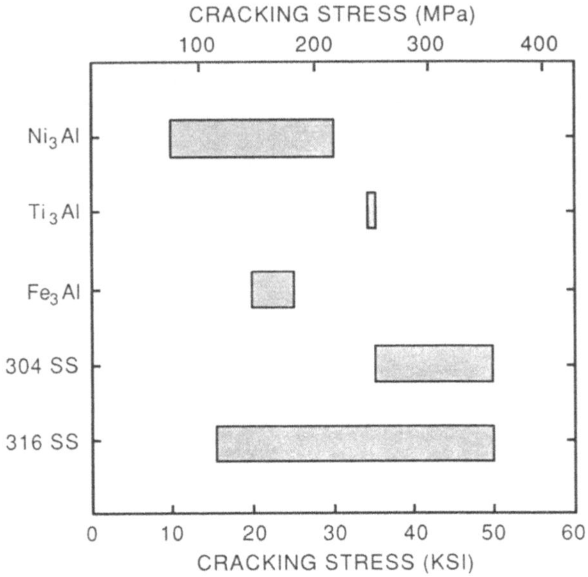

Figure 16-1. Experimentally measured sigmajig test cracking stress for various alloys. (David and Zacharia, 1993)

factor as well. An Fe-containing alloy with 200 ppm of B (IC-103) was much more resistant to HAZ cracking than one containing 500 ppm B (IC-25). A typical HAZ cracking in IC-25 is shown in Figure 16-4. Surface microchemical analyses by Auger spectroscopy were inconclusive regarding the effect of grain boundary composition on the HAZ cracking behavior. However, hot ductility testing (Santella and David 1986) showed that the IC-103 alloy had better ductility at elevated temperature than the IC-25 alloy. It was suggested that the grain boundary cohesive strength of the Fe-containing alloys varied inversely with bulk B concentration. These data indicated that the improved resistance to HAZ cracking of the alloy containing 200 ppm B resulted from the ductility it retained at high temperature. Computer-aided analysis (Santella, Maguire, and David 1989; Maguire, Santella and Damkroger 1992) of the transient longitudinal thermal stresses induced in the HAZ by welding showed that they exceeded the yield strength of the Fe-containing alloys under conditions where cracking was observed during welding. The stress analysis supported the claim that the alloy containing 200 ppm B was resistant to cracking because it could accommodate thermal stresses in excess of the yield strength by local plastic flow.

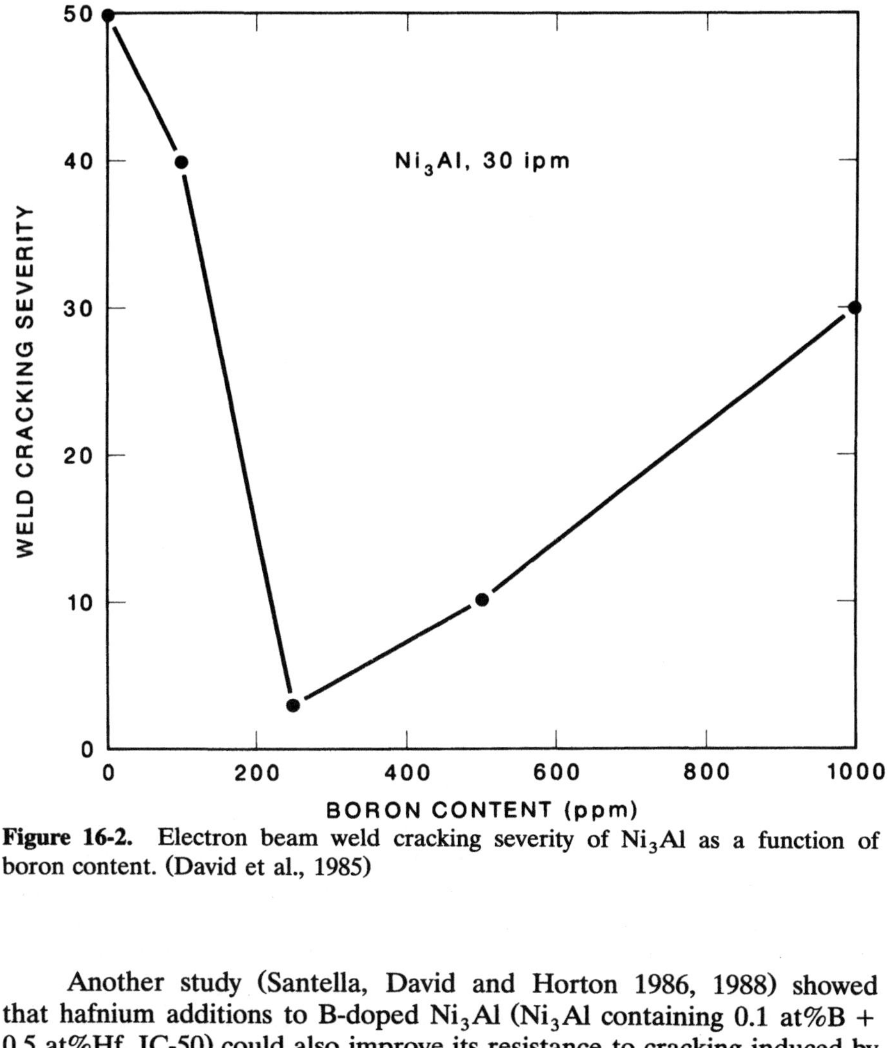

Figure 16-2. Electron beam weld cracking severity of Ni_3Al as a function of boron content. (David et al., 1985)

Another study (Santella, David and Horton 1986, 1988) showed that hafnium additions to B-doped Ni_3Al (Ni_3Al containing 0.1 at%B + 0.5 at%Hf, IC-50) could also improve its resistance to cracking induced by welding processes. This work involved both autogenous welds as well as welds made with filler metal additions. Welds made with and without filler metal on wrought sheet product forms up to 5 mm-thick were free of defects. Gas tungsten arc welds were made on a 2 mm-thick sheet using sheared strips of the IC-50 base material for filler metal. Welding characteristics of the IC-50 alloy as a weld filler metal were acceptable, and tensile testing results suggested that weldment tensile properties were comparable to those of the base material. Metallographic analysis showed that microsegregation produced two-phase microstructures in the weld

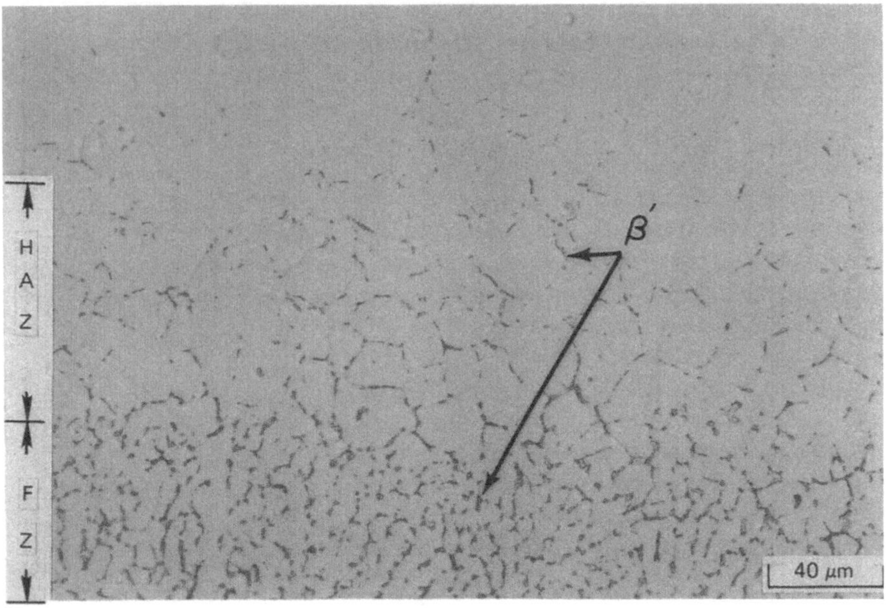

Figure 16-3. The presence of β' in the FZ and along the grain boundaries in the heat-affected zone of alloy IC-25. (David et al., 1985)

fusion zones consisting of Ni-rich dendrite cores comprised of a $\gamma + \gamma'$ mixture and interdendritic regions which were fully ordered. The interdendritic regions were highly enriched in Hf. It was also observed that the cast alloy was not resistant to cracking, and this was attributed to embrittlement of grain boundaries due to microsegregation. Similar behavior was observed for continuous cast sheet of an alloy where zirconium was substituted for Hf (Li and Chaki 1991). In this case, cracking in the cast material was attributed to the formation of low-melting eutectics containing Zr and B. It was also suggested that the relatively large columnar grains found in castings favor fracture and the formation of cracks during welding.

The effect of chromium additions on the weldability of the Ni_3Al + Hf + B alloys has also been studied (Maguire, Edwards, and David 1992). Chromium additions are of interest as they are effective in mitigating an environmental embrittlement problem found in the Ni_3Al alloys at temperatures below about 800° C. It was shown that welding response correlated well with strength and hot ductility. Intergranular HAZ cracking decreased as Cr concentration in the alloys increased, and it was virtually

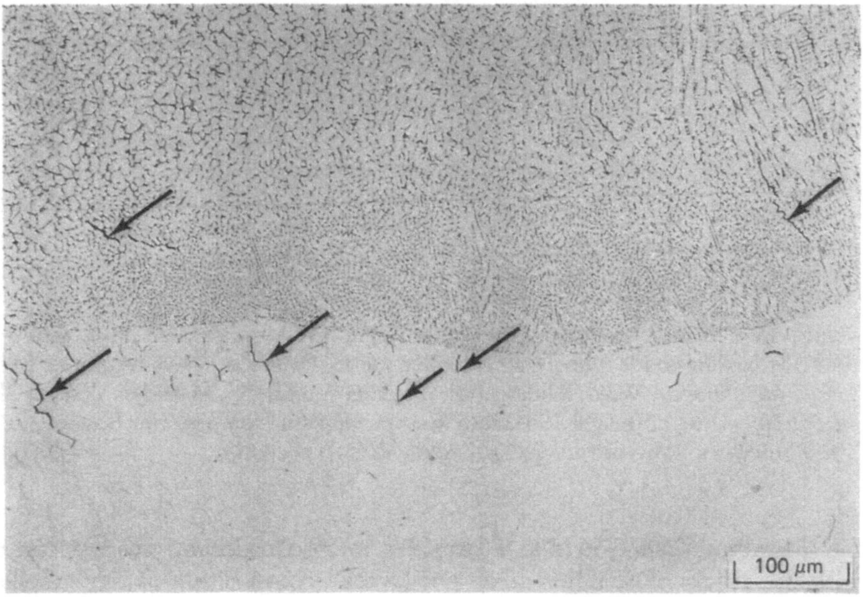

Figure 16-4. Microstructure showing microcracking tendency of alloy IC-25 during electron beam welding. (David et al., 1985)

eliminated at a Cr level of 8 at%. The ability of the alloys to resist cracking was observed to increase with increasing heat input as was found in the earlier studies of other Ni_3Al alloys. The improved hot ductility and resistance to HAZ cracking in the higher Cr alloys was attributed to their increased strength, and to the presence of a fine antiphase domain microstructure.

Currently there is interest in the use of certain Ni_3Al alloys in the cast form for a variety of heat-treating hardware applications. Based on metallographic analysis of solidification cracks in welds of one such alloy, identified as IC221M, modest composition modifications were made to the base alloy to improve solidification-cracking resistance (Santella 1993). The cross-sectional view of a single-pass manual GTA weld made in an autogenous grooved IC221M casting (Figure 16-5) typifies the cracking usually observed in welds of this alloy. The large crack found at the weld centerline extended for nearly the entire length of the weld bead, about 50 mm. The smaller cracks near the edges of the weld bead which did not extend to the weld surface are also often observed.

Metallographic analysis of crack fracture surfaces showed that they occurred during the solidification process, while examination of weld

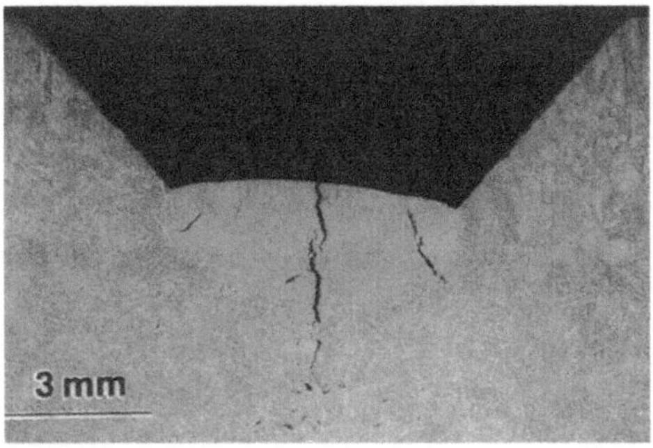

Figure 16-5. Cross-sectional view of a single GTA weld bead in a grooved IC-221M casting made with IC-221M filler metal. (Santella, 1993; reprinted from *Scripta Metallurgica*, Weld solidification cracking in cast Ni_3Al alloys, Volume 28, pages 1305–1310, copyright 1993, with kind permission from Elsevier Science Ltd., The Boulevard, Langford Lane, Kidlington OX5 1GB, UK)

microstructures showed that a lamellar microconstituent was associated with the cracks. The appearance and overall composition of the lamellar microconstituent indicated that it was a variation of the Ni solid solution-Ni_5Zr eutectic found in the Ni-Zr binary system at the composition of Ni-13 Zr wt% and the eutectic temperature of 1170° C (Massalski 1986).

The susceptibility to solidification cracking in welds and castings is related to the amount of liquid phase present when the solidus temperature is reached (Phillips 1980). In binary eutectic systems, there is a critical range between the limit of solid solubility and the eutectic composition where the susceptibility to solidification cracking is maximized. At compositions outside of this critical range, either lower in solute or nearer to the eutectic composition, solidification cracking is much less likely to occur. Based on this theory, and the fact that a Ni-Zr eutectic was associated with solidification cracking in the IC221M welds, it was decided to investigate the importance of Zr concentration in the alloy to solidification cracking. Two additional alloys were melted and evaluated: IC221B, containing no intentional Zr addition; and IC221W, containing twice the nominal Zr concentration as IC221M.

Autogenous bead-on-plate welds in IC221B showed extensive cracking along grain boundaries in the fusion zone, and no further evaluation of this alloy was done. Similar autogenous BOP welds on IC221M and IC221W were crack free. Optical metallography showed that IC221B contained none of the Ni-Zr eutectic, while the microstructures of IC221M and IC221W were similar except that IC221W appeared to contain a slightly higher fraction of interdendritic phases.

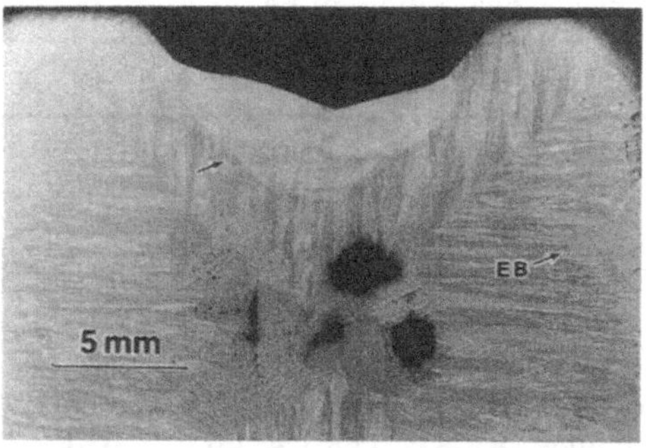

Figure 16-6. Cross-sectional view of GTA weld made in a grooved IC-221W casting with IC-221W filler metal. Weld contains three individual beads. (Santella, 1993; reprinted from *Scripta Metallurgica*, Weld solidification cracking in cast Ni_3Al alloys, Volume 28, pages 1305–1310, copyright 1993, with kind permission from Elsevier Science Ltd., The Boulevard, Langford Lane, Kidlington OX5 1GB, UK)

Autogenous welds on these two alloys were then made in grooves to increase the restraint conditions. Under this condition, only IC221W remained crack free. Also, a weld bead made with a matching filler-metal addition in the grooved IC221W casting was free of cracks. Two additional GTA weld beads were made in this groove, shown in Figure 16-6, and each was acceptable.

The welding results are consistent with the phenomenological theory of solidification cracking (Borland 1960) to the extent that they show the susceptibility of these Ni_3Al alloys to solidification cracking was improved by increasing their Zr concentration. Presumably, the amount of Ni-Zr eutectic in the weld microstructures increased with Zr concentration. On the other hand, eliminating the Zr completely did not eliminate solidification cracking as theory would suggest. This result indicates that factors other than Zr concentration play a role in the solidification cracking behavior of the Ni_3Al alloys.

2.2. Iron Aluminides

Iron aluminides of interest are Fe_3Al- and FeAl-type alloys. Recent renewed interest stems from their unique properties and advantages, such as excellent oxidation and sulfidation resistance, better strength-to-weight ratio compared to stainless steels, and low-cost raw materials. Achieving a reasonable ductility in this class of alloy continues to be a major issue. The weldability of iron aluminides has been investigated using GTA and EB welding processes.

Table 16-2 Chemical Compositions of Fe_3Al Alloys (wt%) Investigated

Alloys	Heat	Al	Cr	Nb	Mn	B	C	TiB_2	Fe
Base alloys	FA-61	15.8							Balance
	FA-39	15.1						0.5	Balance
Cr modified alloys	FA-64	15.8	4.4					0.5	Balance
	FA-72	15.9	4.4						Balance
Nb modified alloys	FA-66	15.5		3.8				0.5	Balance
	FA-79	15.7		1.9					Balance
Mn modified alloy	FA-69	15.8			4.6			0.5	Balance
Cr-Nb-B modified alloy	FA-83	15.8	4.4	1.0		0.01			Balance
	FA-129	15.9	5.5	1.0			0.05		Balance

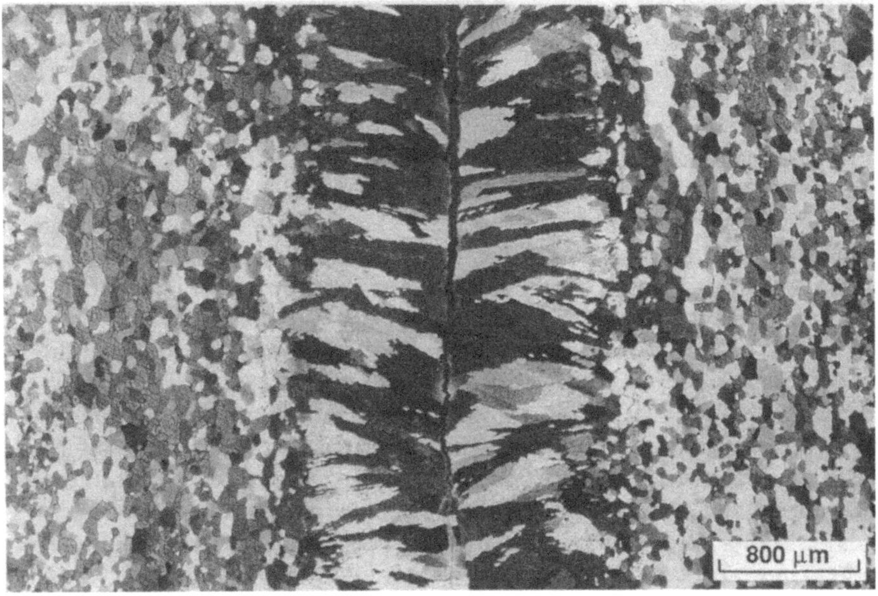

Figure 16-7. Optical micrograph of an Fe_3Al alloy weldment showing a typical centerline crack in the fusion zone. (David et al., 1989)

The composition of the various Fe_3Al-type alloys investigated to date is shown in Table 16-2. Preliminary weldability studies have been carried out by making autogenous welds on thin sheets of these alloys. The weldability of Fe_3Al-type alloys is a strong function of the type of process and composition (Horton et al. 1989). The results of autogenous GTA and EB welds indicate that, although some of these alloys can be successfully welded using EB welding process, some compositions may hot-crack

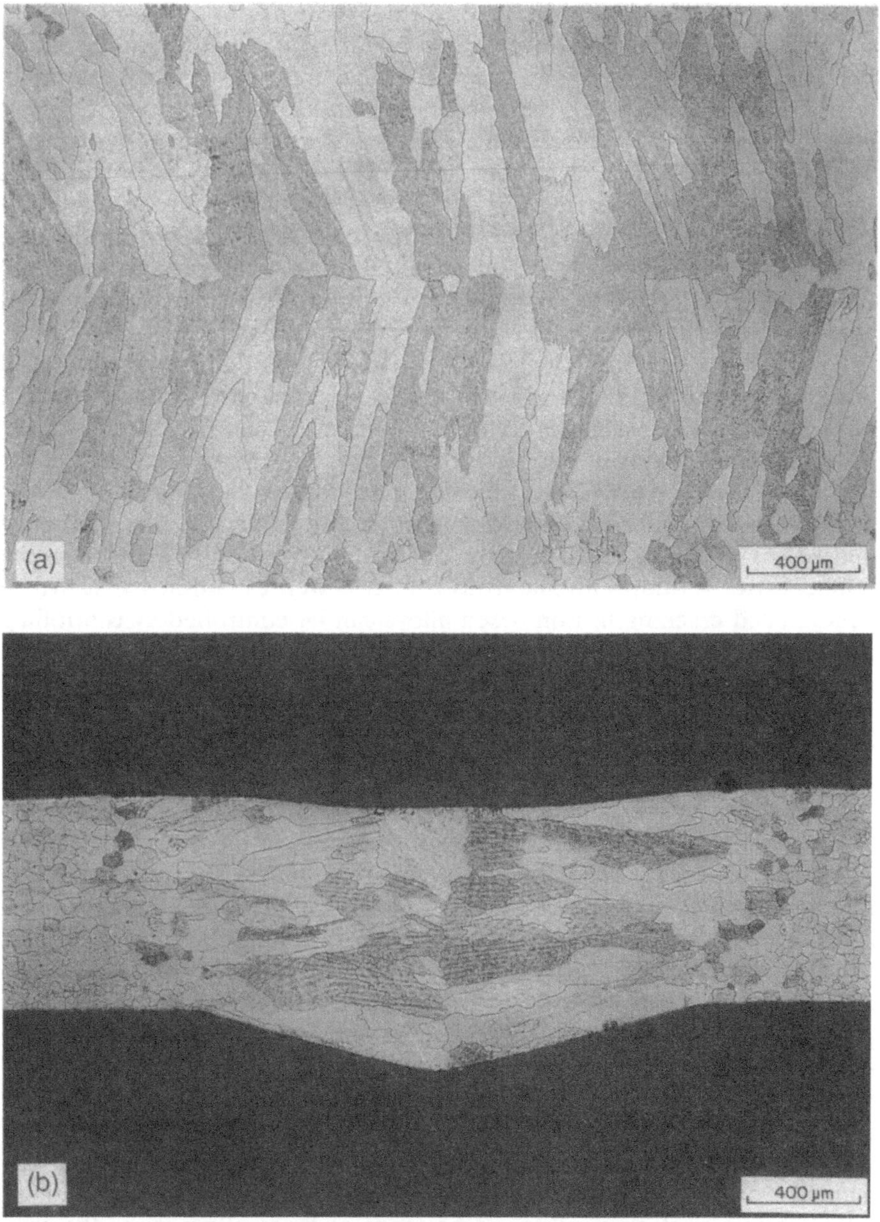

Figure 16-8. Optical photomicrographs of alloy FA-129 gas tungsten arc weld: (a) surface and (b) transverse sections. (David and Zacharia 1993)

during GTA welding. Cracking in most of these alloys has been found to be FZ solidification cracking, as shown in Figure 16-7.

Of the various alloys investigated for weldability, alloy FA-129 modified with chromium, niobium, and carbon has the best weldability characteristics. Successful welds have been produced in this alloy. A typical photomicrograph of a successful weld is shown in Figure 16-8. Sigmajig weldability tests indicated a range of 15–25 ksi (100–175 MPa) threshold stress for FA-129 alloys (see Fig. 16-1). This shows that the weldability of this alloy may be very sensitivite to minor changes in alloy compositions. However, a majority of FA-129 alloys tested had a threshold stress value of 25 ksi (175 MPa). This stress value is within the material range of austenitic stainless steels and some other aluminides (see Fig. 16-1).

Welded FA-129 alloys have been found to be susceptible to cold cracking or delayed cracking. These cracks are commonly found in the FZ or HAZ. Figure 16-9a shows a typical cold crack in the FZ, and Fig. 16-9b shows scanning electron micrographs of the cold-cracked surface indicating a cleavage fracture. This cracking phenomena in iron aluminides is associated with hydrogen embrittlement. Recent studies by Liu, McKamey and Lee (1990) have found that the embrittlement in these alloys, at room temperature, is due to atomic hydrogen. It is well established that hydrogen-induced cracking in iron-based alloys can be controlled by controlling the welding conditions such as the welding environment and pre- and postweld heat treatments. In fact, preliminary work (David and Zacharia 1993) on the effect of pre- and postweld heat treatments at ORNL has shown the benefit of these treatments in eliminating cold cracking in iron aluminides (in particular, FA-129 alloy).

Preliminary weldability studies have been carried out on some FeAl-type developmental alloys. Compositions of some of these alloys are given in Table 16-3. Initial weldability studies were conducted by making autogenous welds on thin sheets of alloys using GTA and EB welding processes. Alloys containing boron and zirconium have been found to be prone to hot cracking. Preliminary results (David, George, and Liu 1990) using auger and secondary ion mass spectroscopy indicate boron and zirconium to be poor activators for weldability. Of the various alloys investigated, FA-385, -386, and -388 containing carbon showed great promise. Successful welds free of hot cracks were produced in these alloys. Carbon in FeAl-type alloys appears to improve weldability. Figure 16-10 shows a typical surface and transverse photomicrograph of alloy FA-386 weld. Further developmental work is under way at ORNL on FeAl-type alloys. Although solidification crack-free welds can be made in these alloys, as in the case of Fe_3Al-type alloys, FeAl-type alloys are also prone to cold cracking. Again, as in the case of Fe_3Al-based alloys, pre- and postweld heat

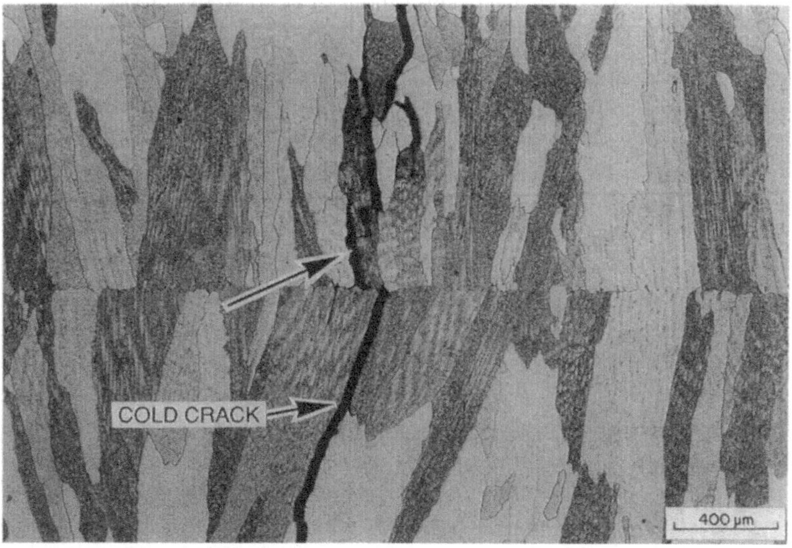

(a)

(b)

Figure 16-9. (a) Optical photomicrograph of delayed cold crack in alloy FA-129 GTA weld; (b) scanning electron micrograph of a typical cold-crack surface. (David and Zacharia 1993)

Table 16-3 Chemical Compositions of FeAl Alloys (at%)

Alloys	Al	Mo	Zr	B*	C*
FA-362	35.8	0.2	0.05	600	
FA-372	35.8	0.2	0.05		
FA-385	35.8	0.2	0.05		300
FA-386	35.8	0.2	0.05		600
FA-387	35.8	0.2		600	
FA-388	35.8	0.2			600

*Parts per million

Source: New Data ORNL (David, unpublished)

treating can be used to overcome or reduce the cold-cracking tendency of FeAl-based alloys.

2.3. Titanium Aluminides

Titanium aluminides of interest for aerospace applications are Ti_3Al (also referred to as α_2 titanium aluminide) and TiAl (referred to as γ titanium aluminide). In recent years, alloys based on these two types of titanium aluminides have been developed. Weldability of titanium-aluminide-based alloys is a critical issue. However, weldability characterization of titanium-based alloys is very limited.

Ti_3Al-based alloys have very limited ductility. The ductility of these alloys has been improved by the addition of β-stabilizing elements such as Nb, V, Mo, and W. An alloy of interest modified with Nb is Ti-13.5% Al-21.5%Nb. Weldability studies (David et al. 1990; Baeslack, Cieslak, and Headley 1988; Baeslack, Mascorella, and Kelly 1989) have been carried out on this particular alloy using GTA, EB, and laser-welding processes. The alloy showed no evidence of hot cracking during welding and had a measured sigmajig threshold stress of greater than 35 ksi (240 MPa). This value is within the range of 10 to 50 ksi (69 to 345 MPa) for crack-resistant austenitic stainless steel and some nickel aluminides (see Fig. 16-1).

During GTA welding, although the alloy showed excellent resistance to solidification cracking, the welds showed a tendency to solid-state cracking well below the solidification temperature, in a very brittle fashion, during cooling subsequent to welding. This may be associated with the presence of predominant α_2 phase in the FZ and HAZ of the weld and limited ductility of the alloy. Microstructural features of this alloy have been found to be a strong function of welding speed and energy input and, thus, the cooling rate. Weldability studies have shown that moderate cooling rates associated with the conventional arc-welding process pro-

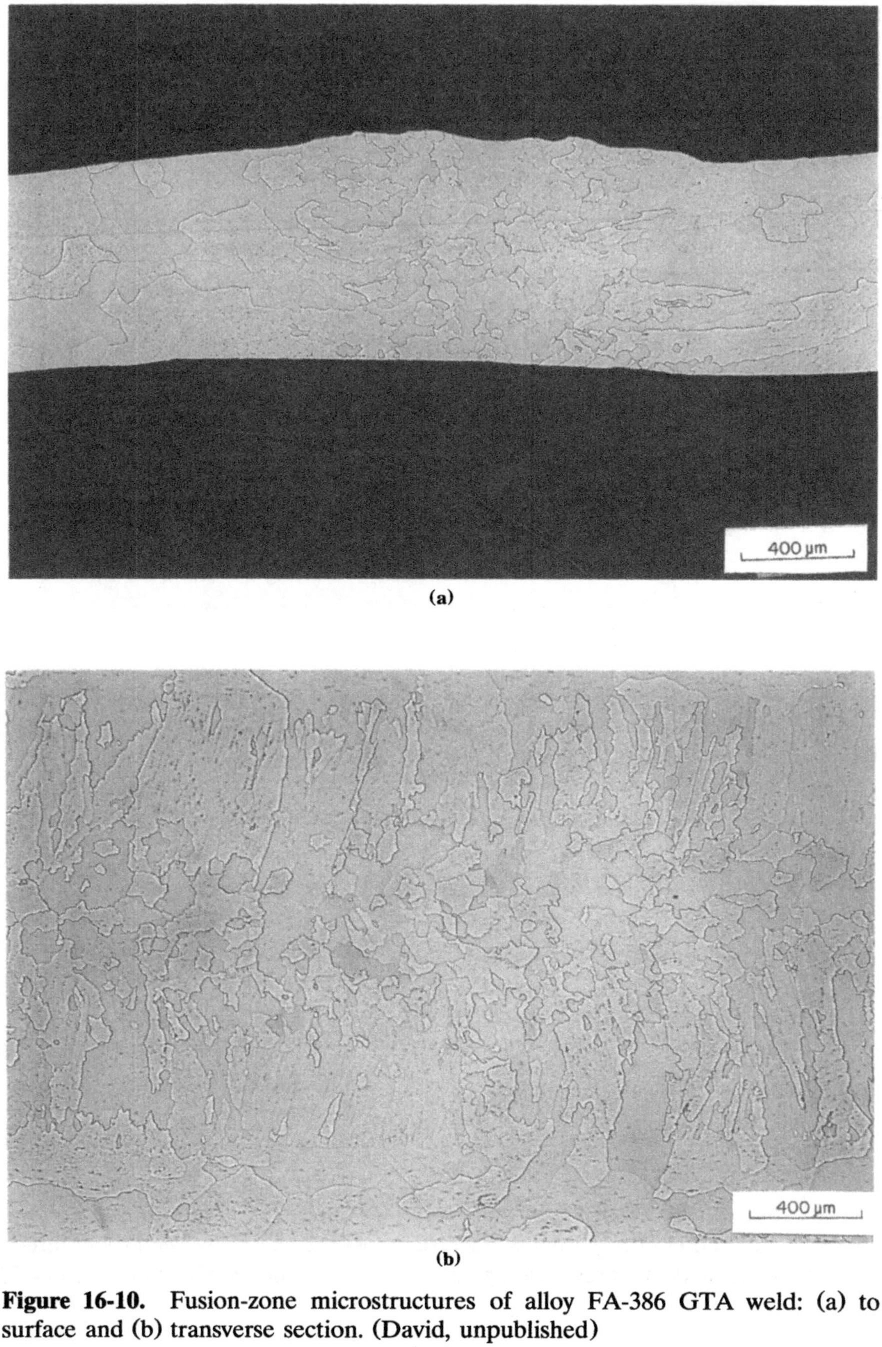

Figure 16-10. Fusion-zone microstructures of alloy FA-386 GTA weld: (a) to surface and (b) transverse section. (David, unpublished)

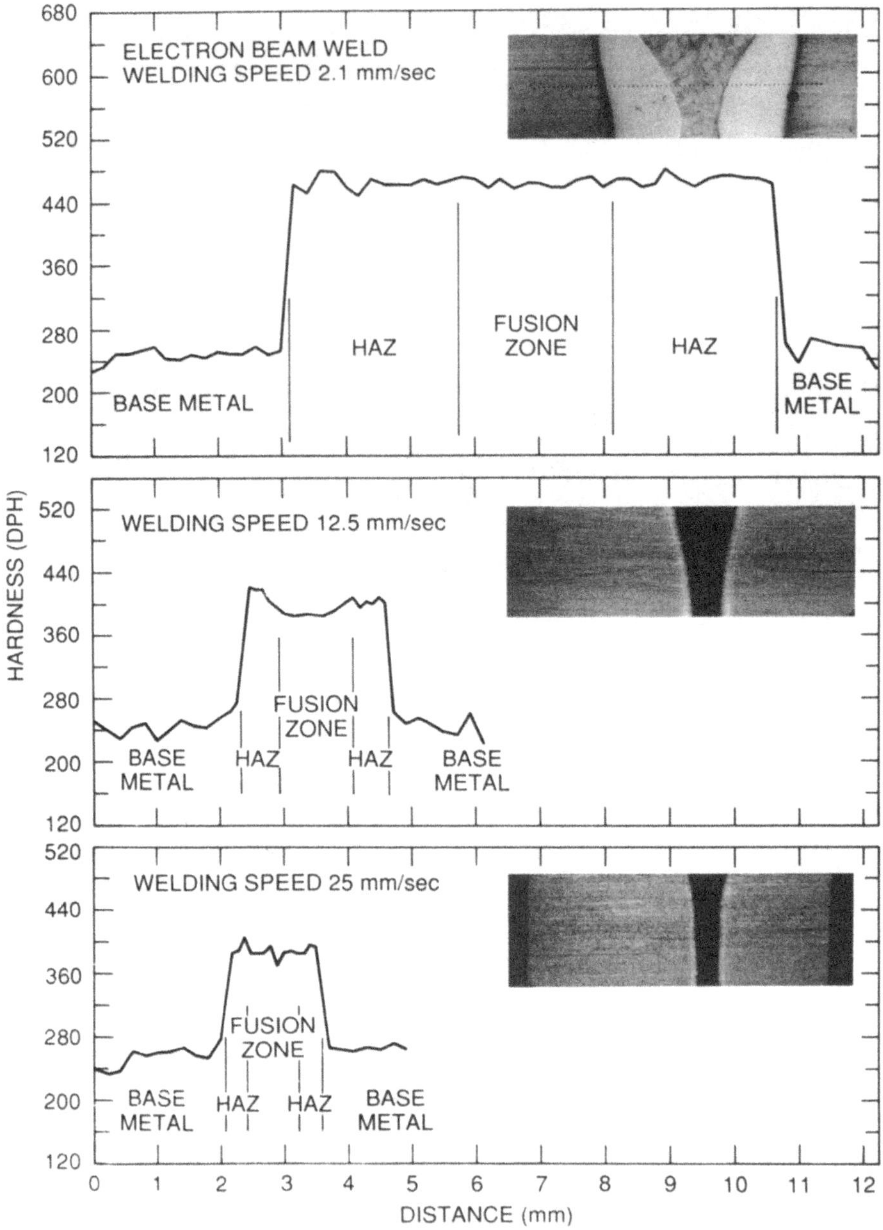

Figure 16-11. DPH microhardness traverse across the transverse sections of three EB weldments made at three welding speeds and heat inputs. (David et al., 1990)

mote the formation of fine acicular α_2 phase with about zero ductility. However, EB welds made over a range of welding speeds and heat inputs have shown that the FZ microstructure of EB welds contained predominantly fine, acicular-order α_2 phase at low welding speeds, and predominantly retained β (metastable) at high welding speeds. In the HAZ there was a gradual variation in microstructure from ordered, acicular α_2 near the fusion line to $\alpha_2 + \beta$ in the far HAZ. Microhardness profiles across the weldment indicated the hardness in the HAZ and FZ to be much higher than the base metal (BM) in all welds (see Fig. 16-11). The hardness profiles across the welds made at higher welding speeds showed a decrease in the weld-centerline hardness relative to those made at lower welding speeds. This may be attributed to retention of metastable β phase. Similar observations (Maguire et al. 1992) have been made on high-speed pulsed laser welds.

Weldability studies (David, George, and Liu 1990; David et al. 1990) related to TiAl-based γ alloys have been minimal. Patterson et al. (1990) have recently characterized EB welds of TiAl-based alloy Ti-48 at% Al-6.5 vol% TiB_2. Weldability of this alloy has been found to be very sensitive to process parameters. Successful welds have been produced without FZ hot cracking. However, a susceptibility to solid-state cracking, similar to the one described above, was observed in this alloy. This cracking tendency can be controlled by adjusting the welding parameters to reduce cooling rate. Preheat also has been found to be very useful in reducing the cracking tendency. Weldability of another alloy of interest, Ti-48 Al-2Cr-2Nb (at%), has been carried out by Mallory et al. (1992). Successful welds in this alloy have been produced using the GTA process and preheating of the alloy.

2.4. Long-Range Ordered Alloys

Long-range ordered (LRO) alloys are also a unique class of material with an atomic arrangement distinctly different from that of conventional or disordered alloys. Again, their tendency to be brittle in the ordered

Table 16-4 Chemical Compositions of Long-Range Ordered Alloys (wt%) Investigated

Alloy	Fe	Ni	Co	V	Ti
$(Fe,Ni)_3(V,Ti)$	37.6	39.6		22.2	0.4
$(Fe,Co)_3(V,Ti)$	16.3		61.1	22.2	0.4

Source: David and Braski (1986)

(a)

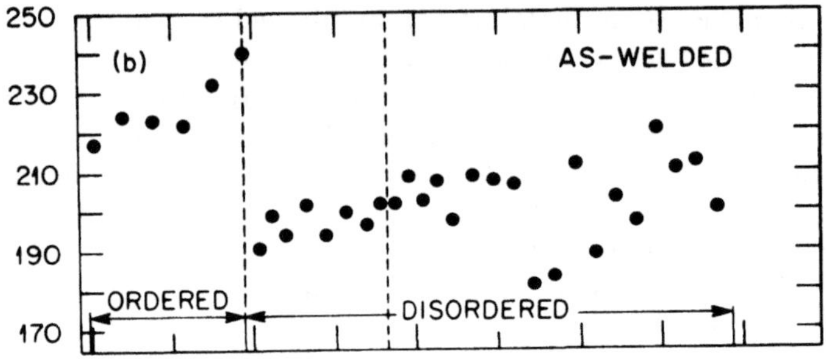

(b)

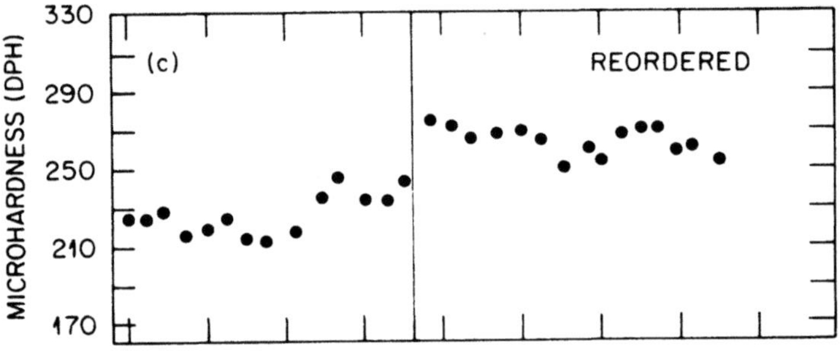

(c)

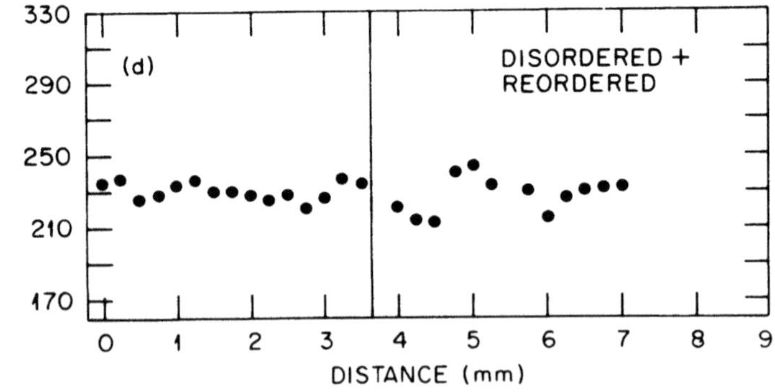

(d)

Figure 16-12. (a) Macrograph of GTA weld of $(Fe,Ni)_3(V,Ti)$ LRO alloy showing base metal HAZ and FZ. Note narrow band between base metal (ordered) and HAZ (disordered). Also shown below are microhardness traverses across the weldments for different conditions: (b) as-welded, (c) welded and heat treated for reordering, and (d) welded and heat treated for complete disordering and reordered (David and Braski, 1986) state has limited their use in structural applications. By adjusting composition and ordered crystal structure, Liu and Stiegler ((1984) have been able to control the ductility of the alloy. This has led to the development of ductile LRO alloys with general compositions $(Fe,Ni)_3(V,Ti)$ and $(Fe,Co)_3(V,Ti)$. The compositions are given in Table 16-4.

Sound GTA or EB welds were produced readily in the ductile LRO alloys (David, Braski, and Liu 1986). However, depending on the order-

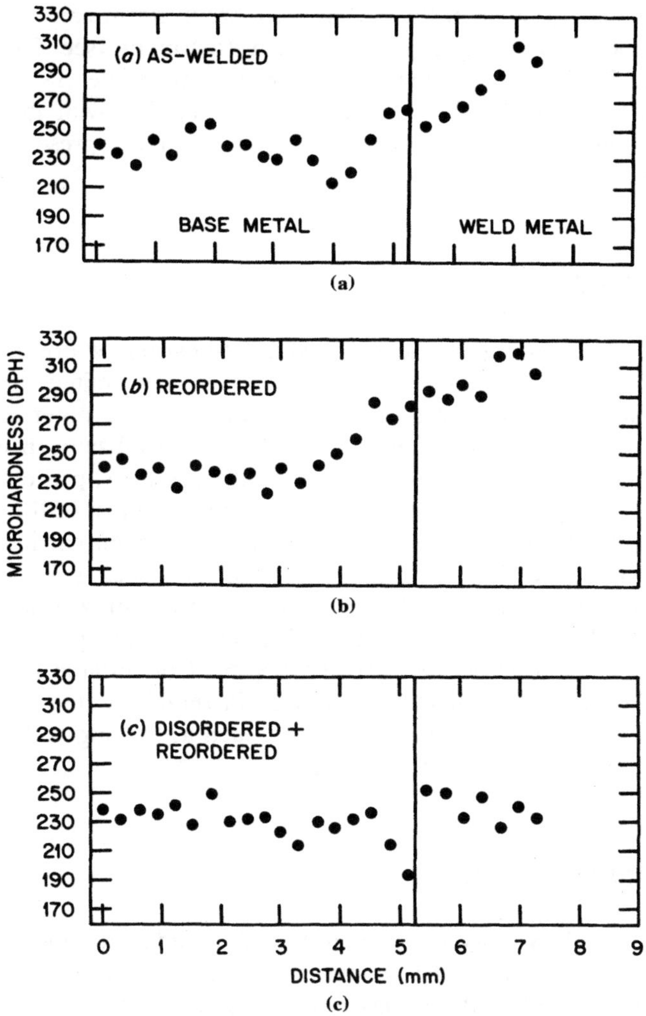

Figure 16-13. Microhardness traverse across the $(Fe,Co)_3(V,Ti)$ LRO alloy weldment for different conditions. (David and Braski, 1986)

disorder transition temperature (T_c), the response of this class of material to weld thermal cycle varies from alloy to alloy. The HAZ and FZ of $(Fe,Ni)_3(V,Ti)$ welds were disordered, with a decreased hardness within the HAZ and the FZ (see Fig. 16-12). The disordered weld was somewhat weaker and less ductile than the BM. Use of a postweld reordering heat treatment restored the tensile properties of the weld to that of the unwelded BM. The HAZ and FZ of the $(Fe,Co)_3(V,Ti)$ welds remained ordered, with increased hardness within the HAZ and the FZ. Tensile properties similar to those of the BM and a more uniform hardness profile can be produced by using a postweld heat treatment (see Fig. 16-13). Welds in the LRO alloys appear to be stronger than the BM when the weldment is ordered. Possible explanations for this behavior include the small size of the ordered domain in the FZ and/or the presence of VC precipitate in the FZ.

3. SUMMARY

Basic knowledge leading to the successful fabrication of ordered intermetallic alloys using fusion-welding processes is very limited. Successful welds have been produced in some of these alloys, while other alloys are prone to cracking during or after welding. Weldability has been found to be a strong function of alloy composition, welding process, process parameter, and weld geometry. It should be mentioned that most of the preliminary weldability studies of intermetallic alloys have been carried out on small test coupons. Welding and weldability of thick-section intermetallic alloys are yet to be carried out. Engineering properties of weldments must be characterized. Weldability of these alloys shows great promise. However, welding of large structures and development of suitable filler alloys to successfully weld this new class of alloys will remain a great challenge. Further, other alternate and novel methods, such as brazing and transient liquid-phase bonding, should be investigated.

ACKNOWLEDGEMENTS

The authors thank S. Viswanathan and G. M. Ludka for their reviews, P. A. Boyd for preparing the manuscript, and K. Spence for editing. Research is sponsored by the Division of Materials Sciences, U.S. Department of Energy, under contract DE-AC05-84OR21400 with Martin Marietta Energy Systems, Inc.

REFERENCES

Aoki, K., and Izumi, D., 1979, *Nippon Kinzatu Gakkaishi* 43:1190.

Baeslack, W. A., Cieslak, M. J., and Headley, T. J., "Structure, Properties and Fracture of Pulsed Nd: YAG Laser Welded Ti-14.8% AL-21.3% Nb Titanium Aluminide," 1988, *Scripta Metall.* 22:1155–1160.

Baeslack, W. A., Mascorella, T. J., and Kelly, T. J., 1989, *Weld. J.* 68(12):483s–498s.

Borland, J. C., 1960, *British Welding Journal* 7:579.

David, S. A., Braski, D. N., and Liu, C. T., 1986, *Weld. J.* 65(4):93s.

David, S. A., Jemian, W. A., Liu, C. T., and Horton, J. A., 1985, *Weld. J.* 64(1):22s.

David, S. A., George, E. P., and Liu, C. T., 1990, Oak Ridge Natl. Lab., Oak Ridge, TN (unpublished).

David, S. A., Horton, J. A., Goodwin, G. M., Phillips, D. H., and Reed, R. W., 1990, *Weld. J.* 69(4):133s–140s.

David, S. A., Horton, J. A., McKamey, C. G., Zacharia, T., and Reed, R. W., 1989 *Weld. J.*, 68(9):372s.

David, S. A., and Zacharia, T., (1993), *Weld. J.* 72(5), 201s.

Goodwin, G. M., 1987, *Weld. J.* 77(2):33s.

Kear, B. H., Sims, C. T., Stoloff, N. S., and Westbrook, J. H., eds., 1970, "Ordered Alloys," Baton Rouge, Louisiana, Claitor's Publishing Division.

Li, H., and Chaki, T. K., 1991, In *High Temperature Ordered Intermetallic Alloys IV MRS Symp. Proc.* 213:919.

Lipsitt, H. A., 1985, In *High Temperature Ordered Intermetallic Alloys, MRS Symp. Proc.* 39:351.

Liu, C. T., 1984, *Int. Metall. Rev.* 29:168.

Liu, C. T., McKamey, C. G., and Lee, E. H., 1990, *Scripta Metall.* 24:385.

Liu, C. T., and Stiegler, J. O., 1984, *Science* 226:636.

Liu, C. T., White, C. L., and Horton, J. A., 1985, *Acta. Metall.* 33:213.

Maguire, M. C., Edwards, G. R., and David, S. A., 1992, *Weld. J.* 71(7):231s.

Maguire, M. C., Santella, M. L., and B. K., Damkroger, 1992, "Factors in Sub-Solidus Heat-Affected Zone Cracking," *The Metal Science of Joining*, TMS, 41.

Mallory, L. C., Baeslack, W. A., Phillips, D., and Kelly, T. J., 1992, *Proceedings of the 7th International Conference on Titanium, San Diego*, TMS, Warrendale, PA.

Massalski, T. B., ed., 1986, *Binary Alloy Phase Diagrams*, p. 1777, ASM.

Patterson, R. A., Martin, P. L., Damkroger, B. K., and Christodoulou, L., 1990, *Weld. J.* 69:39s–44s.

Phillips, R. H., 1980, *Metals Forum* 3:158.

Rhodes, C. G., Hamilton, C. H., and Paton, N. E., 1987, "Titanium Aluminides for Elevated Temperature Application," AFML-TR-78-130, Air Force Materials Laboratory, Wright Patterson AFB, Dayton, Ohio.

Index

D

J

K

L

M

N

O

GPSR Compliance

The European Union's (EU) General Product Safety Regulation (GPSR) is a set of rules that requires consumer products to be safe and our obligations to ensure this.

If you have any concerns about our products, you can contact us on ProductSafety@springernature.com

In case Publisher is established outside the EU, the EU authorized representative is:

Springer Nature Customer Service Center GmbH
Europaplatz 3
69115 Heidelberg, Germany

Batch number: 10755320

Printed by Printforce, the Netherlands